INTRODUCTION TO ELECTRICITY AND ELECTRONICS

CONVENTIONAL-CURRENT VERSION

INTRODUCTION TO ELECTRICITY AND ELECTRONICS

CONVENTIONAL-CURRENT VERSION

ALLEN MOTTERSHEAD

Department of Electronics Technology

Cypress College, California

JOHN WILEY & SONS

NEW YORK CHICHESTER BRISBANE TORONTO SINGAPORE

PHOTO CREDITS

Cover William Hubbell/Woodfin Camp

Part I Tom Carroll/FPG

Part II Fundamental Photographs

Part III Sybil Shackman/Monkmeyer

COVER AND TEXT DESIGN: Judith Fletcher Getman
PRODUCTION SUPERVISOR: Sherry Berg
EDITOR: Judith A. Green

Library of Congress Cataloging in Publication Data:

Mottershead, Allen.
Introduction to electricity and electronics.

Includes index.
1. Electric circuits. I. Title.

TK454.M67 621.319′2 81-10472
ISBN 0-471-05751-7 AACR2

Printed in the United States of America

10 9 8 7 6 5 4 3 2 1

PREFACE

Everyone knows that the world is becoming more complex. From the cars we drive to the calculators, computers, and entertainment equipment familiar to all, we take for granted the advances that electronics has brought. And yet, the fundamental principles that made many of these devices possible first had to be understood before such devices could be developed. This book provides the fundamentals of electricity, magnetism, and electronics to help you understand the world around you. Many everyday applications are described to show these fundamentals at work. Also, completion of the material in this book will allow you to study in detail the advanced theory and applications of electrical and electronic devices, circuits, and systems.

The book is written for the first-year course in an electricity/electronics program at a community college, vocational institute, or technical college. It covers the traditional material of dc and ac circuits, including magnetism, that is required for both electronic and electrical majors. It assumes no previous knowledge of electricity or electronics. Only a basic understanding of algebra is required, and right-angle trigonometry is introduced only when it is necessary to understand alternating current.

There are three parts to the book. In Part One, after introducing calculators, electrical meters, and the basic electrical quantities, series and parallel circuits are covered followed by series-parallel circuits and their applications in loaded voltage dividers. A comprehensive chapter on voltage sources contains up-to-date information on the latest secondary batteries as well as fuel and solar cells. The important effect of internal resistance on maximum power transfer is also considered. A chapter on network analysis covers all the important methods such as branch currents, loop, superposition, Thévenin's, Norton's, and delta-wye transformations. Magnetism is then covered so that its applications in dc ammeters, voltmeters, and other dc measuring instruments can be shown.

Part Two begins with the generation of alternating current (ac) and direct current (dc), and continues with a detailed coverage of alternating voltage and current. This includes a chapter on ac measuring instruments, concentrating on the oscilloscope and its use. The oscilloscope can then be referred to during the following chapters on inductance and capacitance where their effects in both dc and ac circuits are considered. A whole chapter on transformers is included. After treating various combinations of resistance, inductance, and capacitance in alternating circuits, power and resonance in these circuits are also covered. Part Two ends with an optional chapter on complex numbers.

Part Three introduces the broad field of electronics. How a PN junction diode can cause rectification is explained, and its use to convert ac to dc in power supplies is followed by a discussion of filters. Finally, the bipolar junction transistor is introduced, and its use is demonstrated in amplifiers and oscillators.

The book has a number of features. First, theory and principles are made easier to understand by presenting applications whenever possible. For example, following a theoretical discussion of magnetism, we see how loudspeakers and dc motors operate, how solenoids and relays are used in a typical home heating/cooling system, and how the Hall effect is used in a clamp-on ammeter to measure both ac *and* dc.

In the chapter on electromagnetic induction we see the basic principles applied to a microphone, a magnetic tape recorder, practical alternators and dc generators, and a voltage regulator for an automotive battery-alternator system.

In the chapter on transformers, applications to power distribution and residential wiring systems are made, including the safety features of a ground connection and a ground fault circuit interrupter. Other applications include the conventional and electronic automotive ignition system, the role of an inductor in a fluorescent lamp circuit, the use of capacitors in photoflash units, differentiating and delay circuits, and power factor correction applications. The chapter on resonance shows how the principles of resonance are applied to a superheterodyne AM receiver, and the chapter on diodes applies the principles of rectification to the three-phase output of an automobile's alternator.

Second, every opportunity is taken to show how measurements are properly made by the latest in instrumentation. Three chapters are devoted to dc and ac measuring instruments. The conventional meters to measure current, voltage, resistance, and power, both analog and digital, are examined, in addition to the following: Gaussmeter, potentiometer for voltmeter and ammeter calibration, Wheatstone bridge, frequency and period meter, instrument transformers, impedance bridge, capacitance and inductance meters, power factor meter, vector impedance and voltage meters, and curve tracers. The use of the oscilloscope is stressed in the measurement of time constants, inductance, capacitance, phase angle, quality factor, and so on.

Third, every chapter has a summary, self-examination, review questions, and problems. Answers to the self-examination questions and to the odd-numbered problems are at the back of the book. There is also a comprehensive glossary containing all the book's important terms, as well as the symbols and units of each electrical quantity. The appendixes contain 14 sections that cover tables of information on resistors, wire gauges, determinants, trigonometry ratios, derivations, and so on.

Fourth, the text is arranged so that material needed for a concurrent laboratory course is given early in the text. A laboratory manual is available that was written to accompany the text. The laboratory manual has one laboratory experiment with many self-contained sections, for each chapter in this book. It is designed to be used with most standard laboratory equipment and supplies.

Finally, this book is available in both the *conventional-current version* and the *electron-flow version*. Although the SI system recommends the eventual adoption of conventional current, many will prefer to remain with the more familiar electron flow. Either way, it is the hope of the author that this book will make electricity and electronics more understandable and enjoyable to both reader and instructor.

ALLEN MOTTERSHEAD

ACKNOWLEDGMENTS

I am very grateful for the comments and suggestions of the following reviewers: Thomas R. Bertolino, Ventura College; Willie E. Bisson, New England Trade Institute; Robert J. Chenoweth, Chillicothe Area Voc-Tech School; Roy H. DuBose, York Technical College; Thomas C. Harrison, Pickens Area Vocational-Technical School; Robert C. Jones, Gulf Coast Community College; Charles King, New England Institute of Technology; M. B. Rotnem, Kent State University; John C. Schira, Clover Park Vo-Tech Institute; Manuel Stillerman, Bronx Community College; John E. Tillman, Swainsboro Area Voc-Tech School; Victor F. Veley, Los Angeles Trade-Techical College; Donald E. Wallis, Central Missouri State University; and Ulrich E. Zeisler, Utah Technical College at Salt Lake.

I would also like to acknowledge the help, patience and advice of the John Wiley staff who brought this book into production. It has been a pleasure to work with them.

A.M.

TO THE INSTRUCTOR

This book may be used in a number of different ways.

1 For a conventional two-semester dc/ac course:

Semester 1: Part I, DC Circuits: Chapters 1 to 13.

Semester 2: Part II, AC Circuits: Chapters 14 to 27.

2 For a one-semester combination dc/ac course (no analysis) that may include nonelectrical/electronics majors, the following chapters and sections are recommended:

WEEK	DC CIRCUITS
1	Chapter 1—read lightly
	Chapter 2
2	Chapter 3—Omit Sections 3-6 and 3-9
3	Chapter 5
4	Chapter 6
	Section 7-1
5	Chapter 8—read lightly
	Chapter 9
6	Chapter 11—Omit Section 11-14
7	Chapter 12—Omit Section 12-4
	Chapter 13—Omit Sections 13-4 through 13-6
	AC CIRCUITS
8	Chapter 14—All plus Appendix A-6
9	Chapter 15
	Sections 16-1 and 16-6
10	Chapter 17—Omit Sections 17-2, 17-5.3, and 17-5.4
	Chapter 18—Omit Sections 18-1.3 through 18-1.5 and Section 18-6
11	Chapter 19—Omit Sections 19-5.1, 19-6, 19-9, and 19-9.1
	Chapter 20—Omit Sections 20-4, 20-6, 20-6.1, 20-7, 20-8, and 20-9
12	Chapter 21—Omit Sections 21-4, 21-4.1, 21-4.2, 21-4.3, 21-5.1, and 21-7.1
13	Chapter 22—Omit Sections 22-5, 22-7, 22-8, and 22-8.1
	Chapter 23—Omit Sections 23-4 through 23-6
14	Chapter 24—Omit Sections 24-4.1, 24-9, 24-10, and 24-11

WEEK	ELECTRONICS
15	Chapter 28—Omit Sections 28-11.1 through 28-13
16	Chapter 29

3 For a one-semester dc/ac analysis course that concentrates on areas such as network theorems, magnetic circuits, exponential equations for L and C circuits, and complex numbers and is designed to follow the course in No. 2 above, the following chapters, sections, and appendixes are recommended. (It is assumed that coverage of sections omitted in the first course will be accompanied by a general review of pertinent material in the appropriate chapter.)

WEEK	
1	General review plus Chapter 4
2	Chapter 7
3	Chapter 8
4,5	Chapter 10 and Appendix A-4
6	Review plus Section 11-14 plus Appendix A-13
7	Review of Chapter 14 plus Appendixes A-5, A-7, and A-8
8	Chapter 16
9	Review of Chapter 17 plus Sections 17-2, 17-5.3, and 17-5.4
	Review of Chapter 18 plus Sections 18-1.3, 18-1.4, 18-5, and 18-6
10	Sections 19-5.1, 19-6, 19-9, 19-9.1 plus Appendix A-9
	Sections 20-4, 20-6, 20-6.1, 20-7, 20-8, 20-9 plus Appendix A-10
11	Sections 21-4, 21-4.1, 21-4.2, 21-4.3, 21-5.1, 21-7.1 plus Appendix A-11
	Sections 22-5, 22-7, 22-8, 22-8.1 plus Appendix A-12
12	Sections 23-4, through 23-6 plus Appendix A-14
	Review Chapter 24 plus Sections 24-4.1, 24-9, 24-10, and 24-11
13	Chapter 25
14	Chapter 26
15,16	Chapter 27

Individual needs vary from one instructor and course to another so that many variations of the above are possible. For example, Chapter 27 on complex numbers may appear more logically (for some) immediately after Chapter 24 on ac circuits. Others may wish for this material to appear even earlier, before reactance. Or they may feel it is too advanced and omit it altogether. For this reason, Chapter 27 stands on its own as optional material at the end of Part Two.

It is assumed that a laboratory session constitutes part of each semester course, using a minimum of three hours lecture and three hours laboratory per week. Topics have been arranged with this in mind so that, for example, Chapter 16 on the oscilloscope precedes capacitance and inductance. This means that the oscilloscope can be used in the laboratory experiments on these two topics, allowing the observation and measurement of time constants.

Students can check their own progress by doing the review questions, self-examinations, and problems. In general, problems become more difficult toward the end of each chapter, and even more challenging ones are found among the even numbers. A Solutions Manual for all problems is available to instructors.

A.M.

CONTENTS

PART ONE DC CIRCUITS 2

CHAPTER 1 INTRODUCTION TO CALCULATORS, CIRCUITS, AND METERS 4

1-1 The Electronic Calculator 6
1-2 Scientific Notation 6
1-3 Accuracy and Rounding-Off Numbers 8
1-4 The SI System 8
1-5 Conversion Factors 8
1-6 Unit Prefixes 9
1-7 A Simple Electric Circuit 10
1-8 Electrical Symbols for Resistance, Inductance and Capacitance, and Electrical Devices 11
1-9 Pictorials and Schematics 13
1-10 Safety 14
1-11 Reading Scales of Multirange Meters 15
1-11.1 Types of Meters 15
1-11.2 Reading the Scales 16
Summary
Self-Examination
Review Questions
Problems

CHAPTER 2 BASIC ELECTRICAL QUANTITIES 24

2-1 Static Electricity 26
2-2 Coulomb's Law 26
2-3 Applications of Static Electricity (Electrostatics) 28
2-4 Potential Difference 28
2-4.1 The Volt 29
2-4.2 Electromotive Force 30
2-5 Atomic Structure 30
2-5.1 Valence Electrons 31
2-6 Free Electrons 32
2-7 Conductors and Insulators 32
2-8 Current 32
2-9 Conventional Current and Electron Flow 33
2-10 Drift Velocity of Current 34
2-11 Resistance and Conductance 35
Summary
Self-Examination
Review Questions
Problems

CHAPTER 3 OHM'S LAW, POWER, AND RESISTORS 42

3-1 Connection of Circuit Meters 44
3-2 Ohm's Law 44
3-3 Graphical Presentation of Ohm's Law 45
3-4 Other Forms of Ohm's Law 45
3-5 Load Resistance 46
3-6 Mechanical Work and Energy 46
3-7 Electrical Work and Energy 47
3-8 Efficiency 47
3-9 Mechanical Power 48
3-10 Electrical Power 49

3-11 Efficiency Involving Power 49
3-12 Other Equations for Electrical Power 50
3-13 Cost of Electrical Energy 51
3-14 Practical Resistors 51
3-14.1 Fixed Resistors 52
3-14.2 Variable Resistors 53
Summary
Self-Examination
Review Questions
Problems

CHAPTER 4 RESISTANCE 60
4-1 Factors Affecting Resistance 62
4-2 Wire Gauge Table 63
4-3 Insulators and Semiconductors 64
4-4 Temperature Coefficient of Resistance 64
4-5 Linear Resistors 66
4-6 Nonlinear Resistors 67
4-6.1 Thermistors 67
4-6.2 Varistors 69
Summary
Self-Examination
Review Questions
Problems

CHAPTER 5 SERIES CIRCUITS 74
5-1 Voltage Drops in a Series Circuit 76
5-2 Kirchhoff's Voltage Law 77
5-3 Power in a Series Circuit 79
5-4 Open Circuits 80
5-5 Short Circuits 80
5-6 Voltage Division 81
5-7 The Potentiometer 82
5-8 Positive and Negative Grounds 83
5-9 Positive and Negative Potentials 84
5-10 Voltage Sources Connected in Series 85
Summary
Self-Examination
Review Questions
Problems

CHAPTER 6 PARALLEL CIRCUITS 92
6-1 Characteristics of a Parallel Circuit 94
6-2 Kirchhoff's Current Law 94
6-3 Equivalent Resistance for a Parallel Circuit 94
6-4 Other Equivalent Resistance Equations 97
6-5 Current Division 98
6-6 Short Circuits 99
Summary
Self-Examination
Review Questions
Problems

CHAPTER 7 SERIES-PARALLEL AND VOLTAGE-DIVIDER CIRCUITS 104
7-1 Series-Parallel Resistor Circuits 106
7-1.1 Practical Series-Parallel Circuit 109
7-1.2 Thermistor-Resistor Series-Parallel Circuit 110
7-2 Voltage Dividers 111
7-2.1 Series-Dropping Resistor 111
7-2.2 Bleeder Resistor 112
7-2.3 Positive and Negative Potentials 112
Summary
Self-Examination
Review Questions
Problems

CHAPTER 8 VOLTAGE SOURCES 118
8-1 Simple Copper–Zinc Cell 120
8-2 The Carbon–Zinc Cell 121
8-3 The Alkaline–Manganese Cell 122
8-4 Other Primary Cells 124
8-5 Comparison of Primary Cells 124
8-6 Lead–Acid Secondary Storage Cells 125
8-6.1 Chemical Action 126
8-6.2 Construction 127
8-6.3 Specific Gravity 128
8-6.4 Ampere-Hour Capacity 128
8-7 The Nickel–Cadmium Cell 128
8-7.1 Construction 128
8-7.2 Charging 128

8-7.3 Quick-Charging 129
8-7.4 Disadvantages 130
8-8 Other Secondary Cells 130
8-8.1 Gelled-Electrolyte Lead–Acid Cells 130
8-8.2 Silver–Zinc and Silver–Cadmium Cells 130
8-8.3 Cells of the Future 130
8-9 Comparison of Secondary Cells 131
8-10 Other Sources of EMF 132
8-10.1 Fuel Cells 132
8-10.2 Solar Cells 133
8-10.3 Alternators 134
8-10.4 Thermocouples 134
8-10.5 Piezoelectrics 134
Summary
Self-Examination
Review Questions
Problems

CHAPTER 9 INTERNAL RESISTANCE AND MAXIMUM POWER TRANSFER 140

9-1 Internal Resistance of a Cell 142
9-2 Series and Parallel Connections of Cells to Form a Battery 145
9-3 Maximum Power Transfer 147
9-4 Power Measurement—The Wattmeter 148
Summary
Self-Examination
Review Questions
Problems

CHAPTER 10 NETWORK ANALYSIS 154

10-1 Branch Current Method 156
10-2 Loop or Mesh Current Method 158
10-3 The Superposition Theorem 159
10-4 Thévenin's Theorem 161
10-5 Current Sources 164
10-6 Norton's Theorem 166
10-6.1 Converting a Norton Equivalent to a Thévenin Equivalent 169
10-6.2 Combining Current Sources in Series 171
10-7 Delta-Wye Conversions 172
Summary
Self-Examination
Review Questions
Problems

CHAPTER 11 MAGNETISM AND ITS APPLICATIONS 178

11-1 Permanent Magnets 180
11-1.1 Magnetic Field 180
11-1.2 Attraction and Repulsion 180
11-2 Oersted's Discovery 181
11-3 The Electromagnet 182
11-3.1 Solenoids 182
11-3.2 Door Chimes 183
11-3.3 Buzzers and Bells 183
11-3.4 Application of Solenoid Valves in a Home Heating System 183
11-3.5 Relays 184
11-4 Magnetic Units 186
11-4.1 Flux Density 186
11-4.2 Magnetomotive Force 186
11-4.3 Magnetizing Intensity 186
11-4.4 Permeability 187
11-5 Hysteresis 188
11-6 Basic Theory of Magnetism 189
11-6.1 Hysteresis Losses 189
11-7 Materials for Permanent Magnets 189
11-8 Applications of Magnetic Materials 190
11-9 Demagnetization 190
11-10 Force on a Current-Carrying Conductor in a Magnetic Field 192
11-11 Hall Effect and the Gaussmeter 194
11-12 High-Fidelity Loudspeaker 195
11-13 DC Motors 195
11-14 Ohm's Law for a Magnetic Circuit 197
Summary
Self-Examination
Review Questions
Problems

CHAPTER 12 DC AMMETERS AND VOLTMETERS 204

12-1 The D'Arsonval or PMMC Meter Movement 206
12-1.1 Current Sensitivities 207
12-2 Ammeter Shunts 207
12-3 Multirange Ammeters 208
12-4 Ayrton or Universal Shunt 209
12-5 Accuracy and Errors 211
12-6 Ammeter Loading 212
12-6.1 Clamp-On Ammeter 212
12-7 Voltmeter Multipliers 213
12-8 Multirange Voltmeters 214
12-9 Series-Connected Multirange Voltmeter 215
12-10 Voltmeter Sensitivity 216
12-11 Voltmeter Loading 216
Summary
Self-Examination
Review Questions
Problems

CHAPTER 13 OTHER DC MEASURING INSTRUMENTS 224

13-1 Principle of a No-Load Voltmeter 226
13-2 The Potentiometer 226
13-3 Calibration of Voltmeters and Ammeters 228
13-4 Ammeter–Voltmeter Method of Measuring Resistance 229
13-4.1 Measurement of High Resistance 229
13-4.2 Measurement of Low Resistance 230
13-5 Series Type of Ohmmeter 231
13-6 The Shunted-Series Type of Ohmmeter 232
13-7 The Wheatstone Bridge 234
13-8 Digital Multimeters 237
Summary
Self-Examination
Review Questions
Problems

PART TWO AC CIRCUITS 246

CHAPTER 14 GENERATING AC AND DC 248

14-1 Electromagnetic Induction 250
14-2 Lenz's Law 251
14-3 Faraday's Law 251
14-4 Induced Emf Due to Conductor Motion in a Stationary Magnetic Field 255
14-4.1 Effect of Angle on Induced Emf 256
14-5 Elementary AC Generator 257
14-6 The Sine Wave 258
14-7 Factor's Affecting the Amplitude of a Generator's Sine Wave 259
14-8 Factors Affecting the Frequency of a Generator's Sine Wave 259
14-9 Practical Alternators 261
14-9.1 Voltage Regulator for an Automobile Alternator 261
14-9.2 Delta-Wye Connections 262
14-10 Elementary DC Generator 263
14-11 Practical DC Generators 263
Summary
Self-Examination
Review Questions
Problems

CHAPTER 15 ALTERNATING VOLTAGE AND CURRENT 272

15-1 Induced Voltage in Terms of Frequency and Time 274
15-1.1 Angular Frequency 275
15-2 Current and Voltage Waveforms with a Resistive Load 275
15-3 Power in a Resistive Load 277
15-3.1 Equations for Average Power 278
15-4 Rms Value of a Sine Wave 279
15-5 Series and Parallel AC Circuits with Pure Resistance 282
Summary

Self-Examination
Review Questions
Problems

CHAPTER 16 AC MEASURING INSTRUMENTS 288

16-1 The Cathode-Ray Oscilloscope 290
16-1.1 Construction 290
16-1.2 Operation 291
16-1.3 Function of Controls 291
16-1.4 Oscilloscope Probes 293
16-1.5 Measurement of Voltage and Frequency 293
16-1.6 Lissajous Figures 295
16-1.7 Other Oscilloscopes 296
16-2 Ac Meters Using the PMMC Movement 297
16-3 The Electrodynamometer Movement 298
16-4 Moving-Iron Instruments 300
16-5 Frequency Meter 300
16-6 Precautions with Grounded Equipment 300
Summary
Self-Examination
Review Questions
Problems

CHAPTER 17 INDUCTANCE 306

17-1 Self-Inductance 308
17-2 Factors Affecting Inductance 310
17-3 Practical Inductors 311
17-4 Inductances in Series and Parallel 311
17-5 Mutual Inductance 312
17-5.1 Series Connected Coils with Mutual Inductance 313
17-5.2 Dot Notation 313
17-5.3 Coefficient of Coupling 314
17-5.4 Measurement of Mutual Inductance 315
Summary
Self-Examination
Review Questions
Problems

CHAPTER 18 TRANSFORMERS 320

18-1 The Iron Core Transformer 322
18-1.1 Voltage Ratio 322
18-1.2 Current Ratio 323
18-1.3 Power, Volt-Ampere Relations, and Transformer Efficiency 324
18-1.4 Matching Resistances for Maximum Power Transfer 327
18-2 Power Distribution System 328
18-3 Residential Wiring System 330
18-3.1 Safety Features of a Ground Connection 331
18-3.2 Ground-Fault, Circuit-Interrupters (GFIs) 331
18-4 Isolation Transformers 332
18-5 Autotransformers 333
18-6 Instrument Transformers 334
18-6.1 Potential Transformers 334
18-6.2 Current Transformers 334
18-7 Clamp-On AC Ammeter 336
18-8 Checking a Transformer 336
Summary
Self-Examination
Review Questions
Problems

CHAPTER 19 INDUCTANCE IN DC CIRCUITS 342

19-1 Rise of Current in an Inductive Circuit 344
19-2 Time Constant, τ 345
19-3 Universal Time-Constant Curves 347
19-4 Energy Stored in an Inductor 349
19-5 Fall of Current in an Inductive Circuit (with Electronic Switching) 350
19-5.1 Measurement of Coil Inductance 351
19-6 Fall of Current in an Inductive Circuit (with Mechanical Switching) 352
19-7 Conventional Automotive Ignition System 354
19-8 Electronic Automotive Ignition System 355
19-9 Methods of Reducing Inductive Effects 356
19-9.1 Noninductive Resistors 356
Summary

Self-Examination
Review Questions
Problems

CHAPTER 20 INDUCTANCE IN AC CIRCUITS 362

20-1 Effect of Inductance in an AC Circuit 364
20-1.1 Phase Angle Relationship Between Current and Voltage 364
20-2 Inductive Reactance 365
20-3 Factors Affecting Inductive Reactance 366
20-4 Application of Inductors in Filter Circuits 367
20-5 Application of Inductance in a Fluorescent Lamp Circuit 368
20-6 Series Inductive Reactances 370
20-6.1 Series Inductive Reactances with Mutual Reactance 370
20-7 Parallel Inductive Reactances 371
20-8 Quality of a Coil, Q 372
20-9 Effective Resistance, R_{ac} 373
20-10 Measuring Inductance 374
Summary
Self-Examination
Review Questions
Problems

CHAPTER 21 CAPACITANCE 380

21-1 Basic Capacitor Action 382
21-2 Capacitance 382
21-3 Factors Affecting Capacitance 383
21-4 Effect of a Dielectric 385
21-4.1 Electric Field 385
21-4.2 Dielectric Strength 386
21-4.3 Induced Electric Field 387
21-5 Types of Capacitors 388
21-5.1 Other Properties of Capacitors 389
21-6 Capacitors in Parallel 390
21-7 Capacitors in Series 392
21-7.1 Voltage Division Across Capacitors in Series 393
21-8 Energy Stored in a Capacitor 394
Summary
Self-Examination
Review Questions
Problems

CHAPTER 22 CAPACITANCE IN DC CIRCUITS 400

22-1 Current in a Capacitive Circuit 402
22-2 Charging a Capacitor 403
22-3 Time Constant, τ 404
22-3.1 Checking Capacitors with an Ohmmeter 406
22-4 Discharging a Capacitor 407
22-4.1 Application to a Capacitor Photoflash Unit 408
22-5 Universal Time-Constant Curves 409
22-6 Oscilloscope Measurement of Capacitance 411
22-7 Short Time-Constant Differentiating Circuit 412
22-8 Long Time-Constant Circuits 412
22-8.1 *RC* Delay Circuits 414
Summary
Self-Examination
Review Questions
Problems

CHAPTER 23 CAPACITANCE IN AC CIRCUITS 422

23-1 Effect of a Capacitor in an AC Circuit 424
23-1.1 Phase Angle Relationship Between Voltage and Current 424
23-2 Capacitive Reactance 424
23-3 Factors Affecting Capacitive Reactance 425
23-3.1 Comparison of Capacitance in Dc and Ac Circuits 426
23-4 Series Capacitive Reactances 427
23-5 Parallel Capacitive Reactances 428
23-6 Dissipation Factor of a Capacitor 428
Summary
Self-Examination
Review Questions
Problems

CHAPTER 24 ALTERNATING CURRENT CIRCUITS 434

24-1 Representation of a Sine Wave by a Phasor 436
24-2 Voltage and Current Phasors in Pure *RLC* Circuits 437
24-3 Series *RL* Circuit 437
24-4 Impedance of a Series *RL* Circuit 441
24-4.1 Measuring Inductance of a Practical Coil 443
24-5 Series *RC* Circuit 444
24-6 Impedance of a Series *RC* Circuit 446
24-7 Series *RLC* Circuit 447
24-8 Oscilloscope Measurement of Phase Angles 448
24-9 Parallel *RL* Circuit 450
24-10 Parallel *RC* Circuit 451
24-11 Parallel *RLC* Circuit 452
Summary
Self-Examination
Review Questions
Problems

CHAPTER 25 POWER IN AC CIRCUITS 460

25-1 Power in Pure Resistance 462
25-2 Power in Pure Inductance 462
25-3 Power in Pure Capacitance 464
25-4 Power in a Series *RL* Circuit 465
25-4.1 Measuring the Inductance of a Coil 466
25-5 The Power Triangle 466
25-5.1 Power Triangle for a Parallel *RL* Circuit 467
25-5.2 Power Triangle for a Series or Parallel *RC* Circuit 468
25-5.3 Power Triangle for a Series or Parallel *RLC* Circuit 469
25-6 Power Factor 470
25-6.1 Detrimental Effect of a Low Power Factor 473
25-7 Power Factor Correction 473
25-8 Power in Three-Phase Circuits 476
25-9 Maximum Power Transfer 477
Summary
Self-Examination
Review Questions
Problems

CHAPTER 26 RESONANCE 486

26-1 Series Resonance 488
26-2 Resonant Frequency 490
26-3 Selectivity of a Series Circuit 491
26-4 The *Q* of a Series-Resonant Circuit 493
26-5 Resonant Rise of Voltage in a Series *RLC* Circuit 494
26-6 Resonance in a Theoretical Parallel *RLC* Circuit 497
26-6.1 The *Q* of a Theoretical, Parallel Resonant Circuit 498
26-7 Selectivity of a Parallel Resonant Circuit 500
26-8 Resonance in a Practical, Two-Branch Parallel *RLC* Circuit 501
26-8.1 Damping a Parallel Resonant Circuit 503
26-9 Operating Principle of a Superheterodyne AM Receiver 504
Summary
Self-Examination
Review Questions
Problems

CHAPTER 27 COMPLEX NUMBERS 512

27-1 Polar Coordinates 514
27-2 Multiplication and Division Using Polar Coordinates 515
27-2.1 Application of Polar Coordinates to Series Circuits 516
27-3 The *j* Operator 518
27-4 Rectangular Coordinate System 519
27-5 Addition and Subtraction Using Rectangular Coordinates 520
27-5.1 Application of Rectangular Coordinates to Series Circuits 521
27-6 Multiplication and Division Using Rectangular Coordinates 522
27-6.1 Application of Rectangular Coordinates to Parallel Circuits 523
27-7 Conversion from Rectangular to Polar Form 524
27-8 Conversion from Polar to Rectangular Form 526

27-9 Application of Complex Numbers to Series-Parallel Circuits 527
27-10 Conductance, Susceptance, and Admittance 529
27-11 Application of Complex Numbers to Thévenin's Theorem 531
Summary
Self-Examination
Review Questions
Problems

PART THREE BASIC ELECTRONICS 540

CHAPTER 28 DIODES, RECTIFICATION, AND FILTERING 542

28-1 Pure Silicon 544
28-1.1 Pure Silicon at Room Temperature 544
28-1.2 Current in Pure Silicon 544
28-2 *N*-Type Silicon 545
28-2.1 Current in *N*-Type Silicon 546
28-3 *P*-Type Silicon 546
28-3.1 Current in *P*-Type Silicon 546
28-4 *PN* Junctions 547
28-5 Forward-biased *PN* Junction 548
28-6 Reverse-Biased *PN* Junction 548
28-7 Diode VI Characteristics 549
28-7.1 Zener Diodes 550
28-7.2 Checking a Diode with an Ohmmeter 550
28-8 The Germanium Diode 551
28-9 Half-Wave Rectification 552
28-10 Full-Wave Rectification 553
28-10.1 Bridge Rectifier 554
28-10.2 Three-Phase, Full-Wave Rectifier 556
28-11 Capacitor Filtering 557
28-11.1 Ripple Factor 558
28-12 Inductor Filtering 559
28-13 Choke Input or *LC* Filter 560
Summary
Self-Examination
Review Questions
Problems

CHAPTER 29 INTRODUCTION TO TRANSISTORS, AMPLIFIERS, AND OSCILLATORS 568

29-1 Physical Characteristics of Transistors 570
29-1.1 Types 570
29-1.2 Ohmmeter Checking of a Transistor 570
29-1.3 Construction 571
29-1.4 Packaging and Lead Identification 571
29-2 Basic Common-Emitter Amplifying Action 572
29-3 CE Collector Characteristic Curves 574
29-3.1 Forward-Current Transfer Ratio, β 574
29-4 CB Collector Characteristic Curves 576
29-4.1 Forward-Current Transfer Ratio, α 576
29-4.2 Relation Between α and β 577
29-5 Basic Common-Emitter AC Amplifier 577
29-5.1 Waveforms 579
29-5.2 Current, Voltage and Power Gains 579
29-6 Practical CE AC Amplifier with Bias Stabilization 580
29-7 Frequency Response of a CE Amplifier 582
29-8 Negative Feedback 583
29-8.1 Reasons for Using Negative Feedback 584
29-8.2 Introduction of Negative Feedback in a CE Amplifier 585
29-9 Positive Feedback 586
29-9.1 Wien Bridge Oscillator Circuit 587
Summary
Self-Examination
Review Questions
Problems

APPENDIX 597

A-1 Common Symbols used in Schematic Diagrams 598
A-2 Standard Resistance Values for Commercial Resistors 600
A-3 AWG Conductor Sizes and Metric Equivalents 601
A-4 Solving Linear Equations Using Determinants 603

A-5 Effect of Angle on the Voltage Induced in a Moving Conductor 605
A-6 Trigonometric Ratios 606
A-7 Factors Affecting the Amplitude of a Generator's Sine Wave 608
A-8 Induced Voltage in Terms of Frequency and Time 609
A-9 Algebraic Solutions for Instantaneous and Transient Currents and Voltages in a Series *RL* Circuit 610
A-10 Derivation of the Inductive Reactance Formula, X_L 612
A-11 Derivation of the Capacitance Equation with Air Dielectric 614
A-12 Algebraic Solutions for Instantaneous Capacitor Voltage and Current 615
A-13 Observation of a Transformer Core's *B-H* Hysteresis Loop Using an Integrating Circuit 617
A-14 Derivation of the Capacitive Reactance Formula, X_c 619

GLOSSARY 621

ANSWERS TO SELF-EXAMINATIONS 649

ANSWERS TO ODD-NUMBERED PROBLEMS 651

INDEX 665

INTRODUCTION TO ELECTRICITY AND ELECTRONICS

CONVENTIONAL-CURRENT VERSION

PART ONE

DC CIRCUITS

1 INTRODUCTION TO CALCULATORS, CIRCUITS, AND METERS

2 BASIC ELECTRICAL QUANTITIES

3 OHM'S LAW, POWER, AND RESISTORS

4 RESISTANCE

5 SERIES CIRCUITS

6 PARALLEL CIRCUITS

7 SERIES-PARALLEL AND VOLTAGE-DIVIDER CIRCUITS

8 VOLTAGE SOURCES

9 INTERNAL RESISTANCE AND MAXIMUM POWER TRANSFER

10 NETWORK ANALYSIS

11 MAGNETISM AND ITS APPLICATIONS

12 DC AMMETERS AND VOLTMETERS

13 OTHER DC MEASURING INSTRUMENTS

CHAPTER ONE

INTRODUCTION TO CALCULATORS, CIRCUITS, AND METERS

The study of electricity and electronics, in any detail, necessarily involves calculations. Calculations are used in working problems to become familiar with a subject or in predicting results using equations. The electronic calculator has removed much of the tedium in ordinary arithmetic computations and must be considered an essential tool for everyday calculations. The purpose of this chapter is to assist the reader in the choice and use of a calculator. The simple electric circuit, which is also introduced in this chapter, shows how symbols can be used to represent such a circuit in a schematic diagram. Finally, how to read the scales of multirange meters will be covered.

1-1 THE ELECTRONIC CALCULATOR

The four basic arithmetic operations (+, −, ×, and ÷) may be entered and processed on the lowest-cost calculators except for very large or very small numbers. In such cases it is not possible to get an answer. Consider, for example, entering the number 50,000,000 into an eight-digit readout calculator and then multiplying by 2. Of course, the result should be 100,000,000, but the calculator can only display eight of these digits. In some calculators, the 1 disappears off the end and it only displays zeros. In another calculator, the incorrect number 10,000,000 appears, but this answer is shown as incorrect by a flashing readout. Others may show an incorrect answer by decimal points appearing after every digit.

FIGURE 1-1
A typical scientific calculator. (Courtesy of Texas Instruments.)

Most students will find it worthwhile, instead, to invest in a "scientific" calculator or an "electronic slide rule" calculator, perhaps of the type shown in Fig. 1-1. Although it may have some functions that are not immediately used, almost all of them will become understood and valuable as you advance. You certainly should have the trigonometric functions of sin, cos, tan, and their inverse, as well as $\log_{10}$ and ln. Also, if you plan to continue in electronics technology, the ability to convert from rectangular to polar (and vice versa) will be an important asset. Beyond this minimum level, the cost will determine whether you choose a programmable calculator or even one with a printout capability.

1-2 SCIENTIFIC NOTATION

The ability of a scientific calculator to handle very large or very small numbers depends upon its use of scientific notation. This involves writing any given number as a number between 1 and 10 (but not including 10) and multiplied by 10 raised to some suitable exponent, which may be either positive or negative. Thus

$$34{,}275 = 3.4275 \times 10^4$$
$$1000 = 1.0 \times 10^3$$
$$50{,}000{,}000 = 5.0 \times 10^7$$
$$0.006 = 6.0 \times 10^{-3}$$
$$0.000{,}0016 = 1.6 \times 10^{-6}$$

Let us assume that you have a 10-digit-display scientific calculator and you want to multiply the following:

$$5{,}000{,}000{,}000 \times 2.2$$

Using scientific notation and doing it by hand, you obtain

$$
\begin{aligned}
5{,}000{,}000{,}000 \times 2.2 &= 5.0 \times 10^{9} \times 2.2 \\
&= 11.0 \times 10^{9} \\
&= 1.1 \times 10^{10} \\
&= 11{,}000{,}000{,}000
\end{aligned}
$$

This is a result obviously too large to be displayed in normal form.

Now enter 5,000,000,000 into the calculator and multiply by 2.2. The result is shown in scientific notation as 1.1 10 and is understood to mean 1.1×10^{10}. (An E or EE may be displayed between the 1.1 and the 10 on some calculators). Even when the numbers are not fed into the calculator in scientific notation, the result is displayed in this notation to avoid exceeding the display capability.

In order to enter a number into the calculator in scientific notation, we use the scientific exponent or EE button as follows:

Multiply: $(7.6 \times 10^{3}) \times (6.5 \times 10^{4})$.

Enter	*Press*	*Display*	*Purpose*
	C	0	Clears the calculator
7.6	EE 3	7.6 03	Enters the first number in scientific notation
	X	7.6 03	Instruction to perform multiplication
6.5	EE 4	6.5 04	Enters second number in scientific notation
	=	4.94 08	Answer

Answer: 4.94×10^{8}

NOTE If you wish to enter only a power of 10, such as 10^4, it may be necessary to put a 1 in front. That is, enter 1×10^4 (1 EE 4), or the calculator may not respond. See Problem 1-10.

Some calculators are equipped to display the answer in standard decimal notation as follows:

Enter	*Press*	*Display*	*Purpose*
4.94 08	2nd EE	494,000,000	Converts scientific notation to decimal form

If the decimal form is higher or lower than the display capability, there is no change following this sequence or the calculator flashes as described above.

Note that a number may also be converted to scientific notation on some calculators by entering the number and doing the following:

Enter	*Press*	*Display*	*Purpose*
494,000,000		494,000,000	
	EE	494,000,000 00	Converts the decimal number to scientific notation
	=	4.94 08	

Answer: 4.94×10^{8}

Finally, consider $\dfrac{3.04 \times 10^{-2}}{21.6 \times 10^{4}}$.

Enter	*Press*	*Display*	*Purpose*
3.04	EE +/− 2	3.04-02	To multiply 3.04 by 10^{-2}, use +/− key to enter a negative exponent
	÷	3.04-02	Indicates division as the next operation
21.6	EE 4	21.6 04	Enters the denominator in scientific notation
	=	1.407407407-07	
	2nd EE	.0000001407	

Answer: $1.407407407 \times 10^{-7}$ (1.41×10^{-7})

In a similar way, operations involving addition, subtraction, squaring, or square root can be performed on numbers in scientific notation by using the appropriate key.

It should be noted that the method of entering data into the calculator may differ according to the make of calculator. Some use algebraic operation; others use a method referred to as reverse Polish notation. You should consult your calculator's specific instructions to determine how data should be entered.

1-3
ACCURACY AND ROUNDING-OFF NUMBERS

The answer to the last example should raise an important question. How many of the 10 displayed digits are significant in the final answer? This depends upon how many *significant figures* are contained in the original numbers making up the problem. In this case, both 3.04×10^{-2} and 21.6×10^4 contain three significant figures. Thus the answer can be no more accurate than three significant figures and must be "rounded-off" to 1.41×10^{-7}.

If the original numbers had been quoted as 3.039×10^{-2} and 21.58×10^4, these would be accurate to *four* significant figures and the calculator result of $1.408248378 \times 10^{-7}$ would be written as $\mathbf{1.408 \times 10^{-7}}$.

In general, a calculated result is only as accurate as the *least* accurate information fed into the calculator. For example, consider calculating the average miles per gallon obtained by your car when the odometer showed a distance of 221.7 miles for a gas pump reading of 11.5 gallons. A calculator gives

$$\frac{221.7 \text{ mi}}{11.5 \text{ gal}} = 19.27826087 \text{ mi/gal (mpg)}$$

The rounded-off answer is **19.3 mpg** because the gas pump reading contained only three significant figures.

1-4
THE SI SYSTEM

It is well known that the United States is the last major industrial country to adopt the metric system, and then only voluntarily; that is, there is no legislation requiring metric conversion at this time. But to trade with the rest of the world, it is to our economic advantage to standardize measurements using metric quantities. In 1965, the Institute of Electrical and Electronics Engineers (IEEE) adopted the "Système International d'Unites" (SI—International System of Units) founded on the meter, kilogram, second, ampere (or MKSA) system. Table 1-1 shows a comparison of the units used for length, mass, and time in the three systems. Note that the name of each system stems from the first letter of each unit of measurement in Table 1-1.

TABLE 1-1
Physical Units

System	Length	Mass	Time
MKS	Meter (m)	Kilogram (kg)	Second (s)
CGS	Centimeter (cm)	Gram (g)	Second (s)
FPS	Foot (ft)	Pound-mass (lb_m)	Second (s)

Mass, length, and time are known as the *fundamental* units of the systems. We shall later use these and other *derived* units that are based on these fundamental units.

1-5
CONVERSION FACTORS

Although we shall concentrate on the MKSA system, it is sometimes necessary (especially during this transition period of mixed units) to convert units from one system to another. A brief list of some common *conversion factors* is provided in Table 1-2. Note that the unit abbreviations for the conversion factors are also expressed in ratio form. The ratio form is obtained by dividing *both* sides of the equation by the *same* unit. Multiplying or dividing a quantity by a conversion ratio, therefore, *does not* change that quantity because the ratio represents a factor of *1*. (See Example 1-1.)

EXAMPLE 1-1

Convert 3.67 feet to meters.

SOLUTION

Since 1 foot = 0.3048 m, then $\frac{0.3048 \text{ m}}{1 \text{ ft}} = 1$ (see Table 1-2).

$$\begin{aligned} 3.67 \text{ ft} \times 1 &= 3.67 \not{\text{ft}} \times \frac{0.3048 \text{ m}}{1 \not{\text{ft}}} \\ &= 3.67 \times 0.3048 \text{ m} \\ &= 1.118616 \text{ m (by calculator)} \\ &= \mathbf{1.12 \text{ m}} \text{ (rounded to 3 significant figures)} \end{aligned}$$

Note that the units of feet in Example 1-1 cancel leaving the desired units of meters.

TABLE 1-2
Common Conversion Factors, in Equation and Ratio Form

1 inch = 2.54 centimeters; $\frac{2.54\ \text{cm}}{1\ \text{in.}}$

1 foot = 0.3048 meter; $\frac{0.3048\ \text{m}}{1\ \text{ft}}$

1 mile = 1.609 kilometers; $\frac{1.609\ \text{km}}{1\ \text{mi}}$

1 gallon = 3.785 liters; $\frac{3.785\ \text{l}}{1\ \text{gal}}$

1 ounce = 28.35 grams; $\frac{28.35\ \text{g}}{1\ \text{oz}}$

1 pound (mass) = 0.4536 kilogram; $\frac{0.4536\ \text{kg}}{1\ \text{lb}_m}$

1 pound (force) = 4.45 newton; $\frac{4.45\ \text{N}}{1\ \text{lb}_f}$

1 horsepower = 746 watts; $\frac{746\ \text{W}}{1\ \text{hp}}$

Alternatively, some calculators allow conversions to be made directly by using a coded memory system, sometimes given on the back of the calculator. For the SR-51A, the code to use to convert feet to meters is 02:

Enter	*Press*	*Display*
	C	0
3.67		3.67
	2nd	3.67
	02 (Code)	1.118616

Answer: 3.67 ft = 1.12 m

1-6 UNIT PREFIXES

Multiples and submultiples of a basic unit can be indicated by attaching a *prefix* to the *unit,* as shown by the list of prefixes in Table 1-3. Although they also apply to the fundamental units, they are particularly well suited to the electrical units we shall meet soon. In this way any quantity that is much larger or much smaller than the basic unit may be referred to in a compact form that involves the name of the basic unit.

TABLE 1-3
SI Unit Prefixes

Value	Scientific Notation	Prefix	Symbol
1,000,000,000,000	10^{12}	tera	T
1,000,000,000	10^{9}	giga	G
1,000,000	10^{6}	mega	M
1,000	10^{3}	kilo	k
0.001	10^{-3}	milli	m
0.000,001	10^{-6}	micro	μ
0.000,000,001	10^{-9}	nano	n
0.000,000,000,001	10^{-12}	pico	p
	10^{-15}	femto	f
	10^{-18}	atto	a

For example, 1000 meters can be written as 1×10^3 m or 1 kilometer, that is, 1 km. A millionth of a second, 0.000,001 second, can be written as 1×10^{-6} s or 1 μs.

Note in particular the importance of distinguishing between the lower case letter m for milli and the captial letter M for mega.

EXAMPLE 1-2

Convert 0.0005 seconds to:

a. milliseconds (ms)
b. microseconds (μs)

SOLUTION

a. $0.0005\ \text{s} = 0.5 \times 10^{-3}\ \text{s} = \mathbf{0.5\ ms}$
b. $0.0005\ \text{s} = 500 \times 10^{-6}\ \text{s} = \mathbf{500\ \mu s}$

EXAMPLE 1-3

Convert 0.035 nanoseconds to:

a. picoseconds (ps)
b. microseconds (μs)

SOLUTION

a. $0.035\ \text{ns} = 0.035 \times 10^{-9}\ \text{s} = 35 \times 10^{-12}\ \text{s} = \mathbf{35\ ps}$
b. $0.035\ \text{ns} = 0.035 \times 10^{-9}\ \text{s} = 0.000{,}035 \times 10^{-6}\ \text{s} = 0.000{,}035\ \mu\text{s} = \mathbf{3.5 \times 10^{-5}\ \mu s}$

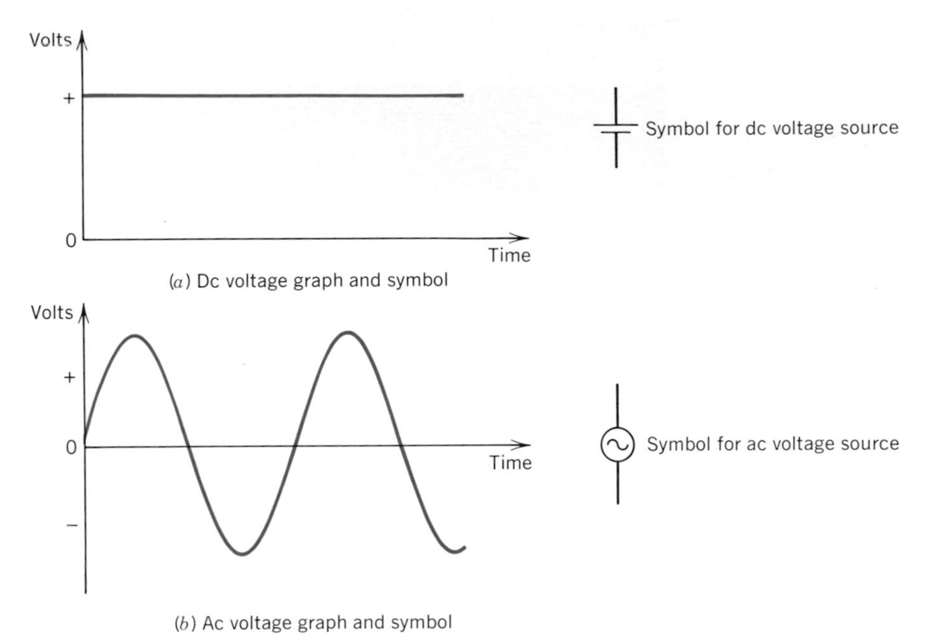

FIGURE 1-2
Graphs showing how voltage varies with time for dc and ac voltages.

1-7 A SIMPLE ELECTRIC CIRCUIT

An electric circuit consists of interconnected electrical components forming a complete path for the flow of electric current. The three basic *passive* electrical circuit *components* are resistors (R), inductors (L), and capacitors (C). But circuits also include electrical *devices* such as buzzers, lamps, relays, transformers, and motors.* Generators and batteries are known as *active* devices. In general, an *electric* "power" circuit involves the transfer of relatively large amounts of energy to produce heat, light, motion, and so on. An *electronic* circuit, in addition to the three basic passive components already mentioned, involves additional active devices such as semiconductors (transistors) and electron tubes.

Most electric circuits have four main parts:

1. A *source* of electric energy such as a chemical battery, generator, or solar cell.
2. A *load* or output device such as a lamp, motor, or loudspeaker.
3. *Conductors* such as copper or aluminum wire, to transport the electrical energy from the source to the load.
4. A *control* device such as a switch, thermostat, or relay, to control the flow of energy to the load.

The energy source may be either dc (direct current—unidirectional flow and constant in value) or ac (alternating current—continuously changing in value and periodically reversing in polarity). The source applies an *electromotive force* (emf) or *potential difference* to the circuit. See Fig. 1-2 for graphical representations of these two sources along with symbols used to represent them in a circuit.

The emf is measured in *volts* (V) and is *related to the work* that a source can do to move electrical *charge* through a circuit. This flow of charge is called *current* and is measured in *amperes* (A). The typical domestic supply voltage in the United States and Canada is 120/240 volts ac at a frequency of 60 hertz. 1 Hz = 1 cycle per second (cps). (See Section 2-4 for a definition of potential; see Section 2-8 for a definition of current.)

* It will later be shown that all passive devices (those devices that do not amplify or do not provide a source of energy) can be reduced to circuit component combinations of R, L, and C.

To illustrate an electric circuit, let us examine briefly a simple *series* dc circuit, namely, the commercial flashlight. To represent a circuit, a *pictorial diagram* may be used which is similar to the physical appearance of the components. However, a method that is preferred by engineers, electricians and technicians is to use a *schematic diagram.* This consists of interconnected symbols, which are drawings that *represent* electrical components. Schematic diagrams are much easier to draw than pictorial diagrams. Both are shown in Fig. 1-3. Note that since symbols are used in Fig. 1-3*b*, no labeling is needed.

Note also that the *long* side of the battery symbol (Fig. 1-3*b*) always represents the *positive* terminal and the *short* side is the *negative* terminal, so the plus and minus signs are optional.

A series circuit is defined as one in which there is only one path for current (to flow). (A parallel circuit has two or more paths.) In Fig. 1-3, the source consists of two series-connected dry cells, each having an emf of $1\frac{1}{2}$ V to supply 3 V to the circuit. A 3-V lamp is the load and a knife-blade switch is connected between the source and load. The conductors in this example are provided by the metallic case of the flashlight body holding the lamphousing and the cells (or batteries).

When the switch is open (OFF), as it is in Fig. 1-3, no current flows and the lamp is unlit. When the switch is closed (ON), a complete current path exists and electrical current flows from the cells to light the lamp. The flow of current in the high resistance lamp filament heats it to a white hot incandescence. The lamp converts electrical energy to light as well as heat energy.

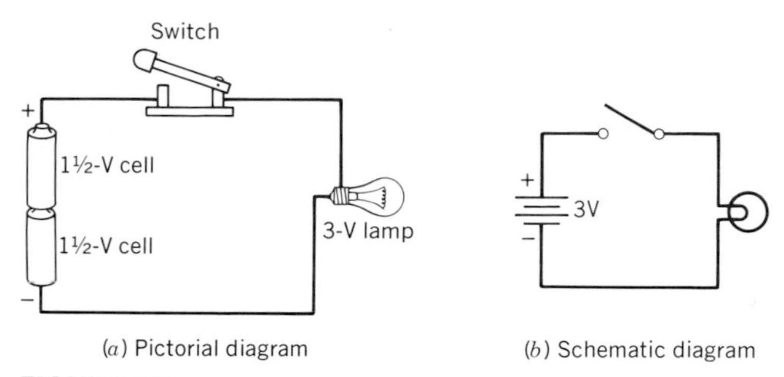

FIGURE 1-3
Simple electric circuit.

1-8 ELECTRICAL SYMBOLS FOR RESISTANCE, INDUCTANCE AND CAPACITANCE, AND ELECTRICAL DEVICES

Table 1-4 shows the symbols for the three basic passive components—resistors, inductors, and capacitors—used in electrical and electronic schematic diagrams. Photo-

TABLE 1-4
Schematic Symbols for Resistors, Inductors, and Capacitors

Type	Symbol	Property	Unit
Fixed resistor	*R*	Limits or opposes current	Ohm
Variable resistor	*R*		
Fixed inductor or coil (air core)	*L*	Opposes *change* of current	Henry
Iron core inductor	*L*		
Variable inductor	*L*		
Fixed capacitor	*C*	Opposes *change* of voltage	Farad
Polarized capacitor	*C*		
Variable capacitor	*C*		

FIGURE 1-4
Typical resistors, inductors, and capacitors. (Components are mounted on bases for breadboarding in a circuit.)

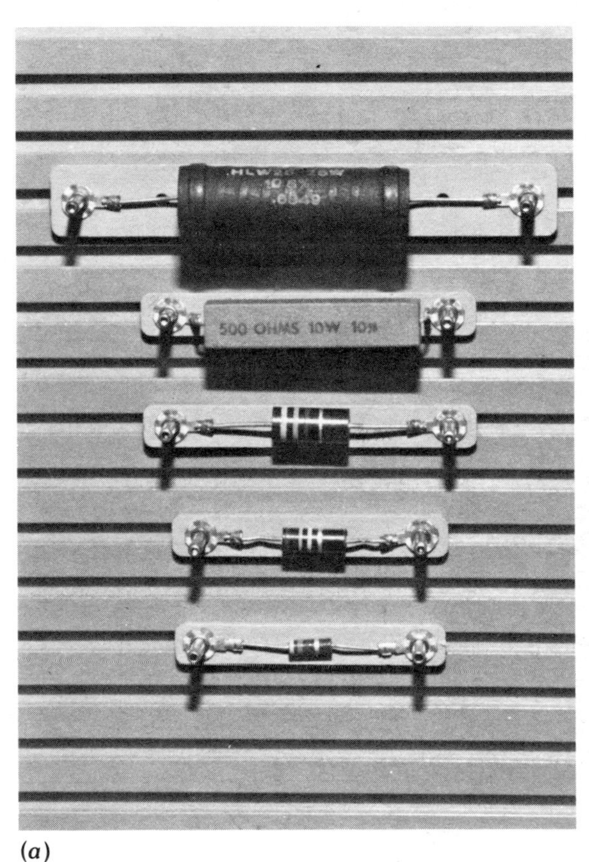

(*a*)

(*b*)

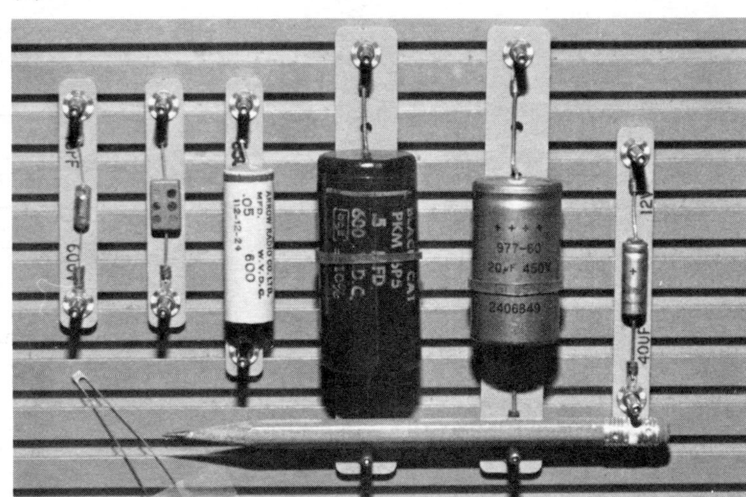

(*c*)

graphs of some representative types of these components are shown in Fig. 1-4. Refer to Appendix A-1 for a more complete list of electrical and electronic components and devices.

We shall see (Section 2-11) that electric current encounters opposition (or resistance) as it flows in any circuit. In some cases, it is desirable to intentionally insert resistance into a circuit for the purpose of *limiting* the current. Resistance is symbolized by R; its unit is the ohm, abbreviated by the Greek letter omega (Ω). Resistors of a wide variety of standard values are commercially available. Their values are indicated in some cases by a series of colored stripes around the body of the resistor (see Section 3-14). Typical values are $R = 220\ \Omega$, $4.7\ \text{k}\Omega$, and $1\ \text{M}\Omega$.

An inductor is a coil of wire that may be wound on an iron core. An inductor has the property of opposing any *change* of current through it. This property is called inductance and is symbolized by L. The basic unit of inductance is the henry (H). Typical measurement values are $L = 100\ \mu\text{H}$, 30 mH, and 7 H.

Capacitors are essentially conductors separated by an insulating material or *dielectric*. Capacitance (C) is the property to store electrical charge and oppose any change in voltage. Capacitance is measured in the basic unit of farads (F). More practical values of measurement are microfarads (μF) and picofarads (pF). Typical values are $C = 270$ pF, 0.1 μF, and 100 μF.

1-9 PICTORIALS AND SCHEMATICS

Figure 1-5 shows a pictorial wiring diagram for a power supply and the corresponding schematic.

There are a number of points to observe:

1. Note how terminal strips are used for solder tie-points and are shown in the pictorial but *do not* appear in the schematic.
2. The connection dots in the schematic *do not* necessarily coincide with tie-points in the pictorial.
3. The schematic is drawn with straight connecting lines either horizontal or vertical. They bear little resemblance in location to the wires in the actual circuit.

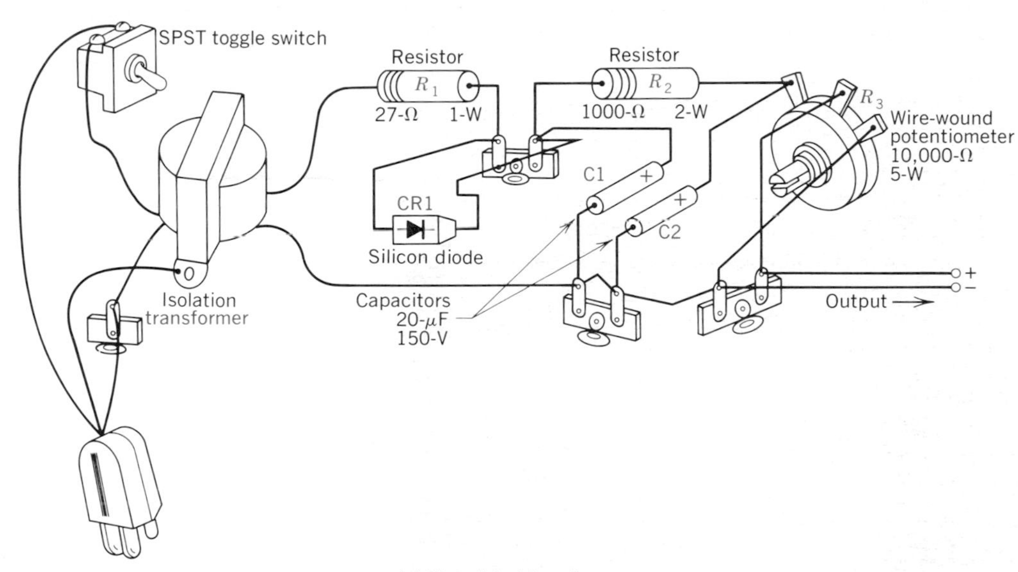

(*a*) Pictorial wiring diagram

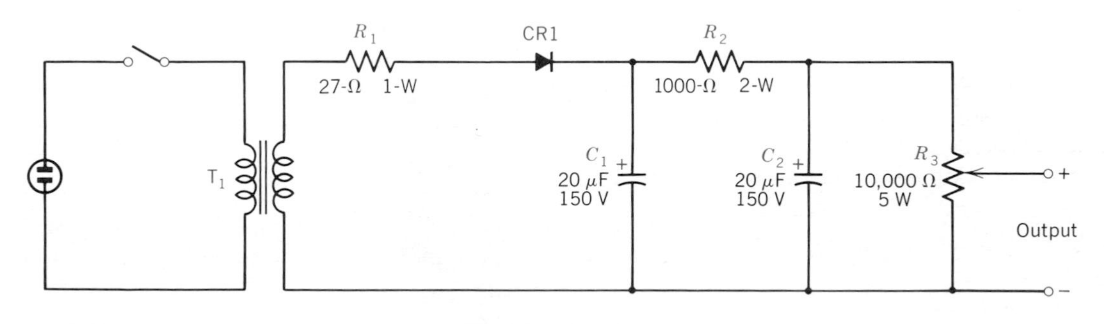

(*b*) Schematic diagram

FIGURE 1-5
Pictorial and schematic diagrams of a power supply. (Courtesy of McGraw-Hill.)

4 The schematic usually "proceeds" from left to right following the current flow.
5 The schematic identifies each component by a letter and number, such as R_1, R_2, C_1, and so on, for reference purposes.
6 Values of the components are shown on the schematic, but no other labeling is used.

It should be obvious that the pictorial diagram is used for component recognition and wiring purposes. The schematic is used to show how a circuit works and to trace signals through a piece of equipment during troubleshooting. In fact, most usually a schematic is the only diagram available. Where a schematic is not available, it is often wise to trace the actual circuit and make your own schematic for troubleshooting purposes. (Practice in doing this is given at the end of the chapter.)

It is also possible to use a schematic to wire a circuit without the use of a pictorial wiring diagram. In the case of laboratory "breadboarding," this is most easily accomplished by laying out the components on the circuit board in the same relative locations as shown in the schematic. As each wire connection is made, that line should be lightly scribbled out or checked off on the schematic to keep track of your work.

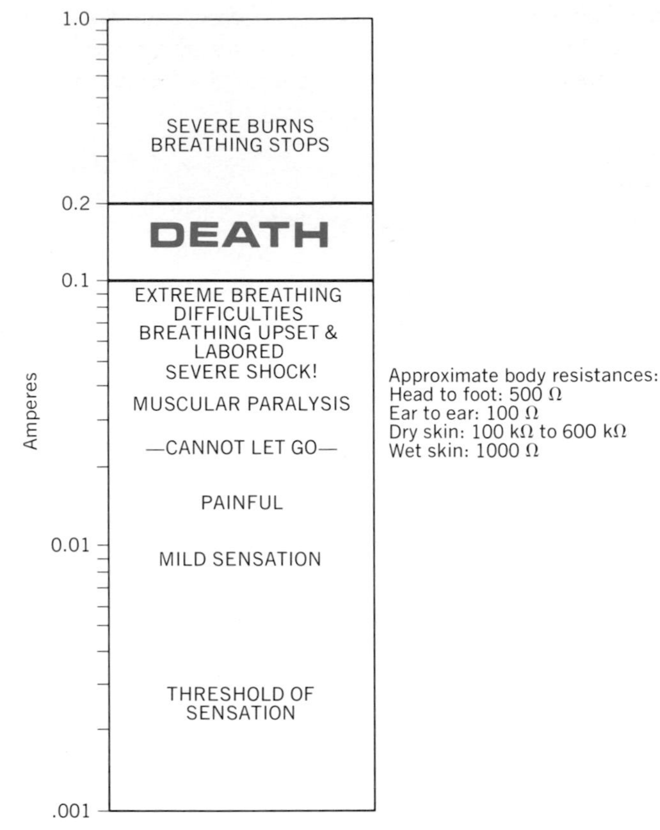

FIGURE 1-6
Diagram showing physiological effects of electric currents. (Courtesy of Graymark International, Inc., Irvine, Calif.)

1-10 SAFETY

One of the reasons for introducing the basic electric circuit and the electrical units that describe emf, current, and resistance is to acquaint you with some precautions in working with electricity.

If you touch a high-voltage source (either ac or dc), your body resistance may be low enough to allow a lethal amount of current to flow through your body. Actually, the amount of current that affects your heart and causes it to stop pumping at its regular rate (a problem called ventricular fibrillation) is in the order of only a few milliamperes. It is, of course, difficult to get exact values, and figures vary widely, but Fig. 1-6 gives you some idea of the effects of current on the body.

If you are in a wet location, or are in contact with a grounded metal water pipe, the domestic supply voltage of 120 V is easily sufficient to kill. The following are some precautions:

1 **Do not work on live circuits if at all possible.**
2 **If you are working on a high-voltage circuit (100 V and up), use only one hand to make a voltage measurement; keep the other hand in your pocket or behind you to avoid a complete path for current to flow through your heart and lungs. (This, however, does not prevent the possibility of being burned on the active hand.)**
3 **Remove rings, wrist watch, and any other metal jewelry that might have a chance of making contact with a conductor.**
4 **Remove the charge from high-voltage electrolytic capacitors by shorting them with a screwdriver. They can hold a charge for hours after the power has been turned off.**

5 Certain components, such a resistors, vacuum tubes, and power transistors, get quite hot. Let them cool before removing them.
6 Do not work with frayed or burned power cords.
7 Know the location of an available fire extinguisher (designed for electrical fires) and know how to use it.
8 Treat all voltages with respect and concentrate on what you are doing.

Fortunately, most circuits involving transistors and integrated circuits seldom operate at voltages in excess of ±15 V. But when these voltages are obtained from a 120-V supply (or higher), precautions are always in order.

1-11 READING SCALES OF MULTIRANGE METERS

1-11.1 Types of Meters

Although digital meters have become quite common, it is very likely that you will probably use deflection (or analogue) meters at some time. Your first experiments will probably involve reading a voltmeter, ammeter, and ohmmeter. All of these readings may be made on a *multimeter*. This is a meter in which a *function* switch selects the mode of operation, and a *range* switch is used to provide a readable deflection for a wide range of measured quantities. Figure 1-7 shows a multimeter in which the selector and range switches have been combined into one switch. This type of meter is known as a volt-ohm-milliammeter, or VOM. It is portable and can measure ac and dc volts, resistance, and dc currents.

Another common multirange instrument is the electronic or vacuum-tube voltmeter, the EVM or VTVM. Although solid-state versions are available, these voltmeters are generally powered from a 120-V outlet. The EVMs can measure ac and dc volts, and resistance. Some solid-state instruments can also measure ac and dc current. See Fig. 1-8.

Also available is the multirange milliammeter, an example of which is shown in Fig. 1-9. These generally have lower resistance than the milliammeter portion of a multimeter.

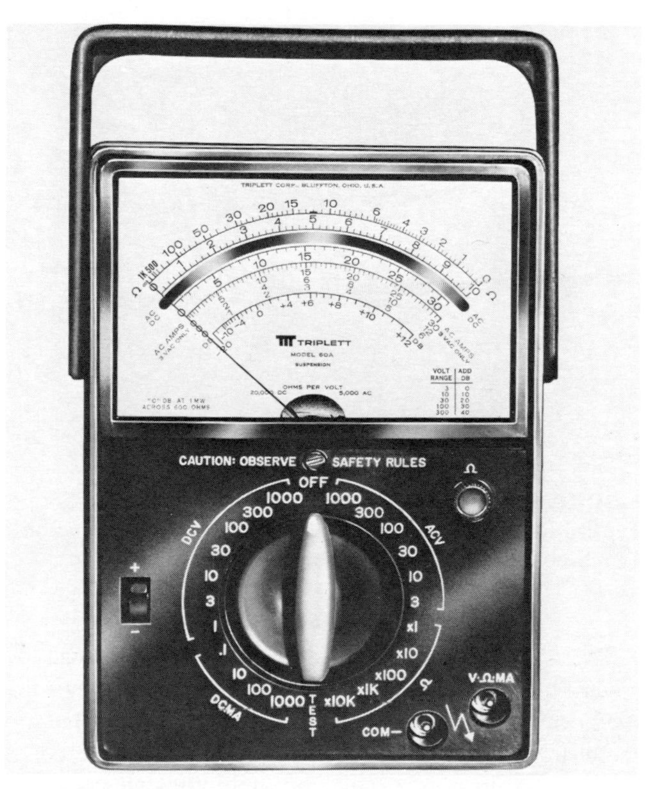

FIGURE 1-7
Typical VOM. (Courtesy of Triplett Corporation.)

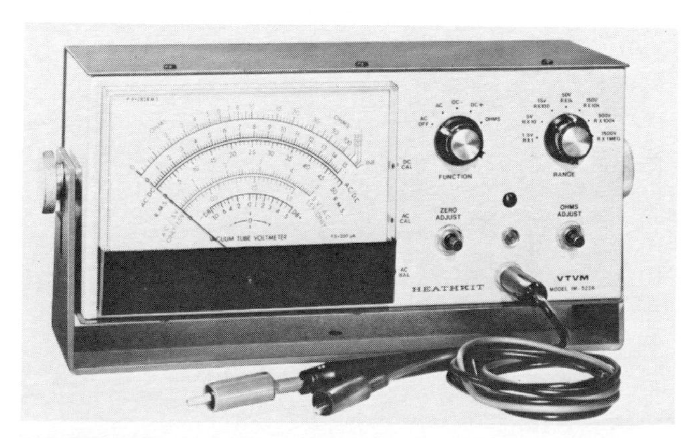

FIGURE 1-8
Typical electronic voltmeter or VTVM. (Courtesy of Heath Company)

FIGURE 1-9
Typical multirange dc milliammeter. (Courtesy of Hickok Teaching Systems, Inc., Woburn, Mass.)

The detailed operation and use of meters will be covered in later chapters. However, Fig. 1-10 shows the method of connecting a voltmeter, ammeter, and ohmmeter in a simple circuit.

Note that in a dc circuit, the ammeter and voltmeter must be connected observing polarity. The ohmmeter can be used to measure resistance in a circuit only if the voltage is first removed from the circuit.

1-11.2 Reading the Scales

When using a multirange ammeter or voltmeter, it is important to know the function of the range switch. The range switch selects the amount of current or voltage that causes full-scale deflection (FSD) of the meter. When measuring an unknown quantity, it is wise to start out on the highest range and reduce to a lower range until you get a deflection somewhere between mid- and full scale if possible.

To avoid placing as many scales on the meter as there are ranges, some scales are used for many different ranges simply by multiplying or dividing the scale numbers by 10 or 100, for instance. To read a multirange meter, find the range being used on the range switch and determine which scale has a full-scale deflection most closely corresponding to the range. Then read off the number from the scale where the pointer has come to rest.

The following examples demonstrate the method.

EXAMPLE 1-4

Determine the readings of the multirange milliammeter in Fig. 1-11 when the range switch is on:

a. 50 mA
b. 100 mA

SOLUTION

a. With the range switch on 50 mA, the FSD must be 50 mA. Therefore, use the 0–5 scale with each number multiplied by 10.

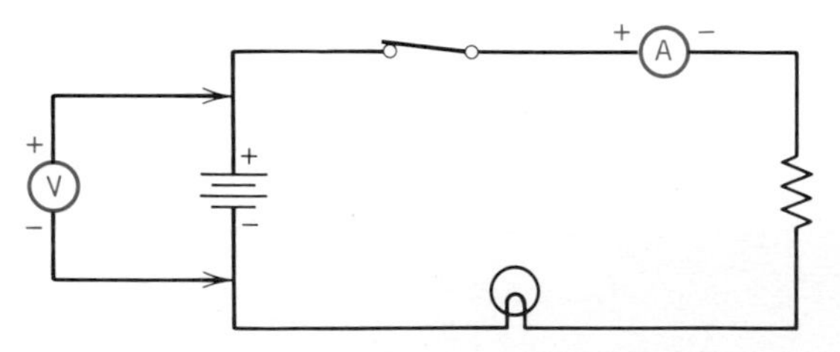

(*a*) Connection of voltmeter in parallel with the source. The ammeter is in series with the load.

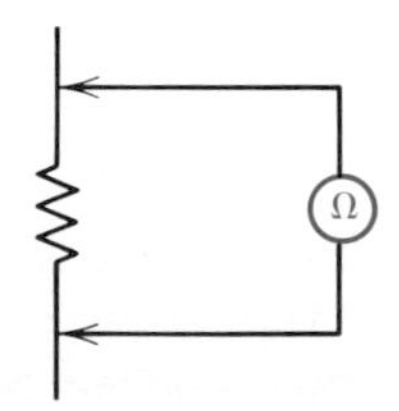

(*b*) The ohmmeter is connected across the component with no applied voltage

FIGURE 1-10
Use of voltmeter, ammeter and ohmmeter to measure voltage, current, and resistance.

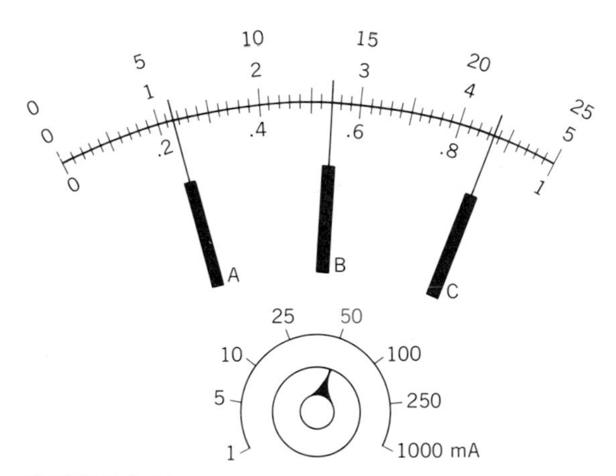

FIGURE 1-11
Multirange dc milliammeter scale with range switch on 50 mA.

At A: reading is between 10 and 20 = **11.5 mA**
At B: reading is between 20 and 30 = **27 mA**
At C: reading is between 40 and 50 = **43.5 mA**

b. With the range switch on 100 mA, the FSD must be 100 mA. Therefore, use the 0–1.0 scale with the decimal point moved to the right two places.

At A: reading is between 20 and 40 = **23 mA**
At B: reading is between 40 and 60 = **54 mA**
At C: reading is between 80 and 100 = **87 mA**

NOTE The only direct reading scales are those corresponding to ranges of 1, 5, and 25 mA.

EXAMPLE 1-5

Determine the readings of the multirange voltmeter in Fig. 1-12 for the following settings:

a. Function: dc; Range: 5 V
b. Function: ac; Range 150 V

SOLUTION

a. With the range switch on 5 V dc, the FSD must be 5 V on the ac/dc scale. Therefore, we must use the 0–50 scale with each number divided by 10.

At A: reading is between 0.5 and 1.0 = **0.72 V**
At B: reading is between 2.0 and 2.5 = **2.37 V**
At C: reading is between 4.0 and 4.5 = **4.30 V**

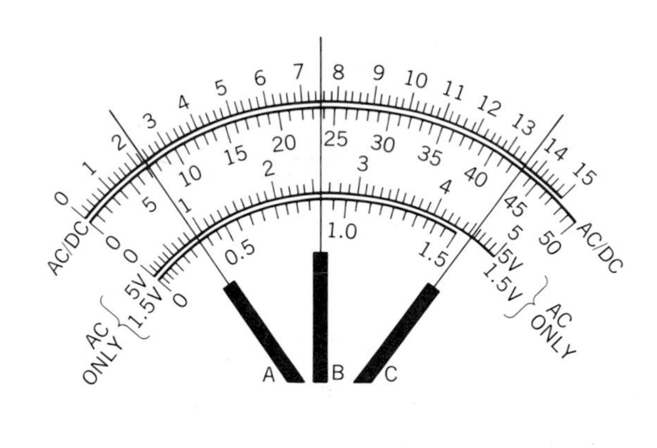

FIGURE 1-12
Multirange ac/dc voltmeter scales with range and function switches set to 5 V dc.

b. With the range switch on 150 V ac, the FSD must be 150 V on the ac/dc scale. Therefore, we must use the 0–15 scale with each number multiplied by 10.

At A: reading is between 20 and 30 = **23 V**
At B: reading is between 70 and 80 = **75 V**
At C: reading is between 130 and 140 = **136 V**

NOTE The direct reading 1.5-V and 5-V scales may be used for ac voltages only.

An ohmmeter scale always has zero at one end and infinity at the other. The scale shown in Fig. 1-13 is the type found in most EVMs while most VOMs will have the zero on the right and infinity at the left. In both cases, the value read from the scale is multiplied by the range factor, $\times 1$, $\times 100$, and so on.

EXAMPLE 1-6

Determine the readings of the multirange ohmmeter in Fig. 1-13 on the following ranges:

a. ×100

b. ×100 k

SOLUTION

a. At A: reading = 1.7 × 100 = **170 Ω**
At B: reading = 7.5 × 100 = **750 Ω**
At C: reading = 35 × 100 = **3.5 kΩ**

b. At A: reading = 1.7 × 100 k = **170 kΩ**
At B: reading = 7.5 × 100 k = **750 kΩ**
At C: reading = 35 × 100 k = **3.5 MΩ**

NOTE Multirange VOMs and EVMs may have all or some combination of the above three sets of scales included on the face of the meter. Great care must be taken to determine which scale and set of numbers must be used according to the settings of the function and range switches.

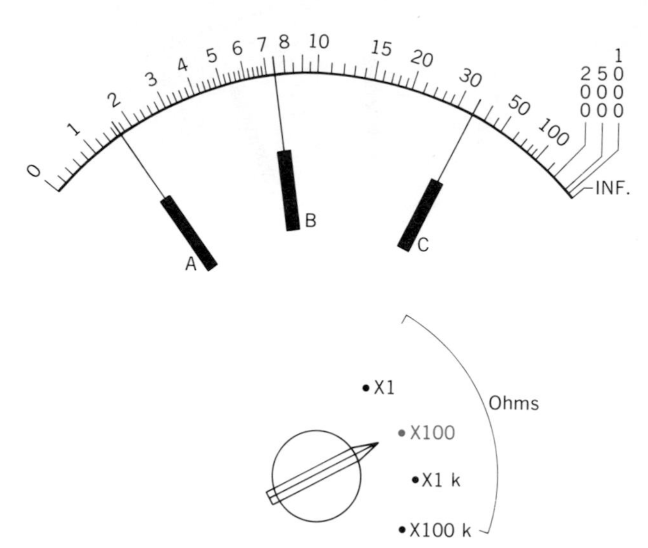

FIGURE 1-13
Multirange ohmmeter set on ×100 range.

SUMMARY

1. Low-cost calculators are limited in the size of numbers they can process to the display capability of the calculator.

2. A scientific calculator uses scientific notation. This requires writing any given number, regardless of size, as a number between 1 and 10 and multiplied by 10 raised to some suitable positive or negative exponent.

3. A calculated result is only as accurate as the least-accurate information fed into the calculator and should be rounded-off accordingly.

4. English or customary units may be converted to the MKS system by multiplying by a conversion ratio given in Table 1-2.

5. Multiples and submultiples of a basic unit can be indicated by attaching a prefix to the unit as listed in Table 1-3.

6. An electric circuit consists of interconnected components forming a complete path for the flow of electric current.

7. An electric circuit consists of a source, load, conductors, and some control device.

8. Symbols for the three basic passive components—resistors, inductors, and capacitors—are listed in Table 1-4.

9. A pictorial diagram is similar to the physical appearance of the components; a schematic diagram uses symbols.

10. Only a small current (a few milliamperes) through the body may be lethal. Follow safety precautions and treat all voltages with respect.

11. Reading a multirange meter involves identifying the function, range, and appropriate scale.

12. For a voltmeter or ammeter, the range switch selects the amount of voltage or current that causes full-scale deflection of the meter.

13. When reading an ohmmeter, the resistance read from the ohms scale must be multiplied by the range setting.

SELF-EXAMINATION

Answer true or false
(Answers at back of book)

1-1. The scientific notation for 1215 is 1.215×10^3. ________

1-2. The scientific notation for 0.0794 is 7.94×10^{-1}. ________

1-3. The standard decimal form for 8.98×10^2 is 898. ________

1-4. The standard decimal form for 3.64×10^{-3} is 0.00364. ________

1-5. The number in question 1 is accurate to four significant figures. ________

1-6. The number in question 2 is accurate to four significant figures. ________

1-7. If 0.494 is rounded-off to two significant figures, the result is 0.49. ________

1-8. If 0.494 is rounded-off to one significant figure, the result is 0.5. ________

1-9. A conversion ratio may be used to convert a quantity given in the FPS system to the MKS system. ________

1-10. One millisecond is one millionth of a second. ________

1-11. A time interval of 27 μs equals 0.027 ms. ________

1-12. A time interval of 300 ps equals 3 ns. ________

1-13. A schematic diagram uses pictures of the components, but a pictorial diagram uses symbols. ________

1-14. Symbols are graphic drawings that represent electrical components. ________

1-15. When a switch is opened in a circuit, the current stops flowing. ________

1-16. A resistor opposes current whereas an inductor opposes a change of current. ________

1-17. The unit of resistance is the henry, and for inductance it is the ohm. ________

1-18. A capacitor opposes change of voltage; its unit of measurement is the farad. ________

1-19. Only a few milliamperes of current through the body is sufficient to be painful. ________

1-20. A multimeter may be of the VOM type that can measure voltage, current, and resistance. ________

1-21. An EVM is portable, but a VOM needs a 120-V ac source. ________

1-22. A range switch in a multirange voltmeter or ammeter selects the quantity required to cause full-scale deflection. ________

1-23. When using an ohmmeter, it is necessary to multiply the ohms scale reading by the range setting. ________

1-24. If, in Fig. 1-11, the range switch is on 1 mA, the current indicated by the pointer at position C is 8.7 mA. ________

1-25. If, in Fig. 1-12, the voltmeter is set to 5 V ac, the voltage indicated by the pointer at position C is 4.7 V. ________

1-26. If, in Fig. 1-13, the ohmmeter range is set to $\times$ 1k, the resistance indicated by the pointer at position C is 350 kΩ. ________

1-27. An ammeter is always connected in series with the component whose current it is going to 'measure. ________

1-28. A voltmeter is always connected in parallel with the component in order to measure the voltage across the component. ________

REVIEW QUESTIONS

1. What is the largest number that a 10-digit read out calculator can display if it does not use scientific notation?

2. What are the largest and smallest (nonzero) numbers that a 10-digit read out calculator can display if it does use scientific notation?

3. Why is scientific notation useful?

4. What does it mean when a given calculator display blinks repeatedly or, on another calculator, decimal points appear after each number?

5. What is one way of converting a decimal number to scientific notation on some calculators?

6. What is the rule for rounding off a calculated number?

7. What are the three systems for measuring physical units?

8. What conversion factor would you use to convert inches to centimeters?

9. Why does multiplying a quantity by a conversion ratio not change that quantity?

10. What do the symbols T, M, m, μ, p stand for when used as unit prefixes?

11. Draw pictorial and schematic diagrams of two electric circuits and indicate the four main parts in each.

12. What is the main difference between dc and ac voltage?

13. Which passive electrical component has the property of opposing any change of current?

14. What is the main use of a pictorial diagram?

15. Give at least three differences between a schematic and a pictorial.

16. Describe how you would "breadboard" a circuit from a schematic diagram and how you would check it.

17. List the important safety precautions in working with electricity.

18. In using a multirange meter, what range should be used when making a measurement of an unknown voltage or current?

19. What precaution *must* be observed when using an ohmmeter to measure resistance in a circuit?

20. Explain the relation between the ranges on a range switch and the various scales on the meter face.

21. What is the main difference between a VOM and an EVM?

22. What other difference is there between a VOM and an EVM as far as the resistance scales are concerned?

23. Which of the two meters in Question 21 generally measures current?

24. What does it mean to say that polarity must be observed when dc voltmeters and ammeters are used to make voltage and current measurements?

25. In what way (series or parallel) are voltmeters and ammeters connected in a circuit?

PROBLEMS

(Answers to odd-numbered problems at back of book)

1-1. Convert to scientific notation:

a. 571,000

b. 0.0000645

c. 2306

1-2. Convert to scientific notation:

a. 49,960

b. 0.007825

c. 1,100,000

1-3. Round-off the numbers in Problem 1-1 to two significant figures.

1-4. Round-off the numbers in Problem 1-2 to three significant figures.

1-5. Round-off the numbers in Problem 1-1 to one significant figure.

1-6. Round-off the numbers in Problem 1-2 to two significant figures.

1-7. Express the following in standard decimal form:

a. 4.79×10^{5}

b. 1.135×10^{-3}

c. 9.079×10^{2}

1-8. Express the following in standard decimal form:

a. 3.57×10^{4}

b. 2.07×10^{-4}

c. 8.064×10^{2}

1-9. Carry out the following computations using a calculator, expressing the answers in scientific notation and rounded-off to show the appropriate significant figures:

a. $(1.357 \times 10^{4}) \times (7.25 \times 10^{-3})$

b. $\dfrac{1.357 \times 10^{4}}{7.25 \times 10^{-3}}$

c. $\dfrac{\sqrt{1.357 \times 10^{4}}}{(7.25 \times 10^{-3})^{2}}$

1-10. Repeat Problem 1-9 for the following:

a. $(8.079 \times 10^{2}) \times (3.574 \times 10^{-7})$

b. $\dfrac{3.574 \times 10^{-7}}{8.079 \times 10^{2}}$

c. $\dfrac{\sqrt{3.574 \times 10^{-7}}}{(8.079 \times 10^{2})^{3}}$

d. $44.3 \times 10^{2} \times 10^{3}$

e. $\dfrac{927 \times 10^{-4} \times 10^{1}}{10^{6} \times 10^{-1}}$

1-11. Using the conversion factors in Table 1-2 (or conversion code on your calculator if available), make the following conversions:

a. 13.5 in. to cm
b. 2.89 ft to m
c. 125 mi to km
d. 8.75 gal to l

1-12. Repeat Problem 1-11 for the following:

a. 32 oz to g
b. 165 lb_m to kg
c. 150 lb_f to N
d. 75 hp to kW

1-13. a. Convert 50 mA to A.
b. Convert 50 mA to μA.

1-14. a. Convert 0.15 A to mA.
b. Convert 0.15 A to μA.

1-15. a. Convert 2500 V to kV.
b. Convert 0.35 V to mV.

1-16. a. Convert 465 kV to V.
b. Convert 50 mV to V.

1-17. a. Convert 1200 Ω to kΩ.
b. Convert 0.75 Ω to mΩ.
c. Convert 1500 kΩ to MΩ.

1-18. a. Convert 5.1 kΩ to Ω.
b. Convert 270 kΩ to MΩ.
c. Convert 2.2 MΩ to kΩ.

1-19. a. Convert 50 ns to ps.
b. Convert 50 ns to μs.

1-20. a. Convert 50,000 μF to F.
b. Convert 30 mH to H.
c. Convert 1500 pF to μF.

1-21. a. Convert 213 ps to ns.
b. Convert 0.001 F to μF.
c. Convert 0.5 mH to μH.

1-22. Given the pictorial diagram of the one-transistor radio in Fig. 1-14, draw the corresponding schematic. Refer to Appendix A-1 for symbols.

1-23. Determine the readings of the multirange milliammeter in Fig. 1-11 when the range switch is on:

a. 1 mA
b. 25 mA
c. 1000 mA

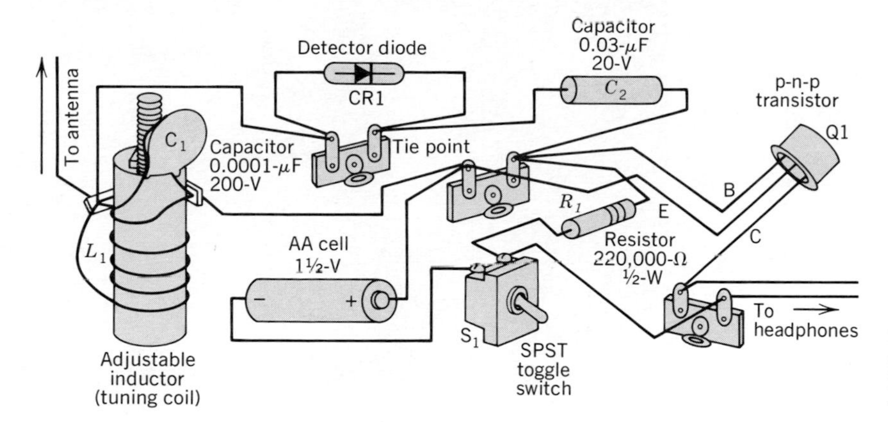

FIGURE 1-14
Pictorial of a transistor radio for Problem 1-22. (Courtesy of McGraw-Hill.)

1-24. Repeat Problem 1-23 for the following:

a. 5 mA
b. 10 mA
c. 250 mA

1-25. Determine the readings of the multirange voltmeter in Fig. 1-12 for the following settings:

a. 1.5 V, ac
b. 1.5 V, dc
c. 50 V, dc
d. 150 V, ac

1-26. Repeat Problem 1-25 for the following:

a. 5 V, ac
b. 15 V, dc
c. 500 V, ac
d. 1500 V, dc

1-27. Determine the readings of the multirange ohmmeter in Fig. 1-13 on the following ranges:

a. ×1
b. ×1 k

1-28. What range (and function) settings would you use to make the following measurements on a multirange multimeter as in Figs. 1-11, 1-12, and 1-13?

a. 6-V battery
b. 120-V ac outlet
c. A current of 28 mA dc
d. 3.15 V ac
e. 10-ohm resistor
f. 8.2-kΩ resistor
g. 4.7-MΩ resistor

CHAPTER TWO

BASIC ELECTRICAL QUANTITIES

The word *electricity* is derived from the Greek word *elektron,* which means amber. When amber is rubbed with fur, the amber becomes capable of attracting lightweight particles such as very small pieces of paper. We now know that this *electrification by friction* is due to the transfer of electrons from the fur to the amber. When this happens, the two materials are said to be *oppositely* charged. In this condition, a *potential difference* is said to exist between the two materials. This potential difference provides an effect similar to that between the two electrodes of a battery. Often mistakenly referred to as a "pressure," this potential difference or electromotive force causes electrons to flow through a metallic conductor connected across the electrodes. This flow of electrons constitutes an electric current, which can be made to perform a useful function such as producing heat or light. In the case of a chemical cell or battery it is chemical action that maintains the potential difference and the flow of current.

In this chapter you will learn the basic units of electricity, including resistance—the opposition to the flow of current.

2-1 STATIC ELECTRICITY

Your first experience with static electricity may have been when, on a dry day, you walked across carpeting with rubber-soled shoes and touched a door handle. The resulting spark and mild shock was because your body became "electrified." You actually accumulated excess electrons, transferred from the carpet by friction. This is called *static electricity* because the charges are at rest.

Specifically, if amber, ebonite, or a hard-rubber rod is rubbed with wool or fur, it accumulates excess electrons at the expense of the fur. It was Benjamin Franklin, in the middle of the eighteenth century, who is credited with stating that the rubber rod is *negatively charged* and the fur *positively charged.* (Actually, not knowing anything about electrons, Franklin suggested the fur *gained* some charge thus becoming *positive,* and the rubber *lost* some charge, becoming *negative*.)

If, on the other hand, a glass rod is rubbed with silk, it is found that the glass rod is positively charged because electrons have been transferred to the silk, making it negative. These effects are summarized in Fig. 2-1 by the use of negative and positive signs placed around the two rods.

Note that the rubber rod, which was initially neutral or uncharged, is said to have a *surplus* of electrons, whereas the glass rod is said to have a *deficiency* of electrons.

That the two rods are in fact oppositely charged may be demonstrated by momentarily *touching* the two rods to two separate lightweight pith balls suspended by fine strings. As shown in Fig. 2-2*c* the pith balls become charged to the same *polarity* as the rods and *attract* each other because they are oppositely charged. If the *same* rubber rod is used to charge both pith balls (Fig. 2-2*a*), repulsion occurs. The same is true if the glass rod is used to charge both balls (Fig. 2-2*b*).

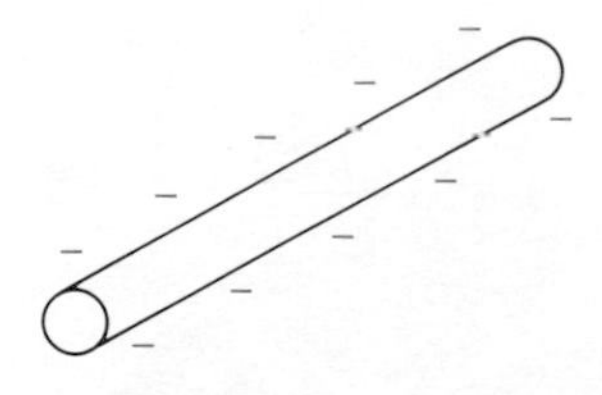

(*a*) Rubber rod rubbed with fur

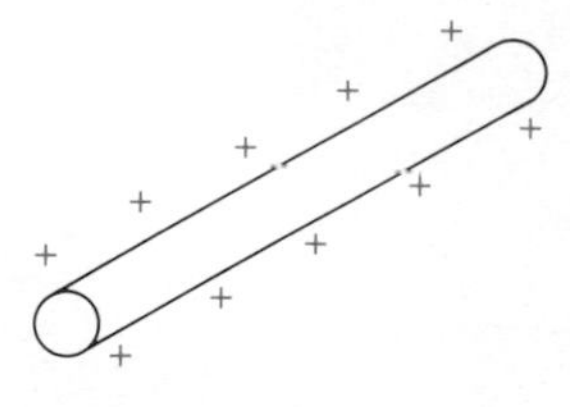

(*b*) Glass rod rubbed with silk

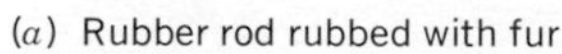

FIGURE 2-1
Two different rods oppositely charged by friction.

This is an important law of electric *forces:*

Like charges repel; unlike charges attract.

2-2 COULOMB'S LAW

In 1785, A French physicist, Charles Coulomb, determined how much force of attraction or repulsion is exerted between two charged bodies separated by a given distance. The force may be calculated using Coulomb's law:

$$F = k\frac{Q_1 Q_2}{r^2} \qquad \text{newtons (N)} \qquad (2\text{-}1)$$

where: F is the force in newtons* (N)

* One newton is the force required to give a mass of 1 kg an acceleration of 1 m/s².

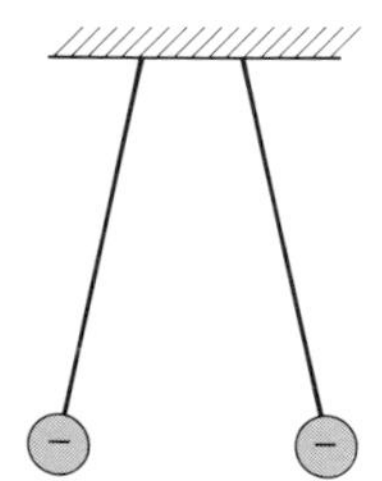

(*a*) Repulsion produced by rubber rod touching both balls

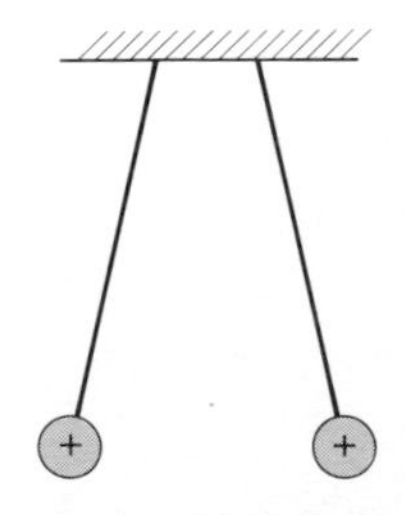

(*b*) Repulsion produced by glass rod touching both balls

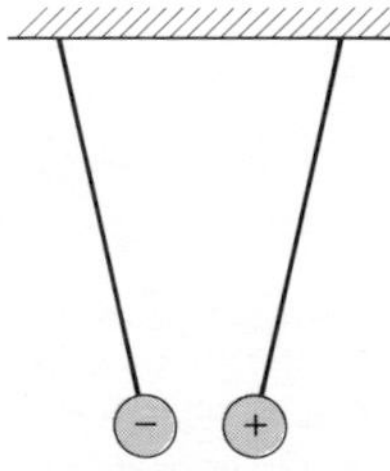

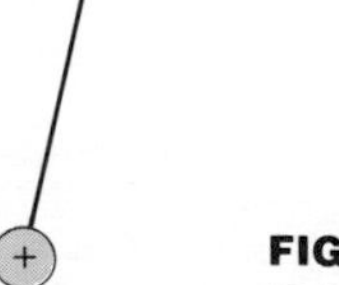

(*c*) Attraction produced by oppositely charged pith balls

FIGURE 2-2
Pith ball demonstration of like charges repelling, unlike charges attracting.

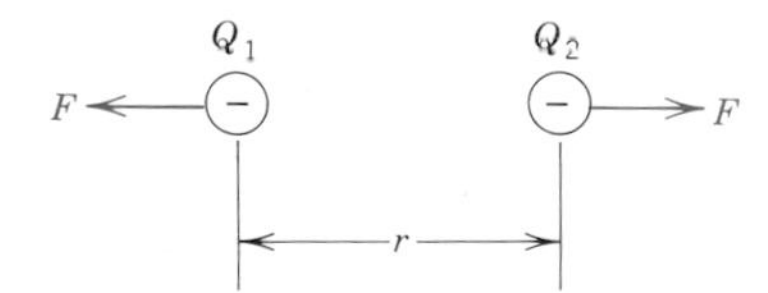

FIGURE 2-3
Quantities and forces involved in Coulomb's law.

Q_1, Q_2 are the charges of the two bodies in coulombs (C)

r is the separation between the two bodies in meters (m)

k is a constant, equal to 9×10^9 Nm²/C²

See Fig. 2-3.

EXAMPLE 2-1

Two pith balls, separated by 3 cm, are each charged positively with 0.25 μC.

a. Calculate the force of repulsion in newtons.

b. Convert the force to pounds.

SOLUTION

a. $$F = k\frac{Q_1 Q_2}{r^2} \qquad (2\text{-}1)$$

$$= 9 \times 10^9 \frac{\text{Nm}^2}{\text{C}^2} \times \frac{(+0.25 \times 10^{-6}\ \text{C}) \times (+0.25 \times 10^{-6}\ \text{C})}{(3 \times 10^{-2}\ \text{m})^2}$$

$$= \mathbf{0.625\ N}$$

b. $$F = 0.625\ \text{N} \times \frac{1\ \text{lb}}{4.45\ \text{N}}$$ (see the conversion factors in Table 1-2.)

$$= \mathbf{+0.14\ lb}\ \text{(repulsion)}$$

Note that all quantities must be expressed in *basic* units in the equation, regardless of the units in which they are given.

Note also that Eq. 2-1 involves the product of the two charges, so that if both are positive or both negative, F will be positive. A *positive* force means *repulsion*. If only one charge is negative, F will be negative. A *negative* force must be interpreted as a force of *attraction*.

The point of considering Coulomb's law at this time is to determine what units are used to measure charge. We have seen how charge depends upon an accumulation (or deficiency) of a number of electrons. It was decided, by experiments carried out by Andrew Millikan during the period 1909–1913, that the combined charge of 6.24×10^{18} electrons should be called 1 coulomb (1C).

We can now see why Q is used to represent charge. **A coulomb is a *quantity* of electrical charge equal to the combined charge of 6.24×10^{18} electrons.** Therefore, 6.24×10^{18} electrons = 1 coulomb, 1 C (of charge).

NOTE A coulomb does not have any mass. It is not a particle, but rather a number. It represents a *quantity* of charge. Note also that Q is the *letter symbol* for charge, but the coulomb, the unit in which charge is measured, is abbreviated C.

Evidently the charge of one electron (which is what Millikan actually determined) is given by:

charge of 1 electron (ε)

$$= \frac{1}{6.24 \times 10^{18}}\ \text{C} = 1.6 \times 10^{-19}\ \text{C}$$

As a conversion factor, we may write the unity ratio of $1.6 \times 10^{-19} \frac{\text{C}}{\epsilon}$ or its reciprocal $6.24 \times 10^{18}\ \epsilon/\text{C}$.

A coulomb is a rather *large* quantity of charge. Typical values obtained in the laboratory by friction seldom exceed a few millionths of a coulomb, so the microcoulomb, μC, is often used. (See Example 2-1.)

EXAMPLE 2-2

Calculate the force of attraction between an electron and the positive nucleus of an atom having a charge 29 times that of an electron if the distance between them is 1×10^{-10} m.

SOLUTION

$$F = k\frac{Q_1 Q_2}{r^2}$$

$$= 9 \times 10^9 \frac{\text{Nm}^2}{\text{C}^2} \times \frac{(-1.6 \times 10^{-19}\ \text{C}) \times (+29 \times 1.6 \times 10^{-19}\ \text{C})}{(1 \times 10^{-10}\ \text{m})^2}$$

$$= \mathbf{-6.68 \times 10^{-7}\ N}\ \text{(attraction)}$$

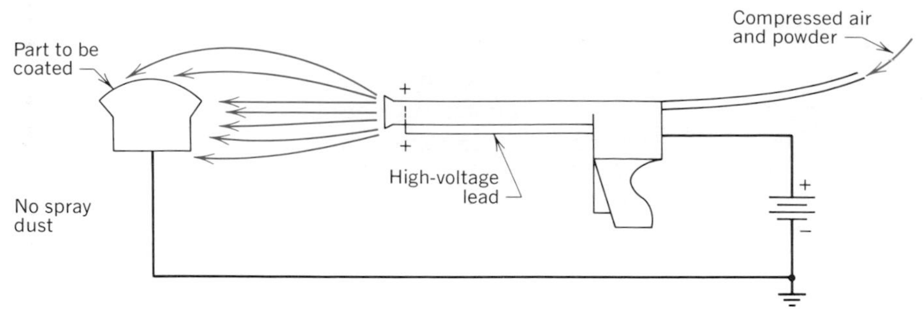

FIGURE 2-4
Electrostatic spray gun.

2-3 APPLICATIONS OF STATIC ELECTRICITY (ELECTROSTATICS)

The attractive force of charged particles for oppositely charged surfaces is used in a number of modern applications. Only two are mentioned here, but they should make you realize that static electricity is more than just a mild shock nuisance.

Anyone who has watched or done ordinary spray painting knows that a lot of wasted paint remains suspended in the air in a vapor cloud. This technique is typically only 65% efficient, and a good ventilation system is needed to clear the room of the paint particles that are so suspended.

An *electrostatic* spray gun (see Fig. 2-4) uses a powder, or resin material, mixed with the paint that is carried to the gun by compressed air. A high-voltage lead (several thousand volts) at the tip of the gun positively charges the powder particles as they exit from it. They are attracted to the part to be painted because it is grounded and therefore at a lower potential. (Effectively, the painted part is connected to the negative side of the high voltage and the gun to the positive side.)

The technique used in Fig. 2-4 produces uniform coatings from 1 mil (0.001 inch) on a cold part up to 20 mils on a preheated part at 98% efficiency. Only 2% of the paint is lost to air. The process is usually followed by heat curing in an oven and is economically competitive with liquid spray painting. Further, personnel using such spray guns can do so safely without the need for a mask.

A second application is in air filtration. An *electrostatic* air cleaner can remove extremely small microscopic particles that normally and easily pass through an ordinary air filter. The air is made to pass through a series of plates that give the particles a positive charge. The air then moves through a negatively charged filter where the positively charged dirt and dust particles are attracted and trapped. A typical air cleaner system used in the home may involve potential differences of about 2 kV.

These two are but a few of the many beneficial uses of electrostatic forces.

2-4 POTENTIAL DIFFERENCE

Let us reconsider the two oppositely charged rods of Fig. 2-1 and see what happens when the two are joined with a conductive metallic wire, as shown in Fig. 2-5.

It is found that soon after the metallic wire (copper, aluminum, etc.) is connected, the two rods are no longer differently charged. Since like charges repel and opposite charges attract, electrons travel from the location of electron surplus (the negatively charged rod) to a location where there is a deficiency of electrons (the positively charged rod). **This flow of electric charges is called *electric current.* And the ability (or *potential*) to cause this flow between the two *differently* charged bodies is called a *potential difference.*** In fact, the negatively charged side is said to have a *negative* potential and the positively charged side a *positive* potential. However, these terms are only relative. It is possible to have two negatively charged bodies, with one more negatively charged than the other. The *less negatively* charged body is considered as having a positive potential with respect to the other *more* negatively charged body. (see Fig. 2-6.)

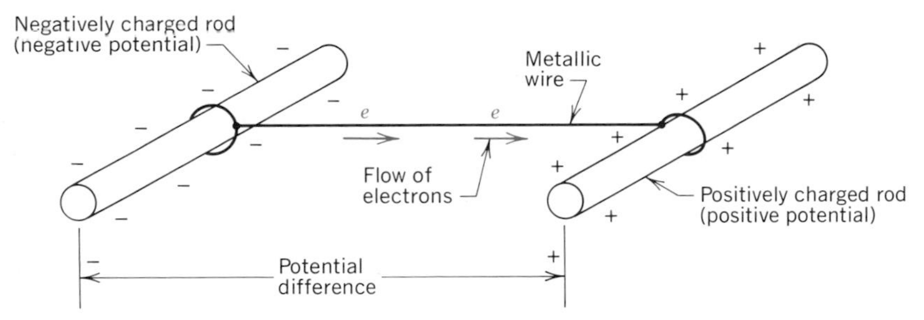

FIGURE 2-5
Two differently charged bodies establish a potential difference that causes electrons to flow between the two. *Note:* The charges actually exist on the *surface* of the rods.

If the two bodies in Fig. 2-6 are connected with a copper wire, electrons flow, from left to right, until equilibrium is established, at which point no potential difference exists.

Prior to connecting the two charged rods, we have a static electricity situation; that is, the electric charges are at rest. We then have a momentary brief current (charge flow) whenever a connection is made between the two differently charged bodies. This discharge current flows only during the time the electrons are rearranging themselves. It is this *discharging* current that we feel when we reach for the door handle after our body has been charged electrostatically by friction with a wool rug. (Incidentally, it is the flow of electrons puncturing our skin that causes the stinging shock sensation.) At night, you can sometimes see a blue spark when the air breaks down and allows the flow of charges or current. If you hold a key tightly in your hand and touch the door handle, the same spark and current occur but without the puncturing sensation.

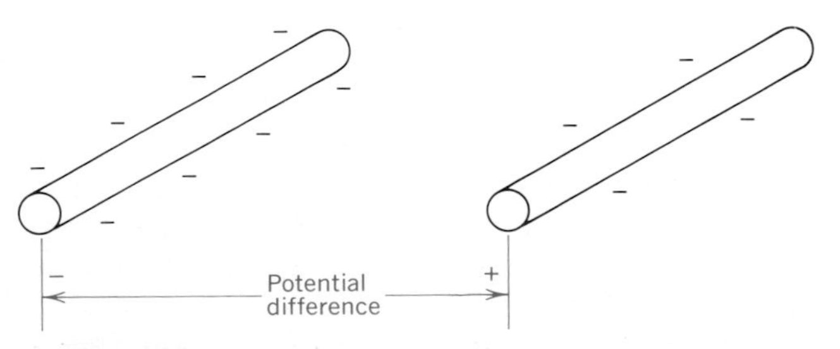

FIGURE 2-6
Two negatively charged bodies establish a potential difference. The rod on the left has more negative charges than the one on the right.

2-4.1 The Volt

As mentioned in Section 1-7, the *potential difference* of a source is related to the work, W, that the source can do to move the charge, Q, around a complete circuit. **The basic unit to measure potential difference is the volt (V). The potential difference, V, is defined as the *ratio* of the work done to the charge transferred.** That is,

$$V = \frac{W}{Q} \quad \text{volts (V)} \tag{2-2}$$

where: W is the amount of work or energy expended in joules,* J

Q is the charge in coulombs, C

From the above Eq. 2-2, we may say:

1 volt = 1 joule/coulomb (1 V = 1 J/C)

A potential difference of 1 V exists when 1 J of work is done in moving 1 C of charge between two points.

* One joule is the work done by a force of 1 N acting through a distance of 1 m.

EXAMPLE 2-3

Calculate the amount of work done by a 12-V battery that causes 50 C of charge to be transported from source to load.

SOLUTION

$$V = \frac{W}{Q} \qquad (2\text{-}2)$$

$$W = V \times Q$$
$$= 12\ \text{V} \times 50\ \text{C}$$
$$= 600\ \text{J/C} \times \text{C} = \mathbf{600\ joules}$$

NOTE *V* is the letter *symbol* for potential difference, whereas V is an abbreviation that stands for the basic *unit* of volt.

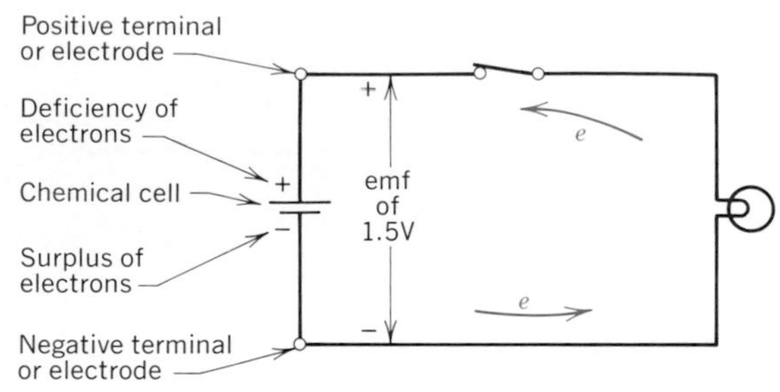

FIGURE 2-7
Electric circuit showing how electrons flow from the negative terminal to the positive terminal.

2-4.2 Electromotive Force

A term used to refer to the potential difference of a source is *electromotive force* or emf. In this way we can distinguish between a source and other potential differences that exist throughout a circuit called voltage drops, which are a consequence of current. Both, however, are measured in volts. In fact, we often refer to the *voltage* between two points and may be referring to either an emf or some other potential difference in the same circuit. But this "voltage" may be in millivolts (mV), microvolts (μV), or kilovolts (kV).

It should be understood very clearly that although we have introduced the concept of potential difference (PD) using electrostatic charges, this PD is the same as the PD produced by a battery. In a carbon-zinc (flashlight) cell that produces 1.5 V or a lead-acid (automotive) cell that produces a potential difference of 2 V, chemical activity makes one electrode have a surplus of electrons, while the other has a deficiency. The only difference is that when electrons are made to flow through an external circuit to light a lamp, for example, the chemical energy in the cell is continuously converted to electrical energy. Thus the cell tries to maintain the potential difference and a continuous flow of current. See Fig. 2-7.

Of course, after a period of continual use, the chemical cell's emf decreases because almost all its chemical energy has been converted to electrical energy. There just is not enough chemical energy left to maintain the potential difference. In the case of a lead–acid cell, *recharging* the cell restores the electrodes to their initial condition. Recharging involves using electrical energy to reconvert the cell electrodes to their original chemical state. Thus, during a recharge, electrical energy is converted to chemical energy in the cell. During a discharge, chemical energy is converted to electrical energy.

In Fig. 2-5 we stipulated that the material used to connect the two charged bodies to cause a charge flow should be a metal wire. **Metallic material, which readily permits the transfer of charges, is called a *conductor*.** If, however, the material used was string or plastic (in fact, almost anything nonmetallic), we would have found no current or charge flow. **Such nonconducting materials are called *insulators*.** To understand these different electrical properties of materials, it is necessary to consider their atomic structure.

2-5 ATOMIC STRUCTURE

In 1913, Niels Bohr proposed that an atom consists of a central nucleus surrounded by orbiting electrons. Although there are more modern theories, the Bohr concept of an atom is well suited for explaining electrical properties. Consider, in particular, an atom of copper, as represented in Fig. 2-8.

The nucleus consists of 29 positively charged protons and 35 neutrons that have no charge, making the charge of the whole nucleus positive. But the charge of a proton

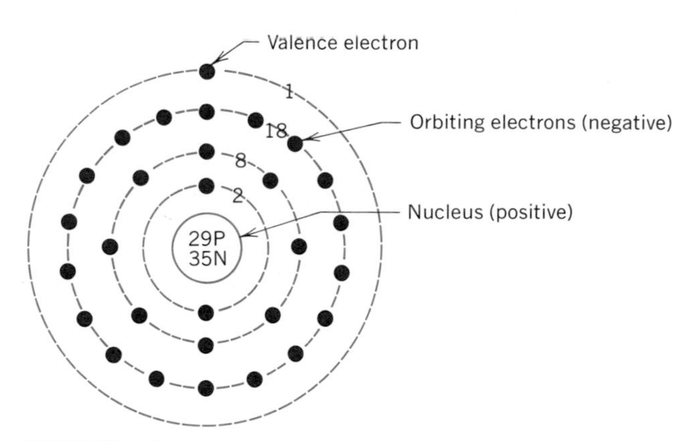

FIGURE 2-8
Bohr concept of an atom of copper.

is the same as that of an electron (1.6×10^{-19} C). Therefore, it requires 29 orbiting electrons to make a *neutral* atom of copper. This is provided by the electrons filling up "shells" from the center outward, until we have 29. That is, the electrons are allowed to exist at only certain energy levels, and all shells or orbits are usually filled before another shell can be started. From Table 2-1 it can be seen that the charge of a whole atom of copper is neutral.

Table 2-2 shows technical data for the mass and charge of protons and electrons.

TABLE 2-1
Atomic Structure of Copper

Shell	Number of Electrons	Charge
1st	2 (Filled)	
2nd	8 (Filled)	
3rd	18 (Filled)	
4th	1 (Could hold 32)	
Total:	29 electrons	Negative
Nucleus:	29 protons	Positive
	35 neutrons	Neutral

TABLE 2-2
Mass and Charge for Atomic Particles

Particle	Mass, kg	Charge, C
Electron	9.11×10^{-31}	-1.6×10^{-19}
Proton	1.67×10^{-27}	$+1.6 \times 10^{-19}$
Neutron	1.67×10^{-27}	0

2-5.1 Valence Electrons

In all atoms of all elements, the outer electron shell is called the *valence* shell. The valence shell determines the chemical and electrical behavior of the material. In copper, at room temperature, the valence electron is very loosely bound to the nucleus, for the following reasons:

1. The outer shell is not completely filled. If it were, a very *stable* geometry or electron configuration would result. Such a geometry tightly binds the electrons together. A filled valence shell does not readily permit any one electron to be removed.
2. The outer single electron is only weakly bound to the nucleus because it is *further away* than the rest of the electrons. Recall Coulomb's law, Section 2-2, that the force of attraction between opposite charges varies inversely as the square of the distance between them.

It is these valence electrons, readily given up by each metallic atom, that account for the ability of copper, and other metallic materials, to conduct current. In an insulating material, such as plastic, mica, glass, and so on, the valence shells are either completely filled or almost full. In these materials, electrons are not readily available, so they do not conduct electricity.

There is a group of materials, between conductors and insulators, called *semiconductors,* which contain four valence-shell electrons. Silicon, for example, will allow the conduction of current but not as well as copper. It is, however, *not* an insulator. Although silicon has four electrons in its valence shell, it shares these, in a manner called *covalent bonding,* with four other silicon atoms. Each atom then effectively has eight electrons in its valence shell but on a *shared* basis. This is a stable situation, since it means that two *subshells* within the third shell are completely filled. The atoms do not *readily* give up electrons for conduction. However, at room temperature, *some* of the electrons do gain enough energy to escape from their atoms, accounting for some *conductivity*. Hence the term *semiconductor*.

2-6 FREE ELECTRONS

The valence electrons that acquire sufficient energy to leave the parent atom are called *free electrons*. That is, the electrons are free to wander from atom to atom in a random nature through the material, colliding with atoms but not attached to any particular atom. The atom that loses an electron becomes a *positively charged ion,* since there are more positively charged protons in the nucleus than there are negatively charged electrons in the atom's shells.

Copper, a good conductor, has approximately 8.5×10^{28} free electrons per cubic meter. In a piece of copper wire, with no emf connected across it, this tremendous number of free electrons is randomly moving within the material. However, the *net* combined average motion of *all* the electrons to the left or right is zero. When a battery is connected across the copper wire, a force acts on these free electrons and moves them. See Fig. 2-9. **It is the motion of these free electrons, in a *definite overall direction,* that constitutes an electric current.** Recall that an electric current is a flow of electrical charges.

2-7 CONDUCTORS AND INSULATORS

The more free electrons a material has, the better it will conduct electricity (allow electrical charge flow through it). Thus silver, which has approximately 5% more free electrons (for the same volume) than annealed copper, is a slightly better conductor than copper. Aluminum, however, has only 61% of the number of free electrons that copper has, and it is a somewhat poorer conductor. Most metals are good conductors of electricity.

Insulators have very few free electrons for reasons explained in Section 2-5. A perfect insulator would have no free electrons. But in practice, a good insulator like polystyrene, may have as many as 6×10^{16} free electrons/cubic meter. Although this is a large number, it is relatively few compared with copper, which has 8.5×10^{28} free electrons per cubic meter. Most of the free electrons in polystyrene come from the impurities in the insulating material or any moisture absorbed by the insulator. It is these free electrons that account for the very small leakage currents through, or on the surface of, most insulators.

2-8 CURRENT

As we have seen, current is the rate of flow of charge and is measured in terms of how much charge flows per

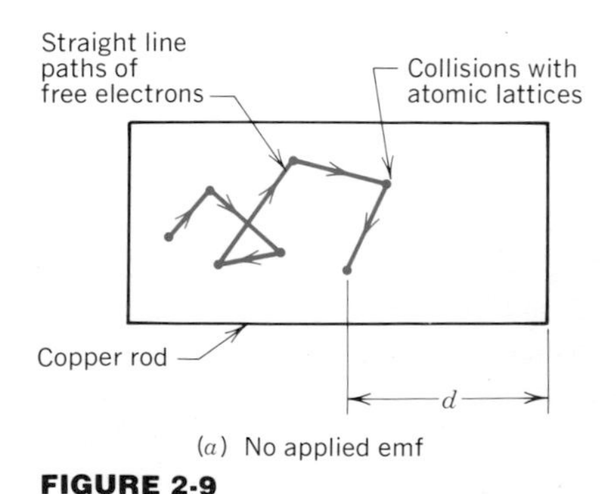

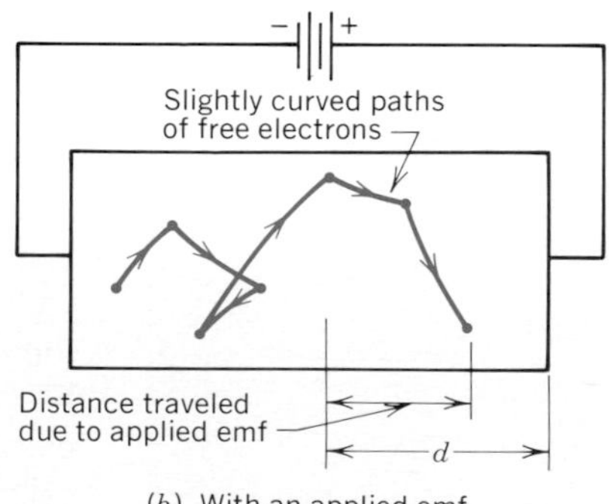

FIGURE 2-9
Schematic representation of a few paths of an electron in copper.

second. (This may be likened to the rate of flow of traffic, measured in cars per second.) The basic unit to measure current is the *ampere* (A), which may be defined as follows:

$$I = \frac{Q}{t} \quad \text{amperes (A)} \qquad (2\text{-}3)$$

where: I is the current in amperes, A

Q is the charge in coulombs, C

t is the time in seconds for the charge to flow

Thus 1 ampere = 1 coulomb/second (1 A = 1 C/s)

EXAMPLE 2-4

In an electroplating circuit it is determined that 60 C of charge have been transferred at a steady rate in 4 min.

a. How much current is flowing in the circuit?

b. At this rate how many coulombs will be transferred in 10 min?

c. How many electrons does this represent?

SOLUTION

a. $$I = \frac{Q}{t} \qquad (2\text{-}3)$$

$$= \frac{60\text{ C}}{4\text{ min} \times \dfrac{60\text{ s}}{\text{min}}} = \frac{60\text{ C}}{240\text{ s}} = 0.25\text{ C/s} = \mathbf{0.25\ A}$$

b. $$Q = It = 0.25\text{ C/s} \times 10\text{ min} \times \frac{60\text{ s}}{\text{min}} = \mathbf{150\ C}$$

c. Since there are 6.24×10^{18} electrons per coulomb

$$150\text{ C} \times 6.24 \times 10^{18}\,\frac{\text{electrons}}{\text{C}}$$

$$= \mathbf{9.36 \times 10^{20}\ electrons}$$

Note from Example 2-3b that 1 coulomb = 1 ampere – second (1C = 1A · s). Since current is easier to measure than charge, one coulomb (6.24×10^{18} electrons) is defined as the charge that flows past a given point in a circuit in 1 second due to a steady current of 1 ampere.

In many circuits, the ampere is too large a unit to be convenient. Recall from Section 1-6:

$$1\text{ mA} = 1 \times 10^{-3}\text{ A} = 0.001\text{ A}$$
$$1\ \mu\text{A} = 1 \times 10^{-6}\text{ A} = 0.000{,}001\text{ A}$$

EXAMPLE 2-5

Convert: a. 0.075 A to mA

b. 0.35 mA to μA

c. 2300 mA to A

SOLUTION

a. $0.075\text{ A} = 75 \times 10^{-3}\text{ A} = \mathbf{75\ mA}$

b. $0.35\text{ mA} = 0.35 \times 10^{-3}\text{ A}$
$= 350 \times 10^{-6}\text{ A} = \mathbf{350\ \mu A}$

c. $2300\text{ mA} = 2300 \times 10^{-3}\text{ A}$
$= 2.3 \times 10^{3} \times 10^{-3}\text{ A} = \mathbf{2.3\ A}$

2-9 CONVENTIONAL CURRENT AND ELECTRON FLOW

Note that current is defined as the rate of charge flow. In metallic conductors, this consists entirely of moving electrons which is a flow of negative charges. **When electrons flow from the negative terminal of the source around the external circuit, toward the positive terminal, this direction is called *electron flow*.**

In semiconductors there is another type of current carrier that is a positive charge (called a *hole*). Holes flow in the opposite direction to electrons. Hole current in this case is the flow of *positive* charges. This is covered in detail in Chapter 28.

The early experimenters in electricity were not aware of the actual mechanism of current flow. They perceived its effect long before they understood its nature. They reasoned, perhaps logically, that current flowed from a point of high potential through a circuit and back to the low potential side of the source in much the same way as water flows from a higher level to a lower level. **Accordingly, the experimenters agreed upon a *convention* for the direction of current that everyone would use. And**

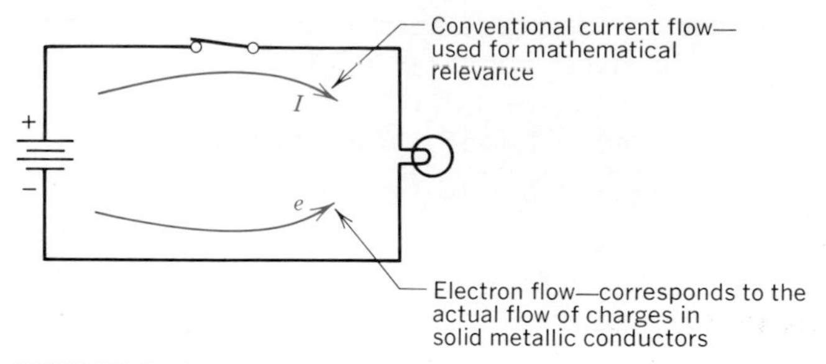

FIGURE 2-10
Opposite directions of electron flow and conventional current flow.

the letter I (after the French word "intensité") was used to designate current, shown flowing from positive to negative as in Fig. 2-10. This is called the *conventional current direction*. It was not until later that it was found that current consisted of the flow of electrons (at least in metallic conductors) from negative to positive. This is usually designated by the letter "e" as in Fig. 2-10.

It should be noted that when an electric current flows in conductive liquids, it is the result of the movement of *both* positive and negative charges. The same is also true of conductive gases or plasmas. The only exception exists in solid metallic conductors where only electrons are in motion. Consequently, there is as much justification for the conventional direction as for the direction of electron flow. (See Fig. 2-10).

Engineers, physicists, and scientists have continued to use *conventional* current flow in developing laws, rules, and terminology. This includes symbols for modern semiconductor devices, *all* of which have arrows on them

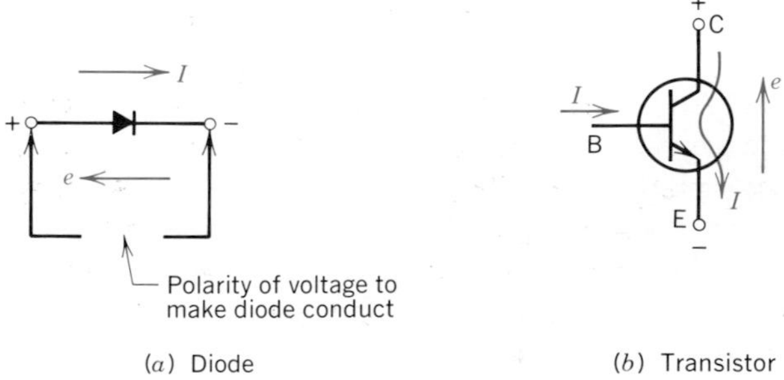

FIGURE 2-11
Graphic symbols for diode and transistor, showing the directions of e, electron flow, and I, conventional current.

pointing in the direction of I or conventional current flow through the devices. See Fig. 2-11.

Finally, the SI notation requires the use of conventional current I as an international standard for all electronic devices. You will find yet another advantage for using this system when we consider the term *voltage drop* and its mathematical meaning in Section 5-1.

2-10
DRIFT VELOCITY OF CURRENT

We have seen how, in a metallic conductor with no emf applied, free electrons are randomly moving about. They move from atom to atom at speeds approaching the speed of light. However, they have no overall definite direction.

When an emf is applied, a steady-state drift speed is *superimposed* upon the random thermal motion of the free electrons. That is, the series of collisions that an electron encounters with the positive ions results in a slow definite progress in the electron toward a point of more positive potential. The relation between the electron drift speed and the current flowing in a metallic conductor is given by the following equation:

$$I = Aenv \qquad \text{amperes (A)} \qquad (2\text{-}4)$$

where: I is the current in amperes, A

A is the conductor's cross-sectional area in square meters, m^2

e is the charge of an electron $= 1.6 \times 10^{-19}$ C

n is the number of free electrons per cubic meter of conductor

v is the velocity of electron drift in m/s

EXAMPLE 2-6

A number 14 gauge copper wire, of diameter 0.064 in., is carrying a current of 15 amperes. Given the number of free electrons per cubic meter is 8.5×10^{28} for copper, calculate:

a. The drift velocity of an individual electron in the conductor.

b. The distance in inches an individual electron moves in the conductor in 1 min.

SOLUTION

a. Diameter $= 0.064$ in.

$$= 0.064 \text{ in.} \times 2.54 \frac{\text{cm}}{\text{in.}} \times \frac{1 \text{ m}}{100 \text{ cm}}$$

$$= 1.63 \times 10^{-3} \text{ m}$$

$$\text{Area, } A = \frac{\pi d^2}{4} = \frac{\pi}{4} \times (1.63 \times 10^{-3} \text{ m})^2$$

$$= 2.09 \times 10^{-6} \text{ m}^2$$

$$I = A\,e\,n\,v \qquad (2\text{-}4)$$

$$\text{Therefore, } v = \frac{I}{A\,e\,n}$$

$$= 15 \text{ A}/(2.09 \times 10^{-6} \text{ m}^2) \times (1.6 \times 10^{-19} \text{ C}) \times (8.5 \times 10^{28} \text{ m}^{-3})$$

$$= \mathbf{5.28 \times 10^{-4} \text{ m/s}}$$

b. $d = vt$

$$= \left(5.28 \times 10^{-4} \frac{\text{m}}{\text{s}}\right) \times 1 \text{ min} \times \frac{60 \text{ s}}{\text{min}}$$

$$= 3.17 \times 10^{-2} \text{ m} \times 10^2 \frac{\text{cm}}{\text{m}} \times \frac{1 \text{ in.}}{2.54 \text{ cm}}$$

$$= \mathbf{1.25 \text{ in.}}$$

The low-charge velocity, as shown in Example 2-6b, is a little over 1 in./min, and it accounts for the use of the term *drift*. Why is the drift velocity so slow? It is because of the tremendous number of free electrons available in the conductor. A current of 15 A corresponds to 15 C/s or approximately 1×10^{20} electrons per s. Since there is a large number of free electrons, they do not have to move very much to transport 15 C of charge every second. Note that if a smaller diameter conductor is used, or a different material with fewer free electrons such as aluminum, the drift velocity of the charges must increase.

If the charges flow so slowly, why does a lamp come on practically immediately after closing the switch in a circuit? Refer to Fig. 2-12.

The answer lies in the fact that free electrons are distributed throughout the entire electric circuit (in the wires and the filament of the lamp) even *before* the switch is closed. Closure of the switch allows the potential of the source ($V = W/Q$) to do work ($W = f \times d$) on all these electrons. A force is exerted on all the electrons and starts them all moving. This is similar to having a trough full of marbles as in Fig. 2-13. A force exerted on the left immediately displaces a marble on the right because all the marbles are set in motion.

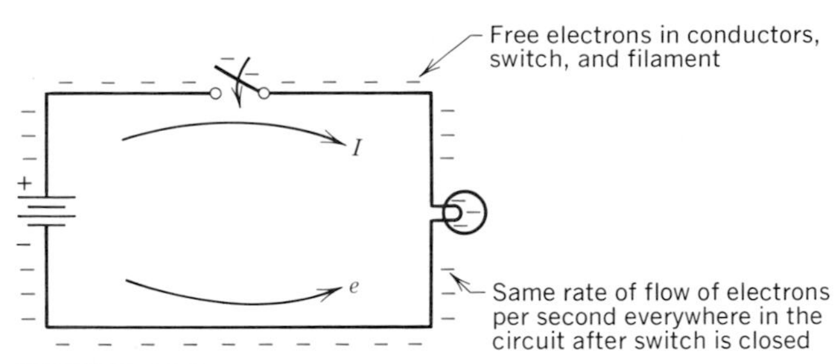

FIGURE 2-12
A circuit has free electrons distributed throughout the circuit even before the switch is closed.

We do not have to wait for an electron from the source to travel to the lamp before the lamp will come on. (This could take many hours.) **Current is not something that squirts out of a battery and gets all used up by the time it reaches the other terminal.** The flow of electrons is the *same* everywhere in the above circuit.

What does move at near the speed of light is the effect of the current starting to move. This is called the impulse or *propagation velocity* and is almost the speed of light or almost 3×10^8 m/s.

2-11 RESISTANCE AND CONDUCTANCE

When current flows in an electric circuit, the free electrons "collide" not only with each other, but also with the bound electrons in the nonmoving atoms of the circuit. **These collisions impede the progress of the free electrons, and the effect is called the *resistance* (*R*) of the circuit.** This opposition to charge flow (current) is measured in the basic unit of **ohms**. The unit symbol used for the ohm is Ω, the Greek capital letter omega.

A circuit is said to have a resistance of 1 ohm whenever a potential difference of 1 volt is applied to the circuit and a current of 1 ampere flows: 1Ω = 1 V/1 A.

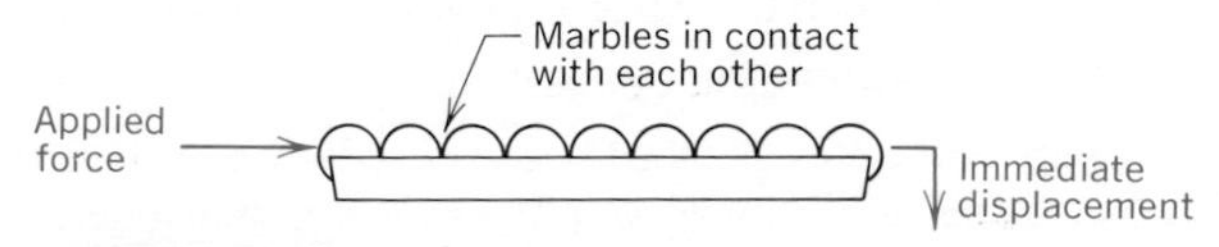

FIGURE 2-13
Marble analogy for immediate effect of current.

To describe larger values of resistance, we use the following:

$$1{,}000\ \Omega = 1\ \text{k}\Omega\ \text{(1 kilohm)}$$
$$1{,}000{,}000\ \Omega = 1\ \text{M}\Omega\ \text{(1 megohm)}$$

Thus 470,000 Ω of resistance can be written as 470 kΩ; similarly, 22,000,000 Ω is written as 22 MΩ.

Small values of resistance may be expressed in milliohms, mΩ, but generally the decimal form is sufficient. For example, 0.030 Ω could be expressed as 30 mΩ.

Details of factors affecting the resistance of conductors will be covered in the next chapter, but we can state here that resistance increases for a longer conductor length or smaller conductor cross-sectional area. This is because *both* of these factors increase the possibility of the number of electron collisions and increase the opposition to the flow of current.

Resistance can be thought of as similar (in some ways) to a frictional effect in a mechanical system. Friction may be undesirable in bearings because of wear and losses that may slow down a rotating shaft, for example. In the same way we can think of resistance in a conductor as detrimental, since it wastes electrical energy in the form of unwanted heat. Thus good conductors should have a very *low* resistance to efficiently transport energy from source to load.

However, the frictional effect found in the braking system of a car is very desirable to the motorist wishing to stop his car. In a similar manner, the heat produced by current flowing through the resistance of a heating element in an electrical appliance (such as a toaster or iron) is very beneficial. Such heating elements purposely are made to have a *higher* resistance, by suitably selecting both the material and cross-sectional area.

Insulators, on the other hand, must have a very high resistance, ideally infinite.

A term that will be useful in our later work is *conductance, G*. **Conductance is the reciprocal of resistance and is an indication of how well a conductor *permits* the flow of current.**

$$G = \frac{1}{R} \quad \text{siemens (S)} \qquad (2\text{-}5)$$

where: G is the conductance in siemens,* S

R is the resistance in ohms, Ω

EXAMPLE 2-7

a. A circuit has a resistance of 20 Ω. What is its conductance?

b. What resistance does a conductance of 1×10^{-6} S represent?

SOLUTION

a. $$G = \frac{1}{R} \qquad (2\text{-}5)$$
$$= \frac{1}{20\ \Omega} = 0.05\ \text{S} = \mathbf{50\ mS}$$

b. $$R = \frac{1}{G}$$
$$= \frac{1}{1 \times 10^{-6}\ \text{S}} = 1 \times 10^{6}\ \Omega = \mathbf{1\ M\Omega}$$

---

* Siemens is the SI unit for conductance. It replaces mho, and its symbol, ℧, for conductance.

SUMMARY

1. Static electricity results from the transfer of electrons from one body to another due to friction.

2. A negatively charged body has a surplus of electrons; a positively charged body, a deficiency.

3. Like charges repel; unlike charges attract.

4. Coulomb's law states that the force exerted between two charges is proportional to the product of the two charges and is inversely proportional to the square of the distance separating the two charges.

5. A coulomb is the combined electrical charge of 6.24×10^{18} electrons.

6. Static electricity has many useful applications, two of which are the electrostatic spray gun and the electrostatic air cleaner.

7. A potential difference exists between two differently charges bodies; it represents the ability to cause a flow of charge between the two bodies.

8. Potential difference is measured in volts. One volt equals one joule per coulomb.

9. The Bohr concept of an atom involves a positive nucleus surrounded by an equally negative charge in the form of electrons, in shells, around the nucleus.

10. A free electron is the outermost or valence electron that leaves the parent atom and is free to wander from atom to atom.

11. Conductors are materials having many free electrons and readily permit the flow of electric charge.

12. Insulators are materials that have very few or no free electrons and do not readily conduct electricity.

13. Current is the rate of flow of charge measured in coulombs per second or amperes. One ampere equals one coulomb per second.

14. In a solid metallic conductor electrons (shown by the letter e) flow from negative to positive. Conventional current (shown by the letter I) has a direction shown flowing from positive to negative.

15. Electron drift is the overall progress made by an electron under the influence of an applied emf; drift velocity is very low, in the order of a few centimeters per minute.

16. The effect of current (in a just completed circuit) moves at almost the speed of light because all the electrons start to move at once. This is called the propagation velocity.

17. Opposition to current, caused by electron collisions, is called resistance and is measured in ohms. One ohm equals one volt per ampere.

18. Conductors have a very low value of resistance; insulators, a very high value.

19. Conductance is the reciprocal of resistance and is measured in siemens.

SELF-EXAMINATION

Answer true or false
(Answers at back of book)

2-1. Friction between two materials causes the transfer of electrons from one material to the other. This is called electrification by friction. ________

2-2. A body is said to be negatively charged if it loses electrons and positively charged if it gains electrons. ________

2-3. Like charges repel; unlike charges attract. ________

2-4. Coulomb's law determines how much force of attraction or repulson occurs between two charged bodies. ________

2-5. The amount of force is directly proportional to the square of the distance and inversely proportional to the product of the charges. ________

2-6. Electrical charge is measured in coulombs. ________

2-7. Applications of static electricity rely upon particles becoming electrically charged and being attracted to oppositely charged surfaces. ________

2-8. A potential difference exists between two differently charged electrodes. ________

2-9. It is necessary to have a positively charged body and a negatively charged body to produce a potential difference. ________

2-10. Potential difference is a measure of the ability to cause charge to flow between two differently charged bodies. ________

2-11. Potential difference is measured in volts. ________

2-12. An electromotive force is a source, measured in volts. ________

2-13. Both emf and voltage drops are measured in volts. ________

2-14. When a cell discharges, chemical energy is converted to electrical energy as the cell tries to maintain the potential difference. ________

2-15. A material that readily permits the flow of charges is called an insulator. ________

2-16. The nucleus of every atom is positive. ________

2-17. There are as many electrons in orbit around the nucleus as there are neutrons in the nucleus. ________

2-18. The outer electron shell is the valence shell of an atom. ________

2-19. In a copper atom the valence electron is loosely bound to the nucleus. At room temperature this electron gains sufficient energy to become a free electron and wanders from atom to atom. ________

2-20. Free electrons have a random thermal motion when no emf is applied to a conductor. ________

2-21. Under the influence of an applied emf the free electrons drift toward the negative terminal of the emf. ________

2-22. A conductor has a large number of free electrons. ________

2-23. Current is the rate of flow of charge, measured in coulombs per second or amperes. ________

2-24. A current of 0.05 A is equal to 5 mA. ________

2-25. A current of 1000 μA is equal to 1 mA. ________

2-26. Conventional current is shown as flowing from positive to negative. ________

2-27. In a solid metallic conductor, current is the flow of electrons from negative to positive. ________

2-28. The arrows in the symbols for solid-state devices indicate the direction of electron flow. ________

2-29. The drift velocity of current in a metallic conductor is very slow because there are a tremendous number of free electrons available. ________

2-30. Propagation velocity refers to the near-instantaneous effect of current when a circuit is first completed. ________

2-31. Resistance in a wire is the opposition to current caused by electron collisions. ________

2-32. A resistance of 1200 Ω equals 12 kΩ. ________

2-33. A resistance of 1200 kΩ equals 1.2 MΩ. ________

2-34. A resistance of 50 Ω has a conductance of 20 mS. ________

REVIEW QUESTIONS

1. a. When two initially neutral bodies, such as rubber and fur, are rubbed together, what can you say about the sign and magnitude of the charge on each?

b. What prevents *all* the electrons from moving from one body to the other?

2. What is the law of electric forces? To what law is this similar?

3. What factors determine the size of forces between charged bodies? Would this same force exist in a vacuum?

4. How many protons are necessary to produce a positive charge of 1 coulomb?

5. If opposite charges attract, why don't the electrons in an atom move toward, and come in contact with, the nucleus?

6. Describe how static electricity may be used in the manufacture of sand paper.

7. What do you understand by the term *potential difference?*

8. Explain how it is possible for two electrodes, both positively charged, to have a potential difference between them.

9. Explain the difference between the potential differences set up by two electrostatically charged rods and the potential difference from a dry cell.

10. Name five conducting materials and five insulators.

11. What is a free electron?

12. How many free electrons would a perfect insulator have? What characteristic would a perfect conductor have?

13. What is covalent bonding? Name a material that has this type of bonding. How does this material's conductance compare with that of copper or an insulator?

14. How is it possible for free electrons to be moving in different directions in a copper wire when there is no emf applied? How does an applied emf change this situation?

15. Is current necessarily the flow of electrons or negative charge?

16. Why is the term *drift* appropriate in describing current in a metallic conductor?

17. What happens to the drift velocity in a series circuit where the cross-sectional area of the conductor decreases? Why is the drift velocity low in a metallic conductor? Would you expect it to be higher or lower (for a given current) in a semiconductor of the same cross-sectional area? Why?

18. What is one ampere-second?

19. Distinguish between electron flow and conventional current. Which direction is shown on solid-state symbols? Why?

20. If the drift velocity of current is in the order of only a few centimeters per minute, explain how a light comes on immediately after closing a switch to complete the circuit.

21. Describe in your own words what resistance to current is in an electric circuit. What factors affect the amount of resistance and why?

22. Why is it desirable for a conductor to have a low resistance? How would you describe the property of an insulator in terms of conductance?

PROBLEMS

(Answers to odd-numbered problems at back of book)

2-1. In an experiment to verify Coulomb's law two metal spheres, separated by a distance of 10 cm, are each charged to 2.5×10^{-8} C. Calculate:

a. The force of repulsion in newtons.
b. The force in pounds.

2-2. What must the separation between an electron and a proton be if the force of attraction between them is to be 1×10^{-6} N?

2-3. How many electrons would have to be removed from each of the spheres in Problem 2-1 if the spheres were initially neutral?

2-4. How many electrons would have to be transferred from one sphere to another, if they were both initially neutral, so that they would experience an attractive force of 1×10^{-6} N when they are 1 cm apart?

2-5. If it takes 30 J of chemical energy to move 20 C of electric charge between the positive and negative terminals of a battery, what is the emf between the terminals?

2-6. What is the emf of a source that moves 600 mC of charge from source to load with the expenditure of 3.6 J of energy?

2-7. If it takes 1 min for the charge in Problem 2-5 to be transported, what is the current?

2-8. What is the current in Problem 2-6 if the time required is 10 s?

2-9. In an electroplating system a steady current of 20 A flows for 10 min. What quantity of charge flows during this period?

2-10. How much charge is transported by a steady current of 50 mA flowing for 15 s?

2-11. How much energy does a 12-V battery supply to a load when the load draws a current of 4.5 A for 5 min?

2-12. A current of 350 mA causes a PD of 20 V across a resistor. How much energy is dissipated in the form of heat in 1 min?

2-13. How long does it take for a source to deliver 50 C of charge to a load if the current is 100 mA?

2-14. How long does it take for 1×10^6 electrons to pass by a given point in a circuit when the current is 10 mA?

2-15. How many electrons are transferred by a current of 10 μA flowing for 10 s?

2-16. What is the current in a circuit if one electron passes by in 1 ns?

2-17. Convert:

a. 0.0025 A to mA

b. 0.075 mA to μA

c. 5000 mA to A

2-18. Convert:

a. 15,000 μA to mA

b. 6500 μA to A

c. 1300 pA to nA

2-19. A copper wire carries a current of 10 A. Determine the electron drift velocity if the wire has a diameter of

a. 1 mm

b. 2 mm

c. 4 mm

2-20. Repeat Problem 2-19 using an aluminum wire with $n = 5.2 \times 10^{28}$ free electrons per cubic meter of aluminum.

2-21. How far does an individual electron move in 1 min for each of the diameters in Problem 2-19?

2-22. How far does an individual electron move in 1 min for each of the diameters in Problem 2-20?

2-23. Convert:

a. 1500 Ω to kΩ

b. 4700 kΩ to MΩ

c. 0.045 Ω to mΩ

2-24. Convert:

a. 100 Ω to kΩ

b. 1.2 kΩ to Ω

c. 2.2 MΩ to kΩ

2-25. Find the conductance of each of the resistances in Problem 2-23.

2-26. Find the conductance of each of the resistances in Problem 2-24.

2-27. Find the resistance represented by the following conductances:

a. 40 mS

b. 0.4545 μS

c. 1 S

CHAPTER THREE

OHM'S LAW, POWER, AND RESISTORS

We have seen that the three basic quantities in an electrical circuit are voltage, current, and resistance. In this chapter, we examine *Ohm's law,* which relates these three variables to each other. Ohm's law permits the determination of any of the three variables if the other two are known.

Power is the *rate* of doing work or consuming energy. One must be careful to distinguish between *power* and *energy,* for they are not the same. The amount of electrical energy consumed depends upon the power *and* the length of the time the power is being used.

Resistors are devices that introduce a known amount of resistance into a circuit. They may be of the *fixed* or *variable* type. Many resistors have their values indicated by a *standard color code* of stripes on the resistor's body. They are available in standard values depending upon the accuracy of their nominal values. Various power ratings are available that determine, through physical size, how much heat can be safely dissipated.

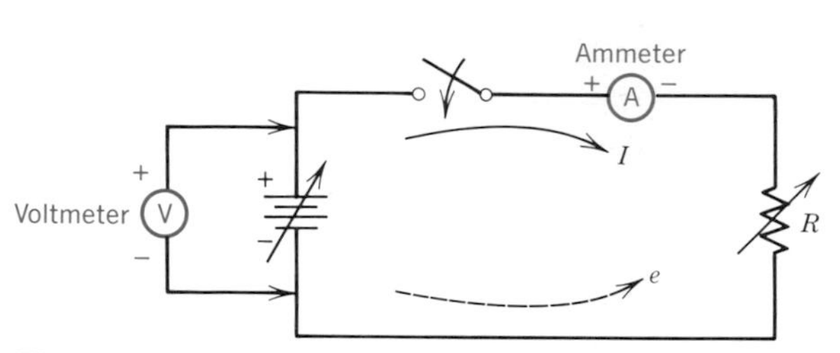

FIGURE 3-1
Circuit showing the use of voltmeter and ammeter for the measurement of applied voltage and current.

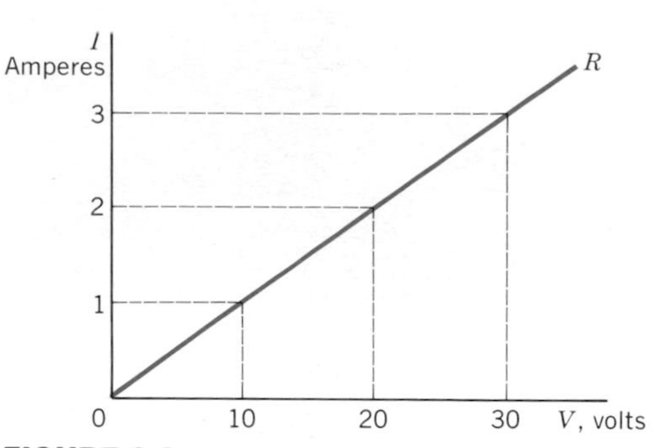

FIGURE 3-2
Ohm's law in graphical form.

3-1 CONNECTION OF CIRCUIT METERS

To consider the variables involved in Ohm's law, refer to Fig. 3-1. This represents a circuit with a variable emf connected to a resistor whose value may be varied. Meters have been included to indicate the amount of voltage (on a *voltmeter*) applied to the circuit and an ammeter to indicate the current flowing in the circuit. If the meters are of the multirange type, their ranges should generally be adjusted to give indications as close to full-scale deflection as possible. This minimizes errors.

Note that the voltmeter, since it is to measure dc volts, is polarity-sensitive. That is, the positive and negative leads of the voltmeter must be connected, respectively, to the corresponding (positive and negative) terminals of the source. If this is not done, the voltmeter reads backward and may be damaged. (It is customary to use a red-colored lead to make connection to the positive terminal and a black lead to the negative terminal.) Note also that the circuit remains intact and functions if the voltmeter is removed. For this reason the voltmeter connections are shown with arrows (rather than dots, which signify a more permanent connection).

The ammeter, however, *is* part of the circuit, and the switch should be opened before the ammeter is inserted or removed. The ammeter is also polarity-sensitive, like the voltmeter. Note the polarity signs on the ammeter to obtain a positive reading. The conventional current must enter the positive terminal or electrons must enter the negative terminal on both instruments.

3-2 OHM'S LAW

In 1827, Georg Simon Ohm observed the relationship between the applied voltage V, the current I, and the resistance R. He found that for a fixed value of resistance, a given current flowed for a certain applied voltage. If this voltage was doubled, the current also doubled. If this voltage was tripled, the current tripled. **That is, the current is *directly proportional* to the voltage if the resistance is maintained *constant*.**

This relationship is presented graphically in Fig. 3-2, where I is plotted against V for a fixed resistance R.

Ohm originally expressed his findings in a form known as Ohm's law and shown in Eq. 3-1.

$$R = \frac{V}{I} = k \qquad \text{ohms } (\Omega) \qquad (3\text{-}1)$$

where: R is the resistance in ohms, Ω

V* is the potential difference in volts, V

I is the current in amperes, A

k is a constant of proportionality, now called ohms.

It should be noted that Fig. 3-2 is the graphical representation of Eq. 3-1, as illustrated by the following example.

* Many texts use the letter symbol E to represent the emf voltage and V for potential differences in the rest of the circuit. However, the SI system uses the symbol V for *all* voltages. The letter E is used to designate electric field intensity.

EXAMPLE 3-1

Calculate the resistance of the resistor in Fig. 3-2 for voltages of:

a. 10 V

b. 20 V

c. 30 V

SOLUTION

a. From the graph, when $V = 10$ V, $I = 1$ A

then
$$R = \frac{V}{I} \tag{3-1}$$
$$= \frac{10\text{ V}}{1\text{ A}} = \mathbf{10\ \Omega}$$

b. When $V = 20$ V, $I = 2$ A
$$R = \frac{V}{I} \tag{3-1}$$
$$= \frac{20\text{ V}}{2\text{ A}} = \mathbf{10\ \Omega}$$

c. When $V = 30$ V, $I = 3$ A
$$R = \frac{V}{I} \tag{3-1}$$
$$= \frac{30\text{ V}}{3\text{ A}} = \mathbf{10\ \Omega}$$

3-3 GRAPHICAL PRESENTATION OF OHM'S LAW

Ohm's law represents a *linear* relationship between current and voltage for any *constant* resistance. This simply means that a graph of a change in current plotted against a change in voltage is a *straight line* for a fixed resistor, as we observed in Fig. 3-2. If graphs of I versus V are plotted for a *family* of fixed resistors, they have the appearance shown in Fig. 3-3. Note that a high resistance has a more horizontal slope. A low resistance has a more vertical slope. But all three are straight lines, representing linear or constant resistance.

If the data for plotting the graphs in Fig. 3-3 were obtained experimentally, from a circuit as in Fig. 3-1, for example, the following points should be noted:

1 The *independent* variable (what we directly vary), voltage, is the abscissa and is plotted horizontally.

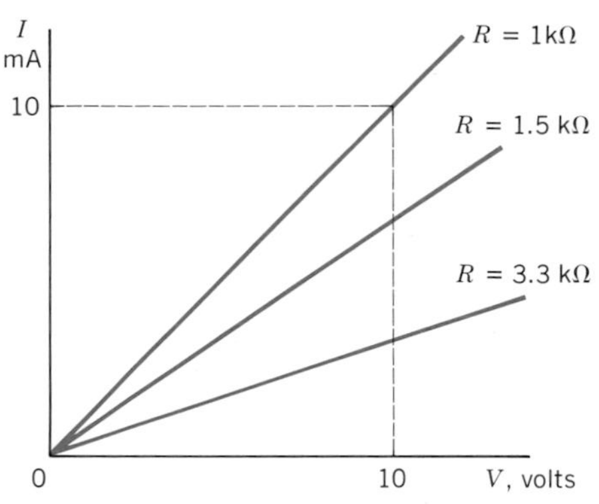

FIGURE 3-3
Graphs of I versus V for three different resistors showing a low slope for high resistance (low conductance) and high slope for high conductance.

2 The *dependent* variable (what changes as a consequence of our actions), current, is the ordinate and is plotted vertically.

3 The graphs are drawn as straight lines through the best average positions of the plotted data and through the origin. (Seldom is plotted data connected directly to each other). This allows for experimental error in reading the meters, and so on.

If the resistance is not constant, such as the filament of a lamp, a graph of I versus V will be curved because the resistance is *nonlinear.* This means that the current does not double for a doubling of the applied voltage, so Ohm's law (Eq. 3-1) is not obeyed. However, it is possible to apply Ohm's law to *any pair* of data for I and V and calculate the resistance R *at those points.*

3-4 OTHER FORMS OF OHM'S LAW

The original statement of Ohm's law (Eq. 3-1) can be algebraically transposed and rewritten in two other forms:

$$\mathbf{I = \frac{V}{R} \quad \text{amperes (A)}} \tag{3-1a}$$

$$\mathbf{V = IR \quad \text{volts (V)}} \tag{3-1b}$$

Equation 3-1a implies that the change in current in a circuit is *inversely* proportional to a change in resistance, and *directly* proportional to a change in voltage. This equation allows us to calculate the current for a given voltage and resistance.

Similarly, Eq. 3-1b can be used to calculate the "voltage drop" that occurs across a resistor R due to a current I flowing through it. The following examples show the use of these equations.

EXAMPLE 3-2

A 150-Ω resistor has an emf of 120 V applied across it.

a. Calculate how much current flows through the resistor.

b. If the resistance is reduced to 50 Ω, calculate the new emf required to maintain the same current as in part (a).

c. To what value must the resistance be increased if a current of 50 mA must flow when the emf is 60 V?

SOLUTION

a. $$I = \frac{V}{R} \qquad (3\text{-}1a)$$

$$= \frac{120\ \text{V}}{150\ \Omega} = \mathbf{0.8\ A}$$

b. $$V = IR \qquad (3\text{-}1b)$$

$$= 0.8\ \text{A} \times 50\ \Omega = \mathbf{40\ V}$$

c. $$R = \frac{V}{I} \qquad (3\text{-}1)$$

$$= \frac{60\ \text{V}}{50\ \text{mA}} = \frac{60\ \text{V}}{50 \times 10^{-3}\ \text{A}}$$

$$= 1.2 \times 10^{3}\ \Omega = \mathbf{1.2\ k\Omega}$$

EXAMPLE 3-3

A 2.2-MΩ resistor has a current of 6 μA flowing through it.

a. Calculate the potential difference across the resistor.

b. The resistor is changed until its conductance is 1×10^{-6} S. What is the new current if the potential difference is now 12 V?

SOLUTION

a. $$V = IR \qquad (3\text{-}1b)$$

$$= 6\ \mu\text{A} \times 2.2\ \text{M}\Omega$$

$$= 6 \times 10^{-6}\ \text{A} \times 2.2 \times 10^{6}\ \Omega = \mathbf{13.2\ V}$$

b. $$I = \frac{V}{R} \qquad (3\text{-}1a)$$

$$= V \times \frac{1}{R} = V \times G$$

$$= 12\ \text{V} \times 1 \times 10^{-6}\ \text{S} = 12 \times 10^{-6}\ \text{A} = \mathbf{12\ \mu A}$$

3-5 LOAD RESISTANCE

The amount of resistance in the circuit of Fig. 3-1 could be measured using an *ohmmeter.* As mentioned in Section 1-11, it would first be necessary to open the switch or completely remove the resistor from the circuit. Then the ohmmeter leads would be connected across the resistor. (In this case it would not matter which way the leads are connected across the resistor, since it is not polarity-sensitive.) **However, an ohmmeter must never be used in a "live" circuit, that is, one with voltage applied to it. The ohmmeter has its own battery and any external voltage will damage the meter and/or the ohmmeter circuitry.**

It is important to realize that the resistor in the circuit of Fig. 3-1 can represent an *electrical load* such as an incandescent lamp or heating element. It is also important to note that an *increase* in load is represented by a *decrease* in resistance. This is because this will result in an increase in load *current,* which is what we are really referring to when we talk about a load. Throughout this text, if we say the load is increased, we mean that the current has increased. In short, load implies current.

3-6 MECHANICAL WORK AND ENERGY

Work is done any time a force is exerted over a distance in the direction of the force. We do work against gravity when we walk up a flight of stairs. We do work in overcoming friction when we slide a heavy box along the floor. And work is done to accelerate a mass such as a car. Thus

work done = (force × (distance moved in the direction of the force)

$$W = Fs \qquad (3\text{-}2)$$

where, in the FPS system:

W is work in foot-pounds, ft lb
F is force in pounds, lb
s is distance moved in feet, ft

and in the MKS system:

W is work in newton-meters, N · m or joules, J
F is force in newtons, N
s is distance moved in meters, m

Note that one newton-meter equals one joule (1 N · m = 1 J)

EXAMPLE 3-4

A 160-lb man walks up a flight of stairs that rises to a vertical height of 20 ft. Calculate the amount of work done against gravity by the man in:

a. ft lb
b. joules

SOLUTION

a. $W = Fs$ (3-2)

$= 160 \text{ lb} \times 20 \text{ ft} = \mathbf{3200 \text{ ft lb}}$

b. $W = Fs$ (3-2)

$$= \left(160 \text{ lb} \times \frac{4.45 \text{ N}}{1 \text{ lb}}\right) \times \left(20 \text{ ft} \times \frac{0.3048 \text{ m}}{1 \text{ ft}}\right)$$

$= 4340 \text{ N} \cdot \text{m} = \mathbf{4340 \text{ J}}$

It should be obvious that doing work requires the expenditure of energy. In fact, work (W) and energy (W) can be thought of as equivalent because they are measured in the *same* units. Thus the amount of energy used by the man in Example 3-4 is 3200 ft lb or 4340 J.

3-7 ELECTRICAL WORK AND ENERGY

We have already seen, from the definition of the volt in Section 2-4 ($V = W/Q$), how to calculate work in an electrical circuit.

$$W = VQ \qquad \text{joules (J)} \qquad (3\text{-}3)$$

where: W is work done or energy expended in joules, J
V is potential difference in volts, V
Q is charge in coulombs, C

Since $Q = It$ (Eq. 2-3), we have the following equation to determine the amount of work done in a length of time, t.

$$W = VIt \qquad \text{joules (J)} \qquad (3\text{-}4)$$

where: W is work done or energy expended in joules, J
V is potential difference in volts, V
I is current in amperes, A
t is time in seconds, s

EXAMPLE 3-5

An emf of 120 V causes a current of 3 A to flow in a circuit. Calculate how much energy is supplied by the source in 12 s.

SOLUTION

$W = VIt$ (3-4)

$= 120 \text{ V} \times 3 \text{ A} \times 12 \text{ s} = \mathbf{4320 \text{ J}}$

Note how the electrical energy calculated in Example 3-5 is practically equal to the amount of mechanical energy calculated in Example 3-4. There is, in fact, no difference between mechanical and electrical energy. The energy exists in different *forms;* they are calculated using different equations, but they are measured in the same units.

3-8 EFFICIENCY

Machines are generally used to convert energy from one form into another. An electric motor, for example, takes electrical energy and converts it into mechanical energy in the form of a rotating shaft. Inevitably, some energy is lost in the process, in the form of heat due to friction, for instance. The ratio of the *useful* output work to the total input work is called the efficiency. This is usually represented by the Greek letter η (eta).

$$\eta = \frac{W_{out}}{W_{in}} \qquad (3\text{-}5)$$

where: η is efficiency (a dimensionless quantity)

W_{out} is useful output work or energy in joules, J

W_{in} is total input work or energy in joules, J

Efficiency is generally expressed as a percentage, as shown in the following example.

EXAMPLE 3-6

A 120-V motor drawing 8 A hoists a 160-lb man through a vertical distance of 20 ft in 6 s. What is the efficiency of the system?

SOLUTION

The useful mechanical work out is

$$W_{out} = Fs \qquad (3\text{-}2)$$
$$= 160 \text{ lb} \times 20 \text{ ft} = 3200 \text{ ft lb}$$
$$= 3200 \text{ ft lb} \times \frac{1.356 \text{ J}}{1 \text{ ft lb}}$$
$$= 4340 \text{ J}$$

The total electrical work in is

$$W_{in} = VIt$$
$$= 120 \text{ V} \times 8 \text{ A} \times 6 \text{ s}$$
$$= 5760 \text{ J}$$

The efficiency is

$$\eta = \frac{W_{out}}{W_{in}} \times 100\% \qquad (3\text{-}5)$$
$$= \frac{4340 \text{ J}}{5760 \text{ J}} \times 100\% = \mathbf{75.3\%}$$

3-9 MECHANICAL POWER

Power is the *rate* at which work is done, or the rate at which energy is converted or consumed. Thus power involves not only *how much* work is done, but also *how long a time* is required to do the work.

$$\text{power} = \frac{\text{work done}}{\text{time taken to do the work}} = \frac{\text{energy}}{\text{time}}$$

$$P = \frac{W}{t} \qquad (3\text{-}6)$$

where, in the FPS system:

P is power in foot-pounds per second, ft lb/s

W is work or energy in foot-pounds, ft lb

t is time in seconds, s

and in the MKS system:

P is power in joules per second, J/s or watts, W

W is work or energy in joules, J

t is time in seconds, s

Note that one joule per second equals one watt. (1 J/s = 1 W) Similarly, it was decided, by James Watt, to call 550 ft lb/s one horsepower. (This was his way of expressing an equivalence between his steam engine and the average power that a horse could develop over some interval of time.)

$$1 \text{ hp} = 550 \text{ ft lb/s}$$

EXAMPLE 3-7

If the 160-lb man in Example 3-4 runs up the 20-ft flight of stairs in 8 seconds, calculate the power he is developing in:

a. hp
b. watts

SOLUTION

a. Work done, W = 3200 ft lb (from Example 3-4)

$$P = \frac{W}{t} \qquad (3\text{-}6)$$
$$= \frac{3200 \text{ ft lb}}{8 \text{ s}} = 400 \text{ ft lb/s}$$
$$= 400 \frac{\text{ft lb}}{\text{s}} \times \frac{1 \text{ hp}}{550 \text{ ft lb/s}} = \mathbf{0.73 \text{ hp}}$$

b. Work done, W = 4340 J (from Example 3-4)

$$P = \frac{W}{t} \qquad (3\text{-}6)$$
$$= \frac{4340 \text{ J}}{8 \text{ s}} = 542.5 \text{ J/s} = \mathbf{543 \text{ W}}$$

It is to be noted from Example 3-7 that if 0.73 hp is equivalent (approximately) to 543 W, then the conversion, 1 hp = 746 W (Sec. 1-5), is verified.

Note also that if the *same* work of 3200 ft lb was accomplished in a *longer* time of 10 s, this would correspond to a *smaller* power of only 320 ft lb/s or 0.58 hp.

3-10 ELECTRICAL POWER

Electrical power is the rate at which work is done, or the rate at which energy is converted or consumed, in an electrical circuit. Thus

$$P = \frac{W}{t} = \frac{V \times Q}{t} = \frac{V \times It}{t}$$

Therefore, $P = VI$ **watts (W)** **(3-7)**

where: P is power in watts, W
V is potential difference in volts, V
I is current in amperes, A

EXAMPLE 3-8

A lamp operating at 120 V draws a current of 0.5 A. Calculate:

a. the rate at which energy is converted into heat and light
b. the amount of power used by the lamp
c. the amount of energy used by the lamp in one minute

SOLUTION

a. The rate at which energy is converted is power. Therefore,

$$P = VI \qquad (3\text{-}7)$$
$$= 120\ \text{V} \times 0.5\ \text{A} = 60\ \text{W or } \mathbf{60\ J/s}$$

b. Power used,

$$P = VI \qquad (3\text{-}7)$$
$$= 120\ \text{V} \times 0.5\ \text{A} = \mathbf{60\ W}$$

c. Energy used,

$$W = VIt \qquad (3\text{-}4)$$
$$= 120\ \text{V} \times 0.5\ \text{A} \times 60\ \text{s}$$
$$= \mathbf{3600\ J}$$

Thus a 60-W lamp uses 60 joules of energy every second it is lit. This example clearly shows how the lamp consumes energy, and power is the *rate* of the energy consumption. The lamp consumes 60 W of power (or 60 J/s of energy) all the time it is on, whether for a tenth of a second or a week.

In some applications the watt is a very small unit of power. Then it is convenient to use the following:

1 kilowatt = 1 kW = 1000 W

EXAMPLE 3-9

a. What maximum power can be delivered from a 120-V outlet if it is fed from a 20-A circuit breaker?
b. What voltage must be applied to a 3.3-kW heating element if the current must not exceed 15 A?
c. What current is drawn by a 40-W lamp operating at 120 V?

SOLUTION

a. $P = VI \qquad (3\text{-}7)$

$$= 120\ \text{V} \times 20\ \text{A} = 2400\ \text{W} = \mathbf{2.4\ kW}$$

b. $V = \frac{P}{I}$

$$= \frac{3.3\ \text{kW}}{15\ \text{A}} = \frac{3300\ \text{W}}{15\ \text{A}} = \mathbf{220\ V}$$

c. $I = \frac{P}{V}$

$$= \frac{40\ \text{W}}{120\ \text{V}} = 0.33\ \text{A} = \mathbf{330\ mA}$$

3-11 EFFICIENCY INVOLVING POWER

We saw in Section 3-8 how the efficiency of a machine or system could be obtained using:

$$\eta = \frac{W_{out}}{W_{in}} \quad (3\text{-}5)$$

But

$$W = Pt \quad (3\text{-}6)$$

Therefore,

$$\eta = \frac{P_{out} \times t}{P_{in} \times t}$$

and

$$\eta = \frac{P_{out}}{P_{in}} \quad \textbf{(3-8)}$$

where: η is efficiency

P_{out} is useful output power in watts, W or hp

P_{in} is total input power in watts, W or hp

Consider a 1-hp electrical motor. If the motor were 100% efficient, it would require 746 W of electrical input power to develop 1 hp of mechanical output power at the shaft. Due to losses, more than 746 W will be required at the input for a practical motor, as shown by the following example.

EXAMPLE 3-10

How much current must be supplied to a 75%-efficient $1\frac{1}{2}$-hp motor operating at 120 V?

SOLUTION

$$\eta = \frac{P_{out}}{P_{in}} \quad (3\text{-}8)$$

Therefore,

$$P_{in} = \frac{P_{out}}{\eta}$$

$$= \frac{1.5 \text{ hp}}{0.75} \times \frac{746 \text{ W}}{1 \text{ hp}} = 1492 \text{ W}$$

$$I = \frac{P}{V}$$

$$= \frac{1492 \text{ W}}{120 \text{ V}} = \mathbf{12.4\ A}$$

Notice how the $1\frac{1}{2}$-hp motor draws almost $1\frac{1}{2}$ kW from the supply. This equivalence of 1 kW per hp is a useful rule of thumb to use for motors of this size. It assumes an average motor efficiency of approximately 75%.

Example 3-7 showed the calculation of mechanical power in both horsepower and watts. Thus Example 3-10 should not suggest that there is anything unique that reserves the watt to measure electrical power and the hp to measure mechanical power. In Europe, it is not unusual to give the power rating of a gasoline engine vehicle as 40 kW, for example. The conversion, 746 W = 1 hp, can be used to identify this vehicle as having a horsepower of:

$$40 \text{ kW} \times \frac{1000 \text{ W}}{1 \text{ kW}} \times \frac{1 \text{ hp}}{746 \text{ W}} = \mathbf{53.6\ hp}$$

3-12 OTHER EQUATIONS FOR ELECTRICAL POWER

When calculating the power dissipated (converted into heat) in a resistor, the quantities known may involve the resistance, R. In this case the following two equations are useful and result when $V = IR$ and $I = V/R$ are separately substituted in Eq. 3-7.

$$\mathbf{P = I^2R \quad watts\ (W)} \quad \textbf{(3-9)}$$

$$\mathbf{P = V^2/R \quad watts\ (W)} \quad \textbf{(3-10)}$$

where: P is power dissipated in watts, W

R is resistance in ohms,

I is current in amperes, A

V is potential difference in volts, V

EXAMPLE 3-11

Calculate:

a. The maximum current that a 100-ohm, 2-W resistor can handle without overheating.

b. How much power is dissipated in a 10-kΩ resistor connected to a 12-V supply?

SOLUTION

a. $P = I^2R$ (3-9)

Therefore,

$$I = \sqrt{\frac{P}{R}}$$

$$= \sqrt{\frac{2 \text{ W}}{100\ \Omega}} = \sqrt{0.02} \text{ A}$$

$$= 0.14 \text{ A} = \mathbf{140\ mA}$$

b. $P = V^2/R$ (3-10)

$$= \frac{(12 \text{ V})^2}{10 \times 10^3\ \Omega} = 14.4 \times 10^{-3} \text{ W} = \mathbf{14.4\ mW}$$

Note again that all substitutions in Eqs. 3-9 and 3-10 are made in basic units. But this is not necessarily the case in Eq. 3-11.

3-13 COST OF ELECTRICAL ENERGY

In Sec. 3-7 we saw how energy could be calculated in joules by the use of Eq. 3-4.

$$W = VIt \qquad \text{joules (J)} \qquad (3\text{-}4)$$

In Example 3-8 we saw how a 60-W lamp, in use for just 1 min, consumed 3600 J of energy. Obviously, the unit of a joule is too small to measure large quantities of energy without obtaining large unwieldy numbers. The preferred unit of electrical energy is the kilowatt-hour, obtained from the following equation.

$$W = Pt \qquad \text{kilowatt-hours (kWh)} \qquad (3\text{-}11)$$

where: W is work or energy, in kilowatt-hours, kWh
P is power in kilowatts, kW
t is time in hours, h

The kilowatt-hour kWh is the basis for calculating the monthly electricity bill. The watt-hour *meter* indicates the amount of energy used (in kWh). This number of kWh is then multiplied by the cost per kWh established by the utility company. Rates vary widely, as shown by the following example.

EXAMPLE 3-12

A 250-W lamp is used 4 h a day for 30 days. Calculate:

a. The amount of energy used by the lamp.
b. The cost of the energy at 5¢/kWh.
c. The cost of the energy at 10¢/kWh.

SOLUTION

a. $W = Pt \qquad (3\text{-}11)$

$$= \left(250\text{ W} \times \frac{1\text{ kW}}{1000\text{ W}}\right) \times \left(4\,\frac{\text{h}}{\text{day}} \times 30\text{ days}\right)$$

$$= 0.25\text{ kW} \times 120\text{ h}$$

$$= \mathbf{30\ kWh}$$

b. $\text{Cost} = \text{kWh} \times ¢/\text{kWh}$

$$= 30\text{ kWh} \times \frac{5¢}{\text{kWh}} = \mathbf{\$1.50}$$

c. $\text{Cost} = 30\text{ kWh} \times \dfrac{10¢}{\text{kWh}}$

$$= \mathbf{\$3.00}$$

EXAMPLE 3-13

How much will it cost to run the $1\frac{1}{2}$-hp motor in Example 3-10 for 1 h a day, for 30 days, at 5¢/kWh?

SOLUTION

$W = Pt \qquad (3\text{-}11)$

$$= \left(1492\text{ W} \times \frac{1\text{ kW}}{1000\text{ W}}\right) \times \left(1\,\frac{\text{h}}{\text{day}} \times 30\text{ days}\right)$$

$$= 44.76\text{ kWh}$$

Therefore, $\text{Cost} = 44.76\text{ kWh} \times \dfrac{5¢}{\text{kWh}} = \mathbf{\$2.24}$

Note how it is possible for a low-powered device to consume a large amount of energy if it is operated for a long time period.

3-14 PRACTICAL RESISTORS

To this point we have considered resistance to be an inherent property of a circuit whose value we do not have much control over. But there are a number of applications when we intentionally want to insert resistance into a circuit. Such devices are called resistors; they may be used to

1. *Limit current* to a safe value.
2. *Drop voltage* to a required value.
3. *Divide* voltage to different values from a single source.
4. *Discharge energy* as from a capacitor.

Resistors are usually described in terms of

1. Their electrical resistance (in ohms).
2. Their ability to dissipate heat (in watts).
3. Their construction (carbon composition, wire-wound, film).

4 Their functional makeup (fixed or variable).

5 The percentage tolerance of their nominal resistive value.

3-14.1 Fixed Resistors

Fixed value, carbon composition resistors are the most common for electronic circuits because of their low cost. They exist in standard values (see Appendix A-2) from 2.2 Ω to 22 MΩ with power ratings from $\frac{1}{8}$ to 2 W. (Also available are resistors ranging in value from 100 MΩ to 1 million MΩ.) As shown by Fig. 3-4*a* this type of resistor consists of powdered carbon and insulating materials molded into a cylindrical shape with axial leads embedded at each end. A bakelite case provides insulation and mechanical protection. Colored bands or stripes on this outer covering indicate the value of resistance using a standard color code shown in Table 3-1.

To read the resistance value, hold the resistor with the colored stripes crowded toward the left-hand side and read from left to right. See Fig. 3-4*b*. The first two colors represent the first two digits, which must then be multiplied by the multiplier value represented by the third color. The fourth color indicates the tolerance, or how close the actual value should come to being equal to the "nominal" value shown by the first three colors. In some resistors, a fifth colored stripe is used to indicate the resistor's reliability or failure rate.

TABLE 3-1
Standard Color Code for Resistors

Color	First and Second Stripes Digit	Third Stripe Multiplier	Fourth Stripe Tolerance
Black	0	$10^0 = 1$	—
Brown	1	$10^1 = 10$	—
Red	2	$10^2 = 100$	—
Orange	3	$10^3 = 1000$	—
Yellow	4	$10^4 = 10{,}000$	—
Green	5	$10^5 = 100{,}000$	—
Blue	6	$10^6 = 1{,}000{,}000$	—
Violet	7	10^7	—
Gray	8	10^8	—
White	9	10^9	—
Gold	—	$10^{-1} = 0.1$	±5%
Silver	—	$10^{-2} = 0.01$	±10%
No color	—	—	±20%

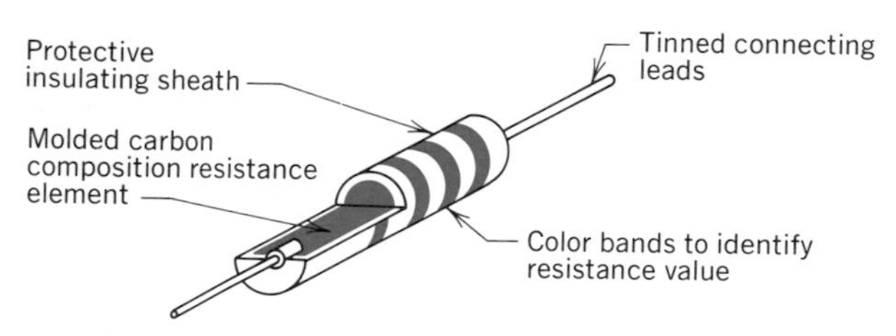

(*a*) Construction of a fixed, carbon composition resistor

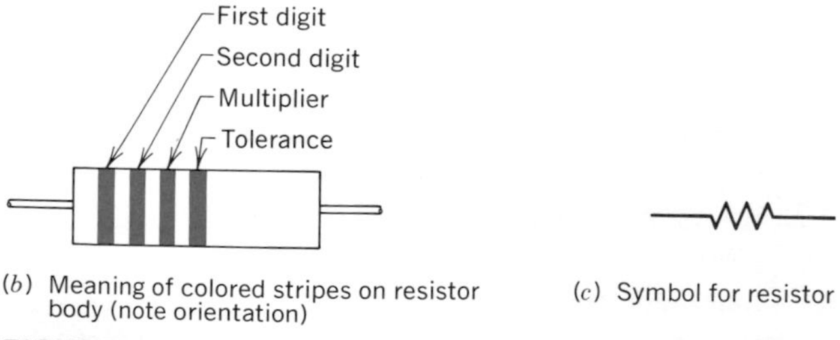

(*b*) Meaning of colored stripes on resistor body (note orientation)

(*c*) Symbol for resistor

FIGURE 3-4
Commercial resistors and resistor color code. *Note:* A wide band for the first digit in (*b*) indicates a wire-wound resistor.

EXAMPLE 3-14

What are the resistance values of resistors coded as follows:

a. Red, violet, brown, silver.
b. Brown, black, yellow, none.
c. Red, red, gold, gold.
d. Blue, gray, green, silver.

SOLUTION

a. red 2, violet 7, brown 10^1, silver ±10% = **270 Ω ± 10%**

b. brown 1, black 0, yellow 10^4, none ±20% = **100 kΩ ± 20%**

c. red 2, red 2, gold 0.1, gold ±5% = **2.2 Ω ± 5%**

d. blue 6, gray 8, green 10^5, silver ±10% = **6.8 MΩ ± 10%**

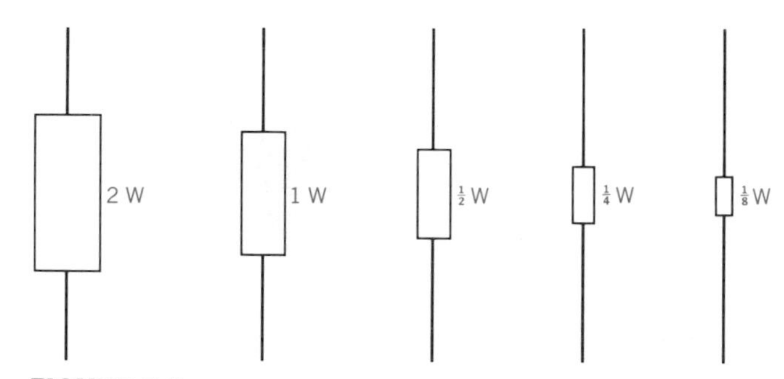

FIGURE 3-5
Full-size drawings of resistors of various power ratings.

If we consider the 270-Ω ± 10% resistor, its actual value may be anywhere between 270 Ω ± 27 Ω; that is, between 297 and 243 Ω. If this is too wide a spread for a given critical application, smaller tolerance resistors may be specified, such as 5%, 2%, and 1%—all at increased cost.

Note that there is no color to specify the power rating. This is determined by physical size (diameter and length). A 2-W resistor is larger and has more surface area than a 1-W resistor in order to dissipate more heat. Experience is necessary to recognize power ratings by size. Figure 3-5 shows full-size pictures of resistors of various power ratings.

Resistors with power ratings above 2 W are generally wire-wound and covered with a porcelain coating. Such power resistors, available up to hundreds of watts, may operate at temperatures as high as 300°C, without damage. Carbon composition resistors, 2 W and under, should not exceed 85°C because excessive heat produces changes in resistance value and may even cause the resistor to open.

Low-powered resistors are also available in packages that resemble integrated circuits. See Fig. 3-6. These may be of the single in-line package (SIP) or dual in-line package (DIP). Fabricated from thin film and thick film metals, often laser-trimmed to close tolerances, these packages may contain eight fully isolated resistors of equal value typically from 100 Ω to 100 kΩ, or some resistor network combination. They are especially useful where matched resistor pairs within ±0.1% to ±0.5% are required.

FIGURE 3-6
Resistors in single in-line package (SIP) and dual in-line package (DIP). (Courtesy of Allen Bradley Co.)

3-14.2 Variable Resistors

Resistors that can be varied in value are generally of the three-terminal type. **Strictly speaking, if all three terminals are actively used, the device is a potentiometer and is used for varying voltage.** This will be explored in the next chapter. **If there are only two active terminals, the device is a variable resistor or *rheostat,* used for varying current.** However, any three-terminal potentiometer can be wired as a rheostat.

Figure 3-7*a* shows a typical wire-wound variable resistor, suitable for power ratings from 2 to 300 W and total resistance values from 1 Ω up to 10 kΩ. They are typically used as rheostats. The ends of the resistance wire are connected to terminals 1 and 3. Terminal 2 is connected to a wiper that slides and makes contact with the wire. The wiper moves when the shaft is rotated.

Low-power variable resistors, often called controls, use a carbon composition for the resistance element. They are available in power ratings from ¼ to 2 W with standard values of 100, 250, 500, 1 k, 2.5 k, 5 k, 10 k, 25 k, 50 k, 100 k, 500 k, and 1 M ohm. Their primary use is as a potentiometer, as in volume controls; or as occasionally adjusted variable resistances called "trimmers." Wire-wound types are also available in slightly higher power ratings.

Figures 3-7*c* and 3-7*d* indicate two methods of wiring a potentiometer as a rheostat. If, for example, the rheostat is 100 Ω, the total resistance from terminal 1 to 3 (in Fig. 3-7*c*) is 100 Ω and does not change. If the shaft is turned, moving terminal 2 "upward" toward terminal 1, the resistance between terminals 1 and 2 will decrease toward zero and the lamp will have maximum current and brightness. Moving the shaft the opposite way will introduce more resistance up to a maximum of 100 Ω, reducing the current. The light will dim. The advantage of the connection in Fig. 3-7*d* is that should a "worn spot" occur between the sliding contact and the resistance, there would be no complete interruption of current, since a path would still exist from 1 to 3.

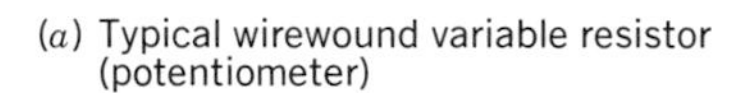

(*a*) Typical wirewound variable resistor (potentiometer)

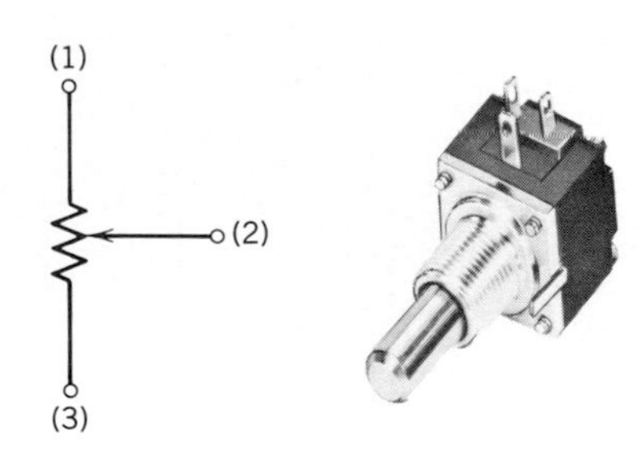

(*b*) Symbol for a potentiometer and photograph

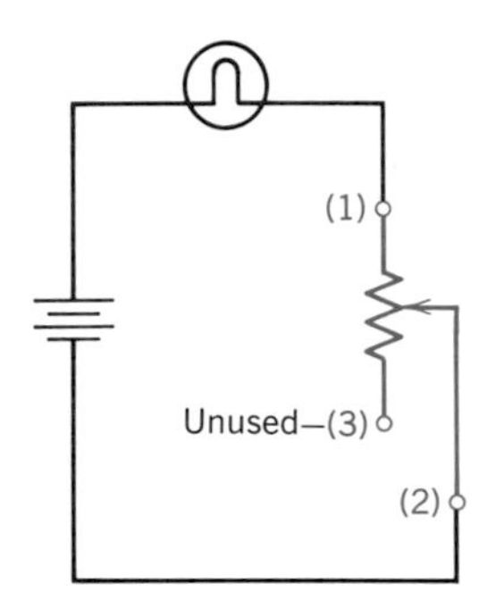

(*c*) Potentiometer used as a rheostat (no connection to terminal 3)

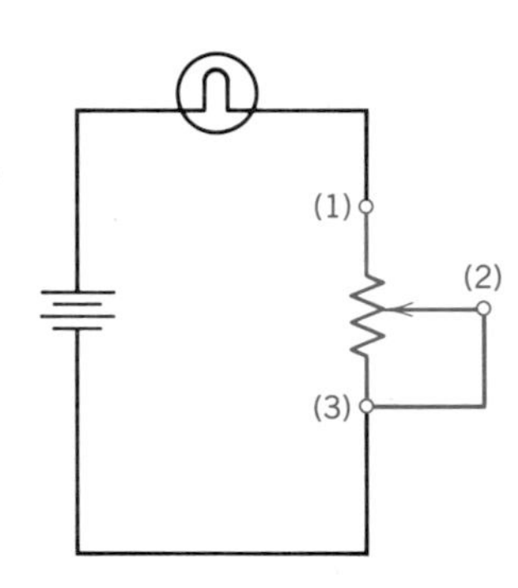

(*d*) Potentiometer used as a rheostat (only two active terminals)

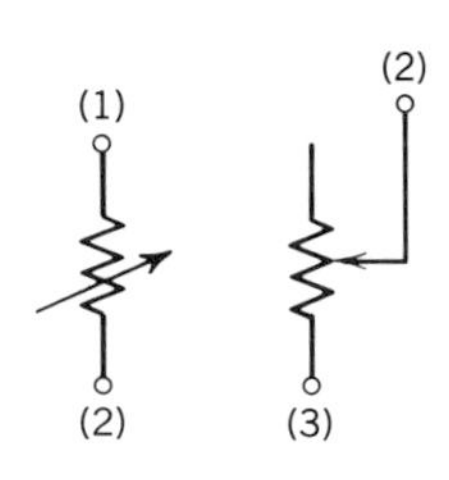

(*e*) Alternate symbols for variable resistor or rheostat

FIGURE 3-7
Variable resistors. [Photograph in (*b*) courtesy of Allen Bradley Co.]

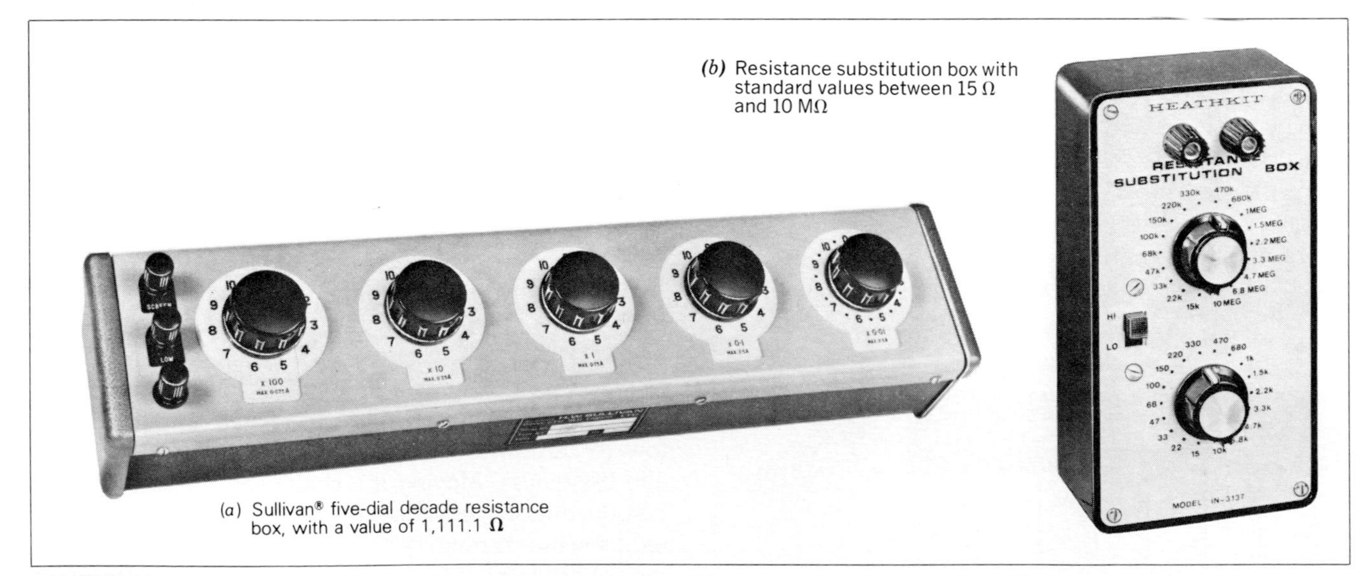

(*a*) Sullivan® five-dial decade resistance box, with a value of 1,111.1 Ω

(*b*) Resistance substitution box with standard values between 15 Ω and 10 MΩ

FIGURE 3-8
Adjustable resistors. [Photograph in (*a*) courtesy of James G. Biddle Co.; Photograph in (*b*) courtesy of Heath Co.]

One type of variable resistor that is very useful in laboratories is a decade resistance box. One type is shown in Fig. 3-8*a*. Some seven-dial decades can produce resistances values from 1 Ω to 10 MΩ in 1-Ω steps, often with an accuracy to within 0.1%, by "dialing" the value required. Each dial controls up to 10 series-connected resistors, which in turn can be connected in series with the other dialed resistors. One precaution is to make sure that the current capability of the highest-valued resistance is not exceeded. These maximum current values are usually shown on each dial.

Another method of obtaining different resistance values is to use a *resistance substitution box*. See Fig. 3-8b for a box that provides standard values of resistance from 15 Ω to 10 MΩ, with a tolerance of ±10%. This is useful when experimenting to determine which standard resistor value to use in a circuit.

SUMMARY

1. A voltmeter connected *across* the source indicates the voltage applied to a circuit, and an ammeter in *series* with the load indicates the load current.

2. Ohm's law gives the relation between voltage, current, and resistance in a circuit: for a fixed resistance, the ratio of potential difference to current is a constant called the resistance, $R = V/I$.

3. The linear relationship between voltage and current for any constant resistance is a statement of Ohm's law and can be shown graphically as a straight line graph of I plotted against V.

4. Other forms of Ohm's law allow the determination of current and voltage given the other two variables, as shown by $I = V/R$ and $V = IR$.

5. A resistor symbol in a circuit can represent an electrical load, such as an incandescent lamp or heating element. A reduction in resistance is an increase in load.

6. Mechanical work can be calculated using $W = Fs$. The unit of work in the FPS system is the ft lb; in the MKS system it is the N · m or joule; 1 N · m = 1 J.

7. Electrical work and energy are also measured in the joule, using either $W = VQ$ or $W = VIt$ for calculations.

8. Efficiency is the ratio of useful work out of a machine compared with the total input work; $\eta = W_{out}/W_{in}$.

9. Power is the *rate* at which work is done or energy is converted, $P = W/t$.

10. The unit of mechanical power in the FPS system is the ft lb/s or the hp, 550 ft lb/s = 1 hp; in the MKS system the unit is the J/s or watt; 1 J/s = 1 W.

11. Electrical power may be calculated using $P = VI$, $P = I^2R$, or $P = V^2/R$. The basic unit of measurement is the watt. 1 kW = 1000 W, 1 hp = 746 W.

12. Efficiency can also be defined as follows: $\eta = P_{out}/P_{in}$. It is usually expressed as a percentage.

13. The practical unit for measuring large quantities of electrical energy is the kilowatt-hour, using the equation $W = Pt$.

14. The cost of electrical energy is the product of the kilowatt-hours used and the cost per kilowatt-hour.

15. Practical resistors are devices that have a known amount of resistance (in certain standard values) and can be used to intentionally introduce resistance to a circuit.

16. Resistors may be fixed in value and identified by a standard color code given in Table 3-1.

17. Variable resistors generally have three terminals and are known as rheostats when used to vary current in a circuit.

18. A decade resistance box is a special type of variable resistance in which almost any desired value can be "dialed" to within 0.1% accuracy.

SELF-EXAMINATION

Answer true or false, or fill in the correct answer
(Answers at back of book)

3-1. A voltmeter is connected in parallel, an ammeter in series. ________

3-2. A voltmeter may be removed from an operating circuit without affecting the circuit, an ammeter cannot. ________

3-3. In a circuit with constant resistance, an increase in voltage results in a proportional increase in current. ________

3-4. If applied voltage is held constant, an increase in resistance results in an increase in current. ________

3-5. If the current through a resistor drops, the voltage drop across the resistor becomes smaller. ________

3-6. A resistor of 2.2 MΩ represents a load that is 1000 times larger than a 2.2 kΩ load. ________

3-7. Work and energy are essentially the same thing. ________

3-8. Power and energy are essentially the same thing. ________

3-9. Mechanical and electrical work are measured in the same units in the MKS system. ________

3-10. Efficiency is the ratio of input work or power to useful output work or power. ________

3-11. Both mechanical and electrical power involve the rate of doing work or converting energy. ________

3-12. Mechanical and electrical power are both measured in watts in the MKS system. ________

3-13. A 100% efficient 2-hp motor would require 1492 W of electrical input power. ________

3-14. Where current flows through a resistor the power dissipated in the resistor depends upon the square of the current. ________

3-15. If the voltage applied to a constant resistance doubles, the power dissipated in the resistance also doubles. ________

3-16. Electrical energy may be measured in joules or kilowatt-hours. ________

3-17. If 500 kWh of energy are used, the total cost at 5¢/kWh is $250. ________

3-18. The resistance of a red, yellow, and green resistor is 2.4 MΩ. ________

3-19. The color stripes for a 100-Ω resistor are brown, black, black. ________

3-20. A rheostat is a two-terminal device that varies resistance and current. ________

3-21. A 120-V appliance that draws 12 A has a resistance of ________.

3-22. A 220-V oven with a resistance of 11 Ω draws a current of ________.

3-23. A 5-mA current through a 10-kΩ resistor causes a voltage drop of ________.

3-24. A 120-V, 360-W television receiver draws a current of ________.

3-25. A 6.6-kW oven that requires a current of 30 A must operate at a voltage of ________.

3-26. A 10-Ω resistor with 2 A of current through it must be able to dissipate a power of ________.

3-27. If a 220-V source must deliver 2.2 kW to a heating element, the resistance of the element must be ________.

3-28. A 200-W lamp used for 6 h consumes an energy of ________.

3-29. If energy costs 20¢/kWh, the cost of running a 100-W lamp for 10 h/day for 10 days is ________.

3-30. A 470-kΩ, ±5% resistor, has color stripes of ________.

REVIEW QUESTIONS

1. State Ohm's law in your own words.

2. Draw a graph of I versus V for a 2-kΩ resistor when the voltage is varied in 2-V steps from 0 to 10 V. Find the slope of this graph (rise/run) using the units on the axes. How is this number representing the slope related to 2 kΩ?

3. On the graph drawn in Question 2 draw a second graph for a 4-kΩ resistor. Does it have a larger or smaller slope than that for the 2 kΩ? Explain why the current through a 3-kΩ resistor at 10 V is *not* given by a point on the above graph located *midway* between the 2- and 4-kΩ lines at 10 V. *Hint:* Draw a graph of I versus R for a constant voltage of 10 V with R varying from 0 to 5 kΩ.

4. Explain why a decrease in load resistance should be considered an increase in load.

5. Explain the difference between work and power.

6. Explain how efficiency can be given in terms of work *or* power ratios.

7. Which is a larger unit of power, the horsepower or the kilowatt?

8. Explain how the *power* used by a 100-W lamp is the same no matter whether the lamp is lit for a second, a minute, or an hour; but the *energy* used is different.

9. Under what conditions would a 100-W lamp use more energy than a 1500-W toaster?

10. Which is the larger unit of energy, the joule or the watt-hour?

11. What is meant by the nominal value of a resistor?

12. Explain why brown, black, black does not represent a 100-ohm resistor. How many stripes would be necessary to represent a 1-million-ohm resistor if a colored stripe were used for each digit? What is the largest theoretical value of resistance the standard color code can represent? What is the smallest?

13. Explain how a three-terminal potentiometer can be wired as a rheostat with two active terminals. What is the advantage of connecting the center terminal to one of the end terminals?

14. What maximum value of resistance can be dialed by a five-dial decade resistance box, whose smallest increments are one tenth of an ohm? Which resistance will determine the current capability of such a value?

PROBLEMS

(Answers to odd-numbered problems at back of book)

3-1. What is the resistance of a circuit that draws 1.5 A from a 220-V source of emf?

3-2. What resistance will limit the current in a circuit to 3.75 A when connected to a 120-V source?

3-3. A voltmeter across a resistor indicates 15 V when an ammeter in series with the resistor indicates 2.5 mA. What is the resistance of the resistor?

3-4. What resistance will allow 25 μA of current when connected across 40 mV?

3-5. A voltmeter connected across a precision 1-kΩ resistor indicates 2.2 V. What is the current through the resistor?

3-6. How much current will flow through a 2.4-MΩ resistor when it is connected across a 1.5-V cell?

3-7. What voltage drop occurs across a 680-Ω resistor that has a current of 450 mA through it?

3-8. What voltage must be applied to a 470-kΩ resistor to obtain a current of 150 μA?

3-9. What work must be done against gravity by the engine in a 3000-lb car to raise it from sea level to 10,000 ft?

3-10. Determine the work done in Problem 3-9 in joules.

3-11. If the work done by the car in Problem 3-9 takes half an hour, calculate the minimum power of the engine in horsepower.

3-12. Determine the power (in kW) of the car in Problem 3-10 if this work is done in half an hour.

3-13. An electric heating element operating at 120 V draws 10 A to boil a kettle of water in 5 min. Calculate how much electrical energy is supplied during this time:

a. in joules

b. in kWh

3-14. An electric clothes dryer operating at 220 V draws 15 A to dry a load of clothes in half an hour. Calculate how much electrical energy is supplied:

a. in joules

b. in kWh

3-15. If the water in the kettle in Problem 3-13 absorbs 300,000 J of energy in the form of heat, determine the efficiency of the kettle.

3-16. If the efficiency of the dryer is 60%, how much energy in the form of heat is absorbed by the clothes in Problem 3-14?

3-17. Calculate the cost of operating the kettle in Problem 3-13 if energy costs 10¢/kWh.

3-18. Calculate the cost of operating the clothes dryer in Problem 3-14 if energy costs 10¢/kWh.

3-19. a. How many joules of energy are equivalent to 1 kWh of energy?

b. What is 1 joule equal to in kWh?

3-20. The annual energy use of the world is 0.3 million billion megajoules (MJ).

a. Express this quantity in kWh.

b. If the total content of proved fossil fuel reserves (oil, coal, and gas) is 25 million billion MJ, how long will this last at the present rate of consumption?

3-21. A steam iron operates at 120 V and draws 12 A. Calculate its power rating in

a. kW

b. hp

3-22. What current is drawn from a 220-V source by a 6-kW heating element?

3-23. At what voltage must a 55-W soldering iron operate if it uses 0.47 A?

3-24. A 220-V motor drawing 12 A develops 3 hp. What is its efficiency?

3-25. What current must be supplied to an 85%-efficient 25-hp motor operating at 550 V?

3-26. Which motor costs more to run, a 60%-efficient $1\frac{1}{2}$-hp motor or an 80%-efficient 2-hp motor?

3-27. What is the maximum current that a 10-kΩ, $\frac{1}{2}$-W resistor can handle without overheating?

3-28. What is the maximum current that a 2.2-MΩ, $\frac{1}{4}$-W resistor can handle without overheating?

3-29. What maximum voltage can safely be connected across a 680-Ω, 1-W resistor?

3-30. What maximum voltage can safely be connected across a 10-Ω, 2-W resistor?

3-31. What must be the nearest standard power rating of a 470-Ω resistor that has a current of 50 mA running through it?

3-32. What is the minimum power rating of a 5.6-kΩ resistor that has a voltage drop of 50 V across it?

3-33. What are the resistance values of resistors coded as follows:

a. Orange, orange, orange, silver
b. Green, brown brown, gold
c. Red, red, green
d. Blue, gray, gold, gold
e. Brown, black black, silver
f. Yellow, violet, red, gold
g. Violet, green, brown, gold
h. Orange, white, green, silver

3-34. What are the color stripes for the following resistors:

a. 11 kΩ, ±5%
b. 820 Ω, ±10%
c. 15 MΩ, ±20%
d. 3 Ω, ±5%
e. 270 kΩ, ±5%
f. 16 Ω, ±5%
g. 4.7 MΩ, ±10%
h. 56 Ω, ±10%

3-35. Select the nearest standard value of 10% resistor for the following:

a. 16 kΩ
b. 290 kΩ
c. 5.9 MΩ

3-36. Repeat Problem 3-35 using 5% resistors.

CHAPTER FOUR
RESISTANCE

Good conductors are those materials whose atomic electronic structures readily provide free electrons. This ability is more easily identified by knowing a property called the material's *resistivity,* a smaller value indicating a better conductor. When combined with other factors of length and cross-sectional area, we have an equation that will enable us to determine the resistance of any conductor, at a specific temperature of 20°C. At any other temperatures, a correction is applied requiring the definition of the *temperature coefficient* of resistance.

In this chapter we also consider *linear resistors*—those that obey the proportionality of V to I in Ohm's law—and *nonlinear* resistors, such as thermistors, in some of their current-surge limiting applications.

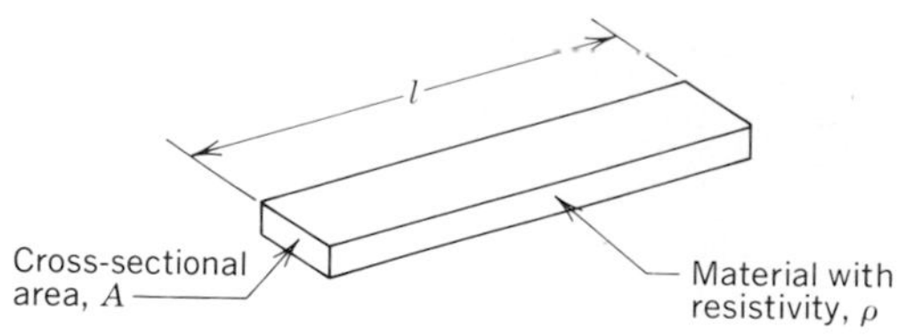

FIGURE 4-1
Factors that determine the resistance of a material at a given temperature.

4-1 FACTORS AFFECTING RESISTANCE

We have seen how resistance to current is caused by the collisions of free electrons within the specific conductive material. Anything that increases such collisions will contribute to an increase in resistance. For a conductor at a given temperature, Fig. 4-1 and Eq. 4-1 show the factors that determine resistance.

$$R = \frac{\rho l}{A} \quad \text{ohms } (\Omega) \qquad (4\text{-}1)$$

where: R is the resistance of a material in ohms, Ω
l is the length of the material in meters, m
A is the cross-sectional area of the material in square meters, m^2
ρ is the resistivity (specific resistance) of the material in ohm meters, $\Omega \cdot m$

Equation 4-1 implies:

1. Resistance is *directly* proportional to length (more collisions for a longer wire).
2. Resistance is *inversely* proportional to cross-sectional area (a smaller cross section will increase the current density and the number of collisions that occur).
3. Resistance depends upon the type of material, as indicated by the specific resistance or *resistivity*, ρ (rho).

Table 4-1 provides a brief list of resistivities at room temperature for some common conductors, semiconductors, and insulators. These numbers actually give the resistance of a specific volume of material. In fact, another name for resistivity is specific resistance. For example, ρ for copper = 1.72×10^{-8} ohm meter ($\Omega \cdot m$) at 20°C. That is, a piece of copper 1 m long and having a

TABLE 4-1
Resistivities at Room Temperature, 20°C

Substance		ρ, $\Omega \cdot m$
Conductors		
Metals	Silver	1.47×10^{-8}
	Copper	1.72×10^{-8}
	Gold	2.45×10^{-8}
	Aluminum	2.63×10^{-8}
	Tungsten	5.51×10^{-8}
	Nickel	7.8×10^{-8}
Alloys	Manganin	44×10^{-8}
	Constantan	49×10^{-8}
	Nichrome	100×10^{-8}
Semiconductors		
Pure	Carbon	3.5×10^{-5}
	Germanium	0.60
	Silicon	2300
Insulators		
	Wood	10^{8}–10^{11}
	Paper	10^{10}
	Nylon	8×10^{12}
	Lucite	10^{13}
	Glass	10^{10}–10^{14}
	Rubber	10^{13}–10^{14}
	Amber	5×10^{14}
	Mica	10^{11}–10^{15}
	Sulfur	10^{15}
	Teflon	10^{15}
	Porcelain	10^{16}
	Polystyrene	10^{16}
	Quartz (fused)	75×10^{16}

cross-sectional area of 1 m^2, has a resistance of $1.72 \times 10^{-8}\ \Omega$. Note in Example 4-1 how the units for resistivity (in $\Omega \cdot m$) give the correct unit for resistance in Eq. 4-1.

EXAMPLE 4-1

The type of copper wire used for house wiring has a diameter of approximately 2 mm.

a. What resistance will a 100-m-long reel of copper wire have at room temperature?
b. What is the resistance if the wire has the same dimensions but is made of aluminum?

SOLUTION

a. Area, $A = \dfrac{\pi d^2}{4}$

$$= \frac{\pi}{4} \times (2 \times 10^{-3}\ \text{m})^2 = 3.14 \times 10^{-6}\ \text{m}^2$$

$$R = \frac{\rho l}{A} \qquad (4\text{-}1)$$

$$= \frac{1.72 \times 10^{-8}\ \Omega \cdot \text{m} \times 100\ \text{m}}{3.14 \times 10^{-6}\ \text{m}^2} = \mathbf{0.55\ \Omega}$$

b. For aluminum, $\rho = 2.83 \times 10^{-8}\ \Omega \cdot$ m. Therefore,

$$R = R_{Cu} \frac{\rho_{Al}}{\rho_{Cu}}$$

$$= 0.55\ \Omega \times \frac{2.83 \times 10^{-8}\ \Omega \cdot \text{m}}{1.72 \times 10^{-8}\ \Omega \cdot \text{m}} = \mathbf{0.90\ \Omega}$$

NOTE In the solution of Example 4-1b, it is not necessary to resubstitute for l and A in solving the problem. Since the only factor that changed is ρ, and the new value of ρ for aluminum is higher than copper, the resistance of copper is multiplied by an *increased* ratio of resistivity.

EXAMPLE 4-2

A 100-W incandescent lamp uses a tungsten filament that is approximately 3 cm long. If its cold (20°C) resistance is 10 ohms, what is its diameter? ($\rho_{\text{tungsten}} = 5.5 \times 10^{-8}\ \Omega \cdot$ m.)

SOLUTION

$$R = \frac{\rho l}{A} \qquad (4\text{-}1)$$

Therefore, $A = \dfrac{\rho l}{R}$

$$= \frac{5.5 \times 10^{-8}\ \Omega \cdot \text{m} \times 3 \times 10^{-2}\ \text{m}}{10\ \Omega}$$

$$= 1.65 \times 10^{-10}\ \text{m}^2$$

$$A = \frac{\pi d^2}{4}$$

Therefore, $d = \sqrt{\dfrac{4A}{\pi}}$

$$= \sqrt{\frac{4 \times 1.65 \times 10^{-10}\ \text{m}^2}{\pi}}$$

$$= 1.45 \times 10^{-5}\ \text{m} = \mathbf{0.0145\ mm}$$

NOTE In both Examples 4-1 and 4-2, regardless of the units in which the information is given, it is first necessary to convert these units to basic units before substitution in the appropriate equation.

4-2 WIRE GAUGE TABLE

Instead of having to calculate resistance (using Eq. 4-1) each time for a given length and diameter of copper wire, a table is available for *standard* wire sizes. This table is given in Appendix A-3. It is based on a system called the American Wire Gauge (AWG), or Brown and Sharpe (B&S) gauge. These are still in use in the United States, although metric equivalents are also given in the table. This system is actually based on measuring the cross-sectional area of the conductors in units called *circular mils*.

If the diameter in mils (1 mil = 0.001 in.) is squared, a measure of the area is obtained in circular mils (e.g., a 0.030-in.-diameter wire has a circular mil area of 900). In fact, some wires larger than 0000 gauge are called 250 MCM, for example, meaning that they have a cross-sectional area of 250,000 circular mils. It can be shown that a decrease in gauge number by 3 (e.g., from 14 to 11) approximately *doubles the area.* Each decrease by a single gauge number increases the area by a factor of 1.26 approximately. **Clearly, smaller gauge numbers are used to describe wires with a larger diameter.**

Wiring in the home varies from No. 12 or 14 for lighting circuits (20 A to 15 A) to No. 8 for stoves and other heavy loads. An electronic circuit may use only No. 22 gauge, good for about 1 A.

4-3 INSULATORS AND SEMICONDUCTORS

As indicated in Table 4-1, insulators have very high values of resistivity. They are used to coat conductors (with rubber, plastic, paper, etc.) to prevent unwanted contact (shorts) with other wires or a metal chassis. However, it is possible for an insulator to *break down* and conduct if a high-enough voltage is connected across it. Under these conditions, an extremely high force acts on the valence electrons to break their bonds and rupture the insulator. The resulting current that flows is usually accompanied by a great deal of heat and in the case of solid insulators leads to their destruction.

Air is a relatively poor insulator compared with mica, for example. A 1-mm air gap will break down with only 4000 V across it whereas mica requires 100,000 V for the same thickness before rupture will occur. Table 4-2 lists some breakdown voltages for various insulating materials. Typical house wiring is insulated for 600 V, but some electronic and power circuits must be able to withstand many thousands of volts.

The resistivity of pure silicon is appreciably higher (in its pure state) than all good conductors but very much less than the resistivity of insulators. Thus silicon is known as a *semiconductor.* But when used in solid-state devices, it is modified or *doped* by adding other materials that increase its conductivity at least a thousand times. It is the way in which this doping changes the conductivity that makes silicon so important in its use in semiconductor devices such as diodes and transistors.

4-4 TEMPERATURE COEFFICIENT OF RESISTANCE

In general, for most metals, an increase in temperature increases the resistance of a conductor in a fairly linear manner over a few hundred degrees Celsius (°C). This increase in resistance is due to the increased molecular activity and number of collisions of free electrons due to heat imparted to the metal. The higher temperature raises the electron energy (thermal agitation) of the electrons in the valence and conduction shells of the atoms. Their more-pronounced vibrations increase the chance of collisions with the free electrons that are trying to move through the material.

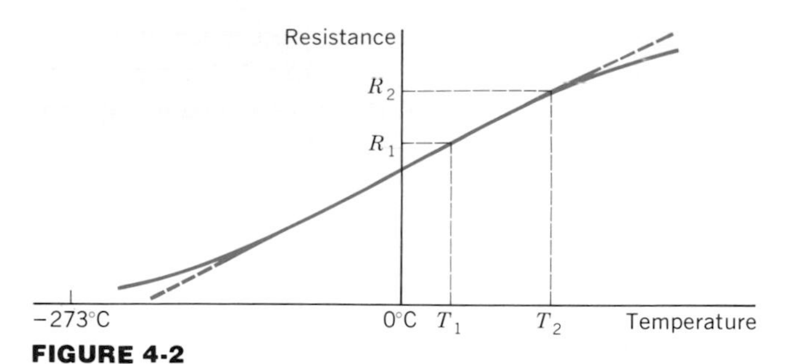

FIGURE 4-2
Variation of resistance with temperature for a metal conductor.

Figure 4-2 shows a typical variation of resistance with temperature for a metal conductor. The graph becomes nonlinear for both very high and very low temperatures, approaching zero resistance at −273°C (absolute zero). Some special alloys exhibit a property called *superconductivity* in which their resistance abruptly drops to zero at temperatures that range between 0.1 and 20°C within absolute zero. A current, once established in a superconducting circuit, will continue almost indefinitely without the need for an emf.

TABLE 4-2
Average Breakdown Voltages for Typical Insulators

Material	Breakdown Voltage (kV/mm)
Air	4
Porcelain	8
Transformer oil	16
Plastic	16
Paper	20
Rubber	28
Glass	28
Teflon	60
Mica	100

The resistance R_2 at a different temperature T_2 may be calculated if the resistance R_1 is known at some temperature T_1 usually taken to be 20°C (room temperature), by using Eq. 4-2.

$$R_2 = R_1 [1 + \alpha_1(T_2 - T_1)] \quad \text{ohms } (\Omega) \quad (4\text{-}2)$$

where: R_2 is the final resistance (at temperature T_2) in ohms, Ω

R_1 is the initial resistance (at temperature T_1) in ohms, Ω

T_2 is the final temperature, in °C

T_1 is the initial temperature, 20°C

α_1 is the temperature coefficient of resistance, per °C, quoted for the material at room temperature, $T_1 = 20°C$

Table 4-3 gives temperature coefficients (α) for some common materials. Note that most are positive temperature coefficients (PTC), meaning that an increase in resistance results from a rise in temperature. Carbon, semiconductors and most insulators have a negative temperature coefficient (NTC), indicating a *drop* in resistance with temperature rise. This is because, in these materials, the current carriers come from electrons *freed* from the valence shells by heat. That is, it requires much more energy to raise an electron from the valence shell and become available for conduction than in a metal conductor. Therefore, any increase in temperature increases the *availability* of current carriers.

TABLE 4-3

Temperature Coefficients of Resistance for Common Materials at 20°C

Material	α_1, per °C
Aluminum	0.0039
Brass	0.0020
Carbon	−0.0005
Constantan (Cu 60, Ni 40)	+0.000002
Copper	0.00393
Gold	0.0034
Iron	0.0050
Lead	0.0043
Manganin (Cu 84, Mn 12, Ni 4)	0.000000
Mercury	0.00088
Nichrome	0.0004
Nichrome II	0.00016
Nickel	0.006
Platinum	0.003
Silver	0.0038
Tungsten	0.0045

EXAMPLE 4-3

A No. 2-gauge copper wire is used in a 10-km-length overhead transmission line. What will be its resistance at:

a. 20°C?

b. 100°F?

c. −40°F?

SOLUTION

a. From Appendix A-3, resistance for No. 2 copper = 0.511 Ω/km, at 20°C. Therefore, resistance for 10 km = 10 km × 0.511 Ω/km = **5.11 Ω** at 20°C.

b. $100°F = (100 - 32)\frac{5}{9}°C = 37.8°C.$

$$R_2 = R_1[1 + \alpha_1(T_2 - T_1)] \quad (4\text{-}2)$$

$$= 5.11\ \Omega \left[1 + \frac{0.00393}{°C}(37.8 - 20)°C\right]$$

$$= 5.11\ \Omega\ [1 + 0.07] = \mathbf{5.47\ \Omega}$$

c. $-40°F = (-40 - 32)\frac{5}{9}°C = -40°C.$

$$R_2 = R_1[1 + \alpha_1(T_2 - T_1)] \quad (4\text{-}2)$$

$$= 5.11\ \Omega \left[1 + \frac{0.00393}{°C}(-40 - 20)°C\right]$$

$$= 5.11\ \Omega[1 - 0.236] = \mathbf{3.90\ \Omega}$$

EXAMPLE 4-4

The resistance of a 100-W tungsten filament lamp increases from 10 Ω at room temperature to 144 Ω when lit. Assuming that this increase in resistance with temperature is linear, calculate the temperature of the hot filament.

SOLUTION

$$R_2 = R_1[1 + \alpha_1(T_2 - T_1)] \qquad (4\text{-}2)$$

Solve for T_2:

$$\alpha_1(T_2 - T_1) = \frac{R_2}{R_1} - 1$$

Hence, $T_2 = \dfrac{1}{\alpha_1}\left(\dfrac{R_2}{R_1} - 1\right) + T_1$

$$= \frac{1°\text{C}}{0.0045}\left(\frac{144\ \Omega}{10\ \Omega} - 1\right) + 20°\text{C}$$

$$= 2978°\text{C} + 20°\text{C} = 2998°\text{C} \approx \mathbf{3000°C}$$

This calculation compares favorably with typical temperatures of 2500°C. Of course, the calculation assumed complete linearity over the wide temperature range which is not the case.

4-5 LINEAR RESISTORS

We saw in Section 3-3 that a resistor that has a constant V/I ratio is known as a *linear* resistor. We now know that when current passes through a resistor, electric energy is converted into heat. This heat has a tendency to cause a slight increase in resistance in most common conductors. However, over the usual temperature range, this change in resistance is so small that materials such as copper, aluminum, and so on, are considered to be linear resistors.

The *carbon composition* resistors (Section 3-14) are also considered to be linear resistors. Figure 4-3 shows that these resistors tend to increase their resistance slightly for any *appreciable* change in temperature from room temperature.*

Metal film resistors, often used for higher precision values of resistance (2% and less), have a very high temperature stability. This is given, for example, as ±50 PPM/°C (50 parts per million per °C, or a change of only 0.005% per °C.) In very critical applications, such as in some accurate measuring instruments, a material called *constantan* is often used. As the name implies, this

* This characteristic is needed in order to prevent "thermal runaway." If the resistance were to decrease (as temperature increases), this would draw more current, increasing the temperature, reducing resistance still further, and the resistor could eventually be destroyed. A *positive* temperature coefficient provides *thermal stability*. A *negative* temperature coefficient leads to *thermal instability*.

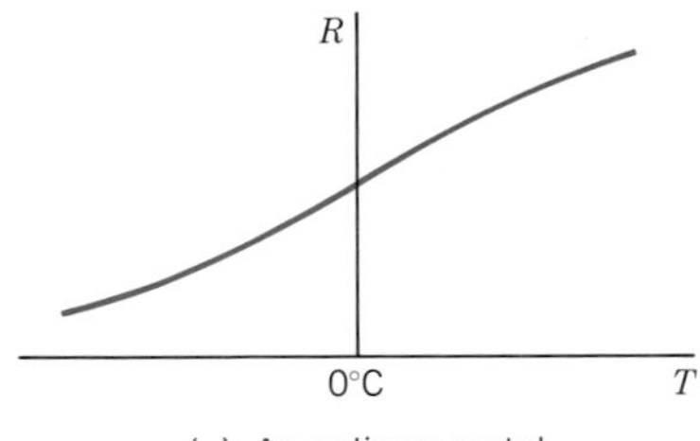

(*a*) An ordinary metal

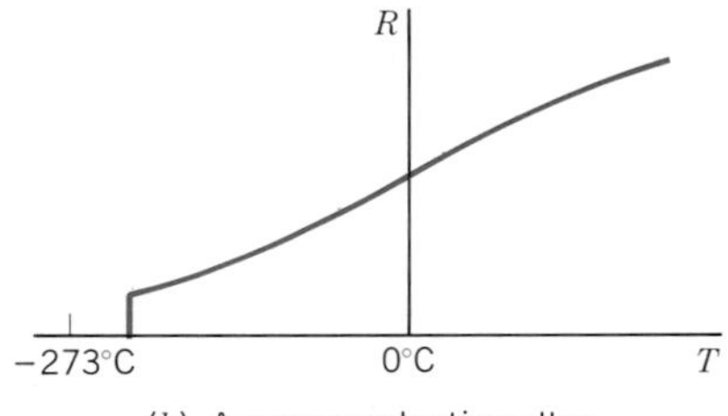

(*b*) A superconducting alloy

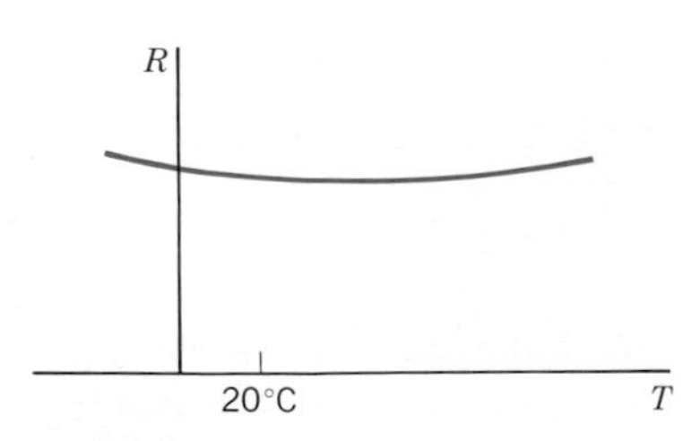

(*c*) A carbon composition resistor

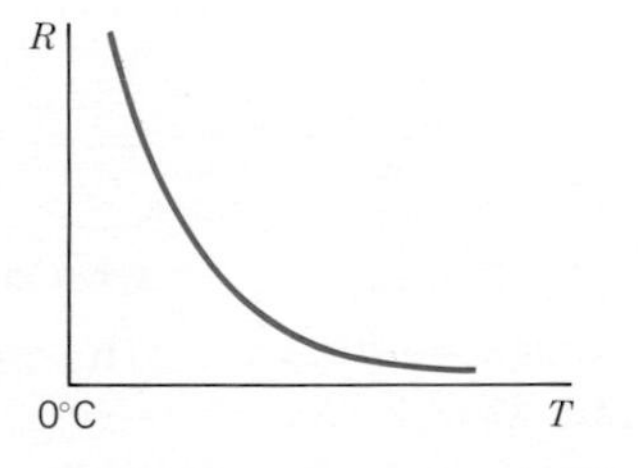

(*d*) A semiconductor (thermistor)

FIGURE 4-3
Variation of resistance with temperature for various materials.

material has an almost zero temperature coefficient and is used to maintain a practically constant resistance when the temperature changes.

4-6 NONLINEAR RESISTORS

The tungsten filament of an incandescent lamp experiences a large change in resistance over its operating range of temperature. This was shown by Example 4-4. A graph of I versus V is not a straight line, so tungsten is a nonlinear resistance in this application. The low initial (cold) resistance of the lamp causes an inrush current when the lamp is first switched on, as shown by Example 4-5.

EXAMPLE 4-5

A 120-V, 60-W incandescent lamp has a cold (20°C) resistance of 18 Ω. Calculate:

a. The initial inrush of current when connected to a 120-V source.

b. The steady operating current of the lamp.

c. The hot resistance of the lamp.

If the lamp takes 1 ms to reach its operating temperature, sketch a graph of the current versus time.

SOLUTION

a. Initial inrush, $I = \dfrac{V}{R}$ (3-1a)

$$= \frac{120 \text{ V}}{18 \ \Omega} = \mathbf{6.7 \ A}$$

b. Final current, $I = \dfrac{P}{V}$

$$= \frac{60 \text{ W}}{120 \text{ V}} = \mathbf{0.5 \ A}$$

c. Hot resistance, $R = \dfrac{V}{I}$ (3-1)

$$= \frac{120 \text{ V}}{0.5 \text{ A}} = \mathbf{240 \ \Omega}$$

See Fig. 4-4.

The lamp's current surge is very short lived because the very small mass of the filament quickly becomes hot enough to reach its hot resistance of 240 ohms. However, this inrush of current does cause a condition of "thermal shock" for the filament. The sudden increase in temperature is accompanied by a sudden expansion of the metal, causing the filament to bend and flex. The repeated flexing of the filament each time the lamp is switched on is partially responsible for the eventual filament rupture and failure of the lamp. (Doesn't a lamp often "blow" or burn out at the instant of switching it on?)

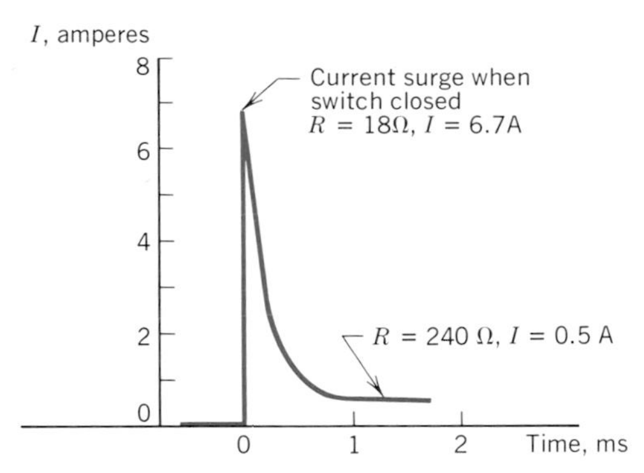

FIGURE 4-4
Graph of the inrush current of an incandescent lamp for Example 4-5.

4-6.1 Thermistors

A method of eliminating current inrush and extending the life of an incandescent lamp is to place a *thermistor* in *series* with the lamp, as shown in Fig. 4-5.

A thermistor is a semiconductor made from metallic oxides. It has a very large, negative temperature coefficient. A thermistor is a nonlinear "thermal resistor" with characteristics shown in Fig. 4-3*d*. Thermistors are manufactured in many different shapes, as shown in Fig. 4-6.

At room temperature the thermistor may have a resistance of approximately 150 Ω so that the total circuit resistance of 168 ohms would allow an initial current of only $\dfrac{120 \text{ V}}{168 \ \Omega} = 0.7$ A. See Fig. 4-5*a*. As this current flows through both thermistor and filament, the thermistor resistance decreases while the filament resistance increases. The thermistor resistance will eventually drop to approximately 1 ohm (Fig. 4-5b), which is negligible compared to the filament. But its characteristic permits

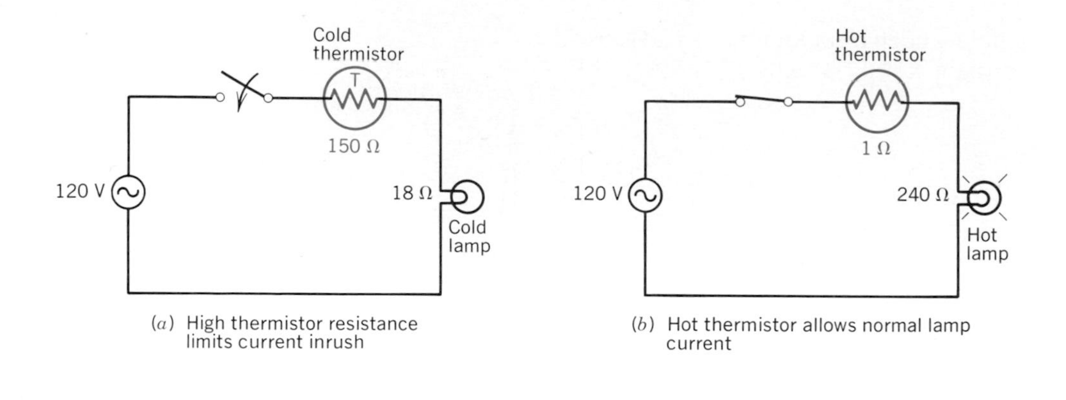

(*a*) High thermistor resistance limits current inrush

(*b*) Hot thermistor allows normal lamp current

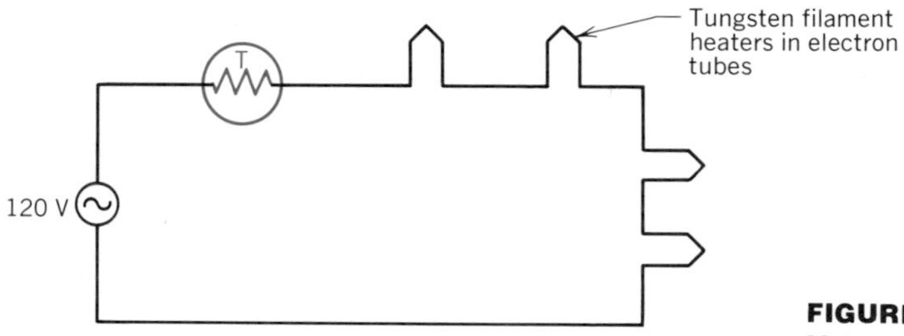

(*c*) Thermistor controls the initial current surge in series-connected, electron-tube heaters

FIGURE 4-5
Using thermistors to limit current inrush.

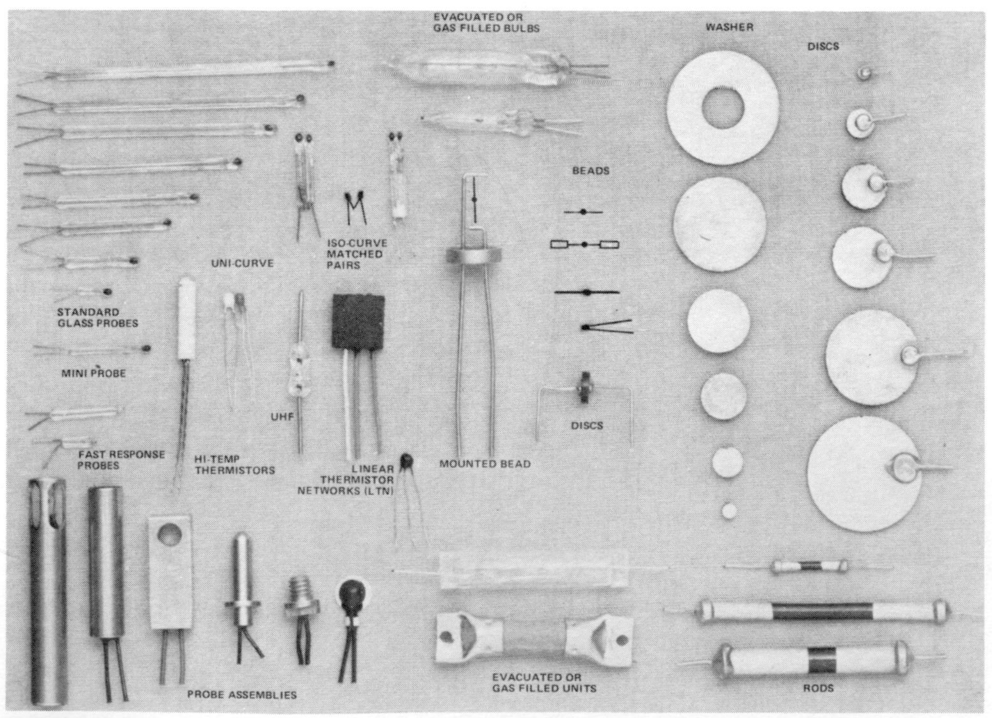

FIGURE 4-6
Various types of thermistors. (Courtesy of Fenwal Electronics.)

the filament to heat up much more gradually, over a period of several seconds, and avoid the "thermal stress." Commercial thermistors are now available that, in disc form, can be dropped into a lamp socket before inserting the lamp. They last the lifetime of the socket and are suitable for one-way lamps up to 300 W.

Thermistors are also used for similar purposes in vacuum-tube-operated radio, TV, and broadcast equipment. In these applications, a number of tungsten heater elements of various tubes are connected in series and operated from a 120-V ac line. See Fig. 4-5*c*. A thermistor reduces the inrush current allowing the heaters to warm up slowly. It should be noted that the cast-iron or other alloy heating elements used in electric stoves take many seconds to achieve their final temperature, so the inrush current would tend to blow fuses or circuit breakers if tungsten were used. Instead, the element is manufactured from an alloy such as Nichrome II because it has a relatively low temperature coefficient. This means there is not a very great change between the hot and cold resistance so that there is no current inrush problem.

4-6.2 Varistors

A *varistor* is a "voltage variable resistor." It consists of semiconductor crystals bound together by vitreous ceramic. When the voltage across a varistor exceeds a given value, dependent upon the wafer thickness, there is an increase in the breaking of valence bonds. This results in a sharp decrease in resistance. Figure 4-7*a* shows that the resulting increase in varistor current maintains the voltage across the device at an almost constant value. Consequently, if a varistor is connected in parallel with a piece of equipment, increases in line voltage are suppressed, with the extra line current passing through the varistor instead of the protected equipment. The trade name *thyrite* is used to identify a common type of varistor. Thyrite was initially developed to provide lightning protection on power transmission lines.

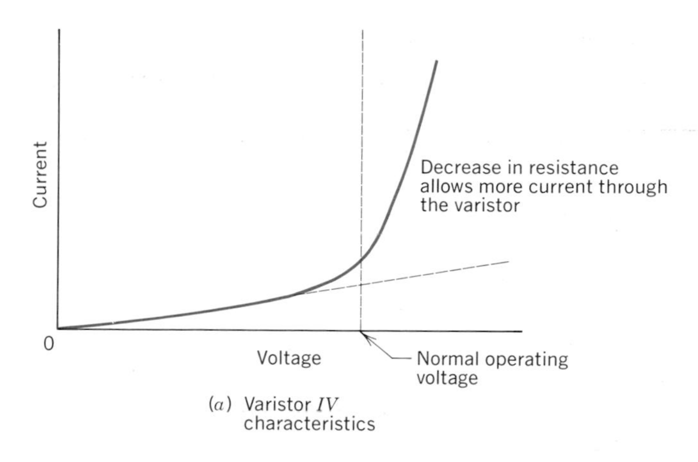

(*a*) Varistor *IV* characteristics

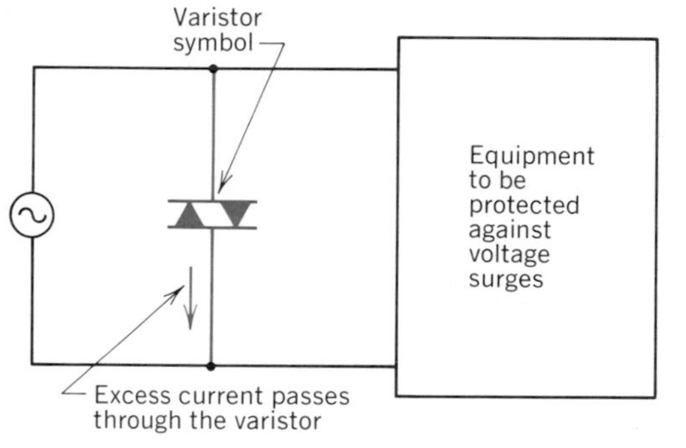

(*b*) Circuit application of a varistor

FIGURE 4-7
How a varistor reduces voltage surges.

SUMMARY

1. For a given temperature, the resistance of a material is directly proportional to the length, inversely proportional to the cross-sectional area and dependent upon the type of material. That is, $R = \rho l/A$.
2. The resistive property of a material is indicated by its resistivity or specific resistance, ρ, measured in $\Omega \cdot$ m, as given in Table 4-1.
3. A low value of resistivity indicates a better conducting material.
4. The circular mil area of a circle is found by squaring the diameter in mils. 1 mil = 0.001 in.
5. The wire gauge table is a list of standard wire sizes based on the circular mil area. A smaller gauge number indicates a larger cross-sectional area and higher current-carrying capacity.

6. Semiconductors have resistivities between conductors and insulators but closer to conductors.

7. Insulators are rated in terms of their breakdown voltages, in volts per meter, as given by Table 4-2.

8. The resistance of most metals increases for an increase in temperature due to increased electron collisions.

9. The change in resistance with temperature is dependent upon the material's temperature coefficient, which has units of per °C, as listed in Table 4-3.

10. Semiconductors (including carbon) and insulators have a negative temperature coefficient because any increase in temperature increases the availability of current carriers.

11. If the resistance and temperature coefficient of resistance are given at one temperature, the resistance at a different temperature may be calculated from $R_2 = R_1[1 + \alpha_1(T_2 - T_1)]$.

12. Linear resistors, those that have a constant V/I ratio, are materials whose resistance changes very little over the range of temperatures involved in the particular application.

13. The nonlinear nature of the resistance of tungsten over its wide temperature range when used as a lamp filament accounts for the large inrush current when the lamp is first switched on.

14. A thermistor is a nonlinear resistor with a very large negative temperature coefficient. Its characteristics enable it to suppress the large current surge through tungsten filaments when connected in series with the filament.

SELF-EXAMINATION

Answer true or false or a, b, c, d
(Answers at back of book)

4-1. If the diameter of a given conductor is doubled with the length and temperature remaining the same, the new resistance will be

a. doubled
b. halved
c. quadrupled
d. reduced to one-fourth

4-2. If the radius of a given conductor is tripled and the length is also tripled, the new resistance at the same temperature will be

a. the same
b. reduced to one-third
c. tripled
d. nine times as large

4-3. The resistivity of a given material is the resistance of a specific volume of the material at a given temperature. ________

4-4. Aluminum is a better conductor than copper because it has a higher resistivity than copper. ________

4-5. 100 ft of No. 18 gauge copper wire has a resistance of approximately

a. 6.5 Ω
b. 0.65 Ω
c. 1.6 Ω
d. 4 Ω

4-6. A 0.020-in.-diameter wire has a circular mil area of

a. 400

b. 40

c. 200

d. 0.0004

4-7. A 2-mm-thick layer of paper will break down when the following voltage is applied:

a. 20 kV

b. 10 kV

c. 40 kV

d. anything above 2 kV

4-8. A material that has an increase in electron collisions at higher temperatures has a positive temperature coefficient. ________

4-9. A negative temperature coefficient material is one in which an increase in heat reduces the number of current carriers. ________

4-10. A reel of copper wire has a resistance of 1 kΩ at 20°C. If its temperature coefficient is 0.004 per °C, the resistance at 120°C will be

a. 1004 Ω

b. 1.4 kΩ

c. 1040 Ω

d. 1000.4 Ω

4-11. A graph of I versus V for a material has a constant slope. This indicates a linear resistance. ________

4-12. A thermistor is an example of a nonlinear resistor, since its resistance decreases significantly with an increase in temperature. ________

REVIEW QUESTIONS

1. What are all the factors that affect the resistance of a conductor?

2. What would be the units for resistivity if the length was measured in feet and the area in CM (circular mils)?

3. Justify, in your own words, why an increase in length increases resistance, but an increase in area decreases resistance.

4. A No. 14 gauge copper wire is recommended to carry a maximum of 15 A in house wiring applications. Why? Does this mean that it is impossible for more than 15 A to pass through a No. 14 wire? Under what conditions, if any, could you allow a No. 14 wire to carry more than 15 A?

5. Describe what happens when an insulator breaks down.

6. Explain why the resistance of common metals increases with temperature, but the resistance of semiconductors and insulators decreases.

7. Given an accurate resistance measuring device and the means to vary temperature, describe how you would determine the temperature coefficient of resistance for a copper coil. Why would α at 30°C be different from the value obtained at 20°C? Why is α usually specified at 20°C?

8. Sketch, on the same set of axes, graphs of R versus T for carbon, constantan, and copper. Which one of carbon and copper is more constant in resistance as the temperature changes? At a given temperature which one has the largest resistance for a given length and cross-sectional area?

9. What causes the current inrush problems in a tungsten filament lamp? Under what conditions could tungsten be considered a linear resistance?

10. Sketch a graph of I versus V for a 100-W incandescent lamp. How does it differ from a graph for a linear resistor? What effect will a series-connected thermistor have upon the $V - I$ graph for the tungsten filament?

PROBLEMS

(Answers to odd-numbered problems at back of book)

4-1. The line cord to a table lamp uses a 2-m (4-m total) length of copper wire that has a diameter of 1.27 mm. Calculate:

a. The resistance of the lamp cord at 20°C.

b. The total voltage drop along the wire when a 120-V, 300-W lamp is in use.

c. The closest standard wire gauge size of the copper wire.

4-2. Repeat Problem 4-1 assuming that aluminum wire is used.

4-3. A relay coil is to be wound using copper magnet wire that has a radius of 0.4 mm. What length of wire is necessary to produce a 100-ohm coil?

4-4. Repeat Problem 4-3 assuming that aluminum wire is used.

4-5. A two-wire copper feeder line is to supply a 1500-W load 500 ft away from a 120-V source. What must be the minimum size of wire if a 6% maximum voltage drop can be tolerated at the load? (Specify the minimum cross-sectional area and nearest wire gauge number.)

4-6. A stove heating element is 1.5 m long and has a cold resistance of 26 ohms. If the element contains Nichrome II wire, determine the diameter of the wire.

4-7. What length of copper wire, 1 mm in diameter, will have the same resistance as 10 cm of gold wire that has a diameter of 1.2 mm?

4-8. A 10-m length of wire having a diameter of 1.8 mm has a resistance of 1.93 ohms. What material is used for this wire?

4-9. a. What air spacing is necessary to provide a minimum breakdown voltage of 2500 V?

b. What spacing would be necessary if Teflon is used instead of air?

4-10. The typical air gap found in an automotive spark plug is 30 thousandths of an inch. What minimum voltage is necessary to cause a spark over this distance?

4-11. The copper winding of a motor has a cold resistance of 0.2 ohm at 20°C. After the motor has been in operation, the winding temperature rises 50°C. What is the new resistance?

4-12. A No. 4 gauge copper wire is used in a 400-km length overhead transmission line.

a. What is the resistance of the line at 20°C?

b. What is the voltage drop along this line when a current of 100 A is carried by the line?

c. Express this voltage as a percentage of the applied voltage of 230 kV.

d. Repeat parts (a), (b), and (c) for a temperature of 120°C.

e. Repeat parts (a), (b), and (c) for a temperature of −30°F.

4-13. If the resistance of a nickel wire is 50 ohms at 300°C, what is its resistance at room temperature, 20°C?

4-14. If the tungsten filament of a 120-V, 250-W lamp is 2800°C, calculate its resistance at room temperature, 20°C.

4-15. A Nichrome II heating element is designed to produce 2100 W at 240 V. If the cold resistance at 20°C is 26 ohms, calculate the temperature of the hot element.

4-16. The inrush current to a 120-V lamp is 15 A. The current drops to 1.25 A after the tungsten filament has reached white heat. Calculate the operating temperature of the filament.

4-17. Early incandescent lamps used carbon filaments and operated at a lower temperature with a more reddish light.

a. If a 120-V, 60-W carbon lamp operated at 1500°C, what was the resistance of the lamp at room temperature, 20°C?

b. What advantage would such a lamp have, compared to a tungsten filament?

4-18. A coil of wire of unknown length and area has a resistance at 20°C of 25 ohms. The resistance increases to 37 Ω when the coil is lowered into boiling water. What is the material of the wire coil?

4-19. At 20°C a 1-kΩ metal film resistor has a temperature coefficient of 60 PPM/°C. What is its resistance at 80°C?

4-20. a. What is the cold resistance of a thermistor required to limit the inrush current of a 120-V, 150-W lamp to a value equal to that of the lamp's final steady current? The cold resistance of the lamp is 8 ohms, and the hot resistance of the thermistor is 1 Ω.

b. What is the power rating of the thermistor?

CHAPTER FIVE

SERIES CIRCUITS

When a number of components are connected in a circuit with only one path for current, they are said to be in *series*. The failure of any one element interrupts current through *all* the elements in the circuit. In this sense, the elements are said to be *dependent* upon each other for the continued flow of current in the circuit.

The relation between the *voltage drops* that occur in such a circuit and the applied emf is governed by *Kirchhoff's voltage law*. This law also provides a way to obtain the total circuit resistance and/or current.

Open circuits (infinite resistance) and *short circuits* (zero or very low resistance) can be located in a series circuit by making voltage measurements and applying Kirchhoff's voltage law.

A series circuit can be used to provide different voltages by a process of *voltage division*. A *potentiometer* provides *continuous* voltage division by means of a sliding contact in a series circuit. The potentiometer is a three-terminal device used to vary the voltage.

Either the positive or negative side of a dc supply may be connected to a metal chassis to provide a common return path in a circuit. This common point is then referred to as *ground*. Also, voltages throughout the circuit may have a positive or negative potential *with respect to* this ground.

5-1
VOLTAGE DROPS IN A SERIES CIRCUIT

In Section 4-6, we considered a circuit (Fig. 4-5*a* and 4-5*b*) in which a thermistor had the *same* current flowing through it as the lamp. Such an arrangement is called a *series* circuit.

The distinguishing feature of a series circuit is that the current is the *same* throughout all the elements in that circuit.

Consider three resistors in series, connected to a supply as in Fig. 5-1*a*.

As the current flows through R_1, a "voltage drop" occurs across R_1, called V_{R_1}. See Fig. 5-1*b*. This means that point *C* is at a lower potential or voltage than *B*. This is shown by the negative sign (−) at *C* with respect to the positive sign (+) at *B*. That is, when the current moves from a point that is positive to a point that is, by comparison, negative, there is a *reduction* in the potential (or voltage) to do work. A voltage drop is said to have taken place. Thus, since point *C* is closer to the negative side of the applied voltage source, point *C* is at a lower voltage than point *B*. This we show by a negative sign (−) at point *C*.

When determining the polarity signs of voltage drops with conventional current, we use the following rule:

Current flows *toward the positive side* of a voltage drop and *away from the negative side* of a voltage drop.

Conversely, if we know from the applied voltage in which direction the current is flowing, we place a *positive* sign (+) at that end of the resistor where current *enters*, and a *negative* sign (−) at that end of the resistor where current *leaves*.

The meaning of the term *voltage drop* may be further understood by the following. Consider Fig. 5-1*b*. If a voltmeter is connected from *B* to *E*, it would indicate the full supply voltage *V*, ignoring any negligible voltage drop

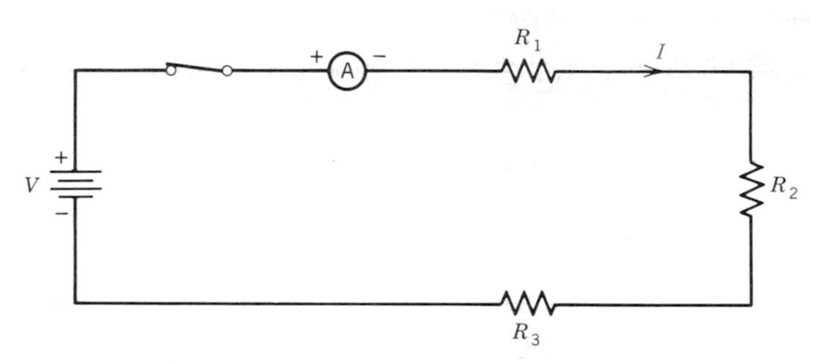

(*a*) Series circuit showing a single path for current

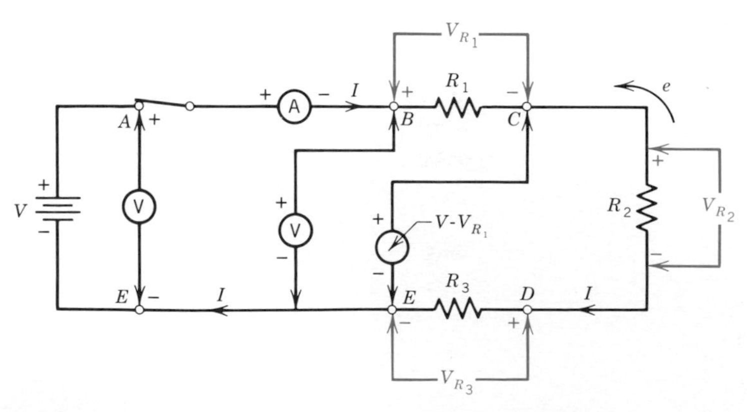

(*b*) Series circuit showing voltage drops across resistors

FIGURE 5-1
Series circuits.

across the extremely low resistance ammeter. Now, if the *positive lead* of the voltmeter is moved from B to C, the voltmeter would indicate some lower voltage, equal to $V - V_{R_1}$. That is, the drop in voltage that occurred across R_1 reduces the voltage (emf) available for the rest of the circuit. Likewise, the voltage from D to E is equal to $V - V_{R_1} - V_{R_2}$ (because of the additional voltage drop across R_2), and this voltage is equal to the voltage drop across R_3.

Note that the voltage drops across the resistors exist only when current actually flows in the circuit. If the switch is open, the voltmeter connected from A to E would still indicate that the source voltage exists, but the resistor voltage drops would be zero, since there is no current flowing. ***A current does not have to flow for an emf to exist.*** (Consider a "live" 120-V outlet with no load connected.) This is why a voltage drop across a resistor is sometimes called an IR (voltage) drop. (See Eq. 3-1b.) This term also helps to distinguish a voltage drop across a resistor from the voltage drop of a battery that is being recharged. In the case of a battery being recharged, current flows *into* the positive terminal instead of *out* of it. Therefore, the battery's voltage is also a voltage drop but not an IR voltage drop. See Fig. 5-2, where V_{R_1} is an IR voltage drop and V_2 is a voltage drop (a "bucking" emf) that must be overcome by supply voltage V_1 to reverse the current through V_2.

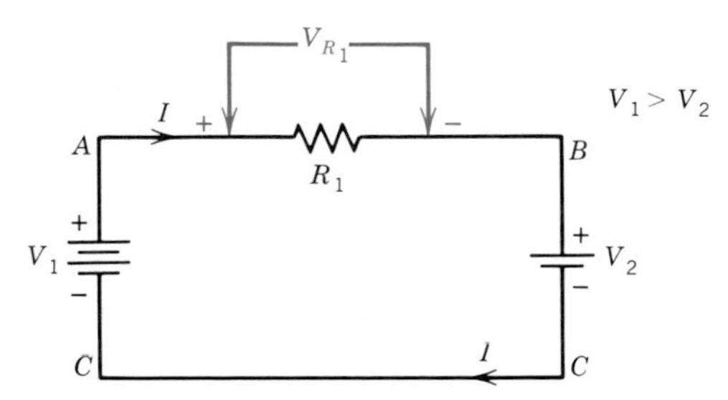

FIGURE 5-2
Battery V_2 is being recharged by V_1 so that V_2 is a voltage drop and V_{R_1} is an IR voltage drop.

5-2 KIRCHHOFF'S VOLTAGE LAW

We saw in Sec. 5-1 that the reading of a voltmeter decreases as we progressively subtract the voltage drops across each resistor. That is, in Fig. 5-1b:

voltage from B to $E = V$

voltage from C to $E = V - V_{R_1}$

voltage from D to $E = V - V_{R_1} - V_{R_2}$

voltage from E to $E = V - V_{R_1} - V_{R_2} - V_{R_3}$

In the last step, we have connected both the negative and positive leads of the voltmeter to the same point, E. Since the voltmeter is now not connected across any source of emf, or across any resistance that could cause a voltage drop, the voltmeter must read zero. Thus

$$V - V_{R_1} - V_{R_2} - V_{R_3} = 0$$

This is one form of Kirchhoff's voltage law:

> **Around any *complete* circuit, the *algebraic* sum of the voltages equals zero.**

By complete circuit we mean that we must end up tracing the current at the point where we started. By algebraic we mean that some voltages are positive, others are negative. Thus V (the voltage source) is a positive voltage while V_{R_1}, V_{R_2}, and V_{R_3} (the voltage drops) are negative.

We can avoid the term *algebraic* if we rearrange the above equation:

$$V = V_{R_1} + V_{R_2} + V_{R_3} \qquad \text{volts (V)} \qquad (5\text{-}1)$$

and restate Kirchhoff's voltage law as follows:

> **Around any complete circuit, the sum of the voltage rises equals the sum of the voltage drops.**

If we use the rule in Section 5-1 to determine the polarity signs of the IR voltage drops, we can use the following to determine whether a voltage is a voltage *rise* or a voltage *drop*.

1. A *voltage rise* occurs if, as we move around a circuit in the direction of the current (conventional current), we proceed from a negative sign (−) to a positive sign (+).
2. A *voltage drop* occurs if, as we move around a circuit in the direction of the current (conventional current)

we proceed from a positive sign (+) to a negative sign (−).

Note that the polarity signs referred to may be those across a resistor or voltage source. For example, if we apply Kirchhoff's voltage law to the circuit in Fig. 5-2, a voltage rise occurs between C and A equal to V_1. As we move from A to B in the direction of the current, we proceed from a positive sign to a negative sign across R_1. Thus V_{R_1} is a voltage drop. As we continue from B to C, we again move from a positive sign to a negative sign across V_2. Thus V_2 is also a voltage drop. We have now returned to our starting point at C, and we have traveled around a *complete* circuit. Therefore, by Kirchhoff's voltage law:

$$\text{sum of voltage rises} = \text{sum of voltage drops}$$

$$V_1 = V_{R_1} + V_2$$

Now let us apply Kirchhoff's voltage law to the circuit in Fig. 5-3*a*.

After determining the polarity signs across the resistors, it is clear that

$$V = V_{R_1} + V_{R_2} + V_{R_3} \qquad (5\text{-}1)$$

But if we apply Ohm's law to each section of the circuit:

$$V_{R_1} = IR_1,\ V_{R_2} = IR_2,\ V_{R_3} = IR_3 \qquad (3\text{-}1b)$$

Thus

$$\begin{aligned} V &= IR_1 + IR_2 + IR_3 \\ &= I(R_1 + R_2 + R_3) \\ &= IR_T \end{aligned}$$

where

$$\mathbf{R_T = R_1 + R_2 + R_3 \qquad ohms\ (\Omega) \qquad (5\text{-}2)}$$

is called the total circuit resistance. Thus, as far as V is concerned, the *same* current I would flow if only one resistor, R_T, is connected across V.

Equation 5-2 proves an important relation:

In a series circuit, the *total* resistance of the circuit is the *sum* of the individual resistances.

EXAMPLE 5-1

A source of 18 V is connected across three series resistors having values of $R_1 = 3.3\ \text{k}\Omega$, $R_2 = 1\ \text{k}\Omega$, and $R_3 = 4.7\ \text{k}\Omega$, as in Fig. 5-4.

Calculate:

a. The total circuit resistance.

b. The current in the circuit.

c. The voltage drops across the resistors to verify Kirchhoff's voltage law.

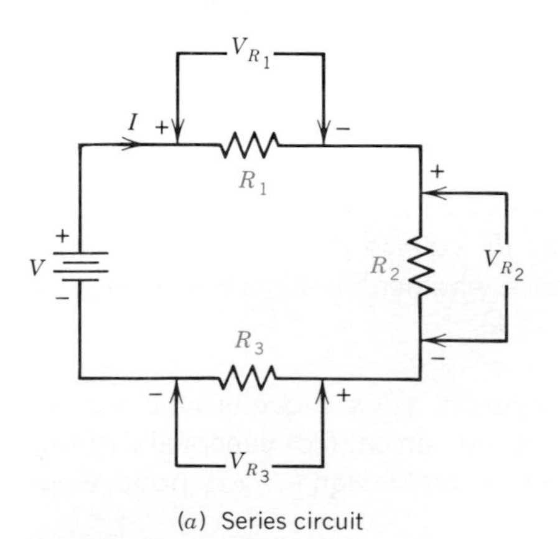

(*a*) Series circuit

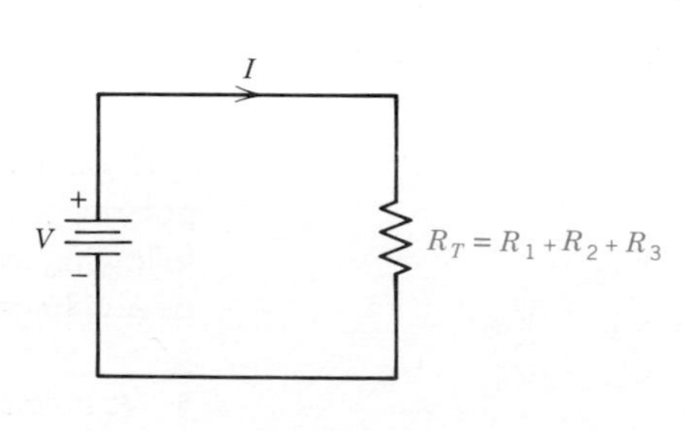

(*b*) Equivalent circuit with same current

FIGURE 5-3
The use of Kirchhoff's voltage law to obtain an equivalent circuit.

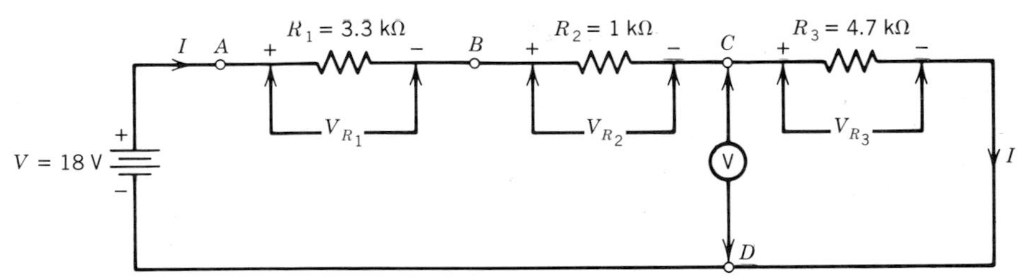

FIGURE 5-4
Circuit for Example 5-1.

d. The reading of a voltmeter connected between points C and D in Fig. 5-4. Show this by two different methods.

SOLUTION

a. Total circuit resistance,

$$R_T = R_1 + R_2 + R_3 \qquad (5\text{-}2)$$
$$= 3.3 \text{ k}\Omega + 1 \text{ k}\Omega + 4.7 \text{ k}\Omega$$
$$= \mathbf{9\ k\Omega}$$

b. Circuit current,

$$I = \frac{V}{R_T} \qquad (3\text{-}1a)$$
$$= \frac{18 \text{ V}}{9 \text{ k}\Omega} = \frac{18 \text{ V}}{9 \times 10^3\ \Omega} = 2 \times 10^{-3} \text{ A} = \mathbf{2\ mA}$$

c. Voltage drops,

$$V_{R_1} - IR_1 \qquad (3\text{-}1b)$$
$$= 2 \text{ mA} \times 3.3 \text{ k}\Omega = \mathbf{6.6\ V}$$
$$V_{R_2} = IR_2$$
$$= 2 \text{ mA} \times 1 \text{ k}\Omega = \mathbf{2\ V}$$
$$V_{R_3} = IR_3$$
$$= 2 \text{ mA} \times 4.7 \text{ k}\Omega = \mathbf{9.4\ V}$$

But $$V_{R_1} + V_{R_2} + V_{R_3} = V \qquad (5\text{-}1)$$
$$6.6 \text{ V} + 2 \text{ V} + 9.4 \text{ V} = 18 \text{ V}$$

Note that the above values verify Kirchhoff's voltage law.

d. The voltmeter indicates $V - V_{R_1} - V_{R_2} = 18 \text{ V} - 6.6 \text{ V} - 2 \text{ V} = \mathbf{9.4\ V}$ or $V_{R_3} = \mathbf{9.4\ V.}$

It should be noted that the wires used to connect resistors R_1, R_2, and R_3 in the circuit of Fig. 5-4 are assumed to have no (zero) resistance and therefore no voltage drops across them. The fact that the wire joining R_2 and R_3, for example, has a negative sign on one end and a positive sign on the other does not mean that there is any voltage difference between these two ends. The negative sign indicates that this wire is at a *lower* potential than the wire on the left side of R_2, which has a positive sign. Similarly, the positive sign indicates that the wire (any point along the wire, in fact) is at a higher potential than the right side of R_3, which has a negative sign. *Polarity markings are only relative to one another.* It depends upon the point of reference whether something in a circuit is positive or negative. For instance, in Fig. 5-4, point B is positive with respect to point C, but it is negative with respect to point A.

5-3 POWER IN A SERIES CIRCUIT

The *total* power dissipated in a series circuit is the *sum* of the individual powers dissipated by each resistor.

$$\textbf{total power } P_T = P_1 + P_2 + P_3 \qquad \text{watts (W)} \qquad (5\text{-}3)$$

where P_1 (etc.) may be calculated using

$$P_1 = I^2 R_1 \qquad (3\text{-}9)$$

or

$$P_1 = V_{R_1} I \qquad (3\text{-}7)$$

or

$$P_1 = \frac{V_{R_1}{}^2}{R_1} \qquad (3\text{-}10)$$

Thus $$P_T = I^2 R_T = VI = \frac{V^2}{R_T} \qquad \text{watts (W)} \qquad (5\text{-}4)$$

EXAMPLE 5-2

For the circuit and data in Example 5-1 calculate:

a. The power dissipated individually in each resistor.

b. The total power delivered by the source, using three different methods.

SOLUTION

a. $P_1 = V_{R_1}I$ (3-7)

$= 6.6 \text{ V} \times 2 \text{ mA} = \mathbf{13.2 \text{ mW}}$

or $P_1 = I^2R_1$ (3-9)

$= (2 \times 10^{-3} \text{ A})^2 \times 3.3 \times 10^3 \ \Omega$

$= 13.2 \times 10^{-3} \text{ W} = \mathbf{13.2 \text{ mW}}$

or $$P_1 = \frac{V_{R_1}{}^2}{R_1} \quad (3\text{-}10)$$

$$= \frac{(6.6 \text{ V})^2}{3.3 \times 10^3 \ \Omega} = \mathbf{13.2 \text{ mW}}$$

$P_2 = V_{R_2}I$ (3-7)

$= 2 \text{ V} \times 2 \text{ mA} = \mathbf{4 \text{ mW}}$

$P_3 = V_{R_3}I$

$= 9.4 \text{ V} \times 2 \text{ mA} = \mathbf{18.8 \text{ mW}}$

b. $P_T = P_1 + P_2 + P_3$ (5-3)

$= 13.2 \text{ mW} + 4 \text{ mW} + 18.8 \text{ mW}$

$= \mathbf{36 \text{ mW}}$

or $P_T = VI$ (5-4)

$= 18 \text{ V} \times 2 \text{ mA} = \mathbf{36 \text{ mW}}$

or $P_T = I^2R_T$ (5-4)

$= (2 \times 10^{-3} \text{ A})^2 \times 9 \times 10^3 \ \Omega$

$= 36 \times 10^{-3} \text{ W} = \mathbf{36 \text{ mW}}$

5-4 OPEN CIRCUITS

An open circuit results whenever a circuit is broken or is incomplete. For a series circuit, this means that there is no path for current and no current flows anywhere. We could say that there is no *continuity* in the circuit.

Let us assume that the circuit from Example 5-1 had been connected in the laboratory, but no current was indicated on the ammeter. (See Fig. 5-5.) How could we determine where a break in the circuit occurred?

Use a voltmeter on a range that can accommodate the supply voltage; connect it across each connecting wire in turn. If one of the wires is open as in Fig. 5-5*a*, the full supply voltage is indicated on the voltmeter. In the absence of current, there is no voltage drop across any of the resistors. Therefore, the voltmeter must be reading the full supply voltage across the open. That is,

$$\begin{aligned}\text{voltmeter reading} &= 18 \text{ V} - V_{R_1} - V_{R_2} - V_{R_3}\\ &= 18 \text{ V} - 0 \text{ V} - 0 \text{ V} - 0 \text{ V}\\ &= 18 \text{ V}\end{aligned}$$

If the circuit were open due to a defective resistor, as shown in Fig. 5-5*b* (resistors usually open when they burn out), the voltmeter would indicate 18 V when connected across this resistor, R_2.

Alternatively, the open circuit may be found using an ohmmeter. With the voltage removed, the ohmmeter will show *no* continuity (infinite resistance), when connected across the broken wire or open resistor.

5-5 SHORT CIRCUITS

A short circuit is a path of zero or very low resistance compared to the normal circuit resistance. All short circuits cause an increase in current. If, in a series circuit, one of the circuit elements is short-circuited, the result is an *increase* in current that may or may not be damaging. (In a simple parallel circuit the result is usually a blown fuse or circuit breaker.)

For example, we know from Example 5-1 that the normal current in the circuit should be 2 mA. Let us assume that the ammeter in Fig. 5-6 reads higher than this and we suspect that one of the components is "shorted-out" or short-circuited. How could we tell which component is shorted by *voltage* measurements?

A voltmeter connected across each of the resistors in turn will eventually indicate 0 volts across the defective resistor. Although current is flowing through this short circuit, no voltage drop occurs if there is zero resistance. That is,

$$I = \frac{V}{R_T} \quad (3\text{-}1a)$$

$$= \frac{18 \text{ V}}{4.3 \text{ k}\Omega} = 4.2 \text{ mA}$$

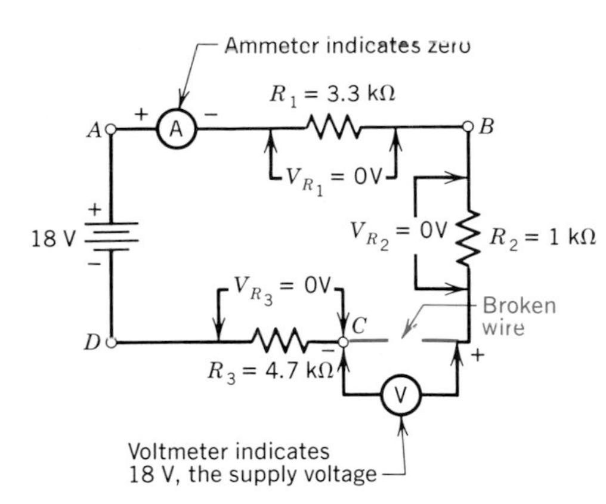

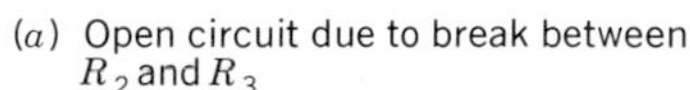
(*a*) Open circuit due to break between R_2 and R_3

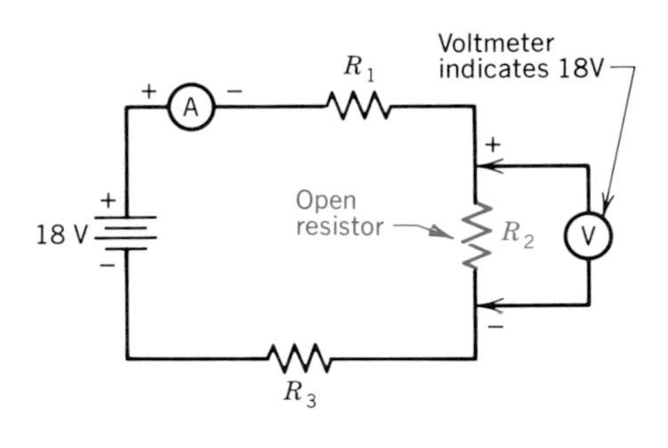

(*b*) Open circuit due to open resistor R_2

FIGURE 5-5
Open circuits.

and
$$V_3 = IR \qquad (3\text{-}1b)$$
$$= 4.2 \text{ mA} \times 0 \ \Omega = 0 \text{ V}$$

Even if the defective resistor, R_3, amounted to a few ohms, say 10 Ω, this would still be effectively a short circuit. This is because R_T would be 4.31 kΩ and the voltmeter would try to indicate

$$V_3 = IR_3 \qquad (3\text{-}1b)$$
$$= 4.2 \text{ mA} \times 10 \ \Omega = 42 \text{ mV} \qquad \text{(essentially zero)}$$

FIGURE 5-6
Short circuit across R_3 results in a current increase.

5-6 VOLTAGE DIVISION

If we refer back to Example 5-1 and examine the relation between voltage drops and resistor values, we find that the *highest* resistance has the *largest* voltage drop; the *lowest* resistance has the *smallest* voltage drop across it. In fact, because the circuit current is common, the voltage drops are in direct proportion to the resistance values.

Consider the circuit of Fig. 5-7.

The voltage across R_2 could be obtained as follows:

$$I = \frac{V}{R_T} = \frac{V}{R_1 + R_2 + R_3} \qquad (3\text{-}1a)$$

$$V_{R_2} = IR_2 = \frac{V}{R_1 + R_2 + R_3} \times R_2 \qquad (3\text{-}1b)$$

Therefore,

$$V_{R_2} = V \times \frac{R_2}{R_1 + R_2 + R_3} = V\frac{R_2}{R_T} \quad \text{volts (V)} \qquad (5\text{-}5)$$

Equation 5-5 is called the voltage division rule. It states that in a series circuit the voltage across the resistors "divides up" in a way that is proportional to the ratio of the resistances.

Similarly,

$$V_{R_1} = V \times \frac{R_1}{R_1 + R_2 + R_3} = V\frac{R_1}{R_T} \quad \text{volts (V)} \qquad (5\text{-}6)$$

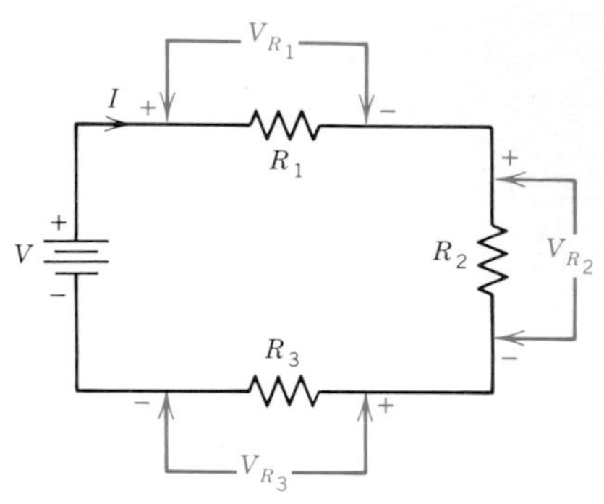

FIGURE 5-7
Voltage division across three series-connected resistors.

and

$$V_{R_3} = V \times \frac{R_3}{R_1 + R_2 + R_3} = V\frac{R_3}{R_T} \quad \text{volts (V)} \qquad (5\text{-}7)$$

The voltage division rule can be applied to any number of series-connected resistors.

Equations 5-5 through 5-7 clearly indicate, since the denominators are equal in size, *that the largest voltage drop occurs across the largest resistance.* For the special case of *equal*-value resistors, each resistor has an *equal* voltage across it.

EXAMPLE 5-3

Resistors of 22 kΩ, 100 kΩ, and 56 kΩ are connected in series across a 12-V supply. What is the voltage drop across the 56-kΩ resistor?

SOLUTION

$$V_{56\,\text{k}\Omega} = V \times \frac{R_3}{R_1 + R_2 + R_3} = V\frac{R_3}{R_T} \qquad (5\text{-}7)$$

$$= 12\text{ V} \times \frac{56\text{ k}\Omega}{22\text{ k}\Omega + 100\text{ k}\Omega + 56\text{ k}\Omega}$$

$$= 12\text{ V} \times \frac{56\text{ k}\Omega}{178\text{ k}\Omega}$$

$$= 12\text{ V} \times 0.315 = \mathbf{3.78\text{ V}}$$

5-7 THE POTENTIOMETER

We have mentioned that a potentiometer is a three-terminal device with a sliding contact that can be moved from one end to the other. It is schematically represented in Fig. 5-8*a* with the upper portion of resistance given by R_1 and the rest of its resistance by R_2.

Consider the voltage division rule:

$$V_{\text{out}} = V_{R_2} = V_{\text{in}} \times \frac{R_2}{R_1 + R_2} \qquad (5\text{-}5)$$

If the shaft of the potentiometer is turned so that B is connected to A, $R_1 = 0$ and $R_2 =$ full resistance of potentiometer. Then

$$V_{\text{out}} = V_{\text{in}} \times \frac{R_2}{R_1 + R_2} = V_{\text{in}} \times \frac{R_2}{0 + R_2} = V_{\text{in}}$$

and the *full supply* voltage will be indicated on the voltmeter.

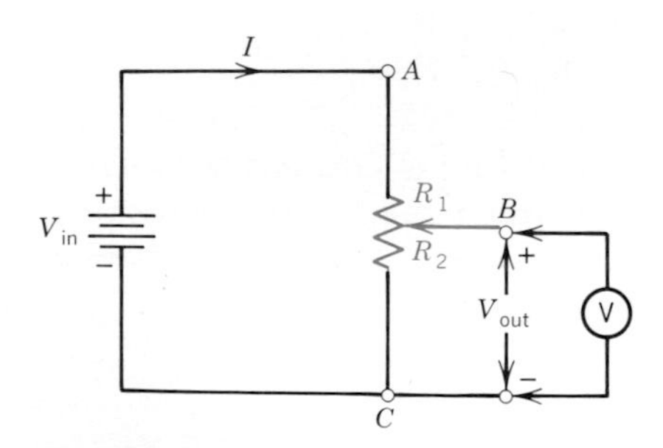

(*a*) Voltage division using a potentiometer

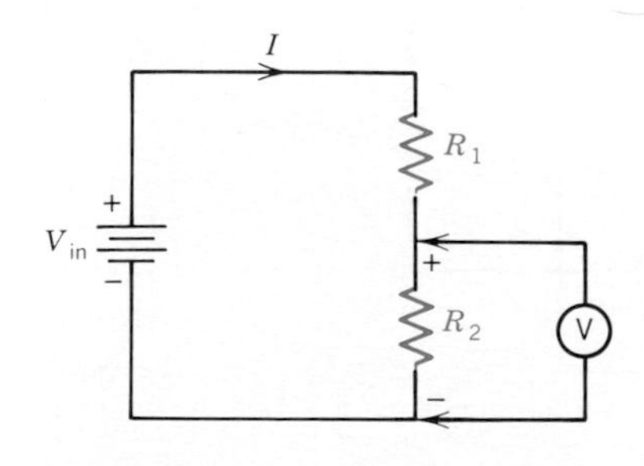

(*b*) Voltage division using two separate resistors

FIGURE 5-8
Circuits providing voltage division.

Conversely, if B is moved to make contact with C, $R_2 = 0$. Then

$$V_{out} = V_{in} \times \frac{R_2}{R_1 + R_2} = V_{in} \times \frac{0}{R_1 + 0} = 0 \text{ V}$$

Clearly, the potentiometer permits a variable voltage to be developed between one of its end terminals and its sliding contact. Note that since the voltmeter is considered ideal (i.e., a zero load having an infinite resistance), there is no load on the potentiometer. This means that no current is drawn from terminal B. Furthermore, there is no change in current I drawn from the supply voltage as the output voltage is varied from minimum to maximum. In practice, potentiometers are used in applications where a load is connected between B and C, but usually this has a very high resistance, as for example, the input resistance to an amplifier. In fact, the potentiometer described is the basis for a volume control in an audio amplifier.

EXAMPLE 5-4

A 5-kΩ potentiometer has 6 V connected across its end terminals. Calculate:

a. The range of output voltage that can be developed.
b. The two possible output voltages when the potentiometer is in its three-quarter position.
c. The current through the potentiometer in this position.
d. The potentiometer's power rating.

SOLUTION

a. Minimum = **0 V**
 Maximum = **6 V**

b. $V_{out} = V_{in} \times \frac{3}{4}$

$= 6 \text{ V} \times \frac{3}{4} = \mathbf{4.5\ V}$

or $V_{out} = V_{in} \times \frac{1}{4}$

$= 6 \text{ V} \times \frac{1}{4} = \mathbf{1.5\ V}$

c. $$I = \frac{V}{R_T} \qquad (3\text{-}1a)$$

$$= \frac{6 \text{ V}}{5 \text{ k}\Omega} = \mathbf{1.2\ mA}$$

d. $$P = I^2R \qquad (3\text{-}9)$$

$= (1.2 \text{ mA})^2 \times 5 \text{ k}\Omega$

$= (1.2 \times 10^{-3} \text{ A})^2 \times 5 \times 10^3 \ \Omega$

$= \mathbf{7.2\ mW}$ (10 mW minimum)

NOTE In this example we assume that the potentiometer's resistance is *linearly* uniform and varies linearly with its position. It would be said to have a linear *taper.* Other variations are possible. An *audio* taper, for example, provides smaller changes in resistance at one end (low-volume settings) than at the other end.

5-8 POSITIVE AND NEGATIVE GROUNDS

In an automotive electrical system it is customary to connect one side of the battery to the metal auto chassis and call it the *ground* side. In this way the metal chassis can be used as the *return path* for any circuit without providing an extra wire. Although most U.S. cars have "negative grounds," some European vehicles have a "positive-ground" system. (Reduced corrosion problems are claimed for the latter system.) See Fig. 5-9.

In a negative ground (Fig. 5-9*a*) system all wiring is at a positive potential with respect to the chassis, whereas in a positive ground system (Fig. 5-9*b*), all potentials are negative. Current flows in opposite directions in the two systems. But in *both* systems, the metal *chassis* is used as a common reference point to state the value of voltage at any point in the system.

Strictly speaking, the use of the word *ground* for the metal chassis is not correct. A better symbol to use for *chassis ground* is shown in Fig. 5-9*c*. This is because *ground* usually implies a connection to *earth ground* such as a cold water pipe. (In a car the chassis is insulated from the ground by its rubber tires.) For example, one side of the domestic 120-V ac outlet, the neutral, is connected to earth ground. Similarly, much electronic equipment is built on a metal chassis, and the chassis may serve as a convenient tie point for components and a common return to one side of a power supply. In some cases this chassis may also be connected to the 60-Hz supply

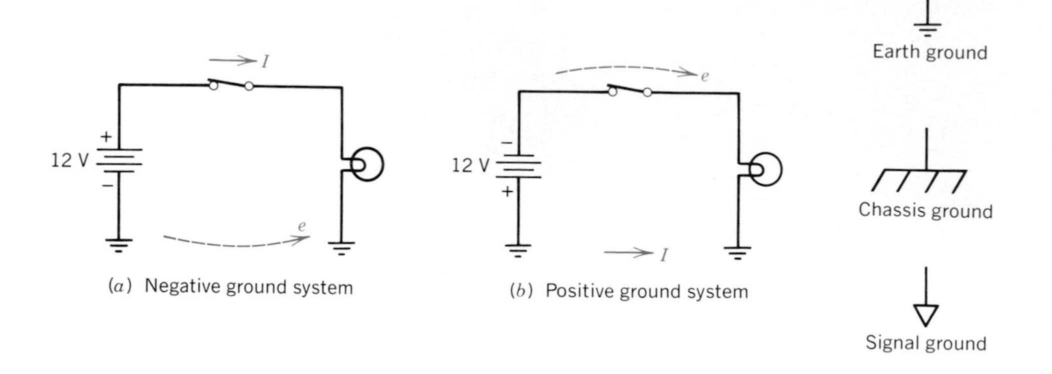

FIGURE 5-9
Possible grounding systems.

(earth) ground and may require polarized plugs to ensure the proper connection of neutral and grounded chassis.

In other cases, a *signal ground* symbol may be used, as shown in Fig. 5-9*c*. This may simply be a conductive strip running around a printed circuit board that serves as a common connection line for a returning signal. It is not necessarily connected to the metal chassis *or* the earth ground. In some equipment, provision is made, by means of a movable shorting strap, to operate the signal ground at a different voltage from the chassis or earth ground.

Note then that if one side of a battery or power supply is to be grounded, there is nothing that requires that it *must* be the negative side. **There is nothing inherently negative about a ground.**

5-9 POSITIVE AND NEGATIVE POTENTIALS

In some cases both positive *and* negative potentials may be required in a circuit. This requires a different placing for the common or ground connection. (See Fig. 5-10.)

In Fig. 5-10*a*:

$$+V_{R_A} = V_{R_1} = 12\text{ V} \times \frac{R_1}{R_1 + R_2} \tag{5-6}$$

$$= 12\text{ V} \times \frac{100\ \Omega}{100\ \Omega + 220\ \Omega} = \mathbf{3.75\ V}$$

$$-V_{R_B} = V_{R_2} = 12\text{ V} \times \frac{R_2}{R_1 + R_2} \tag{5-5}$$

$$= 12\text{ V} \times \frac{220\ \Omega}{100\ \Omega + 220\ \Omega} = \mathbf{8.25\ V}$$

Thus A is positive and 3.75 V higher in potential than C while B is −8.25 V with respect to ground point, C. Since it is understood that all voltages are stated with respect to the common or ground at C, we could simply say $A = +3.75$ V, $B = -8.25$ V.

A very common method of obtaining dual operating voltages for integrated circuits, such as operational amplifiers, is to use two series-connected batteries or power supplies with a common ground as in Fig. 5-10*b*. The advantage of this system is a more constant output voltage when a load is connected than that using a voltage-divider resistor network as in Fig. 5-10*a*. It should be noted that both of these systems require batteries or a "floating" supply. For example, if in Fig. 5-10*b* the two 12-V supplies had their negative sides grounded to the neutral of the 120-V ac, their series interconnection would be impossible.

Figure 5-11 shows a commercial power supply that provides both positive and negative potentials with respect to the common (COM) terminal. In this case, the common is completely isolated from the 120-V ac earth ground (the power supply converts the alternating current to direct current), so no ground symbol is shown. A *tracking control* permits the simultaneous adjustment of both the positive and negative outputs.

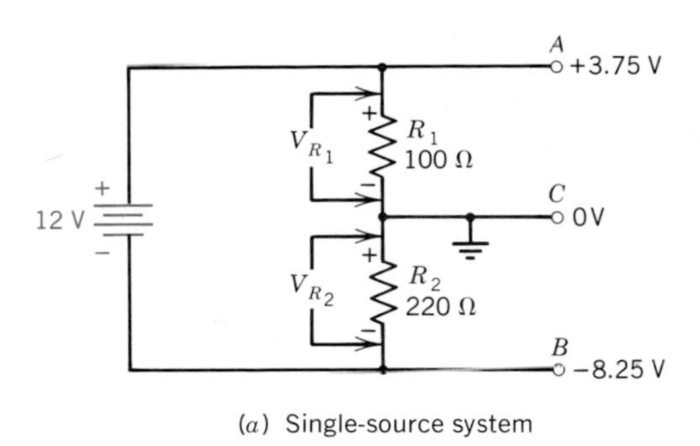

(a) Single-source system

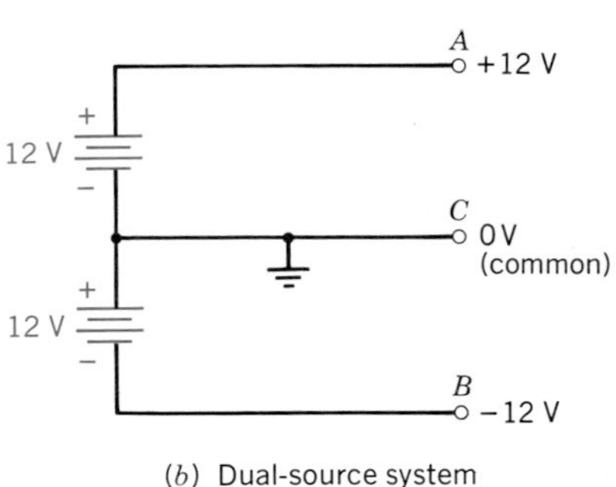

(b) Dual-source system

FIGURE 5-10
Two methods of obtaining positive and negative potentials.

5-10 VOLTAGE SOURCES CONNECTED IN SERIES

The series interconnection of cells (batteries) is covered in detail in Chapter 9. However, we can make two general statements concerning the total or net voltage resulting from two or more series-connected voltage sources.

When the cells are connected as in Fig. 5-12*a*, the cells are said to be *series-aiding* with a total voltage given by

$$V_T = V_1 + V_2 \qquad \text{volts (V)} \tag{5-8}$$

The circuit current, which is the same through both batteries and the resistor, is

$$I = \frac{V_T}{R} = \frac{V_1 + V_2}{R}$$

When the cells are connected as in Fig. 5-12*b* or 5-12*c*, the voltage sources are referred to as being *series-opposing*. The net voltage of the combination is

FIGURE 5-11
Commercial power supply that provides both positive and negative potentials with respect to the common (COM) terminal. (Courtesy of Hewlett-Packard Company.)

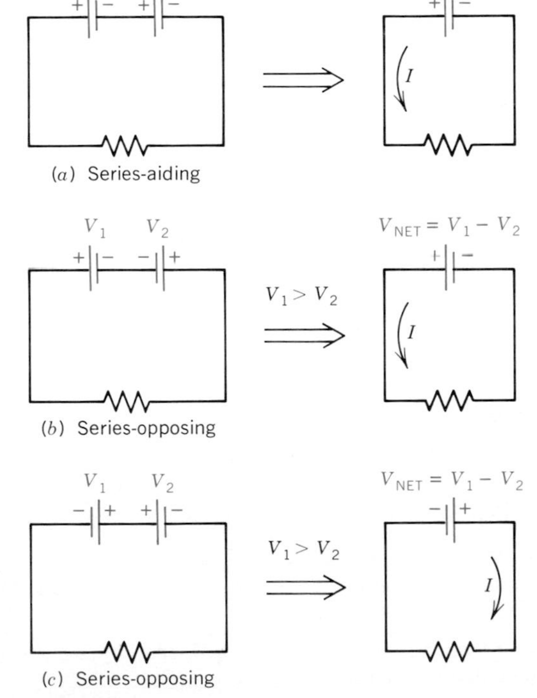

(a) Series-aiding

(b) Series-opposing

(c) Series-opposing

FIGURE 5-12
Series-connected voltage sources.

$$V_{NET} = V_1 - V_2 \qquad \text{volts (V)} \qquad (5\text{-}9)$$

where V_1 is assumed to have the larger emf. Note that the resultant polarity of V_{NET} is the same as the larger emf. Similarly, any number of series-connected emfs may be reduced to a single equivalent voltage. (It should be noted that the internal resistance associated with each cell (Section 9-1) has an additive effect, whether the cells are aiding or opposing.)

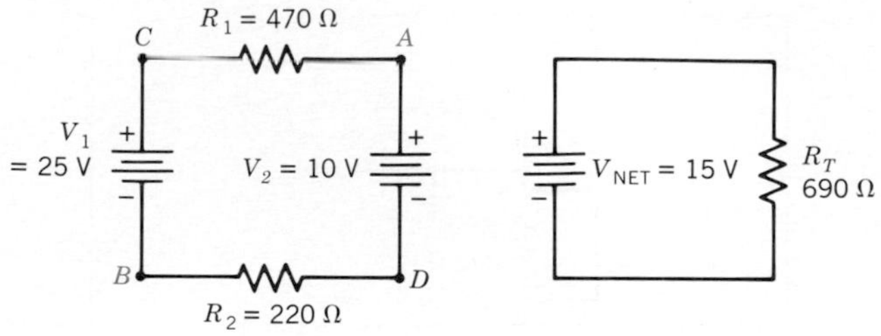

FIGURE 5-13
Circuits for Example 5-5.

EXAMPLE 5-5

For the circuit given in Fig. 5-13*a*, calculate:

a. The current.

b. The voltage at *A* with respect to *B*.

SOLUTION

a. $V_{NET} = V_1 - V_2$ (5-9)

$= 25\ \text{V} - 10\ \text{V} = 15\ \text{V}$

$R_T = R_1 + R_2$ (5-2)

$= 470\ \Omega + 220\ \Omega = 690\ \Omega$

$I = \dfrac{V_{NET}}{R_T}$ (3-1a)

$= \dfrac{15\ \text{V}}{690\ \Omega} = \mathbf{21.7\ mA}$

b. $V_{AB} = V_2 + V_{R_2}$

$= 10\ \text{V} + 21.7 \times 10^{-3}\ \text{A} \times 220\ \Omega$

$= 10\ \text{V} + 4.77\ \text{V} = \mathbf{14.8\ V}$

or $V_{AB} = V_1 - V_{R_1}$

$= 25\ \text{V} - 21.7 \times 10^{-3}\ \text{A} \times 470\ \Omega$

$= 25\ \text{V} - 10.2\ \text{V} = \mathbf{14.8\ V}$

SUMMARY

1. In a series circuit there is only one path for current, so the current is the *same* throughout all the elements.

2. The failure of any one element in a series circuit interrupts the current for all elements in the circuit.

3. An *IR* voltage drop occurs as current passes through a resistor.

4. Kirchhoff's voltage law states that the applied emf equals the sum of the voltage drops around any complete circuit.

5. The total resistance of a series circuit is the sum of the individual resistances.

6. The total power dissipated in a series circuit is the sum of the individual powers dissipated in each resistor.

7. An open circuit creates infinite resistance due to a break in a circuit. The full supply voltage appears across such an open circuit.

8. A short circuit involves zero or very low resistance resulting in an increase in current and zero voltage drop across the short circuit.

9. In a series circuit, the voltage divides with the highest voltage appearing across the largest resistance, which also dissipates the highest power.

10. The potentiometer is a three-terminal device with a sliding contact used to vary the voltage from zero to the full supply voltage.

11. A ground is a common return path for a number of components working from the same supply. Either the positive or negative side may be grounded and may or may not be connected to earth ground.

12. A power supply may consist of two emfs connected in series, with their common interconnection grounded to provide both positive and negative potentials to a circuit.

13. Voltage sources may be connected series-aiding or series-opposing with a total voltage equal to the sum or difference, respectively.

SELF-EXAMINATION

Answer true or false or a, b, c, d
(Answers at back of book)

5-1. No matter where an ammeter is connected in a series circuit the reading is always the same. ________

5-2. Although the current is not "used up" in a series circuit, the voltage is. ________

5-3. The sum of all the voltage drops in a series circuit may slightly exceed the applied voltage. ________

5-4. As electrons pass through a resistor, we still say that there is a voltage drop across the resistor even though electrons enter the negative side and leave the positive side. ________

5-5. An emf can exist only if there is a complete circuit for current to flow. ________

5-6. If there are two emfs in a series circuit, the net applied voltage may be either the sum or the difference of the two emfs. ________

5-7. Three resistors of 1 kΩ, 2 kΩ, and 7 kΩ are connected in series with a 30-V supply. The total resistance and current are

a. 10 kΩ, 3 A
b. 10 kΩ, 300 mA
c. 10 kΩ, 3 mA
d. 5 kΩ, 6 mA

5-8. The voltage drops across the three resistors in Question 5-7 are

a. 1 V, 2 V, 7 V
b. 2 V, 4 V, 14 V
c. 3 V, 6 V, 21 V
d. 3 mV, 6 mV, 21 mV

5-9. The power dissipated in each of the resistors in Question 5-7 are

a. 9 mW, 36 mW, 441 mW
b. 9 mW, 18 mW, 63 mW
c. 9 W, 36 W, 441 W
d. 9 μW, 18 μW, 63 μW

5-10. The total power supplied to the whole circuit in Question 5-7 is

a. 90 mW
b. 90 μW
c. 9 W
d. 900 mW

5-11. If an open circuit occurs between the 2-kΩ and 7-kΩ resistors in Question 5-7, a voltmeter connected across the 7-kΩ resistor will indicate:

a. 30 V
b. 21 V
c. 4 V
d. 0 V

5-12. If an open circuit occurs between the 1-kΩ and 2-kΩ resistors in Question 5-7, a voltmeter connected between these two resistors indicates:

a. 30 V
b. 1 V
c. 4 V
d. 0 V

5-13. If a short circuit occurs across the 7-kΩ resistor in Question 5-7, the current will be

a. 3 mA
b. 10 mA
c. 3 A
d. 10 A

5-14. A voltmeter connected across the shorted 7-kΩ resistor in Question 5-7 indicates:

a. 30 V
b. 21 V
c. 0 V
d. 9 V

5-15. Three resistors are connected in series across a 27-V supply. The second resistor has twice the resistance of the first; the third resistor has three times the resistance of the second. The voltage across the third resistor is

a. 3 V
b. 6 V
c. 9 V
d. 18 V

5-16. A 10-kΩ potentiometer is connected across an 18-V supply. When the shaft is rotated to the $\frac{2}{3}$ position, the output voltage and current through the potentiometer are

a. 6 V and 1.8 mA
b. 12 V and 1.8 mA
c. 12 V and 1.2 mA
d. either (a) or (b).

5-17. Current flows in the opposite direction in a negative ground system compared with a positive ground system. ________

5-18. If two 15-V supplies are series-connected, it is possible to connect loads that operate at +15 V, −15 V, and 30 V. ________

5-19. If two 12-V batteries are series-connected, the total available voltage may be either 24 V or 0 V. ________

5-20. In order to connect two batteries series-opposing, the two negative terminals must be connected. ________

REVIEW QUESTIONS

1. In an experimental verification of the Kirchhoff voltage law, the sum of the measured voltage drops was slightly less than the applied emf. How could you explain this? Why may the sum have been slightly larger? What equipment would you need for an exact verification?

2. Why do the polarity signs across a resistor seem more suggestive of a voltage *drop* when tracing conventional current than when tracing electron flow?

3. What is an *IR* drop? Why is this term useful?

4. Under what conditions would two emfs in a series circuit provide zero current?

5. Give an equation for the total power in a series circuit in terms of the individual resistances and the voltage drops across each.

6. A series circuit is suspected to have an open circuit. How could you verify this to be the case and how would you locate the break?

7. One of the components in a series circuit is shorted-out. How would you locate the fault using the following equipment?

a. A voltmeter.
b. An ohmmeter.

8. Assume a series circuit with a number of resistors connected to a power supply, one side of which is grounded. What difficulty would you have in locating an open resistor if you used a voltmeter that had its negative lead also grounded (as many VTVMs have)? Show by means of a diagram the readings you would obtain for three resistors, the middle one of which is open.

9. Assume a series circuit connected to a 9-V battery. No current is flowing and an open resistor is suspected. It is suggested that an ohmmeter be used to locate the open resistor. Is this a good idea? Will an infinite reading be indicated when connected across the open resistor? If an ohmmeter is to be used, what must be done first?

10. In a series circuit, justify why the largest resistance will have the highest voltage drop.

11. Use the voltage division rule (Eq. 5-6) to justify why a shorted resistor will have zero voltage across it.

12. When an ohmmeter is connected across the outer two terminals of a potentiometer, why is there no change in resistance reading as the shaft is rotated?

13. How is it possible for a potentiometer to reduce the voltage to zero without causing a short circuit and drawing a large current from the supply?

14. If a potentiometer with a *linear* taper varies the resistance linearly as the shaft is rotated, what is a *logarithmic* taper?

15. Why do you think it is unimportant to a headlight in a car whether the electrical system has a positive or negative ground system but very important to a transistor radio?

16. What does it mean to have a "floating" power supply? Draw a diagram to show why it is impossible to obtain positive and negative supply potentials from two power supplies, both of which have their negative sides grounded to the same point.

17. Explain how it would be possible, in a series circuit, for the number of electrons-per-second flowing at two points to be the same but the electron drift velocity at the two points to be different.

18. Under what conditions can two 120-V lamps be safely connected in series across a 240-V supply?

19. What do you understand by the terms *series-aiding* and *series-opposing?*

20. If two batteries are connected in a series circuit so that both batteries are discharging, in what way have the batteries been connected?

PROBLEMS

(Answers to odd-numbered problems at back of book)

5-1. Given $R_1 = 150\ \Omega$, $R_2 = 300\ \Omega$, and $R_3 = 50\ \Omega$, find the total resistance when connected in series.

5-2. Given $R_1 = 5.1\ \text{k}\Omega$, $R_2 = 6.8\ \text{k}\Omega$, and $R_3 = 4.7\ \text{k}\Omega$, find the total resistance when connected in series.

5-3. The three resistors in Problem 5-1 are connected to a 90-V source of emf. Calculate:

a. The circuit current.

b. The voltage drop across each resistor.

5-4. The three resistors in Problem 5-2 are connected to a 24-V source of emf. Calculate:

a. The circuit current.

b. The voltage drop across each resistor.

5-5. Calculate:

a. The power dissipated in each resistor in Problem 5-3.

b. The total power supplied by the source.

5-6. Calculate:

a. The power dissipated in each resistor in Problem 5-4.

b. The total power supplied by the source.

5-7. Refer to Fig. 5-4.

a. What will a voltmeter indicate when connected from B to D?

b. What is the voltage at C with respect to A?

5-8. Refer to Fig. 5-5*a*. Determine the reading of a voltmeter for the following connections:

a. Between A and B.

b. Between B and C.

c. Between C and D.

d. Between B and D.

5-9. Three resistors, when connected in series to a 50-V supply, draw a current of 5 mA. Two resistors are equal in value to each other; the third has a resistance of 2 kΩ. What is the value of each of the other two resistors?

5-10. Three resistors are connected in series to a 20-V source of emf. A current of 2 mA passes through the first resistor; a 10-V drop occurs across the second resistor; and the third resistor has a resistance of 2.2 kΩ. Determine the resistance of the first two resistors.

5-11. Four resistors are connected in series to a 25-V supply that provides 100 mW of power. The first two resistors have equal resistance and a combined voltage drop across the pair of 10 V. If the fourth resistor has half the resistance of the third resistor, calculate the resistance of all four resistors.

5-12. Refer to Fig. 5-6. Determine the reading of a voltmeter for the following connections:

a. Between A and B.

b. Between A and C.

c. Between C and D.

d. Calculate the reading of an ammeter connected between C and D.

5-13. The following three resistors are connected in series: 330 Ω, 2 W; 100 Ω, 1 W; 33 Ω, $\frac{1}{2}$ W. What is the maximum voltage that can safely be connected across this series combination?

5-14. A series Christmas tree light set consists of twenty-five $7\frac{1}{2}$-W lamps operating at 120 V. What is the total hot resistance of the circuit?

5-15. A 14.8 V-source of dc is to be used to recharge a battery as in Fig. 5-2. If the battery is 11.2 V, calculate:

a. The resistance R_1 must have to limit the initial charging current to 6 A.

b. The current that will flow with R_1 in the circuit if the polarity of the battery is mistakenly reversed.

5-16. What resistance must be connected in series with a 6-V car radio if it is to operate from a 13.2-V battery? The radio uses 3 A at 6 V. Determine the power rating of the resistor.

5-17. A 100-W, 120-V lamp, and a 60-W, 120-V lamp are connected in series to a 240-V supply. Assuming that the resistances of the lamps are constant, use the voltage division rule to determine the voltage across each lamp, the power dissipated by each lamp, and the visual effect.

5-18. A 2.2-MΩ resistor is in a series circuit whose total resistance is 8.5 MΩ. How much voltage appears across the 2.2-MΩ resistor when 40 V is applied to the circuit? Calculate the power dissipated in this resistor.

5-19. A 5-kΩ potentiometer is connected in series with a 3.3-kΩ resistor across a 50-mV source as shown in Fig. 5-14. Calculate the voltmeter reading for the following locations for contact *B*:

a. Moved fully upward to *A*.

b. Moved fully downward to *C*.

c. Moved half-way between *A* and *C*.

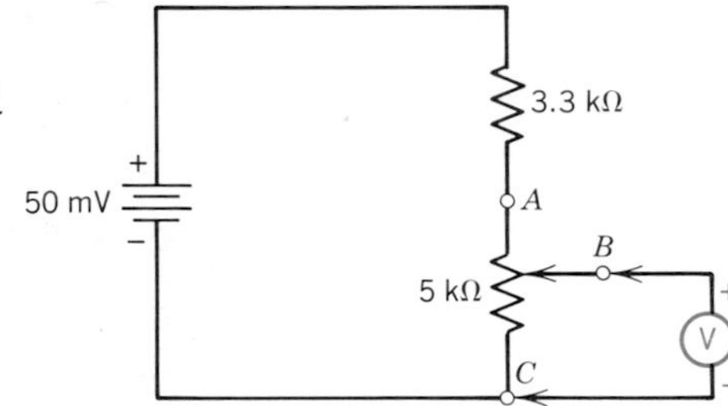

FIGURE 5-14
Circuit for Problems 5-19, 5-21, and 5-23.

5-20. Repeat Problem 5-19 using the values and circuit shown in Fig. 5-15.

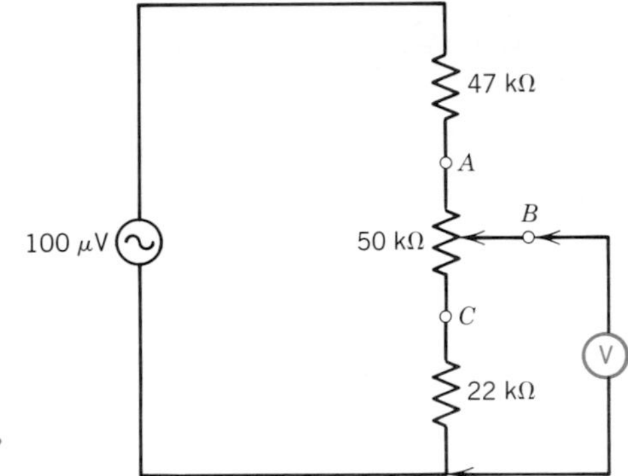

FIGURE 5-15
Circuit for Problems 5-20, 5-22, and 5-24.

5-21. Calculate the current flowing in the circuit of Fig. 5-14 as the potentiometer control is varied.

5-22. Repeat Problem 5-21 using Fig. 5-15.

5-23. If the applied voltage in Fig. 5-14 is 50 V, calculate the required power rating of the potentiometer.

5-24. If the applied voltage in Fig. 5-15 is 100 V, calculate the required power rating of the potentiometer.

5-25. Refer to Fig. 5-10*a*. It is required to produce +9 V at *A* and −6 V at *B* with respect to ground (*C*). Determine:

a. The necessary applied emf.

b. The values of R_1 and R_2 if the no-load current drawn from the applied emf is to be 100 mA.

c. The potential at *B* with respect to *A*.

d. The potential at *C* with respect to *A*.

5-26. Refer to Fig. 5-13, with $V_1 = 20$ V, $V_2 = 45$ V, $R_1 = 15$ kΩ, and $R_2 = 10$ kΩ. Calculate:

a. The current.

b. The voltage at *C* with respect to *D*.

c. The voltage at *B* with respect to *A*.

5-27. Repeat Problem 5-26 using $V_1 = 50$ V and $V_2 = 30$ V.

CHAPTER SIX

PARALLEL CIRCUITS

A *parallel* circuit has two or more current paths but a common applied voltage. This is the method used in wiring homes, schools, and so on, where each load can operate independently of the other loads. *Kirchhoff's current law* applies directly to a parallel circuit. It states, essentially, that the total current supplied by a source equals the sum of the parallel *branch* currents. This law is used to derive a general expression for the total resistance of any number of parallel-connected resistors. We shall also see that the total resistance of such a circuit is always less than the smallest branch resistance. Finally, we shall examine two special equations: one for only *two* resistors in parallel, the other for any number of *equal* resistors in parallel.

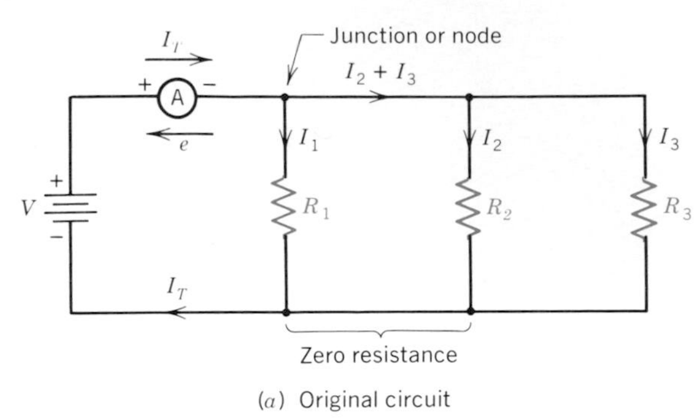

(a) Original circuit

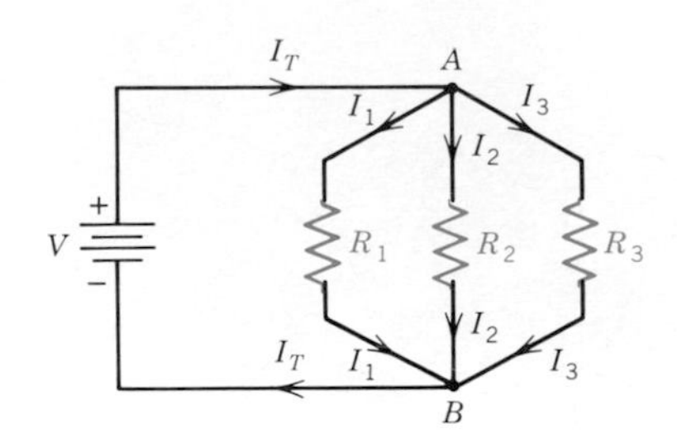

(b) Original circuit redrawn

FIGURE 6-1
Alternate forms of a parallel-connected circuit.

6-1 CHARACTERISTICS OF A PARALLEL CIRCUIT

In a series circuit, the failure of any one element (or the connecting leads) interrupts current for *all* the elements in the circuit (Section 5-1). This is a useful feature when a protective device (such as a fuse or circuit breaker) opens under overload conditions to "clear" the circuit and prevent damage to the wiring in the circuit. But as anyone knows who has had to locate a defective lamp in a series-connected Christmas tree string, this is *not* very suitable for supplying power to a number of different loads. A more practical method is to connect all the various loads in parallel with each other and then connect the combination to the supply.

The distinguishing feature of a parallel circuit is that there is more than one path for current to flow through. (Recall that in a series circuit there is only *one* path for current.) Also, in a true simple parallel circuit, all the components have the *same common* voltage across them.

The current drawn by each component, load or *branch* will be determined by the resistance of that particular branch. Consider the three parallel-connected resistors in Fig. 6-1.

It is clear that in a parallel circuit, the total current I_T from the source divides at each *junction* or *node,* with the largest current passing through the smallest branch resistance.

The currents recombine (as at junction B) to provide the same total current, I_T, back into the source.

6-2 KIRCHHOFF'S CURRENT LAW

In its most simple form Kirchhoff's current law states:

The current flowing into a junction (or node) must equal the current flowing out of the junction.

Kirchhoff's current law strictly applied is the law of conservation of charge. It means that we must be able to account for the flow of charge at any junction. Let us apply Kirchhoff's current law to junction A in the circuit of Fig. 6-2*a*.

Mathematically, Kirchhoff's current law states:

$$I_T = I_1 + I_2 + I_3 \qquad \text{amperes (A)} \qquad (6\text{-}1)$$

6-3 EQUIVALENT RESISTANCE FOR A PARALLEL CIRCUIT

If we apply Ohm's law to each branch resistance, we obtain

$$I_1 = \frac{V}{R_1}, \qquad I_2 = \frac{V}{R_2}, \qquad I_3 = \frac{V}{R_3}$$

Thus Eq. 6-1 may be rewritten as:

$$I_T = \frac{V}{R_1} + \frac{V}{R_2} + \frac{V}{R_3} \qquad (6\text{-}1a)$$

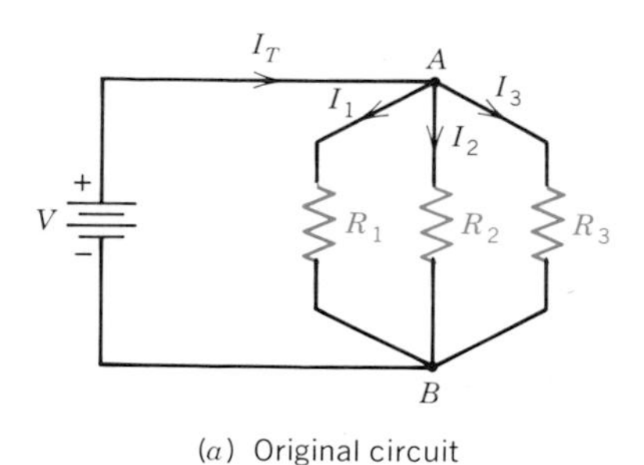

(a) Original circuit

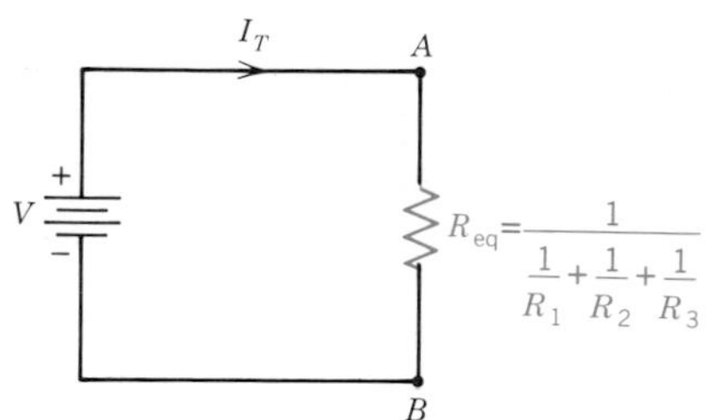

(b) Equivalent circuit

FIGURE 6-2
Parallel resistor circuit.

or

$$I_T = V\left(\frac{1}{R_1} + \frac{1}{R_2} + \frac{1}{R_3}\right) \quad (6\text{-}2)$$

Now let R_{eq} (shown in Fig. 6-2*b*) be the equivalent combined resistance of all three resistors in parallel. That is, R_{eq} could take the place of R_1, R_2, and R_3 and the battery V would supply the same current I_T as it did to the three resistors.

That is

$$I_T = \frac{V}{R_{eq}} = V\left(\frac{1}{R_{eq}}\right) \quad (6\text{-}3)$$

For the current I_T to be the same in Eqs. 6-2 and 6-3 we require:

$$V\left(\frac{1}{R_{eq}}\right) = V\left(\frac{1}{R_1} + \frac{1}{R_2} + \frac{1}{R_3}\right)$$

Therefore,

$$\frac{1}{R_{eq}} = \frac{1}{R_1} + \frac{1}{R_2} + \frac{1}{R_3} \quad \text{siemens (S)} \quad (6\text{-}4)$$

Equation 6-4 may also be written as:

$$R_{eq} = \frac{1}{\frac{1}{R_1} + \frac{1}{R_2} + \frac{1}{R_3}} \quad \text{ohms } (\Omega) \quad (6\text{-}5)$$

Using the concept that conductance is the reciprocal of resistance, that is, $G = \frac{1}{R}$ (Section 2-11), we can also write Eq. 6-4 as:

$$G_T = G_1 + G_2 + G_3 \quad \text{siemens (S)} \quad (6\text{-}6)$$

The total conductance of a parallel circuit is the sum of the individual branch conductances. If G_T is found first, then $R_{eq} = \frac{1}{G_T}$.

Similar to a series circuit, the total power dissipated in a parallel circuit is given by

$$P_T = P_1 + P_2 + P_3 \quad \text{watts (W)} \quad (6\text{-}7)$$

or

$$P_T = I_T^2 R_{eq} \quad (6\text{-}8)$$

or

$$P_T = VI_T \quad (6\text{-}9)$$

Note that the power relations (Eqs. 6-7, 6-8, and 6-9) for a parallel circuit are the same as Eqs. 5-3 and 5-4 for a series circuit.

EXAMPLE 6-1

A source of 24 V is connected across three resistors in parallel having values of $R_1 = 2.2$ kΩ, $R_2 = 1$ kΩ, and $R_3 = 4.7$ kΩ, as shown in Fig. 6-3. Calculate:

a. The current drawn by each resistor.
b. The total current from the source.
c. The combined equivalent circuit resistance using *b*.
d. The equivalent resistance using Eq. 6-4.
e. The total current from the source using the resistance from *d*.
f. The reading of an ammeter between R_1 and R_2.
g. The power dissipated in each resistor.
h. The total power delivered by the source.

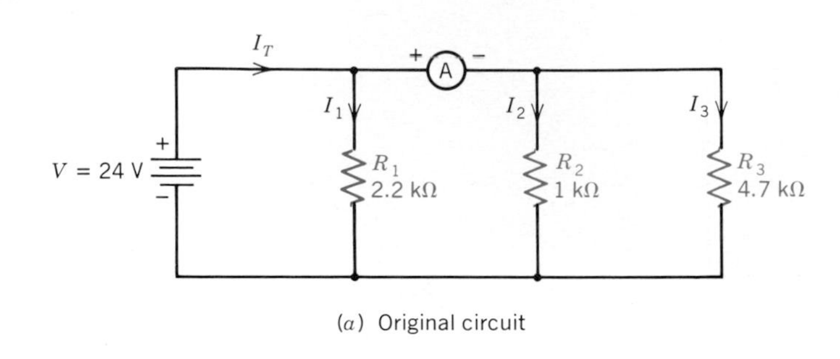

(a) Original circuit

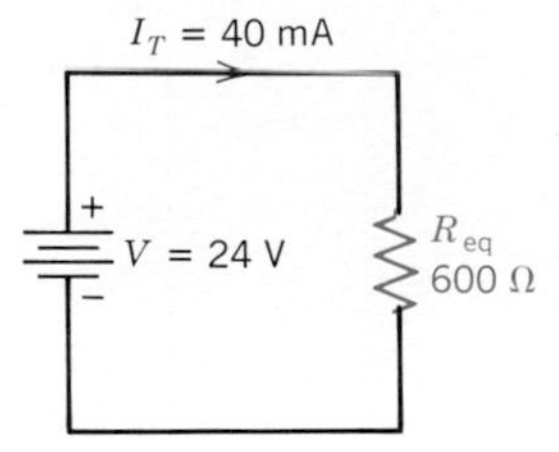

(b) Equivalent circuit

FIGURE 6-3
Circuits for Example 6-1.

i. The voltage across R_3 if an open circuit develops in R_2.

SOLUTION

a. $I_1 = \frac{V}{R_1}$ (3-1a)

$= \frac{24\text{ V}}{2.2\text{ k}\Omega} = \mathbf{10.9\text{ mA}}$

$I_2 = \frac{V}{R_2}$

$= \frac{24\text{ V}}{1\text{ k}\Omega} = \mathbf{24.0\text{ mA}}$

$I_3 = \frac{V}{R_3}$

$= \frac{24\text{ V}}{4.7\text{ k}\Omega} = \mathbf{5.1\text{ mA}}$

b. $I_T = I_1 + I_2 + I_3$ (6-1)

$= 10.9 + 24.0 + 5.1\text{ mA} = \mathbf{40\text{ mA}}$

c. $R_{eq} = \frac{V}{I_T}$ (3-1)

$= \frac{24\text{ V}}{40\text{ mA}} = \frac{24\text{ V}}{0.04\text{ A}} = \mathbf{600\ \Omega}$

d. $\frac{1}{R_{eq}} = \frac{1}{R_1} + \frac{1}{R_2} + \frac{1}{R_3}$ (6-4)

$= \frac{1}{2.2\text{ k}\Omega} + \frac{1}{1\text{ k}\Omega} + \frac{1}{4.7\text{ k}\Omega}$

$= 0.455 \times 10^{-3} + 1 \times 10^{-3} + 0.213 \times 10^{-3}\text{ S}$

$= 1.668 \times 10^{-3}\text{ S}$

Therefore, $R_{eq} = \frac{1}{1.668 \times 10^{-3}\text{ S}} = \mathbf{600\ \Omega}$

Note that this is essentially the same as using Eq. 6-6 with a total circuit conductance of 1.668 millisiemens. See Fig. 6-3b.

e. $I_T = \frac{V}{R_{eq}}$ (3-1a)

$= \frac{24\text{ V}}{600\ \Omega} = 0.04\text{ A} = \mathbf{40\text{ mA}}$

f. The ammeter will indicate:

$I_T - I_1 = 40\text{ mA} - 10.9\text{ mA} = \mathbf{29.1\text{ mA}}$

or $I_2 + I_3 = 24\text{ mA} + 5.1\text{ mA} = \mathbf{29.1\text{ mA}}$

g. $P_1 = \frac{V^2}{R_1}$ (3-10)

$= \frac{(24\text{ V})^2}{2.2 \times 10^3\ \Omega} = \mathbf{262\text{ mW}}$

$P_2 = \frac{V^2}{R_2}$

$= \frac{(24\text{ V})^2}{1 \times 10^3\ \Omega} = \mathbf{576\text{ mW}}$

$P_3 = \frac{V^2}{R_3}$

$= \frac{(24\text{ V})^2}{4.7 \times 10^3\ \Omega} = \mathbf{123\text{ mW}}$

NOTE This is preferable to using $P = I^2R$, since I is a result of a previous calculation and has been "rounded off."

h. $P_T = P_1 + P_2 + P_3$ (6-7)

$= 262 + 576 + 123\text{ mW}$

$= \mathbf{961\text{ mW}}$

or $P_T = I_T^2 R_{eq}$ (6-8)

$= (0.040\ \text{A})^2 \times 600\ \Omega = \mathbf{960\ mW}$

or $P_T = V I_T$ (6-9)

$= 24\ \text{V} \times 0.040\ \text{A} = \mathbf{960\ mW}$

i. If R_2 is removed (due to an open circuit), the voltage across R_3 will remain at 24 V even though the total source current will drop from 40 to 16 mA. This is because all the resistors are connected in parallel with the 24-V source.*

A very important conclusion to draw from Example 6-1 is that *the equivalent resistance R_{eq} of a parallel circuit is always smaller than the smallest branch resistance.* In this case, $R_{eq} = 600\ \Omega$, while the smallest branch resistance was 1000 Ω. This is *always* the case.

In fact, if another resistor, say 600 Ω, is added in parallel with the first three resistors, the equivalent resistance would drop to just one-half the original amount, or 300 Ω. This is because another parallel path permits *more* current to flow meaning that a *drop* in total equivalent resistance has taken place. Another way of looking at it is to consider that some *conductance* has been added, increasing G_T and decreasing $R_{eq} = \dfrac{1}{G_T}$.

6-4 OTHER EQUIVALENT RESISTANCE EQUATIONS

There are two further equations useful for parallel resistance networks.

For the special case of two resistors *only* in parallel:

$$R_{eq} = \frac{R_1 R_2}{R_1 + R_2} \tag{6-10}$$

$$= \frac{\text{product (of 2 } R\text{s in parallel)}}{\text{sum (of the 2 } R\text{s in parallel)}} \text{ ohms } (\Omega)$$

* In practice, the 24-V source would have some internal resistance that would cause the terminal voltage available for the circuit to depend somewhat upon the total load current. A drop in this current would probably cause some increase in voltage across the parallel circuit. In our circuit, in Fig. 6-3, we have assumed this internal resistance to be zero. We shall consider this topic in detail in Chapter 9.

And for N *equal-valued* resistors, each of resistance R, connected in parallel:

$$R_{eq} = \frac{R}{N} = \frac{\text{resistance of one resistor}}{\text{number of resistors}} \tag{6-11}$$

A notation used to show resistors in parallel consists of two parallel lines drawn between the resistors involved. Thus, if three resistors—R_1, R_2, and R_3—are parallel-connected, their equivalent resistance may be represented by

$$R_{eq} = R_1 \| R_2 \| R_3$$

For the special case of two parallel-connected resistors,

$$R_{eq} = R_1 \| R_2 = \frac{R_1 \times R_2}{R_1 + R_2}$$

This notation is especially useful when working with series-parallel circuits, as in Chapter 7.

EXAMPLE 6-2

Calculate the equivalent resistance of a 2.2-kΩ resistor and a 4.7-kΩ resistor in parallel with each other, as in Fig. 6-4*a*.

SOLUTION

$$R_{eq} = R_1 \| R_2 = \frac{R_1 \times R_2}{R_1 + R_2} \tag{6-10}$$

$$= \frac{2.2\ \text{k}\Omega \times 4.7\ \text{k}\Omega}{2.2\ \text{k}\Omega + 4.7\ \text{k}\Omega} = \frac{10.34\ (\text{k}\Omega)^2}{6.9\ \text{k}\Omega} = \mathbf{1.5\ k\Omega}$$

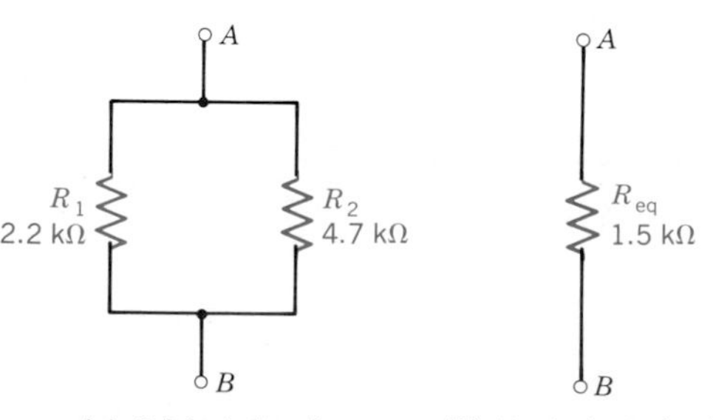

FIGURE 6-4
Circuits for Example 6-2.

Note that if this resistance is now combined with a 1-kΩ resistor in parallel, using the same equation, we obtain:

$$R_{\text{eq}} = \frac{R_1 R_2}{R_1 + R_2} \tag{6-10}$$

$$= \frac{1\text{ k}\Omega \times 1.5\text{ k}\Omega}{1\text{ k}\Omega + 1.5\text{ k}\Omega} = 0.6\text{ k}\Omega = \mathbf{600\ \Omega}$$

This is the same result as in Example 6-1. That is, we can repeatedly use Eq. 6-10 for any number of parallel resistors, provided that we consider them two at a time.

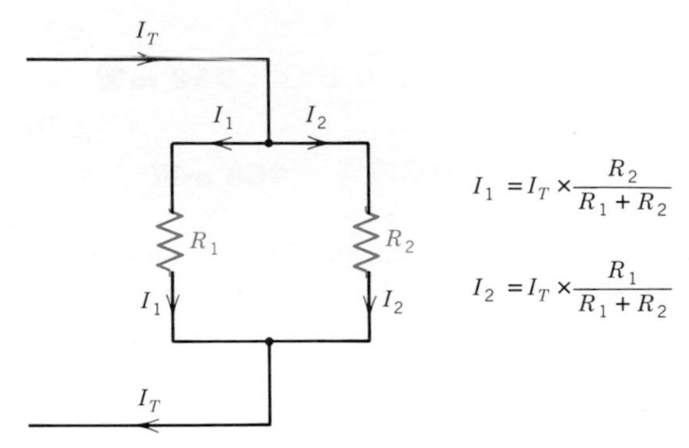

FIGURE 6-6
Current division in two parallel-connected resistors.

EXAMPLE 6-3

Three 600-Ω resistors are parallel-connected across a 60-V source of emf, as in Fig. 6-5a. Calculate:

a. The total circuit resistance.
b. The total current drawn from the supply.
c. The current in each parallel branch.

SOLUTION

a. $$R_{\text{eq}} = \frac{R}{N} \tag{6-11}$$

$$= \frac{600\ \Omega}{3} = \mathbf{200\ \Omega}$$

b. $$I_T = \frac{V}{R_{\text{eq}}} \tag{3-1a}$$

$$= \frac{60\text{ V}}{200\ \Omega} = \mathbf{0.3\ A}$$

c. The total current of 0.3 A must divide equally among the three 600-Ω branches, so **0.1 A** must flow in each branch. See Fig. 6-5a.

6-5 CURRENT DIVISION

If Example 6-1 is re-examined, you will find that the current through the 2.2-kΩ resistor is approximately *double* the current through the 4.7-kΩ resistor with which it is in parallel. This, of course, should not be surprising because of the resistance values of each, but it does suggest a formula by which the current divides in a circuit with two parallel resistances. Consider two resistors in parallel and a total current I_T flowing into the combination as in Fig. 6-6.

Although the source voltage is not shown, it could be obtained using:

$$V = I_T R_{\text{eq}} = I_T \times \frac{R_1 R_2}{R_1 + R_2}$$

Then, $I_1 = \frac{V}{R_1} = \frac{I_T}{\cancel{R_1}} \times \frac{\cancel{R_1} R_2}{R_1 + R_2}$. Therefore,

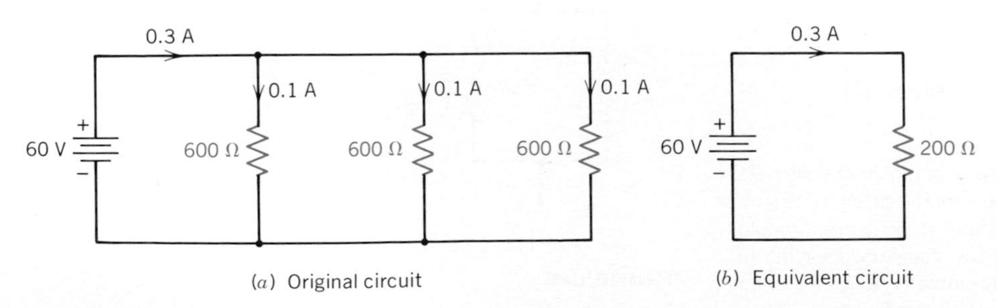

FIGURE 6-5
Circuits for Example 6-3.

$$I_1 = I_T \times \frac{R_2}{R_1 + R_2} \quad \text{amperes (A)} \tag{6-12}$$

and
$$I_2 = I_T \times \frac{R_1}{R_1 + R_2} \quad \text{amperes (A)} \tag{6-13}$$

Equations 6-12 and 6-13 give the current division rule and apply only to two resistors in parallel. Note how the *opposite* subscripts are used for the numerator resistance and the branch current being determined.

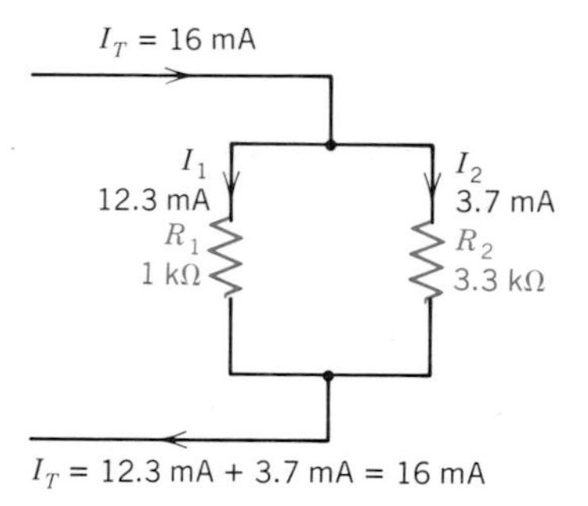

FIGURE 6-7
Circuit for Example 6-4.

EXAMPLE 6-4

A 1-kΩ resistor (R_1) and a 3.3-kΩ resistor (R_2) are in parallel with each other and have a combined current of 16 mA through them. What is the current through each?

SOLUTION

$$I_1 = I_T \times \frac{R_2}{R_1 + R_2} \tag{6-12}$$

$$= 16 \text{ mA} \times \frac{3.3 \text{ k}\Omega}{1 \text{ k}\Omega + 3.3 \text{ k}\Omega}$$

$$= 16 \text{ mA} \times \frac{3.3 \text{ k}\Omega}{4.3 \text{ k}\Omega} = \mathbf{12.3 \text{ mA}}$$

$$I_2 = I_T \times \frac{R_1}{R_1 + R_2} \tag{6-13}$$

$$= 16 \text{ mA} \times \frac{1 \text{ k}\Omega}{4.3 \text{ k}\Omega} = \mathbf{3.7 \text{ mA}}$$

PROOF $I_1 + I_2 = I_T$

$12.3 + 3.7 = \mathbf{16 \text{ mA}}$

Note how the current divides up, so that only approximately one-*quarter* of the current passes through the 3.3-kΩ resistor. See Fig. 6-7.

6-6 SHORT CIRCUITS

In Section 6-5 we saw how the *opposite* subscripts must be used in Eqs. 6-12 and 6-13 for the numerator resistance and the branch current being determined. That this is correct can be seen by assuming $R_1 = 0$ and determining I_1:

$$I_1 = I_T \times \frac{R_2}{R_1 + R_2} \tag{6-12}$$

$$= I_T \times \frac{R_2}{0 + R_2} = I_T \times 1 = I_T$$

That is, all the current I_T will flow through R_1 if its resistance is zero, and the current I_2 through R_2 should be zero:

$$I_2 = I_T \times \frac{R_1}{R_1 + R_2} \tag{6-13}$$

$$= I_T \times \frac{0}{0 + R_2} = \frac{0}{R_2} = 0$$

This condition, of course, represents a short circuit. In a parallel circuit this will always result in no current flowing in the other branches. The current will follow the path of least resistance. However, it is more important to note that the short-circuit current that flows would be damagingly high, limited only by the current capability of the source and its internal resistance. For this reason, where large currents could flow if a short circuit occurred, a fuse or circuit breaker would be used in series with the source, as in Fig. 6-8, for protection.

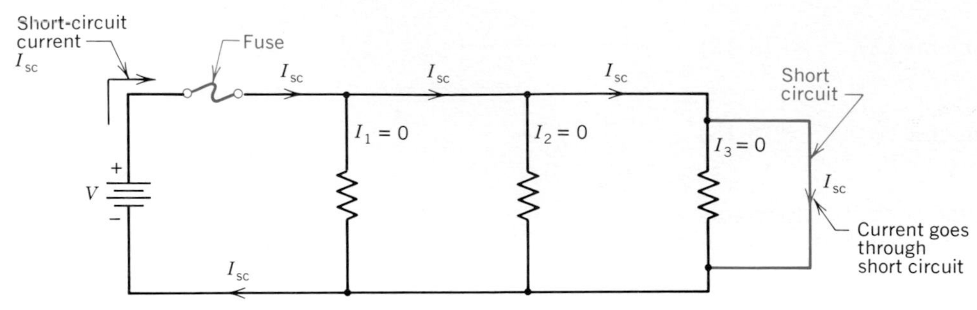

FIGURE 6-8
Short circuit in a parallel network protected by a fuse.

SUMMARY

1. Components connected in parallel have the same voltage across them.

2. The current drawn by each component is determined by the resistance of that particular branch.

3. Each branch is independent of the other branches provided that no short circuit occurs.

4. Kirchhoff's current law states that the current into a junction equals the current out of the junction.

5. The equivalent resistance of a number of parallel-connected resistors may be obtained from $1/R_{eq} = 1/R_1 + 1/R_2 + 1/R_3$.

6. The equivalent resistance of a number of parallel-connected resistors is always less than the smallest individual resistance.

7. For two parallel-connected resistors, the equivalent resistance is given by $R_{eq} = R_1R_2/(R_1 + R_2)$.

8. For N equal resistors connected in parallel, each of resistance R, the equivalent resistance is given by $R_{eq} = R/N$.

9. The total conductance of a parallel circuit is given by $G_T = G_1 + G_2 + G_3$, and the equivalent resistance by $R_{eq} = 1/G_T$.

10. The total power dissipated in a parallel circuit is given by $P_T = P_1 + P_2 + P_3$.

11. For two parallel-connected resistors with a total current I_T, the current division rule gives the current in each resistor by $I_1 = I_T\dfrac{R_2}{R_1 + R_2}$ and $I_2 = I_T\dfrac{R_1}{R_1 + R_2}$.

12. A short circuit in any parallel branch causes a large current to flow from the source and reduces the current through the other branches to zero.

SELF-EXAMINATION

Answer true or false or a, b, c, d
(Answers at back of book)

6-1. A junction or node requires the connection of three or more wires. ________

6-2. Kirchhoff's current law applies only to a parallel circuit. ________

6-3. In Fig. 6-2a, the branch currents will be equal to each other only if all the branch resistances are equal. ________

6-4. In a parallel circuit the combined equivalent resistance is always larger than the smallest resistance. ________

6-5. If, in Fig. 6-3, R_2 is twice as large as R_1 and only half as large as R_3, the smallest current flows in:

a. R_1
b. R_2
c. R_3

6-6. For the same conditions as in Question 6-5, the largest current flows in:

a. R_1
b. R_2
c. R_3

6-7. For the same conditions as in Question 6-5, the ammeter in Fig. 6-3 will have a reading compared with the current through R_1 that is

a. larger
b. smaller
c. the same

6-8. Two 20-kΩ resistors and a 10-kΩ resistor are connected in parallel across a 50-V source. Their combined equivalent resistance is

a. 30 kΩ
b. 50 kΩ
c. 10 kΩ
d. 5 kΩ

6-9. The branch currents in Question 6-8 are

a. All 2½ mA
b. all 1 mA
c. 2½, 2½ mA, 5 mA
d. 1.25 mA, 1.25 mA, 5 mA

6-10. The total source current in Question 6-8 is

a. 7½ ma
b. 3 mA
c. 10 mA
d. 1 mA

6-11. The total power delivered by the source in Question 6-8 is

a. 375 mW
b. 150 mW
c. 0.5 W
d. 50 mW

6-12. One million 1-megohm resistors, if connected in parallel, would have a combined resistance of:

a. 1 MΩ
b. 1 mΩ
c. 1 kΩ
d. 1 Ω

6-13. A 4-Ω and an 8-Ω resistor in parallel have a combined resistance of

a. 6 Ω
b. $2\frac{2}{3}$ Ω
c. $2\frac{1}{3}$ Ω
d. 12 Ω

6-14. If the two resistors in Question 6-13 have a total current of 24 mA flowing through them, the current in the 8-Ω resistor is

a. 8 mA
b. 16 mA
c. 12 mA
d. 24 mA

6-15. If a short circuit develops in the 4-Ω resistor of Question 6-14, the current in the 8-Ω resistor will be

a. 24 mA
b. 0 mA
c. 16 mA
d. 8 mA

REVIEW QUESTIONS

1. What advantages does a parallel circuit have compared with a series circuit?

2. In what ways is a parallel circuit "opposite" to a series circuit? (Name at least three.)

3. In what ways is a parallel circuit similar to a series circuit? (Name at least two.)

4. Justify why the equivalent resistance of a parallel circuit must be less than the smallest branch resistance.

5. Why does "adding" a resistor in a parallel circuit decrease the total combined resistance but increase the total in a series circuit?

6. Justify why it is possible to add conductances to obtain the total conductance in a parallel circuit, but not in a series circuit.

7. What electrical equations are the same for a parallel circuit as for a series circuit? Why is this so?

8. Derive the equation for two resistors in parallel, Eq. 6-10, from the reciprocal equation, Eq. 6-4.

9. Use the current division equations (6-12 and 6-13) to determine what current flows in each resistor when an open circuit occurs in R_1 ($R_1 = \infty$).

10. Why can a short circuit cause damaging currents in a parallel circuit whereas it is less likely to happen in a series circuit?

PROBLEMS

(Answers to odd-numbered problems at back of book)

6-1. What is the equivalent resistance of a parallel circuit consisting of $R_1 = 100\ \Omega$, $R_2 = 220\ \Omega$, and $R_3 = 330\ \Omega$?

6-2. What is the equivalent resistance of a parallel circuit consisting of $R_1 = 100$ kΩ, $R_2 = 220$ kΩ, and $R_3 = 330$ kΩ?

6-3. Repeat Problem 6-1 using $R_1 = 10\ \text{k}\Omega$, $R_2 = 6.8\ \text{k}\Omega$, and $R_3 = 1\ \text{M}\Omega$.

6-4. What is the total conductance of the parallel circuit in Problem 6-2?

6-5. A resistor of conductance 750 μS is connected in parallel with a component having a conductance of 50 mS. What is the equivalent resistance of the combination?

6-6. What is the total current when the parallel circuit of Problem 6-2 is connected across a 40-V source of emf?

6-7. What is the total power dissipated in the three resistors of Problem 6-1 when the parallel circuit is connected across a 120-V supply?

6-8. Three 120-V lamps, rated at 60 W, 100 W, and 200 W are connected in parallel to a 120-V supply. Find the equivalent hot resistance of this combination in the following two ways:

a. Find the individual resistances and combine them.

b. Find the total power and solve for R.

6-9. What resistance must be connected across a 2.2-kΩ resistor to produce a total resistance of 1.5 kΩ?

6-10. What resistance must be placed in parallel with a 33-kΩ resistor to produce a total conductance of 50 μS?

6-11. Three resistors in parallel draw a total current of 7 mA. The first resistor dissipates 50 mW; the second resistor has a voltage drop of 20 V; and the third resistor has a resistance of 10 kΩ. Determine the resistance of the first two resistors.

6-12. Three resistors in parallel draw a total current of 10 mA and dissipate a total power of 500 mW. If the first resistor draws 5 mA and the second and third resistors are equal in value, determine the resistance of all three resistors.

6-13. A 47-kΩ and a 51-kΩ resistor are connected in parallel and draw a total of 12 mA.

a. How much current passes through each resistor?

b. What must be the minimum power rating of each?

6-14. What is the largest resistance that can be connected in parallel with a 10-kΩ, 1-W resistor if the combination must handle a total of 25 mA? What is the power rating of this resistor?

6-15. If the three resistors in Problem 6-1 are each rated at 2 W, what is the highest voltage that can safely be applied to the parallel combination?

6-16. How many parallel-connected 25-W Christmas tree lamps can be supplied from a 120-V source fused at 15 A?

6-17. The total current drawn by the three resistors in Problem 6-2 is 3 mA. What is the current through each resistor?

6-18. Six equal resistors are to be connected in parallel with a 100-kΩ resistor to drop the overall resistance to 40 kΩ. What must be the value of each of the six resistors?

CHAPTER SEVEN

SERIES-PARALLEL AND VOLTAGE-DIVIDER CIRCUITS

Seldom is a practical circuit a truly series or a truly parallel circuit. The insertion of a fuse in series with a power supply and its parallel-connected loads results in a series-parallel circuit. Similarly, the internal resistance of a voltage source or the resistance of feeder lines results in series-parallel circuits. However, all of the rules and laws developed separately for series and parallel circuits can be applied to a series-parallel combination.

An important application of series and parallel principles is in the area of voltage-divider circuits. These permit a load to operate at a voltage different from the supply voltage immediately available. This may be accomplished simply by the use of a *series dropping resistor,* or a *bleeder resistor* may be added to improve the *voltage regulation* if the load should vary. In some transistor and integrated circuit power supplies, it is often necessary to provide both positive and negative potentials with respect to ground. This chapter shows how to select the resistors for such a loaded voltage-divider network.

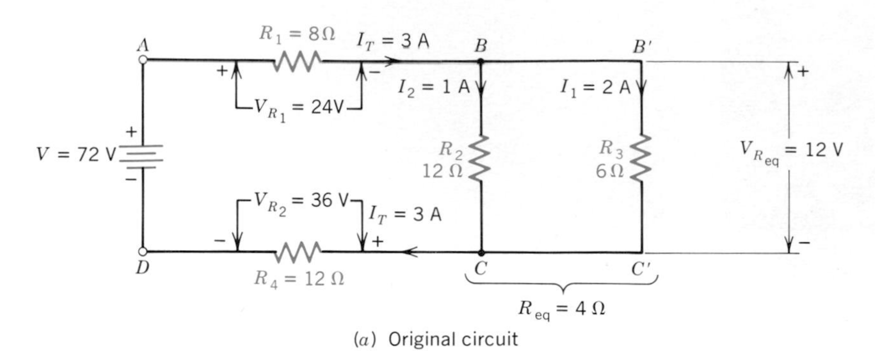

(a) Original circuit

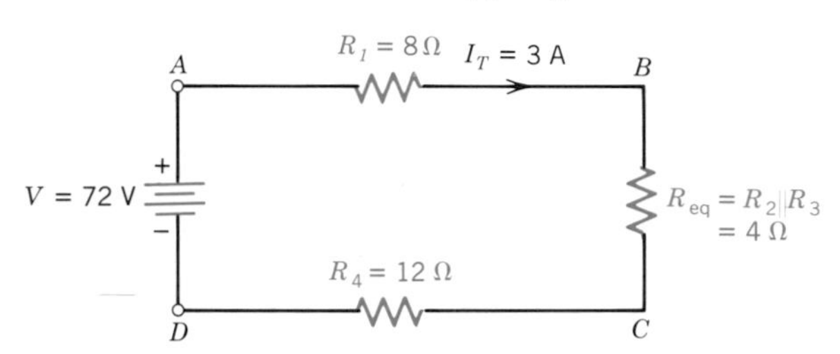

(b) Equivalent circuit for parallel portion = R_{eq}

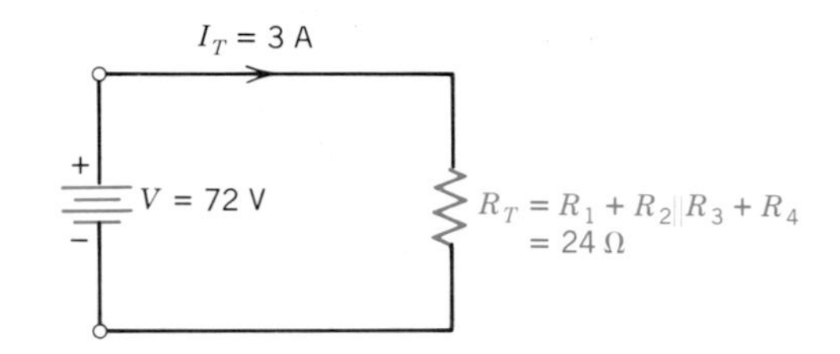

(c) Final equivalent circuit

FIGURE 7-1
Simplification of a series-parallel circuit.

7-1 SERIES-PARALLEL RESISTOR CIRCUITS

When a parallel bank of resistors is connected in series with another resistor or parallel group, we have a *series-parallel* circuit (see Fig. 7-1*a*). Solving for the currents and voltage drops in such a circuit involves the combined methods of the previous chapters. In general, we apply the series circuit rules to the series portions and parallel circuit rules to the parallel portions. Let us make a few general observations about the series-parallel circuit in Fig. 7-1*a*.

1. Resistors R_1 and R_4 are in series with each other because they both have the same current. (The current that leaves the battery and passes through R_1 must return to the battery through R_4.)
2. Resistors R_2 and R_3 are in parallel with each other because there is the same voltage across the two resistors. (A voltmeter connected from B to C indicates the same voltage as from B' to C'.)
3. The parallel *combination* of R_2 and R_3 is in series with R_1 and R_4. (The equivalent resistance of R_2 and R_3 carries the same current as R_1 and R_4. See Fig. 7-1*b*.)
4. The current through either R_2 or R_3 is less than the current through R_1. (This is because the current through R_1 must split-up or *divide* at junction B into two smaller currents.)
5. The sum of the two currents through R_2 and R_3 must equal I_T. That is, the total current equals the sum of the parallel branch currents (Kirchhoff's current law).
6. Only half as much current flows through R_2 as through R_3 because the resistance of R_2 is twice that of R_3 (current division equation).
7. The sum of the voltage drops around either loop *ABCD or AB′C′D* equals the applied voltage (Kirchhoff's voltage law). (**NOTE:** We only count the voltage drop across a parallel circuit *once* when determining the total of the voltage drops around a complete circuit.)

The current and voltages in Fig. 7-1*a* are found as follows.

The total resistance of the whole circuit is given by

$$R_T = R_1 + R_2\|R_3 + R_4$$
$$= 8\ \Omega + \frac{12\ \Omega \times 6\ \Omega}{12\ \Omega + 6\ \Omega} + 12\ \Omega$$
$$= 8\ \Omega + 4\ \Omega + 12\ \Omega = \mathbf{24\ \Omega}$$

See Fig. 7-1*c*.

The current delivered by the battery is

$$I_T = \frac{V}{R_T} \qquad (3\text{-}1a)$$
$$= \frac{72\ V}{24\ \Omega} = \mathbf{3\ A}$$

The voltage drop across any resistance is given by

$$V = IR \qquad (3\text{-}1b)$$

$$V_{R_1} = I_T R_1 = 3\ A \times 8\ \Omega = \mathbf{24\ V}$$
$$V_{R_{eq}} = I_T R_{eq} = 3\ A \times 4\ \Omega = \mathbf{12\ V}$$
$$V_{R_4} = I_T R_4 = 3\ A \times 12\ \Omega = \mathbf{36\ V}$$

Note that $V_{R_1} + V_{R_{eq}} + V_{R_4} = \mathbf{72\ V.}$

The branch currents are

$$I_2 = \frac{V_{R_{eq}}}{R_2} = \frac{12\ V}{12\ \Omega} = \mathbf{1\ A}$$
$$I_3 = \frac{V_{R_{eq}}}{R_3} = \frac{12\ V}{6\ \Omega} = \mathbf{2A}$$

and
$$I_2 + I_3 = \mathbf{3\ A}$$

or
$$I_2 = I_T \times \frac{R_3}{R_2 + R_3} \qquad (6\text{-}12)$$
$$= 3\ A \times \frac{6\ \Omega}{12\ \Omega + 6\ \Omega}$$
$$= 3\ A \times \frac{6}{18} = 3\ A \times \frac{1}{3} = \mathbf{1\ A}$$

$$I_3 = I_T \times \frac{R_2}{R_2 + R_3}$$
$$= 3\ A \times \frac{12\ \Omega}{12\ \Omega + 6\ \Omega}$$
$$= 3\ A \times \frac{12\ \Omega}{18\ \Omega} = 3\ A \times \frac{2}{3} = \mathbf{2\ A}$$

Example 7-1 shows a circuit with two parallel groups that are in series with a resistor and two opposing voltage sources.

EXAMPLE 7-1

Given the series-parallel resistor network in Fig. 7-2*a*, calculate:

a. The total circuit resistance.
b. The total supply current.
c. The current through R_1 and R_2.
d. The voltage drop across R_2.
e. The voltage drop across R_3.
f. The voltage drop across R_5.
g. The current through R_4, R_5, and R_6.
h. The voltage at B with respect to A.
i. The power delivered by V_2.
j. The total power dissipated by all the resistors.
k. The power recharging voltage source V_1.

SOLUTION

a. For R_1 and R_2 in parallel in Fig. 7-2a,

$$R_{eq1} = R_1\|R_2 = \frac{R_1 R_2}{R_1 + R_2} \qquad (6\text{-}10)$$
$$= \frac{1\ k\Omega \times 2.2\ k\Omega}{1\ k\Omega + 2.2\ k\Omega} = 688\ \Omega$$

For R_4, R_5 and R_6 in parallel in Fig. 7-2*a*,

$$\frac{1}{R_{eq2}} = \frac{1}{R_4} + \frac{1}{R_5} + \frac{1}{R_6} \qquad (6\text{-}4)$$
$$= \frac{1}{4.7\ k\Omega} + \frac{1}{5.1\ k\Omega} + \frac{1}{6.8\ k\Omega}$$
$$= 0.213 \times 10^{-3}\ S + 0.196 \times 10^{-3}\ S + 0.147 \times 10^{-3}\ S$$
$$= 0.556 \times 10^{-3}\ S$$

Therefore, $R_{eq2} = \dfrac{1}{0.556 \times 10^{-3}\ S} = 1799\ \Omega$

Using Fig. 7-2*b* and *c*, we obtain

$$R_T = R_{eq1} + R_3 + R_{eq2} \qquad (5\text{-}2)$$
$$= 688 + 3{,}300 + 1{,}799 = \mathbf{5787\ \Omega}$$

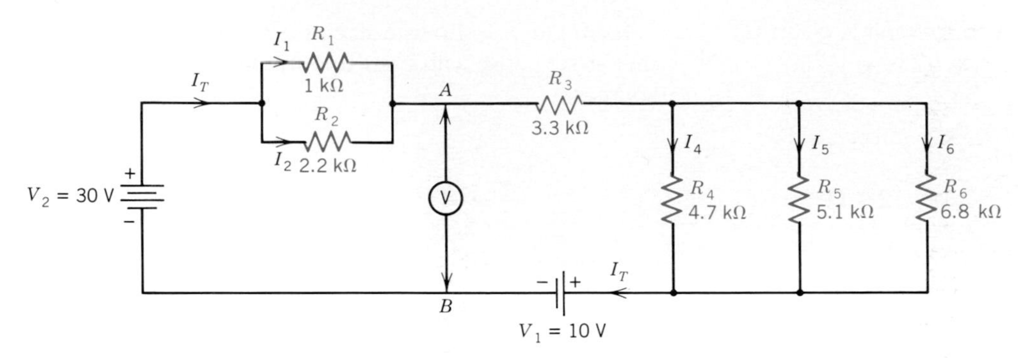

(*a*) Original circuit

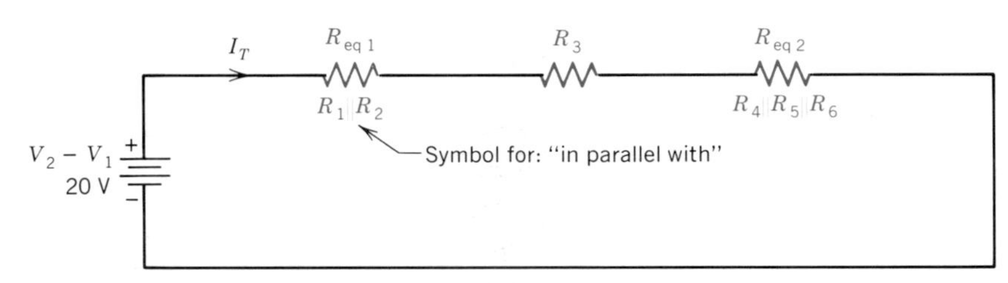

(*b*) Simplified equivalent series circuit

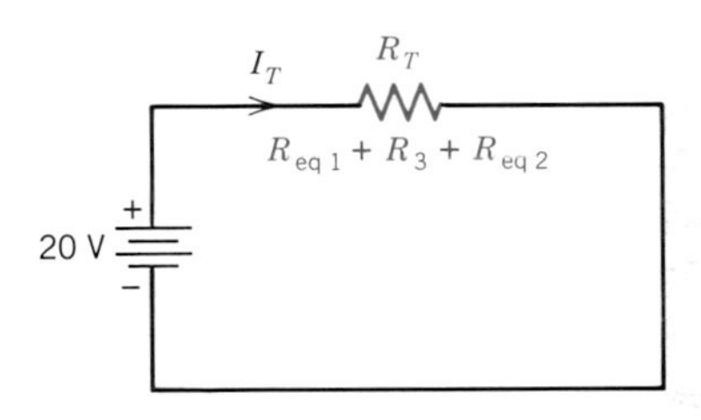

(*c*) Total equivalent resistance R_T

FIGURE 7-2
Circuit for Example 7-1.

b. From Fig. 7-2*c*,

$$I_T = \frac{V}{R_T} \qquad \text{(3-1a)}$$

$$= \frac{V_2 - V_1}{R_T}$$

$$= \frac{30 - 10 \text{ V}}{5.787 \text{ k}\Omega} = \mathbf{3.46 \text{ mA}}$$

This value of I_T is the same as that shown in Fig. 7-2*a* and 7-2*b*.

c. $$I_1 = I_T \times \frac{R_2}{R_1 + R_2} \qquad \text{(6-12)}$$

$$= 3.46 \text{ mA} \times \frac{2.2 \text{ k}\Omega}{1 \text{ k}\Omega + 2.2 \text{ k}\Omega} = \mathbf{2.38 \text{ mA}}$$

$$I_2 = I_T \times \frac{R_1}{R_1 + R_2} \qquad \text{(6-13)}$$

$$= 3.46 \text{ mA} \times \frac{1 \text{ k}\Omega}{1 \text{ k}\Omega + 2.2 \text{ k}\Omega} = \mathbf{1.08 \text{ mA}}$$

CHECK $I_T = I_1 + I_2 = 2.38 + 1.08 = \mathbf{3.46 \text{ mA}}$

d. From Fig. 7-2*a*,

$$V_{R_2} = I_2 R_2 \qquad \text{(3-1b)}$$

$$= 1.08 \text{ mA} \times 2.2 \text{ k}\Omega = \mathbf{2.38 \text{ V}}$$

CHECK $V_{R_1} = I_1 R_1$

$= 2.38 \text{ mA} \times 1 \text{ k}\Omega = \mathbf{2.38\ V}$

e. From Fig. 7-2*a*,

$$V_{R_3} = I_T R_3 \qquad (3\text{-}1b)$$
$$= 3.46 \text{ mA} \times 3.3 \text{ k}\Omega = \mathbf{11.42\ V}$$

f. From Fig. 7-2*b*,

$$V_{R_5} = I_T R_{eq2} \qquad (3\text{-}1b)$$
$$= 3.46 \text{ mA} \times 1799\ \Omega = \mathbf{6.2\ V}$$

g. $$I_4 = \frac{V_{R_4}}{R_4} \qquad (3\text{-}1a)$$
$$= \frac{6.2 \text{ V}}{4.7 \text{ k}\Omega} = \mathbf{1.32\ mA}$$
$$I_5 = \frac{V_{R_5}}{R_5} \qquad (3\text{-}1a)$$
$$= \frac{6.2 \text{ V}}{5.1 \text{ k}\Omega} = \mathbf{1.22\ mA}$$
$$I_6 = \frac{V_{R_6}}{R_6} \qquad (3\text{-}1a)$$
$$= \frac{6.2 \text{ V}}{6.8 \text{ k}\Omega} = \mathbf{0.92\ mA}$$

CHECK $I_T = I_4 + I_5 + I_6 = 1.32 + 1.22 + 0.92 = \mathbf{3.46\ mA}$

h. $$V_{AB} = V_2 - V_{R_2}$$
$$= +30 - 2.38 = +27.62 \text{ V}$$

Thus point B is **−27.62 V** with respect to point A.

i. Power delivered by V_2 is

$$P_2 = V_2 \times I_T \qquad (5\text{-}4b)$$
$$= 30 \text{ V} \times 3.46 \text{ mA} = \mathbf{104\ mW}$$

j. Power dissipated by all the circuit resistors $= I_T^2 R_T$ (5-4a)

$$= (3.46 \times 10^{-3} \text{ A})^2 \times 5787\ \Omega$$
$$= \mathbf{69.3\ mW}$$

k. Power delivered to V_1 in recharging it $= V_1 \times I_T$ (5-4b)

$$= 10 \text{ V} \times 3.46 \times 10^{-3} \text{ A}$$
$$= \mathbf{34.6\ mW}$$

CHECK The power dissipated in all the resistors

$= P_T - P_1 = (104 - 34.6)$ mW

$\approx \mathbf{69.4\ mW}$

Note that there are various ways of checking the validity and accuracy of our calculations, as shown in Example 7-1.

7-1.1 Practical Series-Parallel Circuit

In practice, most circuits are of the series-parallel type. For example, every voltage source has some internal resistance. See Fig. 7-3.

This means that although we may parallel-connect a number of loads right across the terminals of the source, the internal resistance is in *series* with the parallel-connected loads. (The internal resistance is *inside* the source and cannot be separated from the emf of the source.) Only if the internal resistance is negligible, do we have a truly parallel circuit.

Similarly, the circuit is truly parallel only if the connecting leads from source to load also have negligible resistance. When the source and load are well separated, the resistance of the feeder line must be taken into account, as shown in Example 7-2.

EXAMPLE 7-2

An electric motor and an electric heater are to be operated in parallel with each other at a distance of 200 m from a 220-V source of emf. The current drawn by the motor is 16 A and the resistance of the heater is 20 Ω. If the total voltage drop in the line must not exceed 10% of the source voltage, calculate:

a. The minimum wire gauge to be used for the copper feeder line.
b. The power dissipated in the feeder line.
c. The power delivered to the load.
d. The power supplied by the source.

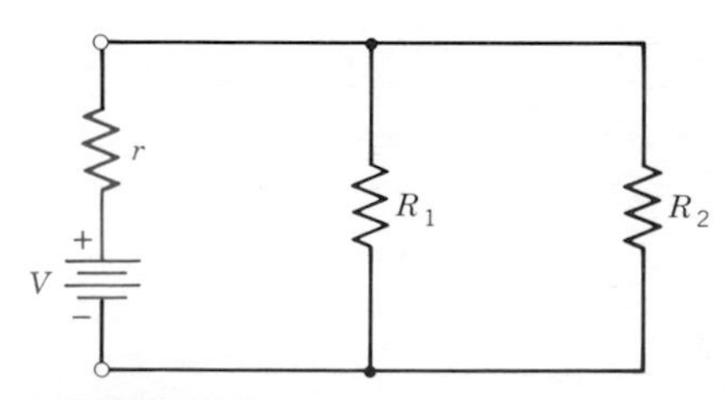

FIGURE 7-3
A series-parallel circuit due to the internal resistance of the voltage source.

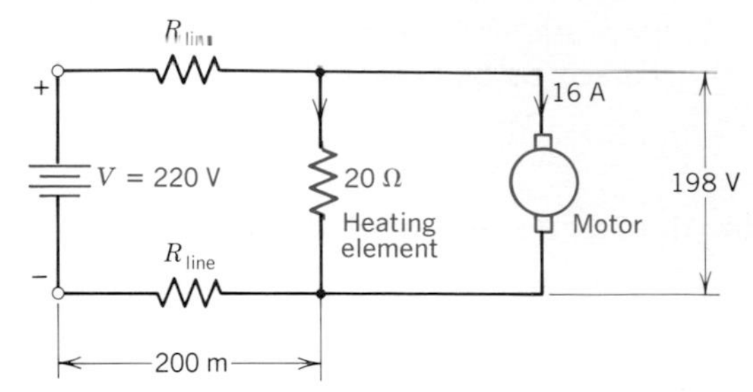

FIGURE 7-4
Circuit for Example 7-2.

Assume that the source has negligible internal resistance. See Fig. 7-4.

SOLUTION

a. Total permissible voltage drop

$$= 10\% \text{ of } 220 \text{ V} = 22 \text{ V}$$

Load voltage = 220 V − 22 V = 198 V

$$\text{Heater current} = \frac{V}{R} \tag{3-1a}$$

$$= \frac{198 \text{ V}}{20 \ \Omega} = 9.9 \text{ A}$$

Total line current = 9.9 A + 16 A = 25.9 A

Maximum resistance of total

$$\text{feeder line} = \frac{V}{I_T} \tag{3-1}$$

$$= \frac{22 \text{ V}}{25.9 \text{ A}} \approx 0.85 \ \Omega$$

Maximum resistance of each line,

$$R_{\text{line}} = \frac{0.85 \ \Omega}{2} = 0.425 \ \Omega$$

$$R_{\text{line}} = \frac{\rho l}{A} \tag{4-1}$$

$$\text{Therefore, } A = \frac{\rho l}{R_{\text{line}}}$$

$$= \frac{1.72 \times 10^{-8} \ \Omega \cdot \text{m} \times 200 \text{ m}}{0.425 \ \Omega}$$

$$= 8.09 \times 10^{-6} \text{ m}^2$$

$$A = \frac{\pi d^2}{4}$$

$$\text{so that } d = \sqrt{\frac{4A}{\pi}}$$

$$= \sqrt{\frac{4 \times 8.09 \times 10^{-6} \text{ m}^2}{\pi}}$$

$$= 3.21 \times 10^{-3} \text{ m}$$

Thus the *minimum* diameter of the wire is **3.21 mm.** Referring to Appendix A-3, we find that the minimum AWG size of wire is **gauge number 8** (3.26 mm).

NOTE If we had selected the wire according to its current-carrying capacity, a number 10 gauge wire (30 A without overheating), would have been too small and would have caused too high a voltage drop.

Also, after determining that the maximum resistance should be 0.425 Ω for 200 m, we know that the resistance is 0.425 Ω × 5 or 2.125 Ω/km, maximum. From the wire gauge table, Appendix A-3, we see that number 8 gauge has a resistance of 2.06 Ω/km, and may be used.

b. Since the diameter of the number 8 gauge wire is practically equal to the calculated value, we can use the voltages and currents calculated in part (a).

Power dissipated in the line

$$= I_T^2 \times 2R_{\text{line}} \tag{3-9}$$

$$= (25.9 \text{ A})^2 \times 2 \times 0.425 \ \Omega$$

$$= \mathbf{570 \ W}$$

c. Power delivered to the load

$$= V \times I_T \tag{3-7}$$

$$= 198 \text{ V} \times 25.9 \text{ A}$$

$$= \mathbf{5128 \ W}$$

d. Power supplied by source = 570 W + 5128 W

$$= \mathbf{5698 \ W}$$

$$\text{or power} = V \times I_T \tag{3-7}$$

$$= 220 \text{ V} \times 25.9 \text{ A} = \mathbf{5698 \ W}$$

Note how 10% of the power supplied by the source is "lost" in the line. This is because we have permitted a 10% voltage drop to occur due to the resistance of the line.

7-1.2 Thermistor-Resistor Series-Parallel Circuit

An interesting series-parallel circuit is a series string of lights each lamp of which has a thermistor *in parallel* with it, as shown in Fig. 7-5.

When the thermistor is cold its resistance is high, so very little current flows through it, giving an essentially series circuit. If a lamp burns out, the current must now pass through the thermistor in parallel with it. The

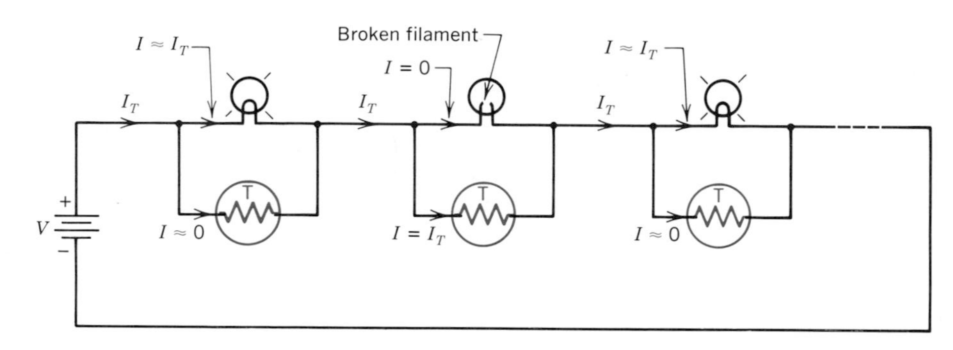

FIGURE 7-5
Series-parallel circuit with a thermistor across each lamp.

resulting increase in temperature of the thermistor will cause its resistance to drop, ideally to a value close to the hot filament resistance. Thus the rest of the lamps will continue to operate at their normal voltage and current. It is now possible to tell which lamp in the series string has burned out in order to replace it. This is *not* possible with an ordinary series circuit.

7-2 VOLTAGE DIVIDERS

A common problem in electronics is to provide power to a load at a voltage different from the available source voltage. In some cases, a number of different loads, all working at different voltages and having various current requirements, may have to operate from a common supply. The following three examples use the principles we have covered in the previous chapters to design three typical divider networks.

7-2.1 Series-dropping Resistor

EXAMPLE 7-3

It is required to operate a 9-V transistor radio from a 12-V automobile battery. If the average load current of the transistor radio is 100 mA, calculate the value of the required series-dropping resistor and its power rating using nearest commercial values.

SOLUTION

Draw the circuit shown in Fig. 7-6 and enter all known values given. The voltage dropped across R must be $V_R = 12\text{ V} - 9\text{ V} = 3\text{ V}$. Using Ohm's law, we obtain

$$R = \frac{V_R}{I} \tag{3-1}$$

$$= \frac{3\text{ V}}{0.1\text{ A}} = 30\ \Omega \qquad \text{(Use a } \mathbf{33\text{-}\Omega} \text{ resistor.)}$$

$$P_R = V_R I \tag{3-7}$$

$$= 3\text{ V} \times 0.1\text{ A} = 0.3\text{ W}$$

(Use a power rating of **½ watt.**)

The nearest standard value of 33 Ω, ±10%, with a power rating of $\frac{1}{2}$ W would be suitable.

The disadvantage of the above circuit is the variation of load voltage with load current. For example, if the volume of the radio is increased, it is possible for the load current to increase to as much as 200 mA. The voltage drop across the series-dropping resistor would increase to 6 V leaving only 6 V for the load. This could cause distortion in the audio section. Similarly, if the current is reduced, the load voltage will increase toward a maximum of 12 V at no load.

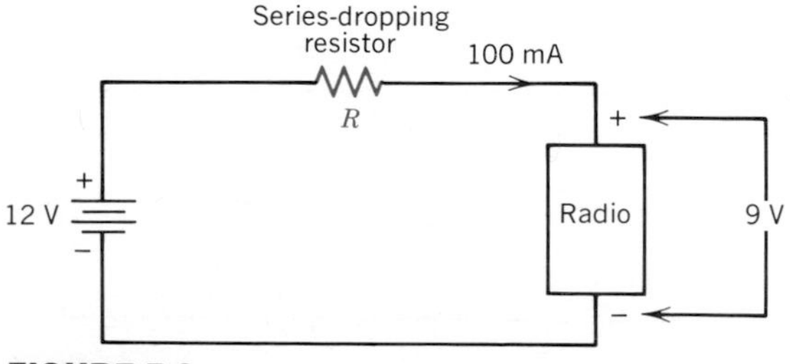

FIGURE 7-6
Circuit for Example 7-3.

7-2.2 Bleeder Resistor

This problem can be overcome to some degree by the use of a *bleeder resistor,* placed in parallel with the load, and selected to draw about 10 to 25% of the total current from the source, as shown in Example 7-4.

EXAMPLE 7-4

Design a simple voltage divider with a series-dropping resistor and a bleeder resistor to provide 9 V at a load current of 100 mA from a 12-V supply. Use a bleeder current of (a) 30 mA, (b) 100 mA, determining the no-load voltage in each case. Use commercial values for all resistors. See Fig. 7-7.

SOLUTION

a. $$I_S = I_B + I_L \qquad (6\text{-}1)$$
$$= 30\text{ mA} + 100\text{ mA} = 130\text{ mA}$$
$$V_{R_S} = 12\text{ V} - 9\text{ V} = 3\text{ V}$$

Therefore, $$R_S = \frac{V_{R_S}}{I_S} \qquad (3\text{-}1)$$
$$= \frac{3\text{ V}}{130\text{ mA}} = \mathbf{23\ \Omega}$$
$$P_{R_S} = V_{R_S} I_S \qquad (3\text{-}7)$$
$$= 3\text{ V} \times 0.13\text{ A} = \mathbf{0.39\ W}$$

Use a 22-Ω, 5%, ½-W resistor, for R_S.

$$V_{R_B} = 9\text{ V}, I_B = 30\text{ mA}$$

Therefore, $$R_B = \frac{V_{R_B}}{I_B} \qquad (3\text{-}1)$$
$$= \frac{9\text{ V}}{30\text{ mA}} = \mathbf{300\ \Omega}$$
$$P_B = V_{R_B} I_B \qquad (3\text{-}7)$$
$$= 9\text{ V} \times 0.03\text{ A} = \mathbf{0.27\ W}$$

Use a 300-Ω, 5%, ½-W resistor for R_B. With the load disconnected

$$V_{R_B} = V \times \frac{R_B}{R_B + R_S} \qquad (5\text{-}5)$$
$$= 12\text{ V} \times \frac{300\ \Omega}{300\ \Omega + 22\ \Omega} = \mathbf{11.2\ V}$$

b. Using $I_B = 100$ mA, the required values are $R_S =$ **15 Ω,** 5%, **0.6 W** (1 W) and $R_B =$ **90 Ω, 0.9 W** (91 **Ω**, 5%, 1 W).

$$\text{unloaded } V_{R_B} = V \times \frac{R_B}{R_B + R_S} \qquad (5\text{-}5)$$
$$= 12\text{ V} \times \frac{91\ \Omega}{91\ \Omega + 15\ \Omega} = \mathbf{10.3\ V}$$

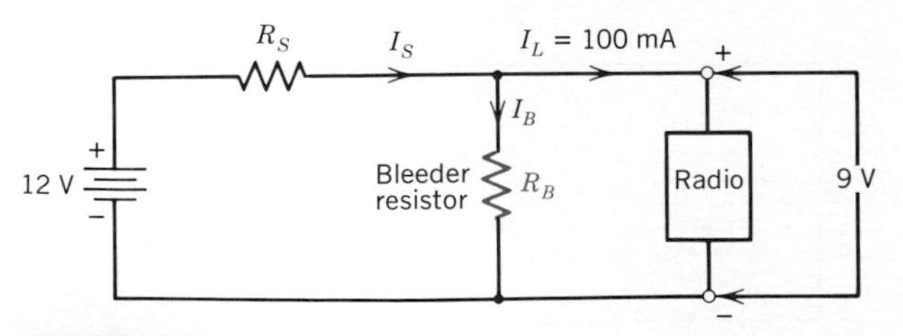

FIGURE 7-7
Circuit for Example 7-4.

Note that a larger bleeder current provides an unloaded output voltage (10.3 V) closer to the loaded value of 9 V. These conditions provide what is called a better *voltage regulation.* But the price paid is more power drawn from the source and wasted in the form of heat in the divider resistors. Even better regulation would be obtained in a practical circuit by the use of a Zener diode connected across R_B. This is covered briefly in Section 28-7.1 and would be included in any application where wide variations in load current might occur. But where the load current is relatively constant, the circuit of Fig. 7-7 will perform quite adequately.

7-2.3 Positive and Negative Potentials

Finally, consider a voltage divider to provide both positive and negative output potentials.

EXAMPLE 7-5

Determine the resistance and power values of R_1, R_2, and R_3 in Fig. 7-8 to provide +25 V at 10 mA for load A, +15 V at 50 mA for load B, and −15 V at 20 mA for load C from a 40-V source of emf with a total current of 100 mA.

SOLUTION

To find the resistances of R_1, R_2, and R_3, we need the voltage drop across each and the current through each.

At junction *W*, $$I_T = I_A + I_{R_1} \qquad (6\text{-}1)$$

Therefore, $$I_{R_1} = I_T - I_A$$
$$= 100\text{ mA} - 10\text{ mA} = 90\text{ mA}$$

At junction *X*, $$I_{R_1} = I_B + I_{R_2} \qquad (6\text{-}1)$$

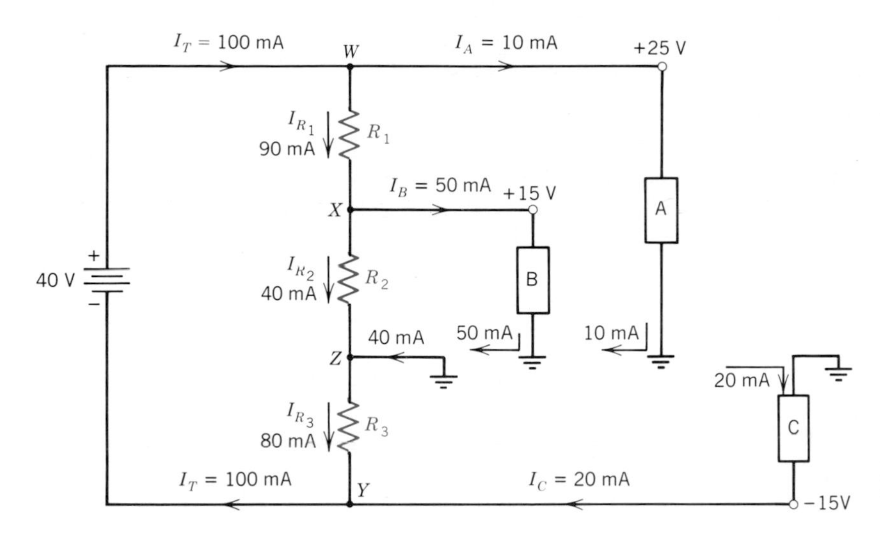

FIGURE 7-8
Voltage-divider circuit for Example 7-5.

Therefore, $I_{R_2} = I_{R_1} - I_B$

$= 90 \text{ mA} - 50 \text{ mA} = 40 \text{ mA}$

At junction Y, $I_T = I_C + I_{R_3}$ (6-1)

Therefore, $I_{R_3} = I_T - I_C$

$= 100 \text{ mA} - 20 \text{ mA} = 80 \text{ mA}$

NOTE At junction Z, 40 mA must be flowing "in" from the ground connection. This is because 60 mA total is "returning" from loads A and B, but 20 mA is flowing "out" to load C. The ground is positive, as far as the top side of load C is concerned, with respect to Y.

Clearly, from Fig. 7-8,

$$V_{R_3} = 15 \text{ V and } V_{R_2} = 15 \text{ V}$$

$$V_{R_1} = \text{load } A \text{ voltage} - \text{load } B \text{ voltage}$$
$$= 25 \text{ V} - 15 \text{ V} = 10 \text{ V}$$

$$R_1 = \frac{V_{R_1}}{I_{R_1}} \quad (3\text{-}1)$$
$$= \frac{10 \text{ V}}{90 \text{ mA}} = \mathbf{111\ \Omega}$$

$$P_1 = V_{R_1} I_{R_1} \quad (3\text{-}7)$$
$$= 10 \text{ V} \times 0.09 \text{ A} = \mathbf{0.9\ W}$$

(110 Ω, 5%, 1 W)

$$R_2 = \frac{V_{R_2}}{I_{R_2}} \quad (3\text{-}1)$$
$$= \frac{15 \text{ V}}{40 \text{ mA}} = \mathbf{375\ \Omega}$$

$$P_2 = V_{R_2} I_{R_2} \quad (3\text{-}7)$$
$$= 15 \text{ V} \times 0.04 \text{ A} = \mathbf{0.6\ W}$$

(390 Ω, 5%, 1 W)

$$R_3 = \frac{V_{R_3}}{I_{R_3}} \quad (3\text{-}1)$$
$$= \frac{15 \text{ V}}{80 \text{ mA}} = \mathbf{187.5\ \Omega}$$

$$P_3 = V_{R_3} I_{R_3} \quad (3\text{-}7)$$
$$= 15 \text{ V} \times 0.08 \text{ A} = \mathbf{1.2\ W}$$

(180 Ω, 5%, 2 W)

SUMMARY

1. A series-parallel circuit consists of some combination of series and parallel-connected components, with series and parallel circuit rules applied to each portion.

2. Most practical circuits consist of some series-parallel combination due to internal resistance of the source or resistance in the connecting leads.

3. A thermistor-resistor combination can be used to maintain a series string of lights in operation, if one lamp fails, by providing a low resistance path through a parallel-connected thermistor.

4. Voltage-divider circuits enable a load to operate at a different voltage from the supply voltage immediately available.

5. A single resistor may be connected between source and load to drop the voltage to a suitable value, depending upon the load current.

6. A series-dropping resistor voltage divider suffers from poor voltage regulation (wide changes in load voltage) if the load current varies widely.

7. A bleeder resistor connected in parallel with the load can improve the voltage regulation of a series-dropping resistor voltage divider.

8. The current drawn by a bleeder resistor should be from 10 to 25% of the total current from the source.

9. The design of a loaded voltage-divider network to provide both positive and negative potentials requires the application of Kirchhoff's current law to determine the current through each of the required resistors.

SELF-EXAMINATION

Answer T or F or, in the case of multiple choice, a, b, c or d
(Answers at back of book)

7-1. In Fig. 7-1*a*, resistors R_1 and R_4 are in series with each other. ______

7-2. In Fig. 7-1*a*, resistors R_1, R_2, and R_4 are in series with each other. ______

7-3. In any series-parallel circuit, the largest current always flows in the smallest resistance. ______

7-4. The total resistance of a series-parallel circuit is always less than the smallest parallel resistor. ______

7-5. In Fig. 7-2*a*, the following resistors are in series:

a. R_1 and R_3 c. Both of the above.
b. R_2 and R_3 d. None of the above.

7.6. Refer to Fig. 7-2*a*.

a. Resistors R_1, R_3, and R_4 are in series. c. Resistors R_4, R_5, R_6 are in parallel.
b. Resistors R_2, R_3, and R_4 are in series. d. All the above.

7-7. Refer to Fig. 7-2. Assume that R_1 and R_2 are each twice as large as R_3 and assume that R_4, R_5, and R_6 are each three times as large as R_3. The total combined circuit resistance is equal to

a. $6R_3$ c. R_3
b. $3R_3$ d. $R_3/3$

7-8. If, in Fig. 7-5, the thermistors are ideally matched with the lamps, there will be no change in the total current drawn from the source regardless of how many lamps burn out. ______

7-9. All complex circuits can be thought of as some combination of series-parallel circuits. ______

7-10. If a load requires one-half of the available supply voltage, a series-dropping resistor must have the same resistance as the load. ______

7-11. A disadvantage of a voltage divider that uses a single-series dropping resistor is the wide change in load voltage with the change in load current. ______

7-12. In a single-series dropping resistor voltage divider, an increase in load causes an increase in load voltage. ______

7-13. A bleeder resistor voltage divider improves voltage regulation because it causes some current to always flow through the series-dropping resistor, even when the load is removed. ______

7-14. The larger the bleeder current, the better is the voltage regulation. ______

7-15. To obtain positive and negative potentials from a single supply in a voltage-divider network requires the selection of 0 V or ground at a point different from the negative or positive side of the single supply. ______

7-16. It is not possible to obtain two different positive potentials and two different negative potentials from a single-supply voltage-divider network. ______

REVIEW QUESTIONS

1. Explain why the circuit in Fig. 7-3 is really an example of a series-parallel circuit.

2. Can you devise any rule concerning the total resistance of a series-parallel circuit?

3. In Fig. 7-2 show all the points at which Kirchhoff's current law can be applied.

4. Apply Kirchhoff's voltage law to the circuit in Fig. 7-2.

5. Draw a circuit that contains two sources of emf and three resistors in which the resistors do not form a series, a parallel, or a series-parallel combination.

6. When is the size of feeder line determined by other factors than the necessary current-carrying capacity?

7. a. Identify which components are parallel-connected and series-connected in Fig. 7-8.

b. Write the equation for the total resistance between W and Y in Fig. 7-8, in terms of the voltage-divider resistors and load resistances.

8. a. What is a disadvantage of the single-series dropping resistor voltage divider?

b. How is this disadvantage overcome?

9. What can be done to improve the voltage regulation of a voltage divider?

10. What is the power rating of a single-series dropping resistor compared with the load if the load operates at half the supply voltage?

11. Explain why a larger bleeder current improves the voltage regulation of a voltage divider when the load varies.

12. Under what conditions might you use a power rating for the voltage-divider resistors double the nearest standard sizes to the calculated values?

13. Would it be possible to obtain +25 V and −15 V supplies from a voltage-divider circuit with a single emf of 45 V? Explain.

PROBLEMS

7-1. Refer to Fig. 7-1. Given $R_1 = 1\ \text{k}\Omega$, $R_2 = 2.2\ \text{k}\Omega$, $R_3 = 3.3\ \text{k}\Omega$, $R_4 = 4.7\ \text{k}\Omega$, and $V = 40$ V, calculate:

a. The total circuit resistance.

b. The total source current.

c. The current through R_3.

d. The voltage drop across R_4.

7-2. For the values given in Problem 7-1 refer to Fig. 7-1 and calculate:

a. The voltage drop across R_2.

b. The total power dissipated in all the resistors.

c. The voltage at C with respect to A.

7-3. Refer to Fig. 7-2. Let $R_1 = 5.6\ \text{k}\Omega$, $R_2 = 3.3\ \text{k}\Omega$, $R_3 = 1\ \text{k}\Omega$, $R_4 = 10\ \text{k}\Omega$, $R_5 = 6.8\ \text{k}\Omega$, $R_6 = 12\ \text{k}\Omega$, and $V_2 = 9\text{V}$, $V_1 = 22.5\ \text{V}$.
Calculate:

a. The total supply current.
b. The current through R_5.
c. The power dissipated in R_3.

7-4. For the values given in Problem 7-3 refer to Fig. 7-2 and calculate:

a. The current through R_1.
b. The power dissipated in R_6.
c. The voltage at A with respect to B.

7-5. An electric motor (effective resistance 12 Ω) and a combined lighting load of resistance 18Ω are to be operated in parallel with each other at a distance of 300 m from a 230-V source of emf. If the total voltage drop in the line must not exceed 8% of the source voltage, calculate:

a. The minimum wire gauge to be used for the copper feeder line.
b. The power dissipated in the feeder line.
c. The power supplied by the source.

Assume that the source has negligible internal resistance.

7-6. Calculate the actual voltage at which the load in Problem 7-5 operates if the selected wire gauge is used.

7-7. Refer to Fig. 7-9. What is the resistance between terminals A and B under the following conditions?

a. The output terminals have no load (open-circuit).
b. The output terminals are short-circuited.
c. A load of 600 Ω is connected at the output.

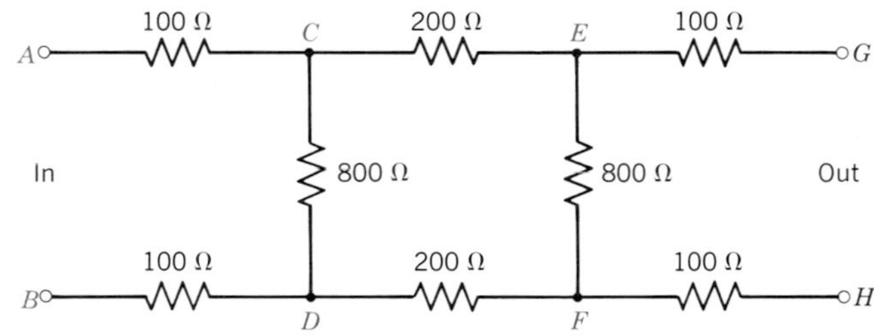

FIGURE 7-9
Circuit for Problems 7-7, 7-8, and 7-9.

7-8. Refer to Fig. 7-9. Assume that a 600-Ω load is connected to the output and a 6-V source at the input. Calculate:

a. The voltage between C and D.
b. The voltage between E and F.
c. The current in the 600-Ω load.

7-9. Refer to Fig. 7-9. Assume that a 600-Ω load is connected to the output and a 10-V source at the input. Calculate:

a. The current and power in the load.
b. Which resistor has the largest power dissipated in it.

7-10. Refer to Fig. 7-10.

a. What voltage exists across the open switch?
b. Find the current through the switch when it is closed.

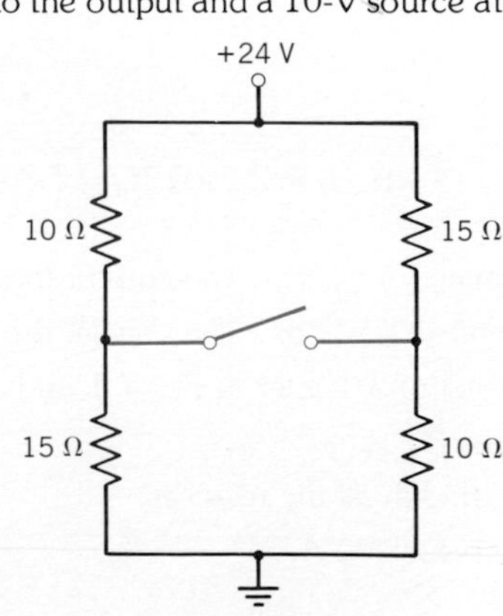

FIGURE 7-10
Circuit for Problem 7-10.

7-11. It is required to operate a 6-V transistor radio from a 12-V automobile battery. If the average load current of the transistor radio is 200 mA calculate:

a. The series dropping resistor and its power rating. (Use commercial values.)

b. The voltage at the radio when the automobile battery is being recharged with a terminal voltage of 14.1 V. (Assume that the load current increases to 250 mA.)

c. The voltage at the terminals of the radio when it is turned off and the battery voltage is 13.2 V.

7-12. A series-dropping resistor had been inserted in the line from a 12-V battery to supply a 9-V tape deck operating at 200 mA. It is required now to replace the tape deck with a 6-V radio operating at 100 mA.

a. What resistance must be connected in parallel with the operating radio to provide 6 V? (Use the nearest commercial value.)

b. What is the power rating of this resistance?

c. How much power is the series-dropping resistor now dissipating? How does this compare with its initial dissipation?

d. What voltage appears at the terminals of the radio when the radio is turned off?

7-13. Design a simple voltage divider with a series-dropping resistor and a bleeder resistor to provide 1.5 kV at a load current of 5 mA from a 2-kV supply. Use a bleeder current of 1 mA and determine the no-load voltage. (Use the nearest commercial value.)

7-14. Repeat Problem 7-13 using a bleeder current of 2 mA.

7-15. a. What nearest commercial bleeder resistor is required to complete a voltage divider to provide 250 V at 40 mA from a 400-V source of emf if the series dropping resistor is 2.2 kΩ?

b. What are the power ratings necessary for these two resistors?

c. What is the no-load voltage?

7-16. Repeat Problem 7-15 using a 3.3-kΩ series-dropping resistor.

7-17. A 12-V power supply is fused at 750 mA.

a. What should be the minimum (commercial value) bleeder resistor to provide a 500 mA load with 9 V?

b. What must be the value of the series-dropping resistor?

c. How much power is dissipated in the series and bleeder resistors?

d. If 5-W power ratings are chosen for the resistors to what maximum voltage can the source increase before either of the resistors overheats? (Assume that the fuse is replaced by a higher value.)

e. What will be the load voltage for the maximum input voltage? (Assume that the load resistance is constant.)

7-18. Repeat Problem 7-17 using an initial power supply of 13 V fused at 1 A.

7-19. Design a voltage divider to provide 90 V at 15 mA and 250 V at 20 mA from a 350-V dc power supply with a bleeder current of 5 mA in parallel with the 90 V load. (Use the nearest commercial values and power ratings.)

7-20. Repeat Problem 7-19 but use a total current from the 350 V supply of 50 mA instead of the 5 mA bleeder current.

7-21. Design a voltage divider to provide ±20 V at 20 mA each and ±12 V at 50 mA each from a 40-V supply with a total current of 100 mA. (Use the nearest commercial values and power ratings.)

7-22. Repeat Problem 7-21 using a 45-V supply.

CHAPTER EIGHT

VOLTAGE SOURCES

Because of the large amount of portable (solid-state) electronic equipment, batteries provide a very important function in everyday applications. Typical applications range from electronic watches to pacemakers; and include toys, calculators, cordless appliances, cameras, cassette recorders, radios, hearing aids, and so on. All rely on a power source using the conversion of chemical energy into electrical energy. This process involves the separation of charges between two terminals. Depending upon the type of chemical activity, two different types of cells (or batteries) are possible: the *primary* cell and *secondary* cell.

The *primary* cell is one in which the chemical materials are used up as electric energy is produced. It must be discarded after its active material is depleted. A *secondary* cell involves a *reversible* chemical reaction to separate charges so that the cell can be repetitively *recharged* to its original state.

The most common primary cell is the *carbon–zinc* cell. It is low in cost and is suited primarily for intermittent use. Where heavier and more continuous loads are experienced, the *alkaline* cell is used. Small, solid-state electronic equipment, such as watches and hearing aids, require higher-energy-density primary cells such as the "button"-type *mercury-oxide* and *silver-oxide* cells.

Where larger amounts of energy at high current are required, a *secondary cell* is most usually specified. The most common example is the (automotive) *lead–acid* cell, which can supply several hundred amperes at 2 V per cell. Many cordless appliances, such as portable garden equipment, use either a gel type of lead–acid cell or a *nickel–cadmium* (nicad) cell, or battery. Strictly speaking, a *battery* consists of a number of interconnected *cells,* each of which has an emf from 1.2 to 2.2 V depending upon the type. However, common usage allows the term of *battery* to be applied to individual AA, C, and D-size cells.

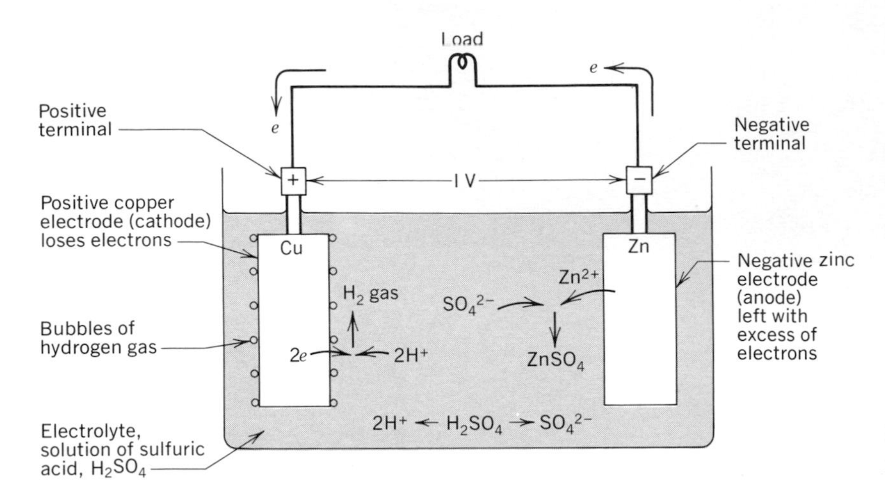

FIGURE 8-1
Simple copper–zinc cell.

8-1 SIMPLE COPPER–ZINC CELL

Although not available in commercial form, the copper–zinc cell illustrates the principles of the chemical reactions in the various types of practical cells. It demonstrates how charges are separated by the motion of ions in a *voltaic cell.*

Consider two *electrodes,* a strip of copper (Cu) and a strip of zinc (Zn), immersed in an *electrolyte* of dilute sulfuric acid (H_2SO_4). See Fig. 8-1. It will be observed that there is a great deal of activity around the metallic zinc electrode as it is "eaten away" by the acid. The metallic copper, however, does *not* dissolve in the acid. But it will be observed that numerous bubbles (hydrogen) cling to the surface of the copper and also rise to the surface. A voltmeter connected between the two metallic terminals indicates approximately 1 V with the copper strip positive and the zinc strip negative. An electrical load (of low resistance) connected between the terminals of the cell allows a current to flow until the zinc is completely used up. The chemical process is described below.

Each molecule of sulfuric acid (H_2SO_4) dissociates (separates), into two positive (hydrogen) ions $2H^+$ and one negative (sulfate) ion SO_4^{2-}.

$$H_2SO_4 \rightarrow 2H^+ + SO_4^{2-} \tag{8-1}$$

Equation 8-1 simply means that a molecule of sulfuric acid is initially neutral, with its molecular components in an ionic bond. When the two hydrogen ions break away, their valence electrons remain with the sulfate molecule giving the sulfate ion a negative charge of two electrons.

Since atomic zinc (Zn^0) is more chemically active than atomic copper (Cu^0), it dissolves more readily in the acid. This means that some zinc atoms (Zn^0) form Zn^{2+} ions and go into solution to replace hydrogen ions and react with the sulfate ions to form zinc sulfate.

$$Zn^{2+} + SO_4^{2-} \rightarrow ZnSO_4 \tag{8-2}$$

The zinc is said to have *replaced* the hydrogen, and so the zinc ion must have the equivalent charge of the two positively charged hydrogen ions. But a doubly charged positive ion of zinc is an atom that is *deficient* in two valence electrons. Thus an ion of zinc departing from the zinc strip leaves behind an *excess* of two electrons on the zinc electrode. The *zinc* electrode, initially neutral, thus becomes *negatively charged.*

The hydrogen (H^+) ions, meanwhile, each lacking one valence electron, pick up an electron from the ready supply of free electrons in the copper strip forming neutral atomic hydrogen (H^0).

$$2H^+ + 2e \rightarrow H_2\uparrow \tag{8-3}$$

This reaction means that each *pair* of hydrogen ions becomes a molecule of hydrogen gas. This gas in the form of visible bubbles coats the copper electrode and also rises to the surface. The copper strip, also initially neutral, having given up two electrons to the hydrogen, becomes *positively charged.*

The overall reaction can be represented by the following chemical equation, although it does not show the manner in which the charge is separated.

$$Zn + H_2SO_4 \rightarrow ZnSO_4 + H_2\uparrow \qquad (8\text{-}4)$$

A potential difference of approximately 1 V is established by the chemical action of *raising* the potential of the copper electrode by 0.5 V and *lowering* the potential of the zinc electrode by 0.5 V.

If a load is connected between the two terminals, electrons flow from the zinc electrode, through the external circuit (Fig. 8-1) to the positive copper electrode. Here electrons are available to combine with hydrogen ions to form more hydrogen gas. Now more positive zinc ions enter solution to replace the lost positive hydrogen ions, until the zinc is completely eaten away, or the acid solution becomes zinc sulfate. Thus we have the motion, in solution, of positive ions toward the positive copper electrode and negative ions toward the zinc electrode.

Although the chemical action tries to maintain the potential difference near 1 V, in practice, the cell is *not* a very successful 1-V source. This is due to *polarization,* the accumulation of (hydrogen) gas around the (positive copper) electrode. Polarization insulates the electrode because hydrogen is a nonconductor. It also reduces the *active surface area* for chemical action and the cell's terminal voltage is soon decreased. Many practical primary cells use a *depolarizer* to prevent this gaseous *insulating* effect which increases the *internal resistance* of the cell and lowers its *terminal voltage.*

The terms *anode* and *cathode* are often used when referring to the electrodes of a battery. It should be noted that the *anode* (Fig. 8-1) is that terminal which *loses* electrons (in the external circuit) during normal operation (discharge). The cathode is the terminal that *gains* electrons. Thus, in the above wet cell, the *zinc* electrode (negative) is the *anode* and the *copper* electrode (positive) is the *cathode* (Fig. 8-1). The above definitions also apply to other electrical and electronic devices that have terminals referred to as anode and cathode. While electron current is from anode to cathode note that conventional current is from cathode to anode, in the external circuit.

8-2 THE CARBON–ZINC CELL

The most common type of primary cell is the carbon–zinc or *Leclanché cell.* This *dry cell* uses a paste instead of a liquid electrolyte so that it can be placed in any position. Figure 8-2 shows the main construction features.

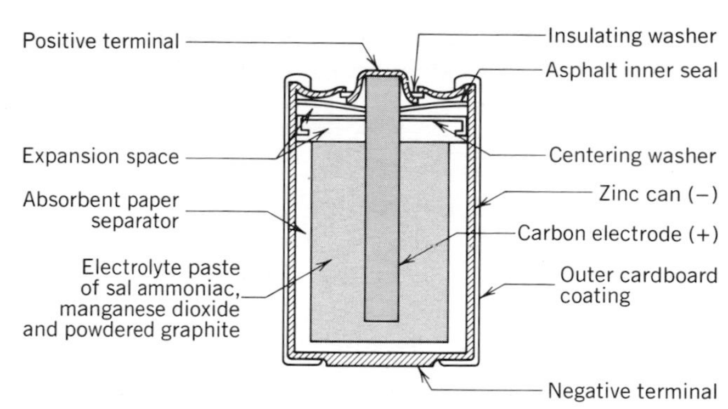

FIGURE 8-2
Construction of a typical carbon–zinc dry cell.

The zinc can is the negative electrode; a carbon rod in the center is the positive electrode. Ammonium chloride (NH_4Cl, also known as "sal ammoniac") and zinc chloride make up the electrolyte. Mixed in with the moist electrolyte is another chemical, manganese dioxide (MnO_2). This is a chemical rich in oxygen that acts as a *depolarizer.* Whenever hydrogen tries to form around the carbon rod, the oxygen reacts with the hydrogen to form water.

$$MnO_2 + 2H_2 \rightarrow Mn + 2H_2O \qquad (8\text{-}5)$$

An absorbent moist paper separator is used to prevent direct contact of the zinc can with the electrolyte. Although the passage of ions is not interrupted, this layer reduces chemical action between the zinc and impurities in the electrolyte. These reactions set up small voltaic cells that would use up the zinc and shorten the life of the cell. This is called *local action.* To reduce local action due to impurities within the zinc itself, mercury is coated on the interior zinc surface during manufacture. This process is called *amalgamation.* (Effectively, it places zinc and its other metallic impurities at the same electrical potential.)

As a result of a motion of ions, similar to the copper-zinc cell, an emf of approximately 1.5 V is produced. This is true no matter what the physical size of the cell. Sizes vary from the small penlight and up through AA, C, D, to No. 6. The physical size determines the amount of current each cell can deliver and the amount of energy stored. The small cells can deliver continuously a maximum current of only a few milliamperes, whereas a 6-inch high (No. 6 cell) can deliver $\frac{1}{4}$ ampere (250 mA). A D-size carbon–zinc flashlight cell can provide a load current of 50 mA for approximately 15 hours. This causes the terminal voltage to drop to approximately 1.3 V. By suitably interconnecting these *cells, batteries* of convenient multiples are obtained, such as 6 V, 9 V, 22.5 V, 45 V, 90 V, and so on.

8-3 THE ALKALINE-MANGANESE CELL

Alkaline–manganese cells are capable of providing heavy currents for long periods. They require no rest periods to recover, as do carbon–zinc cells.

Figure 8-3 shows how the cell consists of two electroplated steel cases. These account for the much lower leakage possibility compared with the zinc–carbon cell, although they do not take part in the chemical action of the cell. The cell incorporates a self-venting system should a sustained short circuit occur and require a release of excessive gas build up.

The positive terminal at the top is in contact with the manganese dioxide through the steel cases. The manganese dioxide plays the part of both cathode and depolarizer. The anode consists of zinc pellets around a central collector that makes contact with the negative terminal at the bottom of the cell. The electrolyte is a paste of potassium hydroxide (KOH), which is in contact with both anode and cathode. A nominal 1.5 V is produced by the cell.

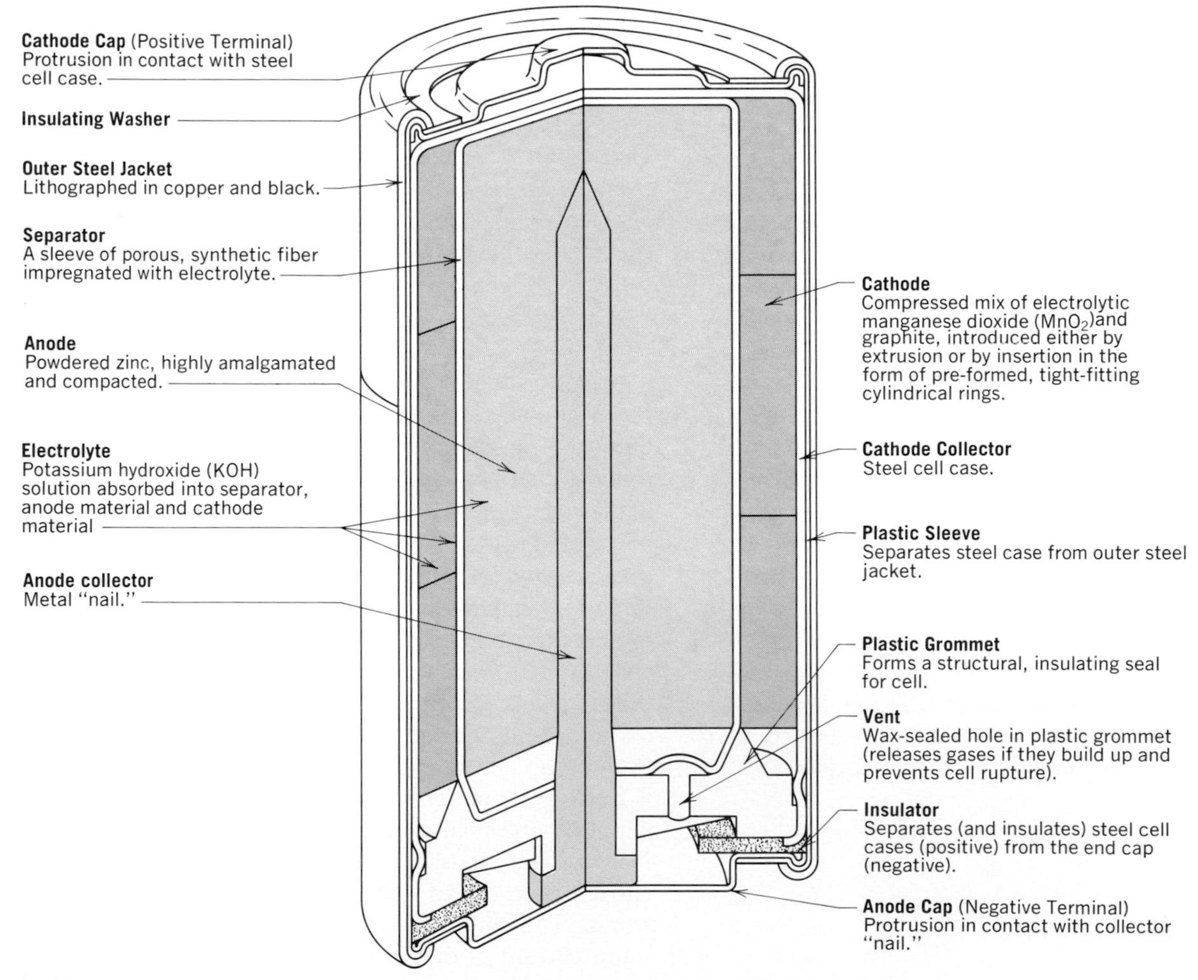

FIGURE 8-3

Cross section of an alkaline manganese cell. (Courtesy of Duracell International Inc.)

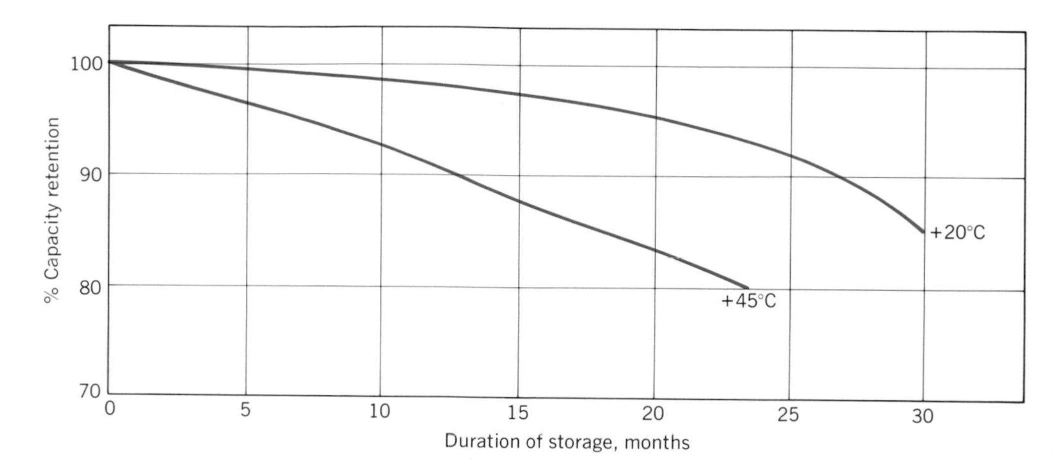

FIGURE 8-4
Typical capacity retention curves for an alkaline cell.

In typical applications, an alkaline cell may last four times as long as a carbon–zinc cell. A D-size alkaline cell can provide a load current of 50 mA for approximately 56 h before the terminal voltage drops to 1.3 V. This corresponds to a *capacity* of:

$$50 \text{ mA} \times 56 \text{ h} = 2800 \text{ mAh (milliampere hours)}$$
$$= 2.8 \text{ Ah (ampere hours)}$$

The capacity of a cell (expressed in ampere hours) is a measure of the total charge (and energy) that a cell can deliver under normal operating conditions. In the example above, if the terminal voltage of 1.3 V is the lowest acceptable value in a given application, the cell would be said to have a capacity of 2.8 Ah.

Another important feature in which the alkaline battery excels is its *shelf life.* This is the length of time the cell retains its energy during storage. Whereas a carbon–zinc cell may retain 80% of its capacity for only 6 to 12 months, an alkaline–manganese cell has a shelf life of 30 to 36 months. Since the loss of energy is due partially to a drying out of the electrolyte, the shelf life can be extended by storing cells at *cooler* temperatures, as shown by curves in Fig. 8-4.

EXAMPLE 8-1

A fresh alkaline D-cell has a capacity of 3 Ah in a given application. Determine the available capacity of the cell after being stored for 20 months at a temperature of

a. 45°C
b. 20°C

SOLUTION

a. From Fig. 8-4, capacity retention after 20 months at 45°C is 84%.

$$\text{capacity} = 3 \text{ Ah} \times 0.84 = \mathbf{2.52 \text{ Ah}}$$

b. After 20 months at 20°C, capacity retention is 96%.

$$\text{capacity} = 3 \text{ Ah} \times 0.96 = \mathbf{2.88 \text{ Ah}}$$

Example 8-1 shows the benefit of storing batteries at reduced temperatures; Example 8-2 shows how this prolongs the active life of the cell.

EXAMPLE 8-2

Determine the length of time the cell in Example 8-1 could deliver a current of 50 mA after being stored for 20 months at a temperature of

a. 45°C
b. 20°C

SOLUTION

a. At 45°C, capacity = 2.52 Ah. Since

$$\text{time} = \frac{\text{capacity}}{\text{current}} \quad \text{or} \quad t = \frac{Q}{I}$$

then $$t = \frac{Q}{I} = \frac{2.52 \text{ Ah}}{50 \times 10^{-3} \text{ A}} = \mathbf{50.4 \text{ h}}$$

b. At 20°C, capacity = 2.88 Ah.

$$t = \frac{Q}{I} = \frac{2.88 \text{ Ah}}{50 \times 10^{-3} \text{ A}} = \mathbf{57.6 \text{ h}}$$

8-4 OTHER PRIMARY CELLS

Many electronic applications would not be possible if it were not for *mercury* cells. These are essentially modifications of the alkaline cell with mercuric oxide (HgO) used as the positive electrode (instead of manganese dioxide). They have 50% more energy for a given volume than the alkaline–manganese cell and three times as much energy as the carbon–zinc.

They are available in 1.35-V cells used mainly for instrumentation applications, and as 1.4-V cells for general use. There are two structures of the 1.4-V cell available. One is suitable for the high surge current, short-duration application in photographic flashes; the other to provide a low continuous current for a year or more in electronic watches. In this form they are usually of the button or flat-pellet type. However, they can be series-connected to produce batteries of 12.6 V, which are popular in smoke alarm circuits.

Another type of primary cell is the *silver oxide* cell. This is similar to the mercury cell but uses silver oxide (Ag_2O) instead of mercuric oxide as the positive electrode. Silver oxide cells develop an emf of 1.5 V and four times as much energy as the carbon–zinc. They are widely used now as button types in electronic watches.

The *heavy-duty zinc–chloride* cell is similar in construction to the carbon–zinc cell, but the electrolyte system is different. In a zinc–chloride cell the ammonium chloride is omitted so that the electrolyte consists only of zinc chloride. This provides vigorous chemical action and minimizes electrode polarization at high currents. Available in all standard sizes with a nominal emf of 1.5 V, the heavy-duty zinc–chloride cell usually lasts twice as long as corresponding standard carbon–zinc cells.

The highest energy density of all primary cells is found in the *lithium* cell. With a nominal emf of 3.0 V, a lithium D-cell has the energy of 30 carbon–zinc D cells. Excellent at low temperatures, they retain 95% of their original capacity after a 5-year period. They are used for continuous or standby power in low-drain integrated circuit memory applications and microprocessor units.

The most recent development is an ultrathin "paper" battery. It operates essentially the same as an ordinary dry cell but uses a stainless steel plate in place of the carbon rod. The electrolyte consists mainly of zinc perchlorate that does not attack the steel plate as would ammonium or zinc chloride. A typical cell 2.75 × 0.78 × 0.03 in. develops 1.5 V, has a capacity of 27 mAh, and weighs only 0.06 ounce. Similar batteries developing 6 V are used in film magazines in cameras to operate the motor mechanism.

8-5 COMPARISON OF PRIMARY CELLS

The basic operating differences between the common primary cells can be seen in the discharge characteristics of Fig. 8-5. The carbon–zinc and alkaline–manganese cells are said to have a *sloping* discharge curve; the mercury, silver oxide, and lithium cells have a *flat* discharge curve. Also, because of the different energy densities, an alkaline battery takes three times as long to drop to 0.8 V as the carbon–zinc, and a mercury cell over four times as long.

EXAMPLE 8-3

Determine the hours of service for a continuous load to reduce the terminal voltage of AA size cells by 20% for the following types:

a. zinc–carbon
b. alkaline
c. mercury

Assume that the discharge curves conform to Fig. 8-5.

SOLUTION

a. Nominal voltage of zinc–carbon cell = 1.5 V.

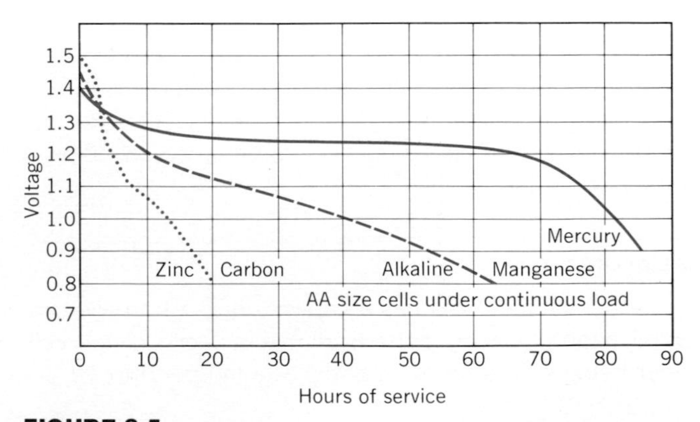

FIGURE 8-5
Comparison of discharge characteristics for primary cells.

$$20\% \text{ reduction} = 1.5 \text{ V} \times \frac{20}{100} = 0.3 \text{ V}$$

final terminal voltage = 1.2 V

From Fig. 8-5, the time required to reach 1.2 V = **5 h.**

b. Final terminal voltage of alkaline = 1.2 V.

From Fig. 8-5, the time required to reach 1.2 V = **10 h.**

c. Nominal voltage of mercury cell = 1.4 V.

$$20\% \text{ reduction} = 1.4 \text{ V} \times \frac{20}{100} = 0.28 \text{ V}$$

final terminal voltage = 1.12 V.

From Fig. 8-5, the time required to reach 1.12 V = **75 h.**

Other electrical characteristics and a summary of the construction materials are contained in Table 8-1. Note how the alkaline cells may be operated at much lower temperatures than the carbon–zinc. It should be mentioned that although shelf life (and capacity) is improved by storing cells at low temperatures, the output of a cell during *operation* is much improved at *higher* temperatures, due to *increased* chemical activity. See Fig. 8-6.

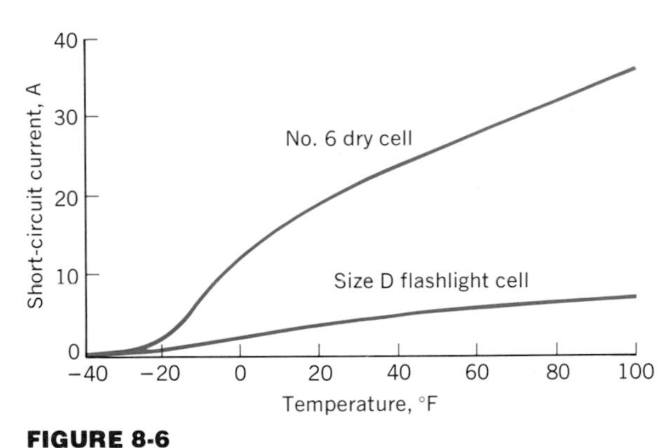

FIGURE 8-6
Typical effects of temperature on cell activity.

8-6 LEAD–ACID SECONDARY STORAGE CELLS

The lead–acid wet cell is the most common *secondary* storage cell. Now found in almost every automobile for

TABLE 8-1
Comparison of Common Primary Cells

	Carbon–Zinc	Alkaline–Manganese	Mercury	Silver Oxide
Negative, anode	Zinc	Zinc	Zinc	Zinc
Positive, cathode	Carbon	Manganese dioxide	Mercuric oxide	Silver oxide
Electrolyte	Ammonium chloride	Potassium hydroxide	Potassium hydroxide	Potassium hydroxide
Nominal voltage, volts	1.5	1.5	1.35 or 1.4	1.5
Maximum rated current amperes	2–30	0.05–20	0.003–3	0.1
Energy output				
Watt-hrs/lb	22	35	46	50
Watt-hrs/in.^{3}	2.0	3.5	6.0	8.0
Temperature range				
Storage, °F	−40 to 120	−40 to 120	−40 to 140	−40 to 140
Operating, °F	20 to 130	−5 to 160	−5 to 160	−5 to 160
Shelf life in months at 68°F to 80% initial capacity	6–12	30–36	30–36	30–36
Shape of discharge curve	Sloping	Sloping	Flat	Flat

starting purposes, the lead–acid cell was first introduced by Gaston Planté in 1860. It is widely used in the telephone industry for standby power and also finds applications in electric vehicles. (The first automobiles were electric and powered by Planté cells). They are capable of being *recharged* several hundred times and last many years with proper use and maintenance.

A fully charged cell has a lead peroxide (PbO_2) *positive* electrode, reddish-brown in color, and a gray spongy lead (Pb) *negative* electrode. With an electrolyte of approximately 27% sulfuric acid (specific gravity of 1.3), the cell produces about 2.2 V. By connecting three or six cells in series, a nominal 6- or 12-V battery is obtained.

8-6.1 Chemical Action

The processes taking place during discharging and charging of the cell are shown in the diagrams of Fig. 8-7. During discharge, the lead in both of the electrodes reacts with the sulfuric acid to displace the hydrogen and form lead sulfate. Since lead sulfate, a whitish material, is somewhat insoluble, both the positive and negative plates become partially coated with this substance (Eq. 8-6). Since both plates approach the same material ($PbSO_4$) chemically, the potential difference begins to decrease. Also, the combining of the oxygen in the lead peroxide with the hydrogen ions of the electrolyte forms water (see Eq. 8-6). Thus the sulfuric acid solution becomes weaker (its specific gravity approaching 1.0) as the cell delivers energy to a load. Both these effects cause the voltage developed by the cell to drop off as the cell loses its charge. Also, the internal resistance of the cell rises due to the sulfate coating on the plates.

Fortunately, the chemical action is reversible. If a battery charger is connected as in Fig. 8-7b (positive-to-positive, negative-to-negative) the direction of current and ionic flow is reversed. Electrical energy causes the recombination of lead sulfate with the hydrogen ions in the electrolyte. Thus, as excess water is removed from the solution, the electrolyte returns to its normal strength of sulfuric acid, and the plates return to their original form of sponge lead and lead peroxide. Since lead sulfate tends to harden into an insoluble salt over a period of time, it is wise to fully recharge a battery if it is not to be used for sometime.

The chemical action involved in the above description can be represented in the following reversible equation:

$$Pb + PbO_2 + 2H_2SO_4 \underset{\text{charge}}{\overset{\text{discharge}}{\rightleftharpoons}} 2PbSO_4 + 2H_2O + \text{electrical energy} \quad (8\text{-}6)$$

$$\text{lead} + \text{lead peroxide} + \text{sulfuric acid} \underset{\text{charge}}{\overset{\text{discharge}}{\rightleftharpoons}} \text{lead sulfate} + \text{water} + \text{electrical energy}$$

It should be noted that an automotive battery charging system requires approximately 14.1 V to recharge a 12-V battery at currents up to 30 A. Recharging at excessively

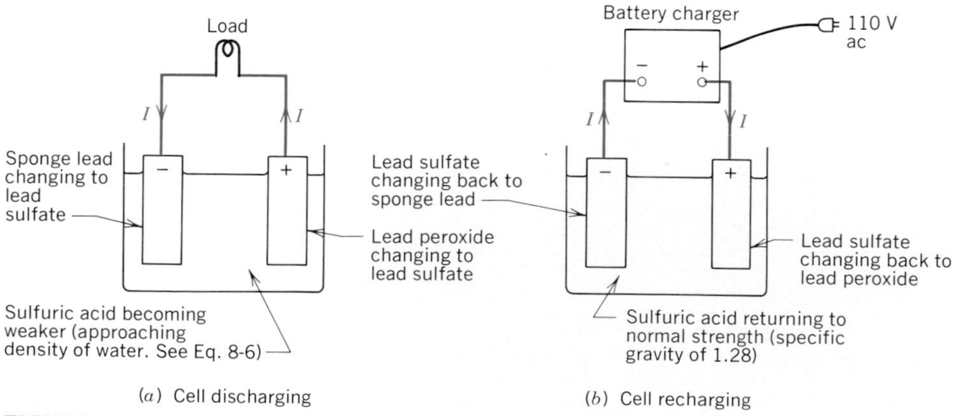

FIGURE 8-7
Chemical actions in a lead–acid cell under conditions of discharge and recharge.

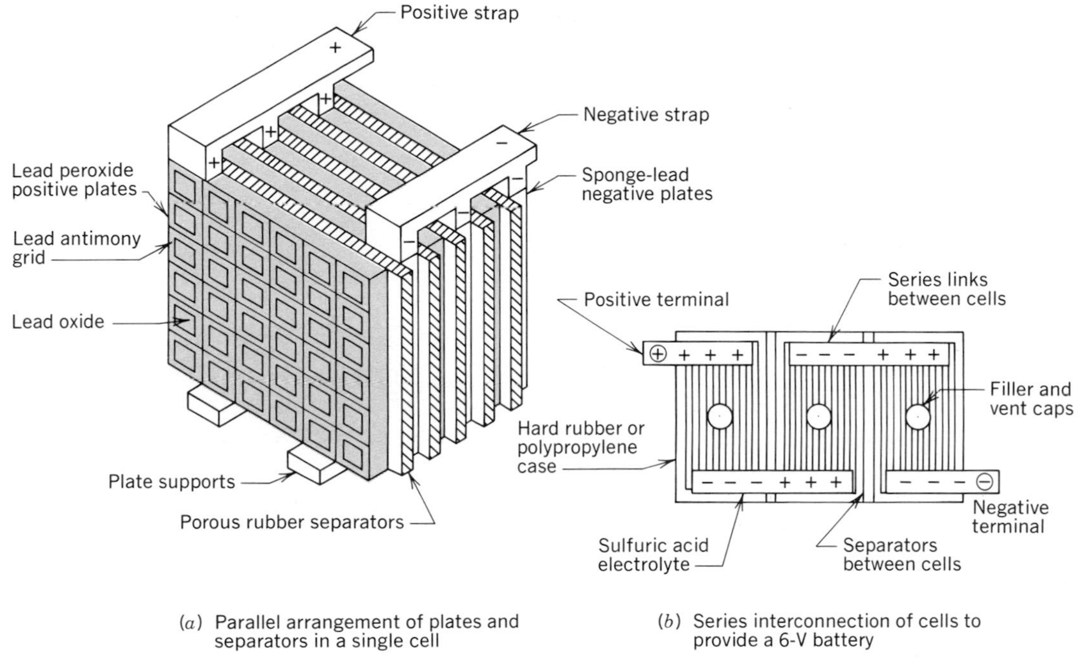

(*a*) Parallel arrangement of plates and separators in a single cell

(*b*) Series interconnection of cells to provide a 6-V battery

FIGURE 8-8
Construction of a lead–acid battery.

higher currents can cause "boiling" of the electrolyte, lowering of the liquid level, and buckling and crumbling of the electrodes, which results in reduced battery life.

8-6.2 Construction

Although the diagram of Fig. 8-7 suggests that each electrode consists of a single plate, this is not the case in a practical cell. To increase the surface area and current availability, a number of positive and negative plates are interleaved and separated by porous rubber sheets, as shown in Fig. 8-8*a*. All the positive plates are electrically connected, and all the negative plates are connected. The connections yield a parallel (higher-current) arrangement for a single cell developing approximately 2 V. Three such cells are series-connected, positive to negative, and positive to negative, to produce a 6-V battery, shown in Fig. 8-8*b*. Since hydrogen gas is produced during recharging, vents are provided to let the hydrogen and water vapor escape. The vents also allow the addition of distilled water to make up for any evaporation of water vapor from the electrolyte.

There are two ways of manufacturing the battery. The plates are made out of lead antimony and cast in the form of grids with holes or spaces in the surface. In the *dry-charge* method, lead peroxide is pressed into the positive plates and sponge lead into the negative plates. A few moments after adding the sulfuric acid, the cell is immediately ready for service without any need for a charge.

In the *wet-charge* process, lead oxide is pressed into both plates. The electrolyte is added and a charging current *forms* the electrodes to lead peroxide and sponge lead.

Recent advances in lead–acid cells have resulted in low-maintenance, and maintenance-free batteries, in which no provision is made, in the latter, for the addition of water. How is this possible?

It has been found that the amount of "gassing" (production of hydrogen) that takes place upon charging a cell can be reduced by lowering the amount of antimony in the lead plates. Antimony is alloyed with the lead to improve the castability of it into grids. Ordinary lead–acid cells have as much as 4% antimony in the lead plates. By reducing this to 2%, *low-maintenance* cells require very little addition of water because very little is "boiled-off" during charging. *Maintenance-free* cells use antimony-free plates (lead–calcium or lead–calcium–tin), allowing complete sealing of the battery, since no vents are necessary to remove gas buildup. Once sealed, no electrolyte can evaporate from the cell. A small vent is provided to relieve the pressure arising from altitude changes.

8-6.3 Specific Gravity

One method of checking a lead–acid cell is to determine the *specific gravity* of the electrolyte. This is a ratio of the weight of a given volume of the acid to the same volume of water. At 68°F a fully charged cell should have a specific gravity (sp. gr.) of exactly 1.28; a fully discharged cell has a specific gravity of 1.13 in comparison to pure water, which has a specific-gravity of unity.

The specific gravity may be measured by means of a *hydrometer.* This consists of a glass tube into which some of the electrolyte may be admitted by squeezing a rubber bulb at the end of the tube. A weighted float inside the tube has a calibrated scale on its side, with numbers reading from 1280 near the bottom to 1120 at the top. Note that the *scale* of the hydrometer multiplies the specific gravity reading by 1000, that is, the specific gravity of H_2O is assumed to be 1000. The lower the density of the electrolyte (closer to water), the lower the float will sink and indicate a smaller number.

A completely discharged cell freezes at 20°F or lower. But if maintained at a fully charged condition, a lead–acid cell will operate between −76 and +140°F. This is because a higher concentration of acid depresses the freezing point. (You probably recall that pure water freezes at 32°F and/or 0°C.)

8-6.4 Ampere-Hour Capacity

The current rating of a battery is usually given in units of the ampere-hour (Ah) capacity based on an 8-h discharge period. During this time, the cell's output voltage must not drop below 1.7 V. For example, a 60-Ah battery (typical for many smaller vehicles), would be able to deliver 7.5 A for 8 h without any cell dropping below 1.7 V. Similarly, it should be able to deliver 5 A for 12 h, but it is unlikely that it could deliver 60 A for 1 h. This is because the cell is less efficient at a higher discharge current. Cell capacities of up to several hundred Ah are available.

Similar to primary cells, the capacity of a lead–acid cell decreases significantly with temperature, losing approximately 0.7% for each decrease of 1°F. At 0°F (−18°C), its capacity is only 60% of the value at 60°F (15.6°C). In automotive applications, a cold cranking "power" is usually specified as the number of amperes available for 30 seconds at 0°F with figures ranging from 300 to 435 A.

8-7 THE NICKEL–CADMIUM CELL

The most widely used rechargeable "dry" cell in electronic calculators, walkie-talkies, portable electric tools, and so on, is the *nickel–cadmium* or nicad battery. A single nickel-cadmium cell develops an open-circuit voltage of approximately 1.2 V and is available in AA, C, and D sizes. They are also available as 6.1, 9.7, and 12.2-V batteries. A nickel–cadmium cell has a lower energy content than a corresponding carbon–zinc cell but greater than a lead–acid cell. Although the initial cost of a nickel–cadmium cell is high, its overall cost is lower than any primary cell because it can be recharged up to 2000 times.

8-7.1 Construction

The base material for the two electrodes is flexible nickel-plated sheets of steel. A rugged porous surface is provided by sintering nickel powder at high temperatures on the sheets. The negative electrode contains *metallic cadmium* and the positive electrode *nickel hydroxide*. The electrolyte is (basic) potassium hydroxide contained in an absorbent material separating the two plates. The two electrodes and the electrolyte separator are spirally rolled up together in a cylindrical form and sealed in a nickel-plated steel container. A resealing safety vent relieves any excess internal pressure resulting from abuse such as overcharging or reverse charging.

8-7.2 Charging

Table 8-2 shows the nominal capacity for some standard size cells. Others are available up to 10 Ah.

The rate at which sealed nickel cadmium cells can be safely charged is the 8-hour or $C/8$ rate, where C is defined as the capacity in Ah. For example, if C is 4 Ah, a maximum charging current of 0.5 A should be used. At this current, it will take 12 h to fully charge a cell that had been fully discharged. Moreover, at this rate, it is permissible to continue charging for an *indefinite period.* Each

TABLE 8-2
Capacities and Maximum
Charge Rates for Nickel–Cadmium Cells

Size	Nominal Capacity, C Ah	Maximum Continuous Charge Rate, mA
AA	0.5	65
CC	2.0	250
D	4.0	500

TABLE 8-3
Normal and Maximum Charge Times
for Various Charge Rates for Nickel Cadmium Cells

Charge Rate, A	Normal Charge Time	Maximum Charge Time
$C/8$[a]	12 h	Indefinite
$C/4$	5 h	6 h
$C/2$	$2\frac{1}{4}$ h	$2\frac{1}{2}$ h
$C/1$	1 h	$1\frac{1}{4}$ h
$2C$	27 min	30 min
$4C$	12 min	12 min
$8C$	5 min	5 min

[a] C is the capacity in ampere hours (Ah).

cell (depending on size) has its own maximum continuous charge current, some of which are given in Table 8-2.

The oxygen that is produced at the positive electrode on overcharge reacts quickly with the cadmium in the charged negative electrode. The oxygen is reused continuously according to the following chemical reaction:

$$O_2 + 2H_2O + 2Cd \rightarrow Cd(OH)_2 + \text{electrical energy} \quad (8\text{-}7)$$

oxygen + water + cadmium →
cadmium hydroxide + electrical energy

The overall chemical reaction in the nickel– cadmium cell is

$$Cd + 2NiOOH + 2H_2O \underset{\text{charge}}{\overset{\text{discharge}}{\rightleftharpoons}} 2Ni(OH)_2 + Cd(OH)_2 + \text{electrical energy} \quad (8\text{-}8)$$

cadmium + nickel hydroxide
+ water $\underset{\text{charge}}{\overset{\text{discharge}}{\rightleftharpoons}}$ nickel hydroxide
+ cadmium hydroxide + electrical energy

Thus the cadmium hydroxide produced by the overcharging is available to be reconverted to cadmium and water to repeat the cycle.

8-7.3 Quick-Charging

For some applications, it is necessary to charge batteries more rapidly. In this case it is important not to exceed the recommended times given in Table 8-3. These times assume that the cell was initially discharged completely and the cell is at a reasonable temperature, 20 to 45°C.

Some automatic chargers sense both the temperature and voltage of the cell; when either reaches its limit, an electronic switch reduces the rate to a trickle charge of $C/8$.

EXAMPLE 8-4

It is required to charge a nickel cadmium C-cell (that has been fully discharged) in approximately 30 min.

a. What is the maximum current that can be used and the maximum time the cell can be left on charge?
b. If the cell is to be left on trickle charge, to what must the current be reduced?

SOLUTION

a. From Table 8-2, a C-cell has a capacity of 2 Ah.

From Table 8-3, a charge rate (in amperes) of $2C$ is required to fully charge in 27 min.

Maximum charge current $= 2C = 2 \times 2\text{ A} =$ **4A**

The maximum charge time at this rate is **30 min.**

b. Trickle charge rate $= C/8 = \frac{2}{8} =$ **0.25 A**

8-7.4 Disadvantages

Nickel–cadmium cells have three disadvantages compared to other secondary cells.

1. *"Memory effect."* If a nickel cadmium battery is operated at certain *low* discharge levels for short, repetitive periods, the cell becomes conditioned to this level and "forgets" its original design capacity. Thus it delivers only its previous output even when the demand for power is increased.
2. *Shelf life.* If a fully charged cell is allowed to stand unused at 20°C, its available capacity will drop to 80% in only 21 days. At 30°C this occurs in only 10 days.
3. Cost. Nickel–cadmium cells cost about twice as much as a lead–acid battery for the same amount of energy.

8-8 OTHER SECONDARY CELLS

8-8.1 Gelled-Electrolyte Lead-Acid Cells

These cells enjoy all the advantages of wet lead–acid cells but avoid the problem of a liquid electrolyte by using a gelled-electrolyte. They use lead–calcium grids, are completely sealed, and can be mounted in any position. A one-way relief valve releases excess pressure if internal pressures rise too high during charging, and automatically recloses. Unlike nickel–cadmium cells a gelled-electrolyte cell should not be left on continuous charge for maximum cell life.

Gelled-electrolyte, lead-acid batteries are available from 2 to 12 V, with capacities ranging from 0.9 to 20 Ah, based on a 20-h discharge rate. The maximum current for these batteries ranges from 40 to 200 A. They are used in portable tools, portable television receivers, and a variety of industrial applications.

8-8.2 Silver–Zinc and Silver–Cadmium Cells

These cells have two to three times the energy output per pound compared with nickel–cadmium cells. They are used to power portable television cameras, video tape recorders, airborne telemetering equipment, guidance systems in missiles, and so on. Battery packs are available with capacities ranging from 0.1 to 750 Ah.

The silver–cadmium cell produces 1.1 V, the silver-zinc produces 1.5 V. Neither suffers from the "memory effect" of the nickel–cadmium cell, but their recharge life cycle is much shorter than the nicads, and they are more expensive.

8-8.3 Cells of the Future

The interest in electric-powered vehicles has promoted the development of a battery with higher energy density and less weight than present lead–acid batteries. The most recent contribution is a zinc–nickel oxide battery. A suitable number of these batteries can provide a vehicle with a range of 100 miles, and a top speed of 50 mph with a battery life of 30,000 miles. The batteries can be recharged in 10 to 12 h from 120 V ac or in 6 to 7 h from 220 V ac. See Fig. 8-9.

FIGURE 8-9
The General Motors breakthrough on electric vehicle battery technology is illustrated here in a size comparison between the conventional lead acid batteries in the foreground and the zinc–nickel oxide batteries in the rear. The battery packs have equal energy, but the zinc–nickel is only half as large and at 900 lb weighs less than half as much as the 2000-lb lead–acid pack. (Courtesy of General Motors Corporation.)

FIGURE 8-10
Comparison of three battery technologies. The three battery cells shown here have the same energy-storage capacity. At the left is a conventional lead–acid battery. The next two are experimental batteries under development at General Motors Research Laboratories, Warren, Mich. In the center is the zinc–nickel oxide battery, about $\frac{1}{3}$ the size and weight of the lead acid; at the right is the lithium–iron sulfide, about $\frac{1}{6}$ the size and weight. (Courtesy of General Motors Corporation.)

The next generation of batteries to be developed includes a lithium sulfide cell, with even greater energy density. See Fig. 8-10. And a sodium-sulfur cell that has four times the energy density of a lead–acid cell, is undergoing tests in vehicles. However, the sodium and sulfur are very corrosive and must be kept at temperatures of 300 to 400°C to maintain the chemical reaction.

8-9 COMPARISON OF SECONDARY CELLS

The differences in the terminal voltage, discharge characteristic, and relative energy content for the four secondary cells discussed above are shown in Fig. 8-11.

Other electrical characteristics are shown in Table 8-4.

TABLE 8-4
Comparison of Common Secondary Cells

	Lead-Acid, Wet and Dry	Nickel-Cadmium	Silver-Cadmium	Silver-Zinc
Negative, anode	Sponge lead	Nickelic hydroxide		Zinc
Positive, cathode	Lead peroxide	Cadmium		Silver oxide
Electrolyte	Sulfuric acid	Potassium hydroxide	Potassium hydroxide	Potassium hydroxide
Nominal voltage, volts	2.2	1.2	1.1	1.5
Maximum rated current	High (100-40,000 A)	Medium (0.1-100 A)	Medium	Medium
Energy output				
Watt-hours/lb	10 to 14	12 to 16	22 to 34	40 to 50
Watt-hours/in.^{3}	0.9 to 1.1	1.2 to 1.5	1.5 to 2.7	2.5 to 3.2
Cycle life	200 to 500	500 to 2,000	150 to 300	80 to 100
Temperature range				
Storage, °F	−76 to 140	−40 to 140	−85 to 165	−85 to 165
Operating, °F	−76 to 140	−20 to 140	−10 to 165	−10 to 165
Shelf life at 68°F to 80% initial capacity, months	8 (with lead–calcium grids)	$\frac{1}{2}$ to 1	3	3
Internal resistance, Ω	Low (0.001 to 0.03)	Low (0.003 to 0.5)	Very low	Very low
Shape of discharge curve	Sloping	Flat	Flat	Flat
Relative cost; 1 = lowest	1	2	4	3

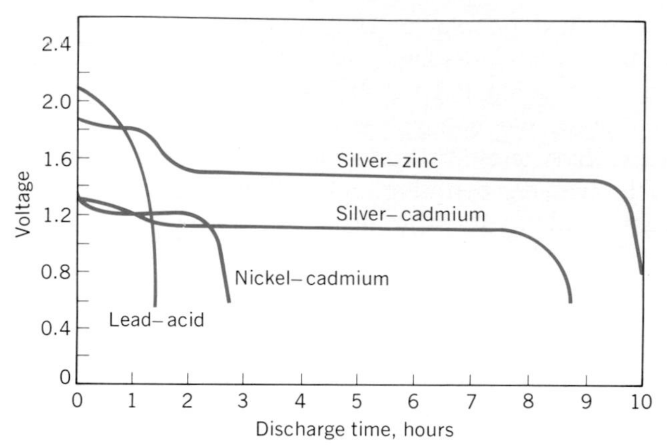

FIGURE 8-11
Discharge characteristics for lead–acid, nickel–cadmium, silver–cadmium and silver–zinc cells of equal weight and current drain.

8-10 OTHER SOURCES OF emf

Table 8-5 shows the different types of energy sources available. All the cells we have considered so far fall in the category of *chemical* sources. They *store* electrical energy in the form of chemical energy. All of the other sources (converters) must produce and deliver the electrical power at the instant it is needed.

8-10.1 Fuel Cells

A *fuel cell* is another example of a chemical source that may provide electrical energy. It generates an emf only as long as the oxygen and hydrogen fuels are fed to it. The arrangement is represented in Fig. 8-12.

TABLE 8-5
Different Sources of emf

Type of Energy Source	Typical Device
Chemical	Voltaic batteries, fuel cell
Photovoltaic (light)	Solar and photocells
Mechanical	Alternators and generators
Thermoelectric (heat)	Thermocouple
Piezoelectric (pressure)	Crystal

A fuel cell is actually the *reverse* process of *electrolysis*. If a dc voltage is applied to two platinum electrodes immersed in water, the current that flows decomposes the water into its basic parts of hydrogen and oxygen. This is called electrolysis. Hydrogen gas accumulates around the negative cathode and oxygen gas around the positive anode. In chemical terms:

$$2H_2O + \text{electricity} \rightleftharpoons 2H_2 + O_2 \tag{8-9}$$

Equation 8-9 is reversible. This means that the combination of hydrogen and oxygen, in some suitable reaction, should produce water *plus* electricity. That is,

$$2H_2 + O_2 \rightleftharpoons 2H_2O + \text{electrical energy} \tag{8-10}$$

When a substance containing hydrogen, such as methyl alcohol or ammonia, comes in contact with a nickel boride catalyst around the anode, the hydrogen ionizes and forms H^+ ions. The anode picks up the electrons and becomes negatively charged. The H^+ ions move through the membrane or potassium hydroxide electrolyte and reacts with the oxygen gas. A silver catalyst at the cathode promotes this reaction, at the expense of electrons, so the cathode becomes positively charged. A by-product of this reaction (Eq. 8-10) is water, which, in the case of manned satellite power supplies, is used for drinking and cooling purposes. Neither the electrodes nor the electrolyte are affected by the operation.

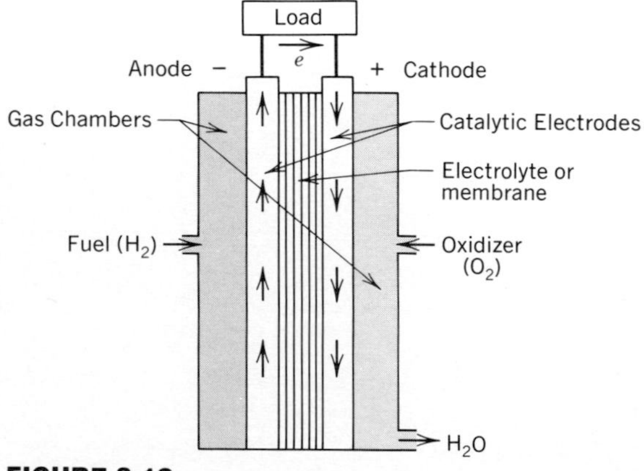

FIGURE 8-12
Arrangement of an oxygen–hydrogen fuel cell.

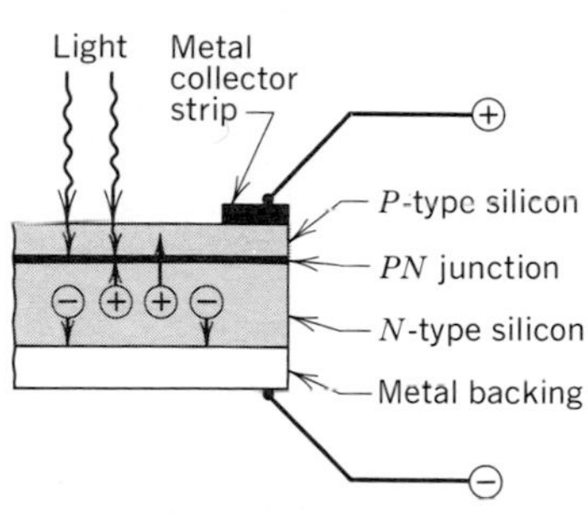

(a) Basic construction of a silicon photovoltaic cell

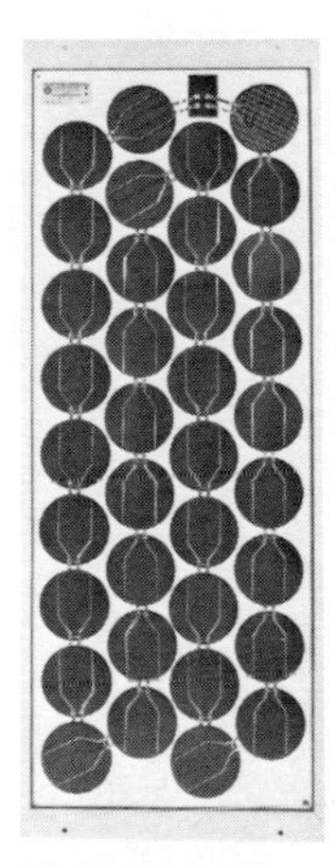

(b) Typical solar module. Thirty-six 90-mm-diameter silicon cells on this 44 × 17 in. module produce 31W of power to charge a 12-V battery at 13.8V and 2.25 A.

FIGURE 8-13
Features of a solar cell. [Photograph in (b) courtesy of the Solar Power Corporation.]

A single fuel cell can generate from 0.5 to 1 V, and by suitable series-parallel interconnection can develop batteries with a typical output of 2 kW for spacecraft applications. Because of the high efficiency of fuel cells (over 40%), low noise, low pollution and waste, the utilities have developed a 40-kW prototype. This is to be followed by 27-MW power plants in the future, since their costs are competitive with conventional and nuclear plants.

8-10.2 Solar Cells

In unmanned spacecraft, electrical power is usually developed by *silicon solar* cells, which convert light or solar energy *directly* into electrical energy.

Basically, the solar cell is a *PN* junction, semiconductor device similar to diodes used for rectification. (See Chapter 28 for a detailed description of *P*-type and *N*-type silicon.) Light energy strikes the solar cell and passes through a thin transparent layer of *P*-type silicon to generate electrons and holes in the *N*-type silicon. These carriers cross over the junction depletion layer, as shown in Fig. 8-13*a*, to produce an emf of approximately 0.5 V on open circuit.

Each solar cell can produce only a few miliamperes, depending upon the surface area. Commercially available cells, 3 in. in diameter, produce approximately 1200 mA at 0.45 V. Solar panels providing 5 W at 12 V result from series-parallel interconnections. Such panels are used to power telephone repeater amplifiers in remote desert locations, 24 h a day, by recharging batteries during daylight hours. A typical solar module is shown in Fig. 8-13*b*.

The most efficient cells to date can convert 28 percent of the light falling on them into electrical energy, although optical losses used in focusing the energy reduce this figure to an overall value of 25 percent. In spite of this relatively low figure (present thermal power plants are approximately 35% efficient), and the high cost (presently $9 per watt of output but expected to be between 50¢ and $2 per watt by 1986), solar cells are being manufactured and installed throughout the world.

One such installation provides the electrical needs for 96 residents in an Indian village in Arizona, which is too far away from existing power lines to make it economically worth bringing in electricity over transmission lines. The installation consists of an array of three 64-ft rows of panels containing a total of 8064 three-in. cells that produce 120 V dc. Excess electrical energy is stored in a bank of lead–acid batteries for use at night and on cloudy days. These 2380-Ah (286-kWh) batteries can deliver full power for 10 days should the need arise. This solar photovoltaic system that powers 47 fluorescent lights, a 2-hp water pump, and 15 small refrigerators, delivers 3.5 kW of peak power at noon on a clear day. Another solar installation array is shown in Fig. 8-14.

FIGURE 8-14
Solar installation at Mt. Laguna, California. (Courtesy of the Solar Power Corporation.)

8-10.3 Alternators

The majority of electrical power used in industry and our homes is generated by using *mechanical* motion to move a conductor through a magnetic field. The process is called *elecromagnetic induction* and is the basic principle of all dc generators and ac alternators. This will be covered in detail in Chapter 14.

8-10.4 Thermocouples

There are a number of *thermoelectric* effects that can utilize temperature differentials to generate an emf. One such device is a *thermocouple.* This consists of two dissimilar metals, such as iron and copper, twisted together at one end and the junction held at some high temperature. A small potential difference of a few millivolts is produced between the wires at the cold end. Thermocouples are used primarily for temperature measurements (from −450°F to over 3000°F) and in heating systems, where the generated voltage is an indication that the pilot light is available to ignite the main gas heating system. See Section 11-3.4.

8-10.5 Piezoelectrics

Finally, an emf may be generated by pressure in the *piezoelectric* effect. Certain quartz crystals, when subjected to a physical force or pressure, generate a voltage across the face of the crystal. In this mode, they are used primarily as *transducers* of mechanical to electrical energy in phonograph pickups, microphones, and sound- and pressure-measuring devices.

SUMMARY

1. A voltaic cell requires two electrodes of different materials, immersed in an electrolyte, to produce an emf.

2. A copper–zinc cell produces approximately 1 V due to the zinc electrode "replacing" the hydrogen in the sulfuric acid and causing a separation of charge through the motion of ions.

3. Polarization is the accumulation of hydrogen gas around a cell's electrode causing an increase in internal resistance and a reduction in generated voltage.

4. The carbon–zinc cell is a dry cell that produces 1.5 V using an ammonium chloride electrolyte. Manganese dioxide is a depolarizer material used to remove accumulated hydrogen gas from the electrode.

5. An alkaline–manganese cell uses a potassium hydroxide electrolyte, zinc anode, and manganese cathode to produce 1.5 V.

6. Shelf-life is the length of time a cell retains 80% of its energy. It can be extended by storage at cool temperatures. An alkaline cell has three to six times the shelf-life of a zinc–carbon cell.

7. Mercury and silver oxide cells have a very high energy density and a flat discharge curve, and are most suitable for low-current applications.

8. A secondary cell, unlike a primary cell, can be recharged numerous times to its initial condition because of a reversible chemical reaction.

9. A fully charged lead–acid cell has a lead peroxide positive plate, a spongy lead negative plate, and an electrolyte of sulfuric acid, and develops approximately 2.2 V.

10. The lead sulfate produced during discharge in a lead–acid cell can be converted back to the original materials by the external connection of a charger, positive-to-positive, negative-to-negative.

11. The state of charge of a lead–acid cell may be determined by measuring the specific gravity of the electrolyte using a hydrometer; 1.13 indicates a discharged cell, 1.28 a fully charged cell.

12. The ampere-hour capacity of a battery is the product of current and time, which determines how much current a cell can deliver based on an 8-h discharge period.

13. A nickel cadmium cell develops 1.2 V, is the most popular rechargeable dry cell for portable equipment, and can be recharged up to two thousand times. It can be left on trickle charge for indefinite periods but loses its charge on standby very quickly.

14. Gelled-electrolyte lead–acid cells are portable versions of the automotive battery but with much smaller capacities. They should not be left on continuous charge.

15. Silver–zinc and silver–cadmium cells are used in lightweight equipment because of their high-energy density and like the nickel–cadmium cells have a flat discharge characteristic.

16. Fuel cells that produce electricity when oxygen and hydrogen combine, and solar cells that directly convert light to electricity in a *PN* junction, are possible sources of the future if economically developed.

SELF-EXAMINATION

Answer T or F or indicate a, b, c, d

8-1. In a copper–zinc cell, both the copper and zinc electrodes are used up. ______

8-2. Current inside a copper–zinc cell consists of the motion of

a. Positive ions.
b. Negative ions.
c. Both positive and negative ions.
d. Electrons.

8-3. The collection of hydrogen gas around the positive electrode of a cell is called

a. Amalgamation.
b. Polarization.
c. Ionization.
d. Hydrogenation.

8-4. A carbon–zinc cell is an example of a dry primary cell. ________

8-5. Local action reduces the shelf-life of a cell. ________

8-6. All cells, no matter what the materials involved, produce 1.5 V. ________

8-7. The electrolyte in an alkaline–manganese cell is

a. Ammonium chloride.
b. Potassium hydroxide.
c. Sulfuric acid.
d. Manganese dioxide.

8-8. Which of the following is false? The advantages of an alkaline cell over a Leclanché cell are

a. Longer shelf-life.
b. More energy.
c. Lower temperature operation.
d. Flat discharge characteristic.

8-9. Mercury and silver oxide cells are secondary cells. ________

8-10. A lead–acid cell is rechargeable because

a. Its electrolyte is sulfuric acid.
b. It is a wet cell.
c. Its chemical reaction is reversible.
d. Its electrolyte has a high specific gravity.

8-11. The cathode of a lead–acid cell is gray spongy lead. ________

8-12. As a lead–acid cell discharges, both electrodes convert to lead sulfate. ________

8-13. At 68°F a fully charged lead–acid cell should have a specific gravity close to

a. 1.13
b. 1.2
c. 1.28
d. 2.2

8-14. A 100-Ah capacity battery should deliver a current of 8 A for approximately

a. 12 h
b. 8 h
c. 20 h
d. 100 h

8-15. The nickel cadmium cell is

a. A wet secondary cell.
b. A dry primary cell.
c. A wet primary cell.
d. A dry secondary cell.

8-16. If a nickel–cadmium cell is to be quick-charged, there is some maximum time that should not be exceeded. ________

8-17. A gelled-electrolyte lead–acid cell is similar in characteristics to a wet cell except that it is portable. ________

8-18. A flat discharge characteristic is one in which the terminal voltage of a cell is relatively constant under use but then decreases rapidly when discharged. ________

8-19. Both fuel cells and solar cells are examples of secondary cells. ________

REVIEW QUESTIONS

1. What is the basic reason for the difference between primary and secondary cells?

2. What two conditions can cause a copper–zinc cell to become fully discharged?

3. a. What do you understand by the words polarization and amalgamation?
   b. What steps are taken in a cell to overcome the first and what condition does the second present?

4. a. How are the words anode and cathode applied to a cell?
   b. Which is which in a copper–zinc cell?

5. How does a depolarizer work?

6. a. Give three advantages that an alkaline–manganese cell has over a carbon–zinc.
   b. What is one disadvantage?

7. a. What are the main applications for mercury and silver oxide cells?
   b. Why?

8. a. Where does the alkaline–manganese cell get its name?
   b. Are the mercury and silver oxide cells also alkaline cells?

9. Why do you think the lead–acid cell is the cheapest of all the secondary cells?

10. Why should a lead–acid cell be fully charged before being stored unused for any length of time?

11. What do you understand by the terms dry-charge and wet-charge, as applied to the construction of lead–acid cells?

12. How is it possible to manufacture a completely sealed lead–acid battery?

13. Why does the specific gravity of a lead–acid cell's electrolyte indicate the state of charge of the cell?

14. What is the ampere-hour capacity an indication of in a battery?

15. What is the reason behind the ability of a nickel–cadmium cell to remain on trickle charge indefinitely?

16. a. Why is it necessary to limit the charge time of a nickel–cadmium cell when it is being recharged quickly?
   b. Where does the charging energy go after the cell becomes charged?

17. a. What other secondary cells are there?
   b. What applications do they have?

18. What do you understand by the term *memory-effect* as applied to a nickel–cadmium battery?

19. a. What are the five types of energy source (as far as producing electrical energy)?

b. Which are not practical for producing large energy quantities for commercial use?

20. a. What is the main advantage of a solar cell?

b. What two disadvantages does it have?

PROBLEMS

(Answers to odd-numbered problems at back of book)

8-1. If a gallon of water weighs 8.3 lb, how much will a gallon of sulfuric acid weigh if a hydrometer gives the following readings?

a. 1130

b. 1280

8-2. A fresh alkaline C-cell has a capacity of 1.5 Ah in a certain application. Determine

a. The available capacity of the cell after being stored for 30 months at 20°C.

b. The length of time the cell in part (a) could deliver 40 mA.

c. The length of time a fresh cell could deliver 40 mA.

8-3. Repeat Example 8-3 for a 40% reduction in terminal voltage. Compare the results with those of Example 8-3. What conclusions can you draw regarding the relative lengths of service of one cell compared with another?

8-4. A battery has a 100-Ah capacity.

a. How many coulombs of charge does this represent?

b. How much current should this battery be able to deliver for 8 hours?

c. If the battery is recharged at the rate of 5 C/s, how long will it take to fully charge the battery?

8-5. It is required to quick-charge a fully depleted 12.2-V, 6-Ah nickel cadmium battery in approximately 12 min.

a. If a 36-V charger with a 0.5-Ω internal resistance is to be used, what additional series resistor is needed and what must be its power rating? (Assume that the voltages remain constant during the charging period.)

b. What must this resistance be increased to if the battery is to remain on trickle charge?

c. If no resistor is used between the charger and battery, what is the maximum length of time charging should be allowed to take place?

d. If a 36-V charger is available that has an internal resistance of 3 Ω, would it be possible to recharge the battery in approximately $\frac{1}{2}$ h? Explain.

8-6. Refer to the solar cell installation in this chapter.

a. How much power must each cell produce?

b. Assuming each cell develops 0.432 V on load, what current is delivered by each cell?

c. Determine a possible series-parallel connection to produce 3.5 kW at 120 V.

d. What is the maximum available current from this combination?

e. If the battery system can supply full power for 10 days, what is the average daily use of energy?

f. Assuming that the cells produce full peak power for 10 hours a day, how long will it take to fully charge the batteries from full discharge while supplying the average daily use of energy?

8-7. A No. 6, 1.5-V dry cell is short-circuited at 70°F.

a. What is the current?

b. What is the cell's internal resistance?

8-8. Repeat Problem 8-7 using a temperature of 20°F.

8-9. a. What would be the short circuit current of a 1.5-V D-size flashlight cell at 70°F?

b. What is the cell's internal resistance?

8-10. Repeat Problem 8-9 using a temperature of 20°F.

CHAPTER NINE

INTERNAL RESISTANCE AND MAXIMUM POWER TRANSFER

***Internal resistance* is an inevitable part of any voltage source.** Although voltage regulators may reduce the effects of internal resistance in a power supply, all batteries suffer a reduction in terminal voltage when a load is connected—because of internal resistance. Also, when cells are interconnected to form batteries, the total internal resistance must be determined using the methods of series and parallel circuits. We also consider the voltage, current, and ampere-hour capacities of such batteries.

The internal resistance of a source determines the maximum power that may be delivered to a load. Where maximum power is to be transferred, the load resistance must be made equal to the internal resistance of the source. This is called a condition of *maximum power transfer,* and the load is said to be *matched* to the source. To illustrate this, we consider the *wattmeter* and its connections to measure power.

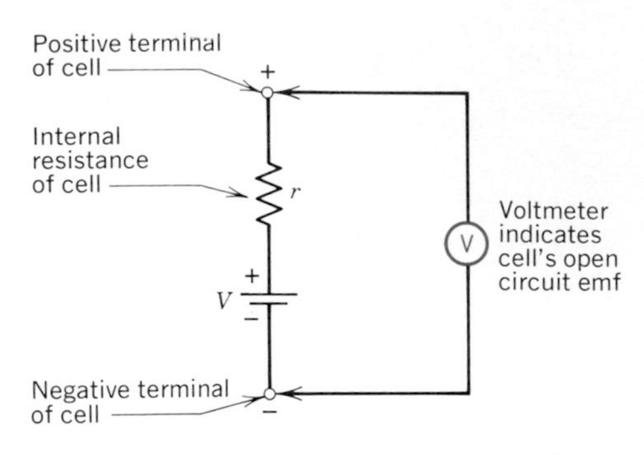

(a) Measuring the open-circuit emf, V, of a cell

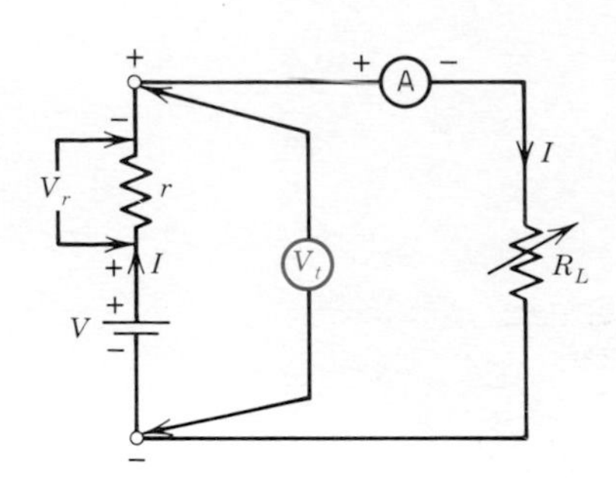

(b) Measuring the terminal voltage, V_t, of a cell under load

FIGURE 9-1
Measuring the open-circuit and terminal voltage of a cell.

9-1 INTERNAL RESISTANCE OF A CELL

We have seen how cells are made up of electrodes and electrolytes, all of which have electrical resistance. **Thus a cell, while generating an emf, has a resistance distributed throughout the cell.**

This *internal* resistance is usually lumped together and is shown as a single equivalent resistance, r, in series with the emf, as in Fig. 9-1*a*. If a voltmeter that draws negligible current is connected across the terminals of the cell (or battery), it will indicate the cell's emf, referred to as the open circuit or no-load voltage, V.

Now consider a load connected to the cell, as shown in Fig. 9-1*b*. The current that flows is, according to Ohm's law,

$$I = \frac{V}{R_L + r} \tag{9-1}$$

This current, flowing through the cell's internal resistance, causes an *internal* voltage drop given by

$$V_r = Ir \tag{9-2}$$

Consequently, as the load current I flows, the terminal voltage V_t of the cell drops below the open circuit voltage, V:

$$V_t = V - Ir \tag{9-3}$$

Equation 9-3 shows that the greater the load current, and the greater the internal resistance, the more the terminal voltage will decrease. In fact, this is how a battery's condition is checked (by noting its terminal voltage under load). When a cell deteriorates, its internal resistance *increases* and, for the same load current, it has a *lower* terminal voltage.

EXAMPLE 9-1

A battery has an open-circuit voltage of 6.5 V and an internal resistance of 2.5 Ω. A load of 10 Ω is connected across the battery. Calculate:

a. The current through the load.
b. The internal voltage drop in the battery.
c. The terminal voltage of the battery under load.
d. The terminal voltage if the battery's internal resistance increases to 5 Ω.

SOLUTION

a. $I = \frac{V}{R_L + r}$ (9-1)

$= \frac{6.5 \text{ V}}{10 + 2.5 \ \Omega}$

$= \mathbf{0.52 \ A}$

b. $V_r = Ir$ (9-2)

$= 0.52 \text{ A} \times 2.5 \ \Omega$

$= \mathbf{1.3 \ V}$

c. $V_t = V - Ir$ (9-3)

$= 6.5 \text{ V} - 1.3 \text{ V}$

$= \mathbf{5.2 \ V}$

d. $I = \dfrac{V}{R_L + r}$ (9-1)

$= \dfrac{6.5\text{ V}}{10 + 5\ \Omega}$

$= 0.43\text{ A}$

$V_t = V - Ir$ (9-3)

$= 6.5\text{ V} - 0.43\text{ A} \times 5\Omega$

$= 6.5\text{ V} - 2.15\text{ V} = \mathbf{4.35\ V}$

It should be clear that if a load resistance R_L is connected so as to drop the terminal voltage to one-half the open circuit voltage, R_L must be equal to the internal resistance, r. This is because the emf is evenly divided between the two resistances. For low internal resistance sources, a more practical way of determining r is to measure the terminal voltage at some known load current and calculate r from Eq. 9-3.

EXAMPLE 9-2

The terminal voltage of a 1.2-V nicad cell drops to 1.1 V when supplying a full-load current of 150 mA. Calculate:

a. The cell's internal resistance.

b. The load resistance that would drop the terminal voltage to 0.6 V.

SOLUTION

a. $V_t = V - Ir$ (9-3)

Therefore, $r = \dfrac{V - V_t}{I}$

$= \dfrac{1.2\text{ V} - 1.1\text{ V}}{0.15\text{ A}}$

$= \dfrac{0.1\text{ V}}{0.15\text{ A}} = \mathbf{0.67\ \Omega}$

b. When the terminal voltage equals one-half the open-circuit voltage, the load resistance must be equal to the internal resistance of **0.67 Ω**.

The change in terminal voltage with load current is shown graphically in Fig. 9-2. An *ideal* voltage source would have no change in terminal voltage. The change that occurs from no load to full load is given by the percentage *voltage regulation:*

$$\%\text{ voltage regulation} = \frac{V_{NL} - V_{FL}}{V_{FL}} \times 100\% \quad (9\text{-}4)$$

where: V_{NL} = the no-load or open-circuit voltage

V_{FL} = the terminal voltage at full-load current

EXAMPLE 9-3

a. Determine the percentage voltage regulation of the 1.2-V cell in Example 9-2.

b. A power supply is known to have a voltage regulation of 40% and a no-load voltage of 21 V. What is the full load voltage?

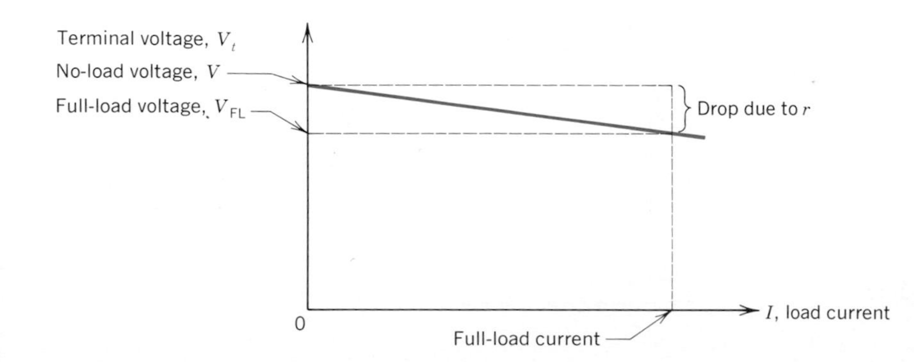

FIGURE 9-2
Effect of internal resistance and load current on terminal voltage, V_t.

SOLUTION

a. % voltage

$$\text{regulation} = \frac{V_{NL} - V_{FL}}{V_{FL}} \times 100\% \quad (9\text{-}4)$$
$$= \frac{1.2\text{ V} - 1.1\text{ V}}{1.1\text{ V}} \times 100\%$$
$$= \frac{0.1\text{ V}}{1.1\text{ V}} \times 100\%$$
$$= \mathbf{9.1\%}$$

b. % voltage

$$\text{regulation} = \frac{V_{NL} - V_{FL}}{V_{FL}} \times 100\% \quad (9\text{-}4)$$
$$40\% = \frac{21\text{ V} - V_{FL}}{V_{FL}} \times 100\%$$
$$0.4 = \frac{21\text{ V} - V_{FL}}{V_{FL}}$$
$$0.4\,V_{FL} = 21\text{ V} - V_{FL}$$
$$1.4\,V_{FL} = 21\text{ V}$$
$$V_{FL} = \frac{21}{1.4} = \mathbf{15\text{ V}}$$

EXAMPLE 9-4

A 12-V, 1.5-Ah gelled-electrolyte lead–acid battery has a terminal voltage of 10.5 V when delivering a full-load current of 10 A. Calculate:

a. The battery's internal resistance.
b. The battery's voltage regulation.
c. The battery's current and terminal voltage when a 2-Ω load is connected.
d. The load resistance that would drop the terminal voltage to 6 V.
e. The terminal voltage of a dc generator to recharge the cell at 5 A.
f. The amount of current the fully charged cell can deliver for 20 h.

SOLUTION

a. $$r = \frac{V - V_t}{I} \quad (9\text{-}3)$$
$$= \frac{12\text{ V} - 10.5\text{ V}}{10\text{ A}} = \frac{1.5\text{ V}}{10\text{ A}} = \mathbf{0.15\ \Omega}$$

b. % voltage regulation

$$= \frac{V_{NL} - V_{FL}}{V_{FL}} \times 100\% \quad (9\text{-}4)$$
$$= \frac{12\text{ V} - 10.5\text{ V}}{10.5\text{ V}} \times 100\% = \mathbf{14.3\%}$$

c. $$I = \frac{V}{R_L + r} \quad (9\text{-}1)$$
$$= \frac{12\text{ V}}{2 + 0.15\ \Omega}$$
$$= \frac{12\text{ V}}{2.15\ \Omega} = \mathbf{5.58\text{ A}}$$
$$V_t = V - Ir \quad (9\text{-}3)$$
$$= 12\text{ V} - 5.58\text{ A} \times 0.15\ \Omega$$
$$= 12\text{ V} - 0.84\text{ V}$$
$$= \mathbf{11.16\text{ V}}$$

d. When the terminal voltage is one-half the no-load voltage, the load resistance equals the internal resistance, **0.15 Ω.**

e. To charge the battery, the generator must be connected across the battery with like polarities, as shown in Fig. 9-3.

Clearly, $V_G = V + Ir$
$$= 12\text{ V} + 5\text{A} \times 0.15\ \Omega$$
$$= 12\text{ V} + 0.75\text{ V}$$
$$= \mathbf{12.75\text{ V}}$$

f. $$\text{current} = \frac{\text{ampere-hour capacity}}{\text{hours}}$$
$$I = \frac{1.5\text{ Ah}}{20\text{ h}} = 0.075\text{ A} = \mathbf{75\text{ mA}}$$

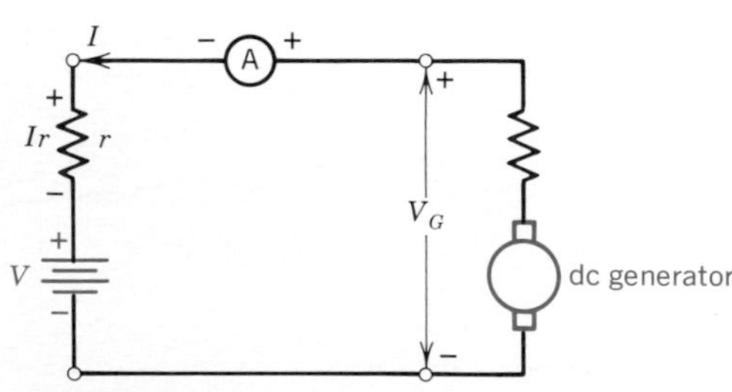

FIGURE 9-3
Charging circuit for Example 9-4.

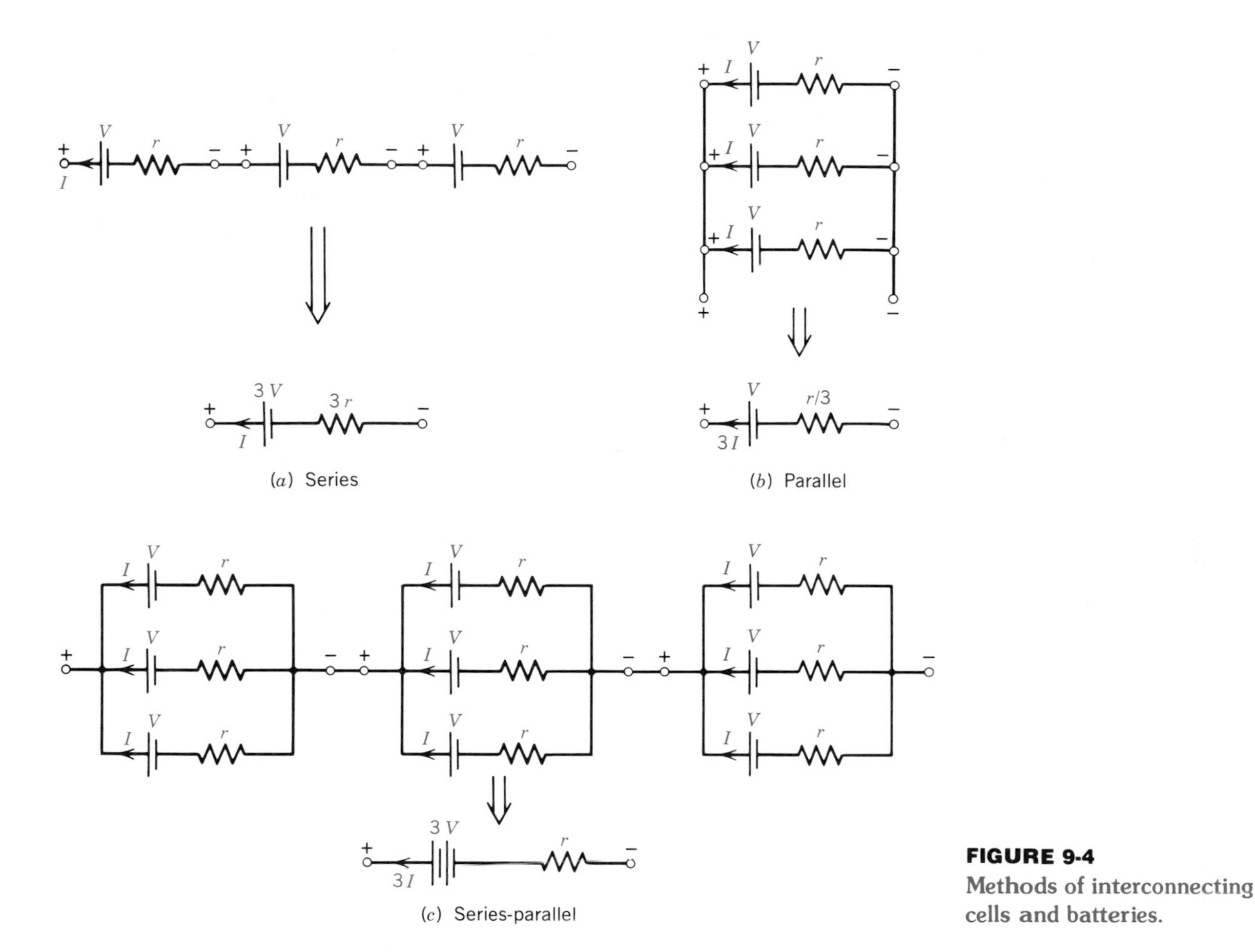

FIGURE 9-4
Methods of interconnecting cells and batteries.

9-2 SERIES AND PARALLEL CONNECTIONS OF CELLS TO FORM A BATTERY

We have seen how the emf developed by individual cells is only 1 to 2 V. To provide higher terminal voltages, *batteries* require a *series* interconnection of cells. As shown by Fig. 9-4*a*, the result is an increase in the overall internal resistance as well. If the cells are identical (not necessarily the case), the total overall voltage and internal resistance are given by

$$V_T = n \cdot V \qquad (9\text{-}5)$$

$$r_T = n \cdot r \qquad (9\text{-}6)$$

where n is the number of cells in series.

Note that the current *capability* is not increased over a single cell. In fact, if the cells are different, the current must not exceed the lowest cell current capability.

The advantages of a *parallel* connection of cells, shown in Fig. 9-4*b*, are to increase the current capability and *reduce* the overall internal resistance. If the cells are identical, or closely matched (which is necessary to avoid large internal circulating currents), the result is

$$I_T = n' \cdot I \qquad (9\text{-}7)$$

$$r_T = \frac{r}{n'} \qquad (9\text{-}8)$$

where n' is the number of cells in parallel.

Note that there is *no* increase in the terminal voltage when cells of equal voltage are connected in parallel.

A *series-parallel* connection is shown in Fig. 9-4*c*, which is typical for a 6-V lead–acid battery and for solar cells. (See Fig. 8-8 to see that each 2-V group consists of three parallel-connected cells.) Both voltage and current capability are increased with this connection. The internal resistances are combined using series-parallel resistance methods.

For cells in series, the total Ah capacity is the same as *each individual* cell's capacity. When the cells are connected in parallel, the total Ah capacity is the *sum* of the individual capacities of each cell. For series-parallel combinations, the total Ah capacity depends upon the particular connection, as shown in Example 9-5.

EXAMPLE 9-5

Six 2-Ah, 1.5-V cells, each with a resistance of 0.1 Ω, are to be interconnected. If each cell is capable of a current of 100 mA determine the terminal voltage, current capability, internal resistance, and Ah capacity of the following arrangements:

a. All connected in series.
b. All connected in parallel.
c. Three series connections of two each in parallel.

SOLUTION

a.
$$V_T = n \cdot V \qquad (9\text{-}5)$$
$$= 6 \times 1.5\ \text{V}$$
$$= \mathbf{9\ V}$$

The current capability is the same as one cell, **100 mA.**

$$r_T = n \cdot r \qquad (9\text{-}6)$$
$$= 6 \times 0.1\ \Omega = \mathbf{0.6\ \Omega}$$

The total Ah capacity is the same as one cell = **2 Ah.**

b. The open-circuit terminal voltage is the same as one cell, **1.5 V.**

$$I_T = n' \cdot I \qquad (9\text{-}7)$$
$$= 6 \times 100\ \text{mA}$$
$$= \mathbf{600\ mA}$$

$$r_T = \frac{r}{n'} \qquad (9\text{-}8)$$
$$= \frac{0.1\ \Omega}{6} = \mathbf{0.016\ \Omega}$$

The total Ah capacity is the sum of the capacities of the six cells = **12 Ah.**

c. Each of the three series connections has a voltage of 1.5 V. The total terminal voltage is given by

$$V_T = n \cdot V \qquad (9\text{-}5)$$
$$= 3 \times 1.5\ \text{V}$$
$$= \mathbf{4.5\ V}$$

The total current capability is the same as each parallel group of two cells, which has a current capability:

$$I_t = n' \cdot I \qquad (9\text{-}7)$$
$$= 2 \times 100\ \text{mA}$$
$$= \mathbf{200\ mA}$$

Each parallel group of two cells has an internal resistance given by

$$r_T = \frac{r}{n'} \qquad (9\text{-}8)$$
$$= \frac{0.1\ \Omega}{2} = 0.05\ \Omega$$

The total resistance of the three series groups is

$$r_T = n \cdot r \qquad (9\text{-}6)$$
$$= 3 \times 0.05\ \Omega$$
$$= \mathbf{0.15\ \Omega}$$

total energy of the cells,

$$W = VIt \qquad (3\text{-}4)$$
$$= 4.5\ \text{V} \times 0.2\ \text{A} \times 20\ \text{h} = 18\ \text{Wh}$$

(Each parallel cell can deliver 0.1 A for 20 h.)

$$\text{total ampere-hour capacity} = \frac{18\ \text{Wh}}{4.5\ \text{V}}$$
$$= \mathbf{4\ Ah}$$

Or each of the two parallel cells (in each series combination) can deliver 0.1 A for 20 h.

Thus total ampere-hour

$$\text{capacity} = 0.1\ \text{A} \times 20\ \text{h} \times 2$$
$$= \mathbf{4\ Ah}$$

Note that although the Ah capacities are different for each connection, the total energy is the same for each connection.

Series: 0.1 A at 9 V for 20 h = 18 Wh of energy.
Parallel: 0.6 A at 1.5 V for 20 h = 18 Wh of energy.
Series-parallel: 0.2 A at 4.5 V for 20 h = 18 Wh of energy.

9-3 MAXIMUM POWER TRANSFER

The internal resistance, r, is not a property that belongs exclusively to cells and batteries. Power sources (ac and dc), amplifiers, microphones, in short every source of emf, has an internal resistance, more generally called internal impedance. In many applications it is desirable to maximize the transfer of power from the source to the load. The question is, given a source of fixed open-circuit voltage and internal resistance, what value of load resistance will have maximum power delivered to it?

Assume that a 40-V open-circuit source of emf with an internal resistance of 8 Ω is connected to a variable load, as in Fig. 9-5*a*. Assume, also, that the load is varied in 1-Ω steps, with the ammeter and voltmeter readings recorded at each step. The product of the two readings gives the power delivered to the load. A graph of the load power against the load resistance will have the shape shown in Fig. 9-5*b*, a definite peak occurring at $R_L = r = 8\ \Omega$.

We can justify the shape of this curve by calculating the load power at several values of load resistance:

1 $R_L = 0$ (short circuit); $P_L = I^2R_L = \mathbf{0}$

2 $R_L = 4\ \Omega$;

$$P_L = I^2R_L = \left(\frac{V}{R_L + r}\right)^2 R_L$$

$$= \left(\frac{40\ \text{V}}{4 + 8\ \Omega}\right)^2 4\ \Omega = \mathbf{44.4\ W}$$

3 $R_L = 8\ \Omega$; $P_L = \mathbf{50\ W}$

4 $R_L = 16\ \Omega$; $P_L = \mathbf{44.4\ W}$

5 $R_L = \infty$ (open circuit); $P_L = I^2R = 0 \times \infty = \mathbf{0}$

The *maximum power transfer theorem* states that maximum power will be delivered to a load from a source when the load resistance equals the internal resistance of the source.

It should not be inferred from the above that the circuit is operating at maximum *efficiency* when $R = r$. This is because we define efficiency as

$$\eta = \%\ \text{efficiency} = \frac{P_L}{P_T} \times 100\% \qquad (9\text{-}9)$$

$$= \frac{P_L}{P_L + P_r} \times 100\%$$

where: P_L = load power

P_r = power dissipated in the source

P_T = total power supplied by the emf

EXAMPLE 9-6

For the above circuit, calculate the load voltage and efficiency at R_L = 4, 8, 16, and 100 Ω.

SOLUTION

$$\text{load voltage } V_L = V \times \frac{R_L}{R_L + r} \qquad (5\text{-}5)$$

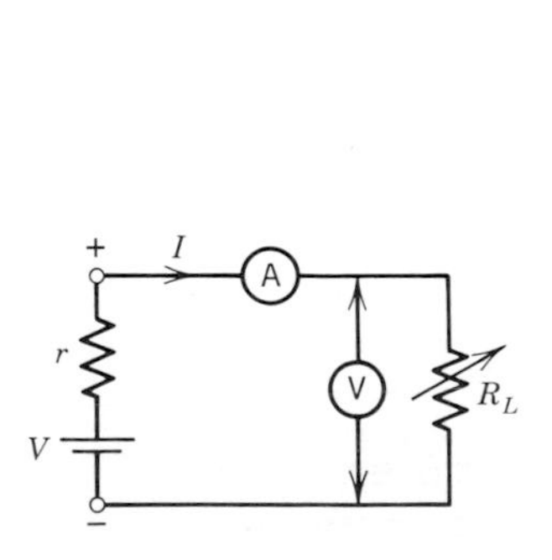

(*a*) Circuit to determine maximum load power

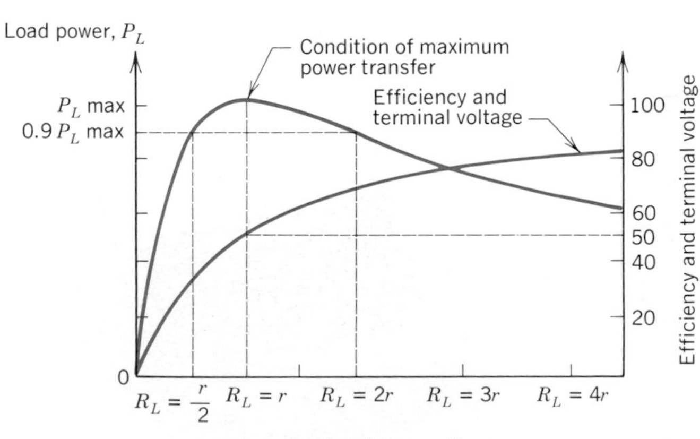

(*b*) Graphs of load power, terminal voltage, and efficiency against load resistance

FIGURE 9-5
Circuit and graphs for maximum power transfer.

When $R_L = 4\ \Omega$, $V_L = 40\ \text{V} \times \dfrac{4\Omega}{4 + 8\ \Omega} = \mathbf{13.3\ V}$,

$$P_L = \frac{V_L^2}{R_L} \qquad (3\text{-}10)$$

$$= \frac{(13.3\ \text{V})^2}{4\ \Omega} = 44.2\ \text{W}$$

$$P_T = \frac{V^2}{R_T} = \frac{V^2}{R_L + r} \qquad (3\text{-}10)$$

$$= \frac{(40\ \text{V})^2}{4 + 8\ \Omega} = 133.3\ \text{W}$$

$$\text{efficiency, } \eta = \frac{P_L}{P_T} \times 100\% \qquad (9\text{-}9)$$

$$= \frac{44.2\ \text{W}}{133.3\ \text{W}} \times 100\% = \mathbf{33.2\%}$$

Similarly, when

$R_L = 8\ \Omega$, $V_L = \mathbf{20\ V}$,
$P_L = 50$ W, $P_T = 100$ W, $\eta = \mathbf{50\%}$

$R_L = 16\ \Omega$, $V_L = \mathbf{26.6\ V}$,
$P_L = 44.2$ W, $P_T = 66.6$ W, $\eta = \mathbf{66.4\%}$

$R_L = 100\ \Omega$, $V_L = \mathbf{37\ V}$,
$P_L = 13.7$ W, $P_T = 14.8$ W, $\eta = \mathbf{92.6\%}$

Thus maximum efficiency, as we have defined it, occurs at loads approaching an open circuit, as shown in Fig. 9-5*b*. This is because the power wasted in the internal resistance, in the form of heat, is less with a smaller current.

The efficiency is (always) 50% at the value of load resistance that provides *maximum power transfer;* that is, when $R_L = r$. Under these conditions, the load voltage is one-half the open-circuit voltage, and the load is said to be *matched* to the source. This is a desirable condition where, for example, a loudspeaker is connected to an amplifier. In the above example, an 8-Ω speaker would have a maximum power of 50 W delivered to it. If either a 4- or a 16-Ω speaker were used with the same 40-V, 8-Ω audio source, only 44.4 W would be transferred. Efficiency is not as important a consideration in *electronic* systems compared to maximum power transfer. But in *power* systems (such as batteries and the 60-Hz ac supply) it is impractical to match the load to the source. In power systems, it is more desirable to aim for maximum *efficiency* (rather than maximum power transfer) to minimize any high-power loss within the source.

9-4 POWER MEASUREMENT—THE WATTMETER

The maximum power transfer theorem may readily be verified experimentally by means of a *wattmeter.* A wattmeter has a deflection that is directly proportional to the product of load current and voltage if the circuit is purely resistive. We are concerned here only with the way in which the wattmeter is *connected* in the circuit to measure power. (The internal operation and construction of the meter is considered in Chapter 16.) See Fig. 9-6.

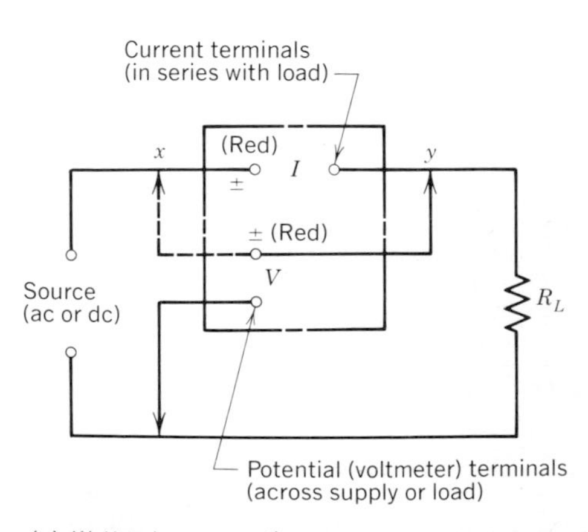

(*a*) Wattmeter connections to measure load power

(*b*) Wattmeter with a 25-W range

FIGURE 9-6
Power measurement. [Photograph in (*b*) courtesy of Hickok Teaching Systems Inc., Woburn, Mass.]

The instrument has a pair of current terminals and a pair of potential (voltage) terminals, with one terminal of each pair marked by a ± sign (or red in color). The current that flows from source to load passes through the current terminals. The voltmeter terminals are usually connected across the load. To avoid any backward reading of the meter pointer, it is necessary to connect the ± voltage terminal to either x or y. This requirement must be met no matter whether the source is ac or dc to make the wattmeter read upscale.

SUMMARY

1. The internal resistance of a cell arises from the cell's materials and causes a reduction in terminal voltage from the open-circuit voltage when a load is connected.

2. The internal resistance of a cell may be determined by noting the change in terminal voltage that occurs when a known current is drawn.

3. The voltage regulation of a source is an indication of the change in terminal voltage under load. An ideal voltage source would have a 0% voltage regulation.

4. If the connection of a load reduces the terminal voltage to one-half of the open-circuit voltage, the internal resistance is equal to the load resistance.

5. A series connection of cells increases the terminal voltage and the internal resistance but not the current capability. A parallel connection increases the current capability, reduces the internal resistance, but does not change the terminal voltage.

6. The total Ah capacity of series-connected cells is the same as each cell's individual capacity.

7. The total Ah capacity of parallel-connected cells is the sum of the individual capacities of each cell.

8. For a series-parallel combination, the total Ah capacity depends upon the particular connection.

9. The total energy of a given number of interconnected cells is independent of the type of connection.

10. Maximum power transfer from a source to a load occurs when the load resistance equals the internal resistance of the source. At this point, the terminal voltage is one-half of the open-circuit voltage and the circuit's efficiency is 50%.

11. A wattmeter is an instrument that automatically multiplies the current and voltage supplied to a load to directly indicate the power delivered to the load. It can measure both ac and dc power.

SELF-EXAMINATION

Answer true or false or a, b, c, d
(Answers at back of book)

9-1. Internal resistance is inevitable in all sources of emf. ________

9-2. If a current of 20 mA is drawn from a source of open circuit voltage, 24 V, and internal resistance, 200 Ω, the terminal voltage is

a. 22 V
b. 24 V
c. 20 V
d. 18 V

9-3. If the connection of a load to a voltage source reduces the terminal voltage to one-half of the open-circuit voltage, the load resistance is one-half of the internal resistance. ________

9-4. If the terminal voltage of a source drops from an open-circuit voltage of 30 V to a value of 20 V when a 60-Ω load is connected, the source's internal resistance is

a. 20 Ω
b. 30 Ω
c. 40 Ω
d. 60 Ω

9-5. The voltage regulation of the source in Question 9-4 is

a. 33.3%
b. 66.6%
c. 50%
d. 10%

9-6. A power supply that has a 100% voltage regulation has an effective internal resistance of 0 Ω. ________

9-7. The terminal voltage of a dc generator required to charge a 24-V battery (whose internal resistance is 0.05 Ω) with a current of 50 A is

a. 21.5 V
b. 24 V
c. 25.5 V
d. 26.5 V

9-8. The total voltage of a battery that consists solely of interconnected series cells is the sum of the emfs of all the cells. ________

9-9. The object of connecting cells in parallel is to produce a battery with increased current capability and decreased internal resistance. ________

9-10. The Ah capacity of a number of interconnected cells is a constant regardless of the method of cell interconnection. ________

9-11. Four $1\frac{1}{2}$-V cells are connected in parallel with each other. This combination is then connected in series with three 2-V cells also connected in series with each other. If each cell has an internal resistance of 1 Ω, the combined voltage and internal resistance is

a. 8 V, 7 Ω
b. 14 V, 7 Ω
c. $3\frac{1}{2}$ V, 3.25 Ω
d. $7\frac{1}{2}$ V, 3.25 Ω

9-12. When a load is *matched* to a source, the load resistance equals the internal resistance of the source. ________

9-13. Maximum power transfer occurs in a circuit when the circuit efficiency is 100%. ________

9-14. The efficiency for the conditions given in Question 9-4 is

a. 33.3%
b. 66.6%
c. 50%
d. 10%

9-15. The graph of load power versus load resistance is symmetrical on either side of the maximum power transfer condition. ________

9-16. A wattmeter in a circuit will read a maximum value if the load resistance is adjusted to one-half the value of the source's internal resistance. ________

REVIEW QUESTIONS

1. What makes up the internal resistance of a battery?

2. a. Why do the headlights of a car dim momentarily when the engine is being started?

b. What visible indication might there be in such a situation when the battery's warranty period has just expired?

3. Is it possible to "jump" a car with a positive ground system using another car with a negative ground? Explain, using a diagram.

4. Describe how you can determine the internal resistance of a supply having:

a. A high internal resistance.

b. A very low internal resistance.

5. a. A power supply is known to have a 100% voltage regulation. Is this good? Explain.

b. An electronic voltage regulator effectively maintains a constant output voltage (in a given power supply), independent of the load current. What can you say about the supply's internal resistance and voltage regulation?

6. What is the result of connecting cells in the following ways, as far as voltage, current, internal resistance, Ah capacity, and total energy are concerned?

a. Series

b. Parallel

c. Series-parallel

7. What is the relation between the power transfer, efficiency, load voltage, internal resistance, and load resistance when a source and load are "matched"?

8. Explain how it is possible for there to be two different load resistances that cause only one-half of the maximum power to be delivered from a source of given internal resistance.

9. When is maximum *power transfer* most desirable and when is maximum *efficiency* most desirable? Why?

10. Describe the proper method for connecting a wattmeter in a circuit.

PROBLEMS

(Answers to odd-numbered problems at back of book)

9-1. A 100-V power supply has an internal resistance of 2.5 kΩ. A 7.5-kΩ load is connected across the power supply terminals. Calculate:

a. The current through the load.

b. The internal voltage drop in the power supply.

c. The terminal voltage of the supply under load.

d. The power dissipated in the power supply when a 5-kΩ load is connected.

e. The current that flows when the power supply terminals are short-circuited.

9-2. Repeat Problem 9-1 using a 5-kΩ internal resistance.

9-3. A 1.5-V cell's terminal voltage drops to 1.25 V when a load is connected that draws a full-load current of 1 A.

a. What is the cell's internal resistance?

b. What is the load resistance?

c. What load resistance would drop the terminal voltage to 1 V?

9-4. A cell has a no-load voltage of 1.2 V. The terminal voltage becomes 1 V when a 2-Ω load is connected. What is the cell's internal resistance?

9-5. A cell has a terminal voltage of 2.2 V with a load of 10 Ω. This drops to 2 V when the load is changed to 2 Ω.

a. Calculate the cell's open-circuit voltage.

b. Calculate the cell's internal resistance.

9-6. A battery has a terminal voltage of 13.6 V when delivering 10 A to a load. When the battery is being charged with 20 A, its terminal voltage is 14.1 V.

a. What is the battery's open-circuit voltage?

b. What is the battery's internal resistance?

9-7. Determine the percent voltage regulation of the cell in Problem 9-3.

9-8. Determine the percent voltage regulation of the cell in Problem 9-4.

9-9. A battery has a no-load voltage of 13.2 V. If the battery is known to have a voltage regulation of 15%, what is its full-load voltage?

9-10. A power supply has a full-load voltage of 22 V and a voltage regulation of 30%. What is its open-circuit voltage?

9-11. Two 6-V cells and one 12-V cell are connected in series. Each of the 6-V cells has an internal resistance of 0.05 Ω, and the 12-V cell has a resistance of 0.1 Ω. How much charging current will flow from a 30-V charger with a 0.3-Ω resistance when connected to the three series-connected cells?

9-12. A 14.1-V charger with an internal resistance of 0.05 Ω is connected across a 12.6-V battery with an internal resistance of 0.1 Ω.

a. Determine the charging current.

b. Determine the terminal voltage.

A second battery, identical to the first, is connected in parallel with the battery and charger.

c. Determine the total current from the charger.

d. Determine the current into each battery.

e. Determine the terminal voltage.

9-13. Four 50-Ah, 6-V cells, each with an internal resistance of 0.01 Ω and a current capability of 10 A are to be interconnected. Determine the terminal voltage, current capability, and internal resistance of the following arrangements:

a. All connected in series.

b. All connected in parallel.

c. Two series connections of two each in parallel.

9-14. For each of the arrangements in Problem 9-13, find:

a. The Ah capacity.

b. The total energy.

9-15. Two batteries are connected in parallel with each other. The first has an emf of 12.2 V and an internal resistance of 0.06 Ω; the second, an emf of 13 V and a resistance of 0.04 Ω.

a. What is the terminal voltage of the combination?

b. What is the internal resistance of the combination?

If the batteries were mistakenly connected with positive-to-negative polarity calculate:

c. The terminal voltage.

d. The current flowing.

e. The power dissipated in each battery in the form of heat.

9-16. A wattmeter is connected between an ac supply and a variable resistive load. The load is adjusted until at a resistance of 10 Ω the wattmeter indicates a maximum reading of 40 W. Determine:

a. The voltage across the load.
b. The internal resistance of the supply.
c. The open circuit voltage of the supply.
d. The voltage regulation of the supply.
e. The efficiency at which the circuit is operating.
f. The load resistance to operate at an efficiency of 60%.

9-17. a. What should be the resistance of an automotive starter motor if it is to be matched to a 12-V battery with an internal resistance of 0.02 Ω?
b. What is the *initial* current that will flow for matched conditions?
c. How much power is delivered to the motor?
d. If the motor is 70% efficient, how much horsepower does the motor develop?
e. How much power is dissipated in heat in the battery?
f. What is the terminal voltage of the battery?
g. What is the efficiency of the circuit?

9-18. a. A power supply is known to have a 50% voltage regulation. If its no-load voltage is 15 V, what will be its full-load voltage?
b. If the current drawn from the above power supply at full load is 0.5 A, what is the internal resistance of the power supply?
c. If the above power supply is connected to a load of 50 Ω, how much power will a wattmeter indicate when connected in series with the load?
d. What value of load resistance will provide a maximum indication of the wattmeter in part (c) and what will this reading be?
e. What will be the load voltage in part (d) ignoring all wattmeter losses?

9-19. An audio power amplifier has an open-circuit voltage of 20 V and an internal resistance of 8 Ω.

a. What two values of load resistance will give one-half of the maximum power possible?
b. What will be the circuit efficiency at these two values?
c. Which of the two loads would be preferable?

CHAPTER TEN
NETWORK ANALYSIS

All the circuits we have considered thus far have been some combination of series, parallel, and series-parallel components. But there are many circuits, especially those involving more than one source of emf, where the components do not have these simple relationships. The analysis of these *networks* requires different *techniques*. There are many methods of analysis. We shall consider some of the more useful ones, which will allow you to solve most ac and dc networks.

The first two, the *branch current* method and the *loop current* method, have similarities and use a combination of Kirchhoff's and Ohm's laws. These methods require the simultaneous solution of two or more equations to solve for the unknown currents.

Circuits that contain more than one source of emf may use the *superposition* theorem to advantage. This considers the currents resulting from each emf applied separately and superimposes these currents to find the overall values when all emfs are in the circuit. This method often permits a solution using simple series-parallel combinations.

The fourth method is called *Thévenin's theorem.* It permits an *equivalent circuit,* consisting of an emf in series with a single resistor, to replace any complex circuit of emfs and linear resistors. It is particularly useful where calculations are to be made for a number of different values of a single component in a given circuit.

Similarly, *Norton's theorem* provides an equivalent circuit consisting of a *constant current source* in *parallel* with a resistance. Both Norton's and Thévenin's equivalent circuits are very useful in the equivalent circuit of a transistor and many other active electronic devices.

Finally, we consider a *delta-wye* conversion technique. Sometimes referred to as a π-T conversion, this method may change a complex network into a series-parallel combination.

10-1
BRANCH CURRENT METHOD

This method involves assigning a designation to each branch current and applying Kirchhoff's voltage law to each complete circuit or loop.

It will be convenient if we repeat Kirchhoff's voltage law in the following way:

> **Around any complete circuit (or loop) the sum of the voltage rises equals the sum of the voltage drops.**

To show the method, let us consider the following example and then make a list of the rules involved in the procedure.

EXAMPLE 10-1

A battery of open-circuit voltage V_B and internal resistance R_B is connected in parallel with a generator of open-circuit voltage V_G and internal resistance R_G. This combination feeds a load resistance R_L. For the following values find the battery current, generator current, load current, and load voltage: $V_B = 13.2$ V, $V_G = 14.5$ V, $R_B = 0.5\ \Omega$, $R_G = 0.1\ \Omega$, and $R_L = 2\ \Omega$.

SOLUTION

1. Draw a circuit diagram. See Fig. 10-1*a*.
2. It can be seen from Fig. 10-1*a* that there are *two* "windows" (loops) in the circuit. This means that we must show *two* currents, one in each loop, of any *arbitrary* direction. (We shall show currents I_B and I_G in the direction we *think* the current might flow.) See Fig. 10-1*b*.
3. Using Kirchhoff's current law, we can identify the current through the load resistor as

$$I_L = I_B + I_G \tag{6-1}$$

Indicate this current in Fig. 10-1*b*.

4. Show the polarity signs of the voltage drops across each resistor using the assumed directions of current. See Fig. 10-1*b*.
5. Indicate, in each window, a loop that goes around a *complete* circuit. The direction is arbitrary, but it is often convenient to use a direction that goes from − to + through an emf. See loops 1 and 2 in Fig. 10-1*b*.
6. Trace around each loop, writing a Kirchhoff's voltage law equation with the voltage rises on one side and the voltage drops on the other.

For loop 1:

$$V_B = I_B R_B + (I_B + I_G) R_L \tag{1}$$

For loop 2:

$$V_G = I_G R_G + (I_B + I_G) R_L \tag{2}$$

7. Insert in Eqs. 1 and 2 the numerical values.

$$13.2 = 0.5 I_B + 2(I_B + I_G) \tag{3}$$

$$14.5 = 0.1 I_G + 2(I_B + I_G) \tag{4}$$

8. Collect together like terms and solve for I_B and I_G:*

$$13.2 = 2.5 I_B + 2.0 I_G \tag{5}$$

$$14.5 = 2.0 I_B + 2.1 I_G \tag{6}$$

Multiply Eq. 5 by four and Eq. 6 by five:

$$52.8 = 10 I_B + 8 I_G \tag{7}$$

$$72.5 = 10 I_B + 10.5 I_G \tag{8}$$

Subtract Eq. 7 from Eq. 8

$$19.7 = 2.5 I_G$$

Therefore, $$I_G = \frac{19.7}{2.5} = \mathbf{7.88\ A}$$

* Linear equations in two or more unknowns may also be solved using the method of determinants. See Appendix A-4.

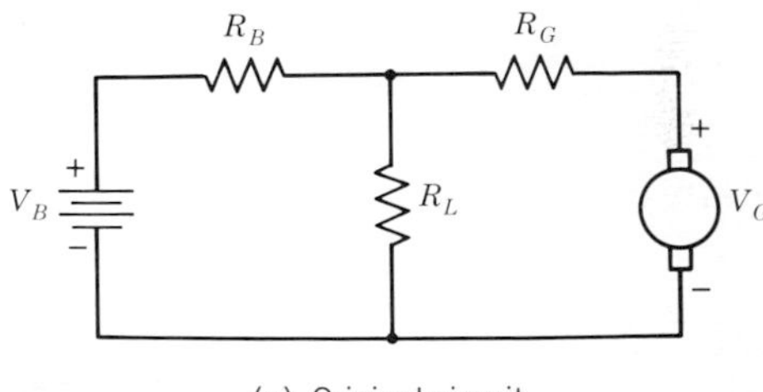

(*a*) Original circuit

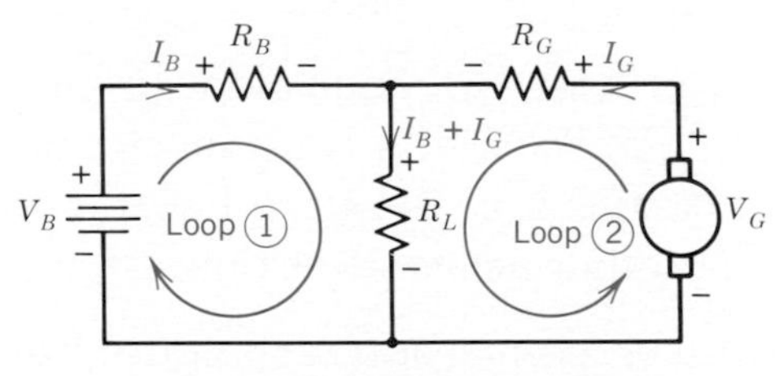

(*b*) Circuit with current designations, voltage drops, and loops

FIGURE 10-1
Circuits for Example 10-1.

Substitute $I_G = 7.88$ A in Eq. 6.

$$2I_B = 14.5 - 2.1I_G$$
$$= 14.5 - 2.1 \times 7.88 = -2.048$$

Therefore, $I_B = \frac{-2.048}{2} = \mathbf{-1.02\ A}$

9. The negative sign with I_B indicates that the *actual* direction of current is opposite to the *assumed* direction. Thus a current of 1.02 A *enters* the positive terminal of V_B and acts to recharge the battery. The load current is given by

$$I_L = I_G + I_B \qquad (6\text{-}1)$$
$$= 7.88 - 1.02 = \mathbf{6.86\ A}$$

and the load voltage is

$$V_L = I_L R_L \qquad (3\text{-}1b)$$
$$= 6.86\ \text{A} \times 2\Omega = \mathbf{13.72\ V}$$

This is a value somewhere between the two values of the voltage sources.

The branch current method is summarized in the following rules:

Rules for Branch Current Procedure

1 Draw a circuit diagram large enough to show all the information clearly.

2 Identify, by the "windows" (or loops) in the circuit, how many *different* currents must be shown (I_1, I_2, I_3, etc.). Show these on the diagram using any arbitrary direction.

3 Use Kirchhoff's current law to identify the remaining currents in the circuit *in terms of* I_1, I_2, and so on.

4 Using the *assumed* directions of current I_1, I_2, and so on, show the *corresponding* polarity signs of the voltages across each component.

5 Indicate on the diagram as many loops as there are windows. The direction of each loop, shown by an arrow, is completely *arbitrary*.

6 Trace around each loop writing a Kirchhoff voltage law equation with *voltage rises* on one side of the equation and *voltage drops* on the other side.

a. A voltage *rise* occurs as you move from − to + in the direction of the loop.

b. A voltage *drop* occurs as you go from + to − in the direction of the loop.

7 Insert in the equations the numerical values of R_1, R_2, and so on, and the emf values (if not done in step 6).

8 Collect like terms and solve for the unknown currents. There must be as many equations as there are unknown currents.

9 Find the current in each component using the current designations in step 3. Interpret any negative currents as having an actual direction opposite to the assumed direction.

To illustrate how the directions of the assigned branch currents and loops can be completely arbitrary, Example 10-2 considers a circuit with two emfs and three loops.

EXAMPLE 10-2

Given the circuit in Fig. 10-2 use the method of branch currents to find the current in each resistor: $V_1 = 10\,V$, $V_2 = 15\,V$, $R_1 = 1\ k\Omega$, $R_2 = 2\ k\Omega$, $R_3 = 3\ k\Omega$, $R_4 = 4\ k\Omega$, and $R_5 = 5\ k\Omega$.

SOLUTION

Since there are three windows, assign currents I_1, I_2, and I_3, and show I_{R_2} and I_{R_4} in terms of these, as shown in Fig. 10-2. Indicate the voltage drop polarity signs across the resistors and three tracing loops, with directions arbitrarily selected. We now write three equations based on Kirchhoff's voltage law.

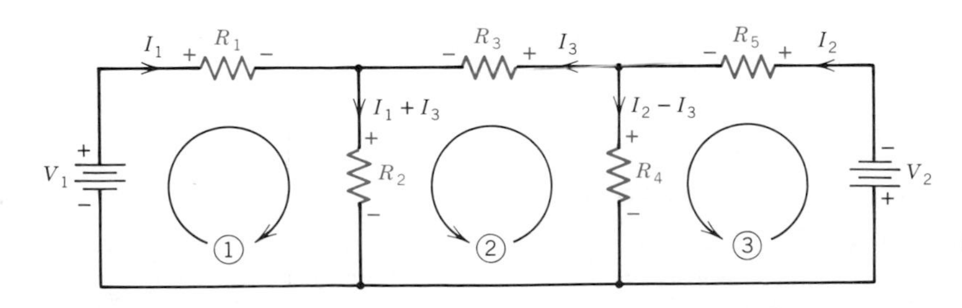

FIGURE 10-2
Circuit for Example 10-2.

Loop 1: $V_1 = I_1R_1 + (I_1 + I_3)R_2$

Loop 2: $(I_2 - I_3)R_4 = I_3R_3 + (I_1 + I_3)R_2$

Loop 3: $0 = V_2 + I_2R_5 + (I_2 - I_3)R_4$

Substituting values and collecting terms:

$$10 = 3I_1 + 0I_2 + 2I_3 \quad (1)$$

$$0 = 2I_1 - 4I_2 + 9I_3 \quad (2)$$

$$-15 = 0I_1 + 9I_2 - 4I_3 \quad (3)$$

Multiply Eq. 1 by two and add to Eq. 3 to obtain:

$$5 = 6I_1 + 9I_2 \quad (4)$$

Multiply Eq. 1 by 9, multiply Eq. 2 by −2 and add to obtain

$$90 = 23I_1 + 8I_2 \quad (5)$$

Multiply Eq. 4 by 8, multiply Eq. 5 by −9 and add to obtain

$$770 = 159\,I_1$$

Therefore, $I_1 = \frac{770}{159} =$ **4.84 mA** (in direction assumed)

NOTE Since resistance values are in kΩ and emfs in volts, the current has units of mA.

By substitution, we obtain

$I_2 = -$**2.67 mA** (opposite to the direction assumed)

$I_3 = -$**2.26 mA** (opposite to the direction assumed)

$I_{R_2} = I_1 + I_3$

$= 4.84 + (-2.26)$mA

$=$ **2.58 mA** (in the direction assumed)

$I_{R_4} = I_2 - I_3$

$= -2.67 - (-2.26)$ mA

$=$ **−0.41 mA** (opposite to the direction assumed)

10-2 LOOP OR MESH CURRENT METHOD

At first sight, this method seems very similar to the branch current method. However, instead of showing a current for each branch, we indicate a current for each loop or window. Only Kirchhoff's voltage law is applied to each loop.

The following rules summarize the procedure:

Rules for Loop Procedure

1. Draw a circuit diagram large enough to show all the information clearly.
2. Indicate, in each window, a loop current of arbitrary direction.
3. Label the loop currents, I_1, I_2, and so on. There must be as many *independent* loop currents as there are windows.
4. Indicate in *each loop* the polarity signs of voltage drops across the resistors for the assumed direction of loop current. Where a resistor has more than one loop current through it, there will be two sets of voltage drops (one in each loop) that may have the same or opposing polarity signs.
5. Apply Kirchhoff's voltage law to each loop, following the assumed direction of the loop current, equating the voltage rises to the voltage drops.
   a. Where a resistor has *opposing* current loops through it, the net voltage drop is the product of the resistance and the *difference* of the two currents, with the larger current assumed to be that of the loop which you are tracing.
   b. If a resistor has current loops through it in the *same* direction, the total voltage drop is the product of the resistance and the *sum* of the two loop currents.
   c. If a loop has no emfs in it, the voltage-rise side of the equation is zero.
6. Insert in the equations the numerical values of all components and sources of emf.
7. Collect like terms and solve for the loop currents. There must be as many equations as there are loop currents.
8. Where a resistor has more than one loop current through it, find the algebraic sum of the loops using the signs (positive or negative) as found in step 7, and the assumed directions as assigned in step 2. Then, interpret any negative currents as having actual directions opposite to the assumed directions.

EXAMPLE 10-3

Given the Wheatstone bridge circuit in Fig. 10-3a, use the loop current method to calculate:

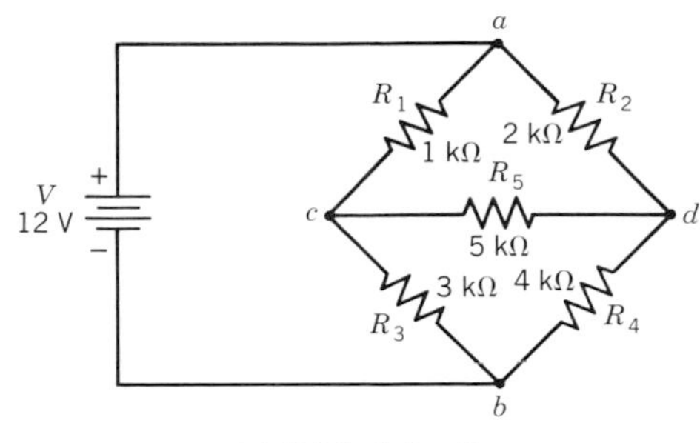

(a) Original circuit

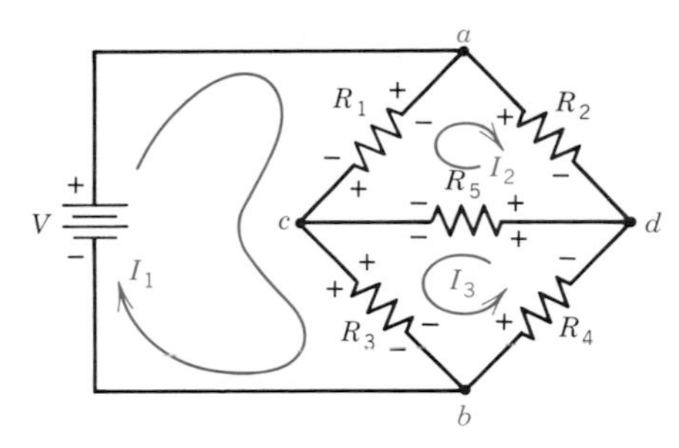

(b) Circuit with current loops and voltage drops

FIGURE 10-3
Circuits for Example 10-3.

a. The current in each resistor.
b. The total resistance of the circuit between a and b given $V = 12$ V, $R_1 = 1$ kΩ, $R_2 = 2$ kΩ, $R_3 = 3$ kΩ, $R_4 = 4$ kΩ, and $R_5 = 5$ kΩ

NOTE It is impossible to solve for the total resistance between a and b using simple series-parallel methods, since there are no resistors either in series or in parallel with each other.

SOLUTION

a. Indicate the three loop currents I_1, I_2, and I_3. Show the voltage drops across the resistors in each loop. See Fig. 10-3b. Write a Kirchhoff's voltage law equation for each loop:

Loop 1: $V = (I_1 - I_2)R_1 + (I_1 + I_3)R_3$
Loop 2: $0 = (I_2 + I_3)R_5 + (I_2 - I_1)R_1 + I_2R_2$
Loop 3: $0 = (I_2 + I_3)R_5 + (I_1 + I_3)R_3 + I_3R_4$

Substituting values and collecting terms, we obtain:

$$12 = 4I_1 - 1I_2 + 3I_3$$
$$0 = -1I_1 + 8I_2 + 5I_3$$
$$0 = 3I_1 + 5I_2 + 12I_3$$

By simultaneous solution we obtain

$$I_1 = 5.03 \text{ mA}, I_2 = 1.92 \text{ mA}, I_3 = -2.07 \text{ mA}$$

Thus

$I_{R_1} = I_1 - I_2 = 5.03 - 1.92$ mA $=$ **3.11 mA**
$I_{R_2} = I_2 =$ **1.92 mA**
$I_{R_3} = I_1 + I_3 = 5.03 + (-2.07)$mA $=$ **2.96mA**
$I_{R_4} = I_3 =$ **−2.07 mA**
(opposite to the direction shown)
$I_{R_5} = I_2 + I_3 = 1.92 + (-2.07)$ mA $=$ **−0.15 mA**
(opposite to the direction assumed)

b. The total current from the supply $= I_1 = 5.03$ mA.
resistance between a and b

$$= \frac{V}{I_1} = \frac{12 \text{ V}}{5.03 \text{ mA}} = \mathbf{2.39\ k\Omega}$$

10-3 THE SUPERPOSITION THEOREM

We have seen that circuits which contain more than one source of emf often cause the resistive components *not* to have a simple series-parallel combination. We then had to write an equation for each loop and solve simultaneously for each current.

The principle of the superposition theorem states that we can superimpose the currents, for any given component, which result from considering *one voltage source at a time.* Since the removal of all voltage sources but one usually results in some series-parallel combination, we can solve for the required current or voltage by the simple methods of Chapters 5, 6, and 7. An *algebraic addition* of each component's voltage or current then yields the total when all voltage sources are in the circuit.

The superposition theorem may be stated as follows:

The total component voltage or current in a multi-emf circuit is the algebraic sum of the individual voltages and currents resulting from each voltage source applied one at a time, with all other voltage sources removed and replaced by their respective internal resistances.

Note that only if a voltage source has zero internal resistance is the removal of a voltage source accomplished by simply "shorting out" the voltage source. This is shown in Example 10-4.

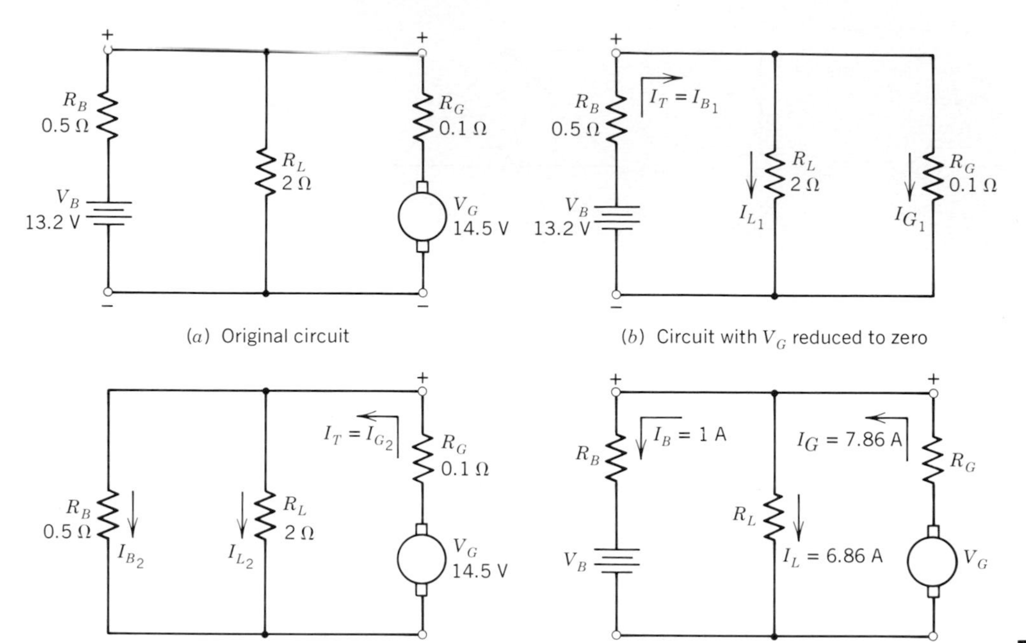

FIGURE 10-4
Circuits for Example 10-4.

EXAMPLE 10-4

A battery of open-circuit voltage 13.2 V and internal resistance 0.5 Ω is connected in parallel with a generator of open-circuit voltage 14.5 V and internal resistance 0.1 Ω. This combination feeds a load having a resistance of 2 Ω. See Fig. 10-4*a*. Using the superposition theorem, calculate:

a. Battery current.
b. Generator current.
c. Load current.
d. Load voltage.

SOLUTION

a. With V_G reduced to zero as in Fig. 10-4*b*, the total circuit resistance is, by series-parallel techniques:

$$R_T = R_B + \frac{R_L \times R_G}{R_L + R_G}$$

$$= 0.5\ \Omega + \frac{2\ \Omega \times 0.1\ \Omega}{2\ \Omega + 0.1\ \Omega} = 0.595\ \Omega$$

$$I_T = I_{B_1} = \frac{V_B}{R_T} \qquad (3\text{-}1a)$$

$$= \frac{13.2\ \text{V}}{0.595\ \Omega} = 22.2\ \text{A}$$

The component of load current due to V_B alone is

$$I_{L_1} = I_T \times \frac{R_G}{R_G + R_L} \qquad (6\text{-}12)$$

$$= 22.2\ \text{A} \times \frac{0.1\ \Omega}{0.1\ \Omega + 2\ \Omega} = 1.057\ \text{A}$$

$$\text{and } I_{G_1} = I_T \times \frac{R_L}{R_L + R_G} \qquad (6\text{-}13)$$

$$= 22.2\ \text{A} \times \frac{2\ \Omega}{2\ \Omega + 0.1\ \Omega} = 21.143\ \text{A}$$

Similarly, for V_B reduced to zero as in Fig. 10-4*c*,

$$R_T = R_G + \frac{R_L \times R_B}{R_L + R_B}$$

$$= 0.1\ \Omega + \frac{2\ \Omega \times 0.5\ \Omega}{2\ \Omega + 0.5\ \Omega} = 0.5\ \Omega$$

$$I_T = I_{G_2} = \frac{V_G}{R_T} \qquad (3\text{-}1a)$$

$$= \frac{14.5\ \text{V}}{0.5\ \Omega} = 29\ \text{A}$$

$$I_{L_2} = I_T \times \frac{R_B}{R_B + R_L} \qquad (6\text{-}12)$$

$$= 29\ \text{A} \times \frac{0.5\ \Omega}{0.5\ \Omega + 2\ \Omega} = 5.8\ \text{A}$$

$$I_{B_2} = 29\ \text{A} - 5.8\ \text{A} = 23.2\ \text{A}$$

The total battery current is the algebraic sum of I_{B_1} and I_{B_2}:

$$I_B = 22.2\ \text{A}\uparrow + 23.2\ \text{A}\downarrow = \mathbf{1\ A}\downarrow$$

This means, because I_{B_2} is greater than I_{B_1}, the actual current through the battery is in the same direction as I_{B_2} in Fig. 10-4c. The battery is being recharged because current is flowing *into* the positive terminal.

b. The total generator current is

$$I_G = I_{G_1} + I_{G_2} \quad (6\text{-}1)$$
$$= 21.143\ \text{A}\downarrow + 29\ \text{A}\uparrow = \mathbf{7.86\ A}\uparrow$$

c. The total load current is

$$I_L = I_{L_1} + I_{L_2} \quad (6\text{-}1)$$
$$= 1.057\ \text{A}\downarrow + 5.8\ \text{A}\downarrow = \mathbf{6.86\ A}\downarrow$$

Note that because I_{L_1} and I_{L_2} are in the same direction, the actual current is the sum of the two.

d. The total load voltage may be found by superimposing the load voltage components due to each voltage source:

$$V_{L_1} = I_{L_1} \times R_L \quad (3\text{-}1b)$$
$$= 1.057\ \text{A} \times 2\ \Omega = 2.114\ \text{V}$$
$$V_{L_2} = I_{L_2} \times R_L \quad (3\text{-}1b)$$
$$= 5.8\ \text{A} \times 2\ \Omega = 11.6\ \text{V}$$

$$V_L = V_{L_1} + V_{L_2} \quad (5\text{-}1)$$
$$= 2.114\ \text{V} + 11.6\ \text{V} = \mathbf{13.7\ V}$$

Actual circuit currents are shown in Fig. 10-4*d*.

The solution of Example 10-4 and the work involved should be compared with Example 10-1, which solved the problem using the method of branch currents. It is doubtful that the superposition method reduces the amount of work (although it allows the use of basic series-parallel circuit theory). This is especially true for a more complex circuit, as in Fig. 10-2.

Finally, it should be noted that the superposition theorem works only for circuits containing linear and bilateral components. For a component to be linear, there must be a directly proportional relationship between voltage and current. To be bilateral, the characteristics of the device must not be dependent upon polarity. That is, the current must be able to flow through the device equally well in either direction. For example, a diode inserted in the battery lead would not allow the application of the superposition theorem, since a diode is neither linear nor bilateral. Also, the total power dissipated in the load resistor of Fig. 10-4*a* is *not* the sum of the two powers dissipated in the load resistor of Fig. 10-4*b* and 10-4*c* because power is *not* linearly related to voltage and current.

10-4 THÉVENIN'S THEOREM

This is a very powerful method of converting a complex circuit containing emfs and linear components into a simple equivalent series circuit of one emf and one resistance. It is particularly useful where repeated calculations are to be made for various values of a single component, since the rest of the circuit to which the component is connected can be represented by an unchanging equivalent circuit.

Thévenin's theorem may be stated in the following manner:

Any linear two-terminal network consisting of fixed resistances and sources of emf can be *replaced* by a single source of emf, V_{Th}, in series with a single resistance, R_{Th}, whose values are given by the following:

1 **V_{Th} is the open-circuit voltage at the two specified terminals of the original circuit.**

2 **R_{Th} is the resistance looking back into the original network from the two specified terminals with all the sources of emf shorted out and replaced by their internal resistances.**

This concept is illustrated in Fig. 10-5. Experimentally, given the circuit in Fig. 10-5*a*, the Thévenin voltage V_{Th} could be measured by a high-resistance voltmeter connected between terminals A and B. Also, in order to measure the equivalent resistance with an ohmmeter between terminals A and B, it would be necessary to remove any emf by shorting the source voltage V.

The Thévenin equivalent circuit shown in Fig. 10-5*b* will then behave, as far as any load connected between terminals A and B is concerned, in the same manner as the original circuit given in Fig. 10-5*a*.

We shall obtain our values of V_{Th} and R_{Th} by calculations, as shown by the following two examples.

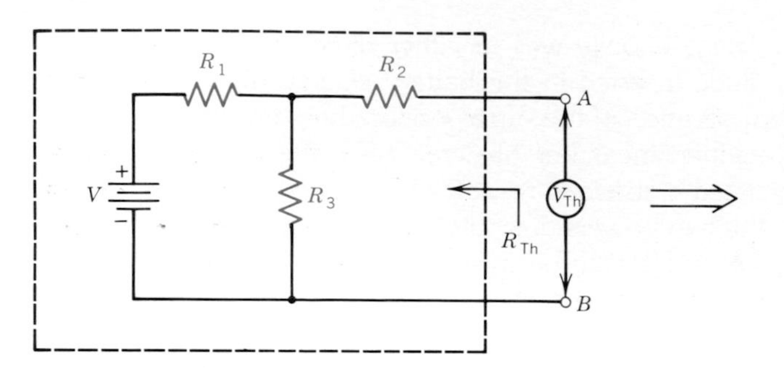

(a) Initial, linear, two-terminal circuit

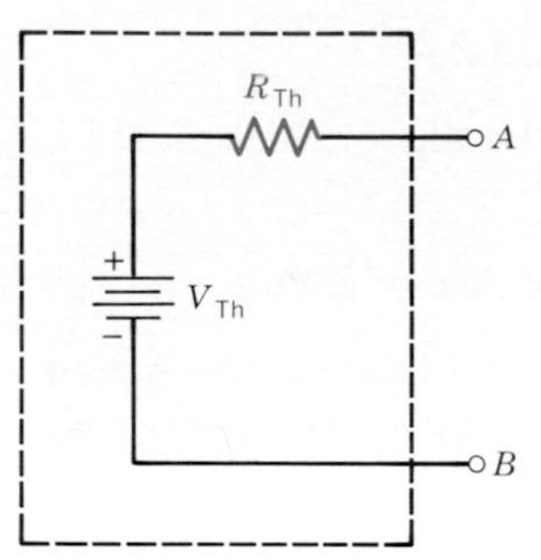

(b) Thévenin equivalent circuit

FIGURE 10-5
Circuits for Thévenin's theorem.

EXAMPLE 10-5

Use Thévenin's theorem to find the current through the load resistor R_L in Fig. 10-6 for the following values of R_L:

a. 1 Ω
b. 2 Ω
c. 3 Ω

SOLUTION

1. Draw the circuit with the load resistor R_L removed. Identify the load terminals as A and B. See Fig. 10-7.
2. We now have a simple series circuit with a current given by

$$I = \frac{V_G - V_B}{R_B + R_G}$$

$$= \frac{14.4 - 13.2 \text{ V}}{0.5 + 0.1 \ \Omega} = \frac{1.2 \text{ V}}{0.6 \ \Omega} = 2 \text{ A}$$

3. The open-circuit voltage between A and B may be found by two methods:

$$V_{R_B} = IR_B \qquad (3\text{-}1b)$$
$$= 2 \text{ A} \times 0.5 \ \Omega = 1 \text{ V}$$

Therefore, $$V_{Th} = V_B + V_{R_B} = V_{AB} \qquad (5\text{-}1)$$
$$= 13.2 + 1 = \mathbf{14.2 \ V}$$

or $$V_{R_G} = IR_G \qquad (3\text{-}1b)$$
$$= 2 \text{ A} \times 0.1 \ \Omega = 0.2 \text{ V}$$

so that $$V_{Th} = V_G - V_{R_G} = V_{AB} \qquad (5\text{-}1)$$
$$= 14.4 - 0.2 = \mathbf{14.2 \ V}$$

Thus 14.2 V is the open-circuit voltage between A and B (with the load removed) and is the Thévenin equivalent circuit voltage.

4. To find the Thévenin resistance, draw the circuit, with the load resistor R_L still removed and then short-out (take out) the emfs V_B and V_G. See Fig. 10-8.
5. Find the resistance between terminals A and B as seen by an ohmmeter connected to these terminals. (Since the voltage sources have been shorted, it would be safe to connect an ohmmeter to $A - B$.) Note that R_B and R_G are now *in parallel* with each other as far as an ohmmeter connected to terminals

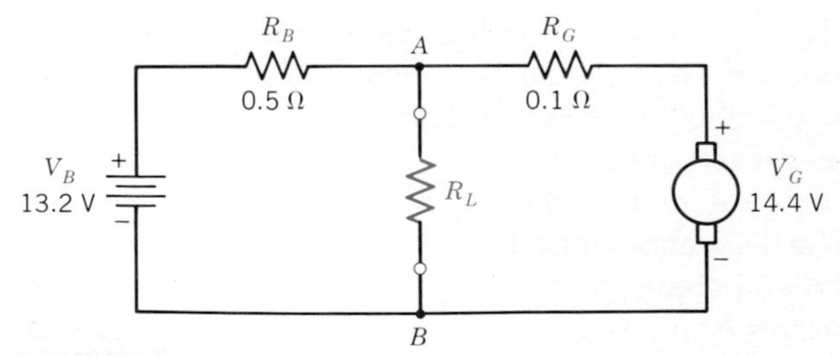

FIGURE 10-6
Circuit for Example 10-5.

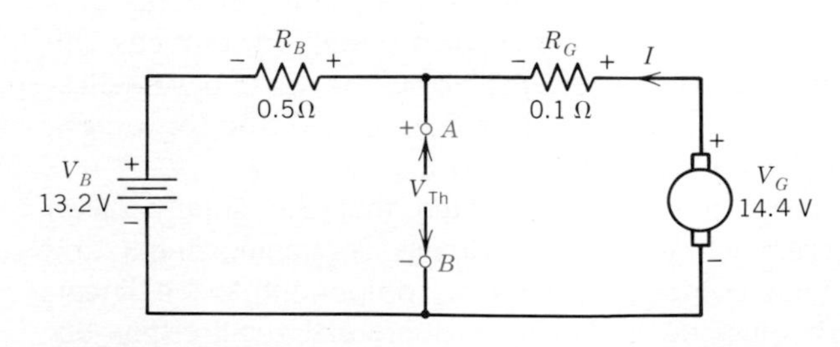

FIGURE 10-7
Open-circuit voltage determination for Example 10-5.

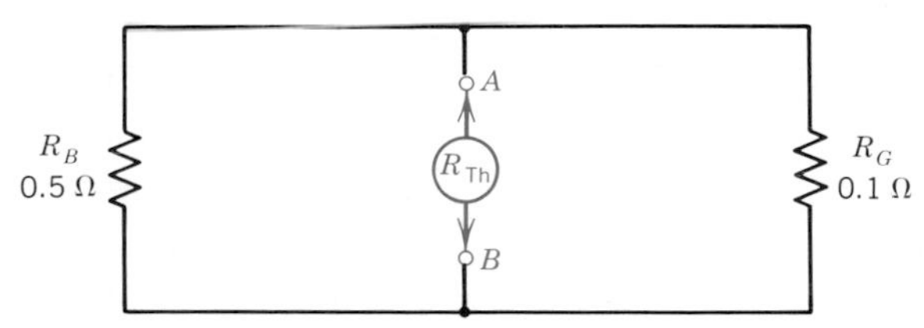

FIGURE 10-8
Resistance determination for Example 10-5.

A and B is concerned. The resistance seen by the ohmmeter is

$$R_{Th} = \frac{R_B \times R_G}{R_B + R_G} = R_{AB} \qquad (6\text{-}10)$$

$$= \frac{0.5 \times 0.1}{0.5 + 0.1} = \frac{0.05}{0.6} = \mathbf{0.083\ \Omega}$$

6. The Thévenin equivalent circuit can now be drawn, as shown in Fig. 10-9.
7. The load resistor can now be reconnected to terminals A and B. It will behave exactly the same as when it was connected to the original circuit in Fig. 10-6.
8. The load currents for the various loads can now be obtained:

$$I = \frac{V}{R_T} = \frac{V_{Th}}{R_{Th} + R_L} \qquad (3\text{-}1a)$$

a. For $R_L = 1\ \Omega$, $I = \dfrac{14.2\ V}{0.083 + 1\ \Omega} = \mathbf{13.1\ A}$

b. For $R_L = 2\ \Omega$, $I = \dfrac{14.2\ V}{0.083 + 2\ \Omega} = \mathbf{6.8\ A}$

c. For $R_L = 3\ \Omega$, $I = \dfrac{14.2\ V}{0.083 + 3\ \Omega} = \mathbf{4.6\ A}$

NOTE If the branch or loop current methods were used, new equations would have to be written and solved for *each* value of load resistance. Since the Thévenin equivalent circuit is found with the variable (load) component *removed,* the *same* Thévenin equivalent circuit can be used with each *different* load.

As a second example of the use of Thévenin's theorem, we now find the current in R_3 in Fig. 10-10. This is similar to the problem solved in Example 10-2.

EXAMPLE 10-6

Given the circuit of Fig. 10-10, find the current in R_3 using Thévenin's theorem.

SOLUTION

1. To find the Thévenin voltage, remove R_3, label the terminals A and B, and solve for the open-circuit voltage, V_{Th}. See Fig. 10-11.

   $V_{AB} = V_{Th}$ is the difference between V_{R_2} and V_{R_4}. By the voltage divider equation:

$$V_{R_2} = V_1 \times \frac{R_2}{R_1 + R_2} \qquad (5\text{-}5)$$

$$= 10\ V \times \frac{2\ k\Omega}{1\ k\Omega + 2\ k\Omega}$$

$$= 10\ V \times \frac{2}{3} = 6.67\ V$$

$$V_{R_4} = V_2 \times \frac{R_4}{R_4 + R_5} \qquad (5\text{-}5)$$

$$= 20\ V \times \frac{4\ k\Omega}{4\ k\Omega + 5\ k\Omega}$$

$$= 20\ V \times \frac{4}{9} = 8.89\ V$$

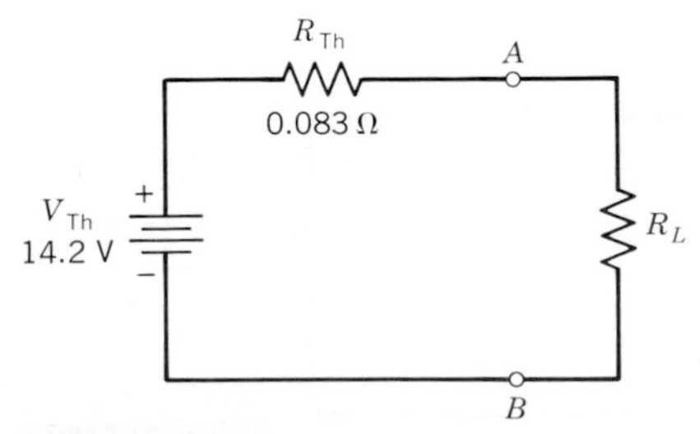

FIGURE 10-9
Thévenin equivalent circuit for Example 10-5.

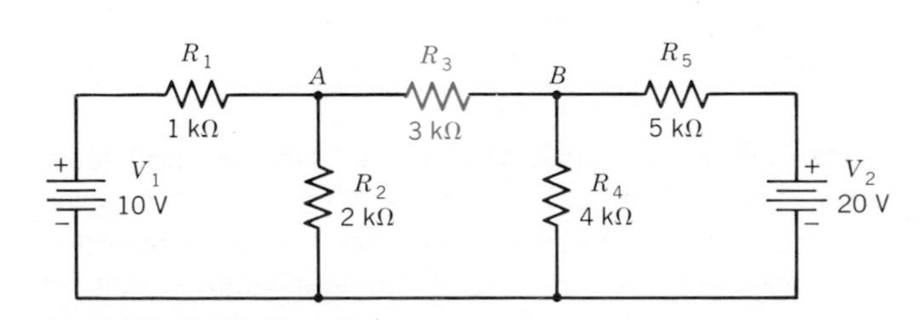

FIGURE 10-10
Circuit for Example 10-6.

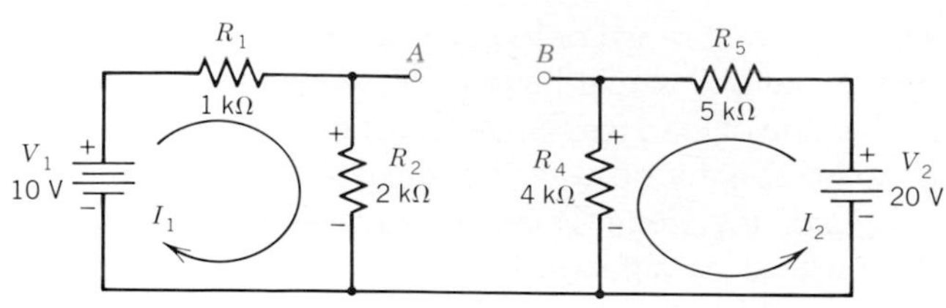

FIGURE 10-11
Open-circuit voltage determination for Example 10-6.

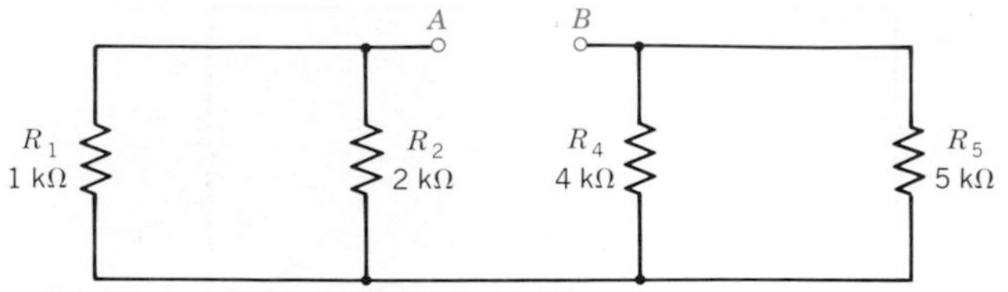

FIGURE 10-12
Resistance determination for Example 10-6.

Therefore, $V_{AB} = V_{R_2} - V_{R_4}$

$$= 6.67 - 8.89 \text{ V} = \mathbf{-2.22\ V}$$

This means that the Thévenin equivalent voltage is 2.22 V with B positive with respect to A.

2. To find the Thévenin resistance, reduce the voltages in Fig. 10-11 to zero and determine the resistance between A and B looking back into the circuit. See Fig. 10-12.

$$\text{resistance } R_{AB} = R_{\text{Th}} = R_1 \| R_2 + R_4 \| R_5$$

$$\text{Therefore, } R_{\text{Th}} = \frac{R_1 R_2}{R_1 + R_2} + \frac{R_4 R_5}{R_4 + R_5}$$

$$= \frac{1\ \text{k}\Omega \times 2\ \text{k}\Omega}{1\ \text{k}\Omega + 2\ \text{k}\Omega} + \frac{4\ \text{k}\Omega \times 5\ \text{k}\Omega}{4\ \text{k}\Omega + 5\ \text{k}\Omega}$$

$$= \frac{2}{3}\ \text{k}\Omega + \frac{20}{9}\ \text{k}\Omega$$

$$= 0.67\ \text{k}\Omega + 2.22\ \text{k}\Omega = \mathbf{2.89\ k\Omega}$$

3. The Thévenin equivalent circuit between terminals A and B with resistor R_3 replaced is now given by Fig. 10-13. Note how B is positive with respect to A.

4. The current in R_3 can now be determined:

$$I_3 = \frac{V}{R_T} = \frac{V_{\text{Th}}}{R_{\text{Th}} + R_3} \qquad \text{(3-1a)}$$

$$= \frac{2.22\ \text{V}}{(2.89 + 3)\text{k}\Omega} = \mathbf{0.38\ mA}$$

flowing from B to A or right to left in the original circuit.

NOTE There is no such thing as the Thévenin equivalent for the *overall* circuit of Fig. 10-10. It depends upon which resistor or portion of the circuit is to be *removed* that determines the nature of the equivalent circuit for what remains. Thus different equivalent circuits would result for R_2, on the one hand, or R_4 on the other.

It should be noted that in some complex circuits, the determination of the open-circuit Thévenin voltage may involve the use of branch or loop current methods. That is, we are not guaranteed that the removal of a given component to determine a Thévenin equivalent will leave a simple series-parallel circuit. For example, determining the Thévenin equivalent for resistor R_4 in Fig. 10-10 will require the use of one of the analysis methods covered in this chapter.

10-5 CURRENT SOURCES

We have seen that a complex circuit may be represented by a constant voltage source in series with a resistance. Let us see if we can obtain an equivalent circuit in terms of a *current source.*

Consider Fig. 10-14*a*. This may be a 12-V battery connected in series with a 4-Ω resistor, or, it may be a Thévenin equivalent of some complex circuit. Either way, an open-circuit voltage of 12 V exists between A and B with an internal resistance of 4 Ω.

If A and B are shorted together, a short-circuit current of 12 V/4 Ω = 3 A flows, as in Fig. 10-14*b*. Now let us assume that it is possible to make a power supply that

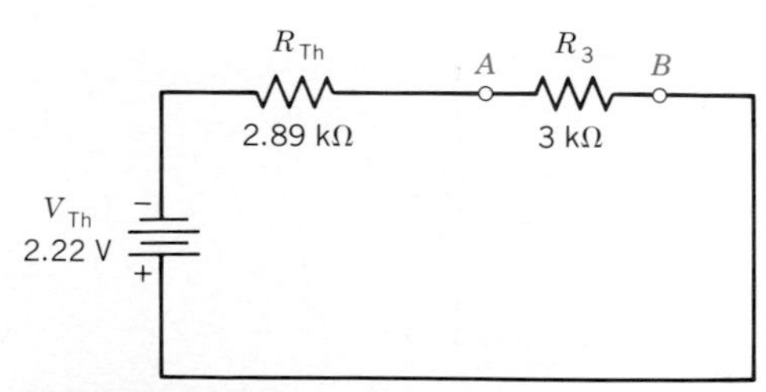

FIGURE 10-13
Thévenin equivalent for Example 10-6.

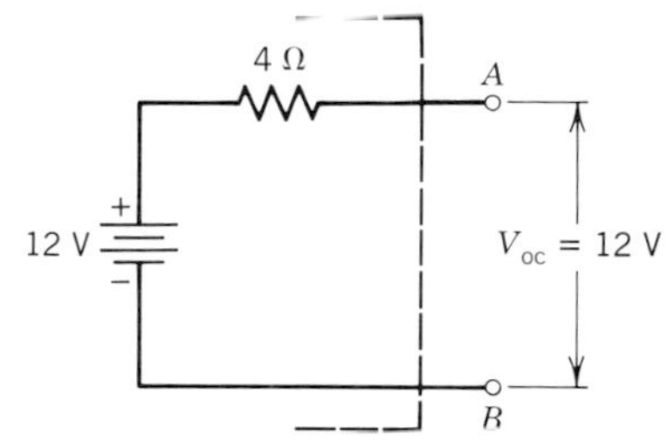

(*a*) Initial constant-voltage-source circuit

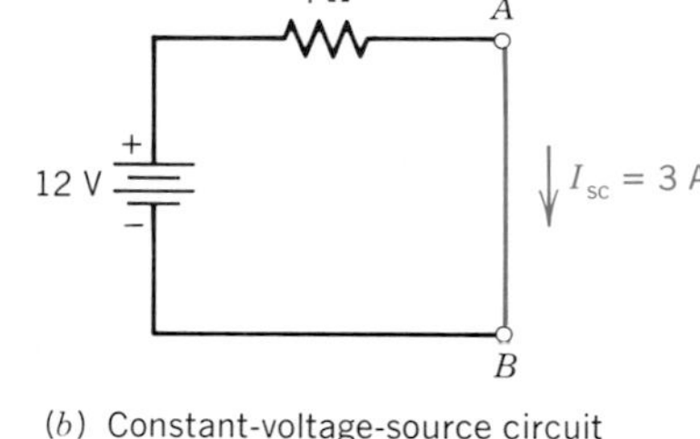

(*b*) Constant-voltage-source circuit provides a short-circuit current of 3 A

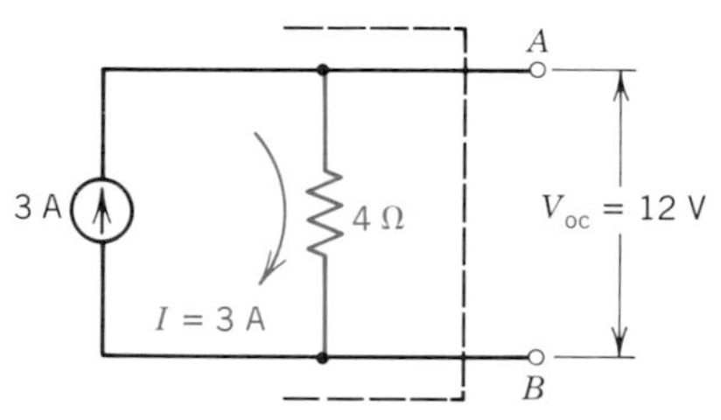

(*c*) Equivalent constant-current-source circuit

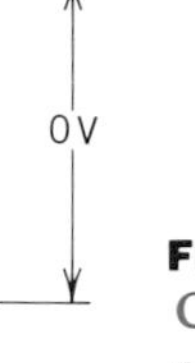
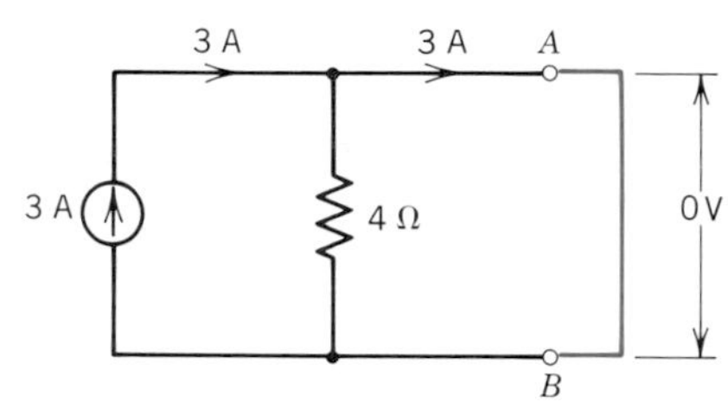

(*d*) Constant-current-source circuit provides a short-circuit current of 3 A

FIGURE 10-14
Constant-voltage-source and equivalent constant-current-source circuits.

provides a constant 3-A output, regardless of the load resistance connected.* This *constant-current source* is shown in Fig. 10-14*c* as a circle with an arrow indicating the direction of the conventional current. Connected in parallel with the current source is a resistance of 4 Ω in this case. This is necessary to provide a complete path for the current source to flow through and develop a voltage of 12 V. That is, our new equivalent circuit in Fig. 10-14*c* has to provide the same open-circuit voltage as the initial voltage source in Fig. 10-14*a*. Also, the required *parallel* resistance is the same as the initial *series* resistance.

Now let us see if our constant-current circuit provides the same short-circuit current as the initial circuit in Fig. 10-14*a*. If terminals *A* and *B* are short-circuited, as in Fig. 10-14*d*, *all* of the current source's 3 A flow between terminals *A* and *B*. This provides the same effect as the circuit in Fig. 10-14*b*.

Another test we can perform to determine equivalency between the two circuits is to measure the circuit's internal resistance. In order to connect an ohmmeter to the circuit of Fig. 10-15*a*, we must remove the constant-current source. See Fig. 10-15*b*. Note that we leave the circuit *open* where we remove the current source, whereas we *short* the leads when removing a voltage source. The ohmmeter now reads 4 Ω as required.

Finally, does the connection of a 4-Ω load, for example, to the equivalent constant-current circuit of Fig. 10-15*a* provide the same load voltage and current as when the load is connected to the initial constant-voltage circuit in Fig. 10-15*c*? See Fig. 10-15*d*. The 3 A from the constant-current source must divide equally between the internal resistance and the load resistance, since both are 4 Ω. As a result, the load voltage is 6 V due to a load current of 1.5 A. These are the same as in Fig. 10-15*c*,

where $I = \dfrac{V}{R} = \dfrac{12\ \text{V}}{8\ \Omega} = 1.5\ \text{A}.$

There are two points to note.

1 Whereas the voltage across the current source on open circuit is 12 V in Fig. 10-15*a*, the voltage is only 6 V when a 4-Ω load is connected in Fig. 10-15*d*. The voltage across a constant-current source varies,

* Using modern IC voltage regulators and current-regulating diodes, it is possible to construct circuits whose current output is substantially constant over very wide ranges of load resistance, from zero to several thousand ohms, depending upon the particular constant current required.

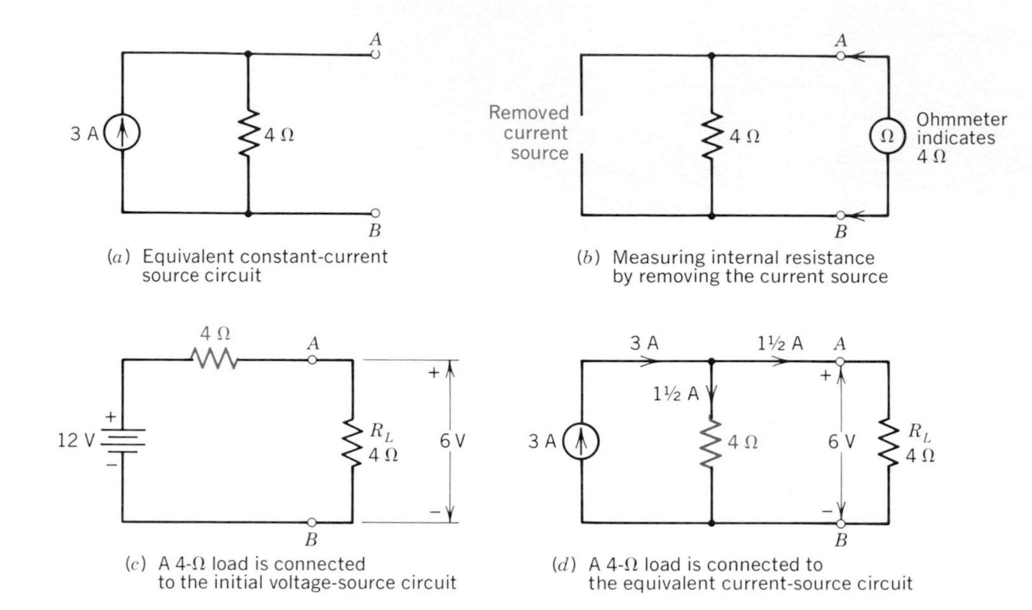

(*a*) Equivalent constant-current source circuit

(*b*) Measuring internal resistance by removing the current source

(*c*) A 4-Ω load is connected to the initial voltage-source circuit

(*d*) A 4-Ω load is connected to the equivalent current-source circuit

FIGURE 10-15
Determining the equivalence of the constant-current-source and constant-voltage-source circuits.

according to the resistance connected across its terminals.

2 Although a constant-current-source circuit is completely equivalent to a constant-voltage-source circuit, as far as any *externally connected load is concerned,* the two circuits *themselves* are not. For example, on open circuit, *no power is dissipated at all* in the initial voltage-source circuit of Fig. 10-14*a*. However, the equivalent current-source circuit of Fig. 10-14*c* dissipates 36 W on open circuit.

10-6 NORTON'S THEOREM

The concept of a constant-current source is required to understand the equivalent circuit proposed by Norton. His theorem may be stated in the following manner:

> **Any linear two-terminal network of fixed resistances and sources of emf, may be *replaced* by a single constant-current source, I_N, in parallel with a single resistance, R_N, whose values are given by the following: I_N is the current that flows between the designated terminals of the original circuit when short-circuited. R_N is the resistance looking back into the original network from the two designated terminals, with all sources of emf shorted-out and replaced by their internal resistances (and any current sources in the network opened up).**

It should be noted that the Norton equivalent resistance is the same as the Thévenin equivalent resistance for any given circuit. The only difference is that this resistance is placed in *parallel* with the constant-current source, I_N.

Experimentally, Norton's equivalent for the circuit given in Fig. 10-16*a* would be found as follows:

1 Connect a low-resistance ammeter between *A* and *B* to measure the short-circuit current, I_N. See Fig. 10-16*b*.

2 Reduce the voltage source to zero and connect an ohmmeter between *A* and *B* to measure the resistance, R_N, looking back into the circuit. See Fig. 10-16*c*.

The Norton equivalent circuit is shown in Fig. 10-16*d*.

Alternatively, the Norton equivalent circuit can be obtained by calculation, as shown by the following examples.

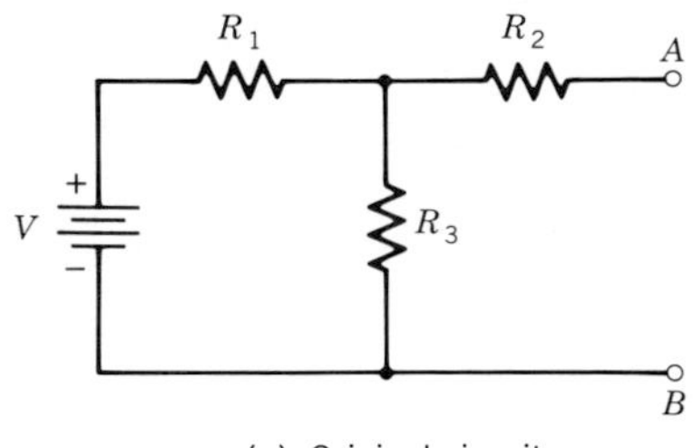

(a) Original circuit

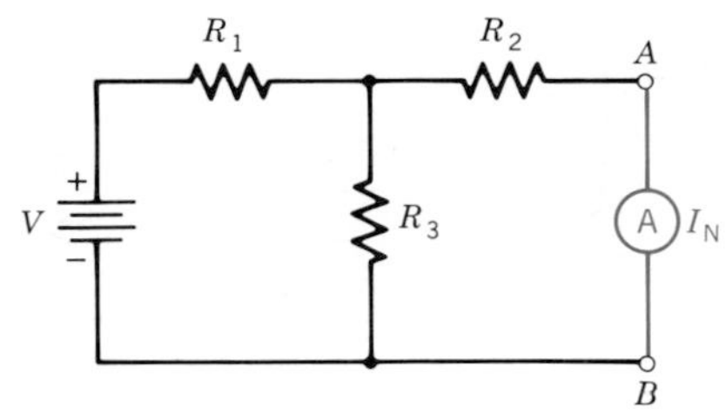

(b) Obtaining I_N. The ammeter indicates short-circuit current I_N

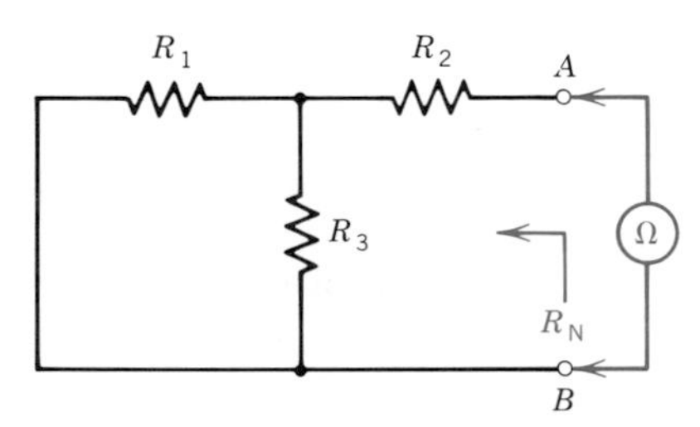

(c) Obtaining R_N. The ohmmeter indicates Norton resistance, R_N

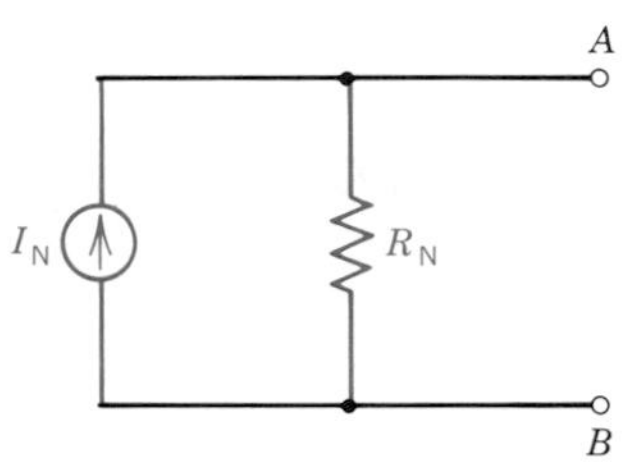

(d) Norton equivalent circuit

FIGURE 10-16
Experimental determination of a Norton equivalent circuit.

EXAMPLE 10-7

Given the circuit in Fig. 10-17*a*, calculate for loads of 55 kΩ and 110 kΩ, using Norton's theorem:

a. Load current.
b. Load voltage and polarity.

SOLUTION

Remove the load, R_L, mark the load terminals A and B, and calculate the short-circuit current between A and B. See Fig. 10-17*b*. The total resistance seen by the 132-V source is given by

$$\begin{aligned} R_T &= R_1 + R_2 \| R_3 \\ &= R_1 + \frac{R_2 \times R_3}{R_2 + R_3} \\ &= 20\ \text{k}\Omega + \frac{40\ \text{k}\Omega \times 60\ \text{k}\Omega}{40\ \text{k}\Omega + 60\ \text{k}\Omega} \\ &= 20\ \text{k}\Omega + 24\ \text{k}\Omega = 44\ \text{k}\Omega \end{aligned}$$

The total current from the 132-V source is

$$I_T = \frac{V}{R_T} \tag{3-1a}$$

$$= \frac{132\ \text{V}}{44\ \text{k}\Omega} = 3\ \text{mA}$$

The current, I_N, between A and B is I_{R_2}:

$$I_N = I_{R_2} = I_T \times \frac{R_3}{R_2 + R_3} \tag{6-12}$$

$$= 3\ \text{mA} \times \frac{60\ \text{k}\Omega}{60\ \text{k}\Omega + 40\ \text{k}\Omega} = \mathbf{1.8\ mA}$$

The Norton resistance is found by shorting the voltage source as in Fig. 10-17*c*.

$$\begin{aligned} R_N &= R_2 + R_1 \| R_3 \\ &= R_2 + \frac{R_1 \times R_3}{R_1 + R_3} \\ &= 40\ \text{k}\Omega + \frac{20\ \text{k}\Omega \times 60\ \text{k}\Omega}{20\ \text{k}\Omega + 60\ \text{k}\Omega} \\ &= 40\ \text{k}\Omega + 15\ \text{k}\Omega = \mathbf{55\ k\Omega} \end{aligned}$$

The Norton equivalent circuit is shown in Fig. 10-17*d*. Note that the current source points "downward," causing the load current to flow from B to A.

a. For a load resistance of 55 kΩ, I_N must divide equally between R_N and R_L, so I_L = **0.9 mA.** When R_L = 110 kΩ, the load current is only one-third of I_N or **0.6 mA.** That is,

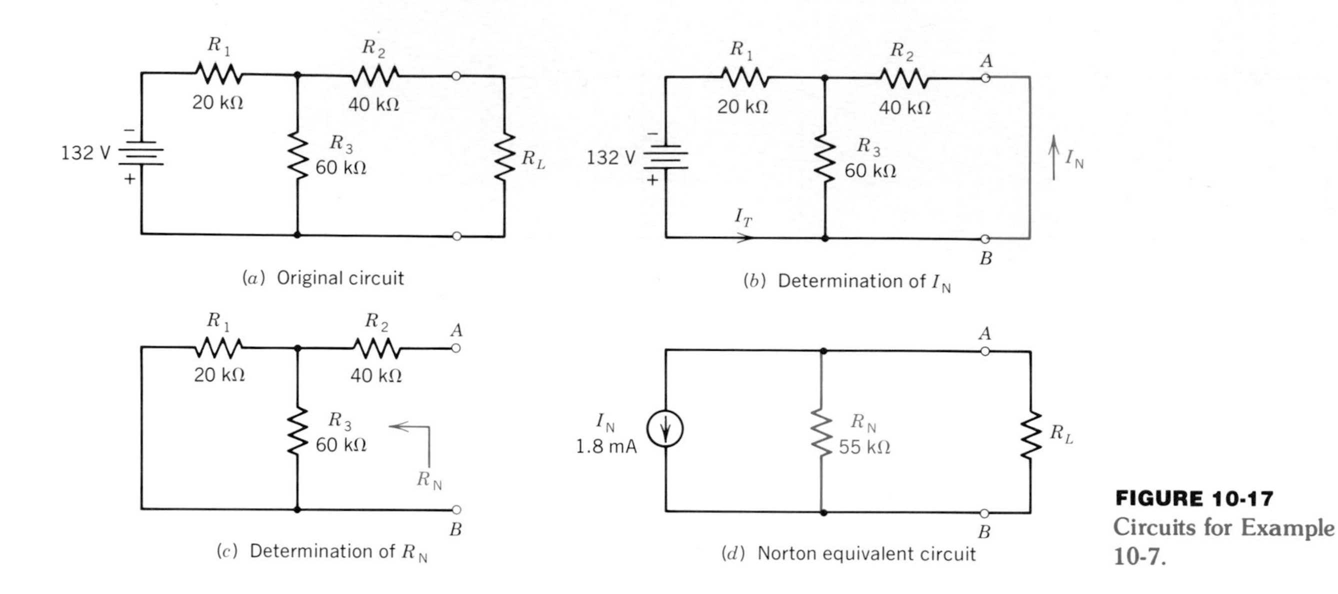

FIGURE 10-17 Circuits for Example 10-7.

$$I_L = I_N \times \frac{R_N}{R_N + R_L} \qquad (6\text{-}12)$$

$$= 1.8\text{ mA} \times \frac{55\text{ k}\Omega}{55\text{ k}\Omega + 110\text{ k}\Omega}$$

$$= 1.8\text{ mA} \times \frac{55\text{ k}\Omega}{165\text{ k}\Omega}$$

$$= 1.8\text{ mA} \times \frac{1}{3} = \mathbf{0.6\ mA}$$

b. For a load resistance of 55 kΩ,

$$V_L = I_L \times R_L \qquad (3\text{-}1b)$$

$$= 0.9\text{ mA} \times 55\text{ k}\Omega = \mathbf{49.5\ V}$$

with B positive with respect to A.

For $R_L = 110$ kΩ,

$$V_L = I_L \times R_L \qquad (3\text{-}1b)$$

$$= 0.6\text{ mA} \times 110\text{ k}\Omega = \mathbf{66\ V}$$

with B positive with respect to A.

As with Thévenin's theorem, note how readily the load voltage and current may be obtained for *any* load without repeating the whole problem. This is *not* the case if simple series-parallel theory had been used.

Now let us consider an example involving two voltage sources that cannot be solved using simple series-parallel theory. We shall discuss the example in two ways. The first is a *direct* application of Norton's theorem.

EXAMPLE 10-8A

Given the battery-generator charging circuit in Fig. 10-18*a*, determine the load current and voltage by obtaining a Norton equivalent for R_L directly.

SOLUTION

Removing R_L and shorting A to B as in Fig. 10-18*b*, we find that

$$I_N = I_B + I_G \qquad (6\text{-}1)$$

$$= \frac{V_B}{R_B} + \frac{V_G}{R_G}$$

$$= \frac{13.2\text{ V}}{0.3\ \Omega} + \frac{14.4\text{ V}}{0.2\ \Omega}$$

$$= 44\text{ A} + 72\text{ A} = \mathbf{116\ A}$$

Shorting-out the voltage sources, but maintaining their internal resistances, as in Fig. 10-18*c*, gives

$$R_N = R_B \| R_G$$

$$= \frac{R_B \times R_G}{R_B + R_G} \qquad (6\text{-}10)$$

$$= \frac{0.3\ \Omega \times 0.2\ \Omega}{0.3\ \Omega + 0.2\ \Omega} = \mathbf{0.12\ \Omega}$$

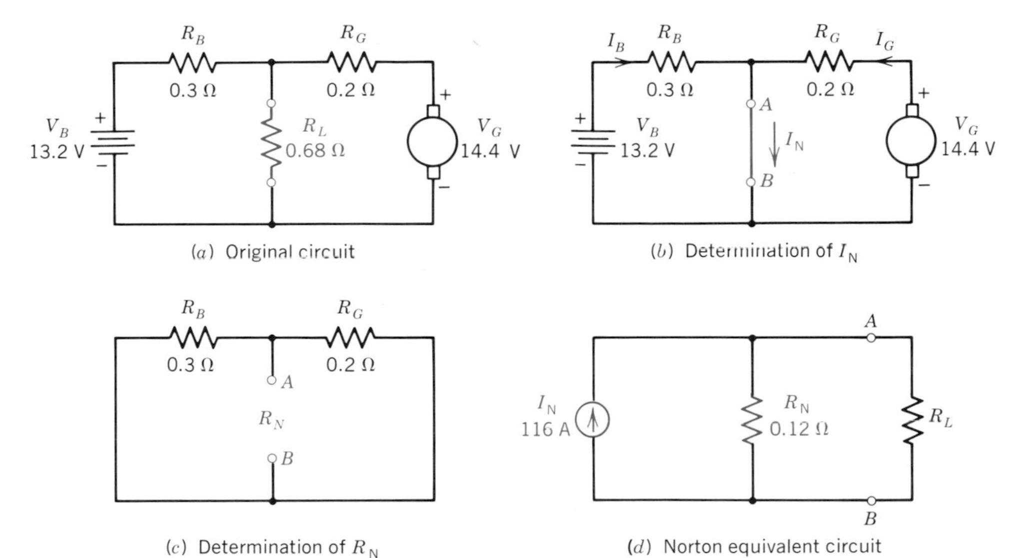

FIGURE 10-18
Circuits for Example 10-8A.

The Norton equivalent circuit is shown in Fig. 10-18*d*.

For $R_L = 0.68\ \Omega$,

$$I_L = I_N \times \frac{R_N}{R_N + R_L} \tag{6-12}$$

$$= 116\text{ A} \times \frac{0.12\ \Omega}{0.12\ \Omega + 0.68\ \Omega} = \mathbf{17.4\ A}$$

and $V_L = 17.4\text{ A} \times 0.68\ \Omega = \mathbf{11.83\ V}$

EXAMPLE 10-8B

Solve Example 10-8A by obtaining a Norton equivalent for each voltage source and combining the result.

SOLUTION

The Norton equivalent for V_B and R_B is shown in Fig. 10-19*b* where:

$$I_{NB} = \frac{V_B}{R_B} \tag{3-1a}$$

$$= \frac{13.2\text{ V}}{0.3\ \Omega} = 44\text{ A}$$

and $\quad R_{NB} = R_B = 0.3\ \Omega$

Similarly, $\quad I_{NG} = \frac{V_G}{R_G} \quad$ (3-1a)

$$= \frac{14.4\text{ V}}{0.2\ \Omega} = 72\text{ A}$$

and $\quad R_{NG} = R_G = 0.2\ \Omega$

Next, the current sources may be combined into one:

$$I_N = I_{NB} + I_{NG} \tag{6-1}$$

$$= 44\text{ A} + 72\text{ A} = \mathbf{116\ A}$$

and the parallel resistances may also be combined:

$$R_N = \frac{R_{NB} \times R_{NG}}{R_{NB} + R_{NG}} \tag{6-10}$$

$$= \frac{0.3\ \Omega \times 0.2\ \Omega}{0.3\ \Omega + 0.2\ \Omega} = \mathbf{0.12\ \Omega}$$

This yields the same Norton equivalent (see Fig. 10-19*c*) with I_L = **17.4 A** and V_L = **11.83 V.**

Although the same calculations are involved in either example, the concept of combining current sources is an important one. For instance, if in a network a current source has a direction opposite to the others, the combined current source is the *algebraic sum* of all the current sources. Also, the combined Norton resistance is the parallel equivalence of all the resistances.

10-6.1 Converting a Norton Equivalent to a Thévenin Equivalent

The method of solution in Example 10-8B illustrates a principle used in *Millman's theorem*. This theorem applies to a multi-emf circuit as in Fig. 10-19*a*, in which

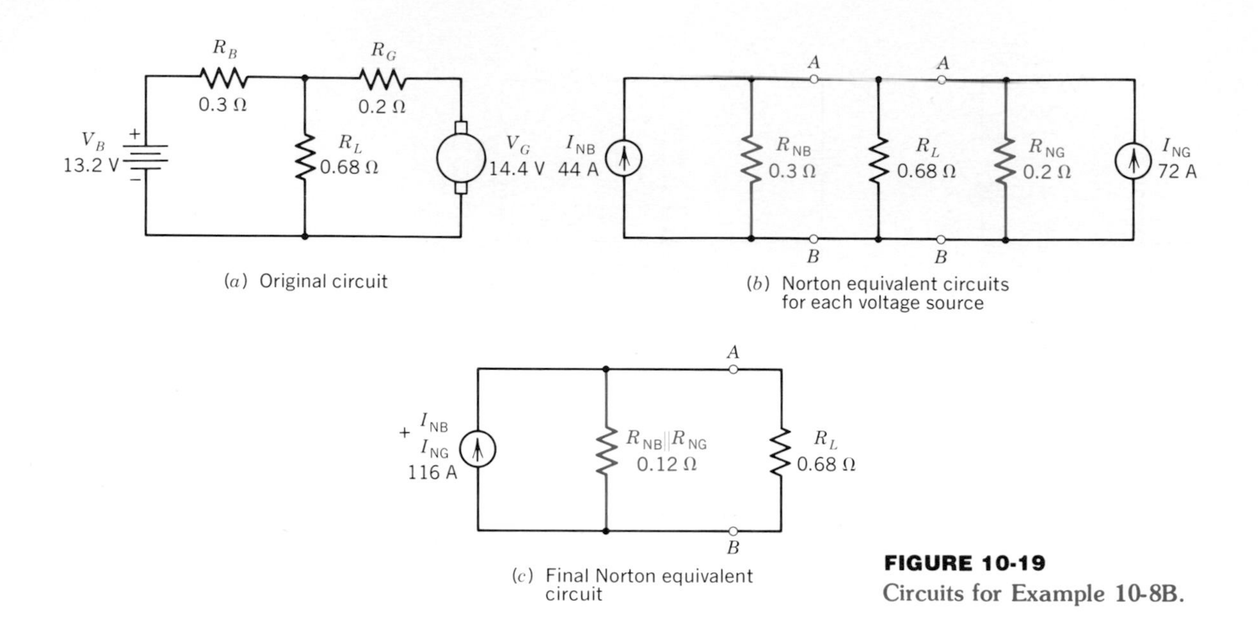

FIGURE 10-19
Circuits for Example 10-8B.

each source is parallel with the other. After finding the Norton equivalent for each and combining as in Fig. 10-19*c*, the last step in Millman's theorem is to convert the constant-current circuit to an equivalent Thévenin circuit.

This is done by removing any load and allowing the constant-current source, I_N, to develop a voltage across the parallel resistor, R_N. This is the Thévenin open-circuit voltage, $V_{Th} = I_N \times R_N$. Also, $R_{Th} = R_N$. See Fig. 10-20.

For example, consider the Norton equivalent in Fig. 10-19*c* with $I_N = 116$ A and $R_N = 0.12\ \Omega$. The Thévenin equivalent voltage is

$$V_{Th} = I_N \times R_N = 116 \text{ A} \times 0.12\ \Omega = \mathbf{13.92\ V}$$

and the Thévenin equivalent resistance is

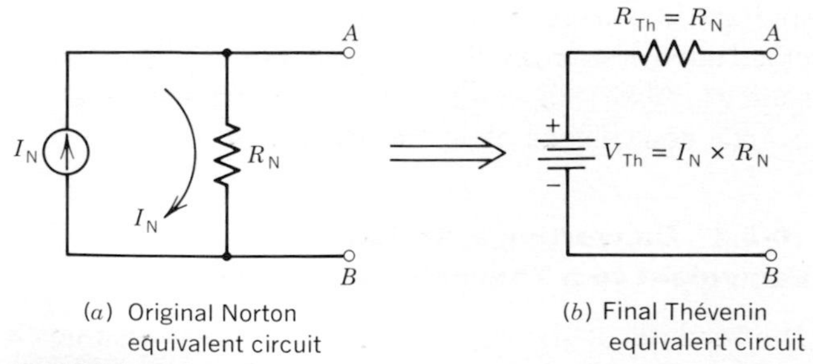

(a) Original Norton equivalent circuit

(b) Final Thévenin equivalent circuit

FIGURE 10-20
Converting a Norton equivalent to a Thévenin equivalent.

$$R_{Th} = R_N = \mathbf{0.12\ \Omega}$$

The load current in Fig. 10-19*c* would then be obtained as

$$I_L = \frac{V_{Th}}{R_{Th} + R_L} \tag{3-1a}$$

$$= \frac{13.92 \text{ V}}{0.12\ \Omega + 0.68\ \Omega} = \mathbf{17.4\ A}$$

as before.

It should be clear from our discussion that a Thévenin equivalent circuit may also be converted to a Norton equivalent, as in Fig. 10-21.

$$I_N = \frac{V_{Th}}{R_{Th}} \tag{3-1a}$$

and $R_N = R_{Th}$

This may raise the question as to which equivalent circuit to use in solving a given problem—Thévenin's or Norton's? This depends upon the particular circuit. For example, it is easier to find the Thévenin equivalent for the circuit in Fig. 10-17*a*. This is because, with the load removed, the open-circuit voltage (which is the Thévenin voltage, V_{Th}) may be obtained *directly* by the voltage-divider equation applied to R_3. On the other hand, if a

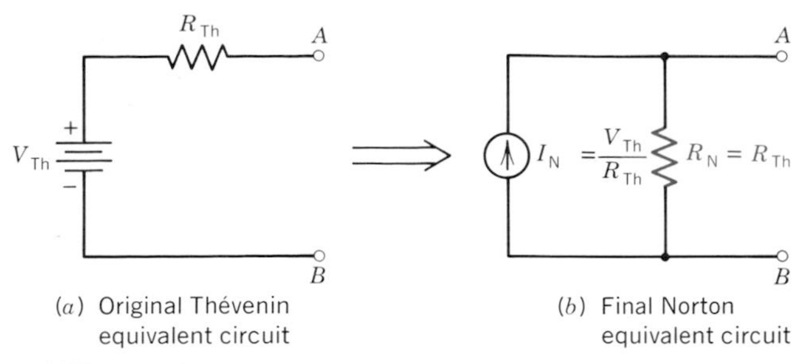

(a) Original Thévenin equivalent circuit

(b) Final Norton equivalent circuit

FIGURE 10-21
Converting a Thévenin equivalent to a Norton equivalent.

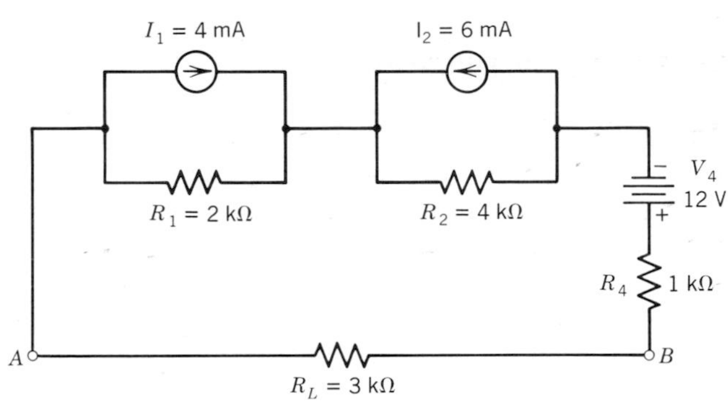

(a) Original circuit with series-current sources

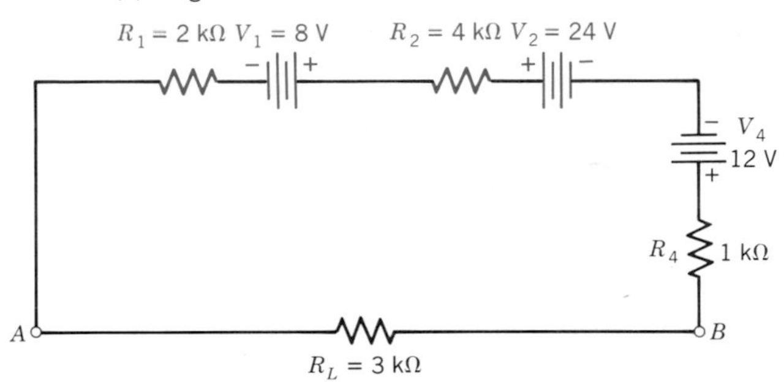

(b) Current sources converted to equivalent Thévenin voltage sources

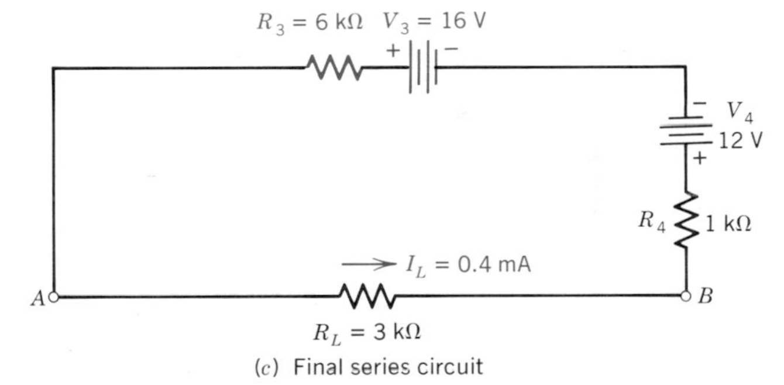

(c) Final series circuit

FIGURE 10-22
Circuits for Example 10-9.

circuit contains two or more parallel sources of emf (as in Fig. 10-18*a*), Norton's theorem may be an advantage. In some circuits, such as one equivalent circuit for a transistor, a Thévenin equivalent is used to represent the input and a Norton equivalent is used to represent the output.

10-6.2 Combining Current Sources in Series

It may be necessary to analyze a circuit in which two or more current sources appear in *series* with each other. In this case, each current source and its associated resistance is converted into its Thévenin equivalent. The voltage sources and resistances are then combined into a single voltage source. This in turn could be converted back to a single current source if this would be an advantage in solving the circuit. See Example 10-9.

EXAMPLE 10-9

Given the circuit in Fig. 10-22*a*, calculate the load current and its direction.

SOLUTION

The Thévenin equivalents for the two current sources are

$$V_1 = I_1 R_1 \quad (3\text{-}1b)$$
$$= 4 \text{ mA} \times 2 \text{ k}\Omega = \mathbf{8\ V}$$

in series with a **2-kΩ** resistance

$$V_2 = I_2 R_2 \quad (3\text{-}1b)$$
$$= 6 \text{ mA} \times 4 \text{ k}\Omega = \mathbf{24\ V}$$

in series with a **4-kΩ** resistance. These two voltage sources are series-connected, as in Fig. 10-22*b*. They have a total value of 16 V in series with 6 kΩ, as shown in Fig. 10-22*c*. Note the opposite polarities for V_1 and V_2 resulting from the opposite directions of the original current sources. The load current is given by

$$I_L = \frac{V_T}{R_T} = \frac{V_3 - V_4}{R_3 + R_4 + R_L} \quad (3\text{-}1a)$$
$$= \frac{16 \text{ V} - 12 \text{ V}}{6 \text{ k}\Omega + 1 \text{ k}\Omega + 3 \text{ k}\Omega}$$
$$= \frac{4 \text{ V}}{10 \text{ k}\Omega} = \mathbf{0.4\ mA}$$

This current flows from A to B because V_3 is larger than V_4.

10-7 DELTA-WYE CONVERSIONS

We have seen that special techniques are necessary to solve complex networks because, very often, the circuit components do not have a simple series-parallel relation. In some cases, a complex network may be reduced to a series-parallel combination by substituting an equivalent network. Figure 10-23 shows two networks, a "delta"—Δ and a "wye"—Y, the names originating in the shapes of the networks.

It is possible, using the appropriate equations, to convert a given delta network to an equivalent wye network. These equations are

$$\Delta \rightarrow \text{Y}$$

$$R_A = \frac{R_2 R_3}{R_1 + R_2 + R_3} \quad \text{ohms } (\Omega) \qquad (10\text{-}1)$$

$$R_B = \frac{R_3 R_1}{R_1 + R_2 + R_3} \quad \text{ohms } (\Omega) \qquad (10\text{-}2)$$

$$R_C = \frac{R_1 R_2}{R_1 + R_2 + R_3} \quad \text{ohms } (\Omega) \qquad (10\text{-}3)$$

where R_A, R_B, and R_C are elements in the Y; and R_1, R_2, and R_3 are in the Δ.

Note how the equation for each resistor in the wye has a numerator equal to the product of the two adjacent resistors in the delta.

Similarly, a wye may be converted to a delta using the following:

$$\text{Y} \rightarrow \Delta$$

$$R_1 = \frac{R_A R_B + R_B R_C + R_C R_A}{R_A} \qquad (10\text{-}4)$$

$$R_2 = \frac{R_A R_B + R_B R_C + R_C R_A}{R_B} \qquad (10\text{-}5)$$

$$R_3 = \frac{R_A R_B + R_B R_C + R_C R_A}{R_C} \qquad (10\text{-}6)$$

where R_1, R_2, and R_3 are elements in the Δ; and R_A, R_B, and R_C are in the Y.

Note how the equation for each resistor in the delta, has a denominator equal to the opposite resistor in the wye.

Let us use a Δ → Y conversion to solve an earlier problem, involving a Wheatstone bridge.

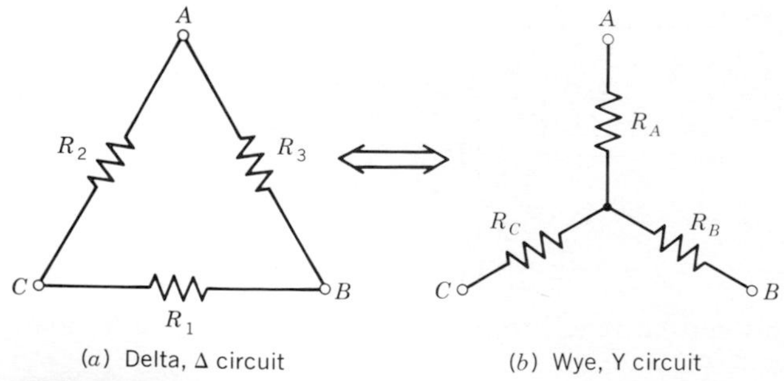

(a) Delta, Δ circuit (b) Wye, Y circuit

FIGURE 10-23
Delta-wye and wye-delta conversions.

EXAMPLE 10-10

Determine the total current supplied by the 12-V source in Fig. 10-24*a* by calculating the total resistance of the network between *a* and *d*.

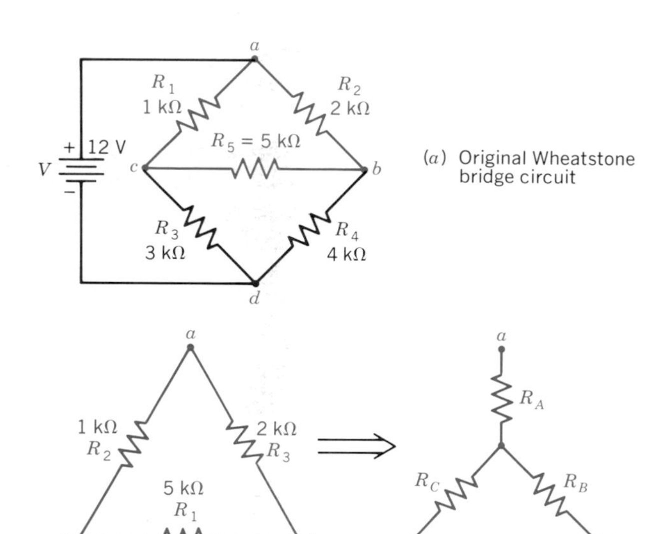

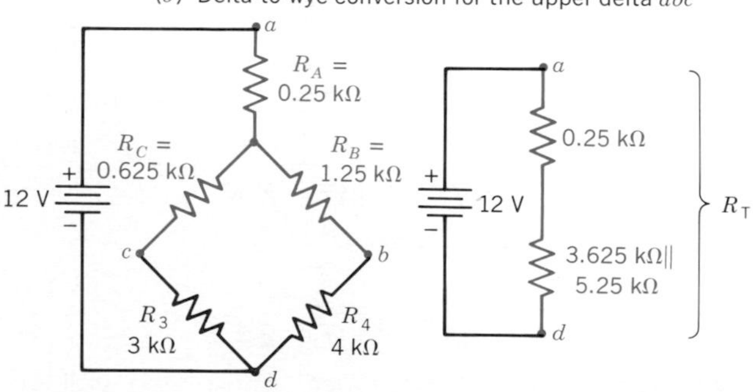

FIGURE 10-24
Circuits for Example 10-10.

SOLUTION

There are four possible conversions in the circuit of Fig. 10-24*a*. We can identify an upper delta—*a, b, c*; a lower delta—*c, b, d*; a left wye—*ca, cb, cd*; and a right wye—*ba, bc, bd*. Let us convert the upper delta into an equivalent wye as in Fig. 10-24*b*. It is suggested that for reference purposes, the delta resistors be renumbered to conform to the Δ-Y conversion equations.

$$R_A = \frac{R_2 R_3}{R_1 + R_2 + R_3} \tag{10-1}$$

$$= \frac{1\ \text{k}\Omega \times 2\ \text{k}\Omega}{5\ \text{k}\Omega + 1\ \text{k}\Omega + 2\ \text{k}\Omega} = 0.25\ \text{k}\Omega$$

$$R_B = \frac{R_3 R_1}{R_1 + R_2 + R_3} \tag{10-2}$$

$$= \frac{2\ \text{k}\Omega \times 5\ \text{k}\Omega}{5\ \text{k}\Omega + 1\ \text{k}\Omega + 2\ \text{k}\Omega} = 1.25\ \text{k}\Omega$$

$$R_C = \frac{R_1 R_2}{R_1 + R_2 + R_3} \tag{10-3}$$

$$= \frac{5\ \text{k}\Omega \times 1\ \text{k}\Omega}{5\ \text{k}\Omega + 1\ \text{k}\Omega + 2\ \text{k}\Omega} = 0.625\ \text{k}\Omega$$

We now see that a series-parallel circuit is formed, as in Fig. 10-24*c*, with a total resistance given by

$$R_T = R_A + (R_C + R_3) \parallel (R_B + R_4)$$

$$= 0.25\ \text{k}\Omega + \frac{3.625\ \text{k}\Omega \times 5.25\ \text{k}\Omega}{3.625\ \text{k}\Omega + 5.25\ \text{k}\Omega}$$

$$= 0.25\ \text{k}\Omega + 2.14\ \text{k}\Omega = \mathbf{2.39\ k\Omega}$$

$$I_T = \frac{V}{R_T} \tag{3-1a}$$

$$= \frac{12\ \text{V}}{2.39\ \text{k}\Omega} = \mathbf{5.02\ mA}$$

These answers and the method of solution should be compared with Example 10-3.

As noted, a Y-Δ conversion could have been used also to reduce the bridge circuit to a series-parallel combination. In this case, the amount of work required would be increased because of more parallel combinations. However, the particular conversion to be used depends upon the circuit to be solved. In some cases, the work may be reduced if it is possible to choose a conversion in which all three resistance values are identical. That is, the "arms" of the equivalent network will all be the same, requiring only one calculation.

Finally, it is interesting to note that delta-wye networks are also referred to as π-T networks, respectively. See Fig. 10-25.

A delta network can be redrawn in the shape of a π (pi) and a wye network can be redrawn in the shape of a T. The π-T designations are common in electronics work, since an "input" and "output" pair of terminals can be identified, as in a filter circuit, for example. The Δ-Y designations are common in electrical machinery, such as ac motors, generators, and transformers, where a three-phase, three-wire (and four-wire) supply is used. (This is discussed further in Chapter 14.)

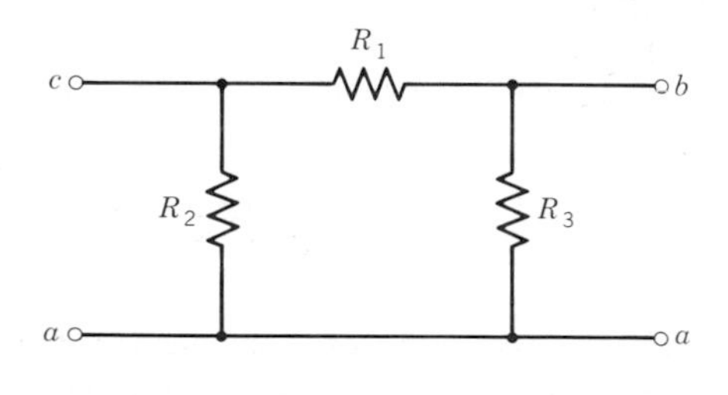

(*a*) A Δ or π network

(*b*) A Y or T network

FIGURE 10-25
Comparison of Δ -π and Y-T networks.

SUMMARY

1. The branch current method of analyzing a circuit involves assigning branch current designations, showing polarity signs of voltage drops across resistors, choosing arbitrary tracing loops, and writing as many independent Kirchhoff's voltage law equations as there are windows in the circuit.

2. After simultaneous solution of the Kirchhoff's voltage law equations, any negative currents must be interpreted as having a direction opposite to that assumed.

3. The loop current method involves indicating a current loop in each window, showing the voltage drops for each loop, and writing a Kirchhoff's voltage law equation for each loop using the algebraic sum of loop currents through each resistor to determine the voltage drops.

4. The superposition theorem applies to a multi-emf circuit and states that the total circuit current is the algebraic sum of the currents resulting from each emf acting independently.

5. The superposition method usually results in a series-parallel combination of components, thereby avoiding the need for writing a set of Kirchhoff's voltage law equations.

6. Thévenin's theorem shows how a complex linear network with at least one emf may be represented by a single source of voltage in series with a single resistor.

7. The Thévenin equivalent voltage is the open-circuit voltage with any load (that may later be connected) removed.

8. The Thévenin equivalent resistance is the resistance looking back into the circuit with all sources removed and replaced by their internal resistance.

9. The Thévenin equivalent of a circuit is very useful for making repetitive calculations for various loads to be connected to the original circuit.

10. Norton's theorem shows how a complex linear network with at least one emf may be represented by a single constant-current source in parallel with a single resistance.

11. The magnitude of the current source is the short-circuit current that flows in the original circuit with the load (which may later be connected) removed.

12. The Norton equivalent resistance is the resistance looking back into the circuit with all sources removed and replaced by their internal resistance. It has the same value as the Thévenin resistance.

13. A Thévenin equivalent circuit can be converted to a Norton equivalent circuit, and vice versa. Each behaves the same as the original circuit as far as any externally connected load is concerned.

14. Millman's theorem applies to a circuit in which two or more emfs, and their internal resistances, are parallel-connected. Each is converted into a Norton equivalent; the respective current sources and resistances are combined; and then a final conversion to a Thévenin equivalent is made.

15. When two or more current sources are series-connected, each is converted into an equivalent Thévenin voltage, which may then be combined into a single voltage source.

16. Three components may be connected in a delta or wye configuration. Using the appropriate set of equations, a delta (or π) may be converted to a wye (or T), and vice versa. These conversions may form some series-parallel combination to simplify the solution of a complex network.

SELF-EXAMINATION

Answer true or false
(Answers at back of book)

10-1. The branch current method of analyzing a circuit involves the use of *both* Kirchhoff's current and voltage laws. ________

10-2. The number of "windows" that a network has determines how many independent Kirchhoff's voltage law equations must be written. ________

10-3. Using the branch current method, it is possible to have more than one set of voltage-drop polarity signs across a given resistor. ________

10-4. The direction of loops shown for tracing currents must always be drawn to make an emf appear as a voltage rise. ________

10-5. The interpretation of a negative current is simply to reverse the direction from the assumed direction. ________

10-6. Whether using the branch current or loop current method, there must *always* be as many independent equations as there are unknown currents. ________

10-7. The loop current method does not involve Kirchhoff's current law in order to write Kirchhoff's voltage law loop equations. ________

10-8. In the loop current method a loop with no emf will always have the voltage-rise side of the equation equal to zero. ________

10-9. The branch current and loop current methods are generally used only when the circuit is not some combination of series and parallel components. ________

10-10. The superposition theorem only applies to a multi-emf network. ________

10-11. There are no restrictions on the type of circuit to which the superposition theorem can be applied. ________

10-12. The advantage of the superposition theorem is that the removal of all emfs but one usually leaves a simple series-parallel circuit. ________

10-13. Thévenin's theorem only applies to nonseries-parallel networks. ________

10-14. The Thévenin equivalent voltage is the open-circuit voltage of the initial circuit, and the Thévenin equivalent resistance is the resistance looking back into the circuit with all voltage sources replaced by their internal resistances. ________

10-15. A Thévenin equivalent circuit acts exactly as the initial circuit, as far as the load is concerned. ________

10-16. The Thévenin equivalent for any given circuit is the same no matter which component in the initial circuit is being removed to find the equivalent. ________

10-17. The resistance in a Norton equivalent is always the same as the resistance in a corresponding Thévenin equivalent circuit. ________

10-18. The current source in a Norton equivalent is always the short circuit current in the initial circuit. ________

10-19. The voltage across a constant current source varies depending upon the load connected to it. ________

10-20. It is impossible to have two current sources of different magnitude in series with each other. ________

10-21. A delta-connected network is found only in three-phase circuits. ________

10-22. A Δ-Y conversion (or vice versa) is made if the result is a simple series-parallel combination with other components in the original circuit. ________

REVIEW QUESTIONS

1. Explain why the direction of the loops in either the branch current method or the loop current method is arbitrary.

2. In the loop current method, what is the rule for writing the Kirchhoff's voltage law equation if a resistor has more than one loop of current through it?

3. What circuit conditions are required to apply the superposition theorem?

4. How would you experimentally apply the superposition theorem?

5. Given a circuit containing a number of emfs (with negligible internal resistance) and resistors, describe how you would experimentally determine the Thévenin equivalent circuit.

6. What is one advantage of solving the current in a component (which takes on different values) using Thévenin's theorem compared with the branch current, loop current, or superposition methods?

7. a. In what way does a Norton equivalent differ from a Thévenin equivalent circuit?

b. If both were contained in two separate black boxes, how may it be possible to determine which one contains the Norton equivalent?

8. Describe how you would experimentally determine the Norton equivalent of a circuit.

9. Draw a network (other than a Wheatstone bridge) in which a Δ-Y conversion would permit a series-parallel solution for total resistance.

PROBLEMS

(Answers to odd-numbered problems at back of book)

10-1. Use the branch current method and Fig. 10-1 with $V_B = 12.6$ V, $V_G = 14.1$ V, $R_B = 0.05\ \Omega$, $R_G = 0.1\ \Omega$, and $R_L = 1\Omega$ to solve for:

a. Battery current. c. Load current.

b. Generator current. d. Load voltage.

10-2. Repeat Problem 10-1 using the loop current method.

10-3. Assume that the battery in Problem 10-1 was installed with reverse polarity. Recalculate the currents and load voltage using either the branch or loop current method.

10-4. Given the circuit and values as in Problem 10-1, to what would the generator open-circuit voltage have to change in order to make the battery current zero?

10-5. Given Fig. 10-10 with values as shown except that $V_2 = 15$ V, find the current in each resistor using the loop current method.

10-6. Repeat Problem 10-5 using the branch current method.

10.7. Given Fig. 10-3*a* with $R_1 = 5.1$ kΩ, $R_2 = 4.7$ kΩ, $R_3 = 3.3$ kΩ, $R_4 = 1$ kΩ, $R_5 = 2.2$ kΩ, and $V = 15$ V, use the branch current method to calculate:

a. The current in each resistor.

b. The voltage from *c* to *d*.

c. The total resistance between *a* and *b*.

10-8. Repeat Problem 10-7 using the loop current method.

10-9. Given the circuit and values in Problem 10-7, determine:

a. The value of resistance that R_4 must be changed to so that no current flows between *c* and *d*.

b. The total resistance between *a* and *b* for this condition.

10-10. Repeat Problem 10-1 using the superposition theorem.

10-11. Given Fig. 10-10 with values as shown, except that $V_2 = 15$ V, find the current in each resistor using the superposition theorem.

10-12. Given Fig. 10-26, find the current through $R_L = 1$ kΩ using the superposition theorem.

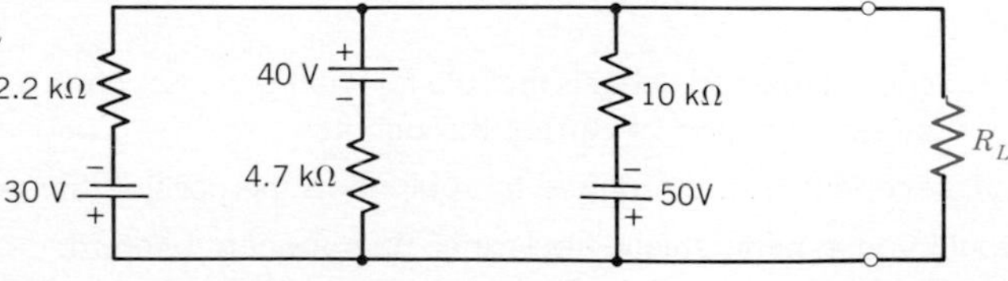

FIGURE 10-26
Circuit for Problems 10-12, 10-20, 10-23, and 10-24.

10-13. Given Problem 10-1, find the Thévenin equivalent circuit for R_L and thus determine all the currents and load voltage.

10-14. Given Problem 10-3, find the Thévenin equivalent circuit for R_L and thus determine all the currents and load voltage.

10-15. Given Fig. 10-2 with $V_1 = 15$ V, $V_2 = 12$ V, $R_1 = 5.1$ kΩ, $R_2 = 4.7$ kΩ, $R_3 = 2.2$ kΩ, $R_4 = 4.7$ kΩ, and $R_5 = 1$ kΩ, find the current through R_3 using Thévenin's theorem.

10-16. Given Fig. 10-10 with values as shown except that $V_2 = 15$ V, use Thévenin's theorem to find the current through R_3.

10-17. Given Fig. 10-3*a* with $V = 15$ V, $R_1 = 5.1$ kΩ, $R_2 = 4.7$ kΩ, $R_3 = 3.3$ kΩ, $R_4 = 1$ kΩ, and $R_5 = 2.2$ kΩ, use a Thévenin equivalent circuit to find the current through R_5.

10-18. Given Fig. 10-10 and the values as shown, use a Thévenin equivalent circuit to find the current through R_4.

10-19. Given Fig. 10-10 and the values as shown, use a Thévenin equivalent circuit to find the current through R_1.

10-20. Given the circuit of Fig. 10-26, use a Thévenin equivalent circuit to find the current through $R_L = 1$ kΩ.

10-21. Given the circuit of Fig. 10-17*a* with $R_1 = 15$ kΩ, $R_2 = 56$ kΩ, $R_3 = 33$ kΩ, and $V = 40$ V, use Norton's theorem to find the current through R_L when

a. $R_L = 82$ kΩ

b. $R_L = 47$ kΩ

10-22. Repeat Problem 10-1 using a Norton equivalent for R_L.

10-23. Given the circuit of Fig. 10-26, use a Norton equivalent circuit to find the current through $R_L = 1$ kΩ.

10-24. Repeat Problem 10-23 using Millman's theorem.

10-25. Given the circuit of Fig. 10-22*a* with $R_1 = 6$ kΩ, and $R_2 = 1$ kΩ, find the current through $R_L = 1$ kΩ.

10-26. Given the delta circuit in Fig. 10-23*a*, with $R_1 = 1$ kΩ, $R_2 = 2.2$ kΩ, and $R_3 = 3.3$ kΩ, determine the equivalent wye circuit in Fig. 10-23*b*.

10-27. Given the wye circuit in Fig. 10-23*b*, with $R_A = R_B = R_C = 10$ kΩ, determine the equivalent delta circuit in Fig. 10-23*a*.

10-28. Given the circuit in Fig. 10-24*a*, find the current in R_2 by converting the lower delta—*cbd*—into an equivalent wye.

10-29. Given the circuit in Fig. 10-24*a*, with $R_1 = 1.5$ kΩ, find the total resistance between *a* and *d* by converting the wye on the right—*ab, cb, db*—into an equivalent delta.

10-30. Given the bridged-T attenuator circuit in Fig. 10-27, find the current through the load and the voltage across the load using a Δ-Y conversion.

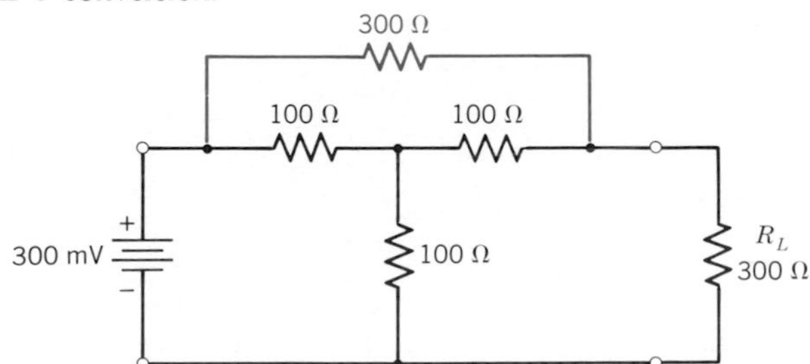

FIGURE 10-27
Circuit for Problem 10-30.

10-31. Solve Problem 10-30 using Thévenin's theorem.

10-32. Solve Problem 10-30 using Norton's theorem.

CHAPTER ELEVEN

MAGNETISM AND ITS APPLICATIONS

Magnetism plays a very important part in the operation of electrical measuring instruments, motors, generators, and so on. Magnetism was used for navigation purposes by the Chinese over 1800 years ago. They found that a natural magnet, called a lodestone, tends to align itself with the earth's magnetic field along a north–south direction. The word *magnet* seems to have been derived from the name of the ancient Greek city Magnesia, where fragments of the iron ore magnetite displayed this peculiar property.

In 1819, Oersted discovered that magnetism could also be produced by an electrical current. This opened the way to "temporary" magnets or *electromagnets.* This principle is responsible for the operation of magnetic relays, solenoid valves, motors, meter movements, and even magnetic-core memory storage in digital computers.

The magnetic properties of a material are revealed by its *hysteresis loop,* a graphical presentation of *flux density* plotted against the *magnetizing force or intensity.* The shape and area of the loop determine the application of the material in dc and ac *magnetic circuits.* Consequently, we consider the system of units used for measurement of magnetic quantities and their relation in *Ohm's law for magnetic circuits.*

11-1 PERMANENT MAGNETS

It is commonly observed that a permanent magnet will attract certain metallic materials. At room temperature, the only common *elements* that are attracted to a magnet are *iron, nickel,* and *cobalt.* These metals are called *magnetic* or ferromagnetic materials. All other elements are termed *nonmagnetic.**

The locations of a bar magnet that have the strongest attraction for a magnetic material are called the *poles* of the magnet. As shown in Fig. 11-1 these are called the *north pole* and *south pole,* after the earth's magnetic poles.

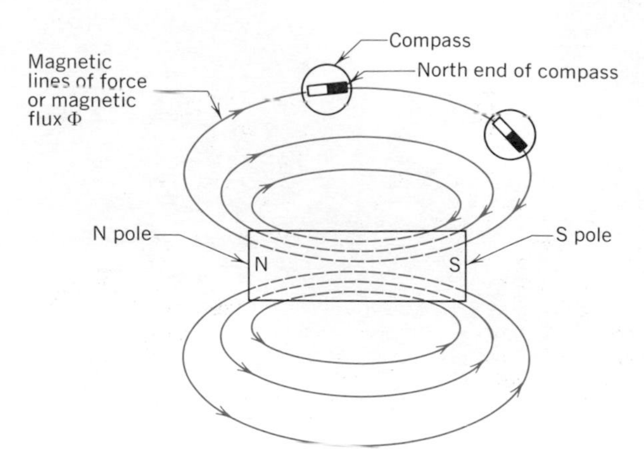

(*a*) Lines of force leave a north pole and enter a south pole

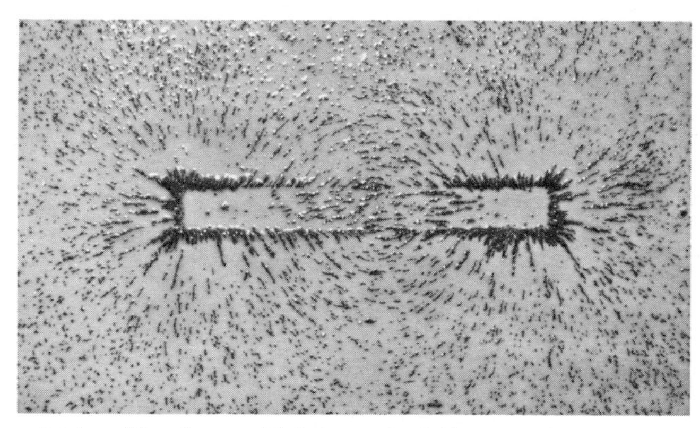

(*b*) Iron filings line up with the magnetic field

FIGURE 11-1
Magnetic field of a permanent magnet.

11-1.1 Magnetic Field

The region surrounding a magnet is called a *magnetic field.* It is useful to *represent* the direction and strength of the field by means of magnetic *lines of force,* or magnetic *flux,* as shown in Fig. 11-1. Although space is shown *between* the lines, this does not mean that there is no magnetic field there. The *spacing* of the lines is used to show the relative strength or magnetic flux *density.* Thus, where the lines of force are crowded together, as at the poles, the magnetic field and its effect are the strongest.

The magnetic lines of force also have *direction,* which is indicated by arrows. They are shown leaving a north pole and entering a south pole. The magnetic field and the lines are *continuous,* completing a closed loop or path from south to north *inside* the magnet.

The *direction* of the magnetic field is simply the direction in which the north end of a compass points when placed in a magnetic field. It should be noted that the lines and arrows do *not* indicate the *flow* of any electrical charge either inside or outside of the magnet. In fact, a magnetic field exists around a permanent magnet if it is placed in a perfect vacuum, as in deep space, where there is no matter to conduct electricity.

11-1.2 Attraction and Repulsion

The attraction and repulsion of two magnets for each other are shown in Fig. 11-2. Since lines of force cannot intersect, the only way we can draw the magnetic fields of two adjacent like poles (both north or both south) is shown in Fig. 11-2*a*. The parallel paths of the lines of force of the two adjacent poles represent a repelling effect. In Fig. 11-2*b*, the lines of force readily pass from the north of one magnet to the nearby south of the other, representing an attracting effect. Thus we have the important rule of magnetism:

Like poles repel; unlike poles attract.

The attraction of an unmagnetized magnetic material by a magnet is shown in Fig. 11-3. Here we see that the field (as shown by lines of force) is distorted to pass through an iron nail, for example. The iron, being a magnetic material, is said to be more *permeable* than air.

* Some materials, such as platinum and aluminum, are very *slightly* magnetic, and are called *paramagnetic* materials. Other elements, such as silver and copper, offer a slight *opposition* to magnetic lines of force compared with free space (air), and are called *diamagnetic* materials. In both cases, however, these effects are so small (approximately 0.001% different) that, for practical purposes, we can consider all elements except iron, nickel, and cobalt to be nonmagnetic.

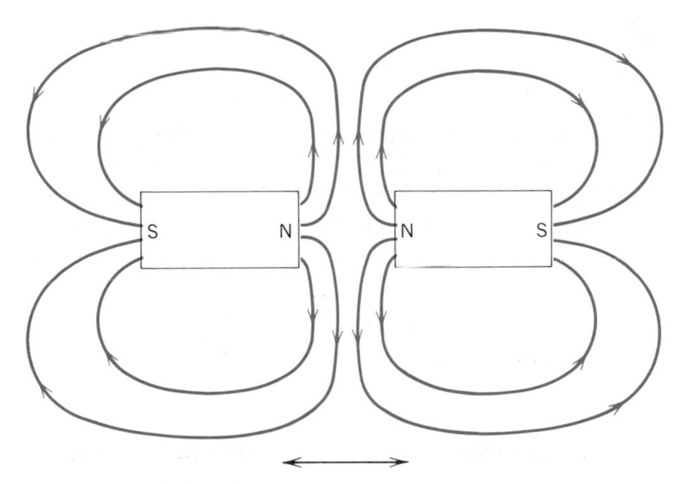

(*a*) Repulsion of like poles

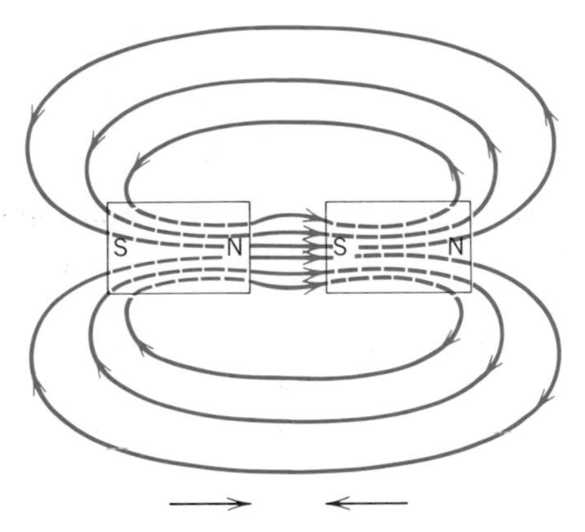

(*b*) Attraction of unlike poles

FIGURE 11-2
Using magnetic lines of force to show why like poles repel and unlike poles attract.

The lines of force would rather pass through iron than through air.

The effect shown in Fig. 11-3 creates a *temporary* magnet by magnetic *induction*. That is, where the field (lines of force) *enters* the nail, the effect is that of a south pole; where the field leaves, a north pole has been induced. Consequently, the nail is attracted toward the magnet. It does not matter which end of the magnet we use; either end will attract an unmagnetized magnetic material.

If left attached to the magnet for some time, the nail will retain some magnetism, even when removed from the magnet. This is one method of magnetizing a material,

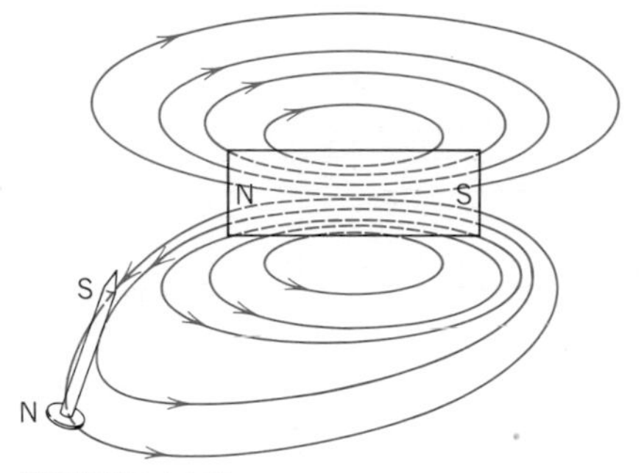

FIGURE 11-3
Attraction of an iron nail by a permanent magnet.

that is, by *induction*. This property of a material, to retain magnetism, is called *magnetic retentivity,* and the magnetism retained is called *residual* magnetism.

11-2 OERSTED'S DISCOVERY

Oersted discovered that a current flowing through a wire had an effect upon a magnetic compass close to the wire. If the region around the wire is explored with a compass, the magnetic field is found to be perfectly circular (concentric) around the wire, as shown in Fig. 11-4. For example, when iron filings are sprinkled on the cardboard through which the wire passes, the iron filings align themselves in the direction of the field and form circular patterns. The magnetic field is strongest at the center of the wire and decreases as the distance from the wire is increased. If the current is increased, the field strength is also increased.

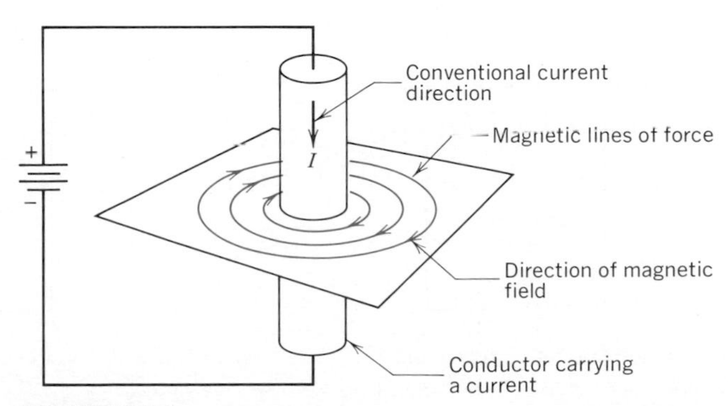

FIGURE 11-4
Magnetic field set up by a current-carrying conductor.

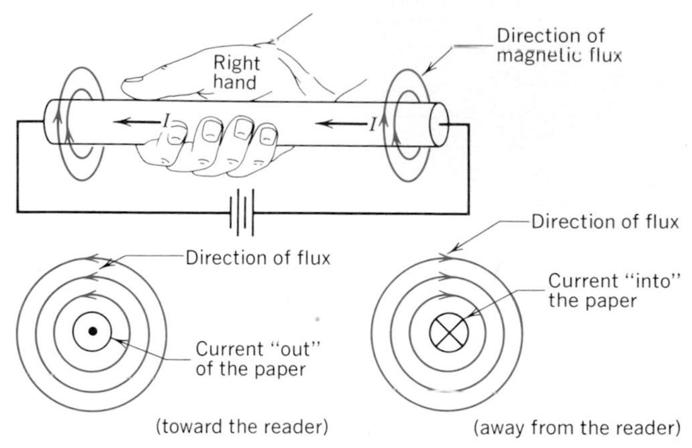

FIGURE 11-5
Right-hand rule to determine the direction of the magnetic lines of force around a straight current-carrying conductor.

The magnetic field has a direction given by the *right-hand rule,* shown in Fig. 11-5. The rule is

> **Grasp the conductor with the thumb of the right hand pointing in the direction of *conventional* current. The fingers will point in the direction of the magnetic field around the conductor.**

The direction of the current with respect to the field around the conductor is sometimes shown using the designation of a dot and a cross, as in Fig. 11-5. The dot represents the head of the current arrow approaching the reader; the cross represents the tail of the current arrow going away from the reader. Applying the right-hand rule gives a counterclockwise or clockwise direction of magnetic field, respectively.

11-3 THE ELECTROMAGNET

If a conductor is wrapped around a hollow cardboard tube (or other nonmagnetic core), the result is a coil (or helix). When current passes through the coil, there is an overall magnetic effect like that of a bar magnet. See Fig. 11-6*a*. The current-carrying coil acts like an *electromagnet.* The strength of the electromagnet can be increased three ways: by inserting a soft iron core into the coil, by using more turns on the coil, and also by increasing the current.

The direction of the field around an electromagnet can also be determined using the *right-hand rule for coils,* as shown in Fig. 11-6*b*.

> **Grasp the coil with the fingers of the right hand pointing in the direction of conventional current. The thumb will point in the direction of the magnetic field through the electromagnet and to the coil's north (N) pole.**

The important point about any electromagnet is that it is a *temporary magnet.* Simply by opening a switch and interrupting the current, we can turn off the magnetic field. It is this property that makes possible many control functions, such as are found in solenoids, relays, bells, buzzers, and door chimes. All rely upon this principle for their operation.

11-3.1 Solenoids

The word *solenoid* is usually applied to those devices that have a *movable* iron core acting *against* an internal spring. Figure 11-7*a* shows a solenoid control mechanism

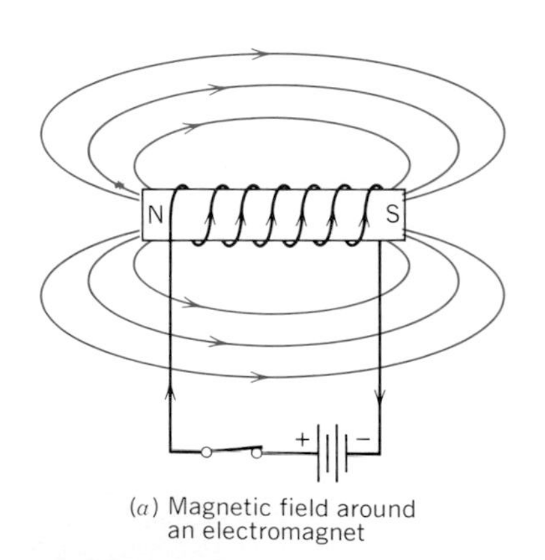

(*a*) Magnetic field around an electromagnet

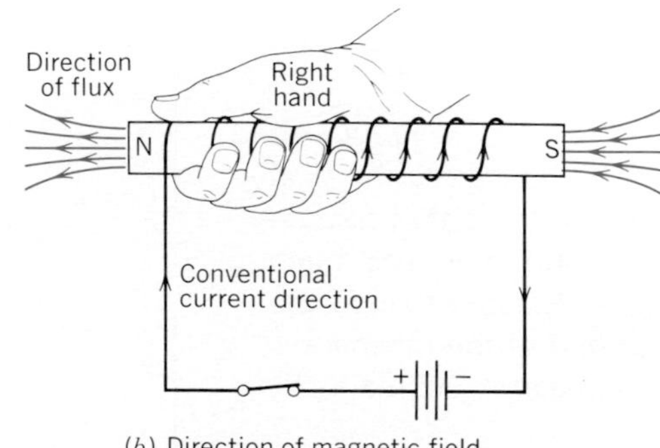

(*b*) Direction of magnetic field given by right-hand rule for coils

FIGURE 11-6
Magnetic effect of an electromagnet.

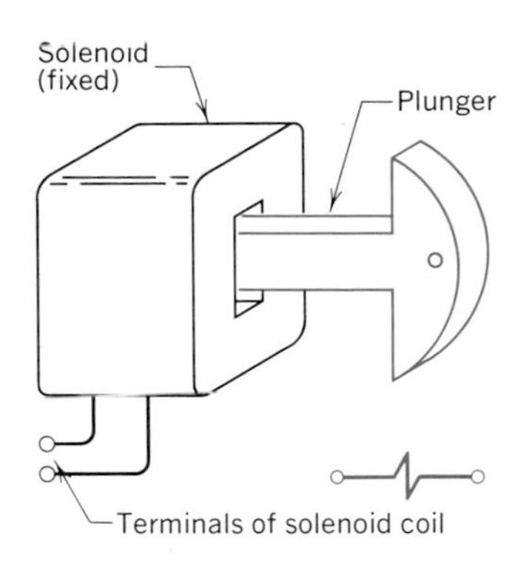

(*a*) Solenoid control mechanism and symbol

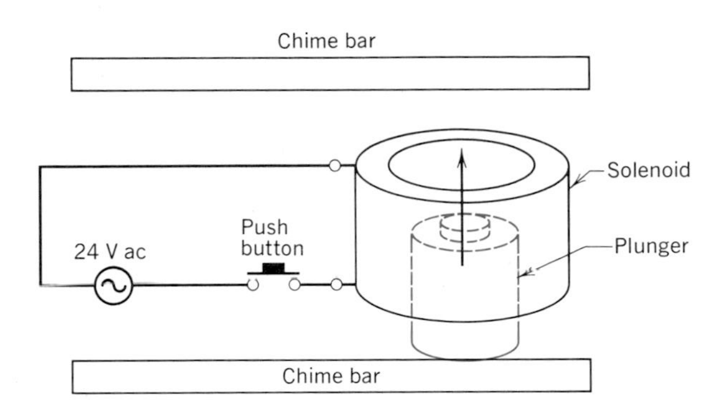

(*b*) Operation of a door chime

FIGURE 11-7
Applications of solenoids.

such as those used in water valves. Solenoids can be operated from ac or dc. A spring acts on the plunger holding it in the open position. Current through the solenoid draws the plunger into the solenoid, actuating the opening or closing of a valve.

11-3.2 Door Chimes

A door chime operates on a similar principle except that it is gravity that makes the plunger rest on one (lower) door chime. See Fig. 11-7*b*. When the button is pushed, the solenoid draws in the plunger, making a sound as it hits the upper chime. When the button is released, another sound is made as it falls back to strike the lower chime bar.

11-3.3 Buzzers and Bells

The operation of a door chime should not be confused with that of a door bell, buzzer, or horn. In these, a *continuous* vibrating effect is obtained by the interruption of current through an electromagnet as an armature is drawn toward the electromagnet. As shown in Fig. 11-8, the armature carries with it a clapper to hit a gong. This motion breaks an electrical contact so that the electromagnet becomes de-energized. A spring pulls the armature back from the face of the electromagnet. But this once again completes the electric circuit for current through the electromagnet, and the cycle is repeated.

11-3.4 Application of Solenoid Valves in a Home Heating System

An interesting application of solenoid valves is found in many domestic gas-fired, forced-air heating systems, as shown in Fig. 11-9. Two solenoid valves (located in the same mechanical unit) are series-connected in the gas line feeding the furnace. A thermocouple controls the safety valve, and the room thermostat controls the main gas valve. A thermocouple consists of two different

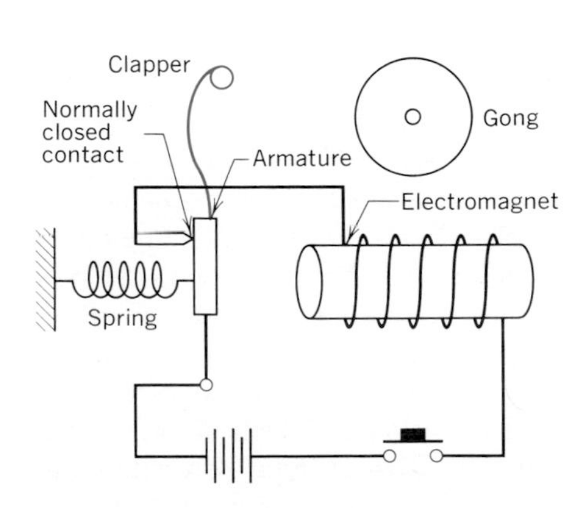

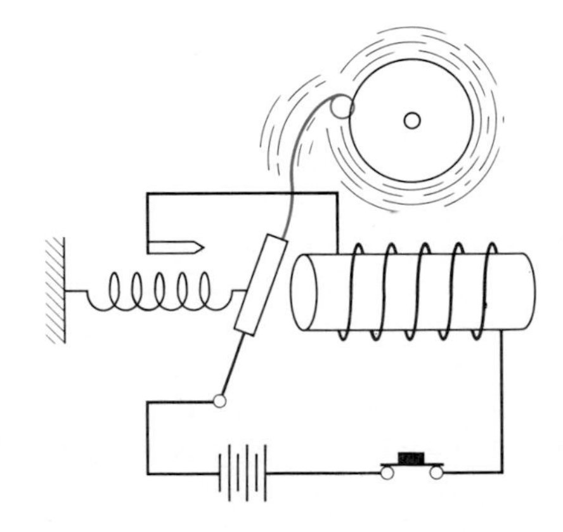

FIGURE 11-8
Operation of a doorbell.

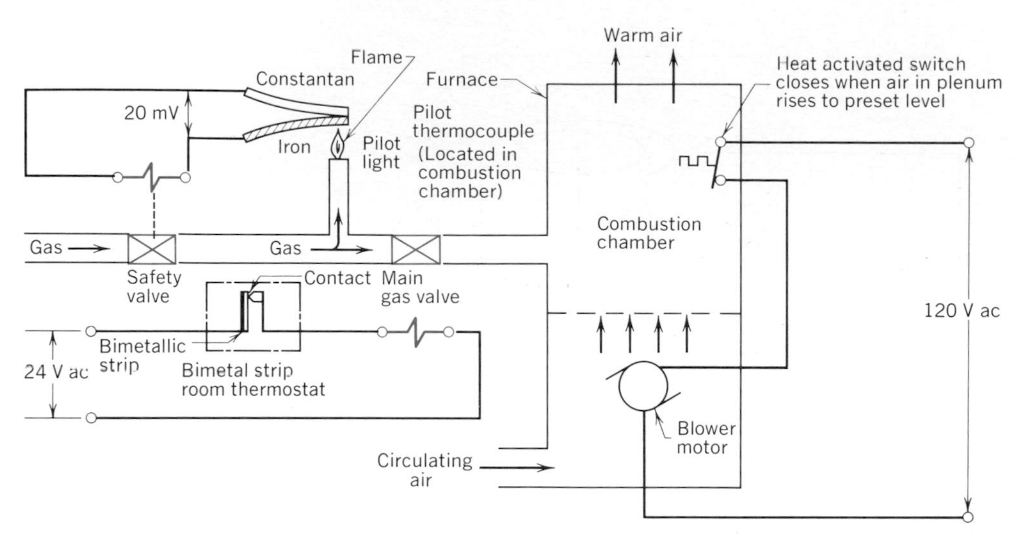

FIGURE 11-9
Application of solenoid valves in a domestic forced-air heating unit.

metals (such as iron and constantan) welded together at one end, and open at the other. When the closed end is located in a pilot light, an emf of approximately 20 mV is generated at the open end. This voltage is used to hold open the safety valve. If the pilot light should be extinguished for any reason, this valve closes and will not admit any gas to the system.

The room thermostat is a bimetallic switch that operates at room temperature. It consists of two different metals, such as iron and brass, bonded together, and adjacent to a stationary contact. Since the brass expands more than the iron as the temperature increases, the bimetal strip bends and opens when the desired temperature of the room is reached. If we assume that the thermostat is closed, calling for heat, we can see from Fig. 11-9 that this applies 24 V ac to the solenoid coil of the main gas valve, opening the valve. If the pilot light is on, the safety valve will be open, and gas will be admitted to the combustion chamber where it is ignited by the pilot flame.

After a few minutes of operation, the temperature in the air plenum causes a switch to be closed, applying 120 V to the furnace blower motor. Warm air is now circulated until the room thermostat once again opens, cutting off the main gas supply to the chamber. The blower motor continues to blow warm air until the plenum is cooled sufficiently to open the heat-activated switch, removing the 120 V ac from the motor. There are, therefore, three separate circuits involved in this type of heating system.

11-3.5 Relays

A very important application of electromagnets is in magnetic relays. A relay is a magnetically operated switch that can control many high-voltage, high-power circuits from a single low-voltage input. By energizing a single electromagnet, the movement of one armature can open and close many contacts at once. These contacts generally control loads operating at higher voltage and current than the input voltage to the coil of the electromagnet. In this sense, a relay can be thought of as an amplifier; a small-power input controls a large-power output. When relays are used to control very large loads, like high horsepower motors, they are generally referred to as *contactors* or starters. Their operating principle is the same, however.

One common type of relay has a normally closed *and also* a normally open contact, as shown in Fig. 11-10. Usually referred to as a form C relay, the normally closed contacts provide continuity between the armature and the upper contact, when the coil is de-energized. A spring holds the armature in this position. When rated voltage is applied to the coil, the armature is drawn downward, "breaking" the normally closed contact and "making" the normally open contact. Continuity now exists between the armature and the lower contact. In a typical relay, a PD of 10 V at just a few milliamperes is sufficient to energize or "pull-in" the relay. The load contacts may be rated at 120 V ac, 1 A, or more. A wide variety of dc and ac relays is available. The ac type can often be

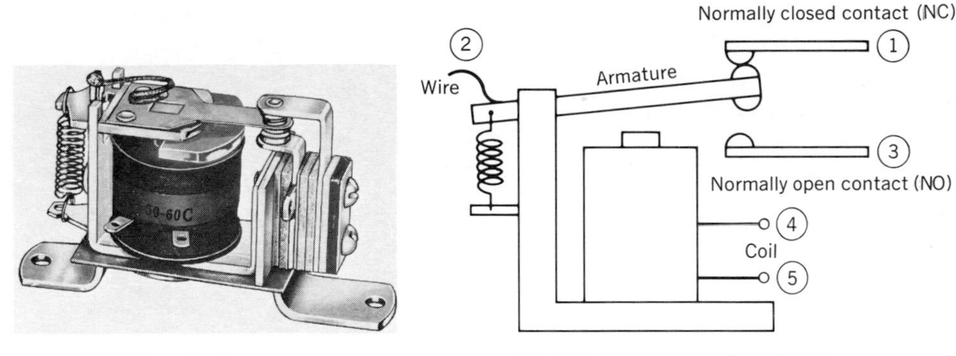

(*a*) A form C relay

(*b*) Pictorial representation of relay

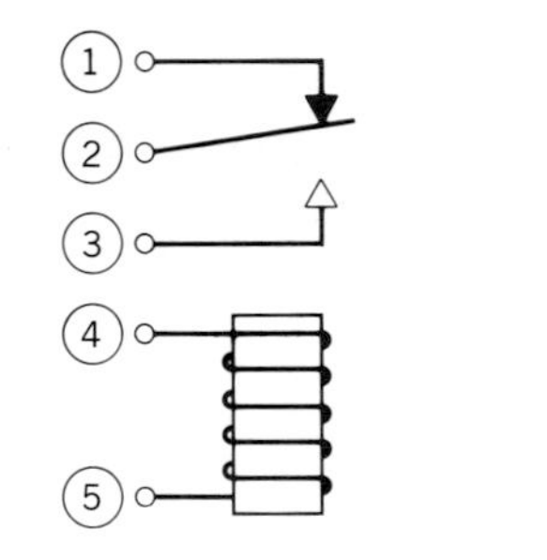

(*c*) Symbol for form C relay

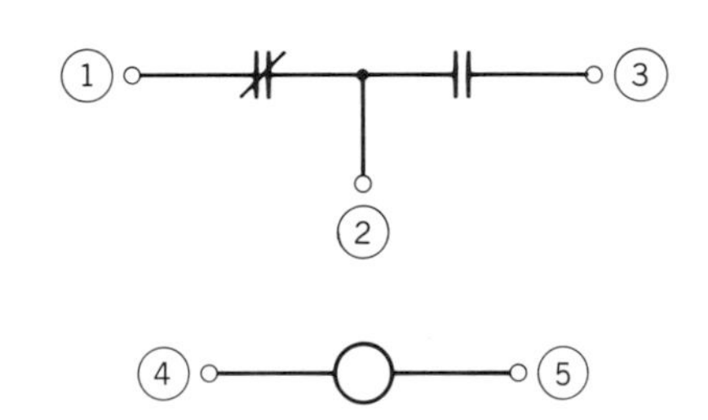

(*d*) Alternate symbol for form C relay

FIGURE 11-10
Magnetic relay and symbols. [Photograph in (*a*) courtesy of Potter and Brumfield Division AMF Incorporated.]

recognized by a copper segment inserted in the face of the core of the electromagnet. This avoids the "chattering" or tendency of the coil to become de-energized on each half-cycle of the ac input voltage to the coil.

We can use the heating system described in Secton 11-3.4 to illustrate the application of an SPDT (single pole double throw) relay. Many forced air heating units are equipped to supply air conditioning through the same ducts used for heating. In order to move the heavier cold air at the proper rate, it is necessary to run the blower motor at a higher speed than that used for heating, where the warm air presents less of a load. The switch-over from low heating speed to high cooling speed is done automatically by a relay, as shown in Fig. 11-11.

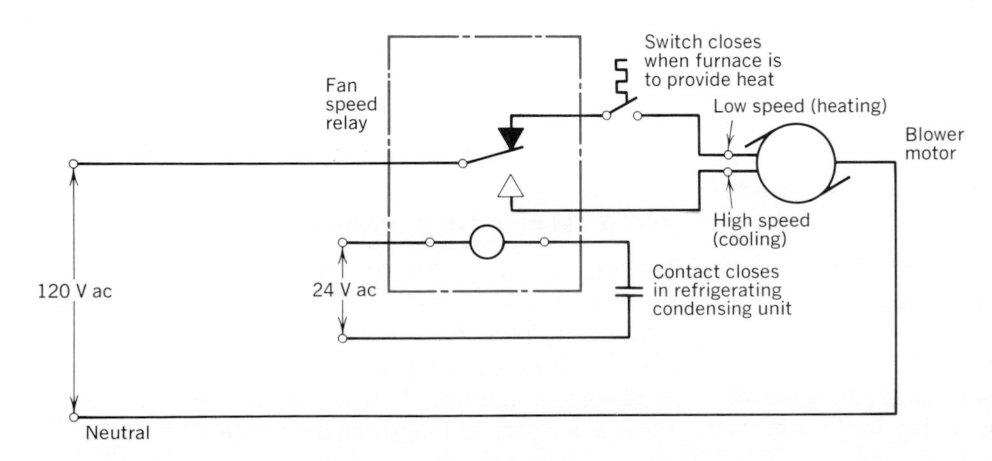

FIGURE 11-11
Relay used to change blower motor from low-speed heating to high-speed cooling.

The relay is in its normally closed position for the heating mode, applying 120 V to the low-speed winding of the motor. (This contact is also in series with the heat-activated switch, as described in Section 11-3.4, as well as a normally closed high-temperature limit switch that opens if temperatures should become dangerously high.) When the room thermostat is switched to cooling, and the air conditioning compressor has been running for a short time, a contact closes in the condensing unit energizing the 24-V fan speed relay. This applies 120 V to the high-speed winding of the blower motor. When the room thermostat is satisfied (open), the contact in the condensing unit eventually opens, de-energizing the speed relay and shutting off the blower motor.

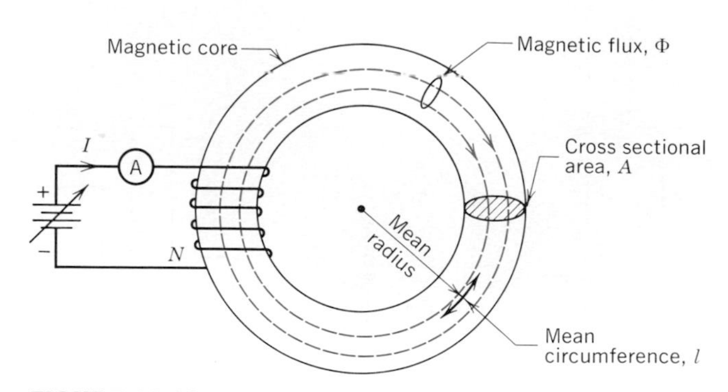

FIGURE 11-12
Quantities for determining B and H in a magnetic core.

11-4 MAGNETIC UNITS

The magnetic properties of a material may be displayed graphically in what is called a *B-H* curve. To explain this term, consider a magnetic core with a given number of turns of wire wrapped around it, as shown in Fig. 11-12. The voltage source is variable allowing us to vary the current in the coil. This in turn determines the amount of magnetic flux in the core.

11-4.1 Flux Density

As the magnetic lines of force in the core are increased, we can think of this as an increase in the *magnetic flux density, B*. We can define magnetic flux density as follows:

$$B = \frac{\Phi}{A} \quad \text{webers per square meter (Wb/m}^2\text{)} \qquad (11\text{-}1)$$

where: B is the flux density, in webers per square meter, Wb/m² or tesla (T)
Φ is the magnetic flux, in webers, Wb
A is the core's cross-sectional area, in square meters, m²

We are not yet in a position to define quantitatively what 1 Wb is, except that it is the MKS unit for measurement of magnetic flux. A flux *density* of 1 Wb/m² is also known as a flux density of 1 tesla, T, that is,

$$1 \text{ T} = 1 \text{ Wb/m}^2$$

11-4.2 Magnetomotive Force

The amount of flux density set up in the core is dependent upon five things: the current, number of turns, material of the magnetic core, length, and cross-sectional area of the core. The more current and the more turns of wire we use, the greater will be the magnetizing effect. We call this product of the turns and current the magnetomotive force (mmf) F_m, similar to the electromotive force (emf).

$$F_m = NI \quad \text{ampere-turns (At)} \qquad (11\text{-}2)$$

where: F_m is the magnetomotive force, mmf, in ampere-turns, At
N is the number of turns wrapped on the core
I is the current in the coil, in amperes, A

11-4.3 Magnetizing Intensity

It should be evident that a given number of ampere-turns on a smaller circumference core represents a greater magnetizing *intensity,* than when used on a larger length of core. **This *magnetizing intensity* is a useful term and is given the symbol H. It is the number of ampere-turns per meter of length of the magnetic circuit:**

$$H = \frac{F_m}{l} = \frac{NI}{l} \text{ ampere-turns/meter (At/m)} \quad (11\text{-}3)$$

where: H is the magnetizing intensity, in ampere-turns/meter, At/m

l is the length of the magnetic core, in meters, m

EXAMPLE 11-1

A cast iron toroid (doughnut) as in Fig. 11-12 has an inside radius of 5 cm and an outside radius of 7 cm. It has a circular cross section. A coil of 400 turns is wrapped around the core and carries a current of 2 A. A flux of 1.5 $\times$ 10^{-4} Wb is set up inside the core. Calculate:

a. The flux density in tesla.
b. The magnetomotive force in ampere-turns (At).
c. The magnetizing intensity in At/m.

SOLUTION

a. Diameter of the circular core cross section,

$$d = 7 - 5 \text{ cm} = 2 \times 10^{-2} \text{ m}$$

area of core's cross section,

$$A = \frac{\pi d^2}{4} = \frac{\pi}{4} \times (2 \times 10^{-2} \text{ m})^2 = \pi \times 10^{-4} \text{ m}^2$$

$$\text{flux density } B = \frac{\Phi}{A} \quad (11\text{-}1)$$

$$= \frac{1.5 \times 10^{-4} \text{ Wb}}{\pi \times 10^{-4} \text{ m}^2} = 0.48 \text{ Wb/m}^2 = \mathbf{0.48\ T}$$

b. Magnetomotive force $F_m = NI$ (11-2)

$$= 400 \text{ turns} \times 2 \text{ A} = \mathbf{800\ At}$$

c. Magnetizing intensity $H = \frac{NI}{l}$ (11-3)

mean length of core $= 2\pi r$

$$= 2\pi \times 6 \text{ cm} = 0.377 \text{ m}$$

$$\text{Therefore, } H = \frac{NI}{l} = \frac{800 \text{ At}}{0.377 \text{ m}} = \mathbf{2122\ At/m}$$

11-4.4 Permeability

We have seen how the flux density B set up in a core is directly proportional to H. (That is, B is directly proportional to N and I, and inversely proportional to l.) A fourth variable is the material of the core itself. **The ability of a material to set up a magnetic field is called its *permeability, μ*.** Thus $B \propto H$ and $B = \mu H$. This provides us with a method of determining a material's permeability.

$$\mu = \frac{B}{H} \quad \text{webers per ampere turn meter (Wb/At-m)} \quad (11\text{-}4)$$

where: μ is the permeability in Wb/At-m

B is the flux density in Wb/m^2

H is the magnetizing intensity in At/m

For *nonmagnetic materials*, $\mu = \mu_0 = 4\pi \times 10^{-7}$ Wb/At-m. But magnetic materials have a much higher permeability that varies widely with the amount of flux density. To show how much more permeable a magnetic material is than air (or any nonmagnetic material), we use the relative permeability, μ_r.

$$\mu_r = \frac{\mu}{\mu_0} \quad (11\text{-}5)$$

where: μ_r is the relative permeability (no units)

μ is the permeability of a given material, Wb/At-m

μ_0 is the permeability of free space (vacuum), $4\pi \times 10^{-7}$ Wb/At-m

NOTE μ_r is dimensionless (much like specific gravity) because it is a ratio of identical units. It can be as high as several thousand in some modern magnetic materials.

EXAMPLE 11-2

a. Determine the permeability of the iron core in Example 11-1.
b. What is the relative permeability of the iron core?
c. If the iron core in Example 11-1 were replaced by a sheet steel core having a relative permeability of 600, what would be the flux density in the core?

SOLUTION

a. $\mu = \dfrac{B}{H}$ (11-4)

$= \dfrac{0.48 \text{ Wb/m}^2}{2122 \text{ At/m}} = \mathbf{2.26 \times 10^{-4}\ Wb/At\text{-}m}$

b. $\mu_r = \dfrac{\mu}{\mu_0}$ (11-5)

$= \dfrac{2.26 \times 10^{-4} \text{ Wb/At-m}}{4\pi \times 10^{-7} \text{ Wb/At-m}} = \mathbf{180}$

c. $B = \mu H$

$= \mu_r \mu_0 H$

$= 600 \times 4\pi \times 10^{-7} \text{ Wb/At-m} \times 2122 \text{ At/m}$

$= \mathbf{1.6\ Wb/m^2 = 1.6T}$

11-5 HYSTERESIS

We are now prepared to consider the graphical relation between B and H for a magnetic material. Since $\mu = B/H$, the graphical relation shows how the permeability of a material *varies* with the magnetizing intensity H.

Assume that the magnetic core in Fig. 11-12 is initially completely demagnetized. As we increase the current, $H = NI/l$ increases, and there will be an increase in the flux density, B. Since the number of turns and the length of core of a coil are fixed, H is directly proportional to the current or ammeter reading. The flux density can be measured by inserting the probe of a flux meter into a small hole drilled in the core.

A plot of the values of B and H gives the *normal magnetization curve*, as shown in Fig. 11-13. There is evidently a linear portion where B is relatively proportional to H. But then a condition of *saturation* occurs when a very large increase in H is required to significantly increase B.

If the current is now gradually reduced toward zero, H returns to zero, but B does not. The core exhibits *retentivity* and retains some residual magnetism. The retentivity is represented by the distance OR.

If the connections to the coil are reversed, and the current is again increased, it is found that a certain amount of H is required to bring the magnetism in the core down to zero. This is called the *coercivity* and is represented by the distance OC. Further increase in the current magnetizes the core in the opposite direction from before, until once again saturation occurs.

Reduction of the current and subsequent reversal of direction will produce a closed figure called a B-H curve or *hysteresis loop*. The name comes from the Greek word *hysteros* meaning "to lag behind." That is, the state of the flux density is always lagging behind the efforts of the magnetizing intensity.

The shape of a B-H loop is an indication of the magnetic properties of the material, which in turn determine the applications for a given material. To understand this, we need to consider briefly one theory of magnetism.

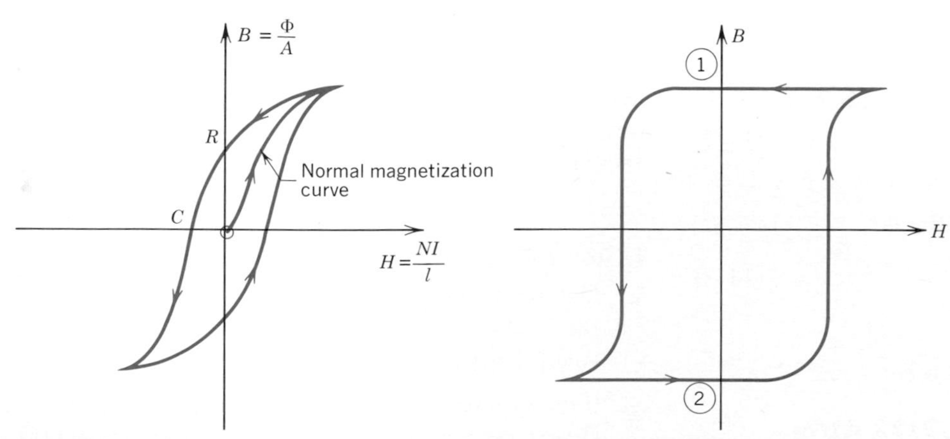

(a) Hysteresis loop for material used in an ac circuit

(b) Hysteresis loop for permanent magnet or ferrite core

FIGURE 11-13
Hysteresis loops.

11-6 BASIC THEORY OF MAGNETISM

We have seen how an electrical current sets up a magnetic field. In the atomic structure of materials, the atoms contain electrons that are not only rotating around the nucleus but also spinning on their own axis. This movement of a negative charge is essentially a very small current, producing a small magnetic field. The combined magnetic effect of between 10^{17} and 10^{21} molecules is called a *magnetic domain*. Each domain acts as a minute bar magnet with its own north and south poles. Only ferromagnetic materials possess magnetic domains because these atoms act *collectively* in their magnetic effect, instead of *singly* as in nonmagnetic materials. In an unmagnetized material, these domains have a random orientation with no *overall* magnetic effect. See Fig. 11-14*a*. When the material is placed in a magnetic field, alignment of the domains takes place producing an overall cumulative magnetic effect, as in Fig. 11-14*b*.

In some materials, those used for making permanent magnets, most of the domains remain aligned after removal of the external field. In materials like soft iron, the domains tend to become disoriented after removing the material from an external field, so the iron quickly loses its magnetism.

11-6.1 Hysteresis Losses

We can see now that saturation is the effect of the domains becoming almost completely aligned. Also, if the magnetic material is subjected to an alternating magnetizing force (as when an ac current flows in a transformer), the magnetic domains in the core undergo a periodic reversal 60 times every second. That is, the hysteresis loop is repeated 60 times per second if the supply frequency is 60 Hz. A magnetic core will become warm due in part to this constant reversal of magnetic field. This means that some energy is lost in the core and is referred to as "hysteresis losses." In a transformer, this results in less power being available at the output compared with the input.

Now it can be shown that the hysteresis losses, as well as being proportional to frequency, are also proportional to the *area* of the *B-H* curve. Consequently, materials used in ac equipment, such as motors, alternators, transformers, and so on, should have low values of

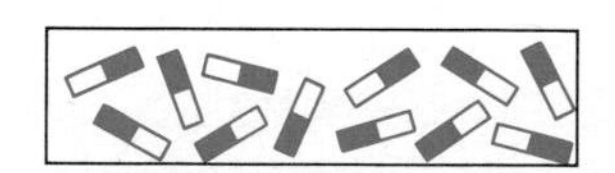

(*a*) Random orientation of domains in unmagnetized material

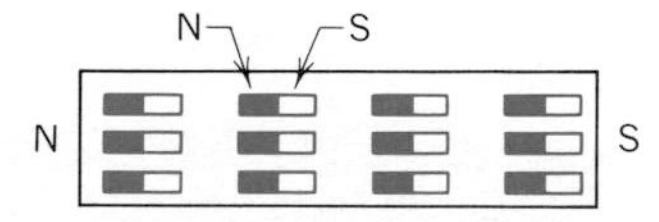

(*b*) Alignment of magnetic domains in magnetized material

FIGURE 11-14
Schematic representation of magnetic domains in unmagnetized and magnetized materials.

coercivity and retentivity, to reduce the area of the hysteresis loop. Soft iron and alloys called Supermalloy and Permalloy meet this need. They have a *B-H* loop more closely resembling that in Fig. 11-13*a*.

11-7 MATERIALS FOR PERMANENT MAGNETS

Permanent magnets require as square a *B-H* loop as possible, similar to that shown in Fig. 11-13*b*. Materials having high values of coercive force (approximately 100,000 At/m) and residual flux density (approximately 1 T) are required.

An alloy of aluminum, nickel, cobalt, iron, and copper (Alnico) is widely used for permanent magnets in loudspeakers. However, due to the increasing cost of cobalt, it is expected that many of these applications will be replaced by *hard ferrites*. These are ceramic materials consisting of barium carbonate with particles of ferric oxides. They have good ferromagnetic properties but are poor electric conductors. This makes them valuable at high frequencies because losses due to *eddy currents* (see Section 16-1.3) are reduced significantly in communication coils. Ferrites are not usually found in power circuits at low frequencies because they tend to saturate at fairly low values of *H*. An advantage of ferrites is that they can be pressed into any desired shape and then fired in an oxygen atmosphere to produce the finished ceramic magnet.

A third group of materials used for permanent magnets is referred to as the rare earth group. This is a group of elements all of which have three electrons in their valence shells. They too are expected to replace Alnico in many applications.

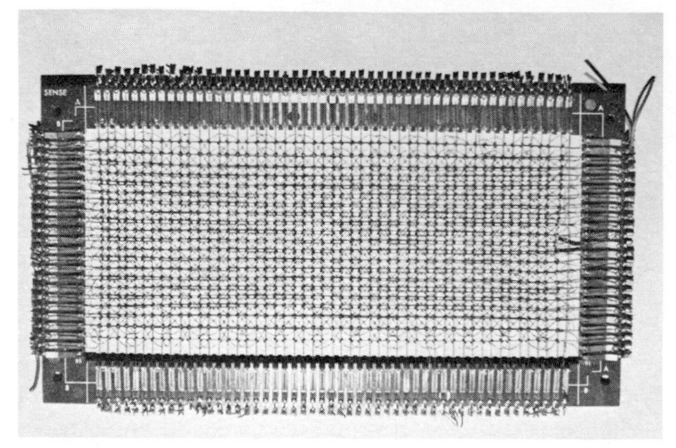

FIGURE 11-15
This 3 × 8 in. plane of magnetic core memory has 1000 ferrite "beads" to store digital information.

11-8
APPLICATIONS OF MAGNETIC MATERIALS

An interesting application for a ferrite core having a square *B-H* loop is the *digital magnetic core memory*. See Fig. 11-15. When the core (having a *B-H* loop as in Fig. 11-13*b*) is magnetized in one direction, it can be considered as storing a digital (1). When a pulse of current is passed through a wire running through the "bead," the core can be magnetized in the opposite direction, storing a digital (0). By using suitable wires and controls, the stored information can be "read" (or "sensed") or new information can be "written" into the core. A series of cores in numerous flat planes stacked one above the other can store "words" of information. This information can be obtained from the core in a very short time by "addressing" the proper location.

Another type of storage used in a computer is *magnetic tape*. The tape consists of a layer of iron oxide coating a transparent mylar film tape. As shown in Fig. 11-16, the tape passes in front of a recording head. This consists of a soft iron magnetic core with a coil wrapped around it. When current passes through the recording head coil in one direction, a certain magnetic polarity is developed. This magnetizes the magnetic coating of the tape in a corresponding direction, representing a digital (1), for example. Current in the opposite direction in the coil reverses the stored magnetic polarity on the tape, representing a (0). Thus bits of information, in the form of a 1 state or a 0 state are stored along the tape. This information can be retrieved later by running the tape under a "read" head. Here voltages of varying polarity will be *induced* in a coil due to the magnetic field stored on the moving tape. (See Section 14-1.)

In the case of an audio tape recording (such as an eight-track cartridge), the frequency of the current flowing in the recording head coil is varying and alternating at an audio rate. The loudness will be determined by the amount of current and thus the strength of field set up on the tape. Some minimum speed of the magnetic tape is necessary to be able to record high fidelity music. This depends in part upon the quality of the tape's magnetic coating and the heads used for magnetizing the tape.

11-9
DEMAGNETIZATION

The magnetic domain theory of magnetism (Sec. 11-6) supports the methods that can be used to demagnetize a magnet. For example, repeatedly hitting a magnet or mechanical vibrations tend to rearrange the aligned

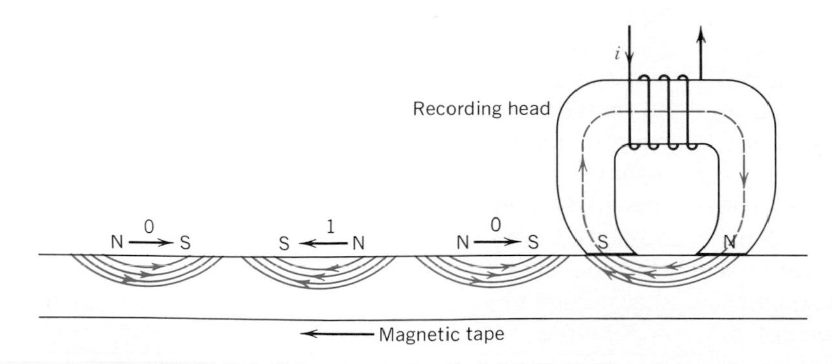

FIGURE 11-16
Magnetic tape recording head principle.

domains, leaving them disoriented. A similar, but much more effective, method is to heat the magnet above what is called the *Curie temperature.* This is the temperature at which ferromagnetism is lost abruptly and completely. Iron has a Curie temperature of 770°C, cobalt 1131°C, nickel 358°C, and alloys various values in between. Apparently, the high temperature causes the atoms and molecules to reach a high state of agitation, disorienting the magnetic domains. Only after the temperature is reduced below the Curie temperature will the material be able to be remagnetized.

In most cases, the above two methods are not suitable for demagnetizing manufactured articles. The most widely used method involves a "demagnetizing" coil with many turns carrying an alternating current. This sets up an alternating magnetic field in the center of the coil. If the part to be demagnetized is held in the center of the coil, the material undergoes a cyclical magnetic reversal around the hysteresis loop. If the part is now *slowly* withdrawn from the coil (or the ac current in the coil is slowly reduced to zero), the magnetic state of the part follows ever-decreasing amplitudes of hysteresis loops. See Fig. 11-17*a*. Eventually, when the coil is completely removed, the coercivity and retentivity are essentially reduced to zero.

There are a number of common examples in which demagnetization is necessary. We have seen how a magnetic recording head must set up a magnetic field only when audio current flows in the coil of the head. After continual use, the recording head retains some residual or static magnetism, which leads to background noise when a desired signal is recorded. One method of removing the unwanted magnetism is to use a demagnetizing tool that operates on the principle outlined above. See Fig. 11-17*b*.

In color TV picture tubes the metal parts around the tube can become permanently magnetized, affecting the quality of the color picture. A coil, operating from the 120-V, 60-Hz supply, is placed near the face of the screen and slowly moved away; or, if the tube is removed, the coil is slowly passed over the entire length of the tube. In most modern color TV sets, a demagnetizing coil is built-in around the tube and automatically energized for a short interval each time the set is turned on. In this way, the TV set can be moved from one location to another without being affected by the variations in the earth's magnetic field.

Another name sometimes used for demagnetizing is

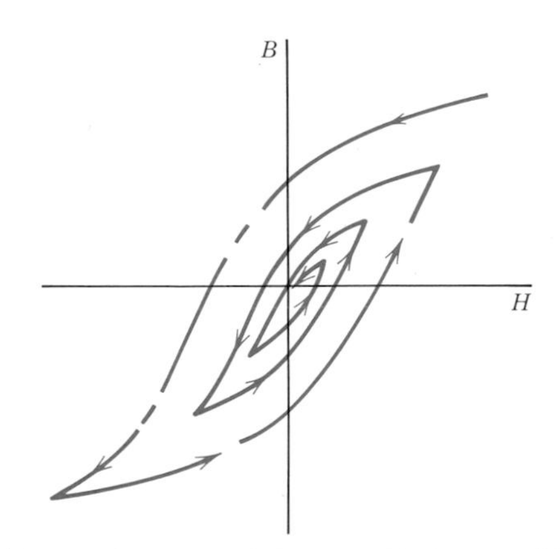

(*a*) Successive hysteresis loops used in demagnetizing a magnetic material

(*b*) A demagnetizing or degaussing coil

FIGURE 11-17

B-H curve and demagnetizer. [Photograph in (*b*) courtesy of the R. B. Annis Co.]

degaussing. Thus the coil used to demagnetize a TV picture tube is sometimes called a degaussing coil. The name comes from the older centimeter-gram-second (CGS) unit of measurement for flux density, the gauss, G. A flux density of 1 T = 10^4 G. The gauss is evidently a small amount of flux density, so the term kilogauss is often used. The earth's magnetic field is approximately one-half a gauss. Research in magnetic fields is often reported in CGS units. For example, the corresponding unit of magnetizing intensity is the oersted; 1 At/m = 0.0126 Oe. Thus a *B-H* curve in the CGS system of units

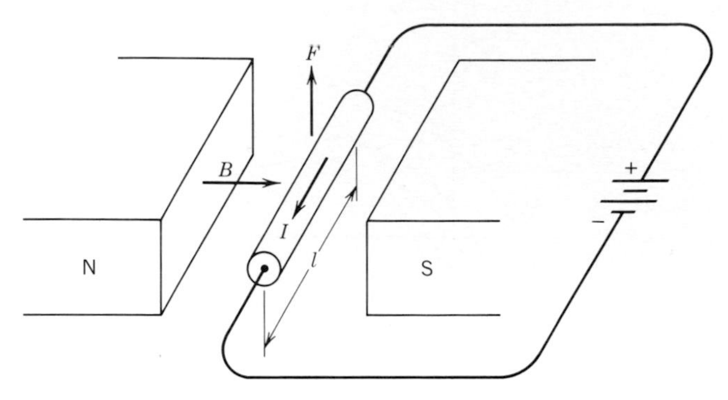

(a) Force F is exerted on a conductor of length l, carrying current I, in a magnetic field B

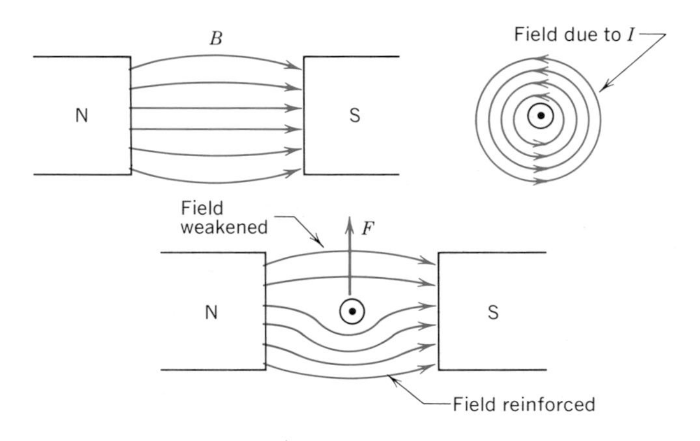

(b) The separate and combined magnetic fields

FIGURE 11-18
Force acting on a current-carrying conductor.

would consist of a graph of gauss versus oersteds. Many meters available to measure flux densities are calibrated in gauss, so they are known as *gaussmeters.* (See Section 11-11.)

11-10 FORCE ON A CURRENT-CARRYING CONDUCTOR IN A MAGNETIC FIELD

It is now time to consider the *interaction* of magnetic fields. In particular, consider the interaction of a permanent magnetic field and that due to a current-carrying conductor placed in the field, as in Fig. 11-18a.

We have seen how the field set up around a wire carrying a current is circular. If the wire (or some part of it) is perpendicular to the field of the permanent magnet, the flux lines on the lower side of the wire will be in the same direction. This means that the field is concentrated in this region due to an additive effect. On the upper side, the two fields oppose each other with a resulting weakened field. The overall field distribution is shown in Fig. 11-18b. The result is a force moving the wire upward from the stronger into the weaker field. If either the current or the permanent magnetic field is reversed, the direction of the force is reversed, and the wire would be moved downward.

The interaction of the two fields has produced a force. In fact, this is how we can determine if a magnetic field exists in a region:

A magnetic field is said to exist at a point if it exerts a force on a moving charge at that point.

Thus we could use a moving beam of charge, such as the electron beam in an oscilloscope, to explore an unknown magnetic field. If the spot of light produced by the beam on a fluorescent screen is always at the same point on the screen, as we move the tube around, no magnetic field exists, or the electron beam is parallel to the field. If we observe a deflection of the spot, upward or downward, we can conclude that a magnetic field exists, and we could determine its direction. In theory, this could be used to examine the earth's magnetic field.

If the beam is placed parallel to the earth's field (roughly north-south), there would be no deflection of the spot. If turned through an angle of 90°, so the electron beam is traveling from west to east, the spot would be deflected downward. This indicates that the earth's magnetic field has a direction in which the lines of force leave a region near the earth's south geographic pole and enter a region near the earth's north geographic pole. We can conclude that a south magnetic pole exists near the earth's north geographic pole, and what we call the north magnetic pole is in fact a south pole. Thus the end of the compass needle that points north (when it aligns itself with the earth's field) is a north pole (and not a north-seeking pole, as it is so often called).

The earth's distribution of magnetic flux is similar to what it would be if the earth's core were considered to be a huge bar magnet, as shown in Fig. 11-19. The angle between the earth's field and the true geographic north-south direction is called the *variation* or *declination.* Approximately 15°W in many parts of the United States,

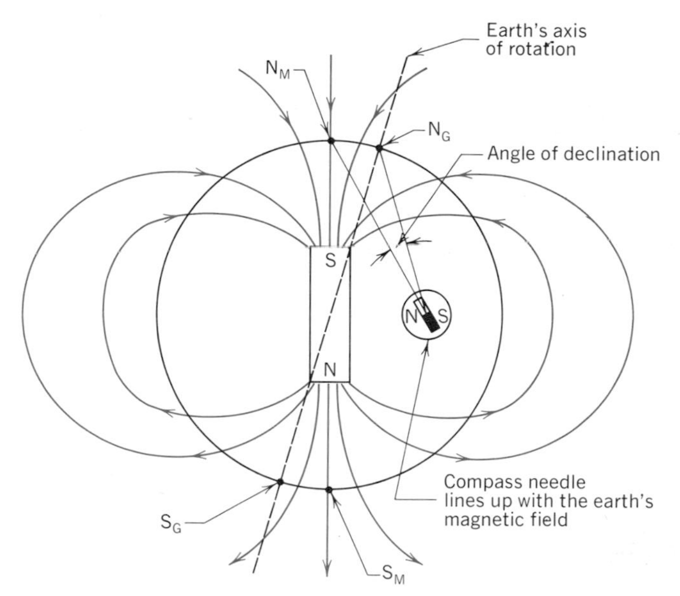

FIGURE 11-19
Simplified diagram of the earth's magnetic field.

the angle of declination varies irregularly over the earth's surface and varies from year to year.

The amount of force acting on a moving charge is given by

$$F = qvB \qquad \text{newtons (N)} \tag{11-6}$$

where: F is the force in newtons, N
- q is the charge of the particle in coulombs, Q
- v is the velocity of the charge perpendicular to B in meters/second, m/s
- B is the flux density of the field in teslas, T

If the particles are moving at some angle with B other than $\theta = 90°$, the force is reduced by a factor of sin θ.

Using the above equation, it is possible to obtain the force acting on a conductor as in Fig. 11-18*a*.

$$F = BIl \qquad \text{newtons (N)} \tag{11-7}$$

where: F is the force on the conductor in newtons, N
- B is the flux density of the field in which the conductor is placed, in teslas, T
- I is the current in the conductor in amperes, A
- l is the length of the conductor perpendicular to the field in meters, m

If the conductor is at some angle with B other than $\theta = 90°$, the force is reduced by a factor of sin θ.

EXAMPLE 11-3

A 1-m length of No. 14 copper wire is located between two pole faces, each of which is 10 cm square. The total flux between the two poles is 1.2×10^{-2} Wb. Calculate:

a. The force acting on the wire when it is carrying 5 A and is perpendicular to the field.

b. The force acting on the wire when it is carrying 5 A and is at an angle of 45° with respect to the field.

c. How much current must pass through the wire when it is perpendicular to the field if it is to support its own weight of 18 g and remain suspended in the magnetic field.

SOLUTION

a. The configuration is as shown in Fig. 11-18*a*, with the *active* length of the conductor in the magnetic field equal to 10 cm. That is, $l = 0.1$ m.

$$B = \frac{\Phi}{A} \tag{11-1}$$

$$= \frac{1.2 \times 10^{-2}\ \text{Wb}}{0.1\ \text{m} \times 0.1\ \text{m}} = 1.2\ \text{T}$$

$$F = BIl \tag{11-7}$$

$$= 1.2\ \text{T} \times 5\ \text{A} \times 0.1\ \text{m}$$

$$= \mathbf{0.6\ N}\ (0.13\ \text{lb})$$

b. $F = BIl \sin\theta$

$$= 0.6\ \text{N} \times \sin 45° = 0.6\ \text{N} \times 0.707 = \mathbf{0.42\ N}$$

c. $w = mg$ (weight of the 1-m length of wire)

$$= 18 \times 10^{-3}\ \text{kg} \times 9.8\,\frac{\text{m}}{\text{s}^2} = 0.176\ \text{N}$$

This must be balanced by the magnetic force

$$F = BIl \tag{11-7}$$

$$= 1.2 \times 0.1 \times I \qquad \text{newtons}$$

Therefore, $0.12\,I = 0.176$

and $$I = \frac{0.176}{0.12}\ \text{A} = \mathbf{1.47\ A}$$

There are numerous applications of a force being exerted on a current-carrying conductor in a magnetic field. We now examine some of them.

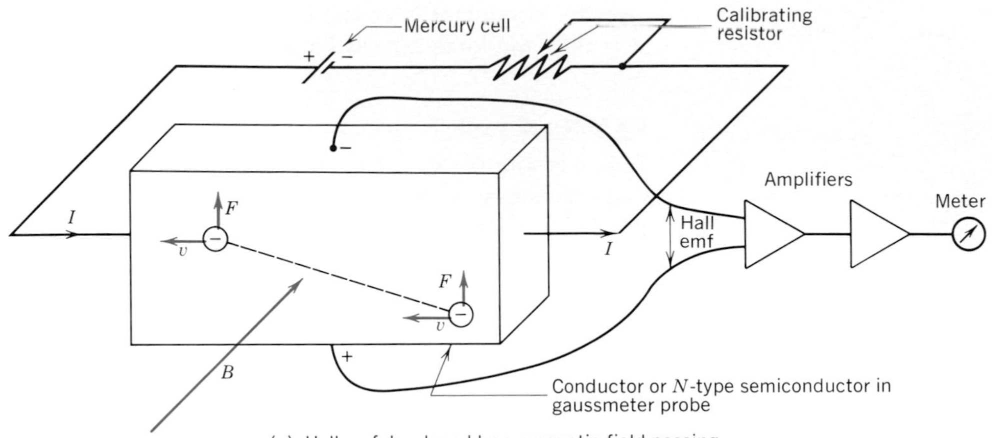

(*a*) Hall emf developed by a magnetic field passing through a current-carrying semiconductor

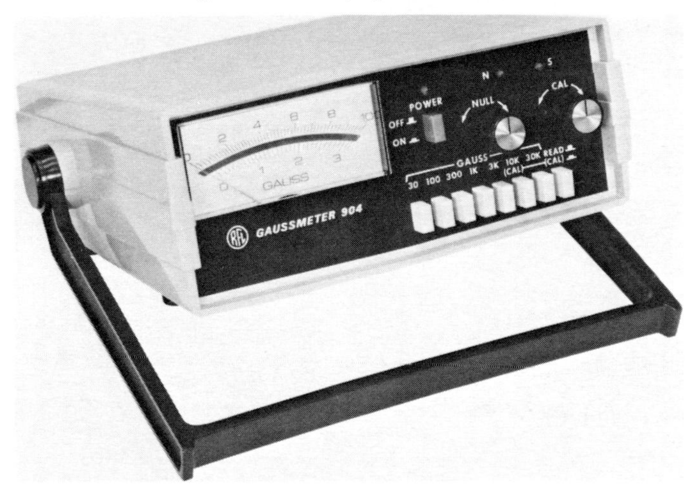

(*b*) Photograph of a gaussmeter

FIGURE 11-20
Hall effect and gaussmeter. [Photograph in (*b*) courtesy of RFL Industries, Inc.]

11-11
HALL EFFECT AND THE GAUSSMETER

Figure 11-20*a* shows a conductor in the form of a flat strip, perpendicular to a magnetic field B. Charges are made to flow through the strip by the external cell. The charges have a magnetic force qvB exerted on them, where v is the drift velocity. If the charge carriers are electrons, they are driven toward the upper edge of the strip. An excess negative charge accumulates at this edge leaving an excess positive charge on the lower edge. A potential difference of a few microvolts is established, which is called the Hall emf or Hall voltage. The Hall emf is always at right angles to both the magnetic field and the current, respectively.

It can be shown that the Hall emf depends directly on the current *and* the magnetic field strength. Thus, if the current is held constant, the emf is a direct indication of the magnitude of the magnetic flux density B. For this reason it is used in instruments to measure the magnetic field strength. The generated Hall emf is much larger if a semiconductor, such as indium arsenide, is used. However, amplification of several hundred times is necessary to drive a dc meter usually calibrated in gauss. Multirange full-scale deflections from 10 to 10^4 G are available on typical portable fluxmeters (or gaussmeters) similar to that shown in Fig. 11-20*b*. The probe that contains the semiconductor strip must be rotated until a maximum reading is obtained.

It is interesting to note that the Hall effect provides a means of distinguishing between N-type and P-type semiconductors. (See Chapter 28.) In the latter type, conduction of current is by means of positive charges called *holes*. Hole flow coincides with the direction of *conventional* current. Thus if, in Fig. 11-20a, the N-type semiconductor is replaced by a P-type, all other variables remaining the same, the Hall emf has an *opposite polarity*. Thus a hole flowing in one direction is *not the same* as an electron moving in the opposite direction.

11-12 HIGH-FIDELITY LOUDSPEAKER

A loudspeaker consists of a fixed permanent magnet (PM) and a moving coil located in an air gap as in Fig. 11-21a. The moving coil, or "voice coil," may have a resistance of from 4 to 20 Ω. It is held in place and centered by a stiffness element called a *spider*. Attached to the spider and suspended from the basket or main frame is the flexible cone.

When current from the audio amplifier flows in the voice coil, a force acts on the coil causing it to move in the magnetic field in the air gap. The cone vibrates back and forth, sending out air pressure waves that have the same frequency as the audio input signal current.

Modern loudspeaker systems, as in Fig. 11-21b, can reproduce frequencies typically from 30 to 18 kHz using three transducers from 1 in. to 12 or 15 in. in diameter. A crossover or dividing network directs the proper frequencies to the respective transducer. The highest frequencies (treble) are handled by a dome radiator or horn that produces very small displacements. The bass frequencies, however, contain more energy, and cause the cone to vibrate much more slowly but with displacements of as much as 1 in. in large "woofers" operating at 20 to 30 Hz.

11-13 DC MOTORS

A dc motor is the direct result of the force acting on a current-carrying conductor in a magnetic field. A motor takes electrical energy and converts it into *rotational*

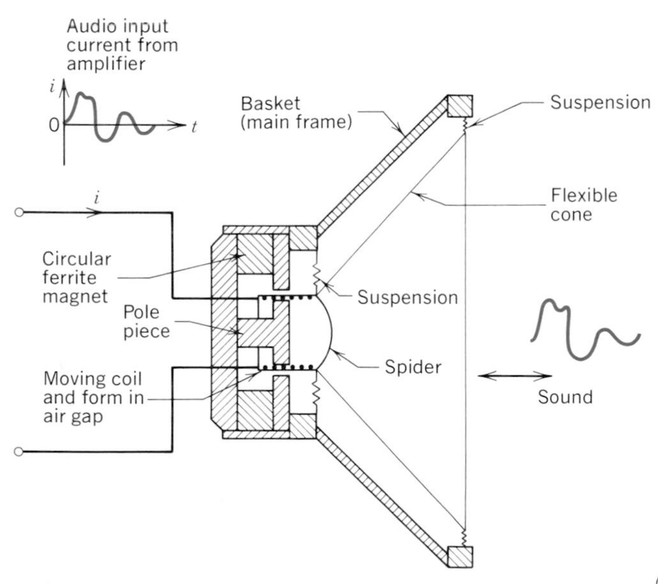

(a) Cross-sectional view of large-cone loudspeaker

(b) A three-way loudspeaker system with dividing network, 1-in. dome radiator, 5-in. midrange speaker, and a 12-in. low-frequency transducer with a 10¼ pound magnet, capable of handling 300 W rms

FIGURE 11-21
Loudspeaker details and system. [Photograph in (b) courtesy of James B. Lansing Sound, Inc.]

mechanical energy. Although some small dc motors use a permanent magnet to set up the stationary or *stator* field, most dc motors use field windings wrapped around pole pieces situated around the frame of the motor, as shown in Fig. 11-22.

The armature or rotor consists of a cylinder of soft steel mounted on a shaft so that it can rotate about its axis. A number of insulated copper wires are placed in slots running the length of the rotor. Current is led into and out of these conductors by means of stationary graphite (carbon) brushes riding on a *commutator* to which the rotor wires are connected. The commutator consists of copper segments insulated from each other and set into a small cylinder on the end of the shaft. The purpose of the commutator is to provide an automatic switching arrangement, to keep the current flowing in the rotor conductors in the direction shown in Fig. 11-22, whatever the position of the rotor. In this way, all the conductors under one field pole experience a side thrust in one direction, and the conductors under the other field pole a thrust in the opposite direction. The result is a counterclockwise torque on the armature for the given current directions. Note how the motor frame serves as a path for the magnetic field that passes through the rotor.

The manner in which the field windings and the armature are connected to the dc supply determines the type and characteristics of the motor. Figure 11-23 shows the four possible connections.

A *shunt* motor has a constant field current because the field winding is in parallel with the supply. The field winding consists of many turns of small-diameter wire to produce a large flux. A shunt motor is considered a constant-speed motor even though its speed does decrease slightly with an increase in load. It is a common motor to use to drive a fan, as in an automobile heating system. The speed is easily controlled by varying a rheostat in series with the field winding. An increase in resistance reduces the field and increases the speed of the motor.

A *series* motor has the field winding connected in series with the armature. Since the field must carry the full armature current, it normally consists of a few turns of comparatively large wire. The primary characteristic of a series motor is its high starting torque. Upon starting the motor, there will be a large inrush of current. This sets up a very strong field in both the stator and the armature causing a large torque. This is the type of motor used in automobile starters. If the load is removed, the current drops and the magnetic field also drops. This causes the motor's speed to increase. (It is trying to generate a counter emf equal to the applied emf with a much lower flux. This it can do only by rotating faster. See Chapter 14.) Thus large-series motors should not be used where the load can become disconnected; otherwise, the dangerously high speed can make the motor fly apart due to the excessive centrifugal force.

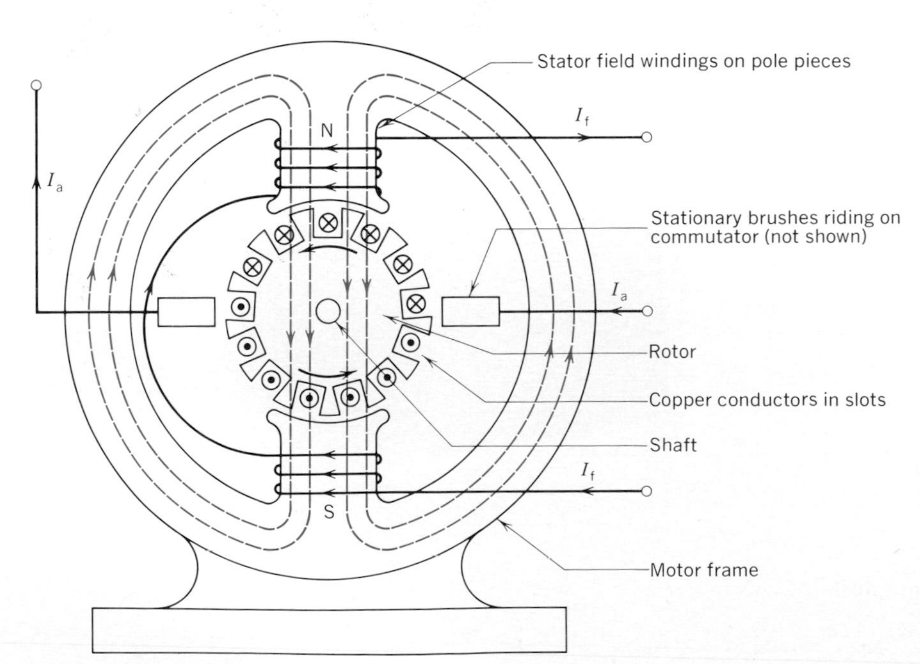

FIGURE 11-22
Pictorial representation of a dc motor.

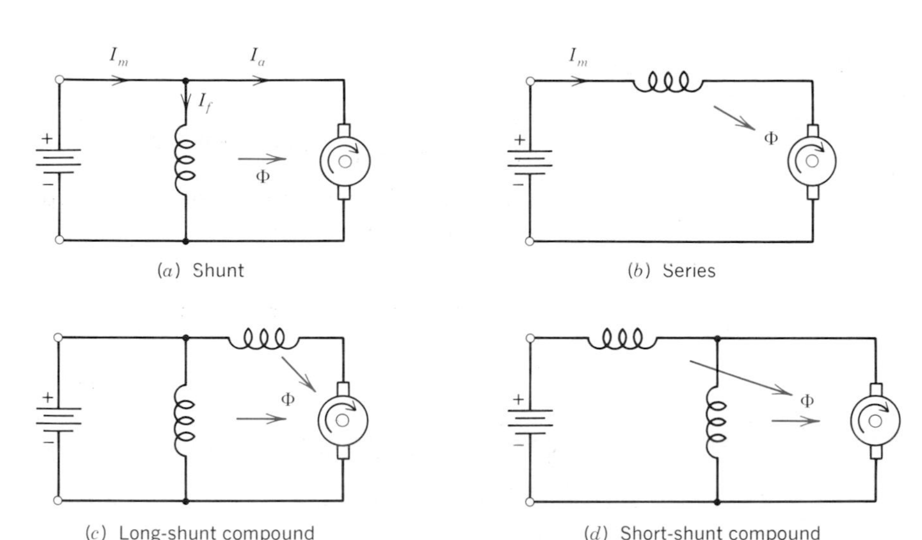

FIGURE 11-23
Schematic diagrams of dc motor connections.

A *compound* motor is built with two field windings, shunt *and* series, which can be connected in either the long- or short-shunt configuration. Compound motors have speed and torque characteristics midway between those for series and shunt motors, as shown in Fig. 11-24.

It should be noted that the direction of rotation of a dc motor is not changed by reversing the applied polarity (unless the motor has permanent magnets for the stator field). The connection to either the field *or* the armature must be reversed to obtain the opposite rotation. This fact makes it possible for a *universal* motor to run on either alternating current or direct current. Universal motors are series motors to develop a high torque at low speeds and are commonly used in hand drills, mixers, and sewing machines. Speed control in these applications is accomplished by means of a solid-state device called a *triac.* This varies how much of the 120-V, 60-Hz sine wave is applied to the motor, without reducing the torque substantially, as is the case in most dc motors with simple rheostat control.

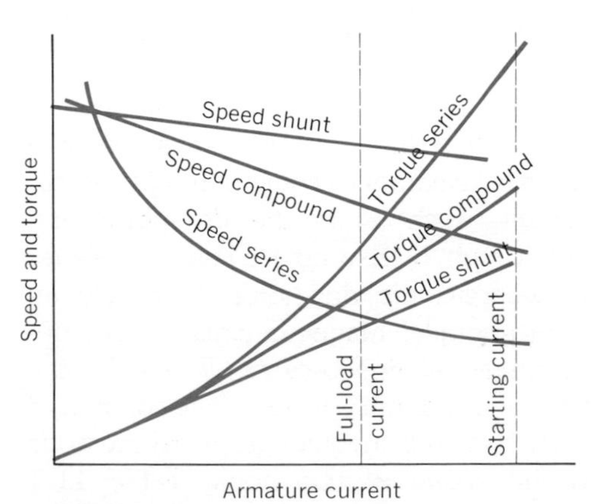

FIGURE 11-24
Comparison of speed and torque characteristics for varying loads on shunt, series, and compound motors.

11-14 OHM's LAW FOR A MAGNETIC CIRCUIT

The amount of magnetic flux in a magnetic circuit can be calculated using a relationship similar to Ohm's law for a dc electric circuit. If we think of magnetic flux Φ corresponding to electric current *I*, and magnetomotive force F_m corresponding to electromotive force *V*, there should be some magnetic quantity corresponding to electric resistance *R*. We call this quantity magnetic reluctance, R_m. It is a measure of a magnetic circuit's opposition to set up a magnetic flux. Thus

$$\Phi = \frac{F_m}{R_m} \quad \text{webers (Wb)} \tag{11-8}$$

where: Φ is the magnetic flux in the magnetic circuit in webers, Wb

F_m is the magnetomotive force NI in ampere turns, At

R_m is the reluctance of the magnetic circuit, in At/Wb

In the same way that we can calculate the resistance of a conductor using $R = \rho l/A$, we can obtain the *reluctance* of a magnetic circuit:

$$R_m = \frac{l}{\mu A} \quad \text{ampere-turns per weber (At/Wb)} \quad (11\text{-}9)$$

where: R_m is the reluctance of the magnetic circuit in At/Wb

l is the length of the magnetic circuit in meters, m

A is the cross-sectional area of the magnetic circuit in square meters, m^2

μ is the permeability of the magnetic material in Wb/At-m

Using Eq. 11-2, $F_m = NI$, we can now obtain an expression for Φ that shows all five factors that affect the magnitude of a magnetic field set up by a current-carrying coil:

$$\Phi = \frac{F_m}{R_m} = \frac{NI}{l/\mu A} = \frac{NI\mu A}{l} \quad (11\text{-}10)$$

where all of the terms are as previously defined.

EXAMPLE 11-4

A cast iron toroid has an inside radius of 5 cm and an outside radius of 7 cm with a circular cross section. A coil of 400 turns is wrapped around the core and carries a current of 2 A. If the permeability of the cast iron is 2.26×10^{-4}, calculate:

a. The reluctance of the cast iron toroid.

b. The flux set up in the toroid.

c. The new value of the current required to maintain the same flux level if a 5-mm air gap is cut in the toroid.

SOLUTION

a. The toroid has the same dimensions as in Example 11-1:
Mean length of the magnetic circuit, $l = 0.377$ m
Cross-sectional area of core, $A = \pi \times 10^{-4}\ m^2$

$$R_m = \frac{l}{\mu A} \quad (11\text{-}9)$$

$$= \frac{0.377\ \text{m}}{2.26 \times 10^{-4}\ \text{Wb/At-m} \times \pi \times 10^{-4}\ \text{m}^2}$$

$$= \mathbf{5.31 \times 10^6\ At/Wb}$$

b. $$\Phi = \frac{F_m}{R_m} \quad (11\text{-}8)$$

$$= \frac{400\ \text{turns} \times 2\ \text{A}}{5.31 \times 10^6\ \text{At/Wb}} = \mathbf{1.5 \times 10^{-4}\ Wb}$$

c. There will be an additional mmf required to set up this flux in the air gap given by

$$F_m = \Phi R_m$$

where R_m is the reluctance of the air gap

$$R_m = \frac{l}{\mu A} \quad (11\text{-}9)$$

$$= \frac{5 \times 10^{-3}\ \text{m}}{4\pi \times 10^{-7}\ \text{Wb/At-m} \times \pi \times 10^{-4}\ \text{m}^2}$$

$$= 1.27 \times 10^7\ \text{At/Wb}$$

$$F_m = \Phi R_m$$

$$= 1.5 \times 10^{-4}\ \text{Wb} \times 1.27 \times 10^7\ \text{At/Wb}$$

$$= 1.9 \times 10^3\ \text{At}$$

The total ampere turns required for the cast iron and the air gap is

$$F_m\ (\text{total}) = F_m\ (\text{cast iron}) + F_m\ (\text{air gap})$$

$$= 800\ \text{At} + 1900\ \text{At}$$

$$= 2700\ \text{At}$$

Therefore, NI total $= 2700$ At

and $$I = \frac{2700\ \text{At}}{400\ \text{t}} = \mathbf{6.75\ A}$$

Note the very large increase in current required. This is because of the very low permeability of the air gap and its high reluctance, even though its length is very small. It should be observed that the air gap in this example represents the practical clearance required in most practical magnetic circuits such as in the dc motor and loudspeaker. The circuit in Example 11-4c is a series magnetic circuit whereas the dc motor is actually an example of a *series-parallel* magnetic circuit. Also, because of the nonlinear relation between B and H, the permeability of a material is not constant. The solution of magnetic flux in such circuits involves graphical methods and is beyond the scope of this book. Table 11-1 summarizes the magnetic quantities.

Other applications of magnetic circuits will be covered in the next chapter.

TABLE 11-1
Summary of Magnetic Quantities

Quantity	Symbol or Equation	Unit
Flux	Φ	weber (Wb)
Flux density	$B = \frac{\Phi}{A}$	Wb/m^2 or tesla (T)
Magnetomotive force	$F_m = NI$	ampere-turn (At)
Magnetizing intensity	$H = \frac{NI}{l}$	ampere-turn per meter (At/m)
Permeability	$\mu = \frac{B}{H}$	weber/At-m $\left(\frac{\text{Wb}}{\text{At-m}}\right)$
Permeability of free space (vacuum)	$\mu_o = 4\pi \times 10^{-7}$	weber/At-m $\left(\frac{\text{Wb}}{\text{At-m}}\right)$
Relative permeability	$\mu_r = \frac{\mu}{\mu_o}$	none, dimensionless number
Reluctance	$R_m = \frac{l}{\mu A}$	ampere-turn per weber (At/Wb)
"Ohm's law"	$\Phi = \frac{F_m}{R_m} = \frac{NI\mu A}{l}$	weber (Wb)

SUMMARY

1. Common magnetic elements are iron, nickel, and cobalt. Most materials are nonmagnetic, such as wood, air, paper, copper, lead, and so on.

2. The attractive force that a magnet has for a magnetic material is shown graphically by lines of magnetic flux shown leaving the north pole and entering the south pole.

3. Like magnetic poles repel each other; unlike poles attract each other.

4. A current in a wire sets up a magnetic field whose direction around the wire is given by the right-hand rule: Point the thumb of the right hand in the direction of conventional current; the fingers point in the direction of the magnetic field.

5. An electromagnet is a coil of wire, or helix, that develops a temporary magnetic field whenever current flows in the coil. If the fingers of the right hand are made to point in the direction of conventional current around the coil, the thumb points to the coil's north pole.

6. Door chimes, doorbells, and solenoids are applications of electromagnets.

7. A relay is an electromagnetically operated switch that can control a load from a remote location by energizing the relay's coil.

8. Permeability is an indication of a material's ability to allow a magnetic field to be set up in the material.

9. Relative permeability shows how much more permeable a material is than air or any other nonmagnetic material.

10. A hysteresis loop shows the nonlinear manner in which B is related to H for a magnetic material.

11. Materials used for permanent magnets should have high values of retentivity (residual magnetism) and coercivity (demagnetizing force), such as the alloy Alnico and ferrites.

12. Materials used for magnetic circuits involving an alternating magnetizing force should have a small area *B-H* curve to minimize hysteresis losses.

13. Hysteresis losses are caused by the continual reversal of magnetic domains, which are groups of a large number of molecules having an overall magnetic effect like a tiny bar magnet.

14. Demagnetization (degaussing) of a magnetized material can be accomplished by repeatedly hitting, heating above the Curie temperature, or using an alternating magnetic field.

15. A charge moving in a magnetic field experiences a force given by $F = qvB$. When the charge is moving in a wire, located in a magnetic field, the force acting on the wire is given by $F = BIl$.

16. The earth's magnetic field is represented by lines of force leaving a north pole in the southern hemisphere and entering a south pole near the north geographic pole.

17. The Hall effect is the generation of an emf across the edges of a conducting strip when charges flowing through the strip are deflected by a magnetic field perpendicular to the current. The principle is used in a gaussmeter to measure flux density.

18. The force acting on a current-carrying conductor in a magnetic field provides a back and forth motion in a loudspeaker, and a rotating motion in a dc motor.

19. Direct current motors may be of the series type (high starting torque), shunt type (constant speed), or the compound type, which has both a series field and a shunt field.

SELF-EXAMINATION

Answer true or false
(Answers at back of book)

11-1. Nickel and cobalt are ferromagnetic materials with a high permeability. ________

11-2. Magnetic lines of flux indicate the paths followed by electrons from a north pole to a south pole. ________

11-3. The end of a compass that points to the earth's magnetic north pole will be attracted to a magnet's south pole. ________

11-4. Like poles attract; unlike poles repel. ________

11-5. A nail is attracted to a magnet because it becomes a temporary magnet of the opposite polarity by induction. ________

11-6. Oersted discovered that a current flowing in a wire affected a compass placed near the wire. ________

11-7. An electromagnet is a temporary magnet most of whose magnetism can be removed simply by turning off the current through the coil. ________

11-8. There are two right-hand rules, one for determining the direction of magnetic flux around a current-carrying wire and another to determine the direction of flux set up in an electromagnet. ________

11-9. A door chime operates in essentially the same way as a doorbell. ________

11-10. A relay is a magnetically operated switch that can only be operated on dc. ________

11-11. Relays can be used to open contacts as well as close contacts on the same relay at the same time. ________

11-12. The unit of flux density is the weber. ________

11-13. A total flux of 2×10^{-3} Wb uniformly distributed over an area of 5×10^{-3} m^2 is a flux density of 0.4 T. ________

11-14. A current of 0.3 A passing through a coil of 500 turns wrapped on a magnetic core of length 10 cm provides a magnetomotive force of 150 At and a magnetizing intensity of 1500 At/m. ________

11-15. If the core in Question 11-14 has a flux density of 0.15 T, the core's permeability is 1×10^4 Wb/At m. ________

11-16. The relative permeability of the core in Question 11-15 is $250/\pi$. ________

11-17. A hysteresis curve results because of the magnetizing intensity H lagging behind the flux density B. ________

11-18. All materials, whether magnetic or nonmagnetic, have their own hysteresis loop. ________

11-19. The magnetic effect of certain materials is believed to be due to tiny clusters of molecules, called domains, that act collectively in their magnetic effect. ________

11-20. Hysteresis losses are due to a dissipation of energy in making the domains switch their alignment back and forth when under the influence of an alternating magnetizing force. ________

11-21. A material used for magnetic core storage should have as thin a B-H loop as possible whereas a material used in a transformer should ideally have a square B-H loop. ________

11-22. Raising a magnetic material above its Curie temperature is just one of the ways of degaussing a magnet. ________

11-23. A degaussing coil is used with an ac current to cycle a material through successively smaller hysteresis loops and to eventually reduce the residual magnetism to zero. ________

11-24. One gauss is the CGS unit of flux. ________

11-25. It is possible for a charge to move perpendicular to a magnetic field and not experience a force. ________

11-26. The force acting on a current-carrying conductor depends only on the field strength, current, and length of conductor. ________

11-27. The Hall effect depends upon the deflection of charge moving perpendicular to a magnetic field to develop a small emf. ________

11-28. Direct current motors use graphite brushes to conduct current through the commutator to the conductors set in slots in the armature. ________

11-29. Shunt motors have field windings in series with the armature to develop a high starting torque. ________

11-30. The flux set up by a magnetomotive force of 500 At applied to a magnetic circuit having a reluctance of 1×10^6 At/Wb is 5×10^{-4} Wb. ________

REVIEW QUESTIONS

1. Name three magnetic materials and four nonmagnetic metals.

2. What do the directions of the lines of force around a magnet indicate?

3. What is magnetic induction?

4. Explain why a nail can be picked up using *either* end of a permanent magnet.

5. What does it mean to say that iron is more permeable than copper?

6. Describe how you would determine the magnetic polarity of an unmarked magnet using a compass.

7. What effect did Oersted discover?

8. How does the right-hand rule for coils differ from the right-hand rule for a single wire?

9. Name four applications of electromagnets.

10. Why does a doorbell provide a continuous vibrating action when a door chime does not?

11. Draw two different symbols for a relay that has two sets of normally open and normally closed contacts.

12. Draw a relay circuit that uses an SPST switch to connect 6 V to a 6-V form C relay. When the switch is open, one 120-V, 100-W lamp is lit. When the switch is closed, the 100-W lamp goes off and a 120-V, 60-W lamp comes on.

13. a. If the core in Fig. 11-12 is of a very high permeability, how much magnetic flux would you expect to find in the circular region of the hole of the doughnut and also immediately outside the core?

b. Why?

c. Would this be different if the core were made of wood?

d. How?

14. Distinguish between magnetomotive force and magnetizing intensity.

15. On the hysteresis loop of Fig. 11-13*a*, sketch a graph showing how the permeability varies with H.

16. In what units are retentivity and coercivity measured? What are typical values for a good permanent magnet?

17. a. Explain in your own words why a material having a narrow B-H loop is desirable for ac applications.

b. Explain how the magnetic domain theory supports your answer.

18. Explain how a ferrite core with a square B-H loop can be used to store digital information.

19. a. What are the three methods of demagnetization?

b. Describe the most practical method of demagnetization as applied to degaussing a carbon steel drill bit.

20. a. What is the operating principle behind a gaussmeter?

b. How can this be used to distinguish between N-type and P-type semiconductors?

21. Assuming that the electron beam of an oscilloscope is sufficiently deflected by the earth's magnetic field to be visible, describe how you would determine the direction of the earth's magnetic field.

22. Describe how a high-fidelity loudspeaker converts audio current variations to sound waves.

23. Explain why a dc motor may be reversed, simply by reversing the applied emf if the motor employs permanent magnets for the field, but not if field windings are used for the stator.

24. Draw a diagram showing how a DPDT switch can be used to reverse the direction of rotation of a series motor.

PROBLEMS

(Answers to odd-numbered problems at back of book)

11-1. A flux of 1.8×10^{-3} Wb is uniformly distributed over a rectangular path of 5 cm × 6 cm. What is the flux density?

11-2. For the flux distribution in Problem 11-1, how much flux passes through a 2-cm-diameter circle?

11-3. The flux density at the pole of a horseshoe magnet is 0.8 T. How much flux passes between the two poles if each pole face is 2 cm × 1 cm?

11-4. If the flux between the two poles in Problem 11-3 is concentrated into a 1-cm × 1-cm region, what is the flux density in this region?

11-5. How much current must flow through a 350-turn coil of wire to provide a mmf of 850 At?

11-6. How many turns of wire are needed to provide a mmf of 150 At if the current to be used is 250 mA?

11-7. If the length of the magnetic circuit in Problem 11-5 is 40 cm, what is the magnetizing intensity?

11-8. How many At are necessary to provide a magnetizing intensity of 2000 At/m in a magnetic circuit of length 60 cm?

11-9. A cast steel toroid (as in Fig. 11-12) has an inside radius of 6 cm and an outside radius of 8 cm with a square cross section. A coil of 300 turns carrying 2.5 A is wrapped around the core. A flux of 5.4×10^{-4} Wb is set up inside the core. Calculate:

a. The flux density.
b. The magnetomotive force.
c. The magnetizing intensity.
d. The permeability of the core.
e. The relative permeability of the core.

11-10. Repeat Problem 11-9 assuming a circular cross-sectional core.

11-11. If a 2-mm air gap is cut into the toroid of Problem 11-9, calculate the new current required to maintain the flux in the core at its initial level.

11-12. Repeat Problem 11-11 using a 3-mm air gap assuming a circular cross-sectional core.

11-13. A cast steel toroid (as in Fig. 11-12) has an inside diameter of 10 cm and an outside diameter of 15 cm with a circular cross section. A coil of 200 turns carrying a current of 750 mA is wrapped around the core. If the permeability of the cast steel is 7.5×10^{-4} Wb/At-m, calculate:

a. The reluctance of the cast steel toroid.
b. The flux set up in the toroid.

11-14. Repeat Problem 11-13 assuming a square cross-sectional core.

11-15. A dc motor has five turns of wire located in a single slot directly under a pole whose field strength is 0.5 T. The length of the armature is 30 cm and the current in each wire is 5 A.

a. Calculate the total force acting on the conductors in this one slot.
b. If current flows in the opposite direction in five turns of wire, located in a single slot on the opposite side of the armature (under the opposite field pole), calculate the torque if the diameter of the armature is 15 cm.

11-16. The electrons in an oscilloscope are accelerated through a potential difference of 2 kV reaching a velocity of 3×10^7 m/s (one-tenth the speed of light). Assuming that the beam is placed in an east-west direction perpendicular to the earth's magnetic field, which has a horizontal component of 0.17 G and a downward vertical component of 0.55 G, calculate the magnitude and direction of the two forces acting on a single electron. Indicate the resultant direction in which the beam should be deflected if it is moving from east to west.

11-17. A dc shunt motor is operating from a 120-V source. The resistance of the field winding is 240 Ω and the resistance of the armature is 2 Ω. When the motor is running, the total current drawn is 3.5 A. Calculate:

a. The field current.
b. The armature current.
c. The voltage drop due to the armature resistance.
d. The amount of back emf generated within the armature.
e. The total I^2R power lost in the field windings and armature.
f. The total power input to the motor.
g. The output power from the motor in watts and horsepower if 60 W are lost in friction and windage.
h. The efficiency of the motor.

CHAPTER TWELVE
DC AMMETERS AND VOLTMETERS

Most analogue electrical instruments (i.e., nondigital), such as ammeters, voltmeters, and so on, use a basic *meter movement* to provide deflection of a pointer. This meter movement relies upon the force exerted on a current-carrying conductor located in a magnetic field. Such a meter movement would cause a full-scale deflection with only a few microamperes of current through it. However, by the use of *parallel* connected resistors, called *shunts,* the range can be extended to much higher values, to produce meters capable of measuring amperes. And if resistors, called *voltmeter multipliers,* are connected in *series* with the movement, a voltmeter of any desired range results.

An ideal ammeter should have zero internal resistance; an ideal voltmeter would have an infinite internal resistance. Practical meters, because of the way they are constructed, have finite resistance and will inevitably have some effect on the circuit they are measuring. If the effect is noticeable, it is called *loading.* In this chapter we shall consider ammeter and voltmeter loading, and the percentage errors of these meters due to what are called full-scale deflection (FSD) accuracies.

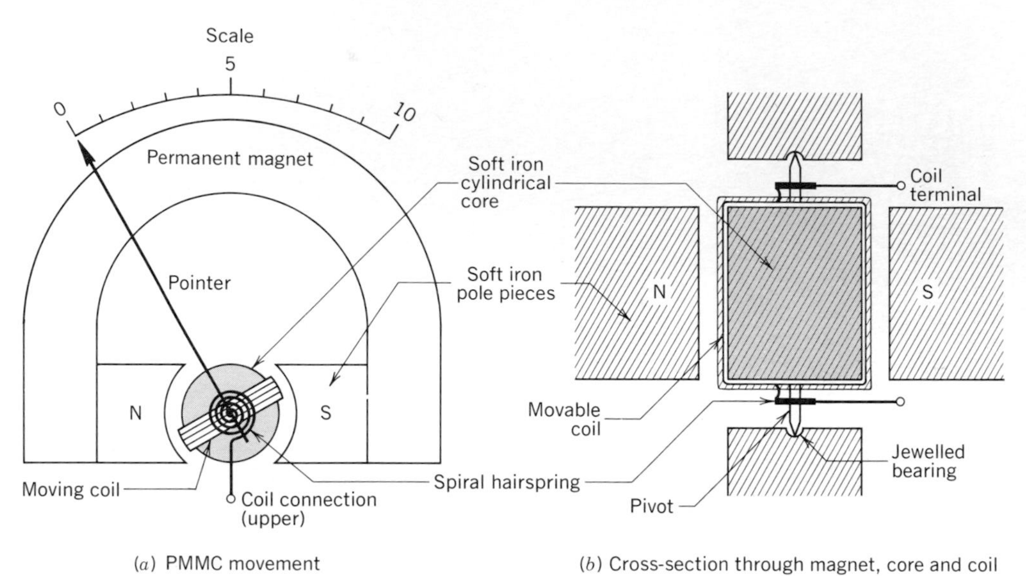

(*a*) PMMC movement (*b*) Cross-section through magnet, core and coil

FIGURE 12-1
The D'Arsonval or permanent-magnet, moving-coil movement.

12-1 THE D'ARSONVAL OR PMMC METER MOVEMENT

The permanent-magnet, moving-coil (PMMC) instrument is the most widely used for dc measurements. It is sometimes referred to as a D'Arsonval movement, after the man who first patented a device based on the moving-coil principle. The modern PMMC movement is also referred to as a Weston movement because of the work done by Edward Weston in converting the initially delicate meter into a rugged portable meter.

As the name implies, and as shown in Fig. 12-1, the PMMC movement consists of a rectangular movable coil suspended in the strong magnetic field of a permanent magnet. The phosphor-bronze hair springs serve to lead the current to and from the coil. The springs also provide a restoring torque. That is, when current flows through the coil, a force acts on the two vertical sides of the coil, causing a turning motion, or torque. A pointer attached to the coil deflects up scale, as the hair springs tighten and resist the motion. An increase in current causes a greater deflection, the pointer coming to rest when the applied torque is balanced by the tightened hair springs. When the current is removed, the springs restore the movable coil to the zero position. Provision is made to zero the pointer mechanically from the outside of the meter by a small screw.

It can be shown from the previous chapter that the torque acting on the movable coil is given by:

$$T = NBIA \sin \alpha \qquad \text{newton-meters (N} \cdot \text{m)} \qquad (12\text{-}1)$$

where: T is the torque acting on the coil in newton-meters, $\text{N} \cdot \text{m}$

N is the number of turns in the coil

B is the flux density of the permanent magnet in webers/m², Wb/m^2

I is the current in the movable coil in amperes, A

A is the area of the coil in square meters, m^2

α is the angle between the pointer and the magnetic field B

This equation (not to be used here for calculations) shows that the torque is linearly related to the coil current if N, B, A, and $\sin \alpha$ are all constant. Also, B and α are kept constant by using curved pole pieces and the soft iron cylindrical core. These ensure a constant *radial* magnetic field in the air gap where the coil moves. See Fig. 12-2. Thus $\alpha = 90°$ and $\sin \alpha = 1$ no matter where

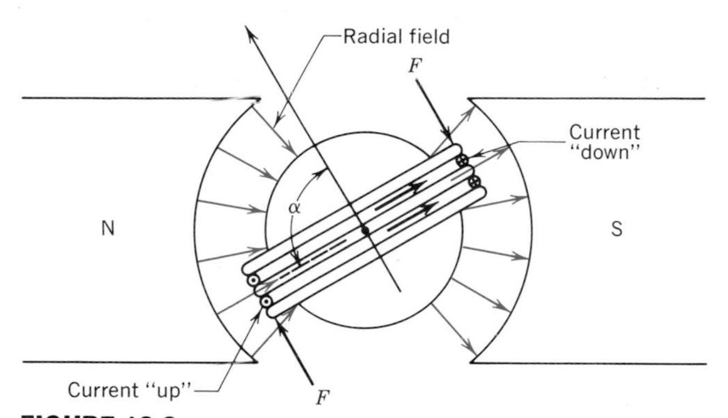

FIGURE 12-2
Detail of radial field in air gap for movable coil of a D'Arsonval or Weston meter movement.

the coil is located as its turns. This is the reason that dc ammeters and voltmeters have *linear* scales. They employ meter movements that have equal divisions for equal current increments. Doubling the current doubles the deflection, and so on.

It should be clear that the permanent-magnet, moving-coil meter movement is polarity-sensitive. If current is allowed through the meter in the opposite direction, the deflection is reversed and damage may occur.

Modifications of the horseshoe magnet shown in Fig. 12-1 include the use of a concentric magnet. This type has a complete outer ring of soft steel, which provides the benefit of magnetic shielding from external fields (as from large current-carrying conductors), which could affect readings. Also, this makes possible an expanded scale covering 240° in a very compact space.

12-1.1 Current Sensitivities

A meter movement's *current sensitivity* is the amount of current through the movable coil to cause full-scale deflection (FSD). In order to make a very sensitive movement, say $I_{FSD} = 50\ \mu A$, the strongest magnetic field and a large number of turns on the coil are desirable. The latter naturally increases the resistance of the coil, and its weight, so that some design compromise is necessary. Table 12-1 shows the typical resistance of one manufacturer's meter movements and ammeters. Note that meter resistance decreases as full-scale deflection current increases. Other movements may be slightly lower in resistance or, in some cases, as much as seven times higher.

TABLE 12-1
Typical Resistances of Meter Movements and Ammeters

Range, I_{FSD}	Approximate Movement Resistance, R_m, Ohms
50 μA	3000
100 μA	2000
200 μA	600
1 mA	20
10 mA	5
50 mA	1
1 A	0.05
10 A	0.005

12-2 AMMETER SHUNTS

In order to extend the range of a basic meter movement, to measure currents larger than the full-scale deflection current, parallel resistors called *shunts* are used to direct most of the current being measured away from the meter movement. Design of the ammeter is based on full-scale deflection currents for the meter movement and the overall ammeter. Refer to Fig. 12-3.

Let I_T be the total current into the ammeter that causes full-scale deflection. This is the *range* of the ammeter. This current divides up with current I_m going through the meter movement R_m, and current I_s going through the shunt resistor R_s. By Kirchhoff's current law:

$$I_T = I_m + I_s$$

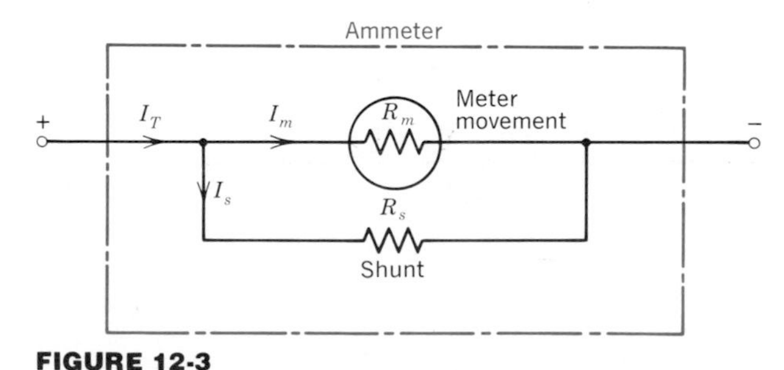

FIGURE 12-3
Basic ammeter shunt circuit.

And by Ohm's law, the voltage across the meter is given by

$$V_m = I_m R_m = I_s R_s$$

Thus $$R_s = R_m \times \frac{I_m}{I_s} \quad \text{ohms } (\Omega) \qquad (12\text{-}1)$$

where: R_s is the shunt resistance in ohms, Ω
R_m is the meter movement resistance in ohms, Ω
I_m is the FSD current of the meter movement in amperes, A
I_s is the shunt current, equal to $I_T - I_m$, in amperes, A

EXAMPLE 12-1

Given a 50-μA meter movement of resistance $R_m = 3$ kΩ, calculate:

a. The value of shunt resistance to make a 1-mA range meter.
b. The voltage drop across the meter at full-scale deflection.
c. The total resistance of the meter.

SOLUTION

Refer to Fig. 12-4*a*.

a. $I_s = I_T - I_m$
$= 1 \text{ mA} - 50\ \mu\text{A}$
$= 1000\ \mu\text{A} - 50\ \mu\text{A} = 950\ \mu\text{A}$

$$R_s = R_m \times \frac{I_m}{I_s} \qquad (12\text{-}1)$$

$$= 3\text{ k}\Omega \times \frac{50\ \mu\text{A}}{950\ \mu\text{A}}$$

$$= 3000\ \Omega \times \frac{1}{19} = \mathbf{158\ \Omega}$$

b. $V_m = I_m R_m$ (3-1b)
At FSD, $V_m = 50\ \mu\text{A} \times 3\text{ k}\Omega = \mathbf{150\ mV}$

c. Total meter resistance,

$$R_T = \frac{V_m}{I_T} \qquad (3\text{-}1)$$

$$= \frac{150 \times 10^{-3}\text{ V}}{1 \times 10^{-3}\text{ A}} = \mathbf{150\ \Omega}$$

Note that this may also be obtained by determining the parallel equivalence of 3 kΩ and 158 Ω.

Note also that if only 0.5 mA is flowing through the meter, only 25 μA will pass through the movement so that it will deflect only to half-scale. This point would be marked as 0.5 mA on the *meter scale,* as shown in Fig. 12-4*b*.

12-3 MULTIRANGE AMMETERS

Many ammeters in the electronic laboratory are multirange meters. That is, the current causing full-scale deflection can be selected by a switch to be close to the value being measured. In this way, larger deflections and more accurate measurements can be made. One method simply consists of selecting the appropriate shunt resistor by means of a *range selector switch,* as illustrated by Example 12-2.

EXAMPLE 12-2

Given a 50-μA, 3-kΩ meter movement, it is desired to construct a multirange milliammeter with ranges of 1 mA, 10 mA, 100 mA, and 1 A. Calculate:

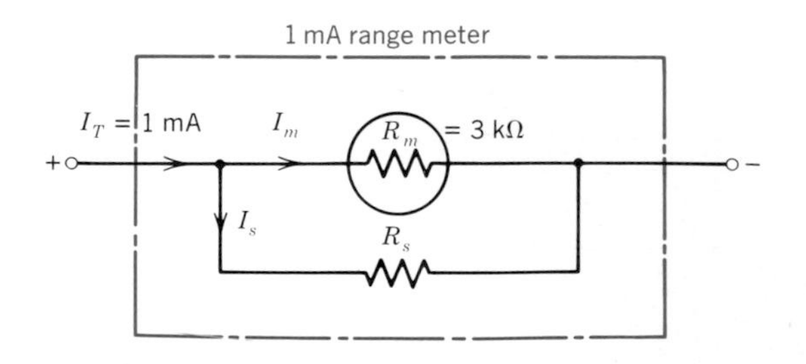

(*a*) Shunt resistor used to convert meter movement to 1-mA range meter

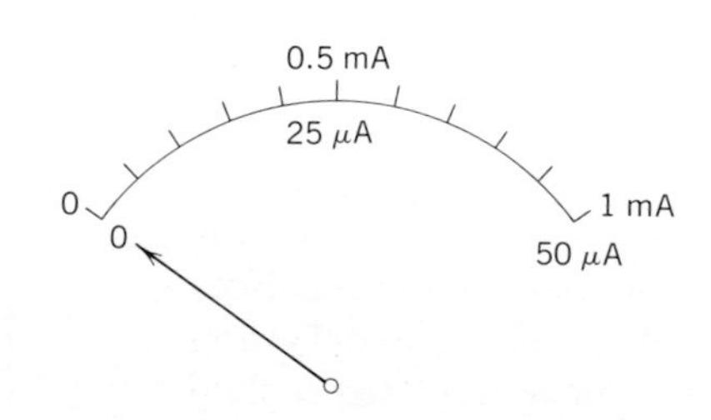

(*b*) 1-mA range meter scale marking corresponding to 50-μA meter movement

FIGURE 12-4
Ammeter shunt circuit and scale for Example 12-1.

a. Appropriate shunt resistor values.
b. The voltage drop across the meter on each range at FSD.
c. The total resistance of the meter on each range.

Draw a circuit diagram of the switching arrangement with all values.

SOLUTION

a. We have already calculated R_s for a 1-mA range in Example 12-1.

$$R_s = 158\ \Omega$$

For 10 mA, I_s = 10 mA − 0.05 mA = 9.95 mA.

$$R_s = R_m \times \frac{I_m}{I_s} \quad (12\text{-}1)$$

$$= 3000\ \Omega \times \frac{0.05\ \text{mA}}{9.95\ \text{mA}} = \mathbf{15.1\ \Omega}$$

Similarly, for 100 mA, $R_s = \mathbf{1.5\ \Omega}$

for 1 A, $R_s = \mathbf{0.15\ \Omega}$

b. Since, at FSD on each range, there will be 50 μA through the 3-kΩ movement, the voltage drop on each range is given by

$$V_m = I_m R_m \quad (3\text{-}1b)$$

$$= 50\ \mu\text{A} \times 3\ \text{k}\Omega = \mathbf{150\ mV}$$

c. Total meter resistance, $R_T = \frac{V_m}{I_T}$ (3-1)

for 1-mA range, $R_T = \mathbf{150\ \Omega}$

for 10-mA range, $R_T = \frac{150 \times 10^{-3}\ \text{V}}{10 \times 10^{-3}\ \text{A}} = \mathbf{15\ \Omega}$

for 100-mA range, $R_T = \frac{150 \times 10^{-3}\ \text{V}}{100 \times 10^{-3}\ \text{A}} = \mathbf{1.5\ \Omega}$

for 1-A range, $R_T = \frac{150 \times 10^{-3}\ \text{V}}{1\ \text{A}} = \mathbf{0.15\ \Omega}$

See Fig. 12-5.

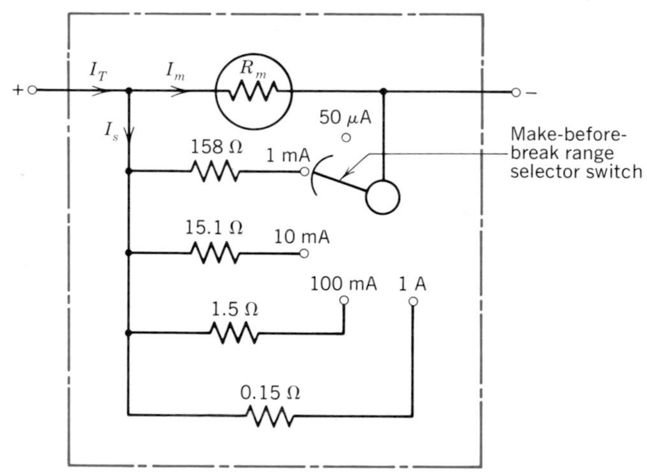

FIGURE 12-5
Multirange ammeter shunt circuit for Example 12-2.

Note how it is necessary to use a *make-before-break* rotary switch for current range selection. This type of switch ensures that while switching between ranges, at no time will the movement be unshunted. With an ordinary switch, there would be a momentary period when *all* the meter current would pass through the movement, causing damage. Many milliammeters use a pair of breakdown diodes connected directly across the meter movement for protection purposes. Then, if the milliammeter is connected into a circuit on too low a range, much of the current is diverted through one of the diodes. On the correct range, however, the diodes have no effect because there is insufficient voltage across the movement to make them conduct.

12-4 AYRTON OR UNIVERSAL SHUNT

Many multirange meters, instead of using a make-before-break switch, use an Ayrton shunt. This type of shunt, also called a *universal shunt,* has the advantage of using resistors that are closer to standard values. This is not the case in the multirange ammeter of Example 12-2, where accurate values of 158 Ω and 15.1 Ω would be necessary. Example 12-3 illustrates the method of selecting resistors for an Ayrton shunt. The application of series-parallel circuit theory is all that is necessary. No special equations are involved.

EXAMPLE 12-3

Given a 50-μA, 3-kΩ meter movement, determine the values of R_1, R_2, and R_3 in Figure 12-6 to design an Ayrton shunt with ranges of 100 μA, 1 mA, and 10 mA.

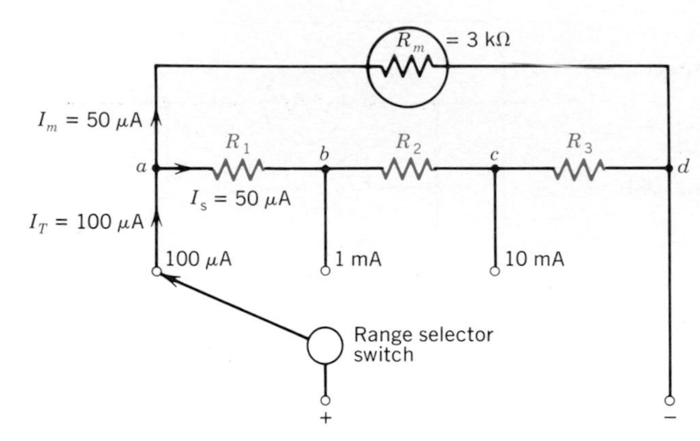

(a) 100-μA range

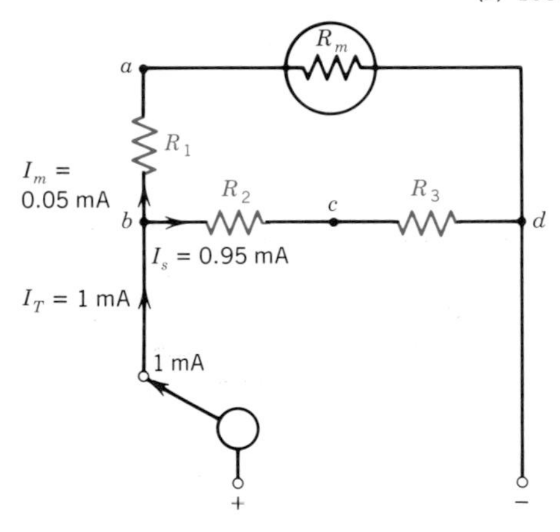

(b) 1-mA range

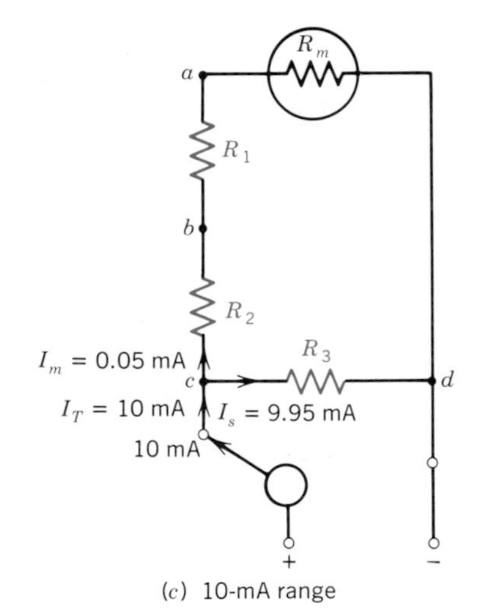

(c) 10-mA range

FIGURE 12-6
Ayrton shunt multirange milliammeter for Example 12-3.

SOLUTION

100 μA range; Fig. 12-6*a*:

The total shunt resistance equals $R_1 + R_2 + R_3$ and must carry 50 μA. In order for the current to divide equally, the resistance of the two paths must be equal, so

$$R_1 + R_2 + R_3 = R_m = 3\ \text{k}\Omega$$

1 mA range; Fig. 12-6*b*:

R_1 is now in series with R_m, with $R_2 + R_3$ shunting these two.

$$0.05\ \text{mA}\ (R_1 + R_m) = 0.95\ \text{mA}\ (R_2 + R_3)$$

But $R_m = 3\ \text{k}\Omega$ and $R_2 + R_3 = 3\ \text{k}\Omega - R_1$

Thus $0.05\ \text{mA}\ (R_1 + 3\ \text{k}\Omega) = 0.95\ \text{mA}\ (3\ \text{k}\Omega - R_1)$

$$R_1 + 3\ \text{k}\Omega = \frac{0.95\ \text{mA}}{0.05\ \text{mA}}(3\ \text{k}\Omega - R_1)$$

$$= 19\ (3\ \text{k}\Omega - R_1)$$

$$20\,R_1 = 54\ \text{k}\Omega$$

$$R_1 = \mathbf{2.7\ k\Omega}$$

10 mA range; Fig. 12-6*c*:

$R_1 + R_2$ are now in series with R_m, with R_3 shunting all three.

$$0.05\ \text{mA}\ (R_1 + R_2 + R_m) = 9.95\ \text{mA} \times R_3$$

But $R_1 = 2.7\ \text{k}\Omega$, $R_m = 3\ \text{k}\Omega$, and $R_2 + R_3 = 300\ \Omega = 0.3\ \text{k}\Omega$, so $R_3 = 0.3\ \text{k}\Omega - R_2$.

Thus $0.05\ \text{mA}\ (2.7\ \text{k}\Omega + R_2 + 3\ \text{k}\Omega)$

$$= 9.95\ \text{mA}(0.3\ \text{k}\Omega - R_2)$$

solving, $R_2 = 0.27\ \text{k}\Omega = \mathbf{270\ \Omega}$

and $R_3 = 300\ \Omega - 270\Omega = \mathbf{30\ \Omega}$

It can be shown that, in general, the total resistance of a meter using an Ayrton shunt on any given range, is slightly higher than the multishunt multirange ammeter. Another disadvantage is that if *one* of the shunt resistors changes value, or opens up, *all* ranges are affected. This is not so for the multishunt ammeter.

12-5 ACCURACY AND ERRORS

If a multirange ammeter is being used, which range should be selected for maximum accuracy? Some errors in making a reading on an instrument are due to the operator. These include:

1. Imprecise interpolation between scale markings when the pointer falls between main divisions.
2. Line-of-sight errors when the operator does not look perpendicularly at the pointer and scale (parallax). Some meters have a mirror to line up the pointer and its image before taking a reading.
3. Not checking the mechanical zero before using the instrument.

Other errors that have a similar *constant* effect are due to the manufacture or use of the instrument:

1. Scale error due to markings not being in exactly the correct place, because mass-produced scales may not fit the actual characteristics of each meter movement. (The instrument needs individual calibration.)
2. Continual use or misuse of the instrument may cause frictional effects that do not allow the pivot to swing freely to its proper location. (This error can be minimized by gently tapping the case before taking a reading.)

All of the above errors are *constant amounts*, independent of the deflection of the pointer across the scale.

There are other errors that *are* proportional to the pointer deflection. They include:

1. Inexact resistances for shunts or multipliers in ammeters and voltmeters.
2. Temperature effects on the resistance values in (1) and the movement coil resistance, as well as the flexibility of springs.

The errors due to the last two effects are relatively small compared with the errors that are constant. Thus the total error due to all effects combined is a relatively constant *amount* at different points along the scale, rather than a *percentage of the reading* being taken. This error is usually expressed, by the manufacturer, as a percentage of the range or full-scale reading. Thus, if a meter is quoted as having an accuracy of ±3%, it means that the error at *any* point of the scale will not exceed 3% of the *full-scale reading*. Thus the percentage accuracy may be very poor for a reading taken on the lower part of the scale. This is illustrated in Example 12-4.

EXAMPLE 12-4

A 10-mA range ammeter is guaranteed to have an accuracy of ±3%, FSD. For readings on the scale of 1 mA, 5 mA, and 10 mA determine:

a. The possible error in mA.
b. The range of values for the true current.
c. The percentage error in each reading.

SOLUTION

a. For all three readings

$$\text{the error} = \pm 3\% \text{ of } 10 \text{ mA}$$

$$= \frac{\pm 3}{100} \times 10 \text{ mA} = \mathbf{\pm 0.3 \text{ mA}}$$

b. The actual reading may be:

For 1 mA indication:

$$1 \pm 0.3 \text{ mA} = \mathbf{0.7 \text{ to } 1.3 \text{ mA}}$$

For 5 mA indication:

$$5 \pm 0.3 \text{ mA} = \mathbf{4.7 \text{ to } 5.3 \text{ mA}}$$

For 10 mA indication:

$$10 \pm 0.3 \text{ mA} = \mathbf{9.7 \text{ to } 10.3 \text{ mA}}$$

c. $\text{Percentage error} = \dfrac{\text{error}}{\text{reading}} \times 100\%$

For 1 mA reading, % error

$$= \frac{0.3 \text{ mA}}{1 \text{ mA}} \times 100\% = \mathbf{30\%}$$

For 5 mA reading, % error

$$= \frac{0.3 \text{ mA}}{5 \text{ mA}} \times 100\% = \mathbf{6\%}$$

For 10 mA reading, % error

$$= \frac{0.3 \text{ mA}}{10 \text{ mA}} \times 100\% = \mathbf{3\%}$$

Thus percentage error due to FSD errors, can be minimized by selecting a range that gives as close a deflection to full scale as possible.

12-6 AMMETER LOADING

When an instrument is used for measurement, it is expected that the application of the instrument to the circuit will *not* change conditions in the circuit.

Since an ammeter is connected in series in a circuit, the *ideal* or perfect ammeter would have *zero* resistance. Then its presence in the circuit would not be noticed under any conditions. But if a practical ammeter's resistance is in any way comparable to the circuit resistance, there is a problem of *loading*, as shown by Example 12-5.

EXAMPLE 12-5

A technician, attempting to determine current in a circuit using Ohm's law, measures the voltage across a 1.2-kΩ resistor and finds it to be exactly 1 V. Not being sure that the resistor is actually 1.2 kΩ, he decides to measure the current by inserting a VOM on the 1-mA range in series with the resistor. If the VOM's resistance is 1 kΩ (typical for many VOMs), calculate:

a. The actual current in the circuit assuming that the resistor is exactly 1.2 kΩ.
b. The indication of the meter on the 1-mA range.
c. The indication of the meter if it is switched to the 10-mA range where its resistance is 100 Ω.
d. The effect of FSD errors on the 10-mA range reading, assuming an accuracy of ±3%.

SOLUTION

a. $$I = \frac{V}{R} \qquad (3\text{-}1a)$$

$$= \frac{1 \text{ V}}{1.2 \text{ k}\Omega} = \mathbf{0.83 \text{ mA}}$$

b. $$I = \frac{V}{R_T}$$

$$= \frac{1 \text{ V}}{1.2 \text{ k}\Omega + 1 \text{ k}\Omega} = \mathbf{0.45 \text{ mA}}$$

c. $$I = \frac{V}{R_T}$$

$$= \frac{1 \text{ V}}{1.2 \text{ k}\Omega + 0.1 \text{ k}\Omega} = \mathbf{0.77 \text{ mA}}$$

d. It is evident that the higher range (10 mA) with its much lower resistance (100 Ω) has reduced loading error compared with the 1-kΩ, 1-mA range meter. However, the 0.77 mA will cause such a small deflection on the 10-mA range that it will be difficult to read accurately.

$$\text{FSD error} = \frac{\pm 3}{100} \times 10 \text{ mA} = \pm 0.3 \text{ mA}$$

Thus the 10-mA ammeter could indicate anything between 0.47 and 1.07 mA. On the 1-mA range,

$$\text{FSD error} = \frac{\pm 3}{100} \times 1 \text{ mA} = \pm 0.03 \text{ mA}$$

Thus the 1-mA ammeter could indicate anything between 0.42 and 0.48 mA. The range of values on the 10-mA scale at least includes the values of 0.77 mA and 0.83 mA whereas the 1-mA scale could never read higher than 0.48 mA. Put another way, on the 10-mA range the readings could be from 43% low to 29% high compared with the true value of 0.83 mA. On the 1-mA range, the readings could be from 42% low to 49% low compared with the true value of 0.83 mA.

Conclusion: Where loading is a problem, switching to the next higher range will reduce loading although the increase in FSD errors may remove much of the increased accuracy.

12-6.1 Clamp-On Ammeter

A meter that causes no loading because it is not connected in series with the circuit is the clamp-around probe shown in Fig. 12-7.

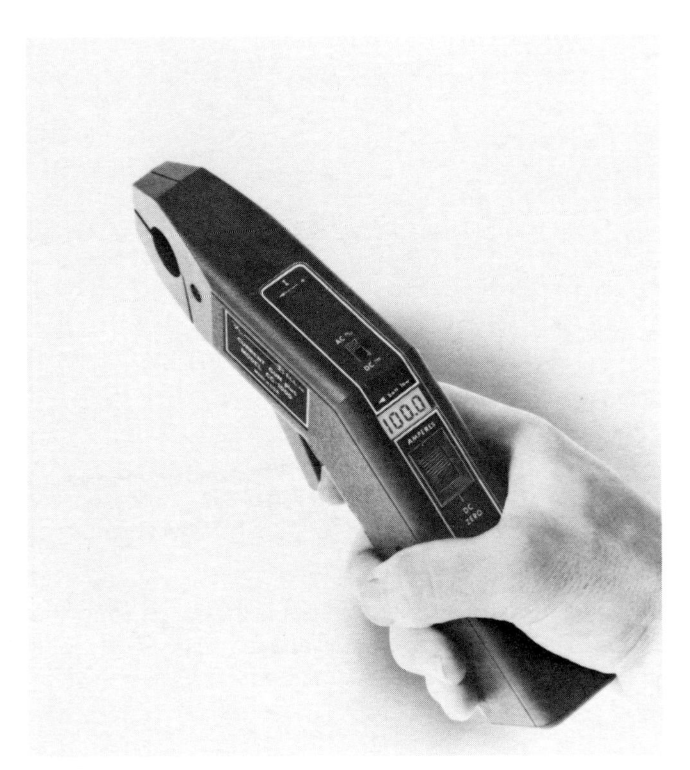

FIGURE 12-7
Clamp-around ammeter probe or "current gun" that measures alternating and direct current. (Courtesy F. W. Bell Inc.)

There is no need to break the circuit because the jaws are opened to clamp the probe around the insulated wire in which the current is flowing. The current in the wire (ac or dc) sets up a magnetic field whose strength is proportional to the current. The probe accurately concentrates this field in a magnetic core surrounding the clamp-around jaws. A Hall-effect generator (Section 11-11) is mounted in an air gap in the core. The Hall emf generated is directly proportional to the current in the conductor.

Four AA cells in the body of the probe power a dc-coupled linear operational amplifier. This drives a three-and-a-half digit LCD readout located in the handle. The "current gun" has a range of ±200 A from dc to 1 kHz with a resolution of 0.1 A.

Currents under 1 A may be read with reasonable accuracy by looping a number of turns of the conductor through the jaws, thus effectively "multiplying" the current.

12-7 VOLTMETER MULTIPLIERS

We have seen how the 50-μA, 3-kΩ meter movement requires a voltage of only 150 mV across it to cause full-scale deflection. This means that the movement may be used to measure voltages up to a maximum of only 150 mV. **It is therefore necessary to use a resistor in *series* with the meter movement to extend the range to higher voltages. Such a resistor is called a voltmeter *multiplier*.**

A multiplier limits the current to the full-scale deflection value of the movement whenever the voltmeter is connected across its rated voltage. The method of determining the required series resistor or multiplier resistor for a given voltage range involves simple series circuit theory. See Fig. 12-8.

As with the ammeter, the design is based on full-scale deflection values.

total voltmeter resistance,

$$R_T = \frac{\text{voltmeter range}}{\text{FSD current of movement}}$$

$$= \frac{V_T}{I_{\text{FSD}}}$$

But $\quad R_T = R_s + R_m$

and $\quad R_s = R_T - R_m$

Therefore, $$R_s = \frac{V_T}{I_{\text{FSD}}} - R_m \quad \text{ohms } (\Omega) \qquad (12\text{-}2)$$

where: R_s is the resistance of the voltmeter multiplier in ohms, Ω

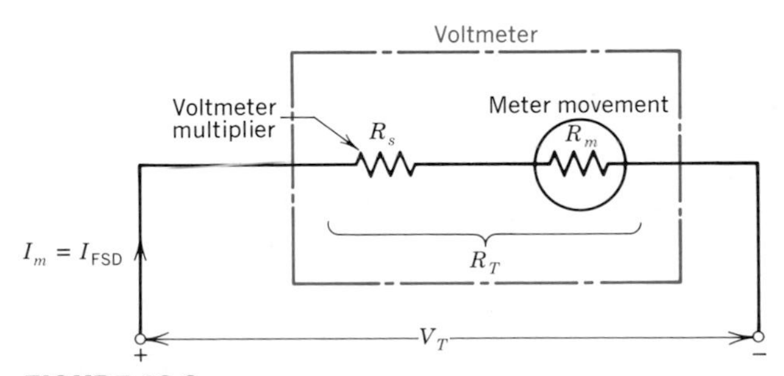

FIGURE 12-8
Basic voltmeter multiplier circuit.

R_m is the resistance of the meter movement in ohms, Ω

V_T is the range of the voltmeter in volts, V

I_{FSD} is the meter movement's full-scale deflection current in amperes, A

If we let

$$\frac{1}{I_{FSD}} = S \qquad (12\text{-}3)$$

Eq. 12-2 can be rewritten as

$$R_s = S \times \text{Range} - R_m \quad \text{ohms } (\Omega) \qquad (12\text{-}4)$$

Since S is the reciprocal of current, it has units of ohms/volt. This is called the voltmeter's sensitivity as further explained in Section 12-10.

EXAMPLE 12-6

It is required to convert a 50-μA, 3-kΩ meter movement to a voltmeter having a range of 5 V. Calculate:

a. The total resistance of the meter.

b. The necessary resistance of the voltmeter multiplier using Eqs. 12-2 and 12-4.

SOLUTION

Refer to Fig. 12-9*a*

a. Total resistance of voltmeter,

$$R_T = \frac{V}{I_m} \qquad (3\text{-}1)$$

$$= \frac{5\text{ V}}{50\ \mu\text{A}} = \mathbf{100\ k\Omega}$$

b. $$R_s = \frac{V_T}{I_{FSD}} - R_m \qquad (12\text{-}2)$$

$$= R_T - R_m$$

$$= 100\text{ k}\Omega - 3\text{ k}\Omega = \mathbf{97\ k\Omega}$$

or, $$S = \frac{1}{I_{FSD}} \qquad (12\text{-}3)$$

$$= \frac{1}{50 \times 10^{-6}\text{A}} = 20\text{ k}\Omega/\text{V}$$

$$R_s = S \times \text{Range} - R_m \qquad (12\text{-}4)$$

$$= 20\text{ k}\Omega/\text{V} \times 5\text{ V} - 3\text{ k}\Omega$$

$$= 100\text{ k}\Omega - 3\text{ k}\Omega = \mathbf{97\ k\Omega}$$

Note that if the voltmeter is connected across only 2.5 V dc, only 25 μA will flow through the meter, and the pointer will deflect halfway toward full scale. This point on the voltmeter scale will be marked 2.5 V, as shown in Fig. 12-9*b*.

12-8 MULTIRANGE VOLTMETERS

One method of making a multirange voltmeter is to use a *range selector switch* to series-connect the appropriate multiplier. See Example 12-7 and Fig. 12-10.

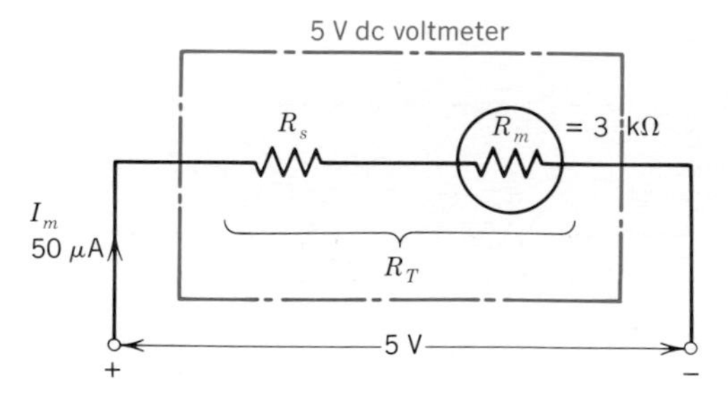

(*a*) Connection of voltmeter multiplier to convert meter movement to 5-V range voltmeter

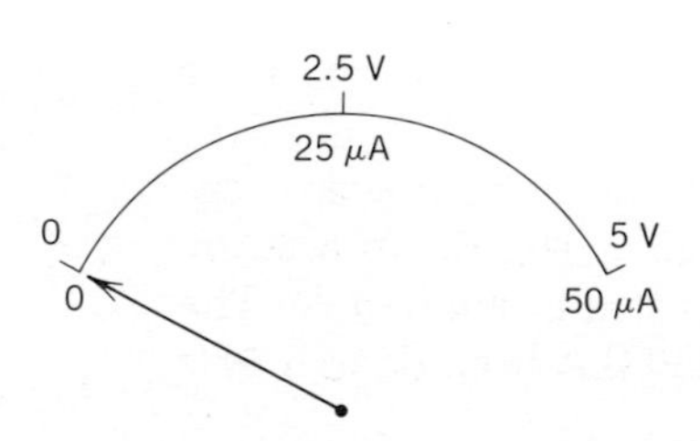

(*b*) Voltmeter markings corresponding to meter movement's current

FIGURE 12-9
Voltmeter multiplier circuit and scale for Example 12-6.

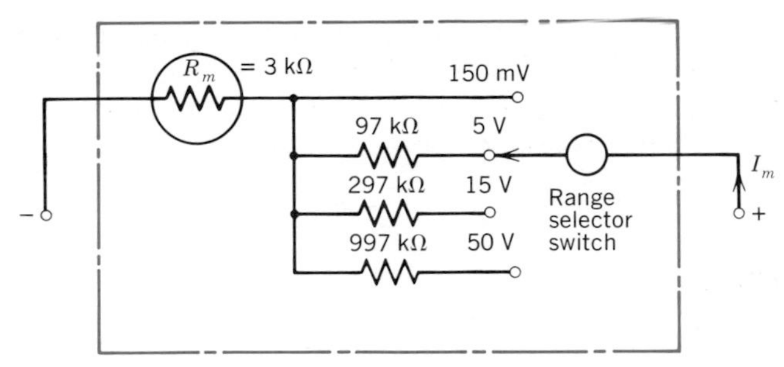

FIGURE 12-10
Multirange voltmeter circuit using individual multipliers for Example 12-7.

EXAMPLE 12-7

Convert a 50-μA, 3-kΩ meter movement into a multirange voltmeter having ranges of 5 V, 15 V, and 50 V.

SOLUTION

For 5-V range, $R_T = 100\ \text{k}\Omega$, $R_s = \mathbf{97\ k\Omega}$

$$\text{For 15-V range, } R_T = \frac{V}{I_m} \tag{3-1}$$

$$= \frac{15\ \text{V}}{50\ \mu\text{A}} = 300\ \text{k}\Omega$$

$$R_s = R_T - R_m \tag{12-2}$$

$$= 300\ \text{k}\Omega - 3\ \text{k}\Omega = \mathbf{297\ k\Omega}$$

For 50-V range,

$$R_s = S \times \text{range} - R_m \tag{12-4}$$

$$= 20\ \text{k}\Omega/\text{V} \times 50\ \text{V} - 3\ \text{k}\Omega$$

$$= 1\ \text{M}\Omega - 3\ \text{k}\Omega = \mathbf{997\ k\Omega}$$

12-9 SERIES-CONNECTED MULTIRANGE VOLTMETER

Instead of using individual multipliers, each having its own nonstandard value of resistance, the multipliers can be connected in series, as in Fig. 12-11.

EXAMPLE 12-8

Determine the values of R_1, R_2, and R_3 in Fig. 12-11 to convert the 50-μA, 3-kΩ movement into a multirange voltmeter having ranges of 150 mV, 5 V, 15 V, and 50 V.

SOLUTION

$$S = \frac{1}{I_{\text{FSD}}} \tag{12-3}$$

$$= \frac{1}{50 \times 10^{-6}\ \text{A}} = 20\ \text{k}\Omega/\text{V}$$

On the 5-V range,

$$R_1 = S \times \text{range} - R_m \tag{12-4}$$

$$= 20\ \text{k}\Omega/\text{V} \times 5\ \text{V} - 3\ \text{k}\Omega$$

$$= 100\ \text{k}\Omega - 3\ \text{k}\Omega = \mathbf{97\ k\Omega}$$

On the 15-V range,

$$R_2 = S \times \text{range} - (R_m + R_1) \tag{12-4}$$

$$= 20\ \text{k}\Omega/\text{V} \times 15\ \text{V} - (3\ \text{k}\Omega + 97\ \text{k}\Omega)$$

$$= 300\ \text{k}\Omega - 100\ \text{k}\Omega = \mathbf{200\ k\Omega}$$

On the 50-V range,

$$R_3 = S \times \text{range} - (R_m + R_1 + R_2) \tag{12-4}$$

$$= 20\ \text{k}\Omega/\text{V} \times 50\ \text{V} - (3\ \text{k}\Omega + 97\ \text{k}\Omega + 200\ \text{k}\Omega)$$

$$= 1000\ \text{k}\Omega - 300\ \text{k}\Omega = \mathbf{700\ k\Omega}$$

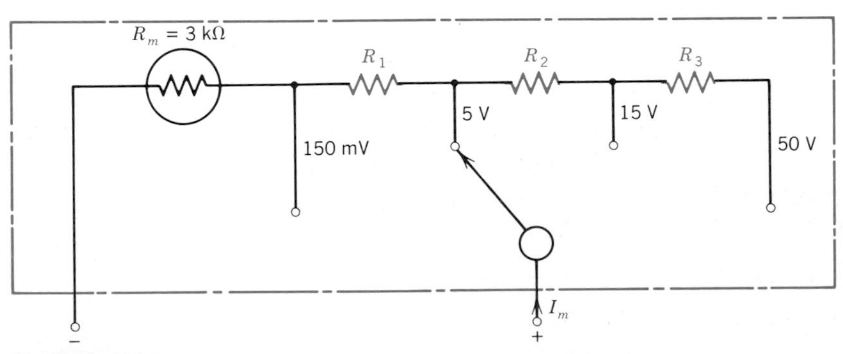

FIGURE 12-11
Series-connected multiplier circuit for Example 12-8.

Note that apart from the first multiplier, R_1, the resistors are different from those used in the individual multiplier multirange voltmeter of Example 12-7 and can be obtained from standard resistance values. A disadvantage of the circuit is the effect that a change in value of one resistor will have on all the other ranges.

12-10 VOLTMETER SENSITIVITY

The sensitivity of a voltmeter may be defined as follows:

$$\text{sensitivity} = \frac{\text{total resistance of voltmeter}}{\text{range of voltmeter}}$$

$$S = \frac{R_T}{V_T} \quad \text{ohms/volt} \qquad (12\text{-}5)$$

For example, determine the sensitivity of the previous two multirange voltmeters.

$$\text{5-V range: sensitivity} = \frac{R_T}{V_T} = \frac{100\ \text{k}\Omega}{5\ \text{V}} = 20\ \text{k}\Omega/\text{V}$$

$$\text{15-V range: sensitivity} = \frac{R_T}{V_T} = \frac{300\ \text{k}\Omega}{15\ \text{V}} = 20\ \text{k}\Omega/\text{V}$$

$$\text{50-V range: sensitivity} = \frac{R_T}{V_T} = \frac{1\ \text{M}\Omega}{50\ \text{V}} = 20\ \text{k}\Omega/\text{V}$$

Evidently, the sensitivity S is a constant for a given meter movement. In fact, the voltmeter sensitivity may also be expressed as the reciprocal of the movement's current sensitivity, as shown in Section 12-7.

$$S = \frac{1}{I_{\text{FSD}}} \qquad (12\text{-}3)$$

In our case, $S = \dfrac{1}{50\ \mu\text{A}} = \dfrac{1}{50 \times 10^{-6}\ \text{V}/\Omega} = 20\ \text{k}\Omega/\text{V}$

The value of knowing a voltmeter's sensitivity is the ability to determine the voltmeter's total resistance for any given *range*. For example, if a voltmeter has a sensitivity of 20 kΩ/V (very typical for many VOMs) the total resistance of the voltmeter on its 150 V range is

$$R_T = S \times V_T$$
$$= 20\ \text{k}\Omega/\text{V} \times 150\ \text{V} = 3000\ \text{k}\Omega = 3\ \text{M}\Omega$$

Note that if this voltmeter, on this range, is measuring a voltage of 75 V, the total resistance of the meter is still 3 MΩ, not 1.5 MΩ. If used on a 1.5-V range, this voltmeter will present a resistance of only 30 kΩ.

12-11 VOLTMETER LOADING

The effect of a voltmeter when it is connected in a circuit is called voltmeter loading. Since a voltmeter is connected across a device, that is, in parallel with it, the voltmeter represents a shunting resistance. If this resistance is low in comparison with the circuit resistance, the voltmeter loading may be extreme, depending upon the circuit conditions. It is evident then that an ideal voltmeter would have an *infinite* resistance. We can now see that this condition can best be approached by a voltmeter having as *high* a sensitivity or as high an ohms/volt rating as possible. (Portable voltmeters found in VOMs may have sensitivities ranging from 1 kΩ/V up to 100 kΩ/V.) Thus the resistance of a given meter varies widely according to the range being used. For many electronic and digital voltmeters (EVMs and DVMs) the resistance of the meter is a constant. For example, the electronic voltmeter shown in Fig. 1-8 has an input (or total) resistance of 11 MΩ on all ranges from 1.5 to 1500 V dc.

EXAMPLE 12-9

Two 1-MΩ resistors are series-connected across a 10-V source. Calculate and compare with initial conditions:

a. The reading of a 20 kΩ/V VOM on the 5-V range connected across one of the resistors.
b. The new reading if the range is increased to 50 V.
c. The reading of an EVM on the 5-V range connected in place of the VOM.
d. Each reading taking into account a ±3% FSD accuracy.

See Fig. 12-12.

SOLUTION

The initial circuit conditions consist of a current of 5 μA causing a voltage drop of 5 V across each resistor. See Fig. 12-12*a*.

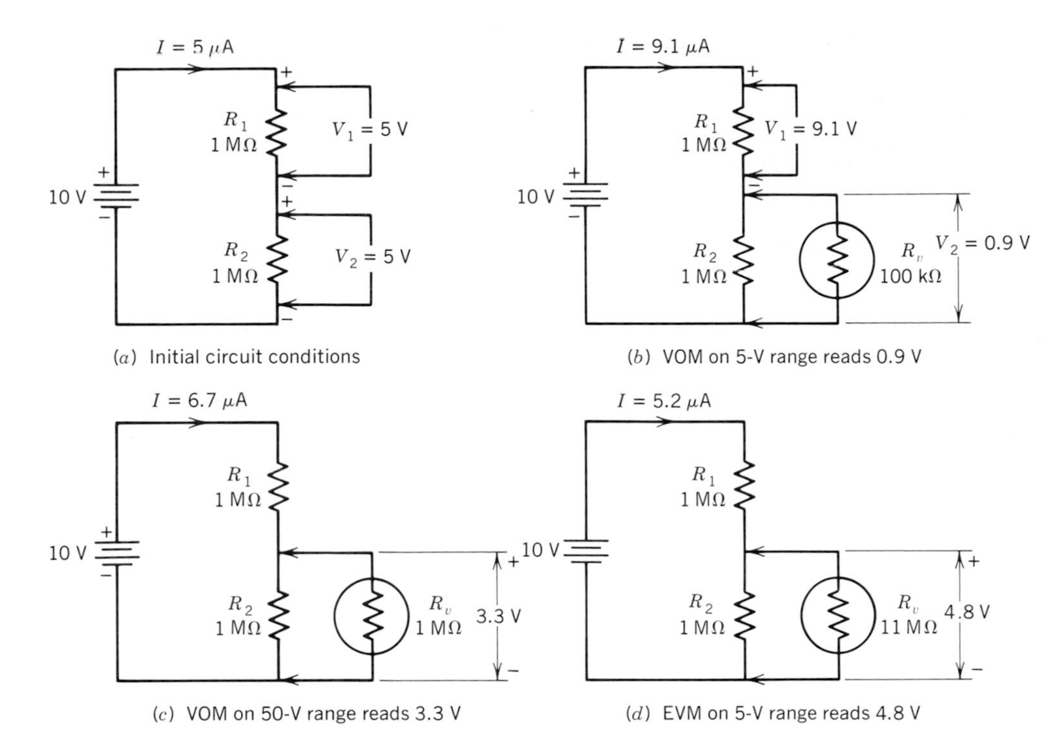

FIGURE 12-12
Loading effects of voltmeters on various ranges for Example 12-9.

a. The resistance of the 5-V range meter = 20 kΩ/V × 5 V = 100 kΩ. Refer to Fig. 12-12*b*.

Total circuit resistance

$$= 1\ \text{M}\Omega + 1\ \text{M}\Omega \| 100\ \text{k}\Omega \approx 1.1\ \text{M}\Omega$$

$$\text{New circuit current, } I = \frac{V}{R_T}$$

$$= \frac{10\ \text{V}}{1.1\ \text{M}\Omega} \approx 9.1\ \mu\text{A}$$

$$\text{Voltage drop across } R_1 = V_1 = IR_1$$

$$= 9.1\ \mu\text{A} \times 1\ \text{M}\Omega = 9.1\ \text{V}$$

Voltage across R_2 and VOM reading

$$= 10\ \text{V} - 9.1\ \text{V} = \mathbf{0.9\ V}$$

Note that the low resistance of the meter has almost doubled the circuit current causing a much larger voltage drop across R_1. This has drastically changed the voltage across R_2 from 5 to 0.9 V.

b. The resistance of the 50-V range meter = 20 kΩ/V × 50 V = 1 MΩ. See Fig. 12-12*c*.

Total circuit resistance

$$= 1\ \text{M}\Omega + 1\ \text{M}\Omega \| 1\ \text{M}\Omega = 1.5\ \text{M}\Omega$$

$$\text{New circuit current, } I = \frac{V}{R_T}$$

$$= \frac{10\ \text{V}}{1.5\ \text{M}\Omega} = 6.7\ \mu\text{A}$$

$$\text{Voltage drop across } R_1 = V_1 = IR_1$$

$$= 6.7\ \mu\text{A} \times 1\ \text{M}\Omega = 6.7\ \text{V}$$

Voltage across R_2 and VOM reading

$$= 10\ \text{V} - 6.7\ \text{V} = \mathbf{3.3\ V}$$

Although perhaps difficult to read on a 50-V range this reading is much closer to the desired 5 V. It results from not as large a voltage drop across R_1 because the current has not increased as much.

c. Refer to Fig. 12-12*d*. Repeating the calculation for an 11-MΩ resistance (the EVM) in parallel with R_2, we obtain:

Total circuit resistance = 1.92 MΩ

New circuit current = 5.2 μA

Voltage drop across R_1 = 5.2 V

VTVM reading across R_2 = **4.8 V**

d. VOM on 5-V range: 0.9 V ±3% of 5 V
= 0.9 V ±0.15 V = **0.75 to 1.05 V**
VOM on 50-V range: 3.3 V ±3% of 50 V
= 3.3 V ±1.5 V = **1.8 to 4.8 V**
VTVM on 5-V range: 4.8 V ±3% of 5 V
= 4.8 V ±0.15 V = **4.65 to 4.95 V**

It is clear, that in spite of increased FSD errors, the 50-V range VOM causes less loading and a closer reading to the initial condition.

Whenever loading is a problem, switching to a *higher* range will in general improve the meter accuracy. However, where voltage measurements are made in high-resistance circuits, the constant high resistance of an electronic voltmeter is essential for accurate readings.

It should be noted that loading occurred in the circuit of Fig. 12-12 because of the high resistance of R_1. If R_1 were 10 kΩ instead of 1 MΩ, the *additional* current that flows when a VOM is connected (although appreciable) will cause very little *additional* voltage drop across R_1. Thus the voltage across R_2 would be relatively unchanged.

Thus voltmeter loading is not caused merely by low-resistance meters connected in parallel with high-resistance devices. It is also dependent upon other high resistances in *series* with the component whose voltage is being measured.

SUMMARY

1. A D'Arsonval or PMMC movement consists of a movable coil pivoted in the strong magnetic field of a permanent magnet.

2. The PMMC movement has a deflection directly proportional to the dc current in the coil, thus providing a linear scale.

3. Current sensitivity is the current required in the movable coil to cause full-scale deflection (FSD).

4. Ammeters use a basic meter movement paralleled by a shunting resistor to extend the current range of the movement.

5. Multirange ammeters use individual shunts, selected by a make-before-break switch, or a series arrangement of shunting resistors called an Ayrton shunt.

6. The higher the current range of an ammeter, the lower the value of the shunt resistance, and the lower the total resistance of the ammeter itself.

7. An ideal ammeter would have zero resistance, regardless of current range.

8. Ammeter loading occurs when the ammeter's resistance is comparable to the circuit's resistance, thus upsetting the circuit's original state.

9. Less ammeter loading occurs on a higher-current range.

10. The accuracy of a meter is quoted as a percentage of its full-scale reading.

11. For maximum percentage accuracy, readings should be taken on a range that most nearly gives full-scale deflection.

12. Voltmeters consist of a multiplier resistor connected in series with the basic meter movement.

13. Multirange voltmeters may consist of individual or series-connected multipliers.

14. A voltmeter's sensitivity is rated in ohms/volt and is the reciprocal of the meter movement's current sensitivity.

15. A high sensitivity voltmeter causes less loading in high-resistance circuits because the voltmeter's large resistance draws negligible additional current.

16. An ideal voltmeter has infinite resistance. Many voltmeters (VOMs) increase in resistance on higher ranges, but electronic voltmeters have a constant high resistance on all ranges.

17. Where voltmeter loading is not a problem, readings should be made on a range that most nearly gives full-scale deflection.

SELF-EXAMINATION

Answer true or false, or in the case of multiple choice, a, b, c or d
(Answers at back of book)

12-1. A PMMC movement responds only to dc current. ________

12-2. The hair springs serve solely to provide a restoring torque. ________

12-3. A radial magnetic field is produced in a PMMC movement to ensure a linear scale. ________

12-4. Current sensitivity is the current required through a movement to produce half of full-scale deflection. ________

12-5. A higher-current sensitivity is usually accompanied by an increase in resistance of the moving coil. ________

12-6. An ammeter shunt has the same voltage drop across it as the meter movement. ________

12-7. Higher-range ammeters have higher shunt resistances. ________

12-8. A 100-μA, 2-kΩ meter movement has a voltage drop across it at half of full-scale deflection of:

a. 100 μV
b. 200 mV
c. 100 mV
d. 200 μV

12-9. A multirange ammeter has the same voltage drop across the meter on each range at FSD. ________

12-10. A multirange ammeter has lower resistance on the higher ranges. ________

12-11. An Ayrton shunt consists of a number of series-connected resistors connected in parallel with the meter movement. ________

12-12. A make-before-break selector switch is required in both the Ayrton shunt and the multishunt multirange ammeter. ________

12-13. A disadvantage of the universal shunt is the change in calibration of all ranges should the resistance of just one resistor change. ________

12-14. The value of shunt resistance required to double the range of a 100-μA, 2-kΩ meter movement is:

a. 1 kΩ
b. 2 kΩ
c. 3 kΩ
d. 4 kΩ

12-15. Ammeter loading occurs when the insertion of the ammeter presents a significant resistance that reduces the current flowing to a value much less than the initial value. ________

12-16. Loading can be minimized by using a higher range. This applies to both voltmeters and ammeters (VOMs). ________

12-17. A voltmeter multiplier is a resistor connected in series with a meter movement to extend the voltage range. ________

12-18. A 50-μA, 2-kΩ movement requires a multiplier to produce a 10-V range voltmeter of:

a. 198 kΩ
b. 19.8 kΩ
c. 18 kΩ
d. 1.98 MΩ

12-19. The lowest range that a multirange voltmeter can have that uses a 50-μA, 2-kΩ movement is:

a. 100 μV
b. 50 mV
c. 100 mV
d. 50 μV

12-20. A multirange voltmeter may use individual multipliers or a number of series-connected multipliers. ______

12-21. A voltmeter that has a total resistance of 1 MΩ on a 25-V range has a sensitivity of:

a. 4 kΩ/V
b. 25 kΩ/V
c. 40 kΩ/V
d. 2.5 kΩ/V

12-22. The higher the sensitivity of a voltmeter, the higher the total resistance of the voltmeter and loading. ______

12-23. Voltmeter loading is due to relatively low-resistance voltmeters connected in high-resistance circuits. ______

12-24. A 50-kΩ/V voltmeter used on a 10-V range indicates 5 V. The total resistance of the voltmeter is

a. 250 kΩ
b. 500 kΩ
c. 10 kΩ
d. 50 kΩ

12-25. A 10-V range voltmeter has an accuracy of ±2% of FSD. The percentage error of a reading of 4 V is

a. 2%
b. 3%
c. 4%
d. 5%

REVIEW QUESTIONS

1. What two functions do the phosphor-bronze hair springs serve in a D'Arsonval meter movement?

2. What is the principle of operation of a PMMC movement?

3. What construction features ensure that the PMMC scale is linear?

4. a. Why would you expect a more-sensitive meter movement to have higher resistance?
b. What would be necessary to achieve higher sensitivity without an increase in resistance?

5. a. What is the primary purpose of a shunt?
b. Why are shunt values selected on the basis of FSD currents in the meter movement?

6. a. Why is it necessary to use a make-before-break selector switch in a multishunt ammeter?
b. Why is it not necessary in an Ayrton shunt?

7. a. Why does the total resistance decrease for higher-range ammeters?

b. Why would an ideal ammeter have zero resistance whereas an ideal voltmeter would have infinite resistance?

8. What factors contribute to manufacturers quoting the accuracy of a meter as a percentage of FSD rather than percentage of reading?

9. Why is the percentage error higher when making a very low reading on a scale than when the reading is near full scale?

10. a. What do you understand by "parallax"?

b. How is this problem reduced in some meters?

11. a. When is ammeter loading likely to occur?

b. If you were using a multirange ammeter, how could you determine if loading is taking place?

12. What is the design principle on which a voltmeter multiplier is selected?

13. Is it necessary to use a make-before-break switch in multirange voltmeters?

14. a. What is an advantage of using series-connected multipliers in multirange voltmeters?

b. What is a disadvantage?

c. Which do you think is most expensive: series-connected or individual multipliers?

15. a. What purpose does knowing a voltmeter's sensitivity serve?

b. Why is it meaningless to refer to an EVM's voltmeter sensitivity?

16. Justify why a voltmeter's sensitivity should also be given by the reciprocal of the meter movement's current sensitivity.

17. Why is voltmeter loading more likely to occur with low-sensitivity voltmeters and high-resistance circuits?

18. a. How can you determine if a VOM is loading a circuit in voltmeter measurements?

b. How can you reduce loading?

c. What penalty do you pay?

19. Will voltmeter loading *always* occur across a high resistance with a low-sensitivity voltmeter? Explain.

PROBLEMS

(Answers to odd-numbered problems at back of book)

12-1. A certain manufacturer's meter movement has a current sensitivity of 100 μA and a resistance of 5 kΩ. In an effort to increase sensitivity, the designer intends to double the number of turns on the coil using wire of half the diameter, use a permanent magnet with a 50% stronger field, and increase the area of the movable coil by $\frac{1}{3}$. Calculate:

a. The sensitivity of the new meter.

b. The resistance of the new meter, assuming that the new coil has the same perimeter as the old one but increased area.

12-2. Repeat Problem 12-1b assuming that the new coil is square like the old one was, increasing the perimeter as well as the area.

12-3. Given a 100-μA, 2-kΩ meter movement, calculate:

a. The shunt resistance needed to make a 5-mA range meter.

b. The voltage drop across the meter at full-scale deflection.

c. The total resistance of the meter.

12-4. Repeat Problem 12-3 for a 10-mA range.

12-5. Given a 100-μA, 2-kΩ meter movement, it is required to construct a multirange milliammeter with ranges of 1 mA, 5 mA, 25 mA, 100 mA, and 500 mA. Calculate:

a. Appropriate shunt values.

b. Voltage drop across the meter on each range at FSD.

c. Total resistance of the meter on each range.

Draw a circuit diagram showing switch arrangements and all values.

12-6. Repeat Problem 12-5 using a 200-μA, 600-Ω movement.

12-7. Given a 100-μA, 2-kΩ meter movement, design an Ayrton shunt with ranges of 200 μA, 1 mA, and 5 mA.

12-8. Repeat Problem 12-7 for ranges of 0.5 mA, 1 mA, and 10 mA.

12-9. a. Given a 50-μA, 3-kΩ meter movement, design an Ayrton shunt with ranges of 1 and 10 mA.

b. Determine the total resistance of the meter on the 10-mA range.

c. Using the above meter movement, determine the single-shunt resistance to make a 10-mA range meter.

d. Determine the total resistance of the 10-mA range meter in part (c) and compare with the answer in part (b).

12-10. Repeat Problem 12-7 for ranges of 1 mA, 5 mA, 10 mA, and 25 mA.

12-11. A 25-mA range ammeter has an accuracy of $\pm 4\%$ of FSD. For readings on the scale of 5 mA, 15 mA, and 25 mA determine:

a. The possible error in mA.

b. The range of values of the true current.

c. The percentage error in each reading.

12-12. Repeat Problem 12-11 for a 50-mA range meter.

12-13. A 2.2-kΩ resistor is connected across a 1.5-V dry cell. Assuming the resistance to be exactly 2.2 kΩ, calculate:

a. The actual current and the power dissipated in the resistor. (Use $P = I^2R$.)

b. The reading of a 1-mA range, 1-kΩ meter connected in series with the resistor, and the power in the resistor. (Use $P = I^2R$.)

c. The range of readings possible in part (b) assuming a $\pm 3\%$ of FSD accuracy.

d. The reading of a 10-mA range, 100-Ω meter in series with the resistor.

e. The range of readings possible in part (d) assuming a $\pm 3\%$ of FSD accuracy.

12-14. Repeat Problem 12-13 using a 1.6-kΩ resistor.

12-15. Given a 100-μA, 2-kΩ meter movement, it is required to produce a multirange voltmeter having ranges of 1 V, 10 V, and 50 V using individual multipliers. Calculate:

a. The total resistance of the meter on each range.

b. The voltmeter multiplier resistors.

c. The voltmeter sensitivity.

d. The resistance of the meter when indicating 20 V on the 50-V range.

Draw a circuit diagram of the switching arrangement with values.

12-16. Repeat Problem 12-15 using a 50-μA, 5-kΩ movement.

12-17. Repeat Problem 12-15 using series-connected multipliers.

12-18. a. What must be the current sensitivity of a meter movement if it is to be used to produce a 50-kΩ/V voltmeter?

b. Assume that the meter movement in part (a) has a resistance of 10 kΩ and is shunted by another 10-kΩ resistor. What series resistor is required to produce a 2-V range voltmeter?

c. What is the voltmeter sensitivity of the voltmeter in part (b)?

d. If the 10-kΩ shunt resistor is removed, what is the new range of the voltmeter?

12-19. A 10-V source of emf with an internal resistance of 100 kΩ is checked by a 10-kΩ/V voltmeter on its 15-V range. Calculate:

a. The current drawn by the voltmeter.

b. The reading of the voltmeter.

c. The reading of the voltmeter if its range is increased to 50 V.

d. The reading of an 11-MΩ EVM on the 15-V range in place of the original meter.

12-20. Repeat Problem 12-19 using a 50-kΩ/V voltmeter.

12-21. A 500-kΩ resistor is series-connected with a 1-MΩ resistor across a 12-V source of negligible internal resistance. Calculate:

a. The voltage across the 0.5-MΩ resistor with no meters connected in the circuit.

b. The reading of an 11-MΩ EVM on the 5-V range connected across the 0.5-MΩ resistor.

c. The reading of a 20-kΩ/V VOM on the 5-V range connected in parallel with the EVM.

d. The new reading of the EVM for the condition in part (c).

e. The reading of both instruments when the VOM is switched to the 50-V range.

12-22. Repeat Problem 12-21 using a 2.2-MΩ resistor instead of a 1-MΩ, and a 25-V source instead of 12 V.

12-23. Repeat Problem 12-21 using a 10-kΩ resistor instead of a 1-MΩ, and a 5-V source instead of 12 V.

12-24. In Problem 12-21, assume that both meters have a ±3% of FSD accuracy. Calculate the possible reading and the percentage error (due to FSD errors—not loading) of the meters in parts (b), (c), and (e).

12-25. A neon lamp that fires (conducts) at 60 V is series-connected with a 100-kΩ resistor (to limit current after firing) across an 80-V source. A milliammeter on the 1-mA range is inserted in the circuit and indicates 200 μA. When a VOM on the 25-V range is connected across the resistor, the current reading increases to 216 μA.

a. What is the sensitivity of the VOM?

b. If the VOM has a FSD accuracy of ±5%, what range of readings could it indicate across the resistor?

c. What would be the milliammeter reading if the VOM were connected across the neon lamp on the 100-V range?

d. What percentage error would the VOM indication in part (c) represent?

CHAPTER THIRTEEN
OTHER DC MEASURING INSTRUMENTS

We have seen how voltmeter loading is caused by current drawn from the circuit by the voltmeter. A potentiometer is an instrument to make voltage measurements with no load at all. This is done by comparing an accurately known and variable voltage with the unknown voltage. When a galvanometer connected between the two shows a "balance," we know that the two voltages are equal. A potentiometer is very useful in calibrating voltmeters and ammeters.

The measurement of resistance is an important operation. Although it can be done simply by an ammeter and a voltmeter, the result is not very accurate. An ohmmeter, especially the shunted-series type, is a convenient instrument for measuring resistances. But where high accuracy is required, a Wheatstone bridge is used. This is a very versatile circuit with four resistance arms, one of which is the unknown. Accuracy is determined only by the resistance values.

Finally, we shall consider the specifications of a digital multimeter, and compare them with the analogue type of VOM and EVM.

13-1 PRINCIPLE OF A NO-LOAD VOLTMETER

Every voltmeter of the type we have examined in Chapter 12 draws some current from the circuit for its operation. Because of this, it *disturbs* the original circuit. Even if this current is very small, the voltmeter reading will be less than the true voltage in any multibranch circuit. In the case of measurement of a source voltage or open-circuit voltage, if the internal resistance of the source is significant, the voltmeter reading is also reduced. See Fig. 13-1*a*. This is especially the case in measuring the small emf (few millivolts) of a thermocouple, where the voltage drop across connecting leads can be comparable to the generated emf.

Figure 13-1*b* shows a measurement principle by which no current is drawn from the source. If the voltages are equal on both sides of a *galvanometer* (a sensitive current indicating device with a zero-center pointer), there is no deflection of the galvanometer. This is called a *null* condition, and no current is drawn from the source. The accurately known voltage of the precision source will then be the same as the voltage of the "unknown" emf, V. Since no current flows in the circuit at all, the measurement is *independent* of the *resistances* of the galvanometer, r_G, the unknown source r_v, and the variable source itself r_k. Also, the calibration accuracy of the galvanometer movement is unimportant as long as it is sensitive enough to clearly show a null condition (zero deflection) at balance. In fact, the greater the sensitivity of the galvanometer the more accurate the measurement. The instrument that provides the variable accurate voltage is called a *potentiometer*.

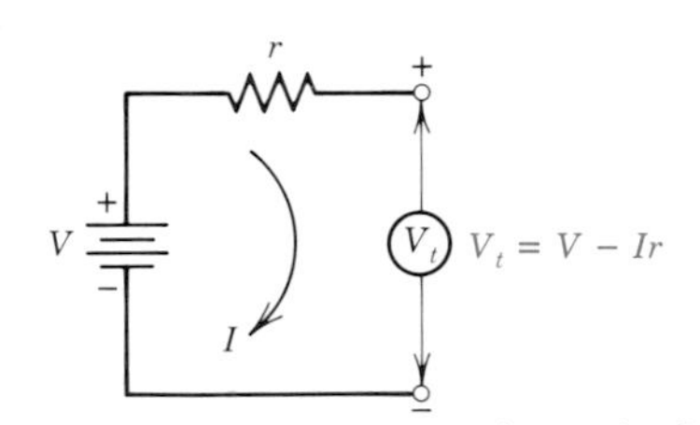

(*a*) Voltmeter resistance causes an internal voltage drop in the source

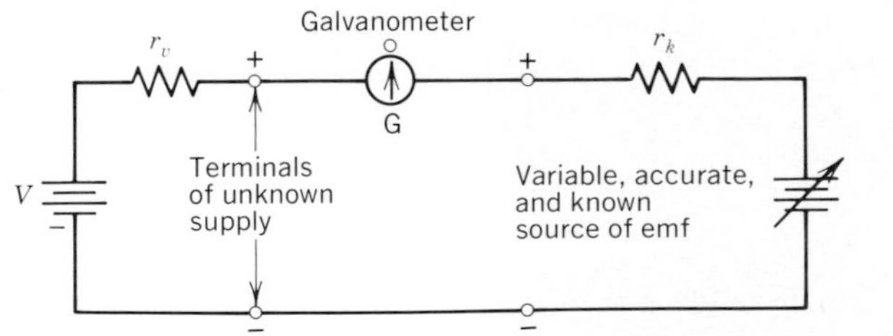

(*b*) Potentiometer principle. Galvanometer indicates "null" (zero) when the two voltages are equal

FIGURE 13-1
Principle of a no-load voltmeter.

13-2 THE POTENTIOMETER

The commercial potentiometer is essentially a *high-resolution, variable voltage-divider.* It uses a sensitive galvanometer (which may be built into the instrument), a *standard cell,* and one (or more) voltage-dividing networks. See Fig. 13-2.

The standard cell is the source of accurately known voltage used only for *calibration* of the potentiometer. Measurement of the unknown emf is accomplished by comparison with known voltage drops caused by current flowing through resistors making up R_1 and R_2.

There are two types of *standard* cells, similar to the one shown in Fig. 13-3. The *saturated* or *normal* cell, manufactured to a high degree of uniformity, develops a known (standard) emf of 1.0183 V. For many years, the National Bureau of Standards (NBS) used this type of cell for calibrating purposes and defined the *international volt* as $\frac{1}{1.01830}$ of the emf produced by this cell at 20°C.

A portable version, without $CdSO_4$ crystals and employing cork washers, is called the *unsaturated* cadmium cell. New cells range between 1.0190 and 1.0194 *absolute volts,* with the actual value determined by calibration with a normal cell. When buying a new cell, it is accompanied by a certificate from the NBS with the actual voltage. (1.01830 international volts equals 1.01864 absolute volts.) The voltage of a standard cell should be checked yearly; its voltage should not be measured with an ordinary voltmeter. If the cell is inverted, it should be allowed to settle for 24 h before use.

The operation of the potentiometer is as follows:

1. Set the dials of the potentiometer to the indicated emf of the standard cell (e.g., R_1 to 1.0 V, R_2 to 19 mV).
2. Throw the DPDT switch to the *calibrate* position. This puts the standard cell into the circuit on one side of the galvanometer.

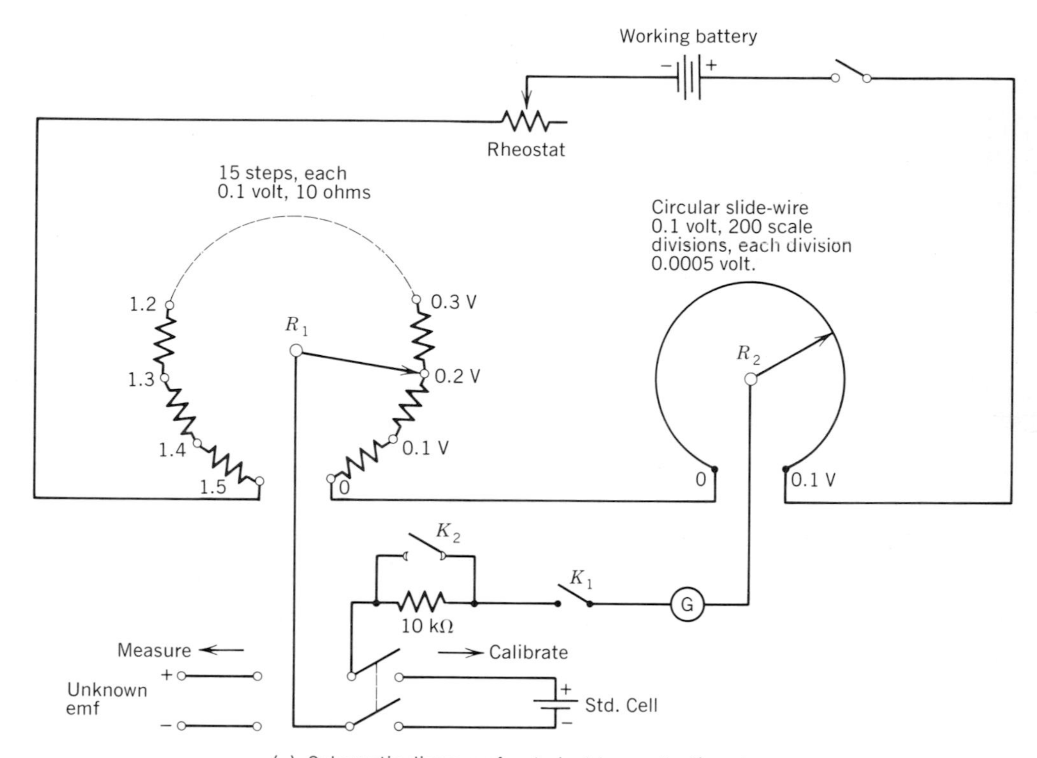

(*a*) Schematic diagram of a student-type potentiometer

(*b*) Photograph of a student-type potentiometer

FIGURE 13-2
Circuit diagram and photograph of a potentiometer (The latter courtesy of the Leeds and Northrup Co.)

(a) External view of mounted cell

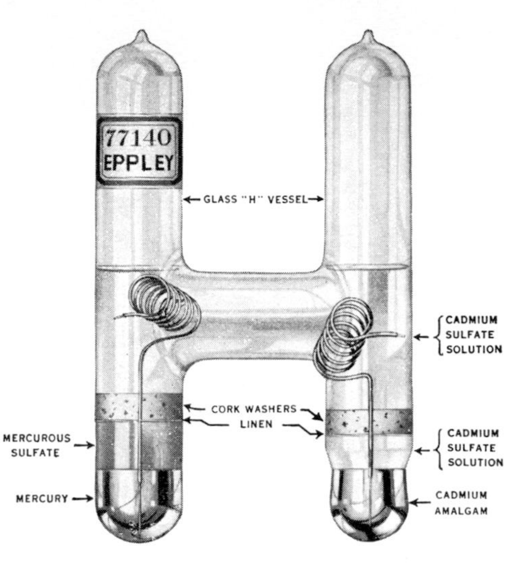

(b) View showing construction of a cell

FIGURE 13-3
Standard cell made by the Eppley Laboratory, Inc. (Courtesy of Eppley Laboratory, Inc.)

3 Close K_1 and adjust the rheostat until the galvanometer shows near zero deflection.

4 Close K_2 as well as K_1 to obtain an accurate null as the rheostat is varied.

The potentiometer is now calibrated or *standardized,* that is, R_1 is calibrated in steps of 0.1 V, up to a maximum of 1.5 V, while R_2 is calibrated up to a maximum of 0.1 V. The current delivered by the working battery causes voltage drops across resistors R_1 and R_2 that correspond exactly to the marked values on the dials. The rheostat should not be changed unless a later calibration check indicates that this needs to be done. Note that no current is drawn from the standard cell at balance so that it cannot be consumed or damaged when checking calibration.

5 Throw the DPDT switch to the measure position. This now permits the unknown voltage to be compared with the potentiometer variable voltage.

6 Adjust the *dials* of the potentiometer (not the rheostat) until the galvanometer shows a null upon closure of both K_1 and K_2.

7 Read the unknown voltage from the two dials, R_1 and R_2.

It should be clear that a maximum voltage of 1.6 V can be measured directly by the potentiometer. Higher voltages are measured by means of a voltage-divider or "volt-box," which reduces the voltage by some known factor to a maximum of 1.5 V. This, however, puts a load on the voltage to be measured, and consequently, may degrade the reading.

13-3 CALIBRATION OF VOLTMETERS AND AMMETERS

Potentiometers are capable of making very accurate and very precise voltage measurements. They are commonly used to calibrate digital voltmeters.

A *calibration curve* for a given voltmeter may be obtained using the circuit of Fig. 13-4*a*, where the voltmeter and potentiometer are in parallel. The *correction* that must be added to (or subtracted from) the voltmeter reading can be obtained at various points on the voltmeter scale and plotted on a graph as in Fig. 13-4*c*.

$$\text{correction} = \text{potentiometer} - \text{voltmeter} \qquad \text{(volts)}$$

On a 1.5-V range voltmeter, this correction would normally be small and perhaps given in millivolts. Note that the correction curve consists of joining the measured points by straight lines, and that a good meter should be correct at three points along the scale as shown. However, many meters will get progressively worse as the reading increases. The FSD accuracy of the meter would be obtained by expressing the largest correction (no matter where it occurred) as a percentage of the full-scale range.

The potentiometer can also be used to calibrate an ammeter using the circuit shown in Fig. 13-4*b*. The ammeter is simply connected in series with a *standard,* highly accurate *precision* resistor (conveniently 1.0000 Ω

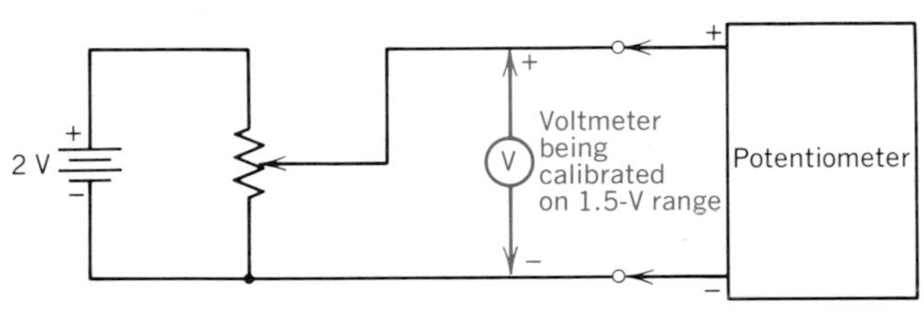

(*a*) Circuit for voltmeter calibration

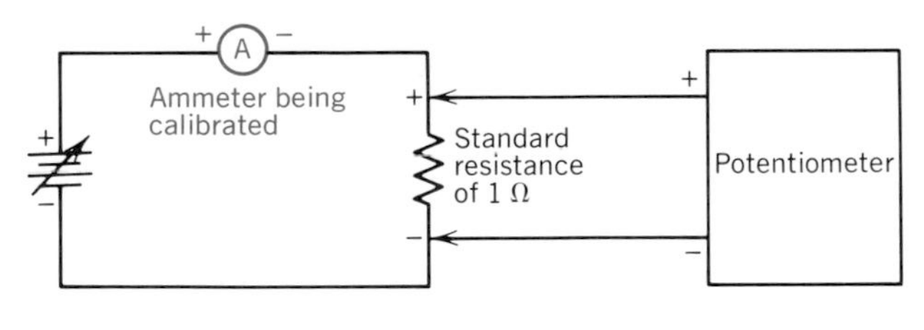

(*b*) Circuit for ammeter calibration

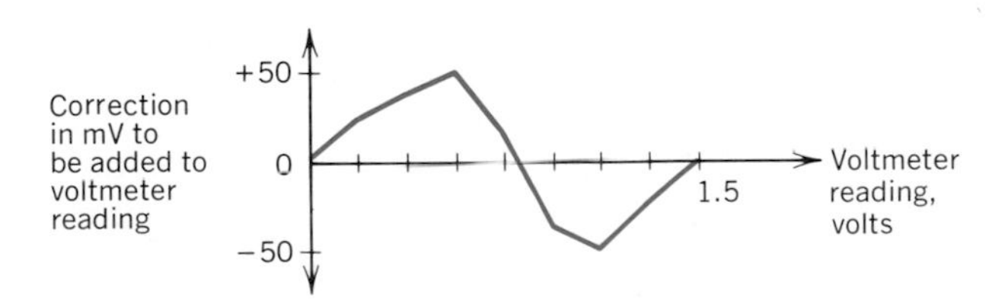

(*c*) Typical correction curve for a voltmeter

FIGURE 13-4
Calibration circuits for the voltmeter and ammeter and a typical calibration curve for a voltmeter.

in many cases to read current directly) and the voltage across the resistance measured by the potentiometer. The numerical voltage readings of the potentiometer by Ohm's law, are numerically the same as the true ammeter current. A correction curve can be drawn, as before, to yield the corrections to be added (or subtracted) when using the instrument.

13-4 AMMETER–VOLTMETER METHOD OF MEASURING RESISTANCE

An apparently simple method of measuring the resistance of a circuit is to divide the voltage across a circuit by the current through the circuit using Ohm's law. That is, $R_x = \frac{V_x}{I_x}$. This simply requires a voltmeter reading and an ammeter reading. But what of the accuracy of such a measurement? There are many possible sources of error.

First, let us ignore the resistances of the meters themselves, and consider only their FSD accuracies.

Let us assume that a 5-V range voltmeter indicates 5 V and a 10-μA range ammeter indicates 10 μA when measuring a circuit's resistance. Also, let us assume that they have ±3% FSD accuracies. Then

$$R_x = \frac{V_x}{I_x} = \frac{5\ \text{V}}{10\ \mu\text{A}} = 500\ \text{k}\Omega$$

The circuit's resistance has been determined to be 500 kΩ with an accuracy of ±6%. That is, the voltmeter *could* be reading 3% high, the ammeter 3% low, (or vice versa), with a maximum possible error of 6%. Of course, it is possible that the two errors could cancel, but we do not know if this is happening without calibration curves for each instrument. We must consider the worst case which is the *sum* of the individual errors.

If the readings of the two meters were not at or near *full scale,* the percentage error could be far worse. If both were read near half-scale, the error could be *double* (12%), and even more at lower deflections.

Now let us consider the additional effect of voltmeter location.

13-4.1 Measurement of High Resistance

If the circuit's resistance is considered high (comparable to the voltmeter's resistance), the voltmeter should be

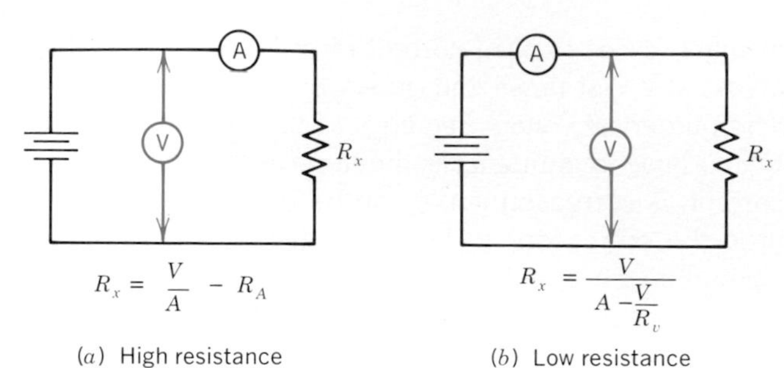

FIGURE 13-5
Showing the voltmeter location to measure an unknown resistance in high- and low-resistance circuits.

connected across the series combination of R_x and the ammeter to avoid loading effects. See Fig. 13-5*a*. Now the ammeter indicates the true resistor current, but the voltmeter reading is a little too high. The following equation makes a correction for the ammeter's resistance.

$$R_x = \frac{V}{A} - R_A \qquad \text{ohms } (\Omega) \qquad (13\text{-}1)$$

where: R_x is the resistance of the unknown in ohms, Ω
V is the voltmeter reading in volts, V
A is the ammeter reading in amperes, A
R_A is the resistance of the ammeter in ohms, Ω

EXAMPLE 13-1

A resistor is being measured using the circuit shown in Fig. 13-5*a*. A 10-V range, 3%-FSD accuracy voltmeter indicates 5 V; a 50-μA, 20-kΩ, 2%-FSD accuracy microammeter indicates 10 μA. Calculate:

a. The resistance.
b. The maximum possible error.

SOLUTION

a. $$R_x = \frac{V}{A} - R_A \qquad (13\text{-}1)$$

$$= \frac{5\text{ V}}{10 \times 10^{-6}\text{ A}} - 20\text{ k}\Omega$$

$$= 500\text{ k}\Omega - 20\text{ k}\Omega = \mathbf{480\text{ k}\Omega}$$

b. Percentage error in voltmeter reading

$$= \frac{\frac{3}{100} \times 10\text{ V}}{5\text{ V}} \times 100\% = 6\%$$

Percentage error in ammeter reading

$$= \frac{\frac{2}{100} \times 50\ \mu\text{A}}{10\ \mu\text{A}} \times 100\% = 10\%$$

Maximum possible error in measurement of R_x = 6% + 10% = **16%**

13-4.2 Measurement of Low Resistance

For *low* resistance measurements, the circuit of Fig. 13-5*b* is used. The voltmeter reading is correct, but the ammeter reading is too high by an amount equal to the current drawn by the voltmeter, V/R_v. That is:

$$R_x = \frac{V}{A - V/R_v} \qquad \text{ohms } (\Omega) \qquad (13\text{-}2)$$

where: R_x is the resistance of the unknown in ohms, Ω
V is the voltmeter reading in volts, V
A is the ammeter reading in amperes, A
R_v is the resistance of the voltmeter in ohms, Ω

EXAMPLE 13-2

The resistance of a circuit is being measured as in Fig. 13-5*b*. The 10-mA range milliammeter indicates 10 mA and the 10-kΩ/V voltmeter indicates 15 V on the 15-V range. If both meters have a 2% FSD accuracy, determine:

a. The resistance.
b. The maximum possible error.

SOLUTION

a. R_v = Ω/V × range = 10 kΩ/V × 15 V = 150 kΩ

$$R_x = \frac{V}{A - V/R_v} \qquad (13\text{-}2)$$

$$= \frac{15\text{ V}}{10 \times 10^{-3}\text{ A} - \frac{15\text{ V}}{150 \times 10^{3}\ \Omega}}$$

$$= \frac{15\text{ V}}{9.9 \times 10^{-3}\text{ A}} = \mathbf{1.52\text{ k}\Omega}$$

b. Maximum possible error = 2% + 2% = **4%**

Note that if an ideal ammeter with $R_A = 0$ is used, and an ideal voltmeter with $R_v = \infty$ is used, both Eqs. 13-1 and 13-2 reduce to $R_x = \frac{V}{A}$, and the voltmeter location is immaterial.

13-5 SERIES TYPE OF OHMMETER

An *ohmmeter* provides a quick, easy method of measuring resistance without taking two readings and making a calculation. Although not of a high accuracy, it is adequate for many commercial applications. In electronics work, most resistors used are only accurate to 5 or 10% anyway.

The most simple type of ohmmeter is the series type shown in Fig. 13-6. **Unlike a voltmeter or ammeter, it requires a battery (1.5 V and up) and is used in circuits with no other power applied. Power must always be turned OFF before the ohmmeter is connected.** The ohmmeter uses a basic meter movement and some provision for *zeroing* the instrument when the two leads of the ohmmeter are shorted, that is, connected.

For the basic circuit shown in Fig. 13-6*a* we can see that, for any value of R_x,

$$V = IR_T \tag{3-1b}$$

and

$$V = I(R_x + R_m + R_z) \tag{13-3}$$

where: V is the emf of the cell in volts, V
I is the current through the resistor and meter movement in amps, A
R_x is the external resistance to be measured in ohms, Ω
R_m is the meter movement resistance in ohms, Ω
R_z is the resistance of the zero control in ohms, Ω

When the leads are shorted together, R_x is zero, and

$$R_m + R_z = \frac{V}{I_{\text{FSD}}} \tag{13-4}$$

Equation 13-4 determines the value of R_z to give full-scale deflection of the meter. This point on the scale is marked zero ohms. The manner in which the rest of the scale of the movement is marked in ohms is shown in Example 13-3. In general, the procedure consists of assuming a convenient value of current and determining the external resistance, R_x, that permits this current to flow.

EXAMPLE 13-3

Given a 0–50-μA, 3-kΩ meter movement, determine the resistance markings at 0, 10, 20, 30, 40, and 50 μA when used in a series-type ohmmeter (Fig. 13-6) with a 1.5-V cell. Also determine a suitable value for the zero ohms control.

SOLUTION

When the leads are shorted together, $R_x = 0$, and

$$R_m + R_z = \frac{V}{I_{\text{FSD}}} \tag{13-4}$$

$$= \frac{1.5\text{ V}}{50 \times 10^{-6}\text{ A}} = 30\text{ k}\Omega$$

therefore, $R_z = 30\text{ k}\Omega - R_m$

$$= 30\text{ k}\Omega - 3\text{ k}\Omega = \mathbf{27\text{ k}\Omega}$$

This is the value to which the zero ohms control (maximum value **30 kΩ**) must be adjusted for a *full-scale deflection* of 50 μA, corresponding to 0 Ω. (This mechanical process is often called *calibrating* the ohmmeter.)

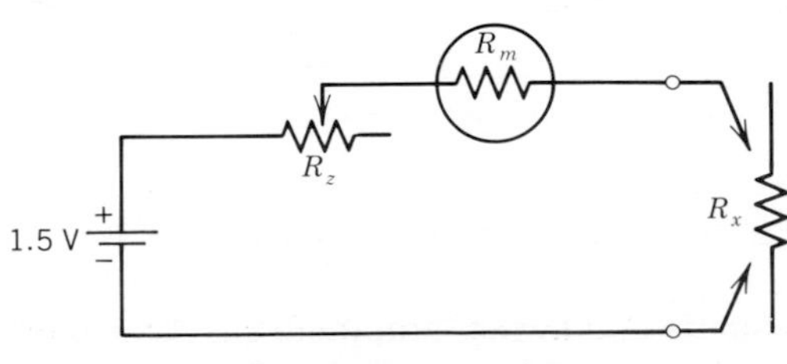

(*a*) Circuit of ohmmeter

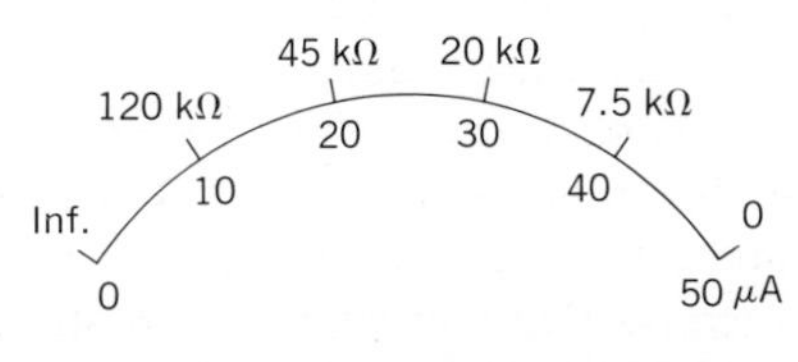

(*b*) Calibration of ohmmeter scale

FIGURE 13-6
Series type of ohmmeter.

When $I = 40\ \mu A$, corresponding to $\frac{4}{5}$ of full-scale deflection,

$$R_x + R_m + R_z = \frac{V}{I} \qquad (1\text{-}3)$$

$$= \frac{1.5\ V}{40\ \mu A} = 37.5\ k\Omega$$

Thus

$$R_x = 37.5\ k\Omega - (R_m + R_z)$$

$$= 37.5\ k\Omega - (3\ k\Omega + 27\ k\Omega)$$

$$= 37.5\ k\Omega - 30\ k\Omega = \mathbf{7.5\ k\Omega}$$

This is the ohmmeter calibration at $\frac{4}{5}$ full-scale deflection. For $I = 30\ \mu A$, corresponding to $\frac{3}{5}$ of full-scale deflection,

$$R_x + R_m + R_z = \frac{V}{I} \qquad (1\text{-}3)$$

$$= \frac{1.5\ V}{30\ \mu A} = 50\ k\Omega$$

Thus

$$R_x = 50\ k\Omega - (R_m + R_z)$$

$$= 50\ k\Omega - 30\ k\Omega = \mathbf{20\ k\Omega}$$

This is the ohmmeter calibration at $\frac{3}{5}$ full-scale deflection. Similarly, for $I = 20\ \mu A$, $R_x = \mathbf{45\ k\Omega}$,

calibration at $\frac{2}{5} I_{FSD}$

$I = 10\ \mu A$, $R_x = \mathbf{120\ k\Omega}$,

calibration at $\frac{1}{5} I_{FSD}$

When the test leads are apart, R_x is infinite, the ohmmeter is on open circuit, and no current flows. Thus $I = 0$ corresponds to infinity, ∞, on the left side of the scale. The markings are shown in Fig. 13-6*b*.

The kind of scale in Example 13-3 is often called a *back-off scale,* being in the opposite direction from a voltmeter or ammeter.

It is evident from the calibration values shown in Fig. 13-6*b* that the scale is *nonlinear,* being very crowded between 120 kΩ and infinity. This is one disadvantage of the series-type ohmmeter in attempting to measure resistance (such as 150 kΩ) accurately.

Another disadvantage of this type of ohmmeter circuit is the large effect on the resistance readings as the battery voltage drops during its aging process. This lower voltage means that a given current (and a given deflection on the ohmmeter) corresponds to a *lower* external resistance. Although a reduction of the zero ohms control, R_z, offsets some of the error, the result is a *reading* that is *higher* than the true external resistance, R_x. This is shown in Example 13-4.

EXAMPLE 13-4

When the ohmmeter battery voltage in Example 13-3 drops to 1.4 V, what value of R_x corresponds to 30 μA on the meter (marked 20 kΩ)?

SOLUTION

When the battery voltage drops, R_z must be readjusted and reduced to be able to zero the ohmmeter. Substituting the reduced battery voltage in Eq. 13-4, we find that

$$R_m + R_z = \frac{V}{I_{FSD}} \qquad (13\text{-}4)$$

$$= \frac{1.4\ V}{50\ \mu A} = 28\ k\Omega$$

Thus $R_z = 28\ k\Omega - 3\ k\Omega = 25\ k\Omega$. (Note that R_z has been reduced from 27 to 25 kΩ because of the reduction in battery voltage.)

For $I = 30\ \mu A$,

$$R_x + R_m + R_z = \frac{V}{I} \qquad (13\text{-}3)$$

$$= \frac{1.4\ V}{30\ \mu A} = 46.7\ k\Omega$$

Thus $R_x = 46.7\ k\Omega - (25\ k\Omega + 3\ k\Omega) = \mathbf{18.7\ k\Omega}$.

Since the calibration scale value is 20 kΩ, the ohmmeter is indicating a value that is 1.3 kΩ too high, an *additional* error of 7%.

This problem due to battery aging can be reduced to less than 1% by connecting the zeroing control in parallel with the meter movement instead of in series with it (see Problem 13-12). However, it is apparent that this type of ohmmeter is *not* capable of measuring either very low (or very high) resistances accurately. This is overcome, in part, in the shunted series type of ohmmeter, which also lends itself readily to multirange capability. The multirange capability enables us to provide an ohmmeter with several scales (from $R \times 1$ for low resistance up to $R \times 1000$ for higher resistances).

13-6 THE SHUNTED-SERIES TYPE OF OHMMETER

A mark of comparison between ohmmeters is the center scale or half-scale deflection resistance. The lower this half-scale resistance, the better the ohmmeter's ability to

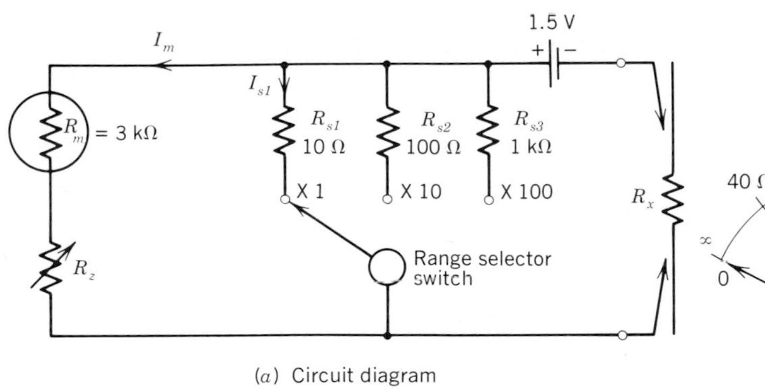

(*a*) Circuit diagram

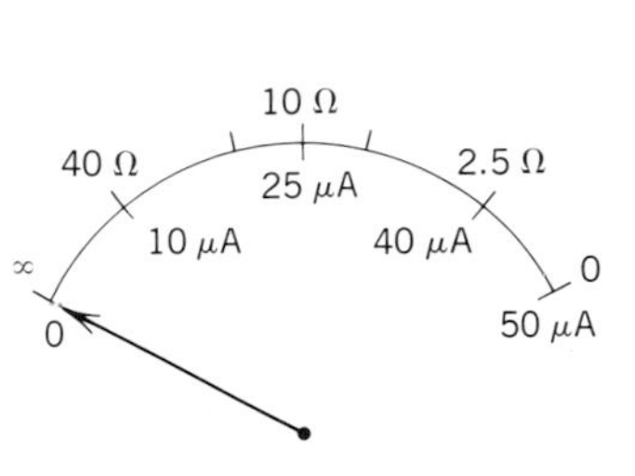

(*b*) Scale markings

FIGURE 13-7
Shunted-series type of multirange ohmmeter.

measure low values of resistance. A value of 10 to 12 Ω is considered an acceptable half-scale calibration. (The value in the series-type ohmmeter of Fig. 13-6*b* is 30 kΩ.) Figure 13-7 shows that a resistor equal to the center scale resistance is used to shunt the movement and the zero control on the $R \times 1$ range.

On any range, the total resistance, R_T, seen by the 1.5-V cell, is given by

$$R_T = R_x + R_s \| (R_m + R_z) \qquad (13\text{-}5)$$

where: R_s is the shunt resistance selected by the range selector switch in ohms, Ω

all other symbols are as in Eq. 13-3.

When the instrument is being zeroed, $R_x = 0$. This connects the cell in parallel with the shunt resistance, R_s, and the series combination of R_m and R_z. The total resistance seen by the cell is now:

$$R_T = R_s \| (R_m + R_z) \qquad (13\text{-}6)$$

Since the current through $R_m + R_z$ is typically only 50 μA, the resistance of $R_m + R_z$ is much larger than R_s, which may be only 10 or a few hundred ohms. Thus, to a close approximation, $R_T \approx R_s$ when $R_x = 0$. This causes a certain current to flow through R_s, and full-scale deflection current I_{FSD} through the movement.

Now, if $R_x = R_s$, the total resistance seen by the cell doubles, so that $R_T \approx 2R_s$. This causes the current through both R_s and the meter movement to be cut in half, and the meter will deflect only to half scale. This half-scale deflection is calibrated with a value equal to R_{s1}, as shown in Example 13-5.

EXAMPLE 13-5

With the range selector switch in Fig. 13-7*a* set on the R × 1 position, calculate:

a. The current through R_{s1} and the movement when $R_x = \infty$.

b. The value of R_z to zero the ohmmeter.

c. The current through R_{s1} when $R_x = 0$.

d. The current through R_{s1} and the movement when $R_x = 10\ \Omega$.

e. The current through R_{s2} and the movement when the range is changed to $R \times 10$ and R_x – 100 Ω. Given $I_m = 50\ \mu A$, $R_m = 3\ k\Omega$, $V = 1.5$ V cell.

SOLUTION

a. When $R_x = \infty$, there is no complete path for the current, so I_m = **0,** I_{s1} = **0,** and the pointer indicates infinity at the extreme left of the scale.

b. To zero the ohmmeter, $R_x = 0$.

This connects the 1.5-V cell in parallel with R_{s1} and the series combination of R_m and R_z. I_{FSD} must flow through the movement for full-scale deflection.

$$R_m + R_z = \frac{V}{I_{FSD}} \qquad (3\text{-}1)$$

$$= \frac{1.5\ V}{50\ \mu A} = 30\ k\Omega$$

Thus $R_z = 30\ k\Omega - 3\ k\Omega =$ **27 kΩ.**

c. $$I_{s1} = \frac{V}{R_{s1}} \qquad (3\text{-}1a)$$

$$= \frac{1.5\ V}{10\ \Omega} = 0.15\ A = \mathbf{150\ mA}$$

d. When $R_x = 10\ \Omega$, the total resistance seen by the 1.5-V cell is

$$R_T = R_x + R_{s1} \| (R_m + R_z) \qquad (13\text{-}5)$$
$$= 10\ \Omega + 10\ \Omega \| (3\ \text{k}\Omega + 27\ \text{k}\Omega)$$
$$= 10\ \Omega + 10\ \Omega \| 30{,}000\ \Omega$$
$$= 10\ \Omega + 9.997\ \Omega$$
$$\approx 20\ \Omega$$

It is to be noted that this is double the resistance seen by the 1.5-V cell when R_x was 0 Ω. Consequently, the battery current will be one-half of what it was when $R_x = 0$.

Thus the current through both R_{s1} and R_m will be cut in half.

$$I_{s1} = \frac{150\ \text{mA}}{2} = \mathbf{75\ mA}$$

$$I_m = \frac{50\ \mu\text{A}}{2} = \mathbf{25\ \mu A}$$

Since the movement current is 25 μA, which is half of the full-scale deflection, the center scale reading of the ohmmeter is 10 Ω.

e. On the $R \times 10$ range, and $R_x = 0\ \Omega$,

$$I_{s2} = \frac{V}{R_{s2}} = \frac{1.5\ \text{V}}{100\ \Omega} = 15\ \text{mA}$$

and $\qquad I_m = 50\ \mu\text{A}$

with $R_T = R_{s2} \| (R_m + R_z) \approx R_{s2} = 100\ \Omega$. When $R_x = 100\ \Omega$, $R_T \approx 200\ \Omega$, so $I_{s2} = \mathbf{7.5\ mA}$ and $I_m = \mathbf{25\ \mu A}$. The ohmmeter will indicate 10 Ω, which must be multiplied by 10 to obtain the deflection scale value of $R_x = 100\ \Omega$.

NOTE On the $R \times 1$ range a large amount of current is required from the cell compared with the $R \times 10$ range. It is because of this that an aging cell may not be able to zero a commercial ohmmeter on the $R \times 1$ range but can on higher ranges. Also, because of the increased internal resistance of the aging cell, the terminal voltage will *not* be constant. This means that the zero ohms control will have to be readjusted, each time the range is changed, to zero the meter.

On very high-resistance ranges, such as $R \times 10$ k, it is necessary (in commercial ohmmeters) to switch in additional cells (as much as 9 V more) to provide enough current to be able to operate the meter movement. See Fig. 13-8. Finally, in many commercial ohmmeters, rather than use independent shunts R_{s1}, R_{s2}, and so on, series-parallel resistor combinations are used based on the Ayrton shunt principle (Section 12-4).

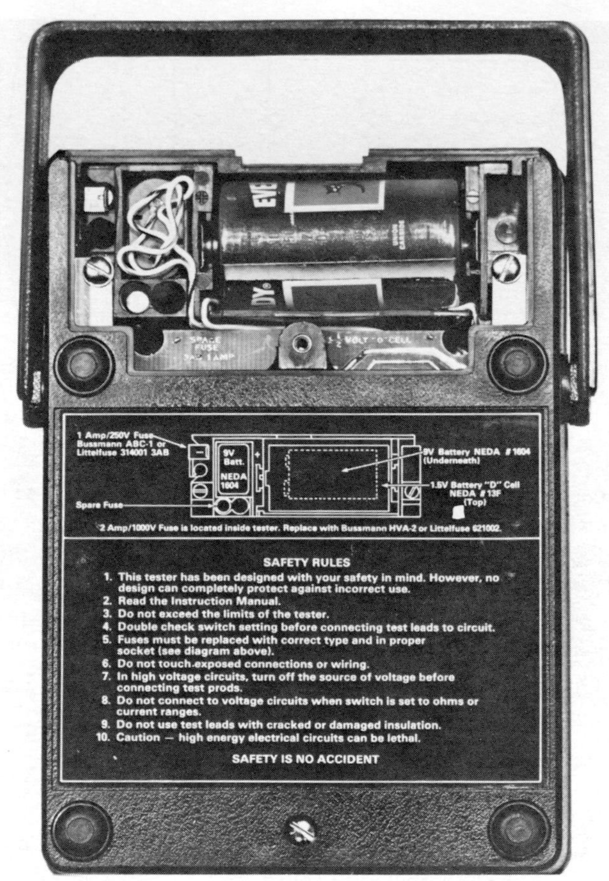

FIGURE 13-8
Model 60A VOM with back removed to show the 1.5-V cell for most ohmmeter ranges and the additional 9-V cell for the high-resistance range. (Courtesy of Triplett Corporation.)

13-7 THE WHEATSTONE BRIDGE

Of all resistance-measuring instruments, the ohmmeter is the least accurate. The Wheatstone bridge is the instrument most often used to make *precision* resistance measurements. Named after its inventor, Charles Wheat-

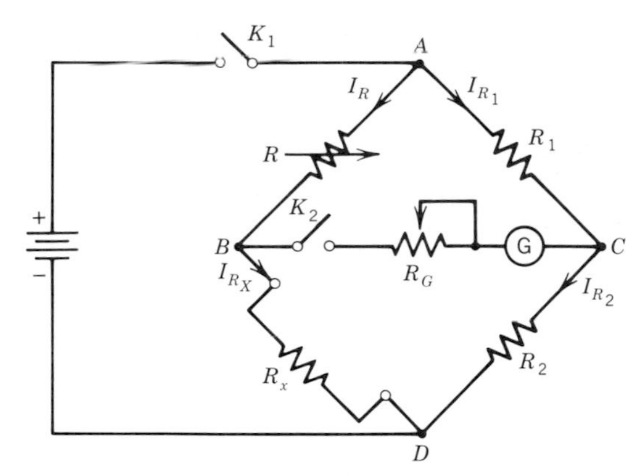

(a) Basic Wheatstone bridge circuit

(b) Commercial Wheatstone bridge

FIGURE 13-9
Wheatstone bridge for measuring resistance. [Photograph in (b) courtesy of Leeds and Northrup Co.]

stone (in 1843), the basic instrument circuit consists of a source, a galvanometer, a key, and a diamond-shaped arrangement of four resistors. A dc voltage is connected across one opposite set of corners, a sensitive galvanometer across the other set, as shown in Fig. 13-9a.

Resistors R_1 and R_2 are usually decade resistors set in value to some accurately known ratio to one another. R is a variable decade resistance with finely divided markings, indicating its value to some high degree of accuracy; R_x is the unknown resistance to be measured.

Keys K_1 and K_2 are closed and R is varied until, with R_G at its minimum value, the galvanometer shows zero deflection or a null. This means that there is no potential difference between B and C. In this condition, the bridge is said to be "balanced."

This null can only happen if

$$V_{BD} = V_{CD}$$

This also implies that

$$V_{AB} = V_{AC}$$

Thus $\quad I_{R_x}R_x = I_{R_2}R_2 \quad$ and $\quad I_R R = I_{R_1}R_1$

Dividing these two equations:

$$\frac{I_{R_x}R_x}{I_R R} = \frac{I_{R_2}R_2}{I_{R_1}R_1}$$

But if there is no current through the galvanometer,

$$I_R = I_{R_x} \quad \text{and} \quad I_{R_1} = I_{R_2}$$

Thus
$$\frac{R_x}{R} = \frac{R_2}{R_1} \tag{13-7}$$

and
$$R_x = \left(\frac{R_2}{R_1}\right)R \tag{13-7a}$$

or
$$R_xR_1 = RR_2 \tag{13-7b}$$

Equation 13-7a shows that R_x can be obtained if the "ratio arms" (the ratio of R_2 to R_1) and R are accurately known.

Equation 13-7b shows a convenient way of remembering the balanced bridge resistance equation: *The cross products of the opposite arms are equal.*

It should be noted that the accuracy of determining R_x in Eq. 13-7a depends upon the *sum* of the precision errors of known resistors: R_2, R_1, and R. Thus, if all three resistors are known to an accuracy of 0.1%, the value of R_x is known to a maximum possible error of 0.3%. This measurement is independent of the applied voltage, the galvanometer accuracy, and the value of R_G in Fig. 13-9a.

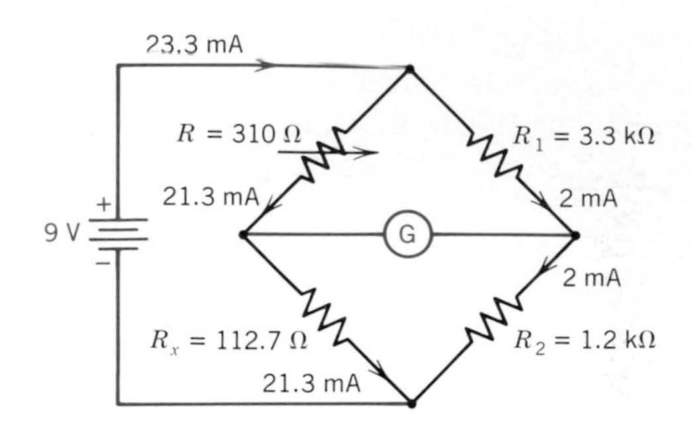

(a) Original bridge circuit

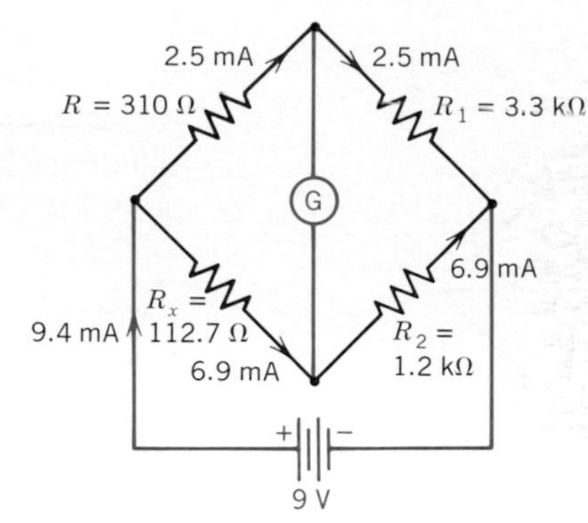

(b) Bridge circuit with galvanometer and battery interchanged

FIGURE 13-10 Wheatstone bridge circuits for Example 13-6.

EXAMPLE 13-6

Using the notation in Fig. 13-10*a*, a student-built Wheatstone bridge consists of 1% ratio arms with $R_1 = 3.3\ \text{k}\Omega$ and $R_2 = 1.2\ \text{k}\Omega$. Bridge balance occurs when R is adjusted to 310 ohms, with an accuracy of 0.1%. Calculate:

a. The unknown resistance and the maximum possible error.

b. The current through R_x if a 9-V battery is used.

c. The power dissipated in R.

SOLUTION

a. $$R_x = \left(\frac{R_2}{R_1}\right)R \qquad (13\text{-}7a)$$

$$= \left(\frac{1.2\ \text{k}\Omega}{3.3\ \text{k}\Omega}\right) 310\ \Omega = \mathbf{112.7\ \Omega}$$

Maximum possible error $= 1\% + 1\% + 0.1\% = \mathbf{2.1\%}$

b. $$I_{R_x} = \frac{V}{R + R_x} \qquad (3\text{-}1a)$$

$$= \frac{9\ \text{V}}{310 + 112.7\ \Omega} = \mathbf{21.3\ mA}$$

c. $$P_R = I_R{}^2 R \qquad (3\text{-}9)$$

$$= (21.3 \times 10^{-3}\ \text{A})^2 \times 310\ \Omega$$

$$= 0.14\ \text{W} = \mathbf{140\ mW}$$

In the commercial Wheatstone bridge, shown in Fig. 13-9*b*, R_2/R_1 is adjustable to provide ratio factors of 1, 10, 100, and so on, and is referred to as the *multiplier*. This multiplies the value of R indicated on the right in Fig. 13-9*b*.

The range of values that may be accurately measured by a Wheatstone bridge extends from 1 to 5 Ω at the lower limit (due to contact resistance) up to 1 MΩ. At high resistances, the sensitivity of the galvanometer to unbalance is severely restricted due to the very small currents in the circuit.

The following example illustrates the effect of changing the locations of the galvanometer and supply voltage.

EXAMPLE 13-7

Repeat Example 13-6 with the positions of the galvanometer and battery interchanged as in Fig. 13-10*b*.

SOLUTION

a. $R_x = \mathbf{112.7\ \Omega}$

Maximum error $= \mathbf{2.1\%}$

b. $$I_{R_x} = \frac{V}{R_x + R_2} \qquad (3\text{-}1a)$$

$$= \frac{9\ \text{V}}{112.7 + 1200\ \Omega} = \mathbf{6.9\ mA}$$

c. $P_R = I_R{}^2 R$ (3-9)

$$I_R = \frac{V}{R + R_1} \quad (3\text{-}1a)$$

$$= \frac{9\ \text{V}}{310 + 3300\ \Omega} = 2.5\ \text{mA}$$

$$P_R = I_R{}^2 R \quad (3\text{-}9)$$

$$= (2.5 \times 10^{-3}\ \text{A})^2 \times 310\ \Omega = \mathbf{1.9\ mW}$$

It is to be noted that although the same value is obtained for R_x in either connection, the latter connection draws much less current and the power dissipation is reduced. (In R_x, the power is reduced from 51 to 5 mW). In most cases, this is the desirable connection if the device being measured is either temperature-sensitive or has a high-temperature coefficient of resistance; that is, its resistance varies greatly as the temperature changes. However, maximum sensitivity of the galvanometer is claimed to occur with the galvanometer as in Fig. 13-10*a*. The battery should be connected with the two low resistances in series. However, the difference in sensitivity is slight, and when using a *commercial* bridge instrument there is no choice in this respect.

It should be mentioned that the Wheatstone bridge is a most versatile circuit and is used in *many* applications other than just measuring resistance. For example, if one of the resistor arms is replaced by a thermistor, which changes resistance with temperature, the bridge will become unbalanced as the temperature changes. The amount of current flowing through the meter is then an indication of temperature. By using the appropriate *transducer* (a device whose resistance changes with a change in a physical quantity), it is possible to use the bridge circuit as a pressure gauge, vacuum gauge, flow meter, pollution indicator, wind speed anemometer, and so on.

13-8 DIGITAL MULTIMETERS

All of the meters we have discussed so far are of the analogue type. That is, they use a pointer to indicate a reading, much as a watch uses hands to show the time. A digital multimeter may use light-emitting diodes (LEDs)

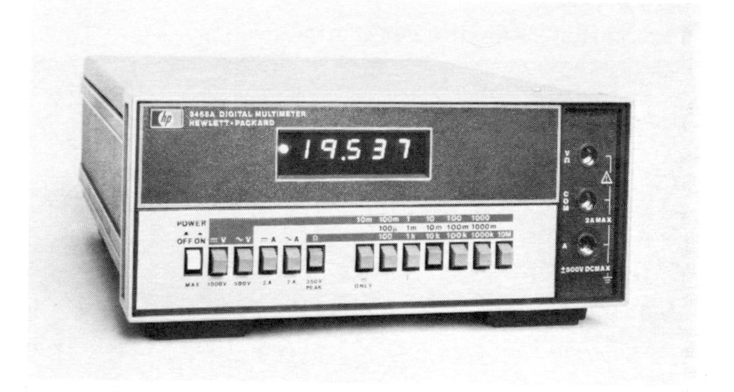

FIGURE 13-11
A $4\frac{1}{2}$-digit five-function multimeter. (Courtesy of Hewlett-Packard.)

or a liquid crystal display (LCD) to show the digital value of the electrical quantity being measured.

Figure 13-11 shows a $4\frac{1}{2}$-digit multimeter. This has a five-digit display with the first digit only able to indicate a 1 when necessary. Both function and range buttons are provided, although some meters automatically select the required range. Polarity is also automatic, a minus sign being displayed when the quantity is negative with respect to common.

A digital multimeter usually has all five functions, being able to measure both ac and dc current, as well as ac and dc volts, and ohms. Its characteristics are compared in Table 13-1 with the analogue type of VOM and electronic voltmeter (EVM). Unlike the latter two, the digital multimeter has no need for an ohms adjustment. However, some may have a zero adjustment when making measurements on the low (10 mV) dc volts range.

In addition to the ease of reading (since no scale and pointer is provided), the primary features of a digital multimeter are its resolution and accuracy. One $3\frac{1}{2}$ digit meter has seven ranges measuring current from 1 pA to 2 mA, with less than 0.2 mV drop caused by the instrument. Some of the accuracies obtainable with the multimeter shown in Fig. 13-11 are indicated in Table 13-2. These are usually in effect for a period of 30 days following calibration, provided that the temperature is within 5°C of 23°C. Note how the accuracy depends upon both the reading and the range in use. Some DMMs may express the accuracy in terms of a percentage of the reading and plus or minus one digit.

TABLE 13-1
Comparison of Multimeter Characteristics

Characteristic	VOM	EVM	DMM
Volts dc	Yes	Yes	Yes
Volts ac	Yes	Yes	Yes
Milliamperes dc	Yes	No	Yes
Milliamperes ac	No	No	Yes
Resistance ranges	$R \times 1$ to $R \times 100$ k	$R \times 1$ to $R \times 1$ M	100 Ω-10 MΩ
Adjustments	$R = 0$	$R = 0$; $R = \infty$	None
Dc volts input resistance	20 kΩ/V	10 MΩ	10 MΩ to 10^{10} Ω
Ac volts input resistance	5–20 kΩ/V	1 MΩ	1 MΩ
Ac volts frequency range	20 Hz–100 kHz	25 Hz–1 MHz	40 Hz–20 kHz
Power	Battery (ohms)	Ac or battery	Ac or battery
Display	Pointer	Pointer	LED or LCD
Accuracy	±2–3% of FSD	±3–5% of FSD	± (% reading + % range) (See Table 13-2)

TABLE 13-2
Typical Digital Multimeter Accuracies

Function	Accuracy
Dc volts (100 mV–100 V)	± (0.02% of reading + 0.01% of range)
Ac volts (100 mV–100 V, 40–10 kHz)	± (0.15% of reading + 0.05% of range)
Dc amperes (100 μA–10 mA)	± (0.1% of reading + 0.01% of range)
Ac amperes (100 μA–10 mA, 1–10 kHz)	± (0.25% of reading + 0.75% of range)
ohms (1 kΩ–1 MΩ)	± (0.02% of reading + 0.01% of range)

EXAMPLE 13-8

A digital multimeter having an accuracy given by Table 13-2 is being used on the 10-V dc range.

a. Determine the possible reading and the percentage error when measuring 5 V dc.
b. Repeat for a measurement of 10 V dc.

SOLUTION

a. From Table 13-2:

$$\text{Possible reading} = 5\text{ V} \pm (0.02\% \text{ of reading} + 0.01\% \text{ of range})$$

$$= 5\text{ V} \pm \left(\frac{0.02}{100} \times 5\text{ V} + \frac{0.01}{100} \times 10\text{ V}\right)$$

$$= 5\text{ V} \pm (0.001\text{ V} + 0.001\text{ V})$$

$$= \mathbf{5\text{ V} \pm 0.002\text{ V}}$$

$$\text{Percentage error} = \frac{0.002\text{ V}}{5\text{ V}} \times 100\% = \mathbf{0.04\%}$$

b. $$\text{Possible reading} = 10\text{ V} \pm (0.02\% \text{ of } 10\text{ V} + 0.01\% \text{ of } 10\text{ V})$$

$$= 10\text{ V} \pm \left(\frac{0.02}{100} \times 10\text{ V} + \frac{0.01}{100} \times 10\text{ V}\right)$$

$$= 10\text{ V} \pm (0.002\text{ V} + 0.001\text{ V})$$

$$= \mathbf{10\text{ V} \pm 0.003\text{ V}}$$

$$\text{Percentage error} = \frac{0.003\text{ V}}{10\text{ V}} \times 100\% = \mathbf{0.03\%}$$

Once again we see the advantage of making readings that are as close to full scale as possible to reduce percentage errors.

SUMMARY

1. A potentiometer is an accurate, high-resolution, variable-voltage divider used to measure unknown voltages without drawing any current.

2. A standard cell is a source of accurately known voltage used for calibrating potentiometers.

3. A potentiometer can be used to calibrate voltmeters and ammeters.

4. A calibration curve consists of a graph of corrections that must be added to (or subtracted from) a voltmeter or ammeter reading to obtain a true reading.

5. Resistance may be measured by applying Ohm's law to actual ammeter and voltmeter measurements.

6. Corrections for measuring circuit resistance taking into account meter resistances can be made using Eqs. 13-1 and 13-2.

7. For calculations involving products and quotients, the total possible error is the sum of the errors of the individual variables.

8. A simple type of ohmmeter consists of a series-connected cell, meter movement, and zero control. Such an ohmmeter has a high center-scale resistance.

9. An ohmmeter scale is nonlinear and of the back-off type, with zero ohms on the right and infinity on the left.

10. When the battery voltage in a simple series-type ohmmeter drops, the indicated readings are too high.

11. The shunted-series type of ohmmeter has a low center-scale resistance and is adaptable to multirange operation.

12. A Wheatstone bridge may be used to measure resistance to a high degree of accuracy.

13. A bridge is balanced and in a "null condition" when the cross products of the opposite arms are equal.

14. Interchanging the location of battery and galvanometer in a Wheatstone bridge does not change the resistance values for balance, but it may have a large effect on power dissipation in the circuit.

15. Digital multimeters have higher resolution, better accuracy, and easier operation than analogue multimeters.

SELF-EXAMINATION

Answer true or false or, in the case of multiple choice, a, b, c, or d
(Answers at back of book)

13-1. A no-load voltmeter measures the open-circuit voltage of a source of emf. ________

13-2. A potentiometer is a source of variable voltage known to a high degree of accuracy for measuring unknown voltages. ________

13-3. A potentiometer compares the voltage from a standard cell with an unknown voltage to determine the unknown voltage. ________

13-4. The calibration of a potentiometer requires setting the dials to the standard cell voltage and adjusting the current from the working battery until a null is indicated on the galvanometer. ________

13-5. A potentiometer can directly measure voltages up to 150 V. ________

13-6. A potentiometer, because it can only measure voltage, can be used to calibrate a voltmeter but not an ammeter. ________

13-7. In calibrating a voltmeter, the correction may be either positive or negative. ________

13-8. The ammeter–voltmeter method of measuring resistance is one of the most accurate methods available. ________

13-9. If a 15-V 3%-accurate voltmeter indicates 7.5 V and a series-connected 10-mA range 4%-accurate milliammeter indicates 10 mA, the resistance and its maximum percentage error are

a. 750 Ω, 7%
b. 75 Ω, 7%
c. 750 Ω, 10%
d. 7.5 kΩ, 10%

13-10. When using the ammeter–voltmeter method to measure high resistance, the voltmeter should be connected directly across the resistor. ________

13-11. A series-type of ohmmeter, like all analogue ohmmeters, has a zero-ohms control. ________

13-12. A series type of ohmmeter has a large change in indicated resistance as the battery voltmeter drops. ________

13-13. A series type of ohmmeter is especially suited for making low-resistance measurements. ________

13-14. A shunted-series type of ohmmeter is used for multirange ohmmeters and has a low center-scale resistance. ________

13-15. Failure to be able to zero an ohmmeter (especially on the X-1 range) is an indication that the main battery needs replacing. ________

13-16. An ohmmeter in a VOM cannot be zeroed on the highest range, but it can on the lowest. This may be an indication that the additional cells were not installed. ________

13-17. In a balanced Wheatstone bridge, the products of the adjacent resistor arms are equal. ________

13-18. At balance, no current flows through the galvanometer in a Wheatstone bridge, so meter accuracy is unimportant. ________

13-19. In measuring an unknown resistance, the choice of location of the battery and galvanometer give the same result. ________

13-20. In a commercial Wheatstone bridge the ratio arms are usually adjusted as multiples or submultiples of 10 and become the multiplier factor. ________

13-21. A Wheatstone bridge can be used only for measuring resistance. ________

13-22. A digital multimeter can also measure ac current while most analogue multimeters do not. ________

REVIEW QUESTIONS

1. Describe the principle of a no-load voltmeter.

2. Describe briefly the complete procedure for using a potentiometer to measure an unknown emf.

3. a. Why doesn't the accuracy of the galvanometer affect the accuracy of voltage determination in a potentiometer or resistance determination in a Wheatstone bridge?

b. What is a very desirable characteristic for a galvanometer in these applications?

4. A standard cell has a high internal resistance of approximately 500 Ω. Why doesn't this affect the accuracy of voltage measurement?

5. a. What is the maximum voltage measurable by a potentiometer?

b. How can this be extended?

c. What disadvantage does this have?

6. Describe the process of calibrating an ammeter.

7. Given an accurately known resistance, suggest how you could use a potentiometer to make an accurate measurement of another resistor similar in value to the first.

8. If you were using the ammeter–voltmeter method for measuring circuit resistance, what *visual* indications would you have on the meters as you tried to decide where the voltmeter should be placed for the following circuits?

a. A high-resistance circuit.

b. A low-resistance circuit.

9. When determining resistance by applying Ohm's law to voltmeter and ammeter readings, under what conditions will the maximum possible error be given by the following sums?

a. The sum of the FSD accuracies of the two meters.

b. Twice the sum of the FSD accuracies of the two meters.

10. Give a simple explanation of why an ohmmeter scale must be nonlinear?

11. Why are most VOM ohmmeter scales of the "back-off" type?

12. Why does a series type of ohmmeter inherently have a high center-scale resistance?

13. Why does a reduction in voltage in a series-type ohmmeter cause the indicated resistances to be high?

14. In a shunted-series type of ohmmeter, why is it more likely to be unable to zero the instrument on the $R \times 1$ range than on a higher range?

15. a. What is often necessary in many VOM ohmmeter circuits to be able to make measurements on the $R \times 10$ k range?

b. Why is this so?

16. a. Why does no current flow through the galvanometer of a balanced Wheatstone bridge?

b. If, in Fig. 13-9*a*, at balance, a connection were made from B directly to C, would this upset the balanced conditions? Explain.

17. Why are two of the resistors in a Wheatstone bridge referred to as the "ratio arms"?

18. Why should interchanging the locations of the dc supply and galvanometer have any effect upon the power dissipations in the Wheatstone bridge circuit?

19. What are four advantages that digital multimeters have over most analogue multimeters?

20. a. How does the error on a digital multimeter measurement differ from that on an analogue instrument?

b. Does minimum percent error occur for readings near full scale on a digital multimeter as it does for an analogue multimeter?

PROBLEMS

(Answers to odd-numbered problems at back of book)

13-1. A potentiometer is being used to calibrate a dc voltmeter on the 1.5-V range. The largest correction occurs when the voltmeter indicates 1.3 V. At this value, the two dials of the potentiometer indicate 1.2 V and 53 mV. Calculate:

a. The true reading of the voltmeter.

b. The correction in mV in magnitude and sign.

c. The percent FSD accuracy of the meter.

13-2. A potentiometer is being used to calibrate a 100-mA range milliammeter. The largest correction occurs when the milliammeter indicates 60 mA. At this value, the two dials of the potentiometer indicate 0.5 V and 76 mV, when connected across a standard 10-Ω resistor connected in series with the ammeter. Calculate:

a. The true reading of the milliammeter.

b. The correction in magnitude and sign.

c. The percent FSD accuracy of the meter.

13-3. A potentiometer is being used to measure resistance. A standard 10-Ω 0.1% resistor is connected in series with the unknown resistor, and the combination is connected to a dc supply. When the potentiometer is connected across the 10-Ω resistor, it indicates 0.105 V at balance; across the unknown resistor it indicates 1.189 V. If the potentiometer is accurate to 0.5% calculate:

a. The unknown resistance.

b. The accuracy of the measured resistance.

13-4. A voltmeter is being calibrated on the 300-V range using a volt-box with a maximum 300-V input and maximum 1.5-V output. When the voltmeter indicates 180 V, the maximum correction occurs for dial readings on the potentiometer of 0.9 V and 35 mV. If the volt-box is 0.1% accurate and the potentiometer 0.2%, calculate:

a. The true reading of the voltmeter.

b. The correction in magnitude and sign.

c. The percent FSD accuracy of the voltmeter.

d. The accuracy to which the determination in part (c) is made.

13-5. A resistor is being measured using the circuit of Fig. 13-5*a*. A 15-V range, 3% FSD accuracy voltmeter indicates 10 V; and a 1-mA, 100-Ω, 4%-FSD accuracy milliammeter indicates 0.5 mA. Calculate:

a. The resistance.

b. The maximum possible error.

13-6. Repeat Problem 13-5 assuming that both meters indicate FSD.

13-7. A resistor is being measured using the circuit of Fig. 13-5*b*. The 100-mA range milliammeter indicates 60 mA, and the 5 kΩ/V voltmeter indicates 3 V on the 5-V range. If both meters have a 3% FSD accuracy calculate:

a. The resistance.

b. The maximum possible error.

13-8. Repeat Problem 13-7 assuming that both meters indicate 50% of FSD.

13-9. Given a 0–100-μA, 2-kΩ meter movement, determine the resistance markings every 20 μA when used in a series-type ohmmeter with two 1.5-V cells.

13-10. Repeat Problem 13-9 using a 0–200-μA, 5-kΩ meter movement, every 40 μA.

13-11. a. Assuming that each cell in Problem 13-9 decreases by 0.1 V, calculate the true resistance being measured when the meter indicates 60 μA.

b. Using the resistance marking for 60 μA from Problem 13-9, determine the additional error introduced by the aging batteries.

c. What could the total battery voltage drop to before being unable to zero the ohmmeter?

13-12. A series-type ohmmeter consists of a variable zeroing rheostat in parallel with the meter movement and this combination in series with a fixed resistor R_1 and a 1.5-V cell. If the meter movement is a 3-kΩ, 50-μA FSD, and the center-scale resistance marking is to be 25 kΩ, calculate:

a. The value of R_1 and the zero-ohms control value.

b. The new value of the zero-ohms control when the voltage drops to 1.4 V.

c. The true resistance being measured when the ohmmeter indicates its center-scale resistance reading.

d. The percentage change in center-scale resistance due to a drop of 0.1 V.

e. Compare the results with the percentage change of center-scale resistance in Examples 13-3 and 13-4.

13-13. Given a shunted-series type of multirange ohmmeter as in Fig. 13-7 with R_{s1} = 12 Ω, R_{s2} = 120 Ω, R_{s3} = 1.2 kΩ, and I_m = 50 μA, R_m = 3 kΩ, V = 3 V, calculate, with the range switch on $R \times 1$:

a. The current through R_{s1} and the movement when $R_x = \infty$.

b. The value of R_z to zero the ohmmeter.

c. The current through R_{s1} when $R_x = 0$.

d. The current through R_{s1} and the movement when $R_x = 12\ \Omega$.

e. The current through R_{s2} and the movement when the range is changed to $R \times 10$ and $R_x = 120\ \Omega$.

f. The resistance markings at 25% and 75% of FSD.

13-14. Repeat Problem 13-13 using a 100-μA, 5-kΩ meter movement and a 1.5-V cell.

13-15. If the range selector switch in Example 13-5 has a fourth position, where no shunt is connected into the circuit, determine the multiplier for this position.

13-16. Repeat Problem 13-15 as applied to Problem 13-13.

13-17. A Wheatstone bridge has the following values at balance, using the notation in Fig. 13-9: $R_2/R_1 = 10$ to an accuracy of 0.5%, $R = 4.52\ \Omega$, $\pm 0.1\%$.
If the dc supply is 9 V, calculate:

a. The value of R_x.

b. The accuracy of the determination of R_x.

c. The current through R_x at balance.

13-18. a. If three of the resistors in a Wheatstone bridge are each 5 kΩ, $\pm 1\%$, what must be the value of the fourth at balance?

b. What range of values could the fourth resistor have?

c. What would be the total resistance seen by the battery at balance?

13-19. Using the notation in Fig. 13-10*a*, a student-built Wheatstone bridge consists of $R_1 = 10$ kΩ, $\pm 2\%$, $R_2 = 220$ kΩ, $\pm 1\%$. Bridge balance occurs when R is adjusted to 4.75 kΩ, $\pm 0.1\%$. Calculate:

a. The unknown resistance.

b. The maximum possible error.

c. The current through R_x if a 6-V battery is used.

d. The power dissipated in R_x.

13-20. Repeat Problem 13-19 with $R_1 = 2.2$ kΩ, $\pm 1\%$.

13-21. Repeat Problem 13-19 with the positions of the galvanometer and battery interchanged.

13-22. Repeat Problem 13-20 with the positions of the galvanometer and battery interchanged.

13-23. A digital multimeter has accuracies given by the data in Table 13-2. It is being used as a dc voltmeter. Determine the possible reading and the percentage error when making a measurement of 1 V on:

a. The 1-V range.

b. The 10-V range.

13-24. Repeat Problem 13-23 for the multimeter being used to make ac voltage measurements at 1 kHz.

13-25. A digital multimeter has accuracies given by the data in Table 13-2. It is being used as a dc milliammeter on the 10-mA range. Determine the possible reading and the percentage error when making a measurement of

a. 2 mA.

b. 9 mA.

13-26. Repeat Problem 13-25 for the multimeter being used to make ac current measurements at 10 kHz.

13-27. a. By how many ohms could the multimeter of Table 13-2 be in error when it indicates 820 kΩ on the 1-MΩ range?

b. What percentage accuracy does this represent?

PART TWO

AC CIRCUITS

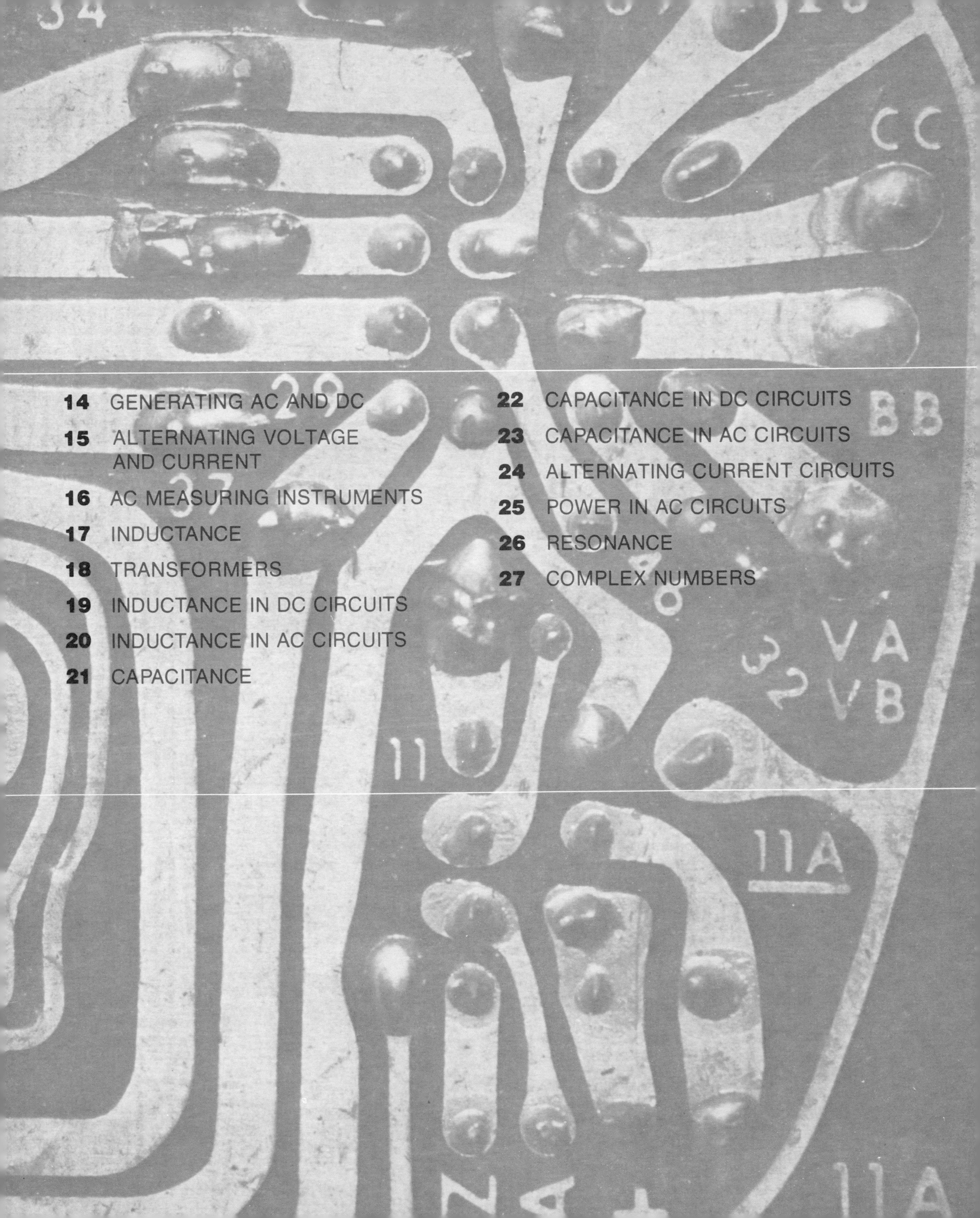

14 GENERATING AC AND DC

15 ALTERNATING VOLTAGE AND CURRENT

16 AC MEASURING INSTRUMENTS

17 INDUCTANCE

18 TRANSFORMERS

19 INDUCTANCE IN DC CIRCUITS

20 INDUCTANCE IN AC CIRCUITS

21 CAPACITANCE

22 CAPACITANCE IN DC CIRCUITS

23 CAPACITANCE IN AC CIRCUITS

24 ALTERNATING CURRENT CIRCUITS

25 POWER IN AC CIRCUITS

26 RESONANCE

27 COMPLEX NUMBERS

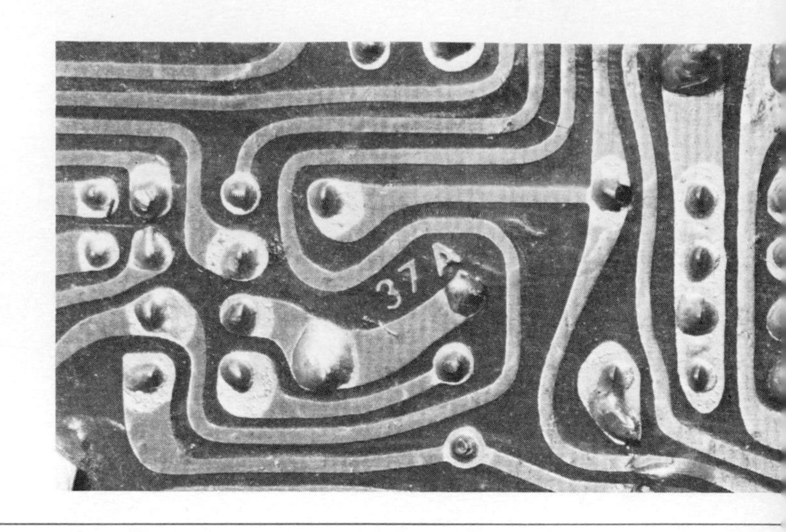

CHAPTER FOURTEEN
GENERATING AC AND DC

We have seen how current in a conductor sets up a magnetic field around the conductor. In this chapter we investigate how a current can be *induced* in a conductor located in a *changing* magnetic field. This process, called *electromagnetic induction,* is the basis of ac alternators and dc generators. The *amount* of voltage induced in a conductor is given by Faraday's law (Section 14-3). The *polarity* of the voltage is given by Lenz's law (Section 14-2).

If a coil is rotated in a fixed magnetic field, a voltage is produced that alternates in polarity. The waveform of this voltage can be made to have a *sinusoidal* variation. That is, its instantaneous value of induced emf depends upon the sine of the angle that the coil makes with the magnetic field. We also consider the factors that determine the amplitude and frequency of the induced sine wave of voltage. Finally, practical dc generators and three-phase alternators are introduced.

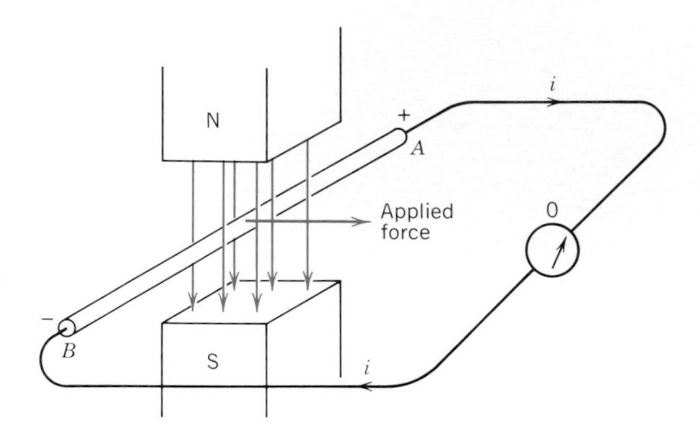

(*a*) Polarities of induced voltage and current are due to moving the conductor to the right

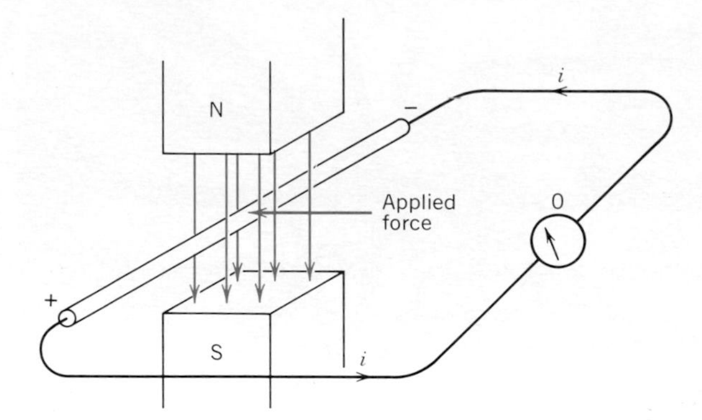

(*b*) Polarities of induced voltage and current are reversed because of the conductor's motion in the opposite direction

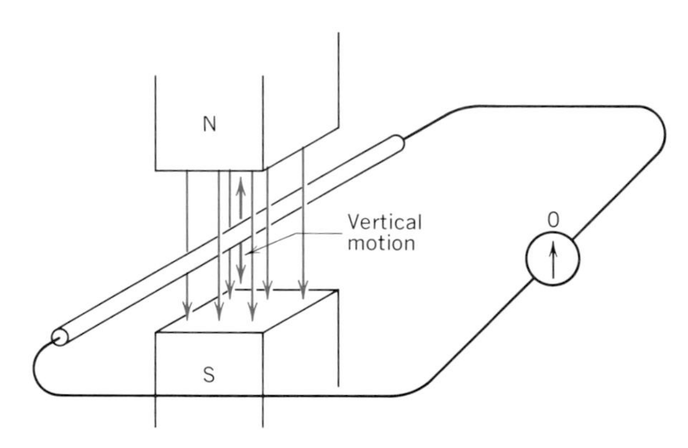

(*c*) No voltage is induced when the conductor is moved parallel to the magnetic field

FIGURE 14-1
Electromagnetic induction.

14-1 ELECTROMAGNETIC INDUCTION

Electromagnetic induction is the process by which a voltage is induced in a conductor whenever there is relative motion between the conductor and a magnetic field. One of the ways of producing this relative motion is shown in Fig. 14-1. Here a conductor is moved through a stationary magnetic field. The direction and strength of the magnetic field is represented by lines of force. In Fig. 14-1*a*, the conductor is said to be "cutting" the lines of the magnetic field. Note the polarity of induced voltage. In Fig. 14-1*b*, the conductor is moved in the opposite direction. This time, the cutting of the field lines produces the opposite polarity. However, when the conductor is moved parallel to the field lines, as in Fig. 14-1*c*, no cutting of field lines occurs and no voltage is induced.

It should be noted that the same effects are obtained if the conductor is held stationary and the magnetic field moved. All that is necessary for electromagnetic induction is relative motion and "cutting" of the field lines. The latter requirement will be modified in Section 14-3.

What is the source of an emf produced in the wire? In Section 11-10 we said that a magnetic field exists at a point if a force is exerted on a moving charge at the point ($F = qvB$). By moving a conductor, we are making the free electrons within the conductor move through the magnetic field. Hence a force is exerted on these free electrons, making them move from one end of the conductor toward the other. This causes a concentration of electrons at one end and a deficiency at the other. In other words, a potential difference has been generated that can cause the current to flow around an externally connected circuit. It is important to note that the polarity signs in Fig. 14-1 indicate a *source* of emf, and not a voltage *drop* caused by the current.

The force required to move the electrons is supplied by the external force applied to the conductor. If the conductor is moved through the field more quickly, a larger force is required inducing a higher voltage. The polarity of this voltage is determined from the application of Lenz's law.

14-2 LENZ'S LAW

The polarity of the induced emf is such that any current resulting from it produces a magnetic flux which opposes the motion or change producing the emf.

Basically, Lenz's law states that the polarity of the induced emf is such that it opposes the cause that produced it. This is shown in Fig. 14-2*a*, where the conductor of Fig. 14-1*a* is represented being moved to the right. The polarity of the induced emf causes current or charge flow through the conductor "into" the paper. This sets up a magnetic field around the conductor that is

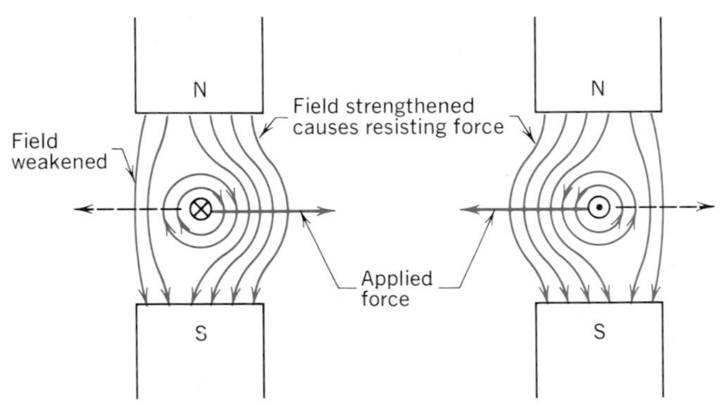

(*a*) Induced current into the paper due to an applied force to the right (*b*) Induced current out of the paper due to an applied force to the left

FIGURE 14-2
Using Lenz's law to determine the polarity of induced emf.

clockwise in direction. The result is a strengthening of the overall field to the right of the conductor, and a weakening on the left. A "motor action" force is developed on the conductor *opposing* the externally applied force. Thus *work* must be done by the external force to develop the induced emf, in line with Lenz's law.

In Fig. 14-2*b*, with the conductor being moved in the opposite direction, a similar magnetic repelling effect is produced by the induced current of the opposite polarity. We shall have many occasions to apply Lenz's law in future applications.

14-3 FARADAY'S LAW

The more quickly a conductor is moved through a magnetic field, the larger is the voltage that is induced. The voltage may also be increased if a stronger magnetic field is used. These two effects are combined in Faraday's law:

$$v_{\text{ind}} = \frac{\Delta\Phi}{\Delta t} \qquad \text{volts (V)} \qquad (14\text{-}1)$$

where: v_{ind} is the emf induced in a conductor, in volts, V

$\Delta\Phi$ is the magnetic flux cut by the conductor, in webers, Wb

Δt is the time required for the conductor to move, in seconds, s

$\frac{\Delta\Phi}{\Delta t}$ is the rate at which magnetic flux lines are cut by the conductor, in webers per second, Wb/s

The symbol Δ means "change." Thus $\Delta\Phi$ really means a *change* in the magnetic flux and Δt means a *change* in time. We use Δt here to refer to an *interval* of time equal to $t_2 - t_1$, where t_2 is the final time and t_1 is the initial time. Similarly, $\Delta\Phi = \Phi_2 - \Phi_1$ is a *quantity* of flux cut by the conductor in the time interval Δt.

Thus Faraday's law states that the voltage induced in a conductor is directly proportional to the rate at which the magnetic flux is cut by the conductor.

EXAMPLE 14-1

The magnetic field strength between the pole faces of a horseshoe magnet is 0.6 T (Wb/m²). The pole faces are 2 cm square. Calculate the voltage induced in a wire if it is moved perpendicularly through the field at a uniform rate in

a. 0.1 second

b. 0.01 second

SOLUTION

a. The flux cut by the wire can be calculated from

$$\Phi = BA \qquad (11\text{-}1)$$

or

$$\Delta\Phi = B \times \Delta A$$

$$= 0.6 \frac{\text{Wb}}{\text{m}^2} \times (2 \times 10^{-2}\text{m})^2$$

$$= 2.4 \times 10^{-4}\ \text{Wb}$$

$$v_{\text{ind}} = \frac{\Delta\Phi}{\Delta t} \qquad (14\text{-}1)$$

$$= \frac{2.4 \times 10^{-4}\ \text{Wb}}{0.1\ \text{s}}$$

$$= 2.4 \times 10^{-3}\ \text{Wb/s} = \mathbf{2.4\ mV}$$

b.

$$v_{\text{ind}} = \frac{\Delta\Phi}{\Delta t} \qquad (14\text{-}1)$$

$$= \frac{2.4 \times 10^{-4}\ \text{Wb}}{0.01\ \text{s}}$$

$$= 2.4 \times 10^{-2}\ \text{Wb/s} = \mathbf{24\ mV}$$

Note that the length of the conducting wire is unimportant, provided that it is long enough to cut through all of the flux. The voltage depends only upon the *rate* at which the flux is cut.

We can now give the formal definition of the unit of magnetic flux, a weber.

The weber is the amount of magnetic flux that, when cut at a constant rate by a conductor in 1 s, will induce an emf of 1 V in the conductor. That is, 1 volt = 1 weber/second.

If, instead of moving a single wire through a magnetic field, a number of wires are series-connected as in the turns of a coil, the induced emf is increased accordingly:

$$v_{\text{ind}} = N\left(\frac{\Delta\Phi}{\Delta t}\right) \qquad \text{volts (V)} \qquad (14\text{-}2)$$

where the symbols are as in Eq. 14-1 and N equals the number of turns in the coil.

EXAMPLE 14-2

The north end of a permanent magnet is moved into a solenoid as in Fig. 14-3. If the coil is cut by magnetic flux lines at the rate of 0.03 Wb/s, determine:

a. The voltage induced in the coil of 100 turns.

b. The polarity of induced voltage in the coil.

c. The effect of withdrawing the magnet at twice the speed.

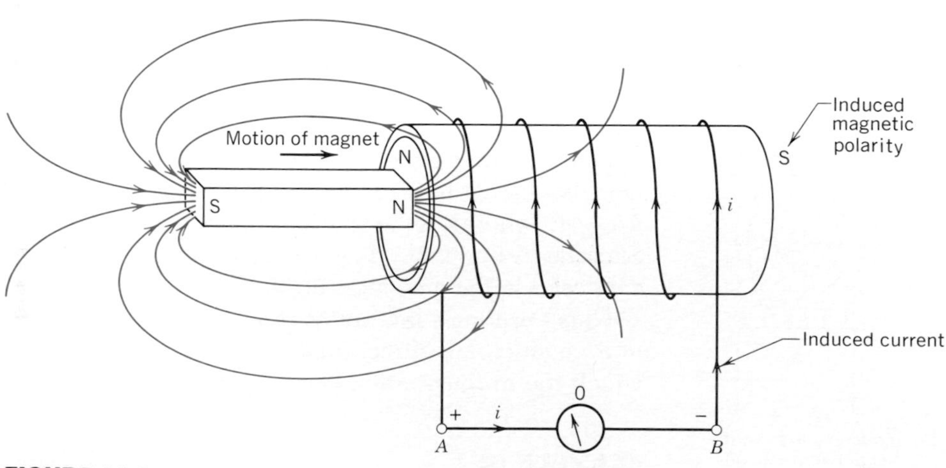

FIGURE 14-3
Using Lenz's law to determine the polarity of induced emf when a permanent magnet is moved toward a solenoid.

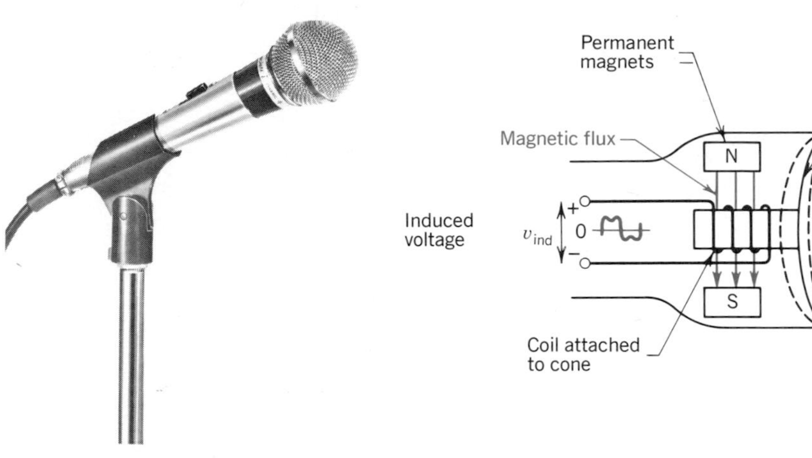

(*a*) Undirectional dynamic microphone

(*b*) Cross-sectional view through a dynamic microphone

FIGURE 14-4
Dynamic microphone. (Photograph courtesy of Shure Brothers Inc.)

SOLUTION

a. $$v_{ind} = N\left(\frac{\Delta\Phi}{\Delta t}\right) \qquad (14\text{-}2)$$
$$= 100 \times 0.03 \text{ Wb/s} = \mathbf{3\ V}$$

b. By Lenz's law, the current must flow in the coil in such a direction as to set up an opposing force. This means that the left side of the solenoid must become a north pole while the permanent magnet is approaching. Applying the right-hand rule, the current must flow upward on the front of the coil. This means that point A is positive with respect to B.

c. If the magnet is removed at twice the speed it entered, the rate of change of flux in the coil, $\frac{\Delta\Phi}{\Delta t}$, is doubled to 0.06 Wb/s, and the induced voltage is doubled to **6 V.**

The polarity of the induced voltage is **reversed** to set up a south pole at the **left** side of the solenoid. This attempts to keep the north pole of the permanent magnet from being removed.

Note that a galvanometer connected between A and B will deflect alternately to the left and to the right as the magnet is moved in and out of the coil. In other words, an alternating current is produced by this back and forth motion. Whenever the motion of the magnet ceases, as at the end of its travel, the induced voltage drops to zero, in accordance with Faraday's law.

This is the principle used in a dynamic microphone. See Fig. 14-4. The sound-pressure waves cause a cone to move in and out. A coil attached to the cone moves within the permanent magnet's field. A voltage is induced in the coil with a frequency dependent upon that of the voice.

After amplification, this voltage can be used to drive a loudspeaker, as described in Section 11-12. The action of the microphone is actually the reverse of a loudspeaker. This principle is used in a number of intercoms and walkie-talkies. Here a single device is used as both a microphone and as a speaker, according to whether the unit is being used to transmit or receive, that is, to convert sound into electricity or to convert electricity into sound.

Faraday's law is also useful for applications where apparently no *cutting* of the magnetic field takes place. This occurs when two coils are wrapped on a common iron core as in Fig. 14-5. Recall that a toroid restricts all the flux to within the core. If the current in the first coil is made to change (by varying the rheostat), a *changing magnetic field* is set up in the core. The magnetic flux *enclosed* by the second coil changes, thus inducing a voltage. (This is the principle of a transformer.) The polarity of this induced voltage can be obtained by applying Lenz's law, as in Example 14-3. In effect, a voltage is induced whenever a change in flux (around a conductor) occurs. This is true if a conductor is moved through a magnetic field or the field moves (changes) about a stationary conductor.

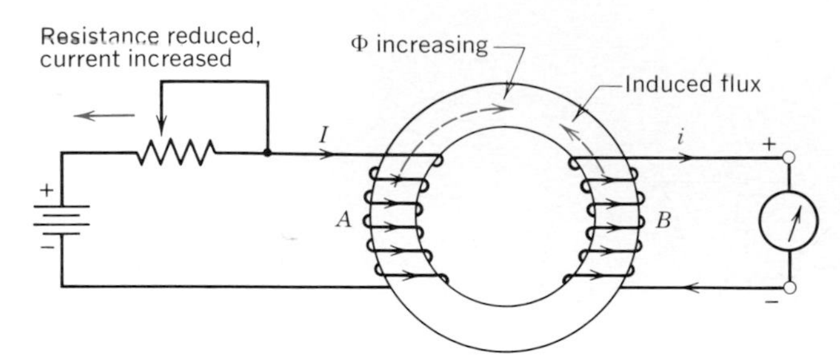

(a) An increasing current in coil A sets up an increasing field in the core. The induced current in coil B sets up a flux that opposes the increase

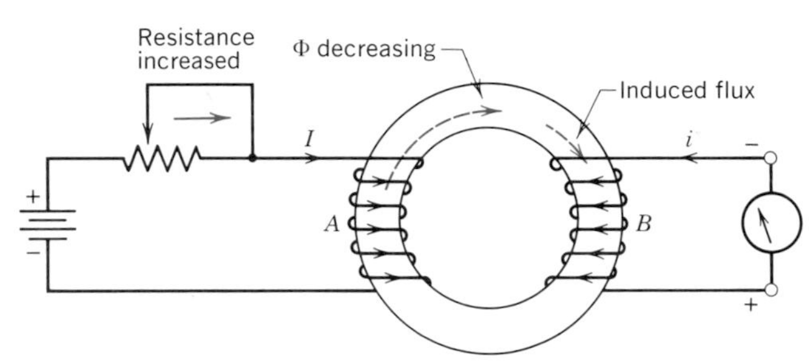

(b) A decreasing current in coil A reduces the field in the core. The induced current in coil B sets up a flux that opposes the decrease

FIGURE 14-5
Application of Faraday's and Lenz's laws to coils on a common core.

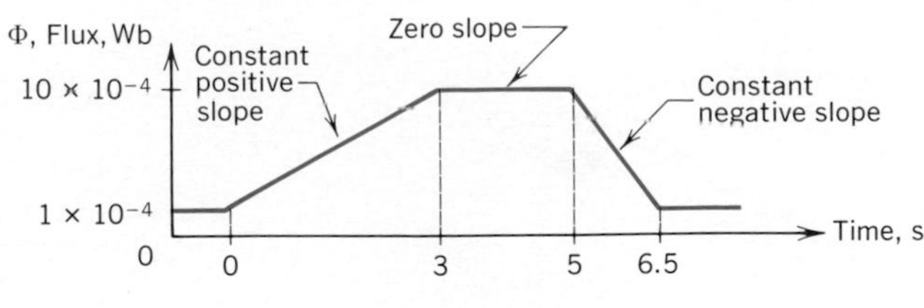

(a) Variation of flux in iron core of Fig. 14-5

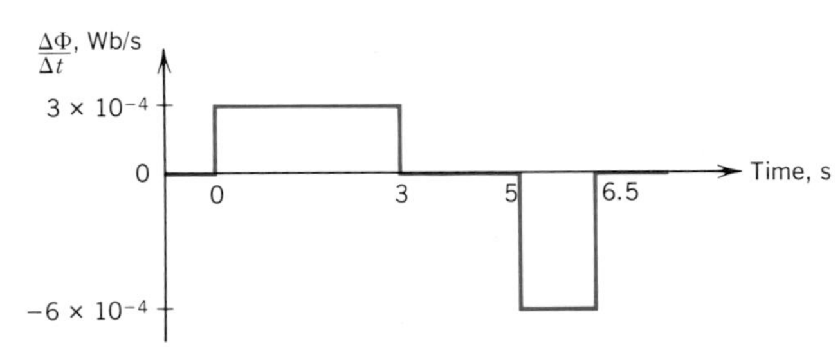

(b) Rate of change of flux in iron core

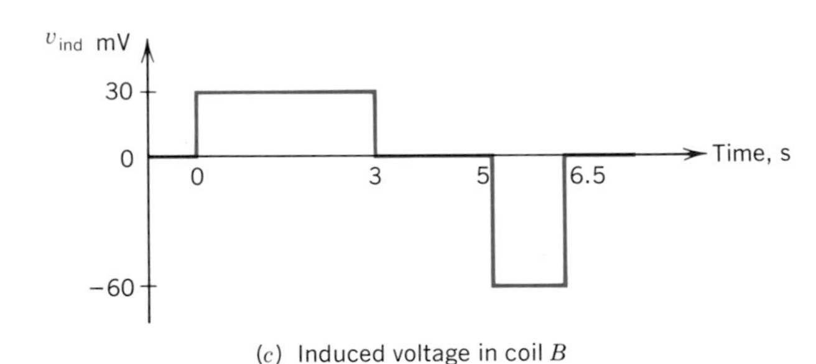

(c) Induced voltage in coil B

FIGURE 14-6
Waveforms for Example 14-3.

EXAMPLE 14-3

Refer to Fig. 14-5. The rheostat is decreased in value in such a way as to cause the magnetic flux in the core to increase uniformly from 1×10^{-4} Wb to 10×10^{-4} Wb in 3 s. After a delay of 2 s, the rheostat is then increased, causing the flux to drop uniformly to its initial value in 1.5 s. Coil B has 100 turns.

a. Draw a graph showing how magnetic flux changes in the core.
b. Underneath the above graph, draw a second graph showing the quantity $\dfrac{\Delta\Phi}{\Delta t}$.
c. Draw a third graph indicating the voltage induced in coil B.
d. Show polarity signs of induced voltage on the second coil for when the flux is increasing and decreasing.

SOLUTION

a. The uniform increase and decrease of the flux in the iron core is shown in Fig. 14-6*a*.
b. The rate of change of flux during the first 3 s is given by

$$\frac{\Delta\Phi}{\Delta t} = \frac{10 \times 10^{-4} - 1 \times 10^{-4}\ \text{Wb}}{3 - 0\ \text{s}}$$
$$= \mathbf{3 \times 10^{-4}\ Wb/s}$$

The flux decreases at a rate given by

$$\frac{\Delta\Phi}{\Delta t} = \frac{1 \times 10^{-4} - 10 \times 10^{-4}\ \text{Wb}}{6.5 - 5\ \text{s}}$$
$$= \mathbf{-6 \times 10^{-4}\ Wb/s}$$

These quantities are actually the slopes of the flux variation graph and are plotted as positive and negative quantities, as in Fig. 14-6*b*.

c. $$v_{\text{ind}} = N\left(\frac{\Delta\Phi}{\Delta t}\right) \qquad (14\text{-}2)$$

When the flux is increasing,

$$v_{\text{ind}} = 100 \times 3 \times 10^{-4}\ \text{V} = \mathbf{30\ mV.}$$

When the flux is decreasing,

$$v_{\text{ind}} = 100 \times (-6 \times 10^{-4})\ \text{V} = \mathbf{-60\ mV.}$$

These voltages are plotted in Fig. 14-6*c*.

d. The polarities of the induced voltages are indicated in Fig. 14-5. When the current increases in coil *A*, the flux enclosed by coil *B* also increases. The induced current in coil *B* flows in such a direction as to set up a flux that opposes the flux *increase* due to coil *A*. When the current in coil *A* decreases, as in Fig. 14-5*b*, the induced voltage has the opposite polarity. The current in coil *B* now tends to *maintain* the decreasing flux. That is, it is opposing the *change* in core flux, not the flux itself.

14-4 INDUCED EMF DUE TO CONDUCTOR MOTION IN A STATIONARY MAGNETIC FIELD

Faraday's law is uniquely suited for the situation in Example 14-3 where the flux is changing within a stationary coil. For the case of a conductor moving through a magnetic field, it is convenient to express the induced voltage in more readily measured quantities. Figure 14-7 shows the quantities involved.

Let us assume that the conductor in Fig. 14-7 is moved through a distance Δs in time Δt. If the magnetic field is uniform, the conductor cuts through a total flux $\Delta\Phi$ given by

$$\Delta\Phi = B \times A \qquad (11\text{-}1)$$

where A is the area $= l \times \Delta s$. Then

$$v_{\text{ind}} = \frac{\Delta\Phi}{\Delta t} \qquad (14\text{-}1)$$

$$= \frac{B \times l \times \Delta s}{\Delta t}$$

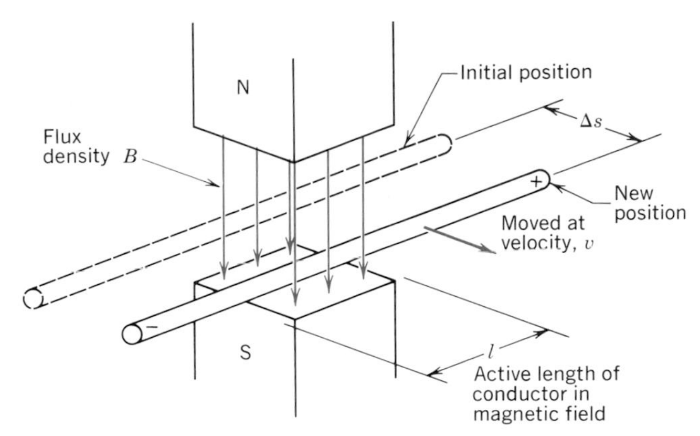

FIGURE 14-7
Quantities involved in $v_{\text{ind}} = Blv$.

$$= B \times l \times \frac{\Delta s}{\Delta t}$$

But $\frac{\Delta s}{\Delta t}$ is the velocity, v, of the conductor. Thus

$$v_{\text{ind}} = Blv \qquad \text{volts (V)} \qquad (14\text{-}3)$$

where: v_{ind} is the emf induced in volts, V
B is the flux density of the field in teslas, T
l is the active length of the conductor in meters, m
v is the velocity of the conductor, in meters per second, m/s

If N conductors are series-connected, the total voltage induced is

$$v_{\text{ind}} = NBlv \qquad \text{volts (V)} \qquad (14\text{-}4)$$

EXAMPLE 14-4

A car with a 0.8-m vertical radio antenna is traveling due west at 90 km/h (approximately 56 mph). If the horizontal component of the earth's magnetic field is 1.7×10^{-5} T at this point, calculate:

a. The voltage induced in the antenna.
b. The voltage if the car travels due north.

SOLUTION

a. $90 \text{ km/h} = 90 \times 10^3 \text{ m/h} \times \dfrac{1 \text{ h}}{3600 \text{ s}} = 25 \text{ m/s}$

$$v_{ind} = Blv \qquad (14\text{-}3)$$
$$= 1.7 \times 10^{-5} \text{ Wb/m}^2 \times 0.8 \text{ m} \times 25 \text{ m/s}$$
$$= 34 \times 10^{-5} \text{ V} = \mathbf{340\ \mu V}$$

b. No voltage is induced when traveling north, since the antenna is traveling *with* the field and *not cutting across* it.

NOTE The voltage induced in part (a) is dc, and while relatively large, it has no effect on the radio, since the radio reacts to RF (radio frequency) or *ac* signals from a transmitting station.

14-4.1 Effect of Angle on Induced Electromotive Force

Equation 14-3 gives the maximum induced voltage V_m when the motion of the conductor is at *right angles* to the field. See Fig. 14-8*a*. We have also seen from Example 14-4 that no voltage is produced when the motion of the conductor is *parallel* to the field. See Fig. 14-8*c*. The voltage produced for an angle between these two extremes, as in Fig. 14-8*b*, is given by (see Appendix A-5):

$$v_{ind} = V_m \sin \theta \qquad \text{volts (V)} \qquad (14\text{-}5)$$

where: v_{ind} is the emf induced in the conductor, in volts, V

B is the flux density of the field in teslas, T

l is the active length of the conductor in meters, m

v is the velocity of the conductor in meters per second, m/s

θ is the angle between the magnetic field and the direction of motion of the conductor

$\sin \theta$ is a trigonometric ratio, between 0 and 1. See Appendix (A-6)

$V_m = Blv$, is the maximum voltage that can be induced in volts, V.

EXAMPLE 14-5

a. What is the maximum voltage that can be induced in a 10-cm-length conductor moving at 20 m/s through a magnetic field of flux density 0.7 T?
b. What voltages are induced in this conductor when moving at angles of 60°, 30°, and 10° through the field?

SOLUTION

a. $V_m = Blv \qquad (14\text{-}3)$

$$= 0.7 \text{ Wb/m}^2 \times 0.1 \text{ m} \times 20 \text{ m/s}$$
$$= \mathbf{1.4\ V}$$

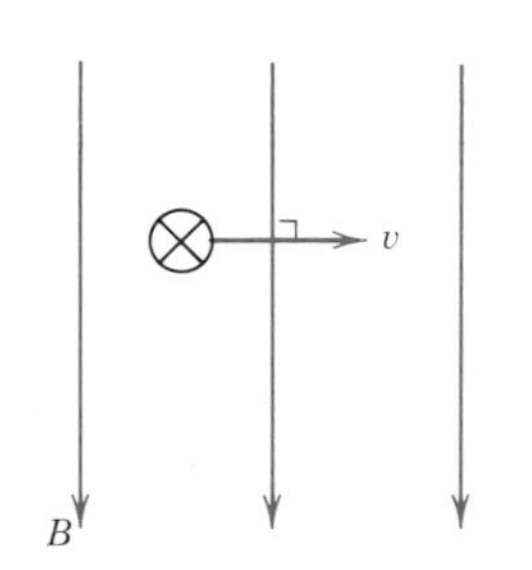

(*a*) Conductor moves at 90° through the field. Maximum voltage induced, V_m

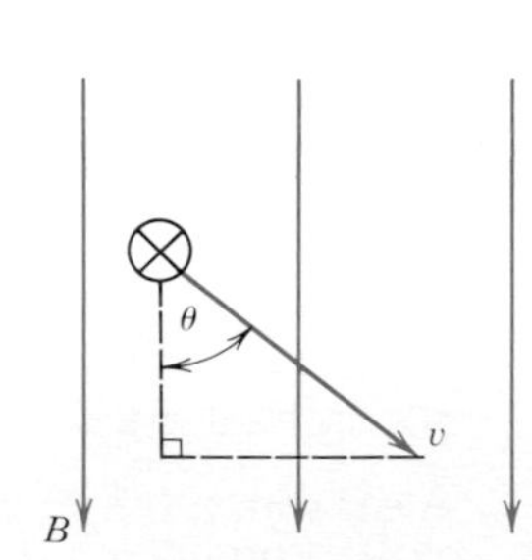

(*b*) Conductor moves at angle Θ through the field. Less than maximum voltage induced, $V_m \sin \Theta$

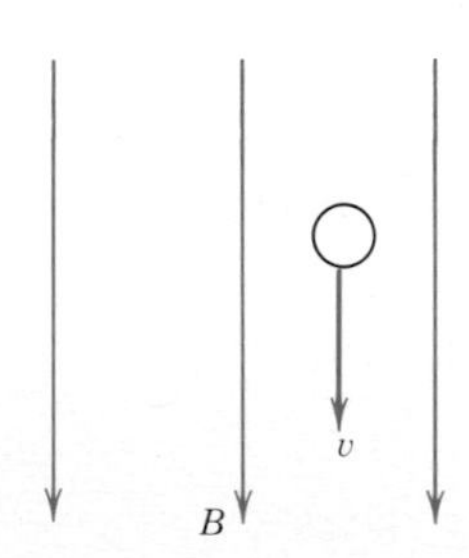

(*c*) Conductor moves at 0° through the field. Zero voltage induced

FIGURE 14-8
Effect of direction angle on induced voltage.

b. $v_{ind} = V_m \sin \theta$ (14-5)

$\theta = 60°$, $v_{ind} = 1.4 \text{ V} \times \sin 60°$
$= 1.4 \text{ V} \times 0.866 = \mathbf{1.21 \text{ V}}$

$\theta = 30°$, $v_{ind} = 1.4 \text{ V} \times \sin 30°$
$= 1.4 \text{ V} \times 0.5 = \mathbf{0.7 \text{ V}}$

$\theta = 10°$, $v_{ind} = 1.4 \text{ V} \times \sin 10°$
$= 1.4 \text{ V} \times 0.174 = \mathbf{0.24 \text{ V}}$

The variation of induced voltage as the angle changes in Example 14-5 occurs continually in alternating current generators. That is, the constantly *changing* angle θ (in rotary motion), is responsible for generating a sinusoidal alternating current.

14-5 ELEMENTARY AC GENERATOR

To continually generate an emf, it is more convenient to *rotate* a conductor in a circular manner than to move a conductor back and forth in a magnetic field. This principle is represented in Fig. 14-9. Mechanical power is used to rotate a coil, having conductors A and B, in a fixed magnetic field. An emf is induced in the two series conductors. This voltage is made available, for external use, by two brushes riding on slip rings connected to the ends of the coil. The slip rings rotate with the coil and provide a means of delivering the generated emf, via the two stationary brushes, to the external circuit and the load.

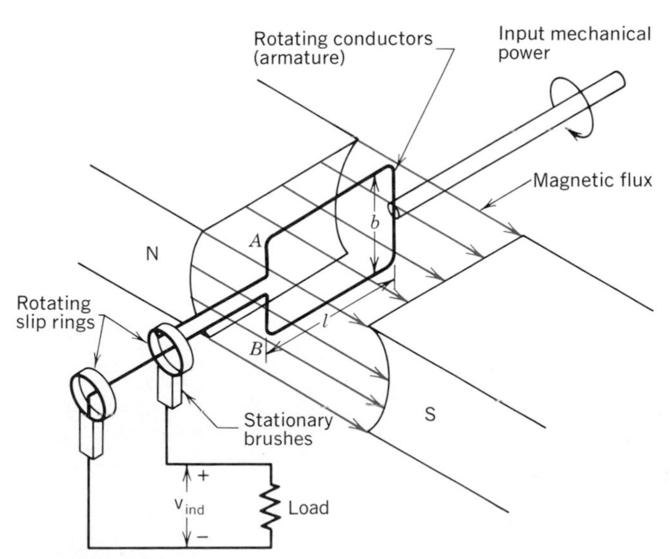

FIGURE 14-9
A simple ac generator.

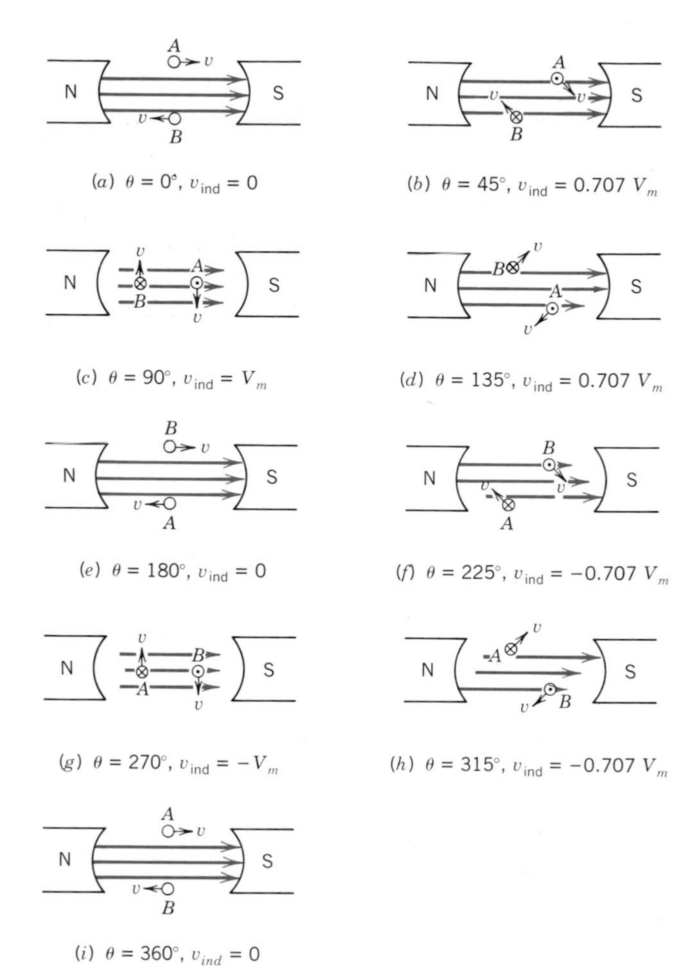

FIGURE 14-10
Various angles and corresponding induced voltages during one revolution of a simple ac generator.

Some of the angular positions assumed by conductors A and B, as the coil makes a complete revolution, are shown in Fig. 14-10. It should be clear that the induced voltage in a given conductor varies instantaneously with the angle between the direction of motion and the magnetic field, as in Section 14-4.1. Thus, when $\theta = 0°$ in

Fig. 14-10*a*, no voltage is induced, since no cutting is taking place. When $\theta = 90°$, the induced voltage is a maximum, since the maximum rate of cutting is taking place. After half a revolution, the induced voltage has dropped to zero. But in Fig. 14-10*f*, when $\theta = 225°$, voltage is again being induced. Note, however, that the direction of induced current in the conductors is opposite to what it was during the first half of a revolution. Thus the voltage generated between the two brushes in Fig. 14-9 *alternates* and follows a *sinusoidal* variation given by $v_{\text{ind}} = V_m \sin \theta$.

14-6 THE SINE WAVE

The shape of the voltage waveform produced by the ac generator may be obtained by graphing the values of v_{ind} against selected values of θ in the equation

$$v_{\text{ind}} = V_m \sin \theta \tag{14-5}$$

The result is shown in Fig. 14-11 for a complete revolution from $\theta = 0°$ to $\theta = 360°$. This is called a cycle, since the variation in voltage will now repeat itself. Note that the sine wave actually consists of four symmetrical segments, each like the first 90° interval.

For example, note that when $\theta = 135°$,

$$v_{\text{ind}} = V_m \sin 135° = 0.707\ V_m$$

the same as when $\theta = 45°$. Also, when $\theta = 225°$,

$$v_{\text{ind}} = V_m \sin 225° = -0.707\ V_m$$

or the negative of the value when $\theta = 45°$.

The maximum value, V_m, of a sine wave, is known as the *amplitude* of the wave and is often referred to as the *peak* value. In oscilloscope measurements it is often more convenient to find the *peak-to-peak* value. As indicated in Fig. 14-11, this is the total voltage from the positive peak to the negative peak. That is, for symmetrical waves such as a pure sine wave,

$$V_{\text{p-p}} = 2V_m \tag{14-6}$$

EXAMPLE 14-6

The peak-to-peak value of a sine wave of voltage is 50 V. Determine:

a. The peak voltage or amplitude.
b. The instantaneous voltages at one-third, three-fifths, and one-and-a-quarter cycles.

SOLUTION

a. $V_{\text{p-p}} = 2\ V_m$ (14-6)

Therefore, $V_m = \dfrac{V_{\text{p-p}}}{2} = \dfrac{50\text{ V}}{2} = \mathbf{25\ V}$

b. At $\frac{1}{3}$ of a cycle, $\theta = \dfrac{360°}{3} = 120°$

$v = V_m \sin \theta$ (14-5)

$= 25\text{ V} \times \sin 120°$

$= 25\text{ V} \times 0.866 = \mathbf{21.7\ V}$

At $\frac{3}{5}$ of a cycle, $\theta = 360° \times \dfrac{3}{5} = 216°$.

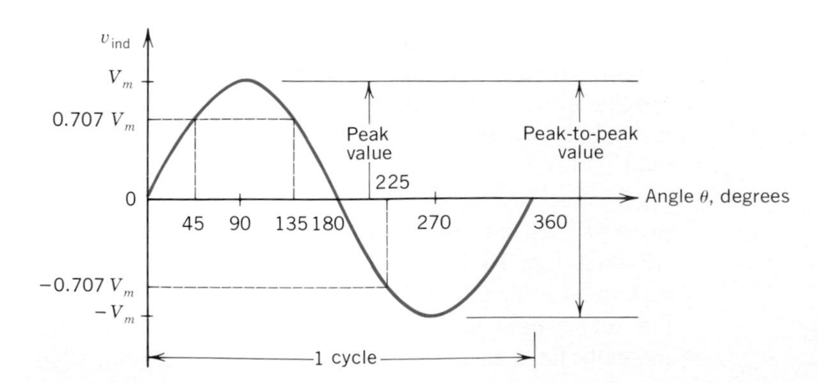

FIGURE 14-11
A sine wave of voltage.

$$\begin{aligned} v &= V_m \sin \theta \quad (14\text{-}5) \\ &= 25 \text{ V} \times \sin 216^\circ \\ &= 25 \text{ V} \times (-0.588) = \mathbf{-14.7 \text{ V}} \end{aligned}$$

At $1\frac{1}{4}$ cycles, $\theta = 360^\circ \times 1\frac{1}{4} = 450^\circ$.

$$\begin{aligned} v &= V_m \sin \theta \quad (14\text{-}5) \\ &= 25 \text{ V} \times \sin 450^\circ \\ &= 25 \text{ V} \times 1 = \mathbf{25 \text{ V}} \end{aligned}$$

NOTE One and a quarter cycles corresponds to a quarter of a cycle or 90° where $v = V_m$. Note also, that the use of a scientific calculator permits the sine of any angle to be obtained directly. It is not necessary to convert the angle to one that lies in the first quadrant (0–90°) unless you are using a set of trigonometric tables that are so limited.

14-7 FACTOR'S AFFECTING THE AMPLITUDE OF A GENERATOR'S SINE WAVE

The more quickly a generator's coil rotates in the magnetic field, the higher will be the induced voltage. Other factors that affect the amplitude are considered in Appendix A-7 with the following equation resulting:

$$V_m = 2\pi nBAN \qquad \text{volts (V)} \qquad (14\text{-}7)$$

where: V_m is the peak voltage generated in volts, V
- n is the speed of rotation of the coil in revolutions/s, rps
- B is the flux density of the magnetic field in teslas, T
- A is the area of the rotating coil in meters, m^2
- N is the number of turns in the rotating coil

For any given machine, A and N are fixed. But n and B may be varied to control the frequency and output voltage.

EXAMPLE 14-7

An ac generator runs at 1200 rpm and uses a flux density of 0.6 T. The length of the rotating coil is 40 cm with a radius of 10 cm. If the coil has 50 turns, calculate the amplitude of the voltage produced.

SOLUTION

$$n = 1200 \text{ rpm} = \frac{1200}{60} \text{ rps} = 20 \text{ rps}$$

$$B = 0.6 \text{ T}$$

$$\begin{aligned} A &= \text{length} \times \text{diameter} \\ &= 40 \times 10^{-2} \times 2 \times 10 \times 10^{-2} \text{ m}^2 = 8 \times 10^{-2} \text{ m}^2 \end{aligned}$$

$$N = 50$$

$$\begin{aligned} V_m &= 2\pi nBAN \quad (14\text{-}7) \\ &= 2\pi \times 20 \text{ rps} \times 0.6 \text{ T} \times 8 \times 10^{-2}\text{m}^2 \times 50 \\ &= \mathbf{301.6 \text{ V}} \end{aligned}$$

14-8 FACTORS AFFECTING THE FREQUENCY OF A GENERATOR'S SINE WAVE

The elementary ac generator of Fig. 14-9 had one pair of poles. It required one mechanical revolution of the coil to induce one cycle of voltage. See Fig. 14-12*a*. If the coil rotates at a speed of n revolutions per second, n complete cycles of voltage will be produced every second. The number of cycles of voltage produced every second is called the frequency, f. Thus, if the coil rotates at a speed of 60 revolutions per second, a voltage having a frequency of 60 cycles per second is generated. The unit of 1 cycle per second, cps, is called 1 hertz (Hz). That is, 1 cps = 1 Hz, and we say f = 60 Hz.

Now consider the addition of a second pair of magnetic poles as in Fig. 14-12*b*. Assuming that the coil rotates at the same speed as before, it will generate a complete cycle of voltage as it turns through just *half* a revolution. That is, a conductor passes under a south and a north pole during this interval, producing a full alternation of voltage. At the end of a complete *mechanical* revolution of 360°, two complete cycles of voltage are produced, amounting to 720 *electrical* degrees. If rotating at 60 rps, this generates a voltage of frequency 120 Hz. See Fig. 14-12.

Clearly then, combining the effects of speed and magnetic poles, we have

$$f = pn \qquad \text{hertz (Hz)} \qquad (14\text{-}8)$$

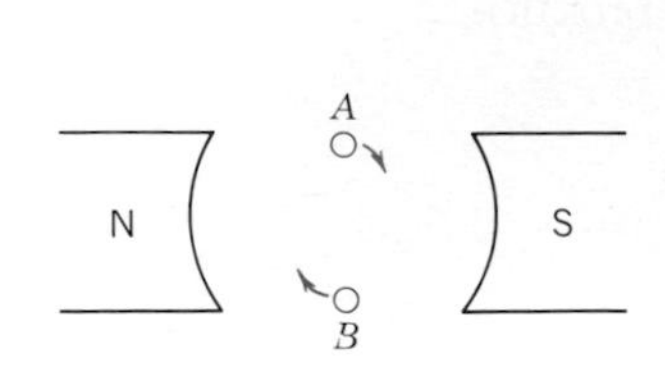

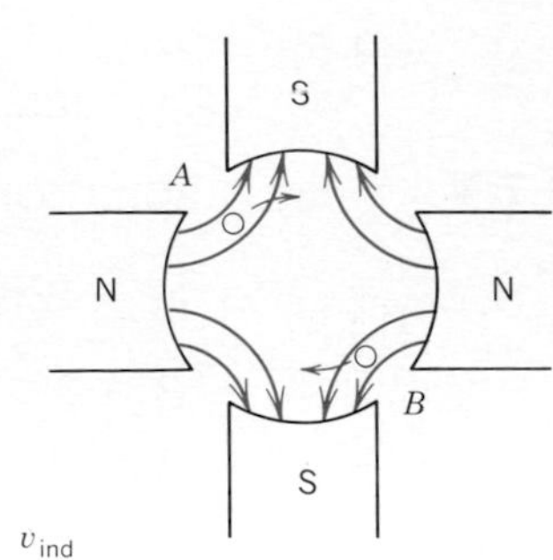

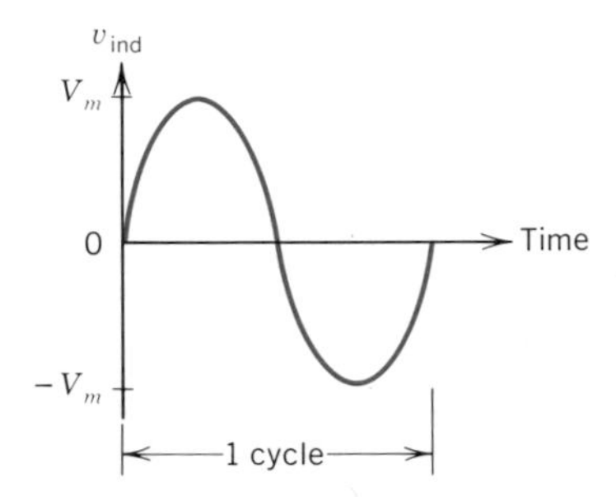

(a) One pair of poles produces 1 cycle in 1 mechanical revolution

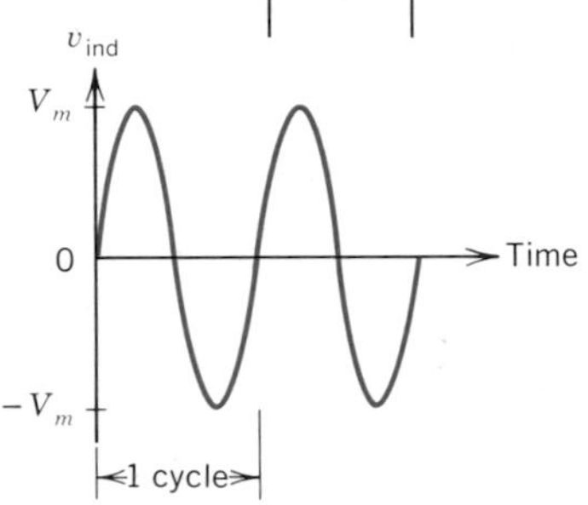

(b) Two pairs of poles produce 2 cycles in 1 mechanical revolution

FIGURE 14-12
Effect of the number of pairs of poles on the frequency of a generator running at a fixed speed.

where: f is the frequency of the induced emf in hertz, Hz
p is the number of *pairs* of poles
n is the speed of rotation in revolutions per second, rps

EXAMPLE 14-8

At what speed in rpm must the rotor of a water turbine turn if it is to produce voltage at a frequency of 60 Hz in a machine having a total number of 30 poles?

SOLUTION

Number of *pairs* of poles, $p = \frac{30}{2} = 15$.

$$n = \frac{f}{p} \qquad (14\text{-}8)$$

$$= \frac{60 \text{ Hz}}{15} = 4 \text{ rps} = \mathbf{240\ rpm}$$

As the frequency of a voltage is increased, the time required for one complete cycle to be generated decreases. The time to generate one cycle is called the *period* of the wave, T. The period is the reciprocal of the frequency, f, or

$$T = \frac{1}{f} \quad \text{seconds (s)} \qquad (14\text{-}9)$$

and

$$f = \frac{1}{T} \quad \text{hertz (Hz or s}^{-1}) \qquad (14\text{-}9a)$$

where: T is the period of the wave in seconds, s
f is the frequency of the wave in hertz, Hz

EXAMPLE 14-9

a. What is the period of a 60-Hz voltage?
b. What is the frequency of a waveform that has a period of 1 ms?

SOLUTION

a. $$T = \frac{1}{f} \qquad (14\text{-}9)$$

$$= \frac{1}{60 \text{ Hz}} = 0.01667 \text{ s} = \mathbf{16.67\ ms}$$

b. $$f = \frac{1}{T} \qquad (14\text{-}9a)$$

$$= \frac{1}{1 \times 10^{-3} \text{ s}} = 1 \times 10^{3} \text{ Hz} = \mathbf{1\ kHz}$$

14-9 PRACTICAL ALTERNATORS

The voltage developed by the simple ac generator is called a single-phase output. Most practical ac generators, or alternators (even those in a modern automobile), produce three-phase ac. They utilize a rotating magnetic field (rotor), which induces voltages in three sets of stationary windings (stator). Direct current is used to excite the field windings in the rotor. The current enters the rotor by means of two stationary brushes riding on two slip rings. See Fig. 14-13*a*.

The armature windings are spaced on the stator surface so that the emfs generated in them reach a peak value 120° apart, as shown in Fig. 14-13*b*. Voltages as high as 18,000 V are produced in large-size alternators.

We have seen how the speed of rotation determines both the amplitude and frequency of the generated voltage. Commercial power stations maintain the frequency of 60 Hz within ±0.02 Hz. This means that the speed must be governed very closely. How much voltage is produced is controlled by varying the field strength B. This is done by varying the direct current in the field windings on the rotor. In case of a need to reduce power output when the electrical demand cannot be met, a reduction in voltage is made by reducing the field strength B but maintaining speed constant to hold the frequency constant. This is often referred to as a "brown out."

14-9.1 Voltage Regulator for an Automobile Alternator

In an automobile alternator, there are wide variations in speed from a few hundred to a few thousand rpm. Here the frequency is not important, since the output is rectified using diodes to produce direct current. But a large variation in output voltage cannot be tolerated. A voltage regulator is used to control the amount of field current and hence the output voltage. Such a system that uses relays (solid-state versions also exist) is shown in Fig. 14-14.

When the ignition switch is turned on, current flows from the battery, through the red warning light and the normally closed contact of the regulator relay, to the field winding. If the engine is started, the alternator turns, generating a voltage at terminal R. This energizes the field relay, providing an alternate path for current to the field winding, and the red warning light goes out. At idle speeds, approximately 14.1 V is generated at the BAT terminal.

If the alternator's output increases due to an increase in speed, the voltage regulator relay is partially energized, opening the normally closed contact. Field current must now pass through resistor R_1, causing a drop in field strength and output voltage. The relay contact will tend to close again. If the speed and output voltage increase far above the required value, the relay's arm is pulled all the way down. This shunts the field current to ground, causing a sudden drop in output voltage. The regulator

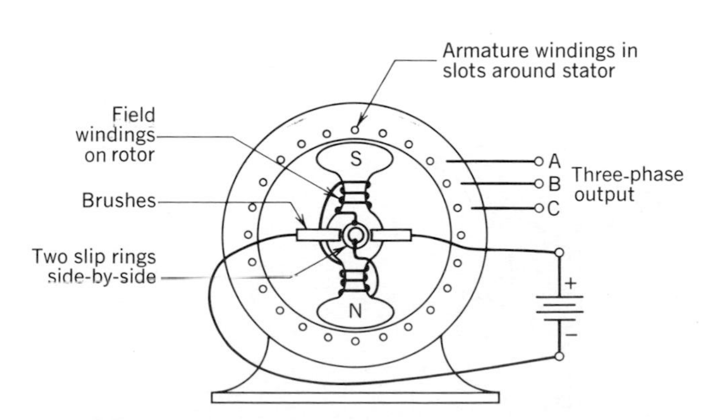

(*a*) Cross-sectional sketch of a three-phase alternator

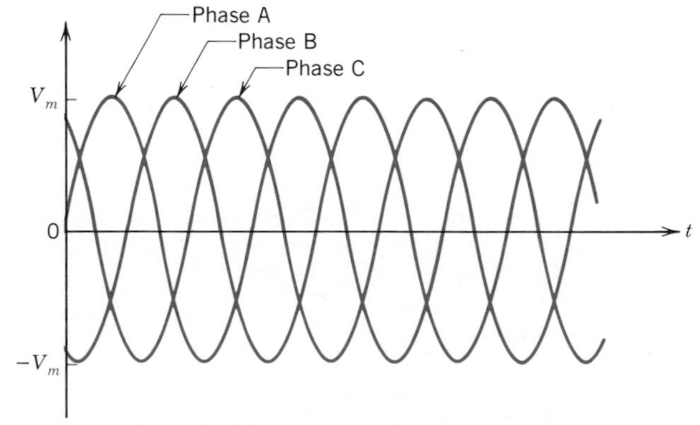

(*b*) Three-phase output voltages are generated 120° apart

FIGURE 14-13
Practical three-phase alternator and its output waveforms.

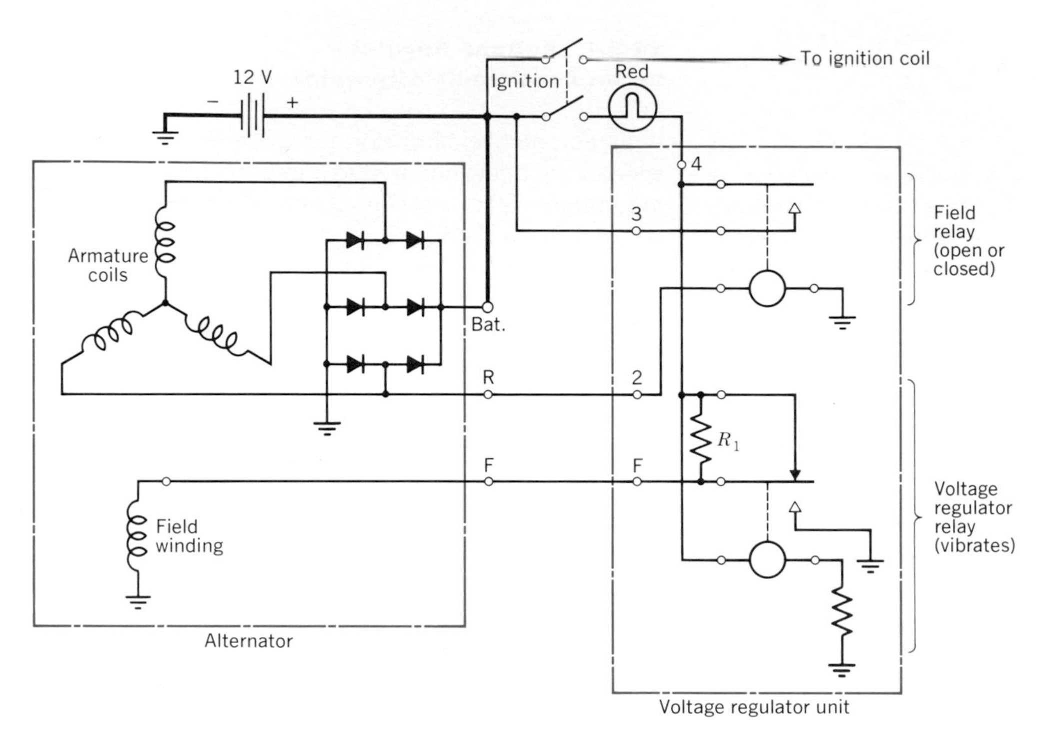

FIGURE 14-14
Schematic of a voltage regulator circuit for an automotive alternator system.

relay tends to vibrate, attempting to maintain the *average* output voltage from the alternator close to 14.1 V. (Other features of the circuit are considered in the review questions at the end of this chapter.)

14-9.2 Delta-Wye Connections

The advantage of a three-phase system is the ability to transmit the *same* power as a single-phase system but use less copper in the conducting wires. Three-phase motors run more smoothly than single-phase motors and are smaller, and more efficient, for the same capacity. Three-phase supplies may be either of the delta or wye connection, as shown in Fig. 14-15.

The delta connection is a three-wire system, primarily for power applications such as three-phase motors. Typical line-to-line voltages may be 220 V, 440 V, 550 V, and so on.

The wye connection is a four-wire, three-phase system having a neutral that is usually grounded. This permits the operation of single-phase equipment such as lights, receptacles and single-phase motors between any phase

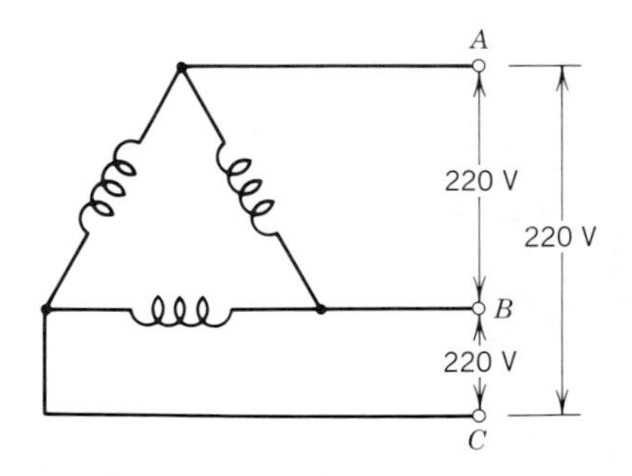

(*a*) 220 V, three-phase, three-wire delta (Δ)

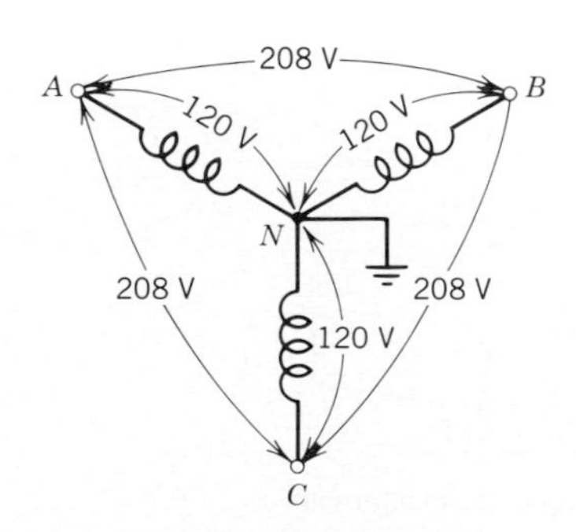

(*b*) 120/208 V, three-phase, four-wire wye (Y)

FIGURE 14-15
Coil connections in three-phase delta and wye systems.

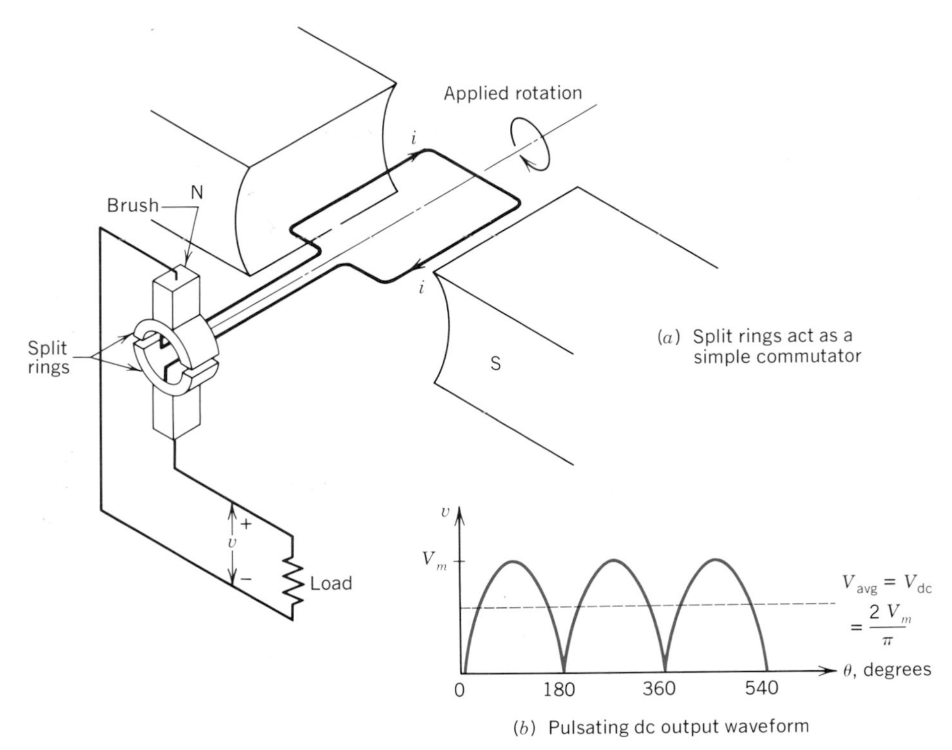

FIGURE 14-16 Elementary dc generator and output voltage waveform.

and the neutral, where 120 V exists. Three-phase motors connected directly to the three lines operate at 208 V. Note that the line-to-line voltage equals $\sqrt{3} \times$ line-to-neutral voltage.

14-10 ELEMENTARY DC GENERATOR

By modifying the brush and slip-ring arrangement of the simple one coil alternator, the negative half-cycles can be "inverted." See Fig. 14-16*a*. This gives a unidirectional, but pulsating, dc output, as shown in Fig. 14-16*b*.

The *split* rings function as a simple *commutator*. The commutator "switches" the armature coil connections to the load at the instant that the induced emf in the coil reverses. This maintains the polarity of one brush at a positive potential with respect to the other, even though the instantaneous value still varies between 0 and V_m.

A dc voltmeter connected across the output of the generator indicates the *average* value of the wave. This can be shown, mathematically, to be

$$V_{dc} = \frac{2 V_m}{\pi} \approx 0.637 V_m \qquad \text{volts (V)} \qquad (14\text{-}9)$$

where: V_{dc} is the dc or average emf in volts, V
V_m is the peak induced emf in volts, V

For example, if a one-coil dc generator produces a peak voltage of 20 V, its dc output is 12.74 V.

14-11 PRACTICAL DC GENERATORS

A practical dc generator uses a large number of coils evenly distributed around the surface of the armature. The coils are connected to a large number of commutator segments. In this way, the output voltage developed across the brushes is nearly always at the peak value produced by each coil. The result is a small "ripple" voltage riding on a dc level of approximately 98% of V_m, as shown in Fig. 14-17.

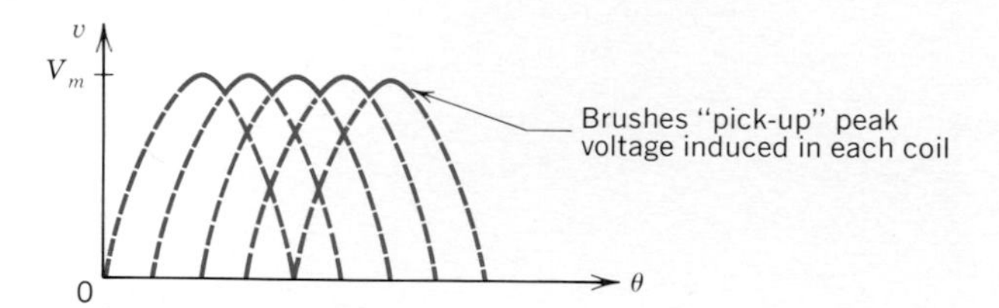

FIGURE 14-17
Output voltage of a practical dc generator.

It is interesting to note that the dc generator's output is practically the same *shape* as the full-wave rectified output of a three-phase alternator. This latter system, used in modern automobiles, is superior to the dc generator because of its ability to deliver a higher voltage at lower speeds. Even at idle speeds, most automobile alternators can supply the electrical load and recharge the battery, which is used solely for starting purposes.

Most dc generators, instead of using a permanent magnet, use field windings to create the magnetic field. These electromagnets may be *separately* excited (from an external source) or *self*-excited. That is, they use the output of the generator itself to supply current to the field windings. In order for a self-excited generator to "build-up," residual magnetism is relied upon to induce a small voltage, which in turn increases the field current, and so on. The field must be properly polarized in order for this cumulative action to occur. That is, the magnetic flux resulting from the initial, small generated voltage must add to the residual flux not oppose it.

Direct current generators may be of the series, shunt, or compound type, the same as the dc motors discussed in Section 11-13. In fact, a dc machine may be run as a motor or as a generator, but it is usually optimized to run as one or the other. The output of a dc generator, like an alternator, may easily be controlled by means of a rheostat in the field-winding circuit.

SUMMARY

1. Electromagnetic induction is the process of generating a voltage in a wire or coil that is located in a changing magnetic field.

2. Lenz's law: The polarity of the induced emf is such that it opposes the *change* that produced it.

3. Faraday's law: When magnetic flux is changing at a rate given by $\frac{\Delta\Phi}{\Delta t}$ Wb/s, the emf induced in N turns is

$$v_{\text{ind}} = N\left(\frac{\Delta\Phi}{\Delta t}\right)$$

4. When flux is increasing with time the quantity $\frac{\Delta\Phi}{\Delta t}$ is positive. When the flux decreases, $\frac{\Delta\Phi}{\Delta t}$ is negative, inducing a voltage of opposite polarity.

5. When a coil of N turns, each turn having an effective length l is moved through a magnetic field B at a velocity v, the total induced emf is given by $v_{\text{ind}} = NBlv$.

6. If a conductor is moved through a magnetic field at an angle θ with the field, the induced emf is $v_{\text{ind}} = V_m \sin \theta$.

7. The sine of an angle is a trigonometric ratio that has a fixed value for every angle. In a right-triangle, $\sin \theta$ is

$$\sin \theta = \frac{\text{opposite side}}{\text{hypotenuse}}$$

8. Maximum voltage, V_m, is induced when a conductor cuts the field at right angles, since $\theta = 90°$ and $\sin 90° = 1$.

9. A simple ac generator consists of a single coil rotating in a fixed magnetic field. The sinusoidal output voltage is available at two stationary brushes rubbing on two rotating slip rings connected to the coil.

10. The sine wave output of an alternator is

$$v = V_m \sin \theta$$

where the amplitude V_m is referred to as the peak value and $2V_m = V_{p-p}$ is the peak-to-peak value.

11. The amplitude of an induced sine wave is determined by the speed of rotation n, flux density B, area of coil A, and the number of turns N in the coil, as given by

$$V_m = 2\pi nBAN$$

12. The frequency of an induced sine wave is determined by the speed of rotation n and the number of *pairs* of poles p,

$$f = pn$$

13. A waveform goes through a complete cycle when it generates all of its possible values that are periodically repeated.

14. The frequency of a waveform is the number of cycles produced in a second. One cps is 1 hertz, Hz.

15. The period T of a wave of frequency f is the time required for a complete cycle:

$$T = \frac{1}{f}$$

16. A practical alternator consists of a rotating magnetic field that induces emf in armature windings around the stator. The frequency is determined by speed and the amount of emf by the field current.

17. Three-phase ac may be developed in either delta- or wye-connected windings. In a wye connection, the line-to-line voltage is $\sqrt{3}$ times the line-to-neutral voltage.

18. An elementary single-coil dc generator differs from an alternator in that split rings act as a simple commutator to invert the negative half-cycle, producing a waveform that has an average or dc value given by

$$V_{dc} = \frac{2V_m}{\pi}$$

19. A practical dc generator has many coils on the armature and many segments on the commutator. It produces an almost constant voltage near the peak value, V_m.

SELF-EXAMINATION

Answer T or F or, in the case of multiple choice, a, b, c, d
(Answers at back of book)

14-1. Electromagnetic induction always requires a conductor to be "cut" by a moving magnetic field. ______

14-2. A potential difference is produced in a wire if there is relative motion between the wire and the field such that a conductor cuts across the magnetic field. ______

14-3. Lenz's law determines the amount of induced voltage. ______

14-4. The induced current always sets up a flux that opposes the change that produced it. ______

14-5. If a coil is moved back and forth across the end of a permanent magnet, an alternating voltage is induced in the coil. ________

14-6. Faraday's law states that the induced emf is inversely proportional to the rate of change of flux. ________

14-7. A dynamic microphone is an application of Faraday's law. ________

14-8. When magnetic flux changes from 3 to 8 mWb in 5 ms, the rate of change of flux, $\frac{\Delta\Phi}{\Delta t}$ is

a. 1 mWb/s

b. 1 μWb/s

c. 1 Wb/s

d. 1×10^3 Wb/s

14-9. If a coil of 10 turns is subject to the flux change in Question 14-8, the emf induced in the coil is

a. 10 mV

b. 10 μV

c. 10 V

d. 0.1 V

14-10. The maximum voltage that can be induced in a 50-cm-length wire moving at 10 m/s through a field of 0.5 T is

a. 25 V

b. 2.5 V

c. 250 V

d. 250 mV

14-11. If the conductor in Question 14-10 is moving at an angle of 30° with the field, the induced voltage is

a. 12.5 V

b. 2.165 V

c. 1.25 V

d. 125 mV

14-12. The sine of 30° is one-half the sine of 60°. ________

14-13. In a simple ac generator with two poles, the induced current drops to zero every half a revolution. ________

14-14. The output voltage of an ac generator is a sine wave because the conductor moves in a circle in a uniform horizontal magnetic field. ________

14-15. If the peak-to-peak voltage of a sine wave is 15 V, the amplitude is 30 V. ________

14-16. If the amplitude of a sine wave is 10 V, the emf after one-sixth of a cycle is

a. 8.66 V

b. 5 V

c. 10 V

d. 7.07 V

14-17. All other factors remaining the same, the output voltage from an alternator coil of length 50 cm and diameter 20 cm is the same as a coil of length 40 cm and diameter 25 cm. ________

14-18. If the speed of an alternator is doubled, the output voltage will double, and so will its frequency. ________

14-19. If the flux density of an alternator is doubled, but the speed is cut in half, the frequency will be half of its original value and the output voltage doubled. ________

14-20. Increasing the number of pairs of poles in an alternator means that the speed can be reduced to produce the same frequency. ______

14-21. Five cycles are counted on an oscilloscope over a time span of 100 μs. The frequency of this waveform is

a. 5 kHz
b. 5 MHz
c. 20 kHz
d. 50 kHz

14-22. The period of the wave in Question 14-21 is

a. 20 ms
b. 20 μs
c. 500 μs
d. 500 ms

14-23. The period of a wave is 0.04 s. Its frequency is

a. 2.5 Hz
b. 25 Hz
c. 2.5 kHz
d. 40 Hz

14-24. In a practical alternator the voltage is always induced in the armature. ______

14-25. The output voltage of a commercial alternator is maintained at its correct value by making slight adjustments in the speed of rotation. ______

14-26. Where three-phase power is to supply both three-phase motors and lighting circuits a three-phase, four-wire wye connection is used. ______

14-27. In an elementary, single-coil dc generator with a peak value of 80 V, the dc output voltage is approximately:

a. 25.5 V
b. 51 V
c. 80 V
d. 56.6 V

14-28. A practical, self-excited dc generator relies upon residual magnetism to build up. ______

REVIEW QUESTIONS

1. What are the required conditions for electromagnetic induction to take place?

2. Give three practical applications where electromagnetic induction is put to use.

3. What is the origin of the emf induced in a wire that is being moved across a magnetic field?

4. State Lenz's law in your own words.

5. a. Give two examples of where Faraday's law is at work in an automobile.

b. What are the fundamental units of $\frac{\Delta\Phi}{\Delta t}$?

c. How is the weber defined?

6. a. If the voltage induced in a conductor is constant, how is the magnetic field changing?

b. If the voltage induced in a conductor is uniformly increasing, how is the magnetic field changing?

7. How can an alternating current be produced using only a solenoid and a permanent magnet?

8. Under what conditions does a magnetic field not have to "cut" through a coil or wire to induce an emf?

9. Under what conditions does a conductor moving through a magnetic field induce:

a. Maximum emf?

b. Zero emf?

c. 0.707 of maximum emf?

10. In Example 14-4 the earth's magnetic field induced 340 μV in a single antenna. Would it be possible to use a number of vertical wires, series interconnected, to increase the total voltage to a higher value? Explain.

11. Why is the sine of 30° not one-half the sine of 60° or one-third the sine of 90°?

12. a. Sketch a right-angled triangle and show why the tangent of a 45° angle is 1. See Appendix A-6.

b. What is the name of this triangle?

c. Why is the tangent of 90° infinite?

13. a. Sketch a right triangle with a 30° angle. Why is the sine of 30° equal to the cosine of 60°?

b. What is the *exact* value of the sine of 45°?

14. Refer to Fig. 14-9. If conductors A and B in the armature coil were held stationary, would the same emf be induced between the brushes if the external permanent magnets were rotated around the coil instead?

15. Refer to Fig. 4-9. If the coil were rotated in the opposite direction from that shown how, if at all, would the output waveform of voltage be different? Explain.

16. a. Why is the output of an alternator sinusoidal?

b. Could we make it have any other shape if we wanted to? How?

c. Why do we choose a sine wave? Why not a triangular shape or semicircular shape?

17. Once an alternator is built, what factors affect the following?

a. The output voltage's amplitude.

b. The output voltage's frequency.

How could you double the output voltage without affecting the frequency?

18. a. What is the relationship between the frequency and period for any alternating wave?

b. Does a change in amplitude change the period?

19. What are the major differences between a practical alternator and a simple single-coil ac generator?

20. a. What is meant by a "brownout"?

b. How is this brought about?

c. How does this affect electrical appliances?

d. Why must the frequency be maintained at 60 Hz, even in a brownout?

21. Refer to Fig. 14-14.

a. If, while the car is running, the alternator fails to develop an output (due to a broken belt perhaps), what visual indication will be noticed?

b. Describe the electrical sequence that takes place following this failure.

c. Does current continue to flow into the field winding?

d. Can the battery discharge through the alternator armature coils?

22. If, after the car engine has been switched off, the field relay remains closed due to a malfunction, what possible effect would be noticed if the car is not used for several days? Explain.

23. What is an advantage of a three-phase, four-wire system over a single-phase system?

24. What is the purpose of a commutator in a dc generator?

25. How does a practical dc generator prevent the output voltage from falling to zero periodically as it does in a single-coil elementary generator?

26. Older automobiles that use a dc generator often had to have the field of the generator "polarized" before it would build up properly.

a. What is meant by this?

b. Why was it necessary?

PROBLEMS

(Answers to odd numbered problems at back of book)

14.1. Refer to Fig. 14-1*a*. What is the polarity of the induced emf at *A* with respect to *B* when

a. The field is moved to the right?

b. The field is moved to the left?

14-2. Repeat Problem 14-1 with the magnetic polarity reversed.

14-3. Refer to Fig. 14-3. What is the polarity of the induced emf at *A* with respect to *B* when

a. The south pole is moved into the coil?

b. The north pole is moved into the coil?

14.4. Repeat Problem 14-3 with the windings wrapped around the solenoid in the opposite direction.

14-5. Refer to Fig. 14-18. What is the polarity of the induced emf at *A* with respect to *B* when

a. The switch is first closed?

b. The switch is just opened?

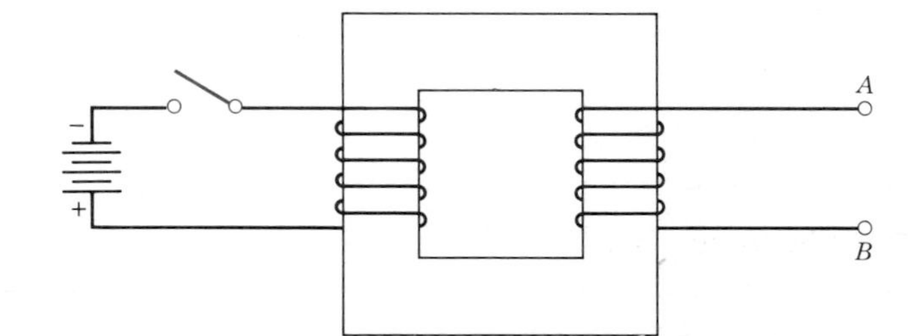

FIGURE 14-18
Circuit for Problems 14-5 and 14-6.

14-6. Repeat Problem 14-5 with the winding connected to the battery wound around the core in the opposite direction.

14-7. What is the rate of change of flux in Wb/s for the following?

a. 2 Wb in 10 min

b. 5 mWb in 10 ms

c. 4.5 μWb in 5 μs

d. 8 mWb in 2 μs

14-8. What emf is induced in a 10-turn coil subjected to the rate of change of flux in Problem 14-7?

14-9. At what rate is the flux changing in a 100-turn coil if 0.5 V is induced in the coil?

14-10. How many turns are there in a coil that has an emf of 80 mV induced when the flux changes from 28 to 30 mWb in one-tenth of a second?

14-11. A wire is moved across the face of a permanent magnet that is 3 cm square, in 0.1 s. If an emf of 9 mV is induced in the wire, what is the flux density of the magnetic field?

14-12. A wire is moved across the face of a permanent magnet that is 4 cm square and 0.8 T in strength. Calculate the voltage induced in the wire if it is moved perpendicular to the field in:

a. 150 ms

b. 5 ms

14-13. Refer to Fig. 14-5. The magnetic flux in the core is increased uniformly from 2 to 8 mWb in 4 ms, held constant for 3 ms, increased again to 10 mWb in 1 ms, then suddenly reduced uniformly to zero in 2 ms. Coil B has 50 turns. Draw graphs showing:

a. How the flux changes with time.

b. The rate of change of flux with time.

c. The voltage induced in coil B.

14-14. Refer to Fig. 14-5. The flux density in the core is increased uniformly from 0.7 to 1 T in 4 ms, held constant for 2 ms, reversed to 0.6 T in the opposite direction in 8 ms, then reduced to zero in 3 ms. If coil B has 40 turns and the toroid has a circular cross section of radius 2 cm, draw graphs showing:

a. How the flux changes with time.

b. The rate of change of flux with time.

c. The emf induced in coil B.

14-15. Five series-connected conductors, each 25 cm long, are moved at 8 m/s perpendicular to a magnetic field of 1.5 T. What is the total induced emf?

14-16. At what velocity must a 40 cm-long conductor be moved perpendicular to a magnetic field of 1.2 T to induce an emf of 5 V?

14-17. What emf will be induced in the five conductors in Problem 14-15 if they move through the field at the following angles with the field?

a. 30°

b. 45°

c. 80°

d. 90°

14-18. At what angle is the conductor in Problem 14-16 moving through the field if the induced emf is as follows?

a. 2.5 V

b. 3.535 V

c. 4.0 V

14-19. The peak-to-peak value of a sine wave is 25 V. Determine:

a. The amplitude.

b. The instantaneous voltages at one-quarter, four-fifths, and two and one-third cycles.

14-20. If the peak value of a sine wave is 20 V, what are the instantaneous voltages at the following angles:

a. 60°

b. 110°

c. 270°

d. 345°

14-21. The armature of an ac generator, running at 3600 rpm, has a length of 50 cm, a diameter of 25 cm, and has 20 turns. If the field strength is 0.5 T, calculate the amplitude of the induced voltage.

14-22. What must be the area of an armature coil rotating at 1500 rpm in a field of 0.8 T if the coil has 30 turns and the induced voltage has a peak-to-peak value of 150 V?

14-23. a. If the generator in Problem 14-21 has 1 pair of poles, what is the frequency of the generated voltage?

b. What is its period?

14-24. a. What is the total number of poles required by the generator in Problem 14-22 if it is to generate a voltage having a frequency of 50 Hz?

b. What is the period of this waveform?

14-25. Consider a three-phase, four-wire supply.

a. If the line-to-line voltage is 10 V, what is the line-to-neutral voltage?

b. What must be the line-to-line voltage of a system if the line-to-neutral voltage is 115 V?

14-26. a. An elementary single-coil dc generator produces a peak voltage of 40 V. What is its dc value?

b. If a single-coil dc generator is used to produce an output voltage of 14.5 V dc, what is the peak value?

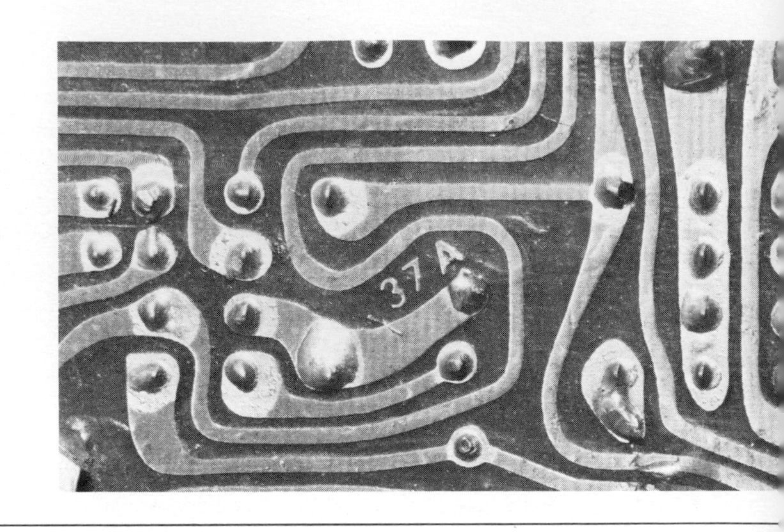

CHAPTER FIFTEEN

ALTERNATING VOLTAGE AND CURRENT

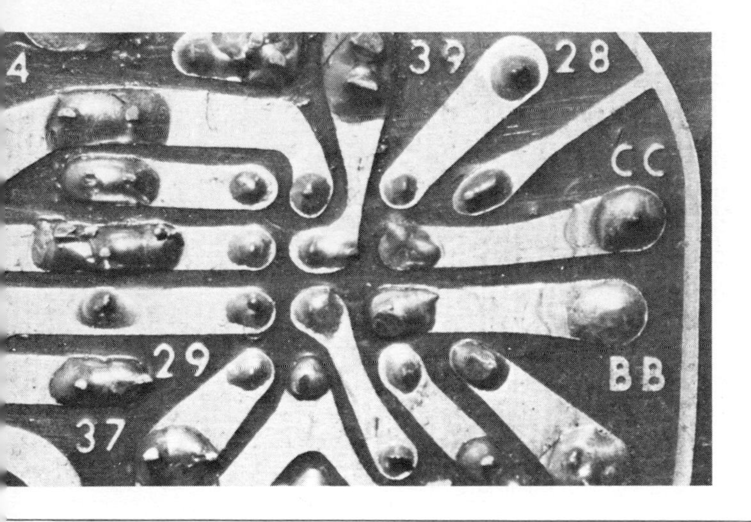

We have seen how the frequency of a sine wave of voltage may be expressed in cycles per second or hertz. In this chapter we introduce the *radian* as a measure of an angle instead of degrees. The frequency can now be given in radians per second and is known as the *angular frequency,* ω. This allows us to express a sinusoidal voltage (or current) in terms of the amplitude, frequency, and time.

One of the questions resolved by this chapter is the relation between an oscilloscope display of a sinusoidal voltage and the single reading of a voltmeter. This requires a consideration of how current varies in a resistor with a sinusoidal applied voltage and the corresponding variation of power. The *average* power dissipated in the resistor is equated to the same amount of power produced by a direct current flowing through the same resistance. The amount of dc that produces the same heating effect as the ac is called the *effective* value of the ac. Because of the steps followed in this process, the effective value is referred to as the root-mean-square or rms value.

Finally, we will see that Ohm's law, and the power equations developed for dc circuits, can be used for an ac circuit *if* rms values of current and voltage are used, *and* only pure resistance is considered in any calculations.

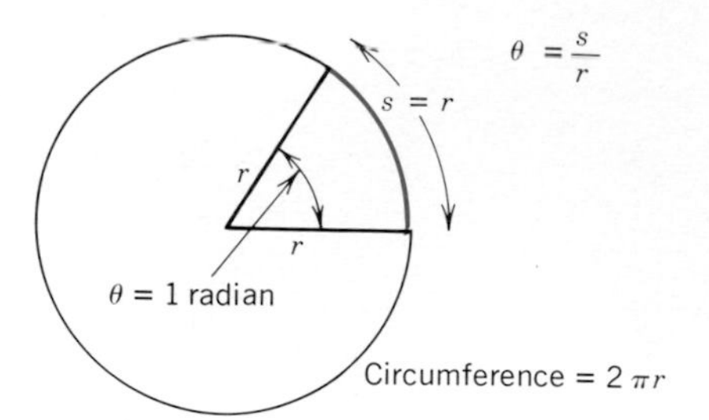

FIGURE 15-1
One radian (approximately 57.3°) is the angle subtended at the center of a circle by an arc equal in length to the radius.

15-1 INDUCED VOLTAGE IN TERMS OF FREQUENCY AND TIME

We have seen how a sine wave of voltage can be expressed by the equation

$$v = V_m \sin \theta \qquad (15\text{-}1)$$

The angle θ can be expressed in terms of the frequency, f, and the instant of time, t, that we wish to know the voltage. See Appendix A-8. The result is

$$\theta = 2\pi ft \qquad \text{radians (rad)} \qquad (15\text{-}2)$$

where: θ is the angle in radians, rad
f is the frequency of the voltage in hertz, Hz
t is the time since the voltage was zero in seconds, s

The 2π term results from expressing the angle in radians instead of degrees. The radian is commonly used in engineering work. As shown by Fig. 15-1, a radian is the angle between two radii of a circle that cut off on the circumference of the circle an arc equal in length to the radius of the circle.

The number of radians in a complete circle is therefore equal to the number of times the radius is contained in the circumference of a circle. The circumference equals $2\pi r$. This means that 2π radians equals 360°.

$$2\pi \text{ radians} = 360°$$
$$\pi \text{ radians} = 180°$$
$$\frac{\pi}{2} \text{ radians} = 90°$$
$$1 \text{ radian} = \frac{180°}{\pi} \approx 57.3°$$

The way in which a sine wave of voltage varies with the angle given in radians is shown in Fig. 15-2.

One of the advantages of using radians is that they are *dimensionless,* being the ratio of two lengths, arc length to radius.

Thus the induced voltage can be written in the following form:

$$v = V_m \sin 2\pi ft \qquad \text{volts (V)} \qquad (15\text{-}3)$$

or

$$v = V_m \sin \omega t \qquad \text{volts (V)} \qquad (15\text{-}4)$$

and

$$\omega = 2\pi f \qquad \text{radians per second (rad/s)} \qquad (15\text{-}5)$$

where: v is the induced emf in volts, V

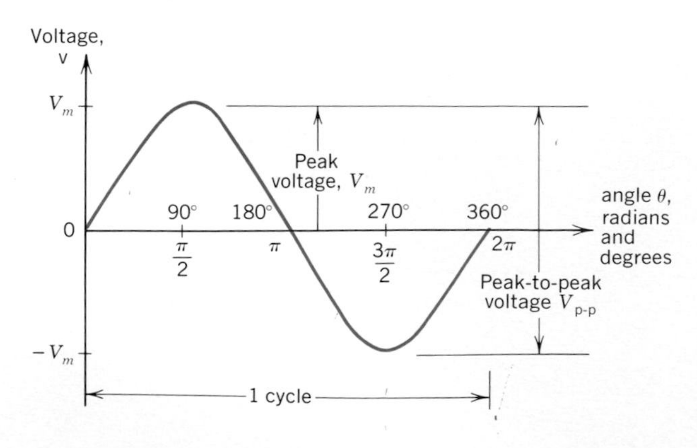

FIGURE 15-2
A sine wave of voltage with the angle given in radians and degrees.

V_m is the maximum induced emf in volts, V

f is the frequency of the emf in hertz, Hz

t is the instant of time under consideration in seconds, s

$\omega = 2\pi f$ is the angular frequency in radians per seconds, rad/s

15-1.1 Angular Frequency

The Greek letter omega, ω, is referred to by some texts as the angular *velocity*. This term is appropriate to describe the speed of a rotating coil or conductor. But as pointed out in Section 14-8, the frequency of the generated voltage depends upon the number of pairs of magnetic poles that the conductor passes under, as well as the speed of rotation.

For example, a coil that makes a full 360° revolution in 1 s has an angular *velocity* of 2π rad/s. If the generator has one pair of poles, the generated voltage has a frequency of 1 cps. For the same *velocity* of 2π rad/s, a generator with two pairs of poles generates a voltage having a *frequency* of 2 cps. Since a cycle of voltage is equivalent to 2π rad, the *angular frequency* of the generated voltage is 2π rad/s in the first case and 4π rad/s in the second case, even though the angular *velocity* of the coil is 2π rad/s in both cases. That is, we are more concerned with the *frequency* of the voltage or current rather than the speed or *velocity* of the conductor that generated the voltage or current.

In summary, if f is the frequency of a voltage in cycles per second, $2\pi f$ is the frequency in radians per second. The term *angular* frequency, ω, is used for $2\pi f$, since an *angle* of 2π rad corresponds to one *cycle*.

Note that the dimension of ω is "per second" or second^{-1}. This is because the radian is a dimensionless quantity. Thus ω has the same dimension as f, and it is consistent to refer to ω as frequency.

EXAMPLE 15-1

A sine wave of voltage has a peak value of 5 V and a frequency of 1 kHz. Determine:

a. The angular frequency.
b. The instantaneous voltage when $t = 400\ \mu$s.

SOLUTION

a. $$\omega = 2\pi f \qquad (15\text{-}5)$$
$$= 2\pi \times 1 \times 10^3 \text{ rad/s}$$
$$= \mathbf{6.28 \times 10^3 \text{ rad/s}}$$

b. $$v = V_m \sin \omega t \qquad (15\text{-}4)$$
$$= 5\text{ V} \sin 6.28 \times 10^3 \frac{\text{rad}}{\text{s}} \times 400 \times 10^{-6}\text{ s}$$
$$= 5\text{ V} \times \sin 2.51 \text{ rad}$$
$$= 5\text{ V} \times 0.59 = \mathbf{2.95\ V}$$

EXAMPLE 15-2

The voltage applied to a circuit is given by $v = 170 \sin 377\,t$ volts. Determine:

a. The amplitude of the voltage.
b. The angular frequency.
c. The frequency in hertz.

SOLUTION

a. By comparison with Eq. 15-4:
$$v = V_m \sin \omega t$$
$$= 170 \sin 377t$$
The amplitude $V_m =$ **170 V.**

b. The angular frequency, $\omega =$ **377 rad/s.**

c. $$\omega = 2\pi f \qquad (15\text{-}5)$$
$$f = \frac{\omega}{2\pi}$$
$$= \frac{377}{2\pi} = \mathbf{60\ Hz}$$

15-2 CURRENT AND VOLTAGE WAVEFORMS WITH A RESISTIVE LOAD

Consider the application of a sinusoidally alternating voltage to a resistor, as shown in Fig. 15-3a. What is the nature of the current variation in the circuit?

Given $$v = V_m \sin \omega t \qquad (15\text{-}4)$$

By Ohm's law, $$i = \frac{v}{R} \qquad (3\text{-}1a)$$

That is, the instantaneous value of current is determined by the instantaneous value of the applied voltage and R. Note the use of small letters to represent quantities varying with time.

$$i = \frac{v}{R} = \frac{V_m}{R} \sin \omega t$$

Therefore, $$i = I_m \sin \omega t \qquad \text{amperes (A)} \qquad (15\text{-}6)$$

and $$I_m = \frac{V_m}{R} \qquad (15\text{-}7)$$

where: i is the instantaneous current in amperes, A
I_m is the peak current in amperes, A
V_m is the peak voltage in volts, V
R is the resistive load in ohms, Ω
$\omega = 2\pi f$ is the angular frequency in radians/second, rad/s
t is the time in seconds, s

Equation 15-6 shows that the current, *i*, varies at the same frequency as the applied voltage, *v*, and *in-phase* with *v*. That is, the current and voltage waveforms pass through their maximum and minimum values at the same instant, as shown in Fig. 15-3*b*. These waveforms may be observed on an oscilloscope, as described in Chapter 16, to identify V_m and I_m.

EXAMPLE 15-3

A sinusoidal voltage of peak-to-peak value, 8 V, is observed on an oscilloscope when connected across a 2.2-kΩ resistor. The period of the voltage waveform is 2 ms. Determine:

a. The peak value of the current.
b. The frequency of the current.
c. The equation representing the current.

SOLUTION

a. $$V_m = \frac{V_{p-p}}{2} \qquad (14\text{-}6)$$

$$= \frac{8\text{ V}}{2} = 4\text{ V}$$

$$I_m = \frac{V_m}{R} \qquad (15\text{-}7)$$

$$= \frac{4\text{ V}}{2.2\text{ k}\Omega} = \mathbf{1.82\ mA}$$

b. $$f = \frac{1}{T} \qquad (14\text{-}9a)$$

$$= \frac{1}{2 \times 10^{-3}\text{ s}} = \mathbf{500\ Hz}$$

c. $$\omega = 2\pi f \qquad (15\text{-}5)$$

$$= 2\pi \times 500$$

$$= 1000\pi \text{ rad/s}$$

Therefore, $$i = I_m \sin \omega t \qquad (15\text{-}6)$$

$$= \mathbf{1.82 \sin 1000\pi t\ mA}$$

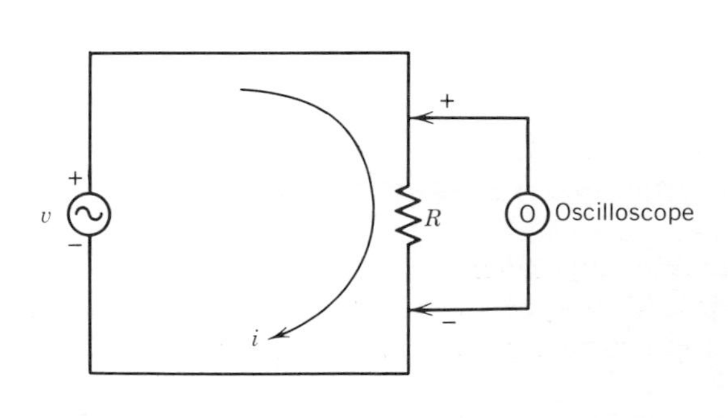

(*a*) Sinusoidal voltage applied to the resistor

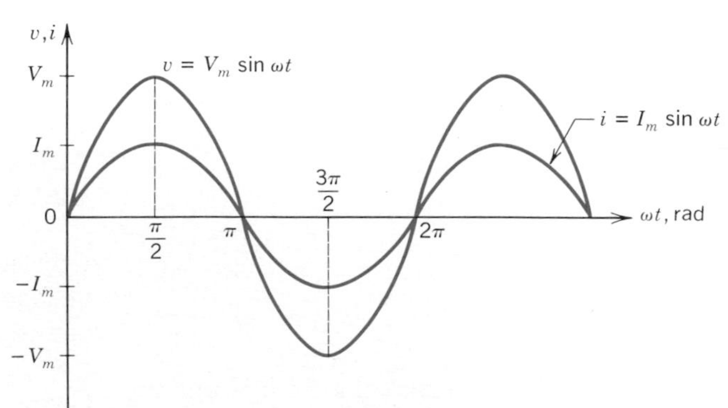

(*b*) Waveforms of voltage and current

FIGURE 15-3
Circuit and waveforms for a resistive load.

NOTE The polarity signs shown in Fig. 15-3 establish a *reference* to interpret the positive and negative values in the graphs for v and i. Also, the graph for i has been drawn arbitrarily smaller than for v.

We can now see that the current variation in the circuit is the same as the voltage variation across a resistor that has the (same) current flowing through it. This is a very important principle. Although an oscilloscope only reacts to voltage, we can use it to show how current varies by displaying the voltage across a resistor in the circuit. And if the resistance is known, the current is also known.

15-3 POWER IN A RESISTIVE LOAD

We know that in a dc circuit

$$\text{the average power } P = I^2R \qquad (3\text{-}9)$$

In an ac circuit, where the instantaneous current i varies, the instantaneous power p must also vary. Thus

$$p = i^2R$$

and since

$$i = I_m \sin \omega t$$

$$p = (I_m \sin \omega t)^2R$$

$$= I_m^2 R \sin^2 \omega t$$

Therefore,

$$p = P_m \sin^2 \omega t \qquad \text{watts (W)} \qquad (15\text{-}8)$$

and

$$P_m = I_m^2 R \qquad (15\text{-}9)$$

where: p is the instantaneous value of power in the resistor, in watts, W

P_m is the peak power in watts, W

I_m is the peak current in amperes, A

R is the resistive load in ohms, Ω

ω is the angular frequency in radians/second, rad/s

t is the time in seconds, s

The power waveform is shown in Fig. 15-4*b*. It evidently has a frequency twice that of the current waveform, varying from 0 to a peak value P_m twice as often as the current.

However, although the power pulsates, in a purely resistive circuit it is always positive. This simply means that an alternating current in a resistor dissipates power in the resistor (in the form of heat), no matter which direction the current flows.

We can also see, by inspection of the symmetrical power curve in Fig. 15-4*b*, that the *average* power, P_{av}, is one-half of the peak power. Thus

$$P_{av} = \frac{P_m}{2} = \frac{I_m^2R}{2} \qquad \text{watts (W)} \qquad (15\text{-}10)$$

where the symbols are as in Eqs. 15-8 and 15-9.

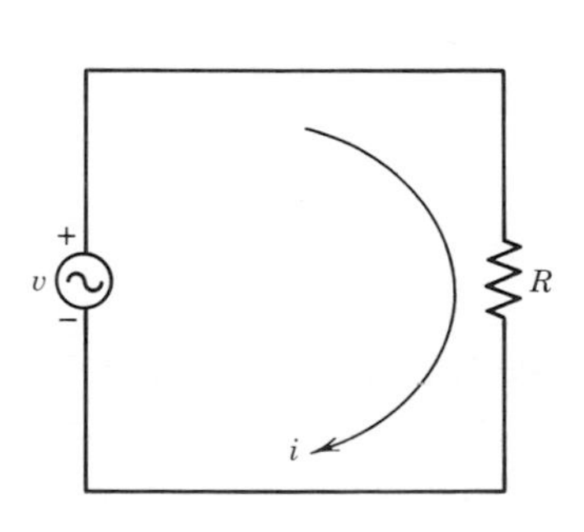

(*a*) Sinusoidal voltage applied to the resistor

(*b*) Waveforms of current and power

FIGURE 15-4
Circuit and waveforms for power in a resistive load.

That is, consider a line drawn through the power curve at one-half its peak value. The average value of a curve is where a horizontal line can be drawn such that it encloses the same area below it as was enclosed by the original curve. In Fig. 15-4*b*, the shaded portion of the power waveform *above* the average power line may be used to "fill-in" the shaded portion *below* the line. This is shown in Example 15-4.

EXAMPLE 15-4

An ac voltage given by equation $v = 50 \sin 800\pi t$ is applied to a 25-Ω resistor. Determine:

a. The peak power dissipated in the resistor.
b. The frequency of the power variation.
c. The average power dissipated in the resistor.

SOLUTION

a. $$I_m = \frac{V_m}{R} \qquad (15\text{-}7)$$

$$= \frac{50\text{ V}}{25\ \Omega} = 2\text{ A}$$

$$P_m = I_m^2 R \qquad (15\text{-}9)$$

$$= (2\text{ A})^2 \times 25\ \Omega = \mathbf{100\ W}$$

b. $$v = V_m \sin 2\pi f t \qquad (15\text{-}3)$$

$$= 50 \sin 800\pi t$$

$$2\pi f = 800\ \pi\text{ rad/s}$$

Therefore, $$f = 400\text{ Hz}$$

Power varies at a frequency of $2f$ or **800 Hz.**

c. $$P_{av} = \frac{P_m}{2} \qquad (15\text{-}10)$$

$$= \frac{100\text{ W}}{2} = \mathbf{50\ W}$$

Example 15-4 shows that a constant power level of 50 W provides the same heating effect as a power waveform varying sinusoidally between 0 and 100 W, no matter what the frequency.

15-3.1 Equations for Average Power

Consider the expression for the average power dissipated in a resistor

$$P_{av} = \frac{I_m^2 R}{2} \qquad (15\text{-}10)$$

The average power determines the heat developed in a resistor and is more important than the peak power. It is therefore convenient to calculate the average power in terms of the current indicated by an ammeter rather than from an oscilloscope measurement. This is done as follows:

$$P_{av} = \frac{I_m^2 R}{2}$$

$$= \left(\frac{I_m}{\sqrt{2}}\right)^2 R$$

and

$$P_{av} = I^2 R \qquad (15\text{-}11)$$

where $I = \frac{I_m}{\sqrt{2}}$ is called the **rms value** of the current. This is the value that an ac ammeter, calibrated for sine waves, will indicate in a sinusoidal circuit. (See Section 15-4 for a fuller explanation of the term rms). It should be noted that the rms value is *not* an average value. (The rms value is sometimes called the *effective* value. The average value of a sine wave over a complete cycle is zero.)

It is important to realize that Eq. 15-11 can be used to calculate power in an ac circuit only if the whole circuit is purely resistive, or if we consider only the current through the resistive portion. To emphasize this fact, the notation I_R should be used. Also, since we can obtain the rms value of voltage indicated by a voltmeter using $V = \frac{V_m}{\sqrt{2}}$ we can state the following equations for calculating power in a resistor:

$$P = I_R^2 R \quad \text{watts (W)} \qquad (15\text{-}12)$$

$$P = \frac{V_R^2}{R} \quad \text{watts (W)} \qquad (15\text{-}13)$$

$$P = V_R I_R \quad \text{watts (W)} \qquad (15\text{-}14)$$

where: P is the average or rms power in watts, W
I_R is the rms current in the resistor, in amperes, A

V_R is the rms voltage across the resistor, in volts, V

R is the resistance in ohms, Ω

These power equations may be used with any shaped waveform, regardless of frequency, including dc, as long as rms values are used.

Equations 15-12 through 15-14 imply that Ohm's law can be applied to an ac circuit. This is true only if we consider the current in the resistor and the voltage across the resistor. That is,

$$R = \frac{V_R}{I_R} \quad \text{ohms } (\Omega) \qquad (15\text{-}15)$$

$$I_R = \frac{V_R}{R} \quad \text{amperes (A)} \qquad (15\text{-}15a)$$

$$V_R = I_R R \quad \text{volts (V)} \qquad (15\text{-}15b)$$

with symbols as above.

EXAMPLE 15-5

A resistor R is connected in a complex ac circuit as shown in Fig. 15-5. An ac voltmeter across the resistor indicates 15 V and an ac ammeter in series with the resistor indicates 50 mA. Calculate:

a. The resistance of R.

b. The average power dissipated in R using Eqs. 15-12 through 15-14.

SOLUTION

a. $$R = \frac{V_R}{I_R} \qquad (15\text{-}15)$$

$$= \frac{15 \text{ V}}{50 \times 10^{-3} \text{ A}} = \mathbf{300\ \Omega}$$

b. $$P = I_R^2 R \qquad (15\text{-}12)$$

$$= (50 \times 10^{-3} \text{ A})^2 \times 300\ \Omega$$

$$= 0.75 \text{ W} = \mathbf{750\ mW}$$

$$P = \frac{V_R^2}{R} \qquad (15\text{-}13)$$

$$= \frac{(15 \text{ V})^2}{300\ \Omega}$$

$$= 0.75 \text{ W} = \mathbf{750\ mW}$$

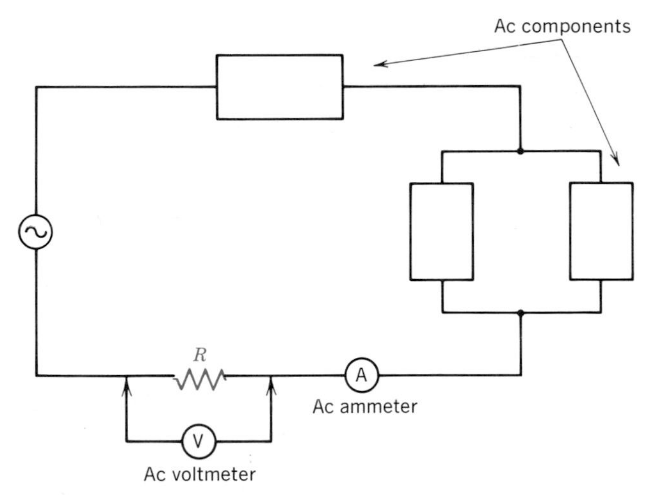

FIGURE 15-5
Circuit for Example 15-5.

$$P = V_R I_R \qquad (15\text{-}14)$$

$$= 15 \text{ V} \times 50 \times 10^{-3} \text{ A}$$

$$= 0.75 \text{ W} = \mathbf{750\ mW}$$

15-4 RMS VALUE OF A SINE WAVE

Let us return to the relation between the readings of an ac voltmeter or ammeter and the instantaneous values of voltage and current in a complete cycle of a sine wave. This approach explains what is meant by the term rms.

If an ac voltmeter is connected across a sine wave of voltage, what should be its reading? The meter can only indicate one value. Which of all the values between $-V_m$ and $+V_m$ should this be? Similarly, what single value should an ac ammeter indicate? Refer to Fig. 15-6.

The answer lies in the comparison of the *average* heating effect of the ac with that produced by some steady dc value. More specifically, **the effective value of an alternating current (or voltage) is that value of dc that is *as effective* in *producing heat* in a given resistance as the given quantity of ac in the same resistance.**

Since heating is dependent upon power, we want the *average power* produced by the ac to be the *same* as the dc.

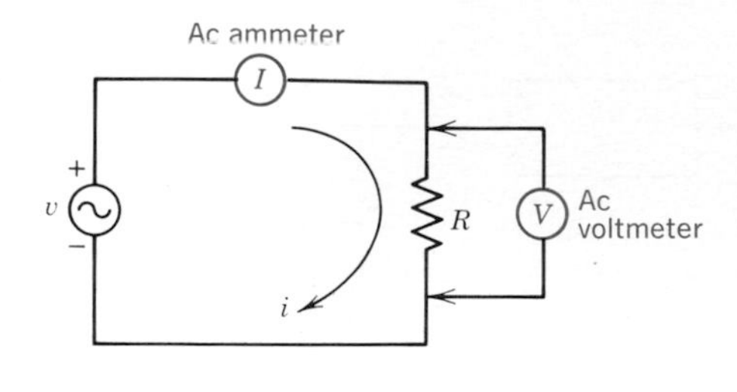

(a) Ac circuit with ammeter and voltmeter

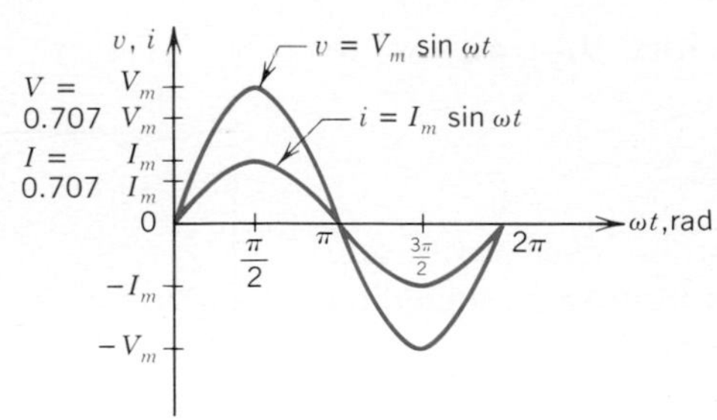

(b) Waveforms of current and voltage with rms values

FIGURE 15-6
Readings of an ac ammeter and ac voltmeter.

The power dissipated by a nonvarying current I_{dc}, flowing through a resistance R, is given by

$$P_{av} = P_{dc} = I_{dc}{}^2R \tag{3-9}$$

The *average* power dissipated by an alternating current of peak value I_m flowing through the same resistance R is given by

$$P_{av} = \frac{I_m{}^2R}{2} \tag{15-10}$$

Since these two must be equal,

$$I_{dc}{}^2R = \frac{I_m{}^2R}{2}$$

Therefore, $I_{dc}{}^2 = \dfrac{I_m{}^2}{2}$

and $$I_{dc} = \sqrt{\frac{I_m{}^2}{2}} = \frac{I_m}{\sqrt{2}} = 0.707\,I_m$$

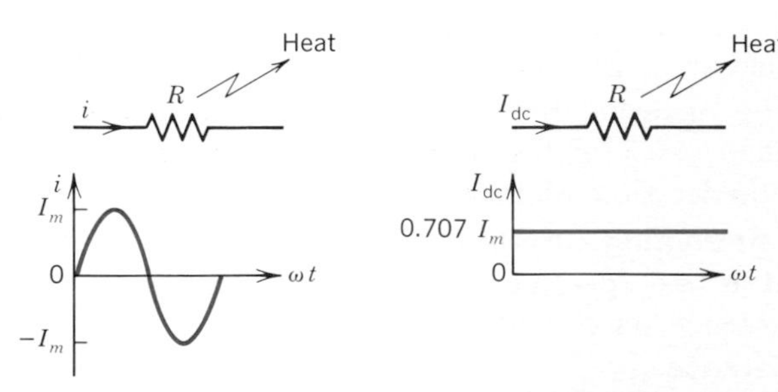

(a) Heat produced by alternating current of peak value I_m

(b) Same heat produced by direct current equal to $0.707\,I_m$

FIGURE 15-7
Effective or rms value of an alternating current.

Thus an alternating current of peak value I_m is as effective in producing heat in a pure resistance as a dc current equal to $0.707\,I_m$ or $\dfrac{I_m}{\sqrt{2}}$. See Fig. 15-7.

Thus $$I_{eff} = \frac{I_m}{\sqrt{2}} = 0.707\,I_m$$

Similarly, $$V_{eff} = \frac{V_m}{\sqrt{2}} = 0.707\,V_m$$

If the steps taken to obtain this result are examined, we first *squared* the current, found the average or *mean* value of the power, and then took the square *root*. This is summarized by referring to the effective value of the ac as I_{rms} (root-mean-square) or V_{rms}. The subscripts are used only for clarification.

Unless otherwise stated, all ac quantities are understood to be rms values. Thus:

$$I = I_{rms} = \frac{I_m}{\sqrt{2}} = 0.707\,I_m \tag{15-16}$$

$$V = V_{rms} = \frac{V_m}{\sqrt{2}} = 0.707\,V_m \tag{15-17}$$

where symbols are as described above.

EXAMPLE 15-6

The domestic ac line voltage, given by the equation $v = 170 \sin 377t$, is applied to a 33-Ω resistor as in Fig. 15-6*a*. Calculate:

a. The reading of an ac voltmeter connected across the resistor.
b. The reading of an ac ammeter connected in series with the resistor.

SOLUTION

a. $V_m = 170\text{ V}$

Voltmeter indicates

$$V = \frac{V_m}{\sqrt{2}} \tag{15-17}$$

$$= \frac{170\text{ V}}{\sqrt{2}} = \mathbf{120\ V}$$

b.

$$I_m = \frac{V_m}{R} \tag{15-15a}$$

$$= \frac{170\text{ V}}{33\ \Omega} = 5.2\text{ A}$$

Ammeter indicates

$$I = \frac{I_m}{\sqrt{2}} \tag{15-16}$$

$$= \frac{5.2\text{ A}}{\sqrt{2}} = \mathbf{3.6\ A}$$

Alternatively, after obtaining the rms value of voltage, the rms value of current can be obtained directly from Ohm's law:

$$I_R = \frac{V_R}{R} \tag{15-15a}$$

$$= \frac{120\text{ V}}{33\ \Omega} = \mathbf{3.6\ A}$$

Thus Ohm's law may be applied to an ac circuit containing *only resistance*, using oscilloscope *or* meter values for I and V.

The above example shows that the 120-V, 60-Hz supply has a peak value of 170 V and a peak-to-peak value of 340 V. However, this ac waveform is just *as effective* in heating the filament of an incandescent lamp (and producing as much light), as 120 V dc.

It should be noted that most ac ammeters and voltmeters are usually calibrated to indicate rms values *only* when used with sine waves. If the meters are connected in a circuit where the waveform is *not* sinusoidal, (such as a square wave or sawtooth wave), the readings will be erroneous, since the 0.707 factor *only* applies to *sine* waves. Although the rms value may be obtained for *any* waveform, a different factor must be used for each. Only a few instruments will show the true rms values regardless of the waveform. These are called *true-reading* rms meters. See Fig. 15-8. Some of them use a thermocouple to provide an indication proportional to the heating effect. It is for this reason that an oscilloscope is so important in electronic measurements. It displays much more information about a voltage or current than does a single-reading meter and is essential in nonsinusoidal circuits such as pulse and digital circuits.

Equations 15-16 and 15-17 can be rearranged so that peak-to-peak oscilloscope readings can be obtained from sinusoidal rms meter values:

$$V_m = \sqrt{2} \times V_{\text{rms}} = 1.414\ V_{\text{rms}} = 1.414\ V \tag{15-18}$$

and $$V_{\text{p-p}} = 2\ V_m = 2\sqrt{2}\ V_{\text{rms}} = 2.828\ V \tag{15-19}$$

Similarly, $$I_m = 1.414\ I \tag{15-20}$$

FIGURE 15-8
True reading rms voltmeter. (Courtesy of Hewlett Packard Company.)

and $$I_{p-p} = 2.828\, I \qquad (15\text{-}21)$$

EXAMPLE 15-7

An ac voltmeter across a resistor indicates 3.5 V and an ammeter in series with the resistor reads 15 mA. Calculate:

a. The peak-to-peak voltage across the resistor.
b. The peak current through the resistor.
c. The resistance of the resistor.

SOLUTION

a. $$V_{p-p} = 2\sqrt{2}\, V \qquad (15\text{-}19)$$
$$= 2 \times \sqrt{2} \times 3.5 \text{ V}$$
$$= \mathbf{9.9\ V}$$

b. $$I_m = \sqrt{2} I \qquad (15\text{-}20)$$
$$= \sqrt{2} \times 15 \text{ mA}$$
$$= \mathbf{21.2\ mA}$$

c. $$R = \frac{V_R}{I_R} \qquad (15\text{-}15)$$
$$= \frac{3.5 \text{ V}}{15 \times 10^{-3} \text{ A}} = \mathbf{233\ \Omega}$$

15-5 SERIES AND PARALLEL AC CIRCUITS WITH PURE RESISTANCE

The solution of series and parallel ac circuits that only involve resistance follows the same principles used for dc. As previously noted, the same equations for power and Ohm's law can be applied if rms quantities are used throughout. In those problems which contain data with peak or peak-to-peak values, care must be taken to make the appropriate conversions, so a consistent set of units results. The same laws of voltage division and current division hold for ac as they do with dc, provided that only pure resistances are involved. Problems 15-19 and 15-20 are designed to give practice in the solution of series- and parallel-resistive ac circuits with a mixed set of units. Examples 15-8 and 15-9 illustrate some of the principles.

EXAMPLE 15-8

Two series-connected resistors, 3.3 kΩ and 4.7 kΩ, have a 36-V peak-to-peak sinusoidal voltage connected to them as shown in Fig. 15-9. Determine:

a. The reading of a series-connected ammeter.
b. The voltage across the 4.7-kΩ resistor.
c. The power dissipated in the 3.3-kΩ resistor.

SOLUTION

a. The rms applied voltage,
$$V_T = \frac{V_{p-p}}{2\sqrt{2}} \qquad (15\text{-}17)$$
$$= \frac{36}{2\sqrt{2}} \text{ V} = 12.73 \text{ V}$$

Total circuit resistance,
$$R_T = R_1 + R_2 \qquad (5\text{-}2)$$
$$= 3.3 \text{ k}\Omega + 4.7 \text{ k}\Omega = 8 \text{ k}\Omega$$

The ammeter indicates
$$I = \frac{V_T}{R_T} \qquad (15\text{-}15a)$$
$$= \frac{12.73 \text{ V}}{8 \text{ k}\Omega} = \mathbf{1.59\ mA}$$

b. Voltage across the 4.7-kΩ resistor,
$$V_2 = V \times \frac{R_2}{R_2 + R_1} \qquad (5\text{-}5)$$
$$= 12.73 \text{ V} \times \frac{4.7 \text{ k}\Omega}{4.7 \text{ k}\Omega + 3.3 \text{ k}\Omega}$$
$$= \mathbf{7.48\ V}$$

c. Power dissipated in the 3.3-kΩ resistor,
$$P = I_R^2 R \qquad (15\text{-}12)$$
$$= (1.59 \times 10^{-3} \text{ A})^2 \times 3.3 \times 10^3\ \Omega$$
$$= \mathbf{8.34\ mW}$$

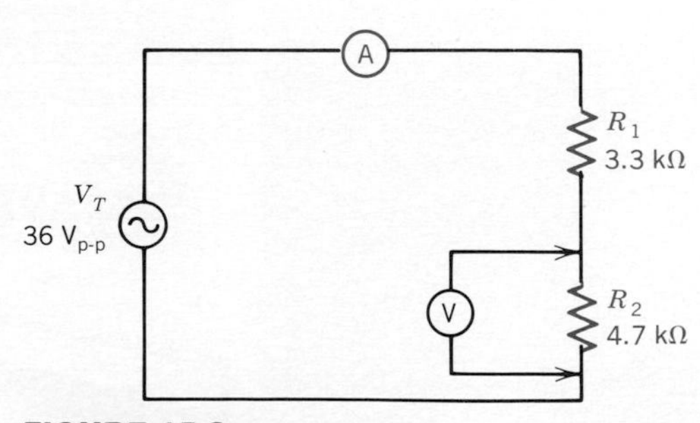

FIGURE 15-9
Circuit for Example 15-8.

EXAMPLE 15-9

Consider $R_1 = 4.7\ \text{k}\Omega$ and $R_2 = 3.3\ \text{k}\Omega$ connected in parallel across a sinusoidal voltage source, as shown in Fig. 15-10. The resistors draw a total of 50 mA. Calculate:

a. The current in the 3.3-kΩ resistor.

b. The peak-to-peak voltage indicated by an oscilloscope across the source.

c. The total power dissipated in the two resistors.

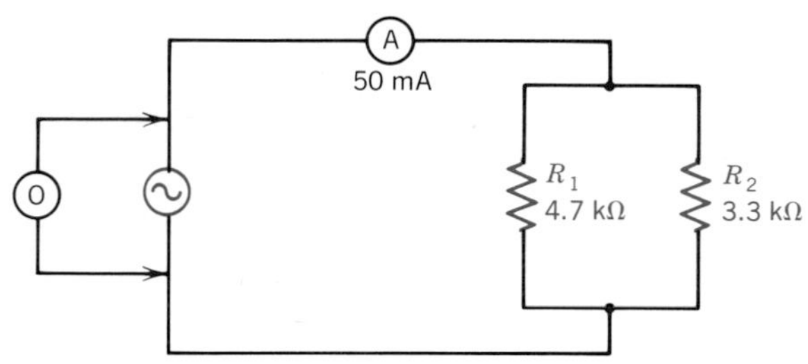

FIGURE 15-10
Circuit for Example 15-9.

SOLUTION

a. Current in the 3.3-kΩ resistor,

$$I_2 = I_T \times \frac{R_1}{R_1 + R_2} \qquad (6\text{-}13)$$

$$= 50\ \text{mA} \times \frac{4.7\ \text{k}\Omega}{4.7\ \text{k}\Omega + 3.3\ \text{k}\Omega}$$

$$= 50\ \text{mA} \times \frac{4.7}{8} = \mathbf{29.4\ mA}$$

b. Voltage across the circuit,

$$V_R = I_R R \qquad (15\text{-}15\text{b})$$

$$= 29.4 \times 10^{-3}\ \text{A} \times 3.3 \times 10^3\ \Omega$$

$$= 97\ \text{V}$$

Oscilloscope indication,

$$V_{\text{p-p}} = 2\sqrt{2}V \qquad (15\text{-}19)$$

$$= 2\sqrt{2} \times 97\ \text{V}$$

$$= \mathbf{274\ V_{p\text{-}p}}$$

(Note that it is not necessary to find the total parallel resistance.)

c. Total power,

$$P = V_R I_T \qquad (15\text{-}14)$$

$$= 97\ \text{V} \times 50 \times 10^{-3}\ \text{A}$$

$$= \mathbf{4.85\ W}$$

SUMMARY

1. The equation of a sinusoidally varying voltage may be given in terms of frequency and time by

$$v = V_m \sin \omega t$$

where $\omega = 2\pi f$ is the angular frequency in rad/s.

2. A sine wave of voltage applied to a resistor causes a sinusoidal current that is in phase with the voltage.

3. The power dissipated in a resistor varies between 0 and a peak value P_m at a frequency twice that of the applied voltage.

4. The average ac power dissipated in a resistor is half the peak power given by

$$P_{\text{av}} = \frac{I_m^2 R}{2}$$

5. Ohm's law and the power equations apply equally well to a resistive ac circuit if rms values are used and care is taken to ensure that voltages and currents are limited to the resistor only.

6. The rms value of an alternating current is that value of dc that is as effective in producing heat in a given resistance as the ac.

7. For sine waves only,

$$I_{rms} = \frac{I_m}{\sqrt{2}}$$

$$V_{rms} = \frac{V_m}{\sqrt{2}}$$

and

$$I_m = \sqrt{2} I_{rms}$$
$$V_m = \sqrt{2} V_{rms}$$

8. Unless otherwise specified all ac quantities indicated without subscripts are understood to be rms values.

SELF-EXAMINATION

Answer T or F or, in the case of multiple choice, a, b, c, d
(Answers at back of book)

15-1. The angular frequency of a 60-Hz waveform is

a. 120 rad/s
b. 240 rad/s
c. 377 rad/s
d. 754 rad/s

15-2. Given $v = 50 \sin 4000\ \pi t$, the amplitude and frequency are

a. 50 V, 4000 Hz
b. 100 V, 4 kHz
c. 100 V, 2000 Hz
d. 50 V, 2kHz

15-3. A sinusiodal voltage causes a sinusoidal current through a resistor of twice the frequency of the applied voltage. ______

15-4. The peak current through a resistor is given by V_m/R and is in phase with the peak of the applied voltage. ______

15-5. When a 60-Hz ac voltage is applied to a resistor, the power varies from 0 to a peak at a rate of 120 Hz. ______

15-6. In a 10-Ω resistor carrying a peak current of 3 A, the peak and average power are

a. 90 W, 45 W
b. 9 W, 4.5 W
c. 30 W, 15 W
d. 45 W, 22.5 W

15-7. The rms value of a sine wave of voltage of peak-to-peak value 10 V is

a. 7.07 V
b. 6.36 V
c. 3.18 V
d. 3.535 V

15-8. The alternating current through a 5-Ω resistor is 2 A. The rms voltage across the resistor is

a. 10 V

b. 5 V
c. 7.07 V
d. 3.535 V

15-9. An ac ammeter with a sinusoidal current indicates 4 mA. The peak-to-peak current is

a. 2.828 mA
b. 11.312 mA
c. 5.656 mA
d. 1.414 mA

15-10. The alternating current flowing through a 1-kΩ resistor is 2 A. A suitably connected wattmeter will indicate:

a. 4 W
b. 400 W
c. 4 kW
d. 4 mW

REVIEW QUESTIONS

1. a. What units are used to measure angular frequency?
b. How is angular frequency related to frequency in Hz?
c. What do you understand by the term angular frequency?

2. a. How many degrees in $\pi/3$ radians?
b. How many radians in 225°?

3. a. If an alternating triangular voltage is applied to a resistor, what is the shape of the current waveform?
b. How can you support this?
c. Is this also true of a square wave?

4. Can you explain qualitatively why power dissipation in a resistor occurs at a rate twice that of the alternating current in the resistor? Given

$$\sin^2 \omega t = \tfrac{1}{2}(1 - \cos 2\omega t)$$

use Eq. 15-8 to show that the power dissipation has an average value of $\frac{P_m}{2}$ and quantitatively why the overall power varies at twice the current frequency.

5. a. Explain why the effective value of a sine wave is *not* the average value of the sine wave.
b. Why is it the average value of the *square* of the sine wave?

6. a. What would be the rms value of a full-wave rectified sine wave as in Fig. 14-16*b*, compared with a pure sine wave?
b. What is the rms value of a square wave that alternates from +10 V to −10 V?
c. What is the rms value of a direct current of 3 A?

7. Under what special conditions can Ohm's law and the power equations be used in ac circuit calculations?

8. What measurement and calculation would you have to make to determine the peak current through a resistor, connected in a sinusoidal circuit, without the aid of an oscilloscope?

9. a. If the power and light in an incandescent lamp are pulsating at a rate of 120 Hz, why is there no visible flicker?

b. What peak ac voltage is required for a 120-V (sinusoidal) lamp?

c. In which circuit do you think a 120-V lamp will last the longer, in a 120-V ac or a 120-V dc circuit? Why?

10. If we state ac voltages and current in terms of effective dc values, why don't we generate, transmit, and use dc instead of ac?

PROBLEMS

(Answers to odd-numbered problems at back of book)

15-1. A sinusoidal voltage has a peak-to-peak value of 10 V and a frequency of 20 kHz.

a. Express this voltage in the form $v = V_m \sin \omega t$.

b. Determine the instantaneous value at $t = 105\ \mu s$.

15-2. Repeat Problem 15-1 for a current waveform with a peak-to-peak value of 50 mA and a frequency of 30 kHz.

15-3. Given the equation $v = 70 \sin 50t$, determine:

a. The amplitude of the voltage.

b. The peak-to-peak value.

c. The angular frequency.

d. The frequency in Hz.

e. The period.

f. The voltage when $t = 1$ s.

15-4. A sinusoidal voltage has a period of 20 μs and has a value of 2 V at $t - 1\ \mu s$. Express this voltage in the form $v = V_m \sin \omega t$.

15-5. A sinusiodal voltage with a peak-to-peak value of 120 V and a period of 2.5 ms is connected across a 33-Ω resistor. Calculate:

a. The peak value of the current.

b. The frequency of the current.

c. The equation representing the voltage.

d. The equation representing the current.

15-6. Repeat Problem 15-5 with a 47-kΩ resistor.

15-7. For the information in Problem 15-5 calculate:

a. The peak power in the resistor.

b. The average power in the resistor.

c. The frequency of the power variation.

15-8. Repeat Problem 15-7 with a 47-kΩ resistor.

15-9. Determine the rms value of the following:

a. 7.5 V peak

b. 14 mA peak-to-peak

c. 20 mV peak-to-peak

d. 17 A peak

15-10. Determine the peak values of the following:

a. 8 V

b. 20 μA rms

c. 120 V effective

d. 10 mA p-p

15-11. Determine the peak-to-peak values of the following:

a. 120 V

b. 240 V rms

c. 15 A

d. 1.6 μV

15-12. Determine the rms values of the following:

a. 340 V p-p

b. 50 mV peak

c. 325 kV peak

d. 100 A p-p

15-13. An ac ammeter connected in series with a 220-Ω resistor indicates 580 mA. Calculate:

a. The applied voltage.

b. The power dissipated in the resistor.

15-14. An electric heater is rated at 1.5 kW at 120 V ac. Calculate:

a. The hot resistance of the heater.

b. The current drawn by the heater.

15-15. What is the maximum permissible reading of an ac ammeter in series with a 33-Ω 2-W resistor?

15-16. What is the minimum resistance that a $\frac{1}{2}$-W resistor may have if it is to be safely connected across an alternating voltage of 6.3 V?

15-17. If an ac voltmeter connected across a resistor indicates 5 V and a series-connected miliammeter reads 2.2 mA, calculate:

a. The power dissipated in the resistor.

b. The resistance of the resistor.

15-18. A voltage having the equation $v = 30 \sin 3000\pi t$ V is applied to a 6.8-kΩ resistor. Calculate:

a. The reading of an ac voltmeter across the resistor.

b. The reading of an ac ammeter connected in series with the resistor.

c. The power dissipated in the resistor.

15-19. Three series-connected resistors have a sinusoidal voltage applied to them. An ac voltmeter across the first resistor indicates 12 V; an ac ammeter in series with the second resistor indicates 2.4 A; and an oscilloscope across the third resistor reads 10 V peak-to-peak. If a wattmeter, connected to read the total power delivered to the series circuit, indicates 55 W, calculate:

a. The resistance of each resistor.

b. The power dissipated in each resistor.

c. The applied voltage.

15-20. Three resistors R_1, R_2, and R_3 are connected in parallel across a sinusoidal voltage. An ammeter in series with the first reads 1.5 A while an oscilloscope across the second indicates a peak voltage of 35 V. If $R_3 = 2R_2$ and a wattmeter indicates a total power of 200 W in the whole circuit, calculate:

a. The resistance of each resistor.

b. The power dissipated in each resistor.

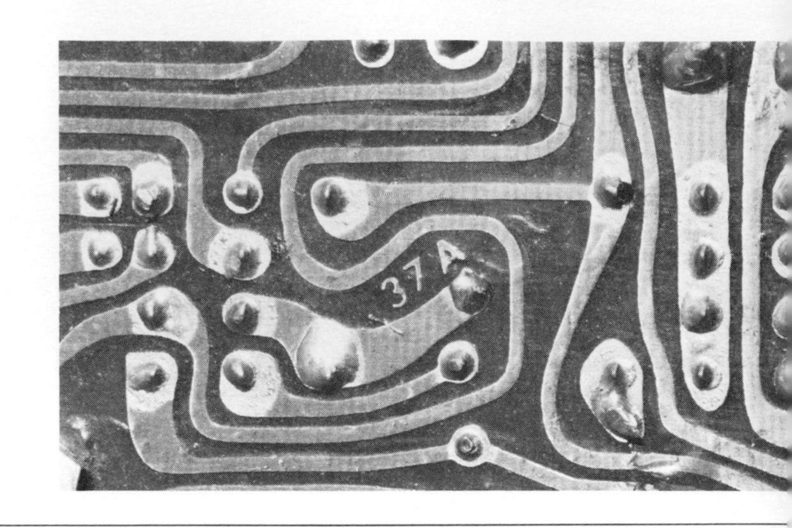

CHAPTER SIXTEEN

AC MEASURING INSTRUMENTS

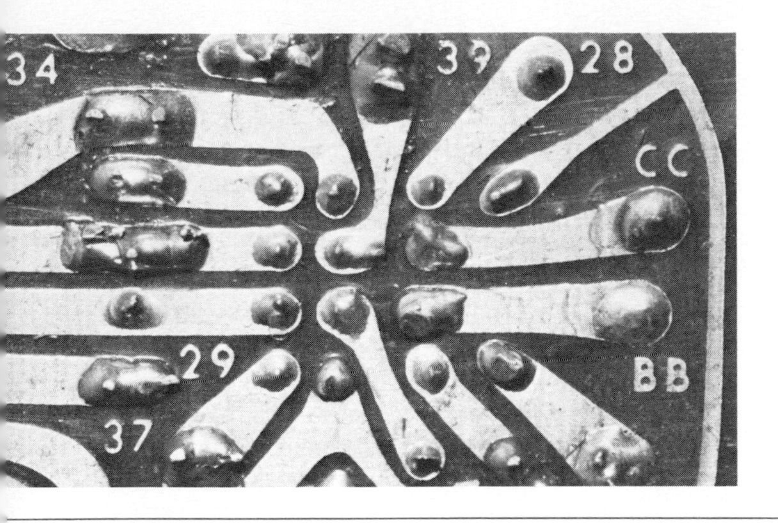

One of the most important instruments used to measure alternating voltage is the cathode-ray oscilloscope (CRO). It measures and displays peak-to-peak voltage, period, frequency, and phase relationships. It is essential to understand the operation of an oscilloscope because it is so widely used in industry and the laboratory. In this chapter we cover construction and operation features that are common to most oscilloscopes and the different probes that are used.

Alternating current voltmeters and ammeters are calibrated most often to indicate rms values. However, many of the meters do so only if used in sine wave applications. These are generally of the PMMC type (Section 12-1) but use additional diodes to rectify or convert the ac to a proportional dc. Other meters show the *true-reading* rms value *regardless of the waveform.* Finally, we consider some precautions when using equipment that is operated from the 120-V, 60-Hz supply, arising from common-ground connections.

16-1 THE CATHODE-RAY OSCILLOSCOPE

The oscilloscope provides us with a means of displaying the manner in which any voltage (or current) varies with time. It relies upon a beam of electrons to strike a fluorescent screen and "write" a waveform of the voltage variation. When viewing a sine wave, it will show all the instantaneous values over a full cycle, by drawing a continuous display of one or more cycles.

16-1.1 Construction

The "heart" of an oscilloscope is the cathode-ray tube or CRT. In a modern solid-state oscilloscope, the CRT is usually the only tube to be found. The CRT consists of a glass envelope that has been highly evacuated. Figure 16-1 shows a schematic representation of the CRT.

A *heater* or *filament* is used to bring the *cathode* to a high temperature to cause the emission of electrons. (These electrons were initially called "cathode rays" before this process of electron emission was fully understood.) The *accelerating anode,* held positive with respect to the cathode by about 2 kV or more, has a small hole at its center. Electrons accelerate toward this anode, pass through the hole, and travel to the screen on the right. The *control grid* regulates the flow of electrons and thus the brightness or intensity of the spot on the screen. The *focusing anode,* as well as providing an accelerating function, acts as a type of lens. This provides a well-defined *trace* on the screen. The cathode, grid, and anodes constitute what is called the *electron gun.* Its sole function is to provide a controlled beam of electrons.

As the electron beam travels to the screen, it passes through two sets of *deflecting plates.* The position of the electron beam on the CRT screen depends solely upon the voltages applied to these plates. This is called an *electrostatic* deflection system. It is quite suitable for the small deflections necessary in oscilloscopes. In TV tubes, deflection of the beam is accomplished through magnetic fields set up by coils in a yoke around the outside of the tube.

When the electron beam strikes the *fluorescent* screen, it excites the *phosphor* coating, causing a bright spot or line. Different phosphors can provide green or yellow displays with varying degrees of *persistence.* That is, the spot will continue to glow for some short time after the electron beam has moved. The electron beam arrives at the screen at approximately one-tenth of the speed of light, or more. This can cause a spot to be burned in the phosphor coating if it is allowed to remain in one position for many hours.

The inside of the tube is painted with a graphite coating called *aquadag,* as shown in Fig. 16-1. This serves as an electrostatic shield against outside fields, an additional

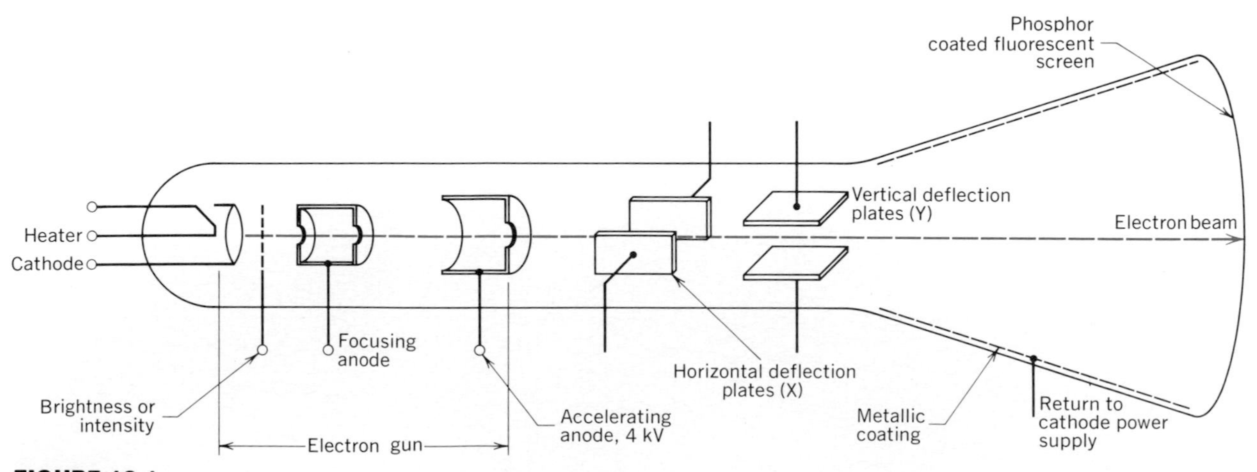

FIGURE 16-1
Schematic cross section through a cathode ray tube (CRT).

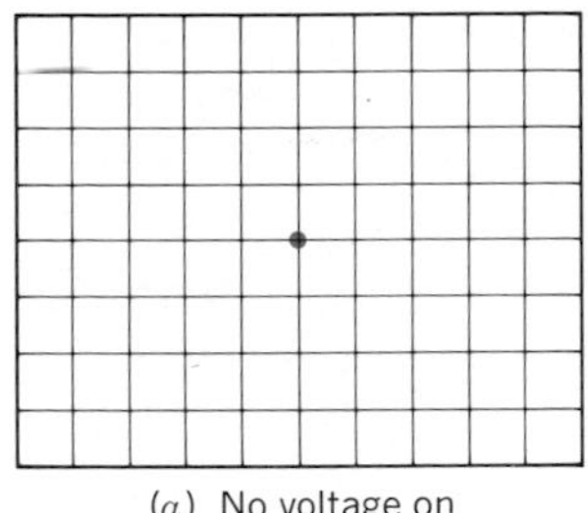

(*a*) No voltage on V or H plates

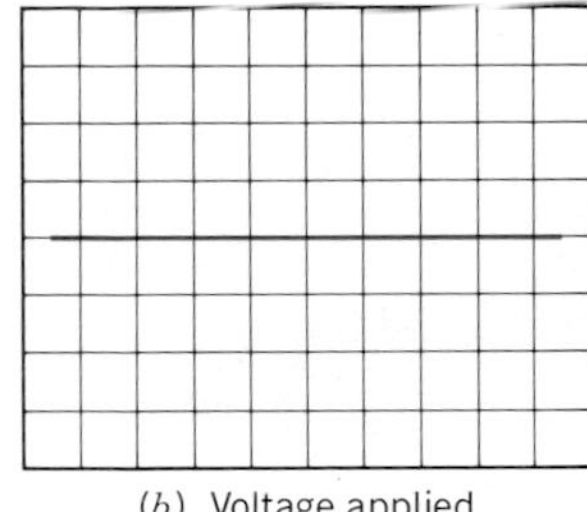

(*b*) Voltage applied to H plates only

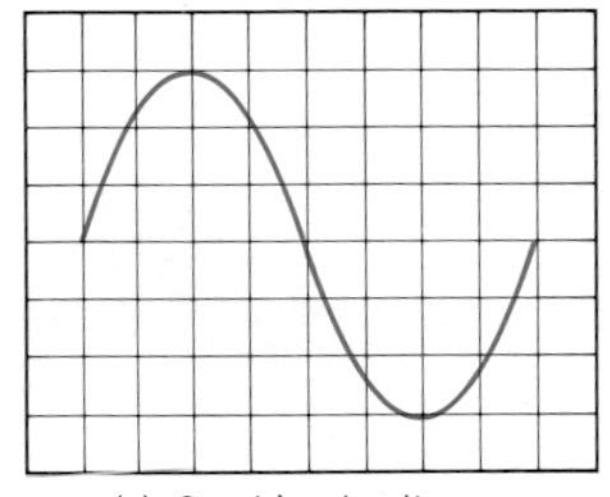

(*c*) Combined voltages on H and V plates

FIGURE 16-2
Screen display for different voltage conditions on deflecting plates.

accelerating anode in some tubes, and a means of avoiding an accumulation of electrons on the screen. When the electron beam arrives at the screen, secondary electrons are emitted, which must be returned to the power supply. They are collected by the graphite coating and make their way back to the cathode through the power supply.

16-1.2 Operation

If no voltage is applied to either set of deflecting plates, the electron beam passes through undeflected, causing a bright spot on the center of the screen, as in Fig. 16-2*a*. If a potential difference is applied to the horizontal (X-axis) deflecting plates, the negatively charged electron stream will move toward the positive potential. This will cause the spot on the screen to move to a new position, horizontally. If a varying (sawtooth) voltage is used to provide this horizontal deflection, at a minimum frequency of about 16 Hz, the movement of the dot from left to right appears as a straight continuous line. See Fig. 16-2*b*. This is usually referred to as the "trace."

Now consider a sine wave of voltage applied to the vertical (Y-axis) deflecting plates, in addition to the above voltage on the horizontal plates. As the electron beam moves from left to right, it is also deflected vertically by the sine wave—upward when the upper plate is positive, downward when it is negative. If the frequency of the voltage on the horizontal plates is suitably adjusted (usually variable from less than 1 Hz to many megahertz), one or two cycles of the voltage on the vertical plates can be seen, as in Fig. 16-2*c*. (Some electronic provision is made to *blank* the electron beam while it is being returned from the right side of the screen to the left. Improper adjustment of the oscilloscope STABILITY control can cause retrace or multiple trace problems.)

16-1.3 Function of Controls

The oscilloscope shown in Fig. 16-3 is representative of many single-beam "scopes." A brief description follows for typical controls found on most oscilloscopes.

BRILLIANCE Adjusted to give convenient intensity.

FOCUS and ASTIG(MATISM) Adjusted to give best definition.

TRIG(GER) LEVEL Controls the point at which the applied input waveform starts to be displayed. Normally left in the counterclockwise **(AUTO)** position where display begins at its average position.

STABILITY Adjusted to give a stable (nonoverlapping) stationary waveform.

Y SHIFT, VERNIER Coarse and fine controls for locating the trace anywhere vertically on the screen.

X SHIFT Shifts the trace horizontally to the left or right.

TRIG SELECTOR

TV FIELD, TV LINE Used in conjunction with **TRIG LEVEL** to provide correct triggering for video pulses. Normally left **OUT.**

HF Used in conjunction with **TRIG LEVEL** for high frequencies from 1 MHz up to 10 MHz.

+/− Used to select triggering of observed input waveform on positive or negative going *slope* of waveform. "Inverts" the waveform.

INT/EXT TRIG On **INTERNAL** (button **IN**) the triggering of the sweep of the beam is controlled by an internally generated voltage.

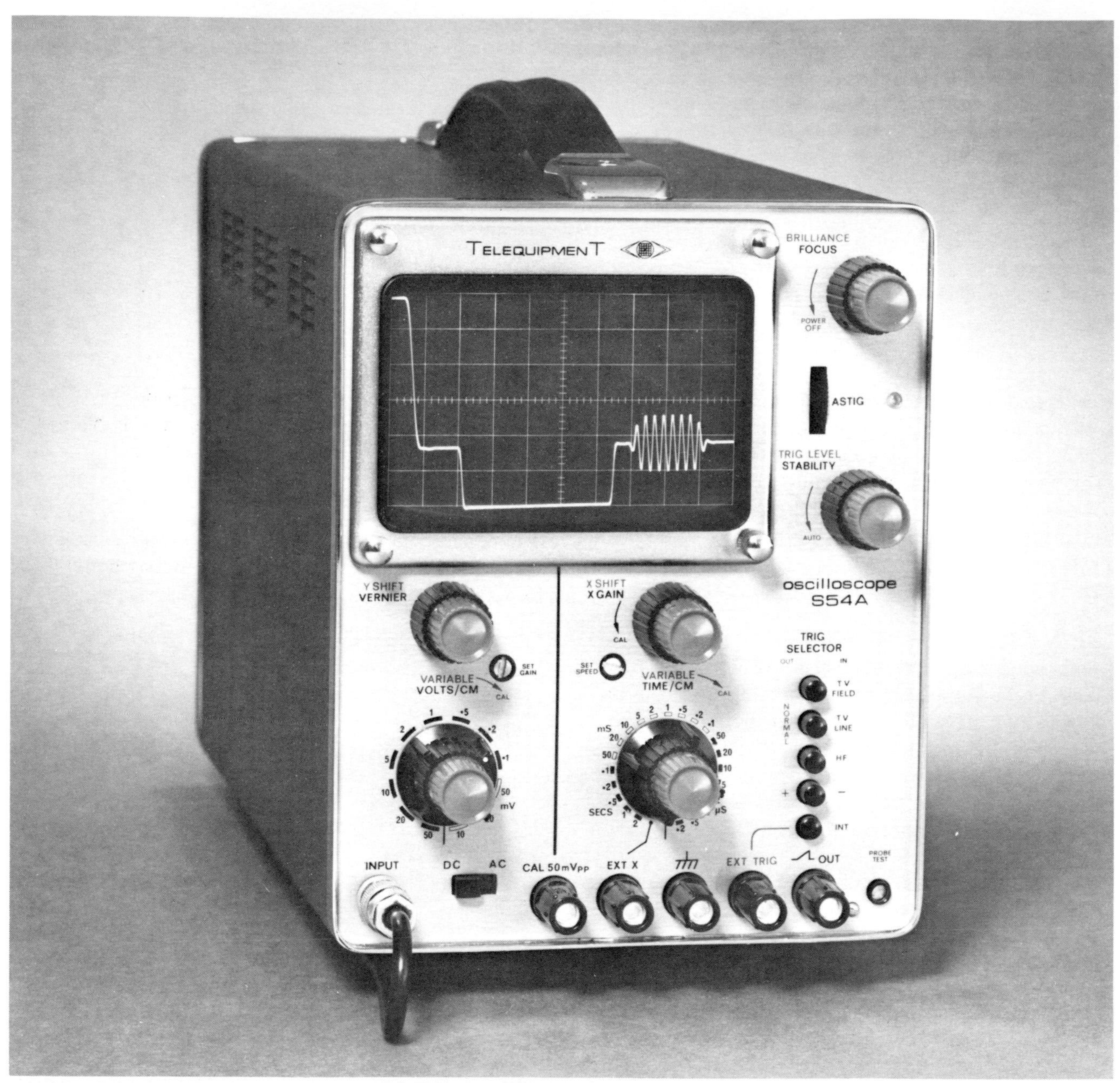

FIGURE 16-3
A single-beam oscilloscope showing the controls. (Courtesy of Tektronix, Inc.)

EXTERNAL TRIGGER An external voltage may be connected for triggering purposes. This gives more stable displays of nonperiodic waveforms.

VOLTS/CM Determines what voltage connected at the **INPUT** will cause a vertical deflection of 1 cm. This knob controls the gain of a vertical amplifier so that a 1-cm deflection on the screen may represent an input of from 10 mV to 50 V. This is only the case if the **VARIABLE** knob in the center is in its fully clockwise or **CALIBRATED** position.

TIME/CM Determines the rate at which the beam sweeps across the screen. When the inner **VARIABLE** knob is in its **CALIBRATED** position (and the **X GAIN** knob is in its fully counterclockwise **CALIBRATED** position) each centimeter horizontally represents the time indicated on the **TIME/CM** knob. This actually varies the frequency of the voltage internally generated and applied to the horizontal deflection plates. It is adjusted to give a convenient number of cycles on the screen.

EXT X If the **TIME/CM** knob is rotated to its fully counterclockwise position, an external voltage may be applied directly to the horizontal plates. (See Section 16-1.6.)

INPUT This is the point at which a voltage to be measured is applied to the oscilloscope by means of a *probe.*

DC/AC In the **DC** (direct coupling) position, a voltage to be observed at the **INPUT** is connected *directly* to the amplifier allowing both dc and ac voltages to be observed. In the **AC** (alternating coupling) position, only the ac portion of an input can be observed. If the input happens to contain a dc level, it is blocked (by a capacitor) and does not appear on the screen. The **DC** (or direct coupling connection) gives a trace of the full information contained in the input.

CAL 50 mV p-p An accurate, internally generated reference voltage used for checking and calibrating the vertical amplifier of the oscilloscope by adjusting the **SET GAIN** control.

PROBE TEST A positive-going, fast-rise pulse used to adjust (compensate) 10X probes. (See Section 16-1.4.)

16-1.4 Oscilloscope Probes

The use of an oscilloscope to measure voltage between two points is initially very similar to using a voltmeter. An oscilloscope probe consists of two leads. The input tip is connected through the body of the probe via a coaxial cable to the input connection on the oscilloscope. The ground lead is connected to the metal sheath of the cable. The tip is considered the positive lead, the ground is the negative side. The ground lead is usually connected to the ground side of the 120-V ac supply that powers the oscilloscope. Care must be exercised in its connection, as described in Section 16-6, to avoid erroneous readings and possible damage.

There are two common types of oscilloscope voltage probes. The most common incorporates a voltage-dividing circuit either in the body of the probe or at the end of the cable, called an *attenuator.* This reduces the input voltage to the oscilloscope to one-tenth the value at the tip. It is referred to as a 10X or 10 : 1 probe. The advantage of such a probe is that it increases the usual input resistance to the oscilloscope from 1 MΩ to 10 MΩ, reducing loading problems. Also, it extends (by a factor of 10) the maximum voltage that can be measured by the oscilloscope. (However, this factor of 10 must be taken into account, when reading the oscilloscope.) Also, the voltage-dividing circuit in the probe needs occasional adjustment to provide the proper flat response to a square wave input. This process, called *probe compensation,* is accomplished by a screwdriver adjustment while the probe is connected to a suitable square wave, as at the **PROBE TEST** terminal. Normally, each probe must be compensated for the oscilloscope with which it is to be used.

Direct probes have no such attenuating circuit. They are used when voltages being measured are only a few millivolts. Such low voltages require the maximum sensitivity of the oscilloscope.

Direct probes are usually marked as 1X; *unmarked* probes are generally 10X.

Current probes are also available that permit direct measurement and display of current waveforms. The current-carrying wire is placed in a slot in the probe, without breaking the circuit under test. A suitable amplifier system, either separate from, or included in, the oscilloscope, can measure dc and ac currents from 1 mA to over 100 A.

16-1.5 Measurement of Voltage and Frequency

Let us assume that the oscilloscope probe has been connected across a sinusoidal voltage and the controls have been adjusted to give the waveform shown in Fig. 16-4.

In our example, the peaks of the sine wave coincide with the *graticule* lines 2 cm above and below the 0-V level. The peak-to-peak displacement thus occupies 4 cm. To convert this to a voltage, we must use the volts/cm calibration—in this example, 0.2 V/cm. If we take into account the 10X factor, the full peak-to-peak voltage is given by

$$V_{p-p} = \textbf{number of cm} \times \textbf{volts/cm} \times \textbf{10} \quad \textbf{(16-1)}$$

Therefore, $V_{p-p} = 4 \text{ cm} \times 0.2 \text{ V/cm} \times 10$
$= \mathbf{8\ V}$

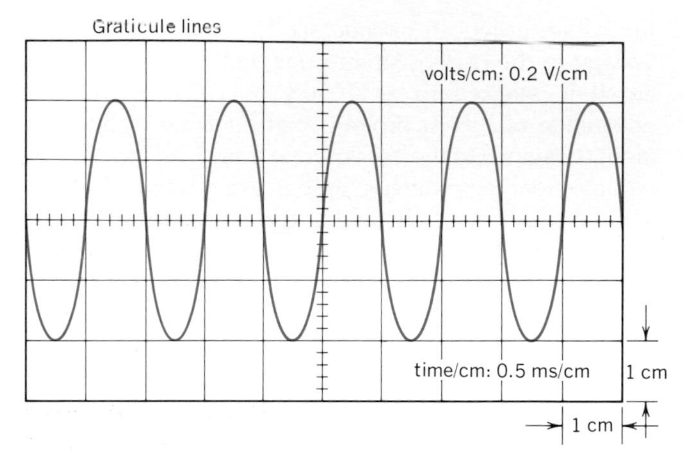

FIGURE 16-4
Typical waveform for measurement of voltage and frequency.

When a direct probe is used the 10X factor is omitted.

The peak voltage that has been measured, V_p, equals 4 V, although only 0.4 V was actually connected to the oscilloscope itself.

The measurement of frequency is done in a similar way except that the 10X factor is *not* used. First, we have to obtain the period, T. From Fig. 16-4, the period is represented by 2 cm, at 0.5 ms/cm.

$$T = \text{number of cm for one cycle} \times \text{time/cm} \quad (16\text{-}2)$$

Therefore,
$$T = 2 \text{ cm} \times 0.5 \text{ ms/cm} = 1 \text{ ms}$$

$$f = \frac{1}{T} \quad (14\text{-}12a)$$

$$= \frac{1}{1 \times 10^{-3} \text{ s}} = \mathbf{1\ kHz}$$

NOTE When making vertical and horizontal measurements, the vertical and horizontal *variable* controls must be in the CALIBRATED position. Also, when making measurements of dc levels, the location of the trace for 0 V (zero volt) must be identified.

In measuring the peak-to-peak voltage, it is not necessary to center the waveform in the middle of the screen. Where the peak-to-peak value is not a whole number of centimeters, the Y-SHIFT is used to make either the positive or negative peaks line up with a graticule line. Then, by suitable adjustment of the X-SHIFT, the finely divided vertical graticule line can be used to give the overall height. A similar method is used to measure period where a cycle does not occupy a whole number of centimeters. This is shown in Example 16-1.

EXAMPLE 16-1

Given the waveform and oscilloscope settings shown in Fig. 16-5, determine, assuming a 10X probe, the following:

a. The peak voltage, V_p.
b. The period, T.
c. The frequency, f.

SOLUTION

a. $$V_{p-p} = \text{number of cm} \times \text{volts/cm} \times 10 \quad (16\text{-}1)$$
$$= 4.6 \text{ cm} \times 50 \times 10^{-3} \text{ V/cm} \times 10$$
$$= 2.3 \text{ V}$$
$$V_p = \frac{V_{p-p}}{2} = \frac{2.3}{2} \text{ V} = \mathbf{1.15\ V}$$

b. $$T = \text{number of cm} \times \text{time/cm} \quad (16\text{-}2)$$
$$= 3.2 \times 0.2 \text{ ms/cm}$$
$$= \mathbf{0.64\ ms}$$

c. $$f = \frac{1}{T} \quad (14\text{-}12a)$$
$$= \frac{1}{0.64 \times 10^{-3} \text{ s}}$$
$$= \mathbf{1.56\ kHz}$$

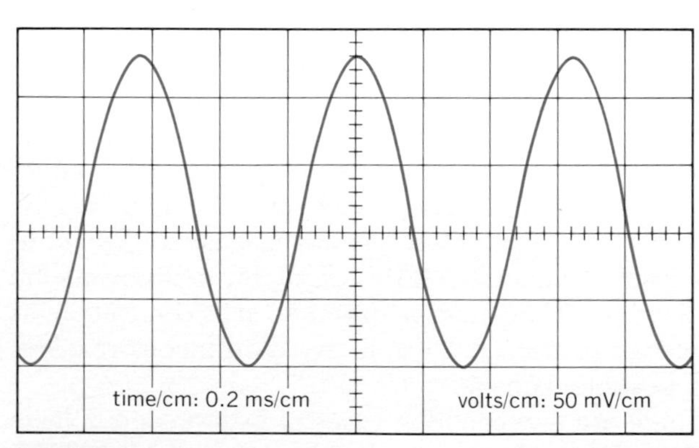

FIGURE 16-5
Waveform for Example 16-1.

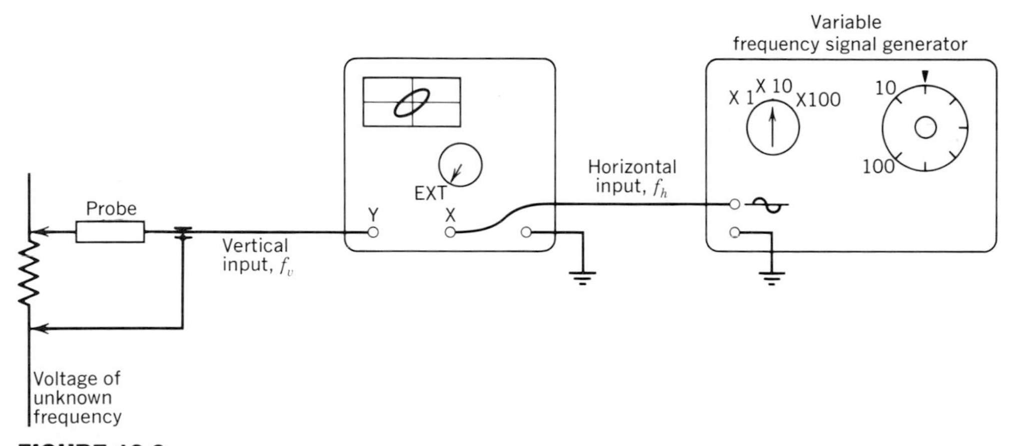

FIGURE 16-6
Measuring frequency using Lissajous figures.

The accuracy of the vertical and horizontal controls ranges from ±3% to ±5% depending upon cost. To this must be added errors arising from reading the vertical and horizontal deflections.

16-1.6 Lissajous Figures

Another method of measuring frequency eliminates the inaccuracy of the **TIME/CM** sweep control. This involves applying the frequency to be measured to the vertical input, as before, but connecting a variable frequency sine wave to the *horizontal* input through the **EXT X** connection. See Fig. 16-6. The beam is now swept from left to right by the external sine wave, instead of the internally generated wave.

The frequency on the horizontal input is varied until a stationary pattern is formed on the oscilloscope. This pattern, called a *Lissajous figure,* takes on different configurations according to the ratio of the two frequencies. Fig 16-7 shows some Lissajous figures for various frequency ratios.

The Lissajous figures obtained on the oscilloscope may not be as symmetrical as those in Fig. 16-7. For example, when the horizontal frequency is the same as the vertical, an ellipse or straight line can be seen changing gradually to a circle. This is because the two waves may differ slightly in time or phase relationship with each other. In fact, this is useful information, and is covered in Section 24-8 under phase angle measurements. Similarly, the other patterns may "drift," but at the ratios shown, the figures will become stationary. The unknown frequency is then known in terms of the variable frequency. In general, the frequency ratio is given by the ratio of the number of tangent points on the horizontal to the vertical. That is,

$$\frac{f_v}{f_h} = \frac{\text{number of horizontal tangent points}}{\text{number of vertical tangent points}} \qquad (16\text{-}3)$$

For example, see Fig. 16-7*d*. A line drawn horizontally on the top of the figure touches the figure at two points. A line drawn vertically on the side of the figure touches the figure at three points. Thus $\frac{f_v}{f_h} = \frac{2}{3}$. If $f_h = 120$ Hz, $f_v = \frac{2}{3} f_h = \frac{2}{3} \times 120\text{ Hz} = \mathbf{80\text{ Hz}}$.

The frequency of the unknown is determined only as accurately as the variable frequency is known. This frequency may be obtained from a *signal generator,* of which there are two common types.

An *audio* signal or function generator (see Fig. 16-8) provides sine and square (and sometimes triangular) output voltages. The frequency is usually variable from approximately 1 Hz to 1 MHz, and the amplitude is also variable, with a maximum value from 10 to 20 V peak-to-peak. Signal generators are used to inject signals into circuits, such as amplifiers, allowing an oscilloscope to make measurements on the gain and frequency characteristics of the circuit. Higher-frequency signal generators, called *radio frequency* (RF) generators, provide frequencies up to hundreds of megahertz for radio, TV, and communications equipment servicing.

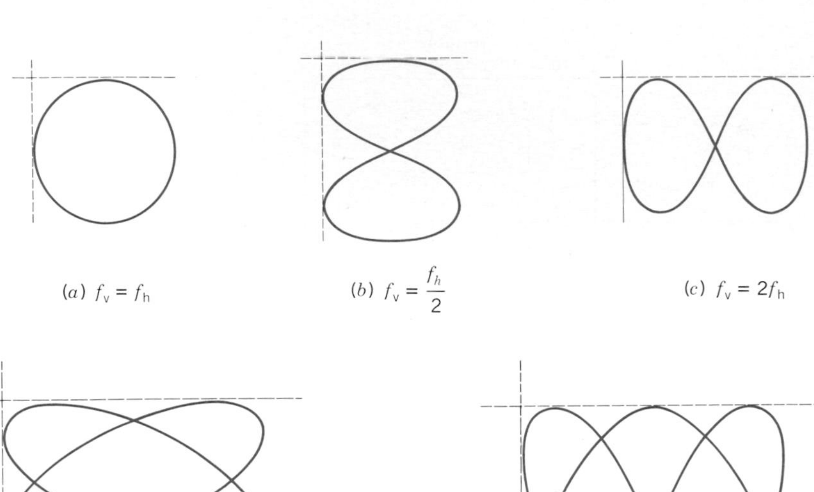

(a) $f_v = f_h$ (b) $f_v = \frac{f_h}{2}$ (c) $f_v = 2f_h$

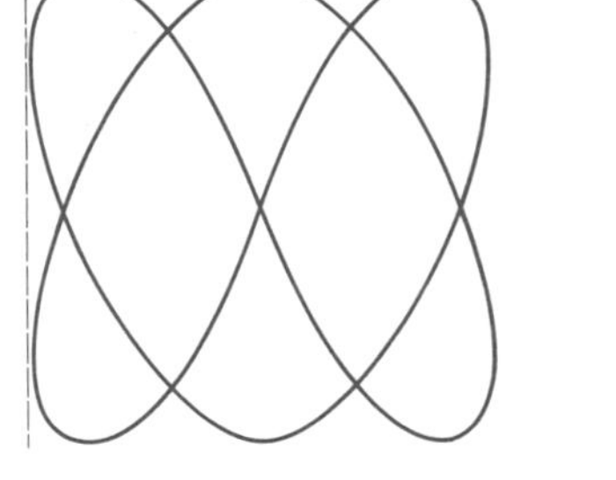

(d) $f_v = \frac{2}{3} f_h$ (e) $f_v = \frac{3}{2} f_h$

FIGURE 16-7
Lissajous figures for various frequency ratios.

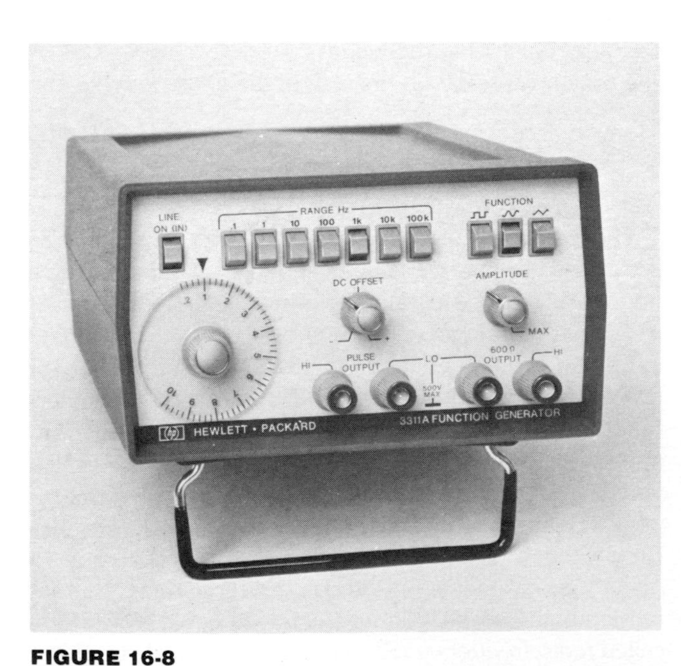

FIGURE 16-8
Typical function generator. (Courtesy of Hewlett-Packard Company.)

The use of Lissajous figures for frequency measurement is very limited where high accuracy is required. A more accurate and convenient method is covered in Section 16-8 using a frequency meter. However, the Lissajous method can be used as a quick check on the calibration of an audio signal generator at low frequencies. This can be done by comparing the line frequency, 60 ± 0.02 Hz, with the 60 Hz setting on the generator. Also, as mentioned, Lissajous figures can be used to make phase angle measurements in circuits at a given (constant) frequency.

16-1.7 Other Oscilloscopes

A common type of oscilloscope has a *dual-beam* or *dual trace* in which two waveforms can be simultaneously displayed. See Fig. 16-9. Channels *A* and *B* control the input signals from two independent probes. In this way, the time and phase relationships of the two waveforms are immediately available. Modern oscilloscopes of this type can display waveforms with frequencies of several GHz.

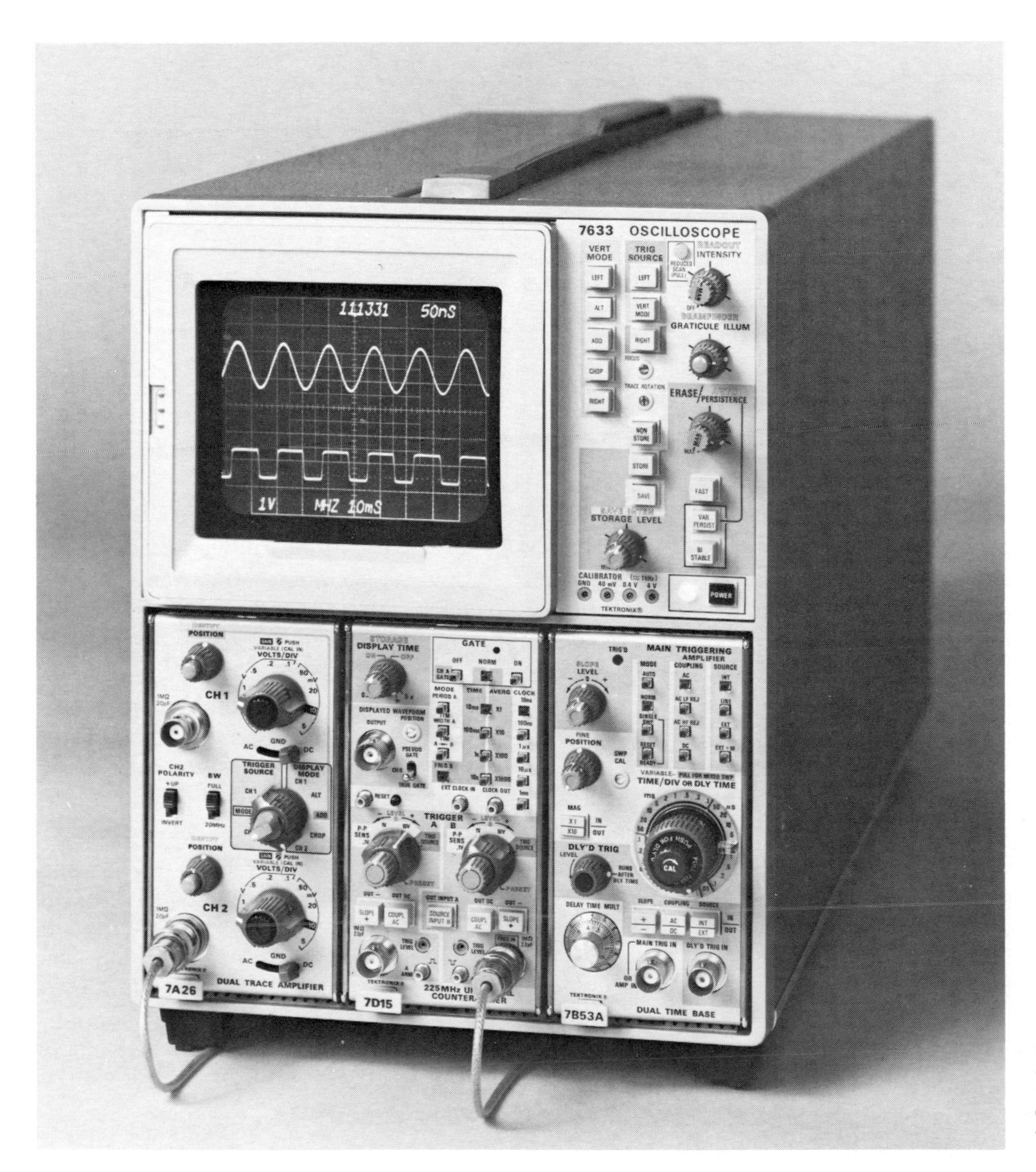

FIGURE 16-9
Dual-beam storage oscilloscope. (Courtesy of Tektronix, Inc.)

Polaroid cameras can be used to make hard copies (or photographs) of waveforms displayed on the screen. Alternatively, a *storage* oscilloscope that can store or "freeze" a given waveform, can be used and then redisplay it a short time later upon demand.

Other modifications of oscilloscopes and their circuitry include *curve tracers* to display characteristics of solid-state devices, *logic-state analyzers* that display digital events in a data stream, automotive *ignition analyzers,* and so on.

We now examine some of the meters designed to measure ac quantities.

16-2 AC METERS USING THE PMMC MOVEMENT

Recall from Section 12-1 that a permanent magnet moving-coil movement is polarity-sensitive and is used for dc measurements. Consider the application of a

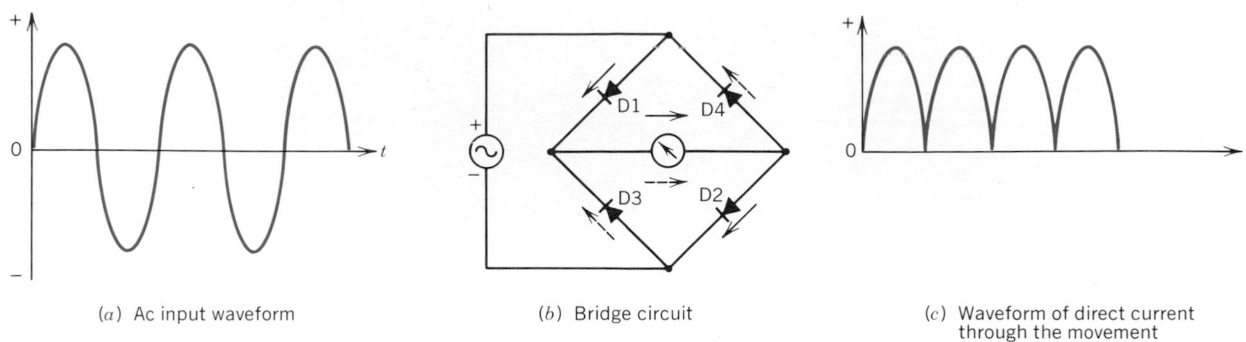

(*a*) Ac input waveform (*b*) Bridge circuit (*c*) Waveform of direct current through the movement

FIGURE 16-10
Operation of an ac meter using a PMMC movement.

sinusoidal voltage to a PMMC movement. During one-half cycle, the PMMC movement experiences a torque in one direction, followed by a torque in the opposite direction during the next half-cycle. At a frequency of 60 Hz the reversal of torque is so rapid that the pointer will not be able to follow the alternations. The pointer will simply stay at zero with a small vibrating motion. The meter is actually indicating the average value of the sine wave, which is zero.

In order to make the pointer deflect in one direction only, we must make the current flow through the moving coil of the movement in one direction only. In some meters (mostly ac milliammeters), this is achieved using *diodes* connected in a bridge circuit as in Fig. 16-10*b*.* (Ac voltmeters usually convert the ac to dc by means of one or two diodes in a shunt rectifier or voltage doubler circuit, followed by a filter.)

A diode is a device that allows current to flow through it in only one direction (in the direction of the arrow) when the polarity across the diode is favorable. (See Section 28-5.) On the positive half-cycle of the input, diodes D1 and D2 conduct; on the negative half-cycle, diodes D3 and D4 conduct. The resultant waveform of current through the PMMC movement is shown in Fig. 16-10*c*. This is a full-wave rectified waveform, which we recognize as the output of a simple dc generator. Its average value is 0.636, $2/\pi$, of the peak value. The pointer turns through an angle proportional to this average value. Where it comes to rest, the scale is marked with the effective value of the applied sine wave. The scale of an ac meter using a PMMC movement is linear, although on very low ranges some nonlinearity occurs due to the characteristics of the diodes. Separate scales are often used for the 1.5- and 5-V ac ranges because of this nonlinearity.

The ac sensitivity (or ohms per volt rating) of ac meters using a PMMC movement is lower than the dc sensitivity because of the necessity of rectification. A typical VOM with a dc sensitivity of 20 kΩ/V will have an ac sensitivity of only 5 kΩ/V. In some electronic voltmeters (VTVM or solid-state) with 11-MΩ resistance on direct current, the input resistance is only 1 MΩ for alternating current. Nevertheless, PMMC movements are widely found in analogue multimeters that can measure alternating current and direct current. Also, they are capable of indicating ac voltages at frequencies of 1 MHz and beyond, something that is not possible with other ac meters described below.

* In addition, a current transformer is used (as described in Section 18-6.2) to provide a low resistance when the meter is inserted in a circuit.

16.3 THE ELECTRODYNAMOMETER MOVEMENT

The limitation of the PMMC movement to direct current is due to the fixed direction of the permanent magnet's field. If we could automatically reverse the field every time the current in the movable coil reversed, the pointer would turn in one direction only. This is achieved by replacing the poles of the permanent magnet with two stationary coils, as in Fig. 16-11*a*. The movable coil is free to turn in the space between the two coils, which act as solenoids. This type of movement is called an *electrodynamometer.*

If the fixed and movable coils are series-connected as shown in Fig. 16-11*a*, the result is an ac ammeter. This basic meter movement can then be used with shunts and multipliers to make any range of ammeter and voltmeter. Because the deflection is proportional to the *square* of the current, the scales of these meters are nonlinear, with markings crowded together at the low end.

The movement will also operate on dc, but its cost is too high to compete with PMMC dc meters. However, the meter can be calibrated on ac by noting its deflection for given values of direct current. In fact, the movement indicates *true-reading* rms values for *all* waveforms. The sensitivity of an electrodynamometer movement is much lower than the PMMC type, having a value of around 50 Ω/V for an ac voltmeter. Also, the frequency range extends from dc up to about only 125 Hz, so they are limited to making measurements in power circuits.

The real value of an electrodynamometer movement lies in its application to measuring average power, either ac or dc, in a wattmeter, as in Fig. 16-11*b*.

We have already considered the external connection of the wattmeter terminals in Section 9-4. We can now see that the current terminals make load current pass through the heavy stationary coils. This sets up a field whose strength depends upon the amount of current. (There is no iron in the coils, so there is a linear relationship between the field and the current.)

The potential terminals connect the movable coil across either the source or the load. The force acting to turn the coil is thus dependent upon both the current and the voltage, with the resulting pointer deflection proportional to the *product* of the two. That is, $P = VI$ for a pure resistor. However, if the circuit contains components other than resistance, the wattmeter reading still gives the true power in watts, although the result is *no* longer given *solely* by the product $P = VI$. (See Section 25-6.) Unlike electrodynamometer voltmeters and ammeters, the scale of a wattmeter is linear.

It should be noted that if the potential coil is connected across the load, there is a slight error due to an additional current drawn by the potential coil. If connected across the supply, the potential coil includes an additional voltage drop across the load-carrying current coils. Consequently, the potential coil is usually connected directly across the load, as shown in Fig. 16-11*b*. In this way, a *constant* error is produced, which can be corrected and compensated for, by opening the load circuit and subtracting the no-load wattmeter reading from all future readings.

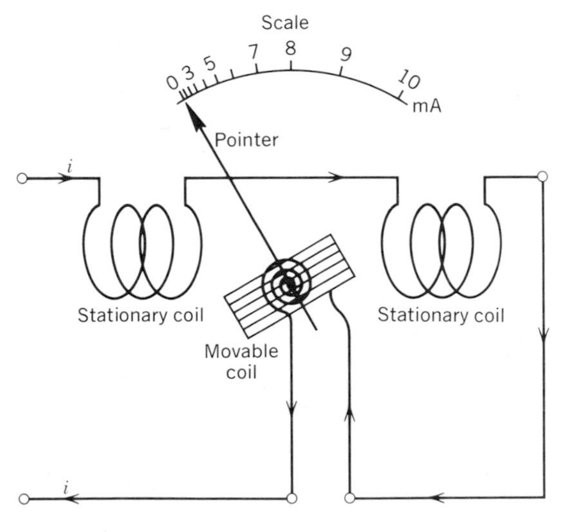

(*a*) Movement used as a milliammeter. (Note the nonlinear scale.)

(*b*) Movement used as a wattmeter

FIGURE 16-11
The electrodynamometer movement.

It is clear that with direct current, polarity must be observed when making wattmeter connections to avoid a backward deflection. But this also applies to alternating current. This is because each half-cycle has its own *instantaneous polarity*. The ± (or red) terminal of the potential coil must be connected to the side of the line where the current coil is located, as in Fig. 16-11*b*. As with any electrodynamometer movement, most wattmeters are *not* suitable for frequencies above 100 Hz, so special ones must be used at 400 Hz, for example.

16-4 MOVING-IRON INSTRUMENTS

There is a group of instruments, primarily for ac measurements, called *moving-iron* instruments. These consist of iron vanes attached to a pointer and free to move inside a current-carrying coil. Since the vanes carry no current (they move because of induced magnetic effects), the movement is simple, rugged and less expensive than the electrodynamometer. It is also an inherent rms indicator for all waveforms. But the moving-iron instrument is less sensitive than the PMMC movement, very limited in frequency, and nonlinear in its scale. Like the electrodynamometer, moving-iron voltmeters and ammeters are found only in low-frequency power circuits.

16-5 FREQUENCY METER

We saw in Section 13-8 how a digital multimeter can be used to measure voltage, current, and resistance to a high degree of accuracy and resolution. Although an oscilloscope, an analogue device, can be used to measure period and frequency, it is seldom more accurate than 2%. An instrument to provide direct digital measurements of these quantities is shown in Fig. 16-12.

Typical of many, this instrument is a multiple-function counter. It presents a six-digit display of frequency and period for waveforms ranging from 5 Hz to 80 MHz. This gives a resolution of 0.1 Hz and 100 ns. It may also be used to totalize input events in the form of pulses and display a maximum count of 999999. The counter requires ony 25-mV input for operation of all functions.

FIGURE 16-12
Multiple-function counter to measure frequency and period. (Courtesy of John Fluke Mfg. Co., Inc.)

16-6 PRECAUTIONS WITH GROUNDED EQUIPMENT

The neutral side of the 120-V, 60-Hz supply is grounded, usually to a cold water pipe where the supply enters the building. The neutral must be a continuous unbroken line through the whole wiring system. The longer side of all 120-V receptacles is the grounded neutral, as shown in Fig. 16-13*a*, with the "hot" or high side connected to the smaller opening.* (The third terminal in 120-V, single-phase polarized receptacles is also connected to ground. The safety feature of this third connection is discussed in Section 18-3.1.)

Most electronic equipment that operates from the 120-V, 60-Hz supply has one side of its output (in the case of a signal generator) or one side of the input leads (in the case of an oscilloscope, voltmeter, frequency meter, etc.) connected to the power line ground. This side is usually designated as the negative or low side. Consider a meter, with its low side grounded, connected across a source that also has one side grounded.

If the grounds are not matched, as in Fig. 16-13*b*, the result is a direct short across the source. Since a signal generator has a rather high internal resistance, from 50 to 600 Ω, the result is merely a 0-V indication. But if the

* For some equipment, it is important that "two-prong" plugs be inserted in the receptacle with the proper orientation. This ensures that the metal cabinet of a TV set, for example, is connected to the grounded neutral, and not the high side. Such two-prong plugs have one side larger than the other to provide this polarization. This is automatically provided for in three-prong plugs that can be inserted in only one way.

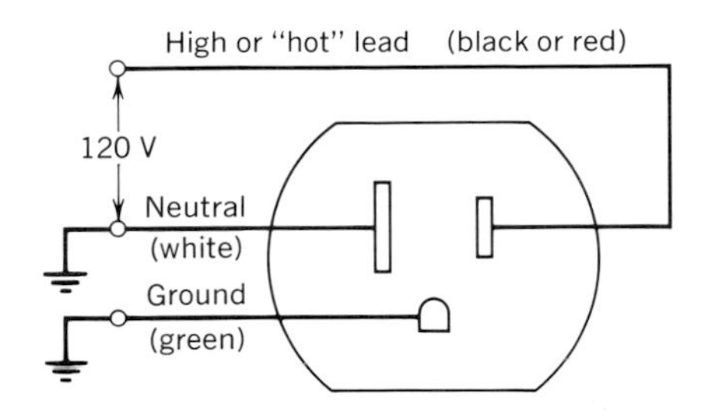

(*a*) Wiring of a 120-V receptacle

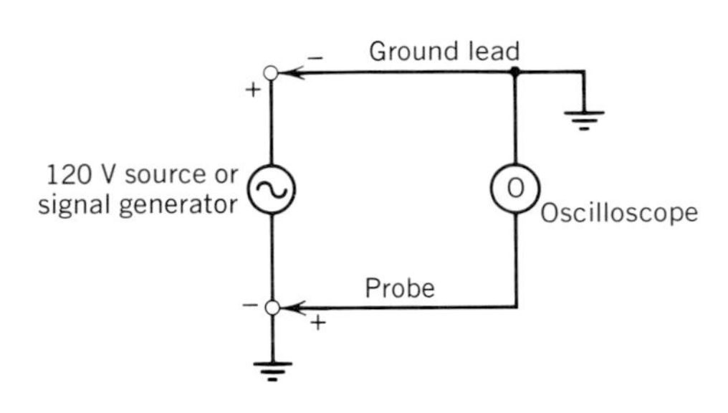

(*b*) Improper connection of grounded oscilloscope causes a short circuit

FIGURE 16-13
Ground connections in a receptacle outlet and oscilloscope.

source is a 120-V outlet, a high current will flow at the *instant* of connecting the ground lead, whether or not the probe has already been connected. In some cases, expensive damage can occur in the measuring instruments. Evidently, the grounds should be connected to each other to avoid this problem. If the identity of the ground is unknown, *only* the probe should be connected. If it is inadvertently connected to ground, no voltage will be indicated, requiring the probe to be moved to the other terminal. In many cases it is not necessary to connect the ground lead at all, since all voltages being measured will be with respect to ground.

Figure 16-14 shows another problem arising from ground loops. When the oscilloscope is connected across R_1 in Fig. 16-14*a*, R_2 is shorted out. The oscilloscope indicates the full voltage from the source, not the true voltage that was across R_1. If it is possible to interchange R_1 and R_2, as in Fig. 16-14*b*, the true value can be obtained if grounds are observed. Otherwise, the voltage across R_2 subtracted from the source voltage will give the voltage across R_1. However, this will only work where the components are resistors, and not for the general case for inductors and capacitors. In this situation, the problem of the grounds can be removed by using an *isolation transformer* to supply power to the source or the measuring instrument. This is discussed further in Section 18-4.

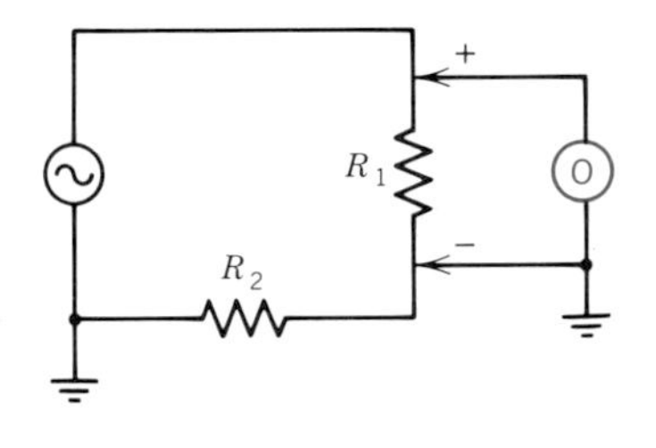

(*a*) Oscilloscope ground shorts out R_2

(*b*) Oscilloscope measures true voltage across R_1

FIGURE 16-14
Oscilloscope measurements taking grounds into account.

SUMMARY

1. A cathode-ray tube provides a beam of electrons that is deflected vertically and horizontally to display a "picture" of a voltage waveform on a fluorescent screen.

2. An oscilloscope measures voltage and period using the calibrated vertical and horizontal controls of VOLTS/CM and TIME/CM as follows:

$$V_{\text{p-p}} = \text{number of cm} \times \text{volts/cm}$$
$$T = \text{number of cm} \times \text{time/cm}$$

3. A 10X probe reduces the input voltage by a factor of 10. This requires the above *voltage* equation to be multiplied by 10.

4. A Lissajous figure results from applying sine waves to the vertical and horizontal deflection plates of an oscilloscope. The resulting pattern determines the ratio of the two frequencies using the method of tangent points:

$$\frac{f_v}{f_h} = \frac{\text{number of horizontal tangent points}}{\text{number of vertical tangent points}}$$

5. A PMMC movement can be used in an ac meter if diodes are used to convert the ac to a proportional dc.

6. An electrodynamometer movement replaces the permanent magnet poles of a PMMC movement with fixed solenoid coils for ac applications.

7. The fixed and movable coils of an electrodynamometer are used to measure power in a wattmeter.

8. Moving-iron instruments use movable iron vanes in a current-carrying coil to cause true-reading rms deflections in primarily ac power circuits.

9. A frequency meter can measure frequency and period of ac waveforms to a high degree of accuracy and resolution.

10. Most electronic equipment operated from the 120-V, 60-Hz supply has one side of its input or output grounded. Ground connections must be closely checked to avoid errors and possible damage.

SELF-EXAMINATION

Answer T or F, or, in the case of multiple choice, a, b, c, or d
(Answers at back of book)

16-1. The electron gun is that part of a CRT that produces the beam of electrons. ________

16-2. An electrostatic deflection method is used in an oscilloscope whereas a magnetic deflection method is used in TV tubes. ________

16-3. The voltage waveform to be examined is connected to the X-axis or horizontal deflection plates. ________

16-4. The Y-SHIFT control can be used to move the display to one side or the other. ________

16-5. The VOLTS/CM control is adjusted for the largest, complete vertical display on the screen. ________

16-6. The TIME/CM control is adjusted for the desired number of cycles or portion of a cycle. ________

16-7. The dc input switch can be used only with pure dc voltages. ________

16-8. When using a 10X probe, both the horizontal and vertical deflections must be multiplied by 10. ________

16-9. A direct probe may be used where small input voltages require the full vertical sensitivity of the oscilloscope. ________

16-10. An oscilloscope displays a peak-to-peak value of 3.2 cm with a 10X probe and a VOLTS/CM setting of 20 V/cm. The peak voltage is

a. 32 V
b. 320 V
c. 640 V
d. 64 V

16-11. An oscilloscope displays 4 cycles over an 8-cm distance with a 10X probe and a TIME/CM setting of 50 μs/cm. The period and frequency are

a. 10 μs, 100 kHz
b. 100 μs, 100 kHz
c. 100 μs, 10 kHz
d. 400 μs, 2.5 kHz

16-12. The frequency of an unknown sinusoidal frequency may be determined to an accuracy of 0.1%, even if the oscilloscope is only 3% accurate, by the use of Lissajous figures. ______

16-13. A Lissajous figure is a stationary pattern only when the two frequencies are exactly equal. ______

16-14. A dual-beam (or dual trace) oscilloscope may simultaneously display two waveforms of the same or related frequency on the same screen. ______

16-15. If a symmetrical ac voltage, at power line frequencies, is applied directly to a PMMC movement, the resultant deflection will be approximately 0.707 of the applied peak voltage. ______

16-16. PMMC movements are used in ac meters because of their high sensitivity. ______

16-17. The electrodynamometer movement may only be used to measure power. ______

16-18. When measuring power line voltages with an ac-operated voltmeter or oscilloscope, it is best not to make any ground connection at all. ______

REVIEW QUESTIONS

1. a. What is an "electron gun"?
b. What polarity voltage must be applied to the control grid to reduce intensity in the CRT trace?
c. What type of deflection system does an oscilloscope use?

2. a. What produces the visible trace on a CRT screen?
b. What might happen if electrons were allowed to accumulate on the screen?
c. How are these excess electrons removed?

3. a. Where does the voltage come from to normally sweep the electron beam across the CRT screen?
b. What would be the visual effect on a displayed waveform if the sweep frequency were reduced to a very low value?
c. When would you want the sweep frequency to be very high?
d. What display would occur for a sine wave input if the TIME/CM control were turned to external?

4. a. What type of probes are commonly available?
b. What advantages does a 10X probe provide?
c. How does it affect the determination of voltage and frequency?

5. a. Describe in your own words the oscilloscope technique to make a peak-to-peak voltage measurement that does not happen to be a whole number of centimeters in height.
b. Repeat part (a) for a frequency determination.
c. Why is it generally more convenient to obtain the peak voltage by first determining the peak-to-peak value?

6. a. Assume that the *same* sine wave is applied to both the vertical and horizontal plates of an oscilloscope. What kind of Lissajous figure will result?
b. Draw two sine waves, one under the other and in phase with each other. Assume that one is connected to the vertical, the other to the horizontal deflection plates. Also assume that positive voltages deflect the electron beam upwards or to the right; negative voltages deflect the beam downward or to the left. Determine what pattern results in a Lissajous figure.
c. Repeat part (b), assuming that one is displaced by 90° or a quarter of a cycle from the other.

7. a. Why does a PMMC movement indicate zero when a pure ac voltage is applied to it?
b. How may a PMMC movement be used to indicate ac values?
c. What effect does this have on the sensitivity of an ac voltmeter in a VOM?

8. a. Explain how an electrodynamometer can be used to measure power.
b. Why is the wattmeter's scale linear but an ammeter's scale nonlinear?

9. Why is it possible for a wattmeter to read backward when measuring ac power?

10. Why is there some slight error in a wattmeter reading according to the way the potential terminals are connected?

11. a. What is a common drawback that moving-iron and electrodynamometer voltmeters have compared with a PMMC voltmeter?
b. What feature do moving-iron and electrodynamometer meters have in common?

12. a. What problems can arise when making ac measurements with power-line-operated equipment in circuits that are also supplied from the 120-V, 60-Hz line?
b. If ac has "no polarity," why does it matter how a voltmeter is connected in a circuit?
c. Does this apply to portable meters such as a VOM?
d. If a circuit is battery-operated, is there any difficulty in using an oscilloscope without regard to grounds?
e. What problems may occur if a line-operated voltmeter and an oscilloscope are used simultaneously? Explain.

PROBLEMS

(Answers to odd-numbered problems at back of book)

16-1. Given a 5.3 cm peak-to-peak oscilloscope deflection, determine the peak voltages for the following VOLTS/CM settings assuming a 1X probe:

a. 10 mV/cm
b. 0.2 V/cm
c. 5 V/cm

16-2. Repeat Problem 16-1 using a 10X probe.

16-3. An oscilloscope displays five complete cycles in 8 cm. Determine the frequency for the following TIME/CM settings assuming a 10X probe:

a. 0.5 s/cm
b. 2 ms/cm
c. 0.1 ms/cm
d. 5 μs/cm

16-4. Repeat Problem 16-3 assuming three complete cycles in 7 cm.

16-5. The frequency of a voltage is being determined using the method of Lissajous figures as in Fig. 16-6. A stationary figure is obtained with five horizontal tangent points and three vertical tangent points when the horizontal frequency is adjusted to 360 Hz. What is the frequency of the unknown?

16-6. Repeat Problem 16-5 with the vertical and horizontal inputs interchanged.

16-7. A sinusoidal voltage having a peak-to-peak deflection of 4.6 cm at 0.2 V/cm is observed on an oscilloscope using a 10X probe across a 4.7-kΩ resistor. If one cycle occupies 2 cm at 0.2 ms/cm, determine:

a. The peak value of the current.
b. The frequency of the current.

c. The equation representing the voltage.

d. The equation representing the current.

16-8. A voltage having the equation $v = 20 \sin 2000\pi t$ volts is applied to a 5.1-kΩ resistor. Calculate:

a. The reading of an ac voltmeter connected across the resistor.

b. The reading of an oscilloscope connected across the resistor in volts.

c. The reading of an ac ammeter connected in series with the resistor.

d. The reading of a frequency meter, in the period mode, connected across the resistor.

16-9. An oscilloscope connected across a resistor indicates 18 V_{p-p} while a series-connected ammeter indicates 2.4 mA. What is the resistance of the resistor?

16-10. What peak-to-peak voltage does an oscilloscope indicate when connected across a 1.2-kΩ resistor when an ac ammeter indicates a resistor current of 10 mA?

16-11. Repeat Problem 16-8 with $v = 60 \sin 10{,}000\pi t$ and $R = 22$ kΩ.

16-12. An oscilloscope indicates 100 V p-p when connected across a resistor that dissipates 100 mW. What is the resistance of the resistor?

16-13. Refer to Fig. 16-15. Given: the oscilloscope across R_2 indicates 28.28 V_{p-p}; the ac ammeter indicates 3 mA; and the ac voltmeter across R_4 indicates 9 V, $R_1 = \frac{R_4}{2}$. Also, $R_3 = 10$ kΩ.

Calculate:

a. The reading of an ac voltmeter connected across V_T.

b. The total resistance of the circuit.

c. The resistance of R_2.

d. The resistance of R_4.

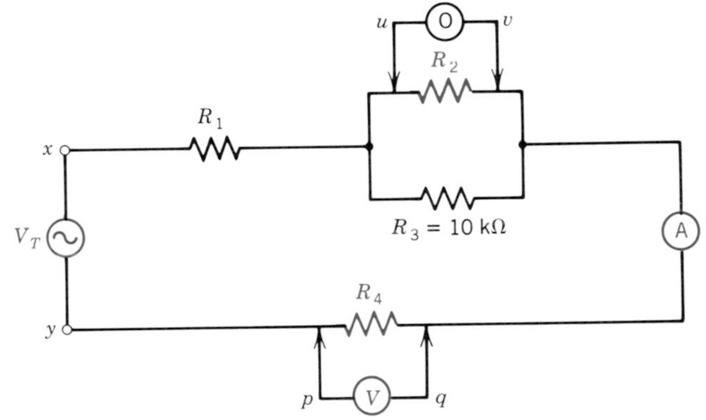

FIGURE 16-15
Circuit for Problems 16-13 through 16-16.

16-14. Repeat Problem 16-13 for $R_3 = 20$ kΩ and a voltmeter reading of 5 V.

16-15. Refer to Fig. 16-15. V_T is a 24-V source with point y grounded. Also, $R_1 = 2$ kΩ, $R_2 = 10$ kΩ, $R_3 = 10$ kΩ, and $R_4 = 5$ kΩ.

a. With neither oscilloscope nor voltmeter connected, what is the reading of the ac ammeter?

b. What are the readings of the ac ammeter and voltmeter when the voltmeter is connected across R_4 with side q grounded?

c. What are the readings of the ammeter, voltmeter, and oscilloscope when the oscilloscope is connected across R_2 with side v grounded?

16-16. Repeat Problem 16-15 with V_T as a 36-V ac source with point x grounded.

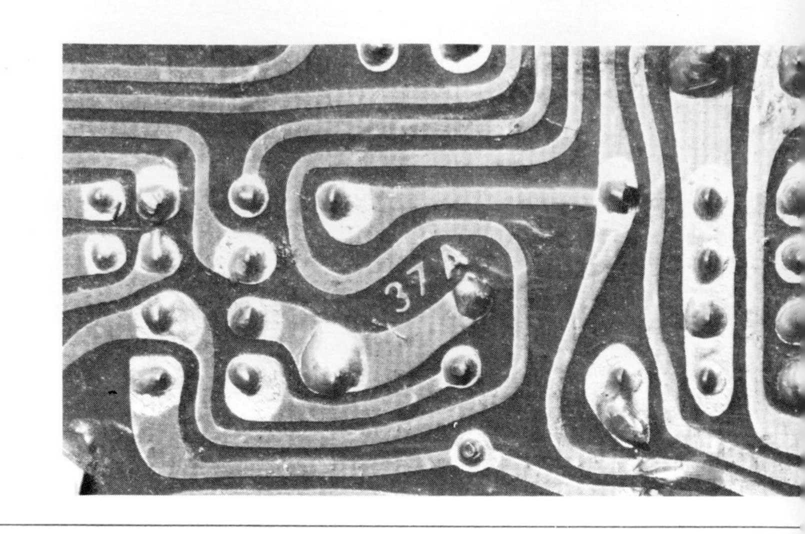

CHAPTER SEVENTEEN
INDUCTANCE

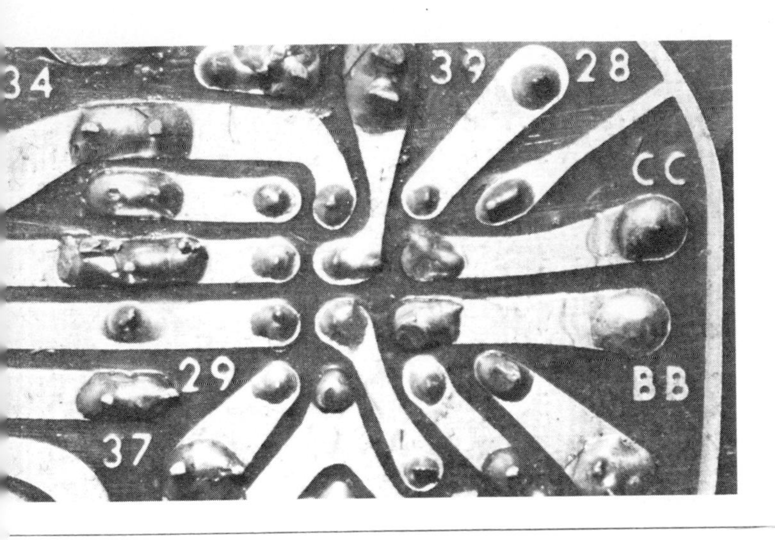

So far, we have considered only one property of an electric circuit: *resistance.* Resistance opposes the *steady* flow of electric current. We now examine a second property called *inductance.* Inductance is the property of a circuit or component to oppose any *change* in current. This property relies upon electromagnetic induction to induce an opposing emf whenever the current is changing. If this emf is induced in the same coil as the changing current, the effect is called *self-inductance.* If the emf is induced in a nearby coil, magnetically coupled to the coil with the changing current, the effect is called *mutual inductance.* This chapter examines the factors that affect inductance and the result of making series and parallel connections of inductors.

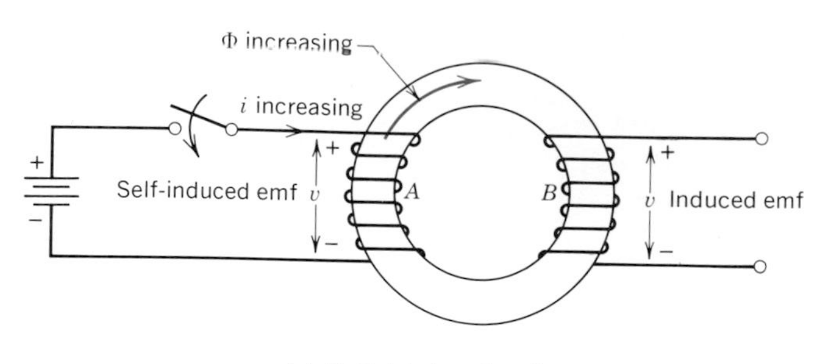

(a) Switch being closed

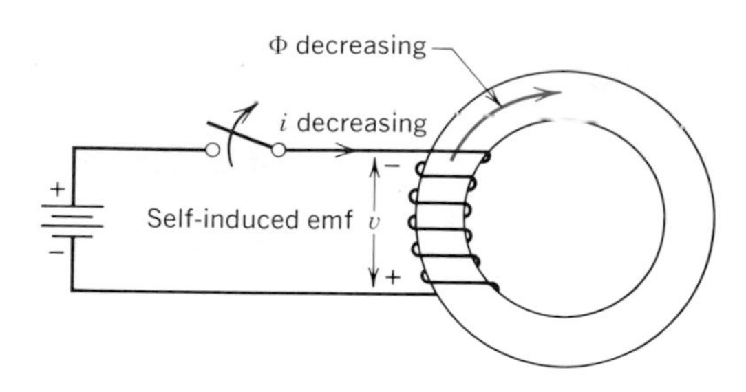

(b) Switch being opened

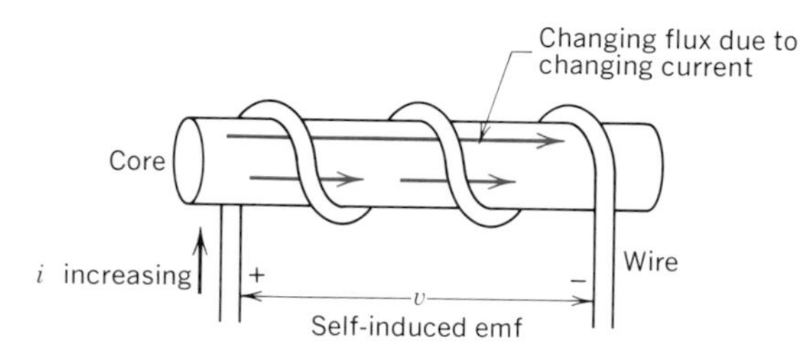

(c) Changing flux from one turn "cuts" adjacent turns in the coil

FIGURE 17-1
Self-induced emf due to inductance and changing current.

17-1 SELF-INDUCTANCE

Inductance is a property of a circuit or coil that arises from electromagnetic induction in the coil itself or in a neighboring coil. See Fig. 17-1*a*.

We saw in Section 14-3 how an emf is induced in coil *B*, due to a changing field *within* coil *B*, at the instant of closing the switch. But the conditions that induce an emf in coil *B* are also present in coil *A*. Thus a *self-induced* emf is developed in coil *A*, within the coil itself, opposing the *applied* emf.

Similarly, when the switch is opened, as in Fig. 17-1*b*, the decreasing current (and flux) in coil *A* self-induce an emf whose polarity tries to keep the current flowing. Once again we see Lenz's law at work.

Figure 17-1*c* shows how the expanding (or collapsing) magnetic field of one turn links and "cuts" adjacent turns in the coil. This is the source of the self-induced emf in a coil with a changing current.

The property of a coil to self-induce an emf when current through it is changing is called the coil's *self-inductance* or simply, *inductance*. The letter symbol for inductance is *L*; its basic unit is the henry, H.

Inductance (by Lenz's law) always causes an induced voltage that *opposes* the effect that created it. But it only does this when the current is *changing* and causing a change in magnetic flux. It is found that the self-induced emf is higher if the current is made to change at a higher rate. We can state these two effects in the following equation:

$$v = L\left(\frac{\Delta i}{\Delta t}\right) \quad \text{volts (V)} \qquad (17\text{-}1)$$

where: v is the emf induced in volts, V

L is the inductance in henrys, H

Δi is the change in current in amperes, A

Δt is the change in time in seconds, s

$\left(\frac{\Delta i}{\Delta t}\right)$ is the rate of change of current in amperes/second, A/s

The polarity of the induced emf is best determined from an application of Lenz's law.*

* Some texts use a negative sign in Eq. 17-1:

$$v = -L\,\frac{\Delta i}{\Delta t} \qquad (17\text{-}1a)$$

to signify that the voltage induced *in* the coil is a *counter* emf. We shall use Eq. 17-1 to give the voltage *across* the coil (either as an emf or voltage drop) and Lenz's law to obtain the polarity from the conditions that caused it.

EXAMPLE 17-1

The current in a 200-mH coil increases from 2 to 5 A in 0.1 s. Calculate:

a. The rate of change of current.

b. The self-induced emf.

SOLUTION

a. $\frac{\Delta i}{\Delta t} = \frac{5\text{ A} - 2\text{ A}}{0.1\text{ s}} = \mathbf{30\ A/s}$

b. $$v = L\left(\frac{\Delta i}{\Delta t}\right) \qquad (17\text{-}1)$$

$$= 200 \times 10^{-3}\text{ H} \times 30\text{ A/s} = \mathbf{6\ V}$$

We can rearrange Eq. 17-1 to obtain the definition for inductance in terms of the induced voltage and rate of change of current:

$$L = \frac{v}{\left(\frac{\Delta i}{\Delta t}\right)} \qquad (17\text{-}1b)$$

From Eq. 17-1b we can see that **An inductance of one henry may be defined as that inductance which, when current through it is changing at the rate of 1 A/s, induces an emf of 1 V.** See Fig. 17-2.

Note the schematic symbol used for inductance shown in Fig. 17-2. The two parallel lines next to the coil indicate a magnetic core (iron). Large values of inductance, 1 H and over, require a magnetic core. However, as explained in Section 17-2, the magnetic core causes the inductance to vary with the current through the coil.

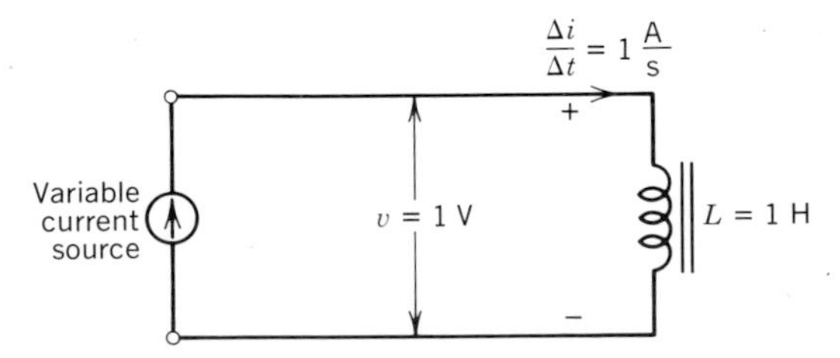

FIGURE 17-2
An inductance of 1 H induces an emf of 1 V due to a changing current of 1 A/s.

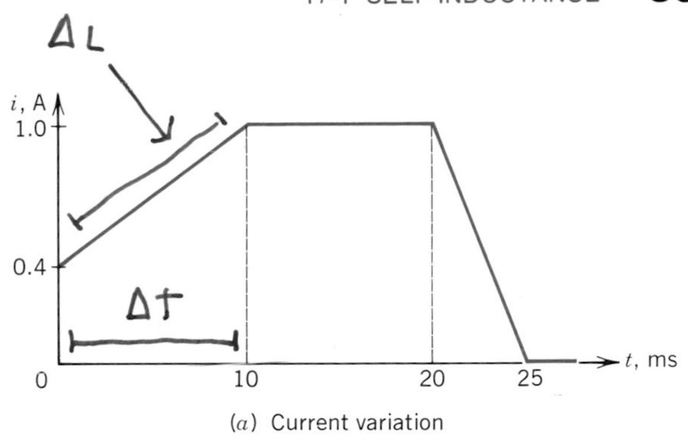

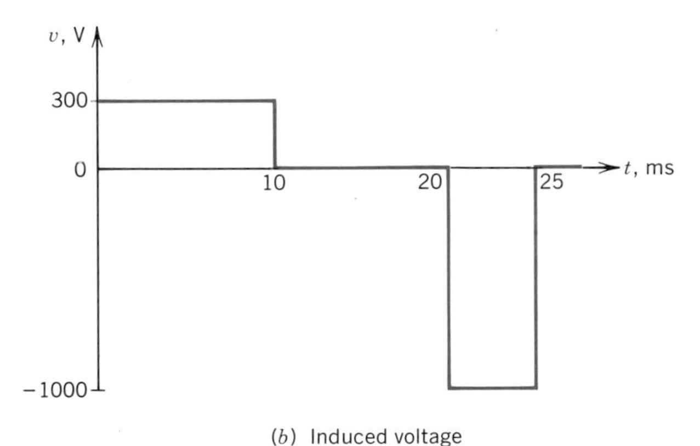

FIGURE 17-3
Waveforms for Example 17-2.

EXAMPLE 17-2

The current in a 5-H coil varies as shown in Fig. 17-3*a*. Draw the waveform of voltage induced in the coil.

SOLUTION

During the first 10 ms:

$$v = L\,\frac{\Delta i}{\Delta t} \qquad (17\text{-}1)$$

$$= 5\text{ H} \times \frac{(1.0 - 0.4)\text{ A}}{10 \times 10^{-3}\text{ s}}$$

$$= 5\text{ H} \times 60\text{ A/s} = \mathbf{300\ V}$$

Note that $\frac{\Delta i}{\Delta t}$ is constant at 60 A/s during this interval (0 to 10 ms), so the induced voltage is constant as shown in Fig. 17-3.

From 10 to 20 ms there is no change in current, so $\frac{\Delta i}{\Delta t} = 0$. As a result, the induced voltage is zero from 10 to 20 ms.

During the interval from 20 to 25 ms, the current is *decreasing* at a constant rate, where:

$$v = L \frac{\Delta i}{\Delta t} \quad (17\text{-}1)$$

$$= 5\ \text{H} \times \frac{(0 - 1)\ \text{A}}{(25 - 20) \times 10^{-3}\ \text{s}}$$

$$= 5\ \text{H} \times (-200\ \text{A/s}) = \mathbf{-1000\ V}$$

Since the current is *decreasing*, the induced voltage has the *opposite*, (negative) polarity, as shown in Fig. 17-3*b*, and is constant at −1000 volts.

17-2 FACTORS AFFECTING INDUCTANCE

The self-induced voltage in a coil with a changing current and flux was given earlier by Faraday's law:

$$v = N\left(\frac{\Delta\Phi}{\Delta t}\right) \quad (14\text{-}2)$$

But

$$\Phi = \frac{Ni}{R_m} \quad (11\text{-}8)$$

where Φ is the instantaneous value of flux due to the instantaneous value of the current i, and

$$R_m = \frac{l}{\mu A} \quad (11\text{-}9)$$

is the reluctance of the coil's magnetic circuit.

Thus

$$v = N \frac{\Delta}{\Delta t}\left(\frac{N\mu A i}{l}\right)$$

But $\frac{N\mu A}{l}$ are constant for a given coil, and so:

$$v = \frac{N^2 \mu A}{l}\left(\frac{\Delta i}{\Delta t}\right)$$

But

$$v = L\left(\frac{\Delta i}{\Delta t}\right) \quad (17\text{-}1)$$

and therefore,

$$L = \frac{N^2 \mu A}{l} \quad \text{henrys (H)} \quad (17\text{-}2)$$

where: L is the inductance of the coil in henrys, H (or Wb/At)

N is the number of turns in the coil (dimensionless)

μ is the permeability of the magnetic core in the coil in Wb/At-m

A is the cross-sectional area of the core in square meters, m^2

l is the length of the core in meters, m (length of coil for air-core coils)

Note that Eq. 17-1 is merely a restatement of Faraday's law, using current variation instead of flux change. Equation 17-2, therefore, assumes that *all the flux* set up by the coil links *all the turns* of the coil. This is reasonably accurate for iron-cored coils. For *long* air-cored coils, however, empirical equations found in handbooks must be used to take into account the *leakage of flux;* that is, some turns do not have the same flux linkage that others have. Air core coils of relatively short length, however, may be treated as shown in Example 17-3.

EXAMPLE 17-3

A coil of 50 turns is wrapped on a hollow nonmagnetic core of cross-sectional area 1 cm^2 and length 2 cm. Calculate:

a. The inductance of the air core coil $\left(\mu_o = 4\pi \times 10^{-7} \frac{\text{Wb}}{\text{At-m}}\right)$.

b. The new inductance when an iron core of *relative* permeability 200 is inserted in the hollow core.

c. The inductance if the number of turns on the iron core is doubled, with no other changes.

SOLUTION

a. $$L = \frac{N^2 \mu_o A}{l} \quad (17\text{-}2)$$

$$= \frac{50^2 \times 4\pi \times 10^{-7}\ \text{Wb/At-m} \times 1 \times 10^{-4}\ \text{m}^2}{2 \times 10^{-2}\ \text{m}}$$

$$= \mathbf{15.7\ \mu H}$$

b. $\mu = \mu_r \mu_o$ (11-5)

$= 200\ \mu_o$

$L = 200 \times 15.7\ \mu\text{H} = \mathbf{3.1\ mH}$

c. If the number of turns is doubled, L will quadruple

$L = 4 \times 3.1\ \text{mH} = \mathbf{12.4\ mH}$

In developing Eq. 17-2, we assumed that the permeability μ is constant. This is true for cores of coils made of nonmagnetic materials (Example 17-3a). But we showed (Section 11-4.4) that the permeability of iron, for example, varies widely with the amount of flux density and thus the current in the coil. This means that the inductance of any coil with a magnetic core is *never* constant. It varies with the amount of current (and flux) in the coil. This is why many coils have their inductance quoted at some specified value of current. For example, a coil may have 8 H of inductance at 85 mA, but 14 H if operated with only 1 mA of current.

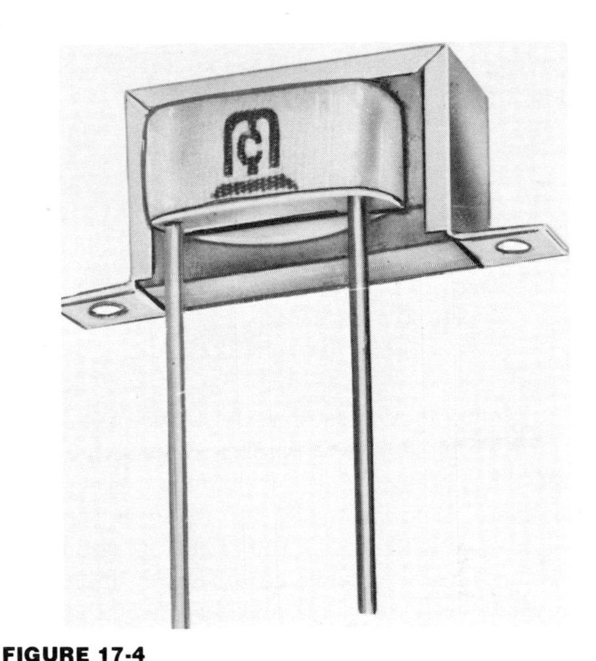

FIGURE 17-4

Typical power frequency filter coil. (Courtesy of Microtran Company, Inc.)

17-3 PRACTICAL INDUCTORS

Coils wound intentionally to produce a specified amount of inductance are known as *inductors,* or *chokes.* They range in value from a few microhenrys in high-frequency communication circuits through millihenrys and up to several henrys in power-supply filter circuits. Figure 17-4 shows a typical filter coil. (See Fig. 1-4*b* for other inductors.)

In some applications, a *variable* inductance is required. This is obtained by using a movable core or *slug* that can be screwed in or out of the coil. Schematic symbols for variable inductors are shown in Appendix A-1.

It should be noted that any conductor, even a straight wire (one turn), has some inductance. Wrapping a wire into a coil makes the flux set up by one turn more readily link the other turns, thus increasing the self-induced voltage and inductance. (See Fig. 17-1*c*).

17-4 INDUCTANCES IN SERIES AND PARALLEL

To obtain a desired value of inductance, some series or parallel combination of inductors can be used.

If the inductances are connected in series as in Fig. 17-5*a*, their self-induced voltages are *additive* and so are their inductive effects. Thus

$$L_T = L_1 + L_2 + L_3 + \cdots + L_n \qquad \text{henrys (H)} \qquad (17\text{-}3)$$

where L_T is the total inductance due to series-connected inductances having individual values of L_1, L_2, and so on.

L_1 L_2

$L_T = L_1 + L_2$

(*a*) Series connection of iron-core coils

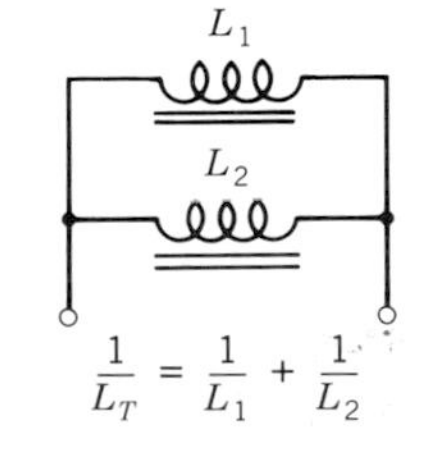

(*b*) Parallel connection of iron-core coils

FIGURE 17-5

Series and parallel-connected inductances.

When connected in parallel, as in Fig. 17-5*b*, the inductance total is given by

$$\frac{1}{L_T} = \frac{1}{L_1} + \frac{1}{L_2} + \frac{1}{L_3} + \cdots + \frac{1}{L_n} \quad \text{henrys (H)} \quad (17\text{-}4)$$

For two inductors in parallel:

$$L_T = \frac{L_1 \times L_2}{L_1 + L_2} \quad \text{henrys (H)} \quad (17\text{-}5)$$

Note that the above equations for inductors have the *same* form as for series and parallel-connected *resistors*. However, they apply only if the coils are well separated so that there is *no* interaction of magnetic fields (or *coupling*) between the coils. (The equations assume that there is no mutual inductance whatever; this occurs whenever two coils are at right angles to each other or far apart.)

EXAMPLE 17-4

Two inductors placed at right angles have values of 8 mH and 12 mH. Calculate the total inductance when they are connected in

a. Series.

b. Parallel.

SOLUTION

a. $L_T = L_1 + L_2$ (17-3)

$= 8 \text{ mH} + 12 \text{ mH}$

$= \mathbf{20\ mH}$

b. $L_T = \dfrac{L_1 \times L_2}{L_1 + L_2}$ (17-5)

$= \dfrac{8 \text{ mH} \times 12 \text{ mH}}{8 \text{ mH} + 12 \text{ mH}} = \dfrac{96}{20} \text{ mH} = \mathbf{4.8\ mH}$

17-5 MUTUAL INDUCTANCE

When two coils are close enough and parallel to each other (or on the same core) so that the field of one interacts with the other, there is an additional inductive effect. Figure 17-6 shows two coils that are not electrically connected but magnetically intercoupled. We have seen how a changing current i_1 not only self-induces an emf in L_1, but also causes a voltage to be induced in L_2. This is due to the changing flux Φ_1. But the voltage induced in L_2 causes a current i_2 that sets up its own changing flux Φ_2. This, in turn, not only self-induces an additional voltage in L_2, but also induces an additional voltage in L_1. That is, a changing current in one coil can induce an emf in the other coil due to a mutual flux. This effect is called mutual induction, and the two coils are said to have a mutual inductance, M, apart from their own self-inductances.

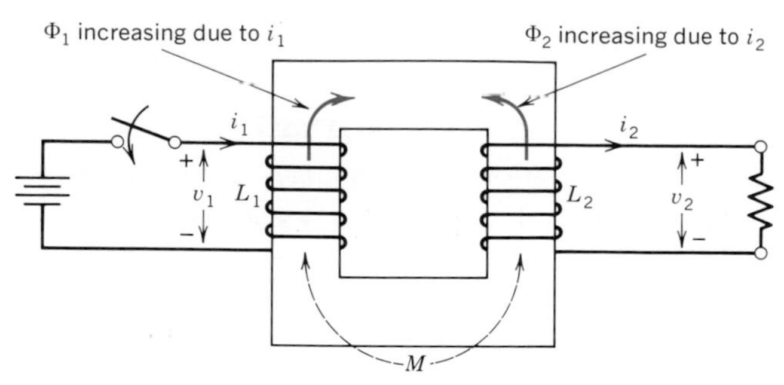

FIGURE 17-6
Mutual inductance M between two coils of self-inductance L_1 and L_2.

Mutual inductance, like self-inductance, is measured in units of henrys.

Two coils have a mutual inductance of 1 henry when a current changing at the rate of 1 A/s in one coil induces an emf of 1 V in the other coil.

Thus the mutually induced emfs in the circuit of Fig. 17-6 are given by

$$v_2 = M\frac{\Delta i_1}{\Delta t} \quad (17\text{-}6)$$

and

$$v_1 = M\frac{\Delta i_2}{\Delta t} \quad (17\text{-}7)$$

These are in addition to the self-induced emfs

$$v_1 = L_1\frac{\Delta i_1}{\Delta t} \quad (17\text{-}1)$$

and

$$v_2 = L_2\frac{\Delta i_2}{\Delta t} \quad (17\text{-}1)$$

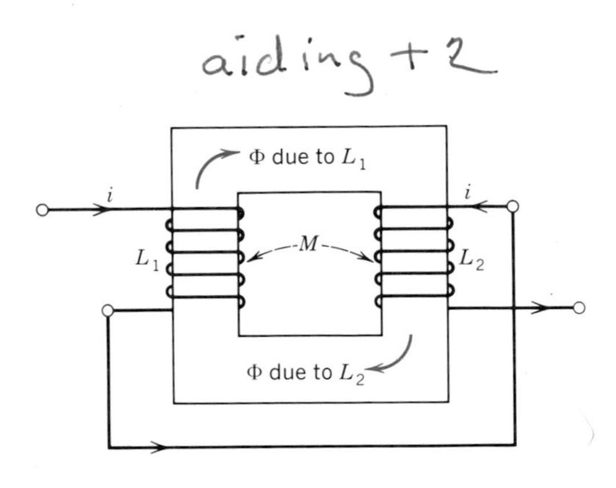

(a) Series-aiding coils set up flux in same direction
$L_T = L_1 + L_2 + 2M$

(b) Series-opposing coils set up flux in opposite directions
$L_T = L_1 + L_2 - 2M$

FIGURE 17-7
Two series-connected coils with mutual coupling.

17-5.1 Series-Connected Coils with Mutual Inductance

Now consider the total inductance when the two coils are series-connected, as in Fig. 17-7*a*.

The total emf induced in coil 1 is due to L_1 and M:

$$v_1 = L_1 \frac{\Delta i}{\Delta t} + M \frac{\Delta i}{\Delta t}$$

The total emf induced in coil 2 is due to L_2 and M:

$$v_2 = L_2 \frac{\Delta i}{\Delta t} + M \frac{\Delta i}{\Delta t}$$

The total emf in both coils $= v_1 + v_2$:

$$v_T = \frac{\Delta i}{\Delta t}(L_1 + M + L_2 + M)$$

$$= \frac{\Delta i}{\Delta t}(L_1 + L_2 + 2M)$$

$$= L_T \frac{\Delta i}{\Delta t}$$

where $L_T = L_1 + L_2 + 2M$ henrys (H) (17-8)

Thus the total inductance is increased above the self-inductances, $L_1 + L_2$, by an amount $2M$ because the coils set up a *series-aiding* flux.

When connected *series-opposing* as in Fig. 17-7*b*, the total inductance is

$$L_T = L_1 + L_2 - 2M \qquad \text{henrys (H)} \qquad (17\text{-}9)$$

In general, the total inductance of two series-connected coils, with mutual inductance M is

$$L_T = L_1 + L_2 \pm 2M \qquad \text{henrys (H)} \qquad (17\text{-}10)$$

where the *positive* sign is used for *series-aiding* coils, and the *negative* sign is used for *series-opposing* coils.

EXAMPLE 17-5

Two coils of inductance 5 and 15 H have a mutual inductance of 4 H. Determine the following inductances available from series-connecting the two coils.

a. The maximum inductance.
b. The minimum inductance.

SOLUTION

a. $L_{T\max} = L_1 + L_2 + 2M$ (17-8)
$= 5\text{ H} + 15\text{ H} + 2 \times 4\text{ H}$
$= 20\text{ H} + 8\text{ H} =$ **28 H**

b. $L_{T\min} = L_1 + L_2 - 2M$ (17-9)
$= 5\text{ H} + 15\text{ H} - 2 \times 4\text{ H}$
$= 20\text{ H} - 8\text{ H} =$ **12 H**

17-5.2 Dot Notation

Whether two coils are to be connected series-aiding or series-opposing is often indicated using a *dot notation,* shown in Fig. 17-8. When current enters both dots (or leaves both dots), the mutual effect is *additive,* as in Fig. 17-8*a*. But when the current enters one dot and leaves

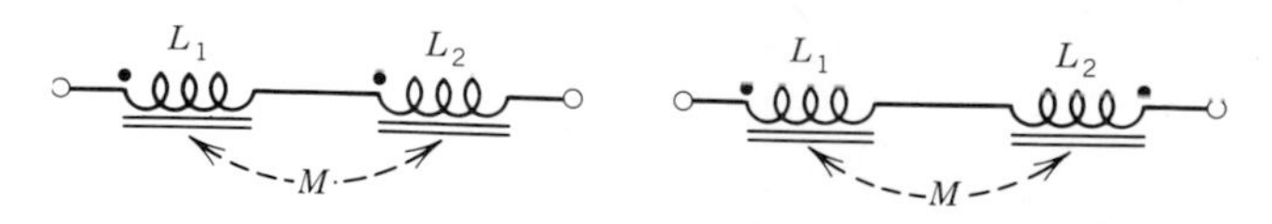

(*a*) Series-aiding coils, M positive $L_T = L_1 + L_2 + 2M$

(*b*) Series-opposing coils, M negative $L_T = L_1 + L_2 - 2M$

FIGURE 17-8
Dot notation for coils with mutual inductance.

the other, as in Fig. 17-8*b*, the mutual inductance is *subtractive*. This dot notation is also used in both transformers and loudspeakers to show correct *phasing*. For example, when paralleling two loudspeakers, we want the two cones to move in the same direction at the same time to provide a reinforcement of sound. Making a common connection to the two dotted inputs will make the two speakers work in phase with each other.

17-5.3 Coefficient of Coupling

The amount of mutual inductance between two coils depends upon the self-inductance of each coil and the amount of mutual flux between the two. The amount of mutual flux—the flux that links both coils—is determined by the physical placement of the two coils and is indicated by the coefficient of coupling, k.

$$k = \frac{\Phi_m}{\Phi} \quad \text{(dimensionless)} \qquad (17\text{-}11)$$

where: k is the coefficient of coupling between the two coils

Φ_m is the mutual flux between the two coils in webers, Wb

Φ is the total flux set up by one coil in webers, Wb

For example, where the coils are wrapped on the same iron toroid, as in Fig. 17-7, there is very little leakage flux and k is practically equal to 1. In Fig. 17-9*a*, if only 30% of the flux set up by coil 1 links coil 2, the coefficient of coupling is only 0.3. If the two coils are perpendicular to each other, as in Fig. 17-9*b*, the coupling is a minimum and close to zero.

It can be shown that the mutual inductance is given by

$$M = k\sqrt{L_1 \cdot L_2} \quad \textbf{henrys (H)} \qquad (17\text{-}12)$$

where: M is the mutual inductance in henrys, H

L_1 is the self-inductance of coil 1 in henrys, H

L_2 is the self-inductance of coil 2 in henrys, H

k is the coefficient of coupling, dimensionless

EXAMPLE 17-6

Two coils having inductances of 5 and 8 H are wound on the same core giving a coefficient of coupling of 0.8. Determine:

a. The mutual inductance between the two coils.
b. The maximum inductance if series-connected.
c. The minimum inductance if series-connected.

SOLUTION

a. $M = k\sqrt{L_1 \cdot L_2}$ (17-12)

$= 0.8 \times \sqrt{5\text{ H} \times 8\text{ H}}$

$= 0.8 \times \sqrt{40\text{ H}}$

$= \mathbf{5.1\ H}$

b. $L_{T\max} = L_1 + L_2 + 2M$ (17-8)

$= 5\text{ H} + 8\text{ H} + 2 \times 5.1\text{ H}$

$= \mathbf{23.2\ H}$

c. $L_{T\min} = L_1 + L_2 - 2M$ (17-9)

$= 5\text{ H} + 8\text{ H} - 2 \times 5.1\text{ H}$

$= \mathbf{2.8\ H}$

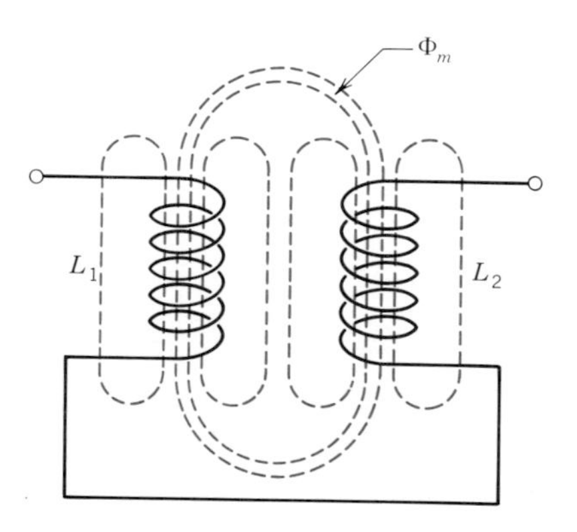

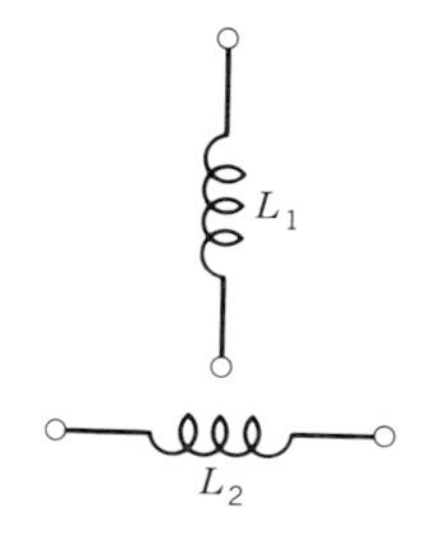

(*a*) Mutual flux, Φ_m between two coils (*b*) Minimum coupling

FIGURE 17-9
Coefficient of coupling between coils.

17-5.4 Measurement of Mutual Inductance

The mutual inductance and coefficient of coupling between two coils can be measured using a technique suggested by Eq. 17-10. If the total inductance is measured with the fields aiding, we obtain:

$$L_{T_a} = L_1 + L_2 + 2M \qquad (17\text{-}8)$$

and when opposing:

$$L_{T_o} = L_1 + L_2 - 2M \qquad (17\text{-}9)$$

Subtracting Eq. 17-9 from Eq. 17-8, we obtain

$$L_{T_a} - L_{T_o} = 4M$$

thus $M = \dfrac{L_{T_a} - L_{T_o}}{4}$ henrys (H) (17-13)

and $k = \dfrac{M}{\sqrt{L_1 \cdot L_2}}$ (dimensionless) (17-14)

where: L_{T_a} is the total inductance with the coils connected series-aiding in henrys, H

L_{T_o} is the total inductance with the coils connected series-opposing in henrys, H

An instrument used to measure inductance is described in Section 20-10. Example 17-7 shows how such an instrument can be used with Eqs. 17-13 and 17-14 to obtain the mutual inductance and coefficient of coupling between two coils.

EXAMPLE 17-7

A coil of 40 mH is placed parallel to a second coil of unknown inductance. When series-connected one way, a total inductance of 100 mH is measured; when the connections are interchanged a total inductance of 36 mH is measured. Calculate:

a. The mutual inductance.
b. The inductance of the second coil.
c. The coefficient of coupling.

SOLUTION

a. $M = \dfrac{L_{T_a} - L_{T_o}}{4}$ (17-13)

$= \dfrac{100 \text{ mH} - 36 \text{ mH}}{4} = \mathbf{16\ mH}$

b. $L_{T_a} = L_1 + L_2 + 2M$ (17-8)

Therefore,

$L_2 = L_{T_a} - L_1 - 2M$
$= 100 \text{ mH} - 40 \text{ mH} - 2 \times 16 \text{ mH}$
$= 60 \text{ mH} - 32 \text{ mH} = \mathbf{28\ mH}$

c. $k = \dfrac{M}{\sqrt{L_1 \cdot L_2}}$ (17-14)

$= \dfrac{16 \text{ mH}}{\sqrt{40 \text{ mH} \times 28 \text{ mH}}}$

$= \dfrac{16 \text{ mH}}{33.5 \text{ mH}} = \mathbf{0.48}$

EXAMPLE 17-8

Two *equal* coils are wrapped on the same core. When connected in series one way, a total inductance of 500 mH is measured; when the connections are interchanged, the total inductance is 50 mH. Calculate:

a. The mutual inductance.
b. The self-inductance of each coil.
c. The coefficient of coupling.

SOLUTION

a. $M = \dfrac{L_{T_a} - L_{T_o}}{4}$ (17-13)

$= \dfrac{500 \text{ mH} - 50 \text{ mH}}{4} = \mathbf{112.5\ mH}$

b. $L_{T_a} = L_1 + L_2 + 2M$ (17-8)

Therefore, $2L = L_{T_a} - 2M$
$= 500 \text{ mH} - 2 \times 112.5 \text{ mH}$
$= 275 \text{ mH}$

and $L = \dfrac{275 \text{ mH}}{2} = \mathbf{137.5\ mH}$

c. $k = \dfrac{M}{\sqrt{L_1 \cdot L_2}}$ (17-14)

$= \dfrac{112.5 \text{ mH}}{\sqrt{137.5 \text{ mH} \times 137.5 \text{ mH}}}$

$= \dfrac{112.5 \text{ mH}}{137.5 \text{ mH}} = \mathbf{0.82}$

Note from Examples 17-7 and 17-8 how two coils wrapped on the same core have a higher coefficient of coupling than when merely placed next to each other.

SUMMARY

1. Self-inductance is the property of a coil to induce an emf within the coil due to a changing current in the coil.

2. A coil's self-induced emf depends upon the inductance L in henrys and the rate of change of current $\frac{\Delta i}{\Delta t}$ in amperes per second as given by

$$v = L\left(\frac{\Delta i}{\Delta t}\right)$$

3. The polarity of the *induced* emf is always such that it opposes any change in the applied current.

4. A coil has a self-inductance of 1 henry when current changing at the rate of 1 A/s causes a self-induced emf of 1 V.

5. The inductance of a given coil depends upon the square of the number of turns in the coil, the permeability and area of the core, and the length of the magnetic circuit as given by:

$$L = \frac{N^2 \mu A}{l}$$

6. Variable inductors use a movable core to vary the permeability and hence the inductance of the coil.

7. The equations for inductances in series and parallel are the same as those for resistances if there is no magnetic interaction (mutual inductance) between the inductors.

8. Two coils have a mutual inductance M of 1 henry if the current changing at the rate of 1 A/s in one coil induces an emf of 1 V in the second coil.

9. When two coils of self-inductance L_1 and L_2 are series-connected with a mutual inductance M, the total inductance is

$$L_T = L_1 + L_2 \pm 2M$$

depending upon whether the mutual inductance of the coils is aiding or opposing.

10. The coefficient of coupling, k, between two coils, determines the amount of mutual inductance as given by

$$M = k\sqrt{L_1 L_2}$$

SELF-EXAMINATION

Answer T or F, or, for multiple choice, a, b, c, or d
(Answers at back of book)

17-1. Self-inductance always induces an emf that opposes the change in current that produced the emf. ________

17-2. A conductor must be wrapped into a coil before it can display the property of inductance. ________

17-3. If the current in a 50 mH coil changes from 10 to 20 mA in 5 ms, the induced emf is

a. 500 mV
b. 0.1 V
c. 10 mV
d. 100 V

17-4. If the number of turns in a coil is tripled, with all other factors constant, the inductance is

a. doubled
b. tripled
c. quadrupled
d. increased by a factor of nine.

17-5. The inductance of a coil is a constant, regardless of its application. ________

17-6. A variable inductor has a movable core that changes the permeability and therefore the inductance. ________

17-7. The same equations can be used for inductances in series and parallel as for resistors, in all applications. ________

17-8. Mutual inductance between coils arises because of an interaction of their magnetic fields. ________

17-9. Mutual inductance always causes an increase in the total inductance when two coils are series-connected. ________

17-10. Two 8-H coils with a mutual inductance of 8 H are connected in series. Their total maximum and minimum possible values are

a. 16 H and 0 H
b. 16 H and 8 H
c. 32 H and 0 H
d. 24 H and 0 H

17-11. The two coils in Question 17-10 must have a coefficient of coupling of:

a. 0
b. 0.5
c. 0.9
d. 1

REVIEW QUESTIONS

1. a. Why should a higher rate of change of current in a coil cause a higher emf to be induced?
b. What other factors will determine how much voltage is induced?
c. How would you determine the polarity of the induced emf?

2. Refer to Fig. 17-1*a*.

a. How will the voltages induced in coils *A* and *B* vary as the switch is closed and then opened?
b. What would have to be done to coil *B* to make its induced voltage have the opposite polarity when the switch is closed and opened compared with part (a)?

3. Define 1 H of self-inductance and 1 H of mutual inductance.

4. a. What causes the inductance of a coil to vary with the current?

b. What would you expect to happen to the inductance at saturation? Why?

c. Why should Eq. 17-2 not hold for long air-core coils?

5. Under what conditions do series- and parallel-connected inductors obey the same equations as those for resistors?

6. a. When two inductors that have mutual inductance are series-connected, how many sources of induced emf are there when the current through the coils changes?

b. What determines whether all these emfs are additive?

c. What is meant by the "dot notation"?

7. a. What is the highest coefficient of coupling that is possible between two coils?

b. How can this be achieved?

c. If the coefficient of coupling is unity, and the two coils have equal self-inductances, what can you say about the value of the mutual inductance?

d. What does this mean about the total inductance when the two coils are connected series-opposing?

PROBLEMS

(Answers to odd numbers at back of book)

17-1. The current in a 7-H coil increased from 0 to 1.4 A in 100 ms. Calculate:

a. The rate of change of current.

b. The self-induced emf.

17-2. The current in a 200-μH coil decreased from 12 to 7 mA in 10 μs. Calculate:

a. The rate of change of the current.

b. The self-induced emf.

17-3. A 50-mH coil of negligible resistance has a voltage across its terminals of 30 V. Calculate the rate at which the current is changing in the coil.

17-4. What is the inductance of a coil if an emf of 300 V is induced when the current changes from 100 to 101 A in 30 ms.

17-5. A coil of 100 turns is wrapped on a hollow nonmagnetic core of diameter 2 cm and length 3 cm. Calculate:

a. The inductance of the coil.

b. The new inductance when an iron core having a relative permeability of 180 is inserted in the hollow core.

17-6. An 8-H inductor is to be made by wrapping wire around a circular cross section iron toroid having inside and outside diameters of 5 and 8 cm, respectively. How many turns are required if the iron's relative permeability is 250?

17-7. The current in a 5-mH coil increases from 0.5 to 1.2 mA in 10 μs, remains constant at this value for 20 μs, and falls to 0 in 2 μs. Draw waveforms of current variation and induced emf in the coil showing all the numerical values.

17-8. A square wave of current of peak-to-peak value 20 mA passes through a 2-H coil. If the current takes 100 μs to change from one level to another and has a frequency of 1 kHz, draw the waveform of induced voltage across the coil as seen by an oscilloscope. Show representative time intervals on the horizontal axis as well as peak voltages on the vertical axis.

17-9. Three inductors have values of 20, 40, and 60 mH. Determine:

a. The total inductance when connected in series.

b. The total inductance when connected in parallel.

c. How they should be connected to give a total inductance of 44 mH.

17-10. Two inductors, with no mutual inductance, have a total inductance of 50 H when connected in series and 5 H when connected in parallel. What are the values of the two inductors?

17-11. a. What are the maximum and minimum possible values of total inductance when two coils of self-inductance, 7 H and 8 H, are series-connected with a mutual inductance of 3 H between them?

b. What is the coefficient of coupling between the two coils?

17-12. Two *equal* coils wound on the same core have a total inductance of 10 H when connected series-aiding and 5 H when connected series-opposing. Calculate:

a. The self-inductance of each coil.

b. The mutual-inductance between the coils.

c. The coefficient of coupling between the two coils.

17-13. If the coefficient of coupling between two coils of self-inductance, 150 μH and 200 μH, is 0.2, find the *maximum* inductance when these two coils are series-connected.

17-14. A coil is known to have twice the self-inductance of a second coil. When connected series-aiding, their total inductance is 100 mH; when connected series-opposing, the total inductance is 10 mH. Calculate:

a. The self-inductance of each coil.

b. The mutual inductance between the coils.

c. The coefficient of coupling between the two coils.

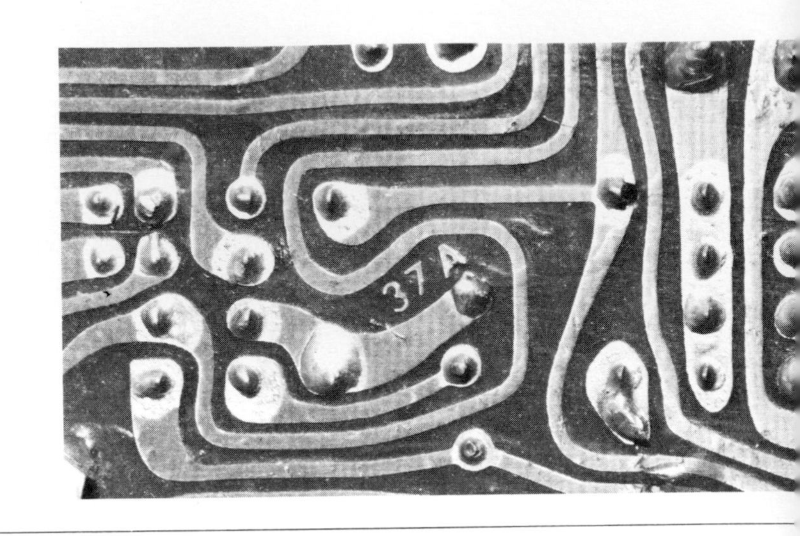

CHAPTER EIGHTEEN
TRANSFORMERS

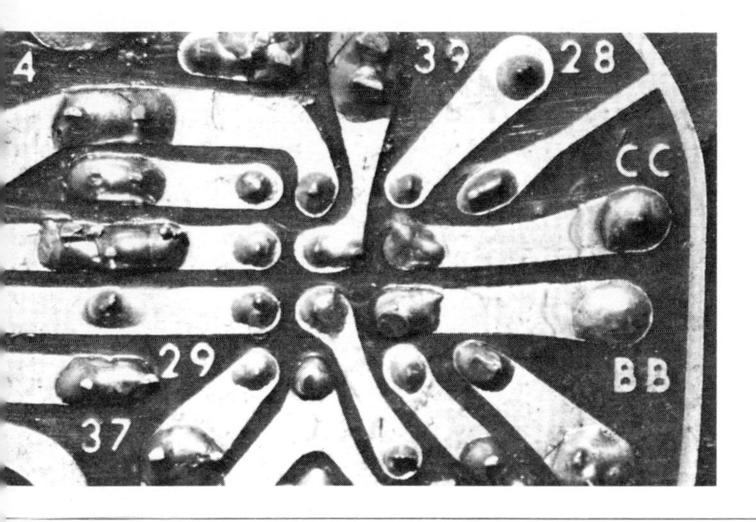

In Chapter 17 we saw that two coils wound on the same iron core have the property called mutual inductance. That is, a changing current in one coil can induce an emf in the second coil. An important application of mutual inductance is found in iron-core *power transformers.* These have the ability to step-up or step-down alternating voltages making possible the economic distribution of electric power between generator and customer. They can also transform load resistances to effect *matching* for maximum power transfer purposes.

A special transformer, with the output voltage the same as the input, is used for *isolation* purposes where supply voltage grounds can cause problems when working with grounded equipment. Another type is the *auto-transformer,* which is particularly adaptable to variable voltage operation. *Instrument* transformers, both the *potential* and *current* types, are designed especially for transforming the high voltage and current in distribution systems for the safe measurement of power. Finally, this chapter considers briefly a residential wiring system, the safety features of a ground system, and the operating principles of a ground-fault, circuit-interrupter.

18-1 THE IRON-CORE TRANSFORMER

We have seen how mutual inductance between two coils can cause an emf to be induced in one coil due to a changing current in the other coil. A transformer is a direct application of mutual inductance. Two coils are wound on the same magnetic core to provide a coefficient of coupling close to unity. An alternating current is applied to one coil, called the *primary* winding; the coil that is connected to the load is called the *secondary* winding. In general, each of these windings has a different number of turns. The ratio of the number of turns on the primary, N_p, to the number of turns on the secondary, N_s, is called the turns ratio, a.

$$a = \frac{N_p}{N_s} \quad \text{(dimensionless)} \qquad (18\text{-}1)$$

For example, if there are 200 primary turns and 50 secondary turns, a equals 4.

18-1.1 Voltage Ratio

Consider a sinusoidal voltage applied to the primary of a transformer as in Fig. 18-1. Since the input voltage is sinusoidal, the flux set up in the magnetic core is also sinusoidal. This means that the flux is always changing, setting up a sinusoidal voltage in the secondary by mutual induction. The primary and secondary voltages are given by Faraday's law:

$$V_p = N_p \frac{\Delta\Phi_p}{\Delta t} \quad \text{volts (V)} \qquad (18\text{-}2)$$

$$V_s = N_s \frac{\Delta\Phi_s}{\Delta t} \quad \text{volts (V)} \qquad (18\text{-}3)$$

In a practical transformer, the primary and secondary coils are wound on top of each other on the same iron core. (See Fig. 18-1c). This means that practically all of the flux produced by the primary links the secondary. Thus

$$\frac{\Delta\Phi_p}{\Delta t} = \frac{\Delta\Phi_s}{\Delta t}$$

in the above equations for V_p and V_s. If, for example, $N_s > N_p$, then $V_s > V_p$, and we see that a transformer can transform one voltage level to another.

Dividing Eq. 18-2 by Eq. 18-3 gives

$$\frac{V_p}{V_s} = \frac{N_p}{N_s} = a \qquad (18\text{-}4)$$

where all terms are as previously defined.

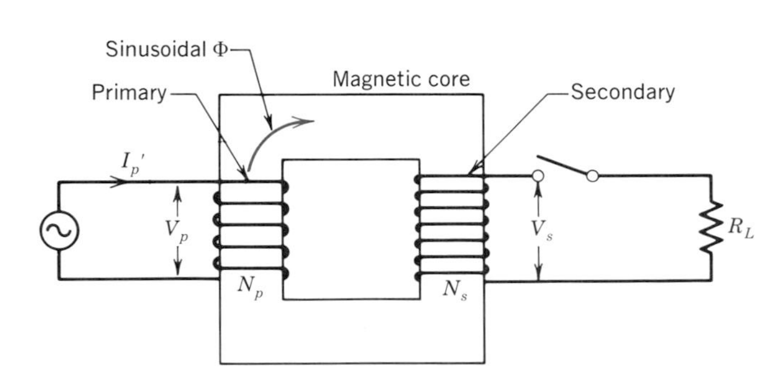

(a) Pictorial diagram showing common flux between the primary and secondary

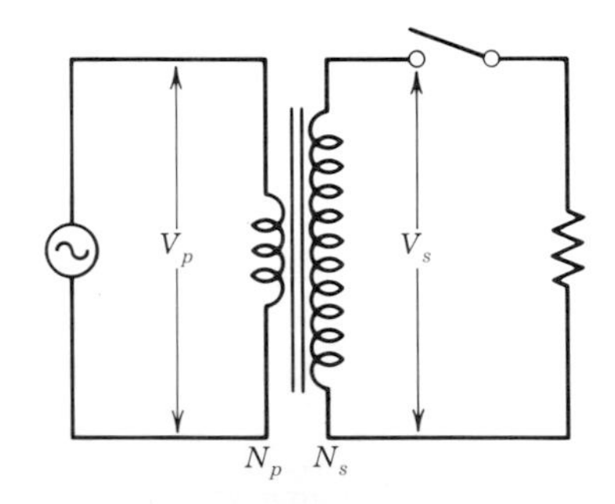

(b) Schematic symbol of a step-up transformer

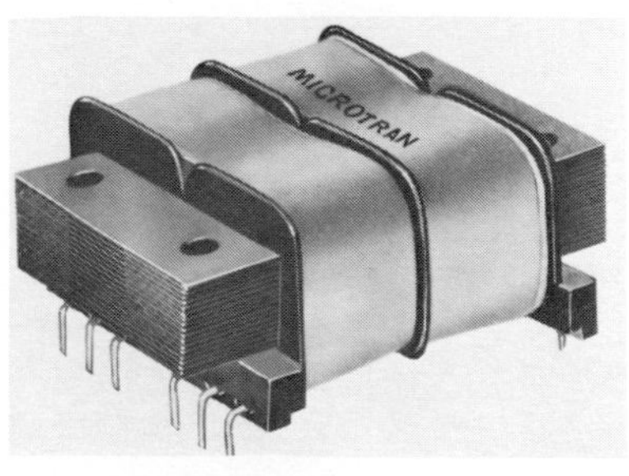

(c) Power line transformer for use in a low voltage dc power supply suitable for mounting on a printed circuit board

FIGURE 18-1
Iron core transformer. [Photograph in (c) courtesy of Microtran Company, Inc.]

Equation 18-4 is often called the *transformation ratio.* It states that the ratio of primary to secondary voltage is the same as the ratio of primary to secondary turns.

Clearly, from Eq. 18-4,

$$V_s = V_p \times \frac{N_s}{N_p}$$

Thus, if $N_s > N_p$, then $V_s > V_p$ and we have a *step-up* transformer. That is, a transformer that increases the output voltage compared to the input voltage is called a *step-up* transformer. It has a transformation ratio, *a, less* than unity.

If $N_s < N_p$, then $V_s < V_p$ and we have a *step-down* transformer. That is, a *step-down* transformer develops a lower output voltage than the applied input voltage. Its transformation ratio, *a,* is *greater* than unity.

Some transformers are designed to develop the *same* output voltage as the input. They are called *isolation* transformers. (An *autotransformer may* develop the same output voltage as the input, but it is used primarily to step voltages up or down. An autotransformer, because of a common electrical connection between input and output, does *not* provide isolation. See Section 18-5.)

EXAMPLE 18-1

A transformer with 50 primary turns and 300 secondary turns has a primary voltage of 120 V. Determine:

a. The transformation ratio.
b. The secondary voltage.
c. The type of transformer.

SOLUTION

a. $$a = \frac{N_p}{N_s} \qquad (18\text{-}1)$$

$$= \frac{50 \text{ turns}}{300 \text{ turns}} = \mathbf{0.167}$$

b. $$\frac{V_p}{V_s} = \frac{N_p}{N_s} \qquad (18\text{-}4)$$

$$V_s = V_p \times \frac{N_s}{N_p}$$

$$= 120 \text{ V} \times \frac{300 \text{ turns}}{50 \text{ turns}}$$

$$= 120 \text{ V} \times 6 = \mathbf{720 \text{ V}}$$

c. Since *a* is less than unity or since $V_s > V_p$, this is a step-up transformer.

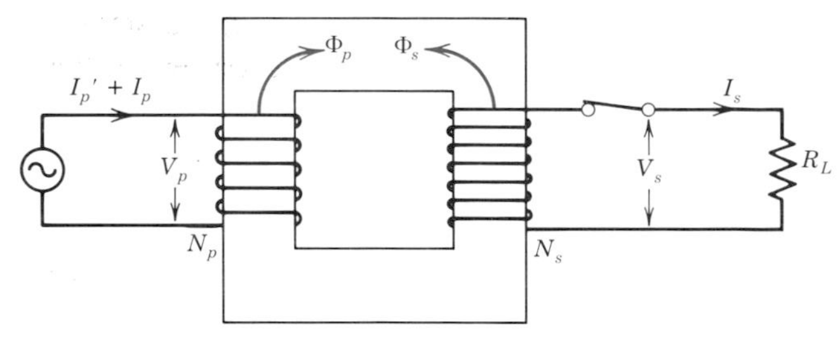

FIGURE 18-2
Transformer with a load sets up a demagnetizing force in the secondary winding.

It should be noted that the ability of a transformer to function depends upon the continual changing of the field $\left(\frac{\Delta\Phi}{\Delta t}\right)$ set up by the alternating current. It is for this reason that transformers cannot operate with pure dc.

18-1.2 Current Ratio

So far, we have not considered any load to be connected to the transformer secondary. Without load, the only current flowing in the primary, I_p', is called the *exciting* or *magnetizing* current. I_p' is quite small but sufficient to set up the changing flux in the core. Now consider what happens when a load is connected, as in Fig. 18-2.

A secondary current I_s flows, determined by V_s and R_L. But I_s must flow through N_s in such a way that, by Lenz's law, it sets up a flux that opposes the changing flux that induced it. This *demagnetizing* force produced by the load current is given by N_sI_s. In order to restore and maintain the flux in the core at its initial level and supply the load, an *additional* current I_p must flow in the primary. This magnetizing force N_pI_p must equal the demagnetizing force N_sI_s:

$$N_pI_p = N_sI_s$$

Thus $$\frac{I_s}{I_p} = \frac{N_p}{N_s} = a \qquad \text{(dimensionless)} \qquad (18\text{-}5)$$

where all terms are as already defined.

Equating the value of *a* in Eqs. 18-4 and 18-5, we see that the current ratio is the inverse of the voltage ratio:

$$\frac{I_s}{I_p} = \frac{V_p}{V_s} = a \qquad \textbf{(dimensionless)} \qquad \textbf{(18-5a)}$$

This implies, for example, that if a transformer steps-*up* a voltage by a given ratio, the current will be stepped-*down* by the same ratio.

EXAMPLE 18-2

A 120-V primary, 12.6-V secondary transformer is connected to a 10-Ω load. Calculate:

a. The turns ratio.

b. The secondary current.

c. The primary current.

SOLUTION

a. $$\frac{N_p}{N_s} = \frac{V_p}{V_s} \qquad (18\text{-}4)$$

$$= \frac{120\text{ V}}{12.6\text{ V}} = \mathbf{9.5}$$

b. $$I_s = \frac{V_s}{R_L} \qquad (3\text{-}1a)$$

$$= \frac{12.6\text{ V}}{10\ \Omega} = \mathbf{1.26\ A}$$

c. $$I_p = I_s \times \frac{V_s}{V_p} \qquad (18\text{-}5a)$$

$$= 1.26\text{ A} \times \frac{12.6\text{ V}}{120\text{ V}} = \mathbf{0.13\ A}$$

Note that the primary current calculated in Example 18-2c is actually the additional current over and above the exciting current I_p'. Therefore, since I_p' is generally only 5% of full-load current, it is small enough to be neglected except in miniature transformers.

18-1.3 Power, Volt-Ampere Relations, and Transformer Efficiency

Equation 18-5a leads directly to a consideration of the input and output power of the transformer.

Since $$\frac{I_s}{I_p} = \frac{V_p}{V_s} \qquad (18\text{-}5a)$$

then $$V_pI_p = V_sI_s \quad \text{volt-amperes (VA)} \qquad (18\text{-}5b)$$

This states that assuming a 100% efficient transformer, the product of the primary volts and amperes equals the product of the secondary volts and amperes. The output *rating* of a transformer is given in volt-amperes (VA) not watts. This is because the load may not be purely resistive, in which case, the current and voltage are *not* in phase with each other. Recall, from Section 15-3.1 that $P = V_RI_R$ watts. That is, only when the voltage and current are in phase with each other, as they are for a pure resistance, is the power in watts given by the product of volts and amperes.*

We shall assume, unless otherwise noted, that all loads are purely resistive. In this case, we can say that the input power V_pI_p equals the output power V_sI_s, for a 100% efficient transformer. That is,

$$P_{\text{in}} = V_pI_p \quad \text{watts} \qquad (18\text{-}6)$$

$$P_{\text{out}} = V_sI_s \quad \text{watts} \qquad (18\text{-}7)$$

The important point to realize is that a transformer, even a (voltage) step-up transformer, does not and cannot step-up power. An *increase in voltage* is accompanied by a similar *decrease* in *current* (or vice versa). At best, only as much power is delivered by the transformer secondary to the load as is drawn from the supply by the primary. The transformer is merely an *energy transfer* device, which transfers energy from the primary to the secondary circuit through magnetic coupling.

In a practical transformer, the output power is always slightly less than the input power because of losses in the transformer itself.

There are three sources of losses in a transformer:

1. *Winding I^2R losses.* The resistance of the primary and secondary coil windings means that some of the input power is converted internally into heat, as given by the equation $P = I^2R$. (These are commonly called copper losses.)

2. *Hysteresis losses.* The magnetic field in the iron core undergoes a complete reversal 60 times each second for power-line frequencies. This expends energy in the form of heat in reversing the magnetism of the

* In general, the product VI is called the *apparent* power in VA. Real or *true* power in watts is obtained by multiplying the apparent power by the *power factor.* The power factor is the cosine of the angle between voltage and current, $\cos\theta$. That is, $P = VI\cos\theta$. Since, for a pure resistor $\theta = 0°$, $\cos\theta = 1$, and the true power in watts is the same as the apparent power in volt amperes. (See Section 25-6.)

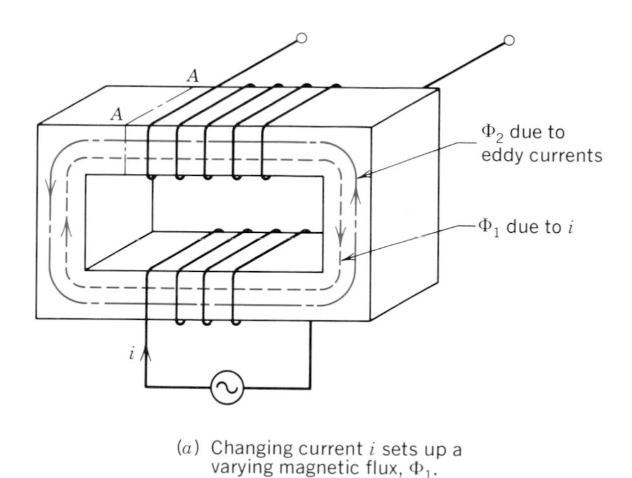

(*a*) Changing current i sets up a varying magnetic flux, Φ_1.

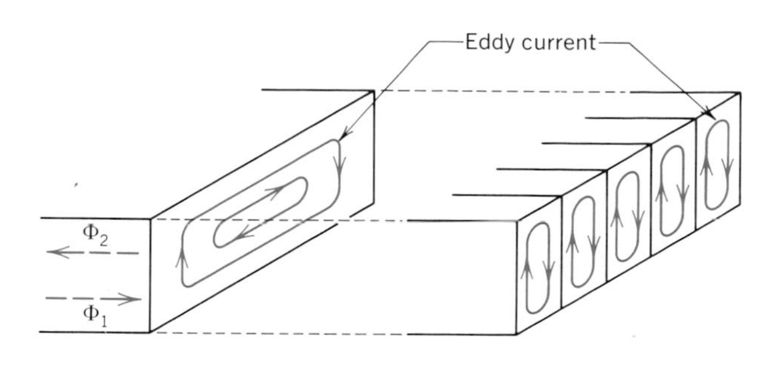

(*b*) Section *A-A* of a solid core showing circulating eddy currents that set up an opposing flux Φ_2 in the core.

(*c*) Section *A-A* of a laminated core showing how longer paths increase the resistance and reduce the eddy currents.

FIGURE 18-3
Eddy currents in solid and laminated cores.

iron core. The heat loss is proportional to the area of the magnetic material's *B-H* loop (Section 11-5).

3 *Eddy currents.* The changing magnetic field in the iron core induces an emf within the iron core, which itself is a conductor. This emf is responsible for setting up currents that circulate in the core perpendicular to the flux through the core. See Fig. 18-3*b*.

These circulating eddy currents generate heat in the core due to the resistance of the core ($P = I^2R$). This means that less power is available at the output of the transformer. In addition, the induced eddy currents set up an opposing flux (Φ_2) in the core. (See Fig. 18-3*a*.) This results in more current flowing in the primary trying to maintain the magnetic field in the core, and increasing losses further.

Note that the core does not have to be a magnetic material for the induction of eddy currents. Any good conductor, such as brass, will become very hot due to the eddy currents. (This is the principle behind an industrial heating method called *induction heating.*)

To reduce eddy current losses, the core is often made from *laminated* sheets of steel, each separated from the other by a thin coating of iron oxide and varnish. See Fig. 18-3*c*. Because of the high resistance of the varnish, the eddy currents must travel longer paths resulting in a greatly increased resistance. Although the induced emf in the core is unchanged, the higher resistance reduces the eddy currents considerably, and the power loss is greatly reduced.

Since the losses increase with frequency, transformers operating at high frequency, called radio frequency (RF) transformers, use a *ferrite* core. As explained in Section 11-7, ferrites are ferromagnetic but are relatively good electric insulators, reducing eddy currents and their heating effects to a minimum.

It is interesting to note that RF transformers may be enclosed in a copper or aluminum cover. This cover provides a *shield* between the *varying* magnetic flux set up in the transformer windings (by the RF current) and any nearby external circuits that may be adversely affected by this field. This is achieved by letting the eddy currents that circulate in the conducting covers set up an opposing magnetic field. These covers also shield the *interior* from the effect of *external,* varying magnetic fields.

Power transformers, on the other hand, usually have a cover that is a good *magnetic* material. This is because the shielding effect resulting from eddy currents is only effective at high frequencies. The magnetic material shields the transformer from external low-frequency and *static* magnetic fields, (as from wires carrying direct current), which would pass through an aluminum cover. See Fig. 18-4. The iron covers provide mechanical protection and also protect external circuits from the transformer's leakage flux.

Large power transformers can have efficiencies as high as 98 to 99.5%. However, the heat generated within the transformer is still appreciable and can be a problem if not removed. Very large transformers have a cooling

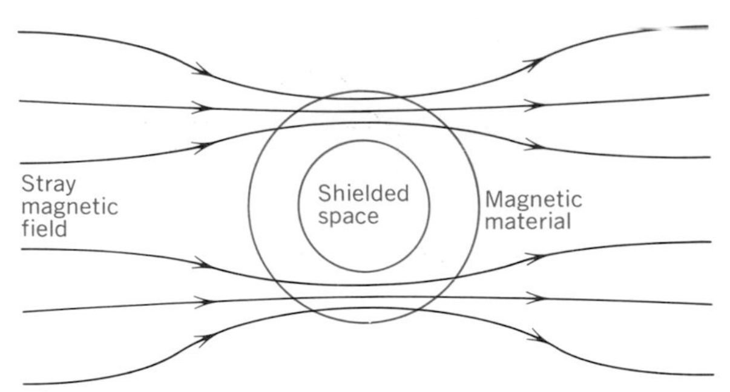

FIGURE 18-4
Magnetic shielding.

system in which nonelectrical-conductive oil is circulated through the windings and core to remove the heat. See Fig. 18-5. Note that

$$P_{in} = P_{out} + \text{losses} \qquad (18\text{-}8)$$

and

$$\text{efficiency } \eta = \frac{P_{out}}{P_{in}} = \frac{P_{out}}{P_{out} + \text{losses}} \qquad (18\text{-}9)$$

EXAMPLE 18-3

A 4160-V/230-V step-down transformer has a load of 4 Ω pure resistance connected to the secondary. Calculate the primary current assuming:

a. 100% efficiency.
b. 98% efficiency.
c. Calculate the losses in the case of part (b).
d. Determine the necessary output volt-ampere (VA) rating of the transformer.
e. On a circuit diagram show the conditions in part (b).

SOLUTION

a. Secondary current, $I_s = \frac{V_s}{R_L}$ (15-15a)

$$= \frac{230\text{ V}}{4\ \Omega} = 57.5\text{ A}$$

For 100% efficiency, $V_pI_p = V_sI_s$ (18-5b)

The primary current,

$$I_p = \frac{V_s}{V_p} \times I_s$$

$$= \frac{230\text{ V}}{4160\text{ V}} \times 57.5\text{ A} = \mathbf{3.18\ A}$$

b. $P_{out} = \frac{V_s^2}{R_L}$ (15-13)

$$= \frac{(230\text{ V})^2}{4\ \Omega} = 13.23\text{ kW}$$

$$\text{efficiency } \eta = \frac{P_{out}}{P_{in}} \qquad (3\text{-}8)$$

$$P_{in} = \frac{P_{out}}{\eta}$$

$$= \frac{13.23\text{ kW}}{0.98} = 13.5\text{ kW}$$

Since the load is pure resistance:

$$P_{in} = V_pI_p \qquad (18\text{-}6)$$

and

$$I_p = \frac{P_{in}}{V_p}$$

$$= \frac{13.5 \times 10^3\text{ W}}{4160\text{ V}} = \mathbf{3.25\ A}$$

c. Losses $= P_{in} - P_{out}$ (18-8)

$$= 13.5\text{ kW} - 13.23\text{ kW}$$

$$= 0.27\text{ kW} = \mathbf{270\ W}$$

FIGURE 18-5
A 15-kV, 3000-kVA, multiple-tapped oil-filled transformer used in an electrostatic precipitator system. (Courtesy of NWL Transformers.)

d. Transformer output rating (minimum)

$$= V_s I_s$$
$$= 230 \text{ V} \times 57.5 \text{ A}$$
$$= 13{,}225 \text{ VA}$$
$$= \mathbf{13.23\ kVA}$$

e. The conditions in part (b) are shown in Fig. 18-6.

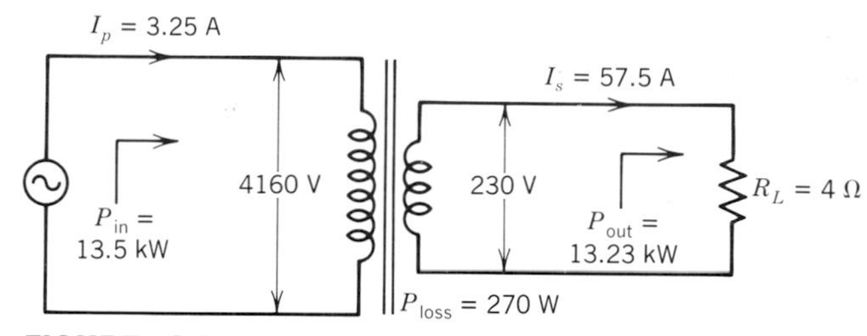

FIGURE 18-6
Circuit for Example 18-3.

18-1.4 Matching Resistances for Maximum Power Transfer

We know from Section 9-3 that maximum dc power is transferred from a dc source to a load if the load resistance, R_L, is equal to the internal resistance, r, of the source. This requirement must also be met to maximize the transfer of ac power for the special case where the source has *only* pure internal resistance and the load is also purely resistive. (The requirement for the more general case that includes inductive and capacitive effects is considered in Section 25-9.)We now consider how a transformer may be used to effect maximum power transfer if the source and load resistances are *not* equal. (Recall that this is only practical in electronic circuits and *not* power circuits.)

We have seen that when a load R_L is connected to the secondary of a transformer, an additional current I_p flows in the primary. See Fig. 18-7*a*.

As far as the input voltage V_p is concerned, it appears that a load R_p has been connected directly in parallel with the input. This load has a value of

$$R_p = \frac{V_p}{I_p} \tag{15-15}$$

since, for pure resistance, the primary voltage and current are in phase with each other.

But
$$V_p = aV_s \tag{18-4}$$

and
$$I_p = \frac{I_s}{a} \tag{18-5}$$

Thus
$$R_p = \frac{V_p}{I_p} = a^2 \frac{V_s}{I_s}$$

But, ignoring the resistance of the secondary winding itself,

$$\frac{V_s}{I_s} = R_L$$

Thus,
$$\mathbf{R_p = a^2 R_L} \quad \textbf{ohms } (\Omega) \tag{18-10}$$

where: R_L is the load resistance on the secondary in ohms, Ω

R_p is the resistance "reflected" into the primary in ohms, Ω

a is the turns ratio of the transformer, $\frac{N_p}{N_s}$, dimensionless

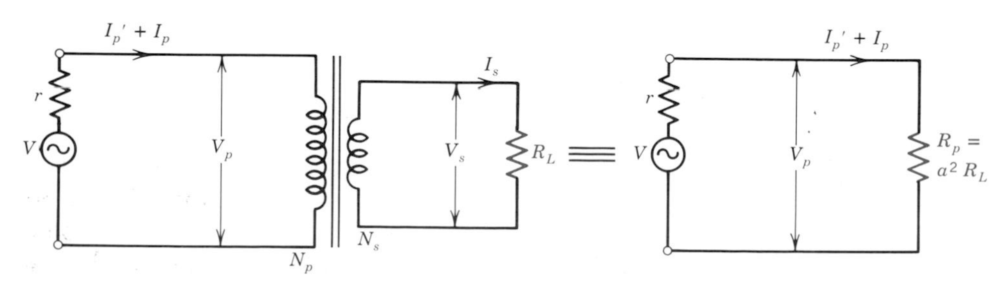

(*a*) Load R_L connected to the secondary

(*b*) Equivalent load a^2R_L connected to the primary

FIGURE 18-7
A load R_L connected to the secondary is reflected into the primary as an equivalent load of a^2R_L.

Equation 18-10 shows that the load R_L on the secondary has been "reflected" into the primary side as an equivalent load equal to a^2R_L. This means that a transformer can transform not only voltage and current, but also resistance.

A transformer of the proper turns ratio a can make a given load resistance R_L appear as a resistance equal to the source's internal resistance, r, as shown in Fig. 18-7b. Whenever $r = a^2R_L$, it will provide maximum power transfer into the transformer and ultimately to the load connected across the secondary. This is shown in Example 18-4.

EXAMPLE 18-4

An audio amplifier has an open-circuit voltage of 20 V and an internal resistance of 125 Ω. A transformer is to be selected to match a 4-Ω loudspeaker to the amplifier. Calculate, assuming 100% efficiency:

a. The necessary turns ratio of the transformer.
b. The primary input voltage and current.
c. The secondary output voltage and current.
d. The output power transferred to the loudspeaker by the matching transformer.
e. The power transferred to the loudspeaker if connected directly to the amplifier.

SOLUTION

a. For matching, $R_p = 125\ \Omega$

$$a^2 = \frac{R_p}{R_L} \qquad (18\text{-}10)$$

$$a^2 = \frac{125\ \Omega}{4\ \Omega} = 31.25$$

Turns ratio $a = \sqrt{31.25} = \mathbf{5.6}$

b. Refer to Fig. 18-7b. When

$$R_p = r$$

$$V_p = \frac{V}{2}$$

$$= \frac{20\ \text{V}}{2} = \mathbf{10\ V}$$

Note that this equal division of voltage only holds for pure resistances. It does *not* apply to the general case where reactances are involved.

$$I_p = \frac{V_p}{R_p} \qquad (15\text{-}15a)$$

$$= \frac{10\ \text{V}}{125\ \Omega} = 0.08\ \text{A} = \mathbf{80\ mA}$$

c. $$\frac{V_p}{V_s} = a \qquad (18\text{-}4)$$

$$V_s = \frac{V_p}{a}$$

$$= \frac{10\ \text{V}}{5.6} = \mathbf{1.8\ V}$$

$$\frac{I_s}{I_p} = a \qquad (18.5)$$

$$I_s = aI_p$$

$$= 5.6 \times 0.08\ \text{A} = \mathbf{0.45\ A}$$

d. $$P_L = I_s^2R_L \qquad (15\text{-}12)$$

$$= (0.45\ \text{A})^2 \times 4\ \Omega$$

$$= \mathbf{0.81\ W}$$

e. $$I = \frac{V}{r + R_L} \qquad (9\text{-}1)$$

$$= \frac{20\ \text{V}}{125\ \Omega + 4\ \Omega} = 0.16\ \text{A}$$

$$P_L = I^2R_L \qquad (15\text{-}12)$$

$$= (0.16\ \text{A})^2 \times 4\ \Omega = \mathbf{0.1\ W}$$

Note that eight times as much power is delivered to the load when the matching transformer is used compared with when the load is connected directly across the source. It should also be noted that matching transformers are generally specified in terms of the resistances that they are to match rather than their turns ratio. An audio output transformer, suitable for mounting on a printed circuit board, is shown in Fig. 18-8. Note the multiple taps available to provide matching for a wide range of loads.

18-2 POWER DISTRIBUTION SYSTEM

The main reason for generating alternating current (instead of direct current) is the ability of a transformer to step-up the voltage and distribute the power economically. (Although some transmission systems use dc, the process of stepping up direct voltage is not simple or economical, whereas it is relatively easy to step-up

FIGURE 18-8
Audio output transformer suitable for printed circuit board mounting. (Courtesy of Microtran Company, Inc.)

alternating voltage using transformers that are highly efficient.) If the alternator's output voltage of approximately 20 kV is increased by a factor of 30, the current is reduced by a factor of 30. This allows smaller-diameter transmission lines to be used and reduces the I^2R power losses that occur. For long-distance transmission lines, it is a general rule of thumb that a potential of 1 kV is necessary for each mile of power transmission. The highest voltage presently in use is 765 kV, but experiments are being conducted in the 1.5 to 2 million volt (megavolt) range.

Figure 18-9 shows that a power system consists of a generating system, a transmission system, and a distribution system. Three-phase transformers, with ratings as high as 1000 MVA, are used to feed the transmission line. At the substations, three single-phase transformers can be interconnected for three-phase operation. The step-up and step-down process is usually carried out in stages,

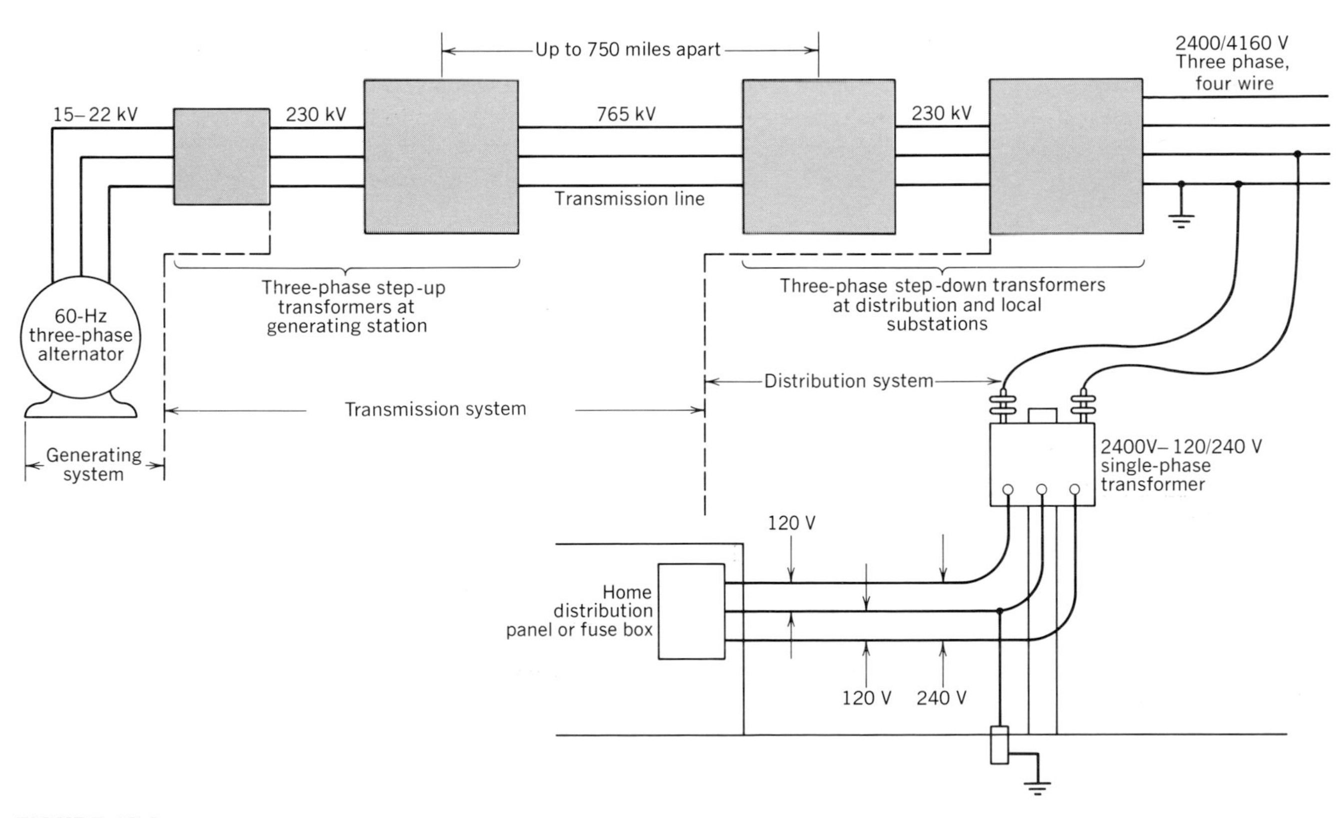

FIGURE 18-9
Use of step-up and step-down transformers in an electric power transmission and distribution system.

rather than in one step. Some industries may use high voltages directly, but residences are fed from a 120/240-V secondary, single-phase transformer.

18-3 RESIDENTIAL WIRING SYSTEM

Figure 18-10 shows a typical, modern residential wiring system. It is important to note that the local transformer (above or below ground) is single-phase with a *center-tapped* secondary that is grounded. This produces a 120/240-V, three-wire single-phase system. It is *not* a three-phase system. After passing through the watthour meter to show the energy used, the three wires feed the fuse box or circuit-breaker panel for the branch circuits. The grounded center tap is connected to the neutral and ground bus in the panel. It is also connected to the cold water pipe in the house to ensure a good earth-connected ground.

Note that some circuit breakers are *ganged* together in pairs to provide a 240-V circuit for large appliances like electric driers and ovens. Lights only require a 120-V "live" and a neutral, with the switch required (by National and local codes) to be in the live side. (See Fig. 18-10.)

Figure 18-11 shows how a light can be controlled from two locations using what are called *three-way* switches. These switches are actually of the single-pole, double-throw type. Thus, if in the circuit of Fig. 18-11, either S1 is thrown to position *A* or S2 is thrown to position *B*, the lamp will light. It is possible, by adding *four-way* switches between the two three-way switches, to control the light from any number of locations, as in a stairwell feeding many floors.

Receptacles require a third (grounding) wire that runs separately to the neutral or ground bus, for reasons explained in Section 18-3.1. (Some local codes may also require that a ground be run to *all* outlets, whether for lighting or receptacles.)

It should be noted that the size of wire to be used in wiring a home is determined by the load and the fuse or circuit-breaker rating. Thus a minimum of a No. 14 gauge should be used with a 15-A circuit breaker, a No. 12 gauge for 20 A, No. 10 for 30 A, and No. 8 for 40 A. (These are based on household types of wire commonly available.) In this way, a fuse will "blow" or a circuit breaker will "trip" if the current in the wire exceeds the safe current-carrying capacity of the wire, avoiding overheating that may lead to fire hazards.

It should also be noted that where local codes permit the use of aluminum wire, special care must be taken to

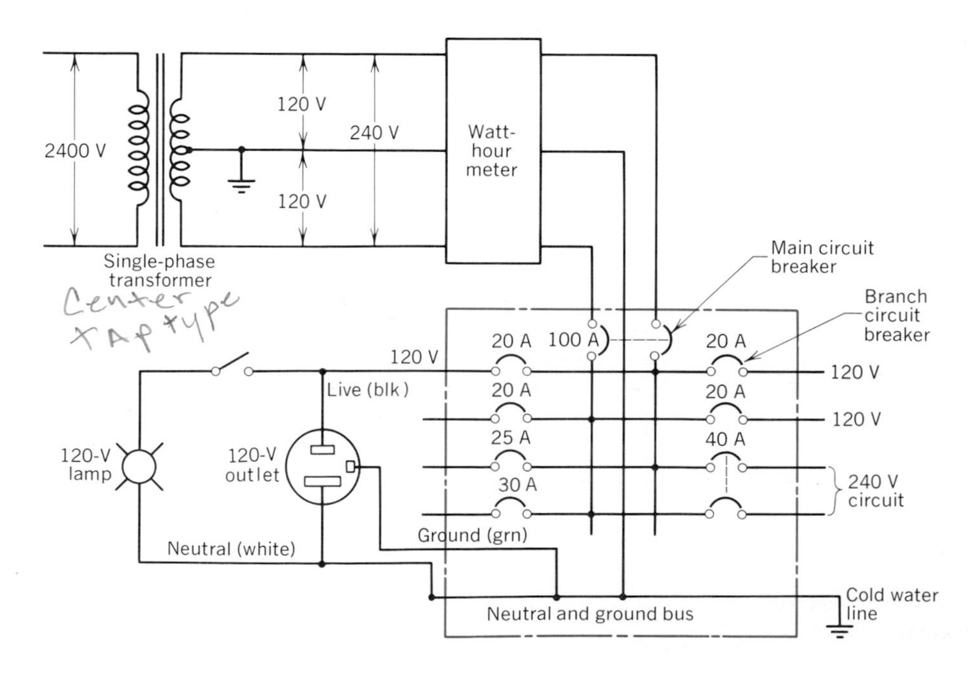

FIGURE 18-10
Residential wiring system.

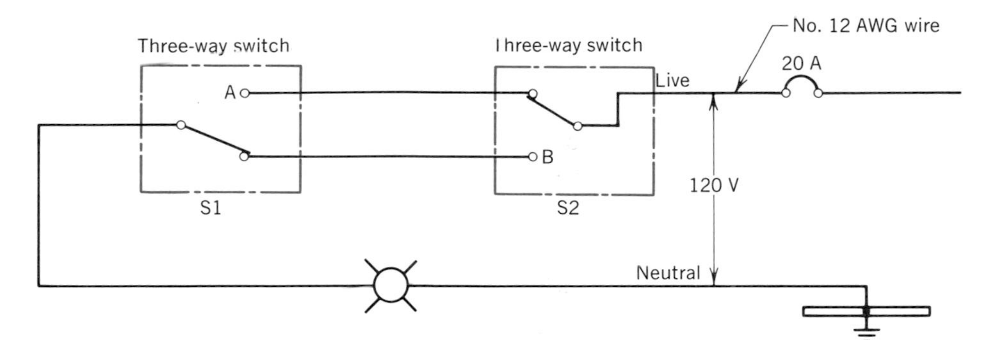

FIGURE 18-11
Three-way switched circuit that controls a light from two locations.

make sure connections are secure, since loose connections can develop enough heat to ignite the aluminum.

18-3.1 Safety Features of a Ground Connection

Consider an appliance, such as that in Fig. 18-12, fed from a 120-V receptacle using only two wires for the live and neutral. Assume that where the live wire passes through the metal case, a short develops due to a frayed wire. The whole metal case now becomes "hot" or "live." Someone touching the metal frame and ground (simultaneously) will now complete a path for current to flow to ground. This can easily occur in the case of a dishwasher or clothes washer or especially if there is a water pipe near the appliance. The person touching the "live" case and ground receives a shock that can be fatal.

Now consider that a separate wire is used to connect the ground in the receptacle to the metal frame of the appliance (the dotted line in Fig. 18-12). Should a short develop as before, a path is immediately available for current to flow and in most cases is enough to cause the circuit breaker to open, disconnecting the power. Even if the current were not sufficient to trip the circuit breaker, it is now impossible for the metal frame to be 120 V above ground. No dangerous potential now exists for anyone touching the metal frame.

18-3.2 Ground-Fault, Circuit-Interrupters (GFCIs)

The ground protection described above is not sufficient in some locations. For example, circuit breakers are thermally operated and may not open quickly enough if a fault occurs in equipment feeding a swimming pool where short circuit current is instantly fatal. In such

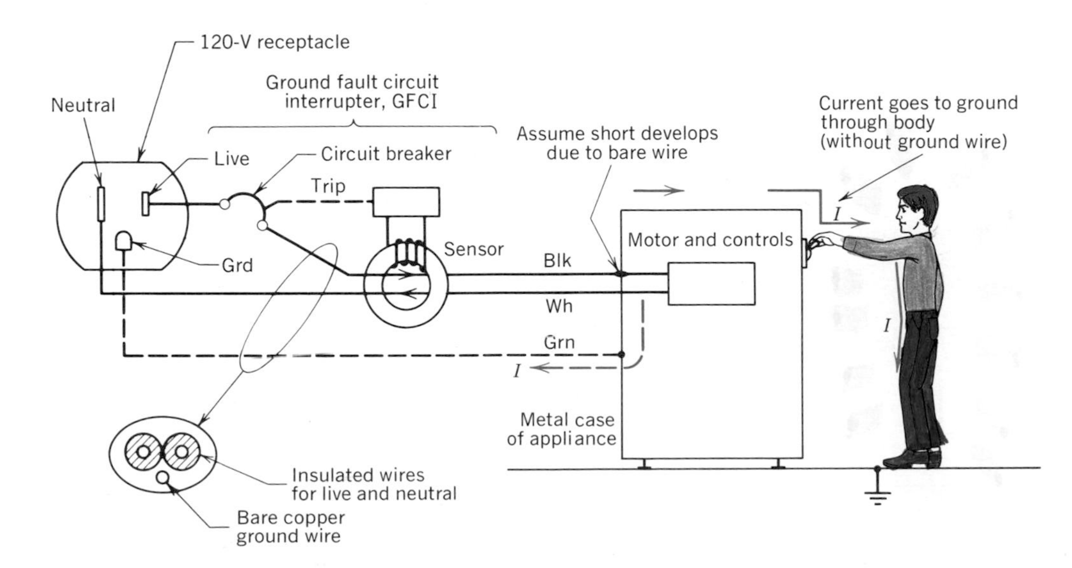

FIGURE 18-12
Current paths with and without ground-return wire.

locations, and where portable equipment is operated, the National Electrical Code requires the installation of ground-fault circuit-interrupters, GFCIs. As shown in Fig. 18-12, a toroidal sensing transformer has both the live and neutral wires passing through the center. Under normal operation with no fault, the two equal currents in opposite directions set up equal opposing magnetic fields. If a fault occurs, as above, current returns by an alternate path. The imbalance of currents through the toroid develops a voltage in the coil of the sensor which opens the contacts. A differential of only 5 mA causes the current to be interrupted in just 25 ms.

18-4 ISOLATION TRANSFORMERS

In Section 16-6, we saw that precautions must be taken when using grounded equipment. This is especially important if measurements are being made in a circuit to be operated from the 120-V, 60-Hz supply. Figure 18-13*a* shows that a short circuit will occur if the equipment ground of the measuring instrument is connected to the "live" side of the supply. A large current can flow through the low-resistance path and can cause damage to the circuit and measuring instrument. This problem is avoided by using an *isolation transformer.* As shown in Fig. 18-13*b*, this takes 120 V at the input and develops 120 V at the output. There is no *electrical* connection, however, between the input and output. Thus the secondary side is not grounded on either side. This means that the equipment ground of an instrument on the secondary side can provide no low-resistance return path to the ungrounded side of the primary. However, if two grounded instruments are used *simultaneously* on the secondary side, as in Fig. 18-13*c*, their common grounds can cause a short circuit on the transformer secondary. It is then necessary to make sure that the instrument grounds are connected to each other.

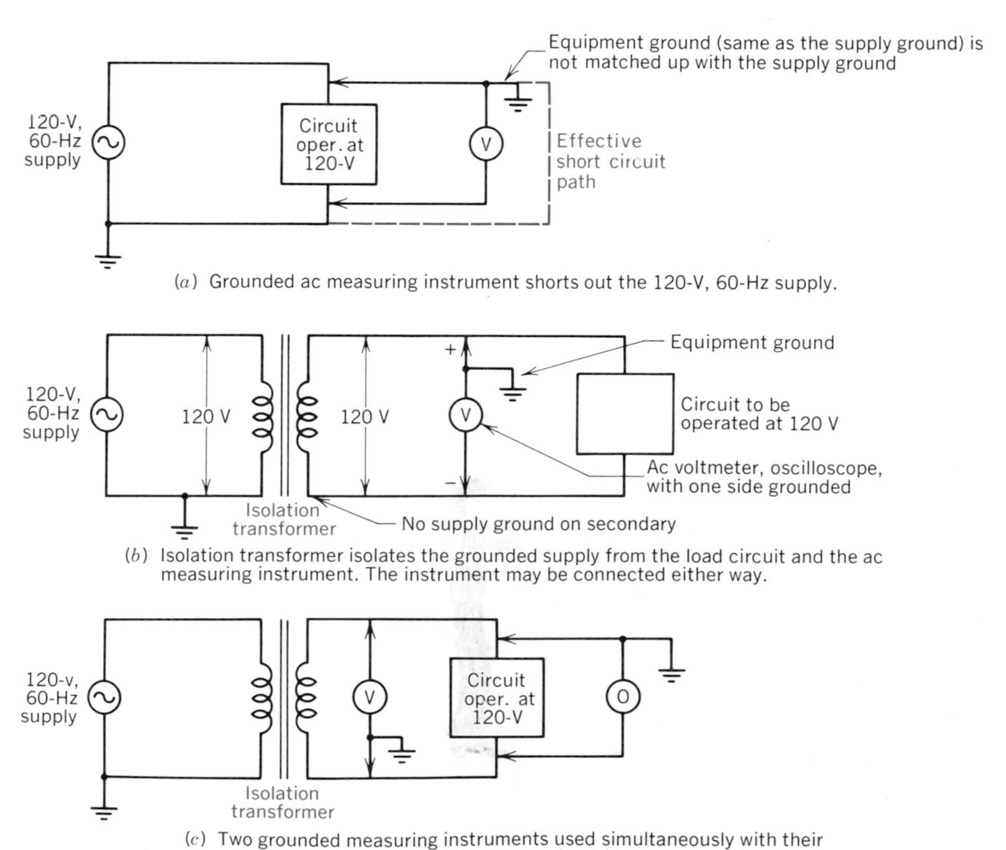

(*a*) Grounded ac measuring instrument shorts out the 120-V, 60-Hz supply.

(*b*) Isolation transformer isolates the grounded supply from the load circuit and the ac measuring instrument. The instrument may be connected either way.

(*c*) Two grounded measuring instruments used simultaneously with their grounds not matched cause the secondary side to be shorted.

FIGURE 18-13
Use of an isolation transformer and precautions with grounded equipment.

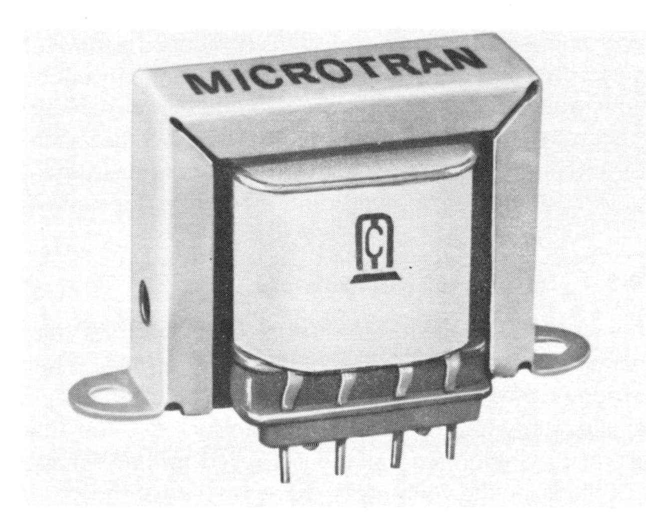

FIGURE 18-14
Isolation transformer suitable for printed circuit board mounting. (Courtesy of Microtran Company, Inc.)

It should be noted that the isolation transformer can be used either to operate the circuit under measurement or the measuring instrument. In either event, the transformer volt-ampere rating must be checked to see that it can handle the load. Figure 18-14 shows a low-powered isolation transformer, suitable for mounting on a printed circuit board.

18-5 AUTOTRANSFORMERS

Where isolation between input and output is unimportant, an autotransformer can be used with considerable savings over a conventional transformer. Figure 18-15 shows that this transformer has a common electrical connection between input and output. It is also very adaptable to variable voltage operation by using a sliding contact like a potentiometer. However, an autotransformer does *not* function as a simple voltage divider.

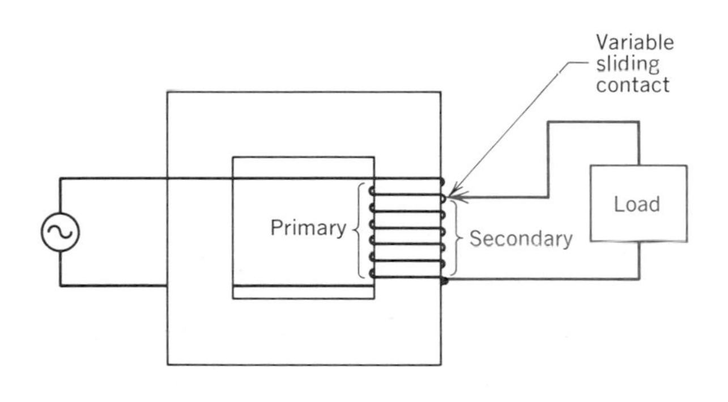

(*a*) Arrangement of windings on core.

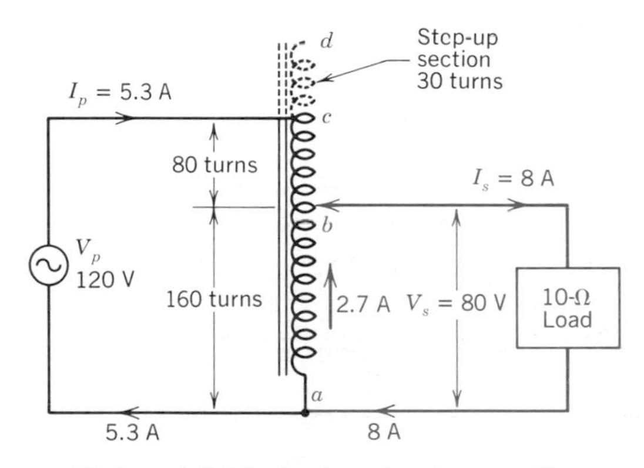

(*b*) Current distribution for a step-down condition

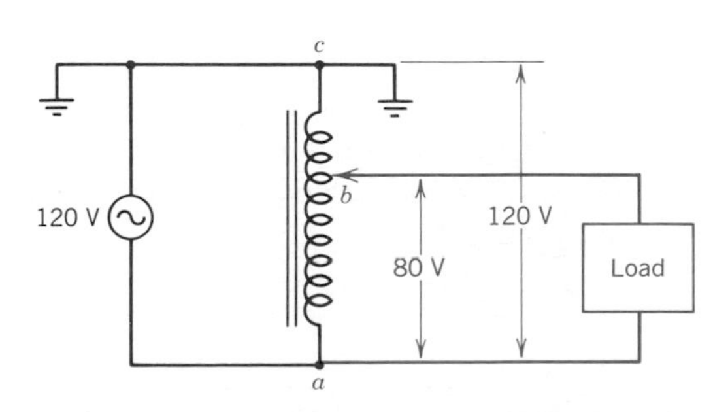

(*c*) Reversing the input voltage connections makes output terminal *a* 120 V above ground at all times.

(*d*) A variable autotransformer.

FIGURE 18-15
Variable autotransformer. [Photograph in (*d*) courtesy of Staco Energy Products Co.]

Assume an autotransformer with a total of 240 turns from point a to c in Fig. 18-15b. With a primary voltage of 120 V, 80 V will be across the 160 turns from point a to b and 40 V across the 80 turn-section from point b to c. A secondary output voltage of 80 V across a 10-Ω pure resistance load causes a secondary current of 8 A. The power delivered to the load is 80 V × 8 A = 640 W.

If we neglect transformer losses, the input power from the 120-V source must also be 640 W, requiring a primary input current of 640 W/120 V or 5.3 A. Applying Kirchhoff's current law at point a shows that 8 A-5.3 A or 2.7 A must flow from a to b.

The actual power transformed by the 160-turn secondary is 80 V × 2.7 A or 216 W, compared with a total load power of 640 W, or only 33.8%. Much of the load power (424 W) is conducted directly from the input through the 80-turn section of the primary winding. Consequently, an autotransformer is smaller and uses less iron than a conventional two-winding transformer of the same rating.

Some autotransformers, used for variable voltage operation (sometimes referred to by the trade name of Variac®) incorporate a small step-up section, as shown dotted in Fig. 18-15b. This enables the transformer to develop a variable voltage output from 0 to 135 V from a 120-V input supply.

It should be noted that if a two-prong, unpolarized plug is used to connect the autotransformer to the 120-V supply, the grounded neutral may be connected to point c instead of being connected to point a, as shown in Fig. 18-15c. In this case, the output voltage between a and b may be only 80 V, but the output terminal a will be at 120 V with respect to ground (point c).

A variable autotransformer is shown in Fig. 18-15d. This 14 × 15 × 8-in variable transformer can deliver a maximum of 8.4 kVA and 60 A at 0– 140 V from a 120-V input.

It should be noted that *variable* transformers are now available which have separate primary and secondary windings, with no electrical connection between the two, thus providing isolation. However, their current and volt-ampere ratings are approximately one-half of the values that are in effect when operating in a nonisolated mode.

18-6 INSTRUMENT TRANSFORMERS

The measurement of voltage, current, and power in high-voltage ac power circuits is not generally made by the direct connection of measuring instruments. *Instrument transformers* are used to drop the voltage and current by known ratios so that standard ac voltmeters and ammeters can be used. This also reduces the shock hazard to instruments and operators. The two kinds of instrument transformer are the *potential* and *current* transformer.

18-6.1 Potential Transformers

These are small transformers (from 50 to 600 VA) that operate on the principle of power transformers. They step-down the voltage connected across the primary by a definite ratio. They are designed to produce 120 V on the secondary when rated voltage is applied to the primary. By multiplying the voltmeter reading on the secondary by the known ratio, the high voltage on the primary is obtained.

18-6.2 Current Transformers

Current transformers are designed to be connected in *series* with the line carrying the high current to be measured. (Some current transformers now use the pass-through, clamp-on method, as discussed in Section 18-7.) In order to keep the resistance inserted in the line as low as possible, only a few turns are used in the primary. Since the current must be stepped-down, the number of secondary turns must be greater than the primary. The turns ratio is usually selected so that 5 A will flow in the secondary when rated current flows in the primary.

Since the connection of an ammeter on the secondary provides a very low resistance, very little voltage appears across the secondary in spite of the transformer being of a voltage step-up type. However, if the current transformer is operated on open circuit, there is no secondary demagnetizing mmf set up so that the flux density in the core becomes very high. This can cause dangerously high voltages to be induced in the secondary on open circuit. The secondary should always be short-circuited before removing the ammeter or making any other changes in the secondary circuit.

EXAMPLE 18-5

Determine the load current, voltage, and power for the given instrument readings and the instrument transformer ratios shown in Fig. 18-16.

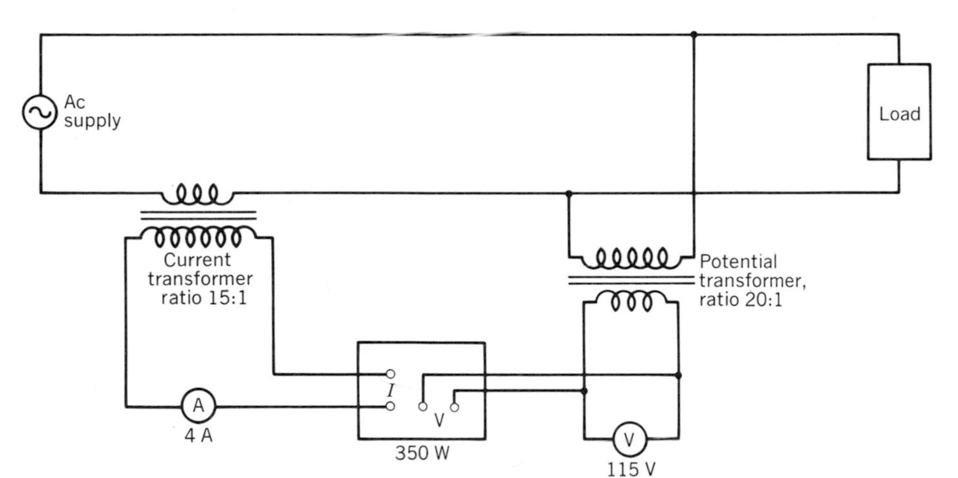

FIGURE 18-16
Instrument transformers to measure the line current, voltage, and power for Example 18-5.

SOLUTION

load current = ammeter reading × ratio

$= 4\text{ A} \times 15 =$ **60 A**

load voltage = voltmeter reading × ratio

$= 115\text{ V} \times 20 =$ **2300 V**

load power = wattmeter reading × ammeter ratio × voltmeter ratio

$= 350\text{ W} \times 15 \times 20$

$= 105{,}000\text{ W} =$ **105 kW**

Note how the load power is *not* the product of the load voltage and load current. This means that the load is not purely resistive, so the current and voltage are not in phase with each other (i.e., the power factor, cos θ, is less than unity).

A current transformer is also used in ac milliammeters that use a PMMC type of meter movement. As explained in Section 16-2, diodes are used to convert the alternating current to be measured to a proportional direct current to operate the dc meter movement. However, the voltage drop across the diodes would cause an excessive voltage drop across the meter. The current transformer, with just a few turns in the primary, introduces a very small voltage drop. Figure 18-17 shows how taps on the primary winding can be used to give a multirange ac milliammeter. Diodes connected to the secondary in a bridge configuration rectify the alternating current to operate the PMMC meter movement.

Note how fewer turns are used on the primary for the 1000 mA range than for any other current range. This makes the turns ratio, a, *smaller* and causes a *greater* step-down of current.

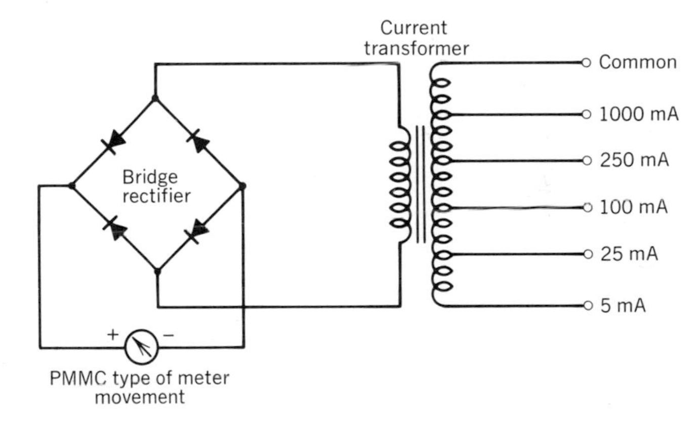

(*a*) Current transformer and bridge rectifier

(*b*) Ac milliammeter

FIGURE 18-17
Use of a current transformer to operate a PMMC meter movement in an ac milliammeter. [Photograph in (*b*) courtesy of Hickok Teaching Systems Inc., Woburn, Mass.]

FIGURE 18-18
Digital clamp-on ac ammeter, voltmeter, and ohmmeter. (Courtesy of TIF Instruments, Inc.)

18-7 CLAMP-ON AC AMMETER

The ac clamp-on ammeter shown in Figure 18-18 also operates on the principle of a current transformer. A wire carrying an alternating current, when placed inside the jaws, functions as a single-turn primary. The current in the wire sets up an alternating magnetic field in the iron core (jaws) and induces current in a secondary winding. This secondary current, proportional to the primary current, is used (after suitable processing) to operate an LED display. Because of the *very* small turns ratio, current up to 1000 A can be measured. By using suitable probes, this multimeter instrument can also measure ac volts and ohms.

Note, however, that this type of instrument is not suitable for measuring direct current. Although similar in appearance to the clamp-on ammeter in Figure 12-7, which uses a Hall effect device and *can* measure direct current (and alternating current), the ac clamp-on ammeter is essentially a transformer and requires an alternating field for its operation.

18-8 CHECKING A TRANSFORMER

If a transformer fails to develop an output voltage, either an open primary or secondary could be responsible. If the primary is open, no current flows in either the primary or secondary. If the secondary is open, only a very small magnetizing current flows in the primary. If a transformer has multiple secondary windings (see Fig. 18-19), an open state in one will not affect the operation of the others. An ohmmeter will show an infinite reading when connected across the open winding.

If some of the windings on the secondary short to each other, or to the core, the additional load will draw an

FIGURE 18-19
A small power transformer that has a center-tapped high voltage secondary on one side (next to the 117-V primary terminals), and low voltage secondary outputs on the other side. (Note the iron cover as discussed in Section 18-1.3.)

increased primary current, often burning out the primary winding. Seldom is it practical to repair open or shorted windings, making it necessary to replace the whole transformer. One word of caution. Before deciding to replace a transformer because of a very low ohmmeter reading across a winding, remember that the dc resistance of most windings is very low. This is especially true for power transformers, where the windings that carry large currents have wire with large cross-sectional area, producing very little dc resistance.

SUMMARY

1. The primary winding of a transformer is connected to the source; the secondary winding is connected to the load. The two are magnetically coupled to each other on the same core.

2. The turns ratio, voltage ratio, and current ratio are related by

$$a = \frac{N_p}{N_s} = \frac{V_p}{V_s} = \frac{I_s}{I_p}$$

3. A step-up transformer increases output voltage compared with the input but decreases the output current by the same ratio. The output power can never be greater than the input power.

4. A transformer can be used to match resistances for maximum power transfer because a load resistance R_L is reflected into the primary as a load resistance R_p given by

$$R_p = a^2 R_L$$

5. Transformers depend upon the continual change in magnetic field set up by the alternating current in the primary winding. They cannot operate from pure direct current.

6. A power distribution system may step up generator outputs to as much as 765 kV for transmission lines to carry the power to distant customers.

7. A residential wiring system is fed from a single-phase, center-tapped secondary transformer to provide a 120/240-V, three wire system with a grounded neutral.

8. A third-wire grounded receptacle allows metal frames of appliances and power tools to be grounded, avoiding shock hazards in the event of a short between the "live" wire and the frame.

9. A ground-fault circuit interrupter (GFI) is used with a 120-V outlet to sense when an imbalance in neutral return current requires the interrupting of the supply as a safety measure.

10. An isolation transformer has two separate windings with the output voltage equal to the input but electrically isolated from any ground on the primary.

11. An autotransformer shares a single-tapped coil between the input and output to transform voltages using a much smaller core than a conventional two-winding transformer.

12. Instrument transformers step down voltage or current by known ratios so that standard meters can be used to make ac measurements.

SELF-EXAMINATION

Answer true or false, or, for multiple choice, a, b, c, or d
(Answers at back of book)

18-1. An iron-core transformer is an application of mutual inductance in which the coefficient of coupling is close to unity. ______

18-2. A transformer will not operate with pure direct current applied because there is no continual changing flux set up to induce a secondary voltage. ________

18-3. The voltage and current ratios have the same value as the turns ratio. ________

18-4. A 1 : 5 step-up transformer with a primary voltage of 120 V has a pure resistance load of 300 Ω. The primary current and secondary voltage are

a. 10 A and 600 V

b. 0.4 A and 600 V

c. 0.4 A and 24 V

d. 10 A and 24 V

18-5. Assuming an 80% efficiency, the input power to the transformer in Question 18-4 is

a. 6 kW

b. 7 kW

c. 7.5 kW

d. 8 kW

18-6. The volt-ampere rating of a transformer is the same as the power rating in watts. ________

18-7. A transformer is used to match an 8-Ω load to a 288-Ω source. The necessary turns ratio is

a. 36

b. 6

c. 8

d. 288

18-8. A power distribution system needs step-up transformers to allow smaller diameter conductors to be used for the transmission line. ________

18-9. A residential 120/240-V, three-wire system is an example of a three-phase system. ________

18-10. A variable autotransformer can be used as an isolation transformer simply by adjusting the output voltage to be the same as the input voltage. ________

18-11. When an isolation transformer is used, there is never any problem with using grounded measuring equipment on the secondary side. ________

18-12. A GFI relies upon equal but opposite currents flowing through a toroidal transformer to indicate when a fault occurs. ________

18-13. A current instrument transformer is connected across the line like a potential transformer. ________

REVIEW QUESTIONS

1. a. If a transformer has a 120-V primary and a 6.3-V secondary, could it be operated in reverse to produce 120 V with a 6.3-V input?

b. What would determine the maximum input voltage with this mode of operation? (Could you apply 120 V to the 6.3-V winding?)

2. What general principle can you use to justify that in a step-up transformer the current must be reduced by at least the same factor?

3. If a transformer cannot operate on pure direct current, how is it possible for an automotive coil (transformer) to induce 25 kV on the secondary from a 12-V dc supply?

4. a. What are the factors contributing to less than 100% operating efficiency in a transformer?
   b. How are these effects minimized?
5. a. What part does a transformer play in providing maximum power transfer?
   b. How does it do this?
6. a. Under what special condition is the volt-ampere rating of a transformer the same as the power rating?
   b. Why is this not *generally* the case?
7. a. What maximum distance could power be economically transmitted if transmission lines were operated at 2 million V?
   b. Why is the step-down process in substations accomplished in stages rather than by dropping immediately to 120/240 V?
8. a. How is it possible for the 120/240 V residential system to be single phase with three wires?
   b. Why is the neutral grounded?
   c. In a two-prong receptacle with no separate ground why, if the neutral is grounded, is the neutral not tied to the metal frame of the appliance?
9. a. Describe the operation of a ground-fault circuit interrupter.
   b. Where is a GFCI likely to be used? Why?
10. a. Explain how an isolation transformer can be used to avoid dangerous short circuits when using grounded test equipment on line-operated circuits.
    b. What precautions must still be observed if more than one grounded instrument is being used?
11. a. Why is an autotransformer smaller than a conventional two-winding transformer for a given load?
    b. What is a Variac?
    c. Can it be used to step-up the voltage?
    d. What is the disadvantage of an autotransformer?
12. a. How does a current transformer differ from a potential transformer?
    b. What precaution should be taken when removing an ammeter from the secondary of a current transformer?
    c. Why is this necessary?

PROBLEMS

(Answers to odd-numbered problems at back of book)
(Unless otherwise stated, neglect transformer losses.)

18-1. A transformer with 200 primary turns and 65 secondary turns has a primary voltage of 50 V. What is the secondary voltage?

18-2. A transformer with a turns ratio of 0.4 has a secondary voltage of 230 V. What is the primary voltage?

18-3. A 4160/230-V single-phase distribution transformer has an output rating of 7.5 kVA. Calculate:

a. The maximum current that can safely be drawn from the secondary.
b. The minimum resistance that can be connected to the secondary.

18-4. A 2300/120-V single-phase transformer is connected to an effective load resistance of 2 Ω. Calculate:

a. The minimum output rating of the transformer.

b. The input power if the transformer is 92% efficient.

18-5. A 230-V primary, 24-V secondary transformer is connected to a 15-Ω resistive load. Calculate:

a. The turns ratio.

b. The secondary current.

c. The primary current.

d. The volt-ampere rating of the transformer.

18-6. A 115-V primary transformer has a turns ratio of 6.5 and a secondary current of 2 A. Calculate:

a. The secondary voltage.

b. The primary current.

c. The load resistance.

d. The volt-ampere rating of the transformer.

18-7. A 2400 V/240-V step-down transformer has a load of 5 Ω pure resistance. Calculate the primary current assuming:

a. 100% efficiency.

b. 90% efficiency.

18-8. A 4160 V/240-V step-down transformer has a load of 2.5 Ω pure resistance. If the core losses amount to 2 kW, determine:

a. The efficiency of the transformer.

b. The primary current.

c. The volt-ampere rating of the transformer.

18-9. An audio amplifier has an open-circuit voltage of 100 V p-p and an output resistance of 64 Ω. A transformer is to be selected to match an 8-Ω loudspeaker to the amplifier. Calculate, assuming 100% efficiency:

a. The necessary turns ratio of the transformer.

b. The primary input voltage and current.

c. The secondary output voltage and current.

d. The output power in the loudspeaker.

e. The power in the loudspeaker if connected directly to the amplifier.

18-10. A voltage source with an internal resistance of 80 Ω is connected to the primary of a step-down transformer that has a turns ratio of 5. Calculate:

a. The load to be connected to the secondary for maximum power transfer.

b. The open-circuit voltage of the source if 10 W of power are developed in the load.

18-11. The total power in a three-phase system is given by $P = \sqrt{3}\, VI \cos \theta$, where V is the line-to-line voltage and I is the line current. Consider a 230/765-kV 1000-MW step-up transformer. Calculate, assuming a purely resistive load:

a. The secondary current when operating at its rated value.

b. The input current if 99.5% efficient.

c. The losses in the transformer in the form of heat.

18-12. The variable autotransformer in Fig. 18-15*b* is adjusted to provide an output voltage of 90 V with an 8-Ω load. Neglecting transformer losses, determine:

a. The power delivered to the load.
b. The load current.
c. The source current.
d. The current in the secondary winding.
e. The percentage of power transformed by the secondary compared with the total circuit power.
f. The instantaneous directions of current in the circuit.

18-13. Repeat Problem 18-12 with the output adjusted to 110 V and a 10-Ω load.

18-14. Repeat Problem 18-12 with the output adjusted to 135 V utilizing the 30-turn step-up section, and a 12-Ω load.

18-15. Instrument transformers are connected as in Fig. 18-16 to measure line current, voltage, and power. If the current transformer ratio is 10 : 1 and the potential transformer ratio is 15 : 1, determine:

a. The line current.
b. The line voltage.
c. The line power.
d. The apparent line power (volt-amperes).

18-16. For the instrument transformer ratios given in Fig. 18-16 determine the wattmeter reading if the ammeter indicates 5 A, the voltmeter indicates 120 V, and

a. The load is purely resistive.
b. The power factor is 0.9.
c. The power factor is zero.

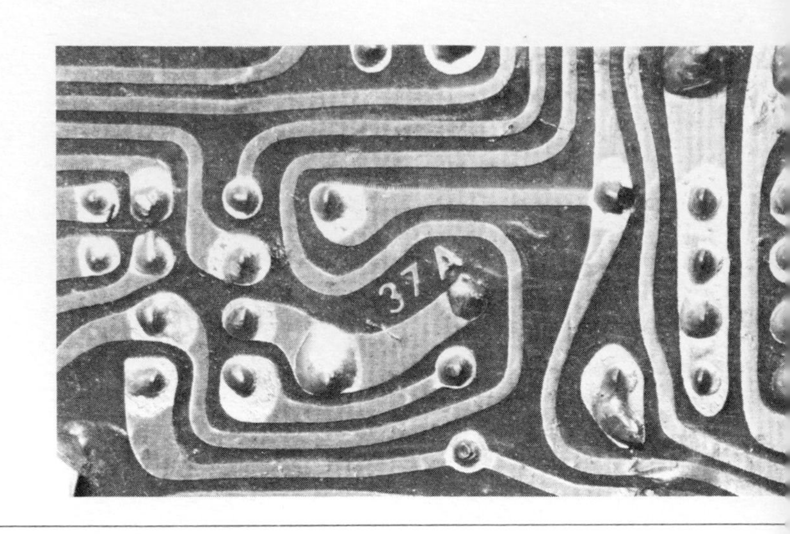

CHAPTER NINETEEN

INDUCTANCE IN DC CIRCUITS

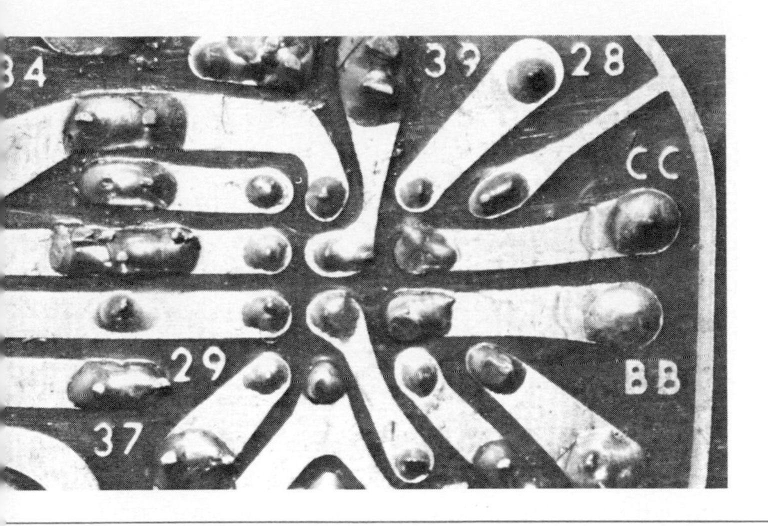

We have seen that the property of inductance is to oppose change of current. In this chapter we examine in detail how current increases (and decreases) when a dc voltage is applied to a series circuit of resistance and inductance. This requires the definition of inductive *time constant* (L/R) of the circuit, which depends *only* upon the inductance and resistance. The time constant is not affected by voltage or current or any other factors. This time constant allows us to examine the *transient responses* in the L-R circuit, including the temporary changes in current and voltage that take place prior to the establishment of *steady-state* conditions. These transients may be analyzed using a graphical method of universal time-constant curves.

If mechanical switching is used to interrupt current in an inductive circuit, the energy stored in the coil's magnetic field is capable of inducing high voltages. This high voltage is used to advantage in a fluorescent lamp circuit and also in an automotive ignition system with the help of a step-up autotransformer. However, methods of reducing inductive effects are necessary where these high voltages could cause damage.

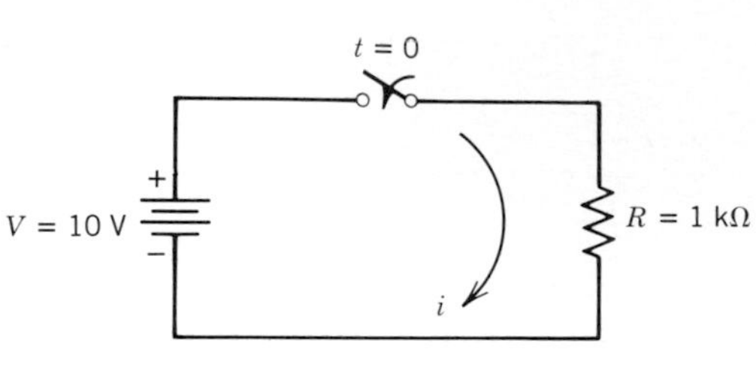

(a) Dc voltage applied to a resistor

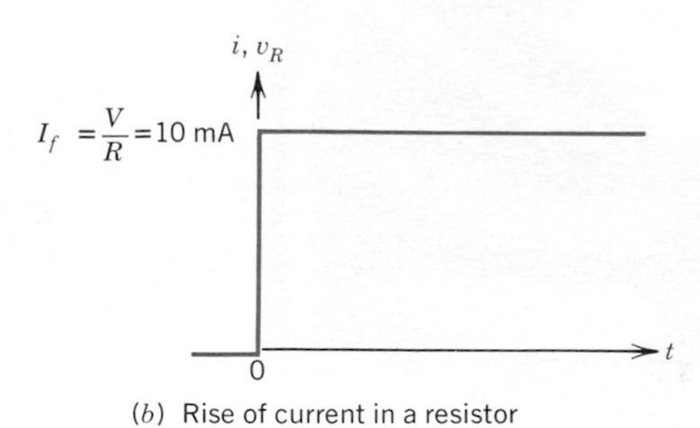

(b) Rise of current in a resistor

FIGURE 19-1
Rise of current in a purely resistive dc circuit.

19-1 RISE OF CURRENT IN AN INDUCTIVE CIRCUIT

The rise of current in a *purely* resistive circuit is instantaneous, as shown in Fig. 19-1. The current increases instantly to its final Ohm's law value immediately after closing the switch. For a given applied voltage, resistance (by itself) only determines *how much* current flows—not the rate at which current *changes*. The final steady current is determined by $I_f = V/R$ as shown in Fig. 19-1*b*.

Now consider the circuit of Fig. 19-2*a*. This circuit shows a coil of negligible resistance in series with a 1-kΩ resistor. Or we can think of the resistance of the coil as being "lumped" into a separate series-connected resistance of 1 kΩ. The effect on the rise of current is the same in either case and is shown in Fig. 19-2*b*. Note that the current does *not* rise instantly as in a pure resistive circuit but rises exponentially. Let us determine why.

When the switch is closed, current tries to rise instantaneously as in the case of the resistive circuit. But we know from Lenz's law that a voltage, v_L, will be induced across the coil opposing any increase (change) in current in the circuit. This results in the polarity shown across v_L in Fig. 19-2*a*, which can be thought of as an *opposing* emf or a *counter*voltage v_L. The applied voltage must overcome both v_L and v_R, as shown in Fig. 19-2*a*. By Kirchhoff's voltage law, therefore,

$$V = v_R + v_L = iR + v_L$$

Now, at $t = 0^+$ (an instant after closing the switch), the instantaneous current i in the above equation must

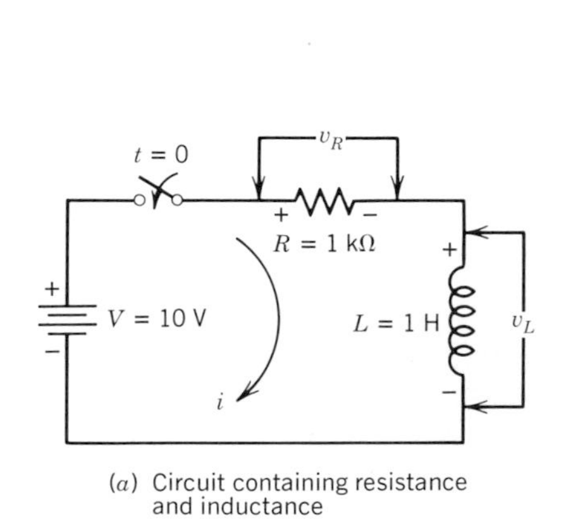

(a) Circuit containing resistance and inductance

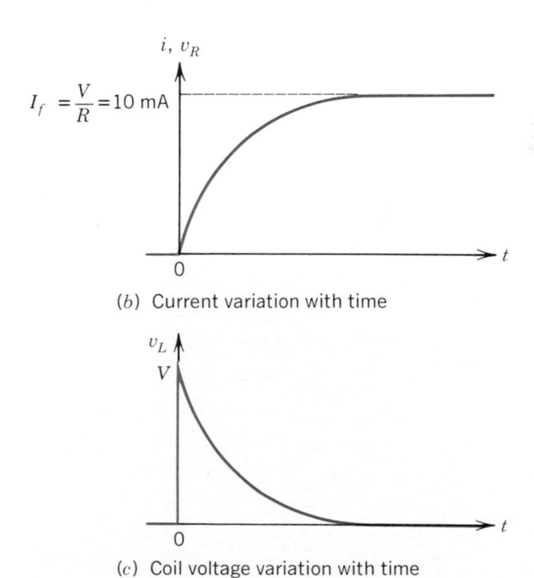

(b) Current variation with time

(c) Coil voltage variation with time

FIGURE 19-2
Current and voltage waveforms in a dc circuit containing resistance and inductance.

continue to be zero. This is because inductance is the property of a coil to oppose any change in current, and, for an initial instant, the coil is completely successful in doing this. This means that there is no voltage drop across the resistor, so:

$$v_R = iR = 0$$

and since $$V = v_R + v_L$$

we have $$V = 0 + v_L$$

This means that at the instant of closing the switch, the rapid rate of change of current in the circuit is sufficient to induce a voltage across the inductance equal to the applied voltage, or $v_L = V$. See Fig. 19-2c.

However, the current *is* changing in the circuit so that some short time later, when the current has increased above zero, there must be a voltage drop v_R across the resistor. See Fig. 19-2b. For Kirchhoff's voltage law to be obeyed, the voltage across the coil at any instant is now given by

$$v_L = V - v_R = V - iR$$

The voltage across the coil is now less than before as a result of the iR drop across the resistor. But the voltage across the coil is

$$v_L = L\left(\frac{\Delta i}{\Delta t}\right) \tag{17-1}$$

so that the rate of change of current with time, $\frac{\Delta i}{\Delta t}$, must be less than the initial value and must decrease as time goes on and the current rises.

This implies that although the current is *increasing*, it is doing so at a slower rate. This is shown in Fig. 19-2b, where the slope of the curve at any instant represents $\frac{\Delta i}{\Delta t}$. The rate continues to decrease until eventually, as shown in Fig. 19-2b, the current reaches a final steady-state value, I_f, where the slope is horizontal and no change occurs. Thus, when $i = I_f$, no voltage is induced across the inductance, as shown in Fig. 19-2c, and all the applied voltage appears across the resistance.

The final steady-state current is now given by Ohm's law:

$$I_f = \frac{V}{R} \tag{3-1a}$$

Using the values given in Fig. 19-2a, $I_f = \frac{10\text{ V}}{1\text{ k}\Omega} = 10$ mA. The inductance has no effect in the circuit at this point as long as no change takes place, such as opening the switch, changing the resistance or changing the applied voltage. The time interval after closing the switch for the current to rise to the steady-state value is called a *transient*. Also, the changes occurring in the current (Fig. 19-2b) and voltage (Fig. 19-2c) are called *transient responses*.

How long does it take to reach the steady-state value or how long does the transient last? This is answered by determining the *time constant*, τ, (lower case Greek letter tau).

19-2 TIME CONSTANT, τ

We have seen that the rate of rise of current (and its slope) is the highest when the switch is first closed because at this instant

$$v_L = V$$

and $$v_L = L\left(\frac{\Delta i}{\Delta t_0}\right) \tag{17-1}$$

Therefore, $$L\left(\frac{\Delta i}{\Delta t_0}\right) = V$$

and the initial rate of rise of current is

$$\frac{\Delta i}{\Delta t_0} = \frac{V}{L} \tag{19-1}$$

where: $\frac{\Delta i}{\Delta t_0}$ is the initial rate of rise of current in amperes per second, A/s

V is the applied dc voltage, in volts, V

L is the inductance of the coil in henrys, H

For the numerical values given in Fig. 19-2a, with $V = 10$ V and $L = 1$ H, the initial rate of rise of current is given by

$$\frac{\Delta i}{\Delta t_0} = \frac{V}{L}$$

$$= \frac{10\text{ V}}{1\text{ H}} = \mathbf{10\ A/s}$$

Now let us consider how long it would take the current to reach its final value *if* it continued to increase at its *initial* rate of change of $\frac{V}{L}$ amperes per second. Let us call this length of time the *time constant,* τ. (See Fig. 19-3.) We know from Fig. 19-3 that the current does not increase at this rate. In fact, when $t = \tau$, the actual current is only 0.63 of its final value. This is also used as a definition of the time constant. (See Section 19-3.)

Let the dotted line that is tangent to the current curve reach the final value of current in a time interval τ. This line represents the initial rate of rise of current. Its slope is given by

$$\text{initial slope} = \frac{\text{rise}}{\text{run}} = \frac{\text{vertical intercept}}{\text{horizontal intercept}}$$

$$= \frac{V/R}{\tau} \text{ amperes per second}$$

but the initial slope $= \frac{\Delta i}{\Delta t_0} = \frac{V}{L}$ amperes per second.

Therefore, $$\frac{V/R}{\tau} = \frac{V}{L}$$

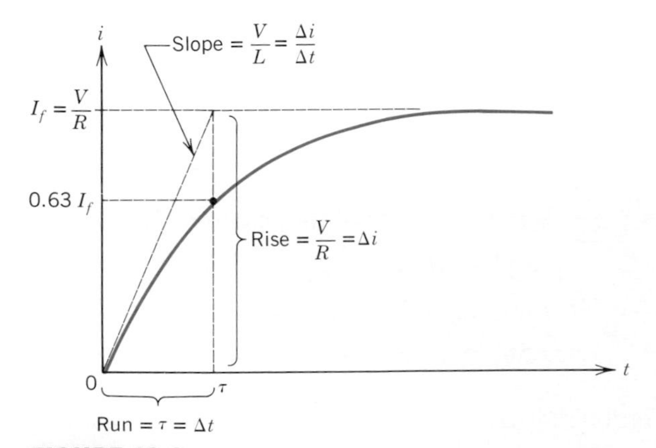

FIGURE 19-3
Determination of the time constant, τ.

and $$\tau = \frac{L}{R} \quad \text{seconds (s)} \qquad \textbf{(19-2)}$$

where: τ is the time constant of the circuit in seconds, s
L is the inductance of the circuit in henrys, H
R is the resistance of the circuit in ohms, Ω

Note that τ only depends on the values of L and R. If the circuit is extremely inductive (L is high), the time to reach steady state is *long.* If the circuit is extremely resistive (R is high), the time to reach steady state is *short.* This is shown in Examples 19-1 and 19-2.

EXAMPLE 19-1

If a circuit's resistance is 1 kΩ and its inductance is 1 H, what is the circuit's time constant?

SOLUTION

Time constant $$\tau = \frac{L}{R} \qquad (19\text{-}2)$$

$$= \frac{1\text{ H}}{1 \times 10^3\ \Omega} = \mathbf{1\ ms}$$

EXAMPLE 19-2

If the circuit of Example 19-1 has its inductance doubled and its resistance cut in half, what is the new time constant?

SOLUTION

time constant $$\tau = \frac{L}{R} \qquad (19\text{-}2)$$

$$= \frac{2\text{ H}}{500\ \Omega} = \mathbf{4\ ms}$$

EXAMPLE 19-3

A coil has an inductance of 8 H and a resistance of 400 Ω. A 20-V dc supply is connected across the coil by means of a switch. Calculate:

a. The time constant of the coil.
b. The final value of current through the coil.
c. The initial rate of rise of current in the coil after the switch is closed.
d. The time it would take for the current to reach its final value *if* the current continued to increase at its initial rate of rise calculated in part (c).

SOLUTION

a. $\tau = \dfrac{L}{R}$ (19-2)

$= \dfrac{8\text{ H}}{400\ \Omega} = 0.02\text{ s} = \mathbf{20\ ms}$

b. $I_f = \dfrac{V}{R}$ (3-1a)

$= \dfrac{20\text{ V}}{400\ \Omega} = 0.05\text{ A} = \mathbf{50\ mA}$

c. $\dfrac{\Delta i}{\Delta t_0} = \dfrac{V}{L}$ (19-1)

$= \dfrac{20\text{ V}}{8\text{ H}} = \mathbf{2.5\ A/s}$

d. If the current is changing at 2500 mA/s, it will change 50 mA in

$$\frac{50\text{ mA}}{2500\text{ mA/s}} = \frac{1}{50}\text{ s} = \mathbf{20\ ms}$$

Note that this is the same time as the time constant in part (a).

19-3 UNIVERSAL TIME-CONSTANT CURVES

It can be shown that the rate of increase of current at *any* instant on the curve is such that if the current continued increasing at *that* rate, the current would reach its final value of V/R amperes in L/R seconds. Thus the time constant is not a characteristic that applies only to the initial conditions, it applies to the whole curve. This means that the manner in which current changes in an inductive circuit can be graphed in terms of time constants. Such a graph is called a *universal time*-constant curve. Figure 19-4 shows two such curves.

Curve *a* gives the percentage of the final steady-state current when the switch is closed and the current is rising. Since $v_R = iR$, curve *a* also represents the change in the voltage across the resistance when the current is rising. Curve *b* gives the voltage across the inductance as a percentage of its initial value whenever the switch is closed and the current is rising.

The curves may also be used when the switch is being opened. In this case, i, v_R, and v_L all decay and are

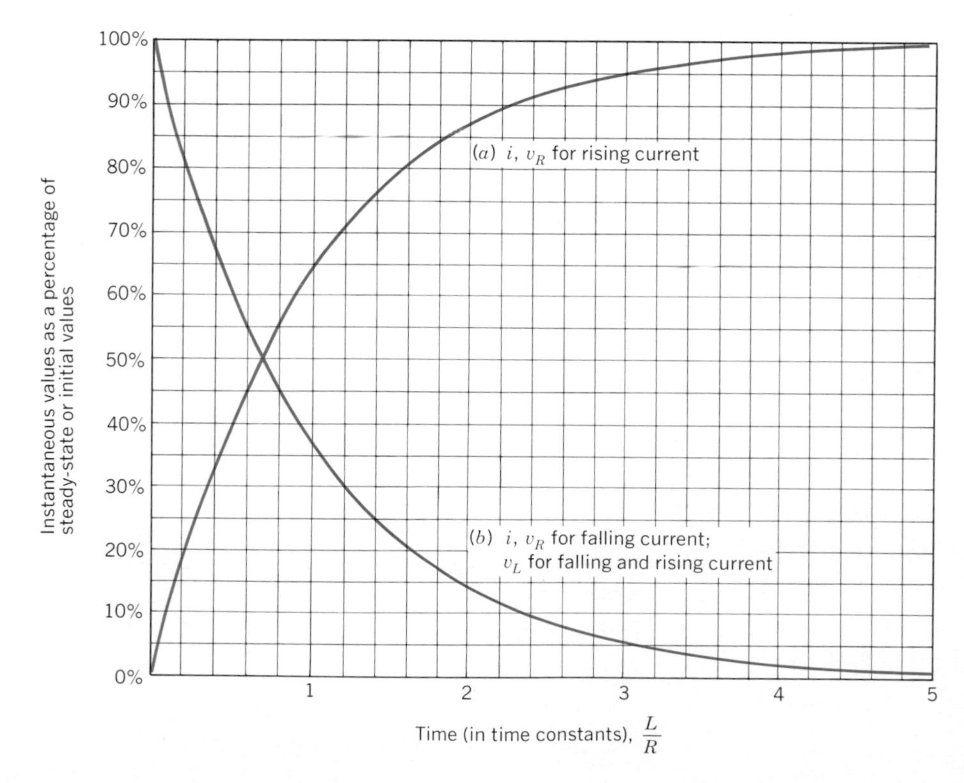

FIGURE 19-4
Universal time-constant curves for the rise and fall of current and voltage in an inductive circuit with direct current.

represented by curve *b*. (The *polarity* of v_L is opposite when the current is falling but the *magnitude* of the voltage v_L is given by curve *b* whether the current is rising or falling).

It can be seen from Fig. 19-4*b* that the current reaches 63% of its steady-state current in one time constant, L/R s.

In fact, the time constant of an inductive circuit is more usually defined as the time required for the current (or coil voltage) to change 63% from its original value to its final value.

Specifically, assume that the time constant is 1 ms and the circuit's steady-state (final) current is 10 mA. Then the current reaches 63% of 10 or 6.3 mA in 1 ms.

At the end of two time constants or 2 ms the current is

$$\text{new value} = \text{original value} + 63\% \text{ of (final value} - \text{previous value)} \tag{19-3}$$

$$= 6.3 \text{ mA} + 0.63 \times (10 \text{ mA} - 6.3 \text{ mA})$$

$$= 6.3 \text{ mA} + 0.63 \times 3.7 \text{ mA}$$

$$= 6.3 \text{ mA} + 2.33 \text{ mA}$$

$$= 8.63 \text{ mA at the end of 2 ms or } 2\,\tau$$

This current corresponds to 86.3% of the steady-state current, and it could have been read directly from the universal time constant curve (Fig. 19-4*a*) after 2 τ.

NOTE After five time constants (5 τ), the increasing curve has reached 99.3% of its final value. This time of 5 τ is usually considered to be the end of the transient.

For all practical purposes, an inductive circuit reaches its steady-state after 5 τ has elapsed, equal to $5L/R$.

Thus, in the above, the current will reach its steady-state value of 10 mA in approximately 5 τ or 5 ms.

EXAMPLE 19-4

Consider the coil in Example 19-3; $L = 8$ H, $R = 400\ \Omega$, $V = 20$ V, $\tau = 20$ ms, and $I_f = 50$ mA. Use the universal time constant curves (Fig. 19-4) to find:

a. The current 46 ms after closing the switch.

b. How long it takes for the current to reach 27.5 mA.

SOLUTION

a. $t = 46$ ms

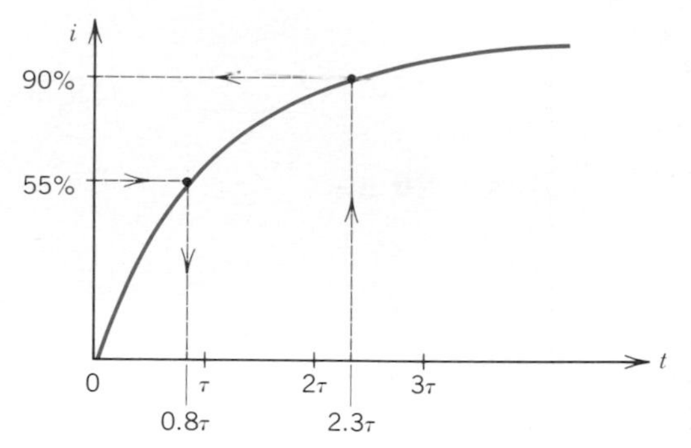

FIGURE 19-5
Sketch of *i* versus *t* for Example 19-4.

$$= \frac{46 \text{ ms}}{20 \text{ ms}/\tau} = 2.3\,\tau$$

From Fig. 19-4, when $t = 2.3\,\tau$, $i = 90\%$ of I_f (see Fig. 19-5). Therefore,

$$i = 90\% \times I_f$$

$$= \frac{90}{100} \times 50 \text{ mA} = \mathbf{45\ mA}$$

b. $i = 27.5$ mA

$$= \frac{27.5 \text{ mA}}{50 \text{ mA}} \times 100\%$$

$$= 55\% \text{ of final current, } I_f$$

From Fig. 19-4, when $i = 55\%$, $t = 0.8\,\tau$ (see Fig. 19-5). Therefore,

$$t = 0.8\,\tau$$

$$= 0.8 \times 20 \text{ ms} = \mathbf{16\ ms}$$

EXAMPLE 19-5

A circuit consists of a 30-mH coil (of negligible resistance) in series with a 2-kΩ resistor. A switch is closed to apply 50 V dc. Using the universal time constant curves (Fig. 19-4), find:

a. The time constant of the circuit.

b. The initial voltage across the coil when the switch is closed.

c. The voltage across the coil 7.5 μs after closing the switch.

d. The steady-state current in the circuit.

e. The current 45 μs after closing the switch.

f. How long it takes for the voltage across the resistor to reach 37.5 V.

g. Approximately how long it takes for the voltage across the coil to drop to zero.

SOLUTION

a. $$\tau = \frac{L}{R} \qquad (19\text{-}2)$$

$$= \frac{30 \times 10^{-3}\ \text{H}}{2 \times 10^{3}\ \Omega} = 15 \times 10^{-6}\ \text{s} = \mathbf{15\ \mu s}$$

b. Initially, $v_L = V =$ **50 V**

c. $$t = 7.5\ \mu\text{s} = \frac{7.5\ \mu\text{s}}{15\ \mu\text{s}/\tau} = 0.5\ \tau$$

From Fig. 19-4, after 0.5 time constant, the decreasing v_L curve is at 60% of its initial value.

Therefore, $$v_L = 60\% \text{ of } 50\ \text{V} = \mathbf{30\ V}$$

d. Steady-state current,

$$I_f = \frac{V}{R} \qquad (3\text{-}1a)$$

$$= \frac{50\ \text{V}}{2 \times 10^{3}\ \Omega} = \mathbf{25\ mA}$$

e. $$t = 45\ \mu\text{s} = \frac{45\ \mu\text{s}}{15\ \mu\text{s}/\tau} = 3\ \tau$$

From Fig. 19-4, after three time constants, the increasing i curve is at 95% of its final value.

Therefore, $$i = 95\% \text{ of } 25\ \text{mA} = 0.95 \times 25\ \text{mA} = \mathbf{23.8\ mA}$$

f. The resistor voltage $v_R = iR$ has the same percent variation with time as the current, with a final value of 50 V.

$$v_R = 37.5\ \text{V} = \frac{37.5\ \text{V}}{50\ \text{V}} \times 100\% = 75\%$$

of the final value. From the increasing i, v_R curve in Fig. 19-4, the time required for the curve to reach 75% is 1.4 τ. Therefore,

$$t = 1.4\ \tau = 1.4 \times 15\ \mu\text{s} = \mathbf{21\ \mu s}$$

g. The coil voltage is approximately zero after 5 time constants. This time is

$$t = 5\ \tau = 5 \times 15\ \mu\text{s} = \mathbf{75\ \mu s}$$

Instead of using a *graphical* method to obtain the circuit current and coil voltage, an *algebraic* solution may be used. Equations are given in Appendix A-9 for both rising and falling current in a series *R-L* circuit with an applied dc voltage. Examples 19-4 and 19-5 are solved using these equations.

19-4 ENERGY STORED IN AN INDUCTOR

During the transient period, as the current rises in a coil, energy is being stored in the growing magnetic field around the coil. The total amount of energy when the transient ceases may be determined from the following:

$$W = \tfrac{1}{2} L I_f^2 \quad \text{joules (J)} \qquad (19\text{-}4)$$

where: W is the stored energy in joules, J
L is the inductance in henrys, H
I_f is the final current in amperes, A

EXAMPLE 19-6

How much energy is finally stored in a 1-H coil that is connected in series with a 1-kΩ resistor and a 10-V dc source?

SOLUTION

Steady-state current through coil,

$$I_f = \frac{V}{R} \qquad (3\text{-}1a)$$

$$= \frac{10\ \text{V}}{1\ \text{k}\Omega} = 10\ \text{mA}$$

$$W = \tfrac{1}{2} L I_f^2 \qquad (19\text{-}4)$$

$$= \tfrac{1}{2} \times 1\ \text{H} \times (10 \times 10^{-3}\ \text{A})^2$$

$$= 50 \times 10^{-6} \quad \text{joules}$$

$$= \mathbf{50\ \mu J}$$

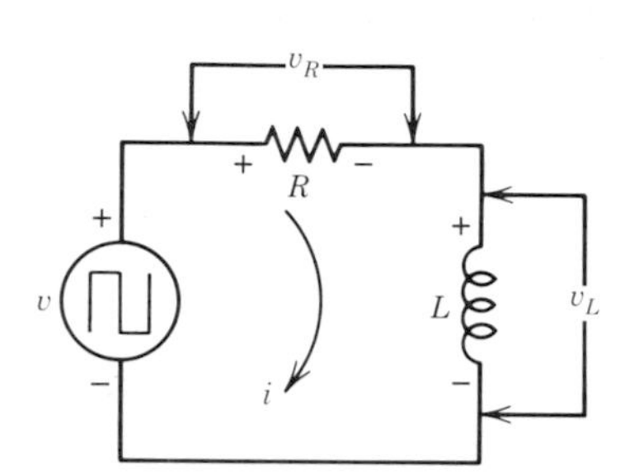

(*a*) Square wave generator connected to an inductive circuit

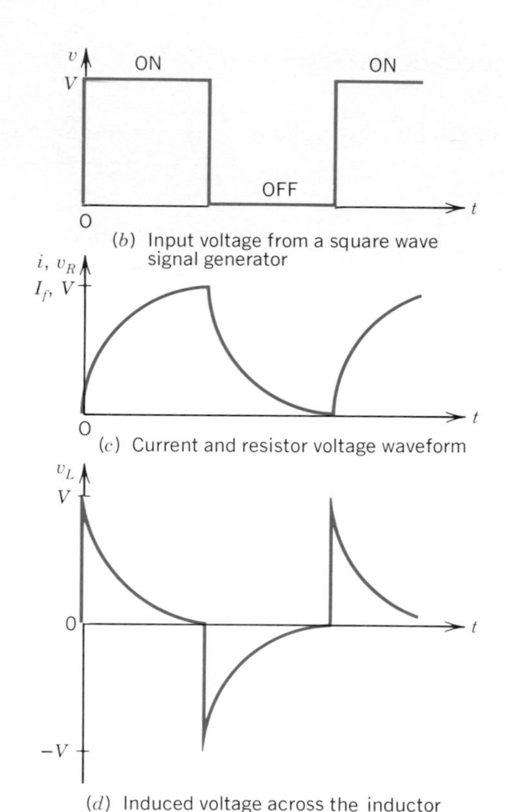

(*b*) Input voltage from a square wave signal generator

(*c*) Current and resistor voltage waveform

(*d*) Induced voltage across the inductor

FIGURE 19-6
Current and voltage waveforms in an inductive circuit with electronic switching of the applied voltage.

This energy is stored in the coil as long as the current in the coil is maintained at its final, steady-state value.

19-5 FALL OF CURRENT IN AN INDUCTIVE CIRCUIT (WITH ELECTRONIC SWITCHING)

Instead of applying a voltage to an inductive circuit by closing a mechanical switch, we can use the output of a square wave signal generator. Depending upon the frequency, this will apply a voltage for a given period and then reduce the voltage to zero for an equal period of time. The variation in current and voltage in the circuit may then be examined on an oscilloscope, since the switching from ON to OFF can be made repetitively at a suitably high frequency. Figure 19-6 shows the circuit arrangement and the resulting waveforms.

When the applied input voltage goes positive, the current builds up gradually as in Fig. 19-4*a*, and the inductive voltage v_L decays with time, following curve *b* in Fig. 19-4.

When the input voltage drops to zero, the current must also drop. The collapsing magnetic field around the coil induces a voltage v_L, which, by Lenz's law, has a polarity opposite to that induced when the current was rising. This voltage, shown negative in Fig. 19-6*d*, is responsible for the gradual decay of current in the circuit toward zero. The energy stored in the magnetic field is dissipated in the resistor in the form of heat.

Note that the polarity signs (+, −) shown across the coil (and the square wave generator) in Fig. 19-6*a* are used for *reference* purposes only. Thus, when v_L is shown positive in Fig. 19-6*d*, this means that v_L has an instantaneous polarity equal to that shown in Fig. 19-6*a*. When v_L goes negative in Fig. 19-6*d*, the opposite polarity is induced across the coil in Fig. 19-6*a*. Without the reference polarity signs, we would not know how to interpret a positive or negative voltage in a graph. Thus polarity signs *are* used, even where a voltage alternates, to provide a reference to interpret graphical values.

With electronic switching, the falling time constant is the same as the rising time constant. We can therefore use the universal time-constant curves of Fig. 19-4 to solve for instantaneous current and voltage values, as in Example 19-7.

EXAMPLE 19-7

A square wave signal generator develops a 12-V peak output at a frequency of 2.5 kHz. It is connected across a series circuit consisting of a 20-mH coil of negligible resistance and a 500-Ω resistor. Find:

a. The current 0.1 ms after the input voltage goes to zero.

b. The voltage across the coil 20 μs after the input voltage goes to zero.

c. From a consideration of the generator's frequency and the time constant of the circuit, the circuit current and inductor voltage 0.2 ms after the input voltage goes to zero.

SOLUTION

a. When the input voltage goes to zero, the current decays following curve *b* in Fig. 19-4. The initial value of current was

$$I_f = \frac{V}{R} \quad (3\text{-}1a)$$

$$= \frac{12\ \text{V}}{500\ \Omega} = 24\ \text{mA}$$

The time constant of the circuit

$$\tau = \frac{L}{R} \quad (19\text{-}2)$$

$$= \frac{20 \times 10^{-3}\ \text{H}}{500\ \Omega} = 40\ \mu\text{s}$$

$$t = 0.1\ \text{ms} = 100\ \mu\text{s} = \frac{100\ \mu\text{s}}{40\ \mu\text{s}/\tau} = 2.5\ \tau$$

when $t = 2.5\ \tau$, $i = 8\%$ of I_f. Therefore,

$$i = 8\% \times 24\ \text{mA}$$

$$= 0.08 \times 24\ \text{mA} = \mathbf{1.92\ mA}$$

b. The initial coil voltage is 12 V; it also follows the decaying curve *b* in Fig. 19-4.

$$t = 20\ \mu\text{s} = \frac{20\ \mu\text{s}}{40\ \mu\text{s}/\tau} = 0.5\ \tau$$

when $t = 0.5\ \tau$, $v_L =$ 60% of the initial voltage.

Therefore, $v_L = 60\% \times 12\ \text{V} = \mathbf{7.2\ V}$

c. $$T = \frac{1}{f} \quad (14\text{-}9)$$

$$= \frac{1}{2.5 \times 10^3\ \text{Hz}} = 0.4\ \text{ms}$$

The duration of time that the input voltage is zero $= T/2 = 0.2$ ms. This corresponds to $\frac{200\ \mu\text{s}}{40\ \mu\text{s}/\tau} = 5\ \tau$. Thus both i and v_L are approximately zero after 0.2 ms.

This problem is also solved using exponential equations in Appendix A-9.

19-5.1 Measurement of Coil Inductance

The waveforms of Fig. 19-6 suggest that we can determine the inductance of a coil if we measure the time constant of a circuit in which the coil is series-connected with a known resistance. Figure 19-7*a* shows an oscilloscope connected across the resistance and a square wave generator used to apply a voltage that varies between 0 and V_m. The frequency of the signal generator is adjusted until the oscilloscope displays the transient response of current having a duration of at least five time constants. See Fig. 19-7*c*. The horizontal distance required for the voltage waveform across *R* to reach 63% of its final value is the time constant. From this, the inductance may be determined as shown in Example 19-8.

EXAMPLE 19-8

A coil of dc resistance of 50 Ω is connected in series with a 10-Ω resistor and a square wave signal generator with a 10-V p-p output. An oscilloscope across the resistor displays a waveform as in Fig. 19-7*c*, reaching 63% of its peak value in a horizontal distance of 0.8 cm. If the horizontal calibration is 0.5 ms/cm, determine:

a. The time constant of the whole circuit.

b. The inductance of the coil.

SOLUTION

a. The time constant of the whole circuit is the time required for the current to reach 63% of the final value = 0.8 cm × 0.5 ms/cm

$$= \mathbf{0.4\ ms}$$

b. $$\tau = \frac{L}{R_T} \quad (19\text{-}2)$$

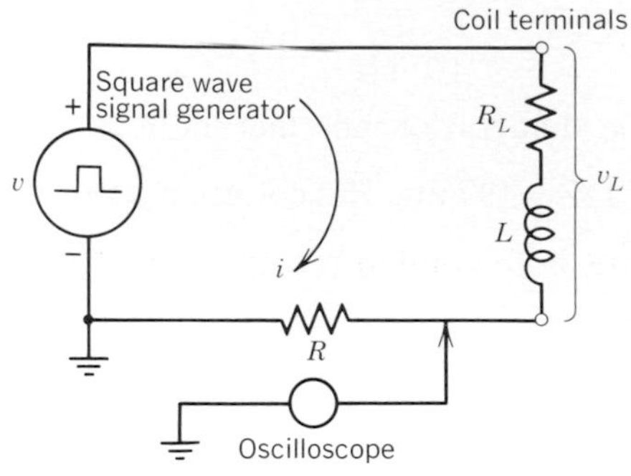

(a) Connection of an oscilloscope to measure v_R

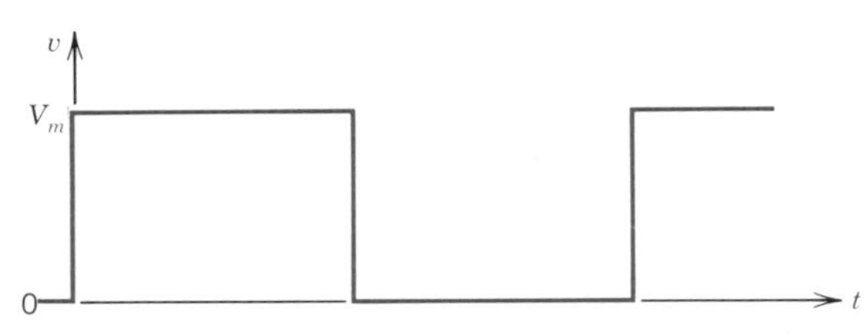

(b) Square wave input voltage from a signal generator

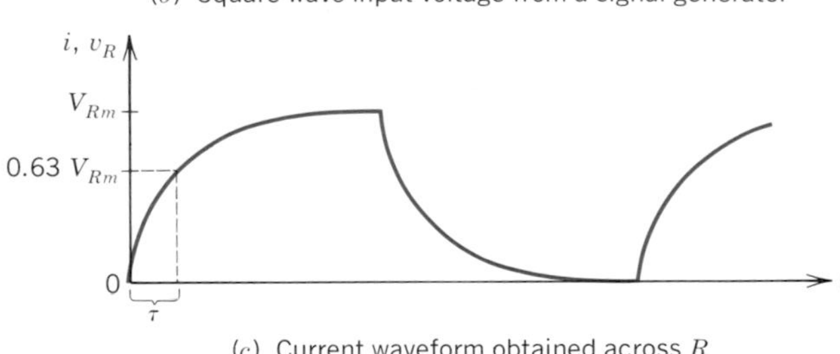

(c) Current waveform obtained across R

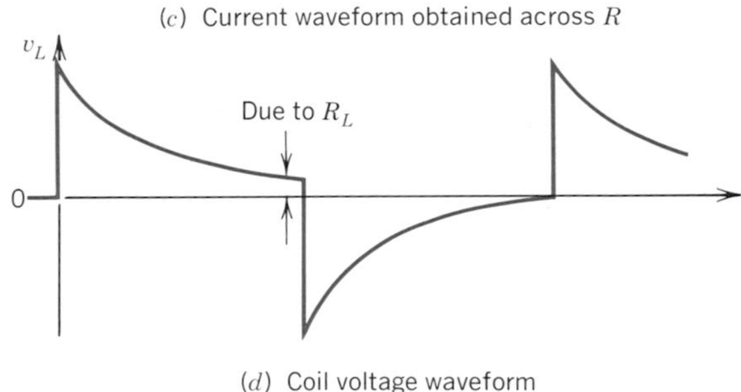

(d) Coil voltage waveform

FIGURE 19-7
Using the method of time-constant measurement to determine the inductance L.

$$\text{Therefore,}\quad L = \tau R_T = 0.4 \times 10^{-3}\text{ s} \times (50\ \Omega + 10\ \Omega) = 24 \times 10^{-3}\text{ H} = \mathbf{24\ mH}$$

Note that the voltage waveform across the coil in Fig. 19-7*d*, does not drop to zero on the positive half-cycle because of the voltage drop across the resistive portion, R_L. This also means that the final value of voltage across R in Fig. 19-7*c* is not V_m, but V_{Rm}, where:

$$V_{Rm} = V_m \times \frac{R}{R + R_L}$$

Thus the sum of the voltage drops across the coil v_L and across the resistor v_R is always equal to the input voltage V_m, *at all times*, as required by Kirchhoff's voltage law.

19-6 FALL OF CURRENT IN AN INDUCTIVE CIRCUIT (WITH MECHANICAL SWITCHING)

We have seen that current builds up gradually when a dc voltage is applied to a series circuit of resistance and inductance, the coil's induced voltage opposing the increase in current through the coil. Also, energy is stored in the magnetic field around the coil.

Now consider the mechanical opening of a switch to interrupt the current through the coil as shown in Fig. 19-8*a*.

The immediate effect of the inductance of the coil is to induce an emf whose polarity is such that the emf tries to *maintain* the current at its initial level. That is, v_L in Fig. 19-8*a* acts in series with the applied voltage (10 V, in this case) to try to keep 10 mA flowing, even though the switch blade is being opened. The induced voltage is sufficiently high to make the current of 10 mA "jump the gap" across the contacts of the opening switch. This effect, called *arcing,* results from the air breaking down as an insulator and ionizing to conduct current for a brief instant. A blue spark is usually visible when this happens. Why is such a high voltage induced?

Let us assume that the resistance of the ionized air between the contacts of the opening switch, R_{sw}, is 9 kΩ. If a current of 10 mA is flowing through the switch, there must be a voltage drop across the switch given by

$$V_{sw} = IR_{sw} = 10\text{ mA} \times 9\text{ k}\Omega = \mathbf{90\ V}$$

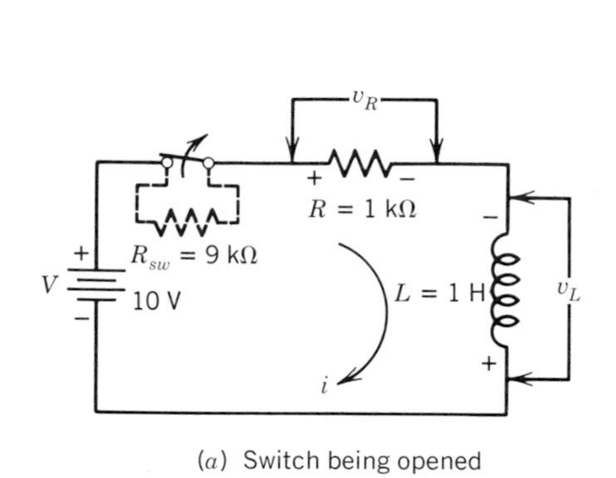

(a) Switch being opened

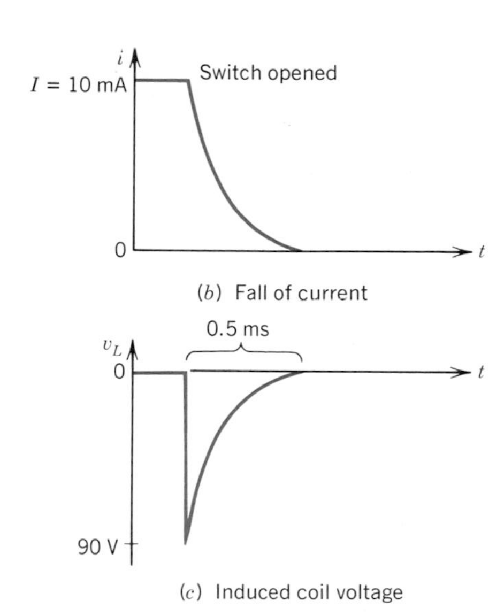

(b) Fall of current

(c) Induced coil voltage

FIGURE 19-8
Current and voltage waveforms resulting from interruption of current in an inductive circuit with dc.

This voltage must have been developed across the coil by the magnetic field's collapsing when the current *started* to decrease. See Fig. 19-8c. We can determine the initial rate at which the current decreases as follows:

$$v_L = L\left(\frac{\Delta i}{\Delta t_0}\right) \tag{17-1}$$

and

$$\frac{\Delta i}{\Delta t_0} = \frac{v_L}{L}$$

$$= \frac{90 \text{ V}}{1 \text{ H}} = \mathbf{90\ A/s}$$

Note that this rate of change of decreasing current is 9 times greater than when the switch was initially closed and current was rising. Why is this so?

Let us calculate the time constant for the circuit when the switch is being opened. Note that we must now use the *total* series resistance in the circuit:

$$\tau = \frac{L}{R_T} = \frac{L}{R + R_{sw}} \tag{19-2}$$

$$= \frac{1 \text{ H}}{1 \text{ k}\Omega + 9 \text{ k}\Omega} = \mathbf{0.1\ ms}$$

This is only one-tenth the time constant for the interval when the switch was being closed. This means that the current decreases much more quickly than it increases. (Actually, the time "constant" is *not constant* because the switch resistance increases as it opens. The current will be essentially zero in less than 0.5 ms.) It is this very rapid rate of change of current that is responsible for the high induced voltage across the coil. And it is the high resistance introduced by the switch that makes the circuit more resistive and causes this rapid change in current.

The ability of a coil to induce this high voltage is readily seen by observing a neon lamp connected across the coil as in Fig. 19-9a. A neon lamp, often used as an indicator in amplifiers, requires approximately 70 V to "fire" or light.*

It is observed that closing the switch in Fig. 19-9a does not light the lamp; a peak of only 10 V exists across the coil and lamp. But when the switch is opened, the lamp flashes because of at least 90 V induced. The waveforms of current and voltage resulting from a repeated opening and closing of the switch are shown in Fig. 19-9b, 19-9c, and 19-9d. The large negative voltage "spike" responsible for lighting the neon lamp may also be observed on an oscilloscope.

A major application of the high voltage induced in a coil by interrupting the current through a coil is found in a fluorescent lamp circuit. (See Section 20-5.) The momentary self-induced voltage of several hundred volts is sufficient to "fire" the fluorescent lamp. The coil (or

* Only the electrode that is at a negative potential glows. In this way a neon lamp can be used to determine the polarity of direct current, and distinguish it from alternating current which causes both electrodes to glow.

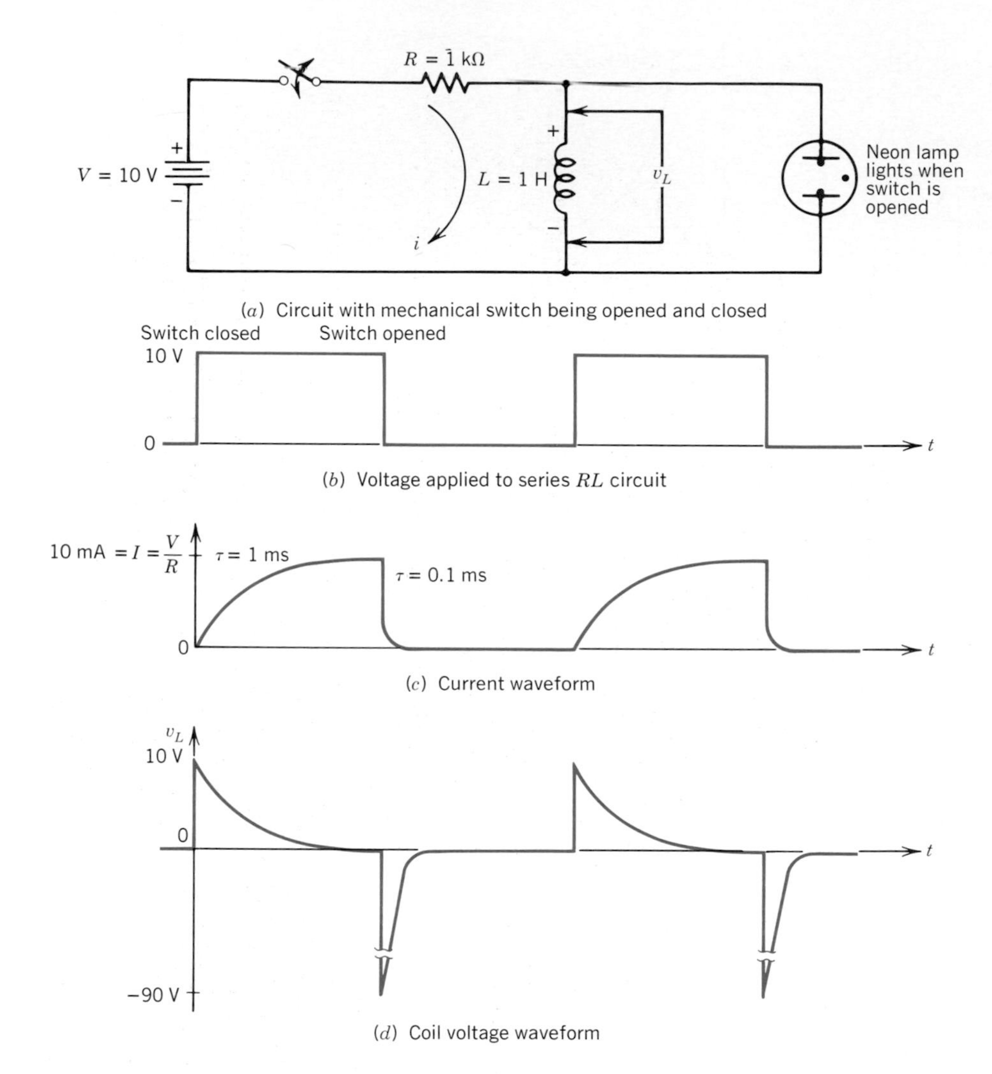

FIGURE 19-9
Waveforms resulting from mechanical switching of current in an inductive circuit (switch resistance assumed to be 9 kΩ).

ballast) then limits the current in the circuit after the lamp is lit.

19-7 CONVENTIONAL AUTOMOTIVE IGNITION SYSTEM

Another application of interrupting current through a coil is found in a conventional ignition system that uses a distributor and points. In this case, it is not the self-induced voltage that is of value, but the much higher mutual induced voltage. Figure 19-10 shows that the automotive ignition coil is actually a step-up autotransformer. The primary consists of several hundred turns of wire capable of carrying several amperes. The secondary has thousands of turns of very fine wire.

If the ignition switch is ON and the engine is cranked by the starter motor, a shaft driven by the engine turns a cam in the distributor to alternately open and close the breaker points. When the points are closed, a direct current flows through the primary winding setting up a strong magnetic field in the "coil." When the cam rotates and opens the points, the primary current is interrupted. The magnetic field collapses and induces a very high voltage in the secondary winding, as much as 25 kV. This voltage is directed to the proper spark plug by the rotor

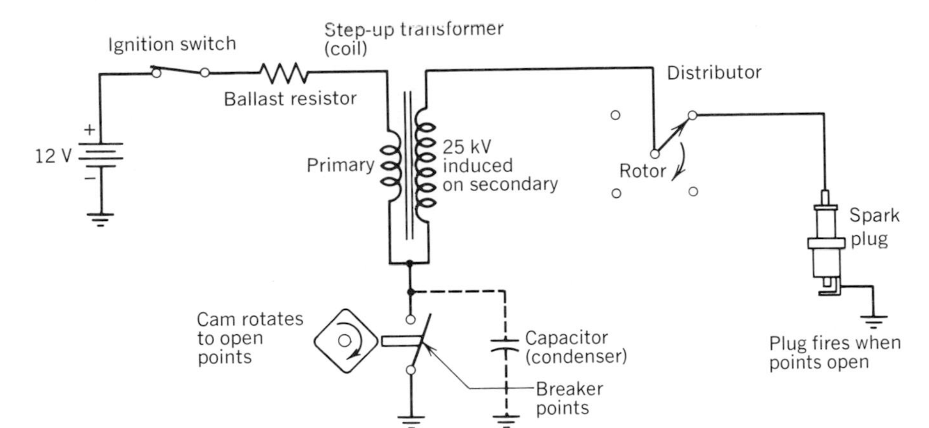

FIGURE 19-10
Conventional ignition system.

(rotating switch also driven by the same shaft that turns the cam), in the top of the distributor. The timing is set to fire the spark plug at the correct position of the associated piston. When the points close again, the field builds up once more, ready to fire the next plug in sequence.

In order to produce the very high voltage on the secondary, the magnetic field must collapse very quickly. This can happen only if the primary current is interrupted very abruptly. But we have seen that the self-inductance of the primary induces a high voltage, which tends to keep the current flowing. This results in arcing across the contacts, but, even worse, it drags out the decay of current. To speed up the decrease of current (and make the system work), it is necessary to connect a capacitor (once called a condenser) across the points. (See Section 21-1.)

When the points now open, the current flows into the capacitor instead of jumping the gap at the opening points. (This practically eliminates arcing.) Very quickly, the capacitor becomes charged and acts like an open circuit with an effective infinite resistance. The time constant for current decay on the primary is now extremely small ($\tau = L/R_T$), so the current and field decrease rapidly. There is also a voltage induced on the secondary at the instant the points close. But because the time constant is relatively long for the buildup of current, this voltage is relatively small and is not used. It does, however, provide a useful indication of point closure on an ignition analyzer oscilloscope, when connected in the system for servicing. This instrument displays a primary or a secondary "parade" pattern, in which the firing characteristics of all engine cylinders are shown from left to right across the screen in their normal firing order.

19-8 ELECTRONIC AUTOMOTIVE IGNITION SYSTEM

The induced voltage on the secondary of the ignition coil can be increased to 35 or 40 kV by using electronic switching. As shown by Fig. 19-11, this involves using a transistor as a solid-state switch in place of the breaker points. Since there is no mechanical switching of current, there is no arcing or need for a capacitor.

The switch is turned ON and OFF by a signal derived from a pick-up coil on a stationary permanent magnet core. An emf is induced in this coil every time one of the

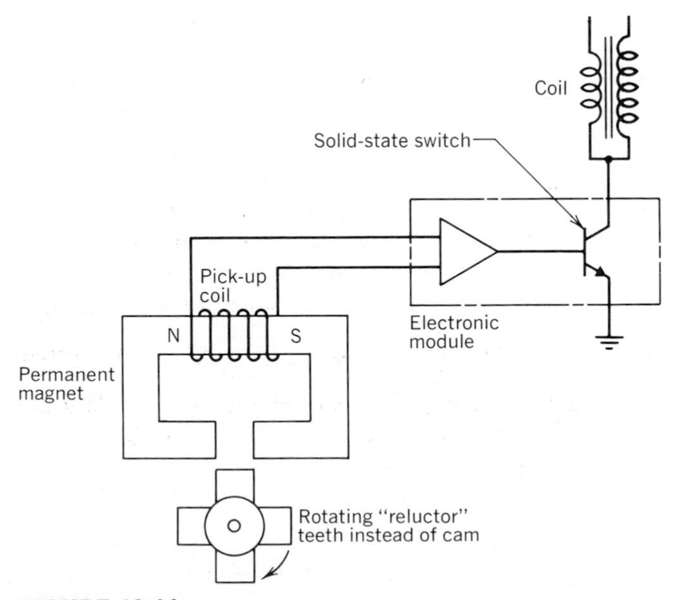

FIGURE 19-11
One type of electronic ignition system.

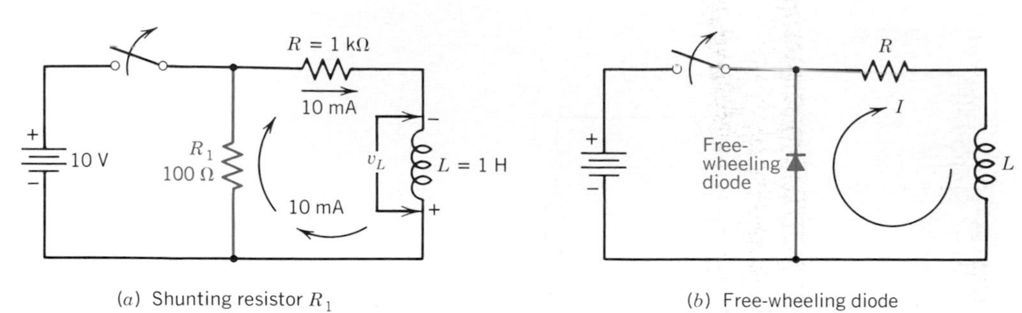

(*a*) Shunting resistor R_1 (*b*) Free-wheeling diode

FIGURE 19-12
Methods of reducing high inductive voltage.

rotating teeth passes between the gap changing the field strength. This signal is suitably amplified in the electronic module and "conditioned" by other information such as dwell time, engine speed, atmospheric pressure, and so on, to maximize performance consistent with pollution emission controls.

19-9 METHODS OF REDUCING INDUCTIVE EFFECTS

Although the interruption of current through an inductance may have useful applications, as already discussed, it is often necessary to reduce the inductive "kick." Dangerously high voltages that may break down insulation occur in both direct current and alternating current circuits, unless countermeasures are taken. The problem is particularly noticeable in dc circuits and can be minimized in a number of ways.

One method is to use a relatively low resistance to provide current an alternate path to flow through, instead of through the opening switch. See Fig. 19-12*a*.

For the values shown, the steady-state current through the coil with the switch closed is $\frac{10\text{ V}}{1\text{ k}\Omega}$ or 10 mA. When the switch is opened, this current continues to flow but through R_1, assumed here to be 100 Ω. Thus the voltage across the coil is

$$\begin{aligned} v_L &= IR_1 + IR \\ &= 10 \times 10^{-3}\text{ A} \times 100\ \Omega + 10 \times 10^{-3}\text{ A} \times 1 \times 10^{3}\ \Omega \\ &= 1\text{ V} + 10\text{ V} \\ &= \mathbf{11\ V} \end{aligned}$$

This compares with 90 V without R_1 and an assumed switch resistance of 9 kΩ.

The disadvantage is the high current (100 mA) drawn by R_1 all the time the switch is closed. This can be overcome by the use of a diode, connected as in Fig. 19-12*b*. When the switch is closed, no current can flow through the diode because it is reverse-biased and effectively open. Current flows normally through the coil.

Upon opening the switch, the polarity of the induced emf causes the diode to conduct, shunting the coil with a very low resistance. The time constant $\frac{L}{R_T}$ is large, and the current drops off gradually. Very little voltage is induced across the coil.

When used in this manner, the diode is often referred to as a "free-wheeling" diode. They are often used across dc motors and relay coils to suppress high induced voltages when the equipment is turned off.

19-9.1 Noninductive Resistors

Wire-wound resistors, used for higher-power applications, inevitably possess some inductance. Where this inductance cannot be tolerated, *special* wire-wound resistors are available. As illustrated in Fig. 19-13 a *bifilar* winding is used. That is, the resistor is wound with a double winding. In this way, the magnetic fields set up by current flowing side by side in opposite directions cancel.

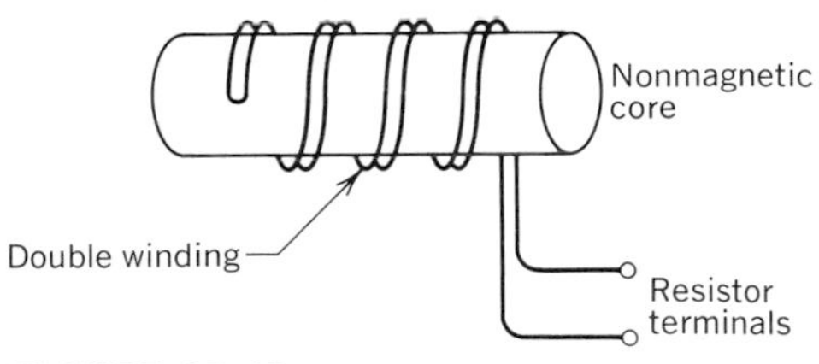

FIGURE 19-13
Bifilar noninductive resistor.

The result is very little induced voltage, and inductance is minimized.

Conversely, a bifilar winding can be used to provide a coefficient of coupling between two coils that is very close to unity. This is necessary where iron cores are not practical, as in radio frequency transformers. In this case, two terminals are available at each end of the bifilar winding.

SUMMARY

1. When a dc voltage is applied to an inductive circuit, the current builds up gradually due to the opposing induced emf.

2. The current has its greatest rate of change when the voltage is first applied and is given by

$$\frac{\Delta i}{\Delta t_0} = \frac{V}{L}$$

3. The final value of current is determined only by the resistance and the applied voltage as given by Ohm's law.

4. The time constant is the time that would be required for the current to reach its final value if it continued to increase at its initial rate.

5. Alternatively, the time constant is the time required for the current to change 63% from its original value to its final value.

6. The time constant is given by

$$\tau = \frac{L}{R}$$

7. Universal time-constant curves give the current and voltage values as a percentage of maximum in terms of the time constant.

8. The final energy stored in an inductor's magnetic field is

$$W = \tfrac{1}{2} L I_f^2$$

9. Mechanical switching in an inductive circuit causes current to fall at a much faster rate than it rises, due to the resistance introduced by the switch, which causes a much smaller time constant.

10. The rapid rate of the fall of current in a coil causes high induced voltages.

11. The high voltage produced, on the secondary side of an automotive coil, results from the use of a step-up transformer and also the sudden interruption of current in the primary by breaker points or an electronic module switch.

12. Inductive effects can be minimized by shunting the inductance with a low resistance or a free-wheeling diode.

SELF-EXAMINATION

Answer T or F, or, for multiple choice, a, b, c, or d
(Answers at back of book)

19-1. In a dc circuit, inductance determines how the current changes and not the final value of current. ________

19-2. In a dc circuit with inductance and resistance, the initial voltage across the inductance is equal to the applied voltage. ________

19-3. Interrupting the current in an inductive circuit always induces the same voltage across the inductance as when the voltage is first applied. ________

19-4. The current builds up in an inductive circuit containing resistance at the same rate until the final steady state is reached. ________

19-5. The time constant determines how long it takes for steady-state conditions to be reached. ________

19-6. The time constant applies only to the first 63% change in the current. ________

19-7. If a 12-V dc supply is connected to a coil having an inductance of 100 mH and a resistance of 1 kΩ, the initial rate of rise of the current and the final current are

a. 12 A/s and 12 mA
b. 1.2 A/s and 120 mA
c. 120 A/s and 12 mA
d. 12 mA/s and 120 A

19-8. The time constant in Question 19-7 is

a. 100 ms
b. 100 μs
c. 12 ms
d. 10 s

19-9. Energy stored in an inductor is in the form of an electric field. ________

19-10. Arcing across the contacts of an opening switch is due to dirty contacts. ________

19-11. High voltage is induced across an inductor when the current through it is interrupted because of a reduction in the time constant because of switch resistance. ________

19-12. An automotive ignition system induces a high voltage as a result of mutual inductance. ________

19-13. A capacitor is used across the breaker points solely to reduce arcing and avoid the constant replacement of the points. ________

19-14. A higher voltage is induced in an electronic ignition system because the current can be interrupted much more suddenly. ________

19-15. The high voltage induced across a dc relay coil when it is shut off can be minimized by using a diode across the coil. ________

REVIEW QUESTIONS

1. a. What is the basic reason for the gradual buildup of current in an inductive circuit with an applied dc voltage?

b. Why doesn't this happen in a pure resistance?

c. If the voltage induced across an inductance is initially equal to the applied dc voltage, why does any current flow at any time?

d. Why does the current eventually reach a final value?

2. a. What do you understand by the term *time constant?*

b. Give two definitions.

c. How long does a transient last?

3. Prove that the time constant, given by L/R, has units of seconds.

4. a. Under what conditions does the voltage across a coil not drop to zero at the end of a transient with a dc voltage?

b. Under what conditions does the voltage across a coil exceed the input dc voltage?

c. Why doesn't this happen when a square wave signal generator is used?

5. a. What factors determine the energy stored by an inductor?

b. What prevents an inductor being used as a portable energy storage device like a battery?

c. In an inductive circuit with a mechanical switch, what evidence is there of the energy stored in the coil?

6. a. What causes arcing?

b. Why does it not occur, as badly, when closing a switch?

c. Describe a demonstration to illustrate the high-voltage-inducing capability of an inductance.

7. a. Describe how it is possible for a 12-V dc source to produce 25 kV if a transformer cannot be operated from dc.

b. If the capacitor were removed from a conventional ignition system, why would the engine be unlikely to start?

c. Why is high voltage produced by the *opening* of the breaker points but not by the *closing?*

8. a. What basic difference exists between a standard and an electronic ignition system?

b. What causes a higher secondary voltage to be induced in the latter?

c. Why is no capacitor required?

9. a. Describe two methods of reducing inductive effects in a dc circuit.

b. What is the advantage of each?

c. Which system can be used for alternating current and direct current?

d. How does a bifilar winding reduce the inductance in a wire-wound resistor?

10. Describe how you would measure the inductance of a coil.

PROBLEMS

(Answers to odd numbers at back of book)

19-1. A 200-μH coil with a resistance of 150 ohms is connected across a 300-mV dc supply. Calculate:

a. The initial rate of rise of current.

b. The final value of current.

c. The time constant.

d. The time required for the current to reach its final value.

19-2. Repeat Problem 19-1 with a 7-H choke, a resistance of 15 Ω, and a dc supply of 120 V.

19-3. The current through a coil is observed to increase from 0 to a final value of 0.25 A in 0.15 s when a 12-V battery is connected across it. Calculate:

a. The resistance of the coil.

b. The time constant.

c. The inductance of the coil.

19-4. The current in a coil of 20-Ω resistance increases from 0 to 0.315 A in 2 s when connected across 10 V dc.

a. What is the inductance of the coil?

b. How long will it take the current to reach its final value?

19-5. An 8-H inductance is connected in series with a 320-Ω resistance. A switch is closed to apply 120 V dc. Use the universal time constant curves to calculate:

a. The current after 50 ms.

b. The voltage across the inductance after 0.1 s.

c. How long it takes for the current to reach 0.3 A.

d. The time for the voltage across the inductance to become 60 V.

19-6. Repeat Problem 19-5 using $L = 12$ H, $R = 200$ Ω, and $V = 150$ V.

19-7. Determine the energy stored in the coil of Problem 19-5 after five time constants.

19-8. Determine the energy stored in the coil of Problem 19-6 after seven time constants.

19-9. A square wave signal generator develops a peak output of 20 V at a frequency of 50 kHz when connected across a 200-μH coil of negligible resistance in series with a 100-Ω resistor.

a. Draw, in the proper time relationship, the input voltage, the current waveform, and the voltage waveform across the coil. Show peak values of voltage and current on all three waveforms.

b. Use Fig. 19-4 to find the current 3 μs after the input voltage goes to zero.

c. Use Fig. 19-4 to find the voltage across the coil 1 μs after the input voltage goes to zero.

d. Use Fig. 19-4 to find how long it takes for the current to fall to 50 mA after the input voltage goes to zero.

19-10. Repeat Problem 19-9 assuming a peak voltage of 10 V.

19-11. A coil of dc resistance 150 Ω is connected in series with a 22-Ω resistor and a square wave signal generator. An oscilloscope across the resistor indicates that the current reaches 63% of its final value in a horizontal distance of 3.2 cm. If the horizontal calibration is 20 μs/cm, determine:

a. The time constant of the whole circuit.

b. The inductance of the coil.

c. What the coil voltage drops to if the signal generator has an internal resistance of 600 Ω and an open circuit voltage of 10-V peak.

19-12. What is the maximum frequency the signal generator can be set to in Problem 19-11 and still maintain the ability to measure the coil's inductance?

19-13. Refer to Fig. 19-9*a*; $R = 3.3$ kΩ, $L = 8$ H, and $V = 20$ V. When the switch is closed calculate:

a. The initial rate of rise of current.

b. The time constant.

c. The final current.

d. The time for the current to reach its final steady-state value.

e. The peak voltage across the inductance.

When the switch is opened, assume an initial switch resistance of 20 kΩ and calculate:

f. The peak voltage across the inductance.

g. The initial rate of fall of the current.

h. The time constant.

i. The approximate time for the current to drop to zero.

19-14. The circuit in Problem 19-13 has a 2.2-kΩ resistor connected across the 3.3-kΩ resistor and 8-H inductance as in Fig. 19-12*a* to reduce the inductive effect.

a. What voltage will be induced across the coil when the switch is opened? Will the lamp light?

b. What is the total current drawn from the supply when the switch is closed?

The following problems are based on exponential equations in Appendix A-9.

19-15. Repeat Problem 19-5 using Eqs. A9-1 and A9-2 instead of the universal time-constant curves.

19-16. Repeat Problem 19-6 using Eqs. A9-1 and A9-2.

19-17. Repeat Problem 19-9 assuming a peak voltage of 10 V and using Eqs. A9-3 and A9-4.

CHAPTER TWENTY

INDUCTANCE IN AC CIRCUITS

In this chapter we consider the current and voltage relationships of an inductance in a sinusoidal ac circuit. Because the current is always changing, there is some effective opposition to the current drawn by a coil at a given frequency. This is called the *inductive reactance* (X_L), which is determined both by inductance and frequency.

Because of inductive reactance, a coil has a greater opposition to alternating current than to direct current. This is made use of in a dc *filter* circuit to reduce the ac ripple component. An inductor is also used in a fluorescent lamp circuit, where, as a *ballast,* it is used to induce a momentary high voltage and then limit the current through the lamp after it is lit.

The series and parallel connection of inductive reactances can be treated similarly to resistances except when *mutual reactance,* X_m, is involved. In this case, the coefficient of coupling permits a determination of mutual inductance and the total inductive reactance.

The *quality of a coil,* Q, is expressed in terms of inductive reactance, and is a "figure of merit" to show how inductive a coil is compared with its *ac resistance.* The ac resistance of a coil is shown to be different from its dc ohmic resistance due partially to the *skin effect.*

20-1 EFFECT OF INDUCTANCE IN AN AC CIRCUIT

We have seen how an inductance induces an emf whenever the current through it changes. If we cause a sinusoidal current to flow through an inductance, we should expect an emf to be induced almost continually. Since this emf opposes any change in current, the inductance must have some effective opposition to the flow of alternating current. This opposition is over and above any resistance that the coil may have.

20-1.1 Phase Angle Relationship Between Current and Voltage

Let us assume that an inductance of negligible resistance has a sinusoidal current flowing through it as in Fig. 20-1. The current waveform is given by

$$i = I_m \sin \omega t$$

The voltage across the coil, which must also be the voltage applied from the source, is

$$v = L\left(\frac{\Delta i}{\Delta t}\right) \qquad (17\text{-}1)$$

Recall that $\frac{\Delta i}{\Delta t}$ is the *slope* of the current waveform. Only when the current is at its peak values of I_m and $-I_m$ is the slope zero. Only at these instants, then, is the voltage across the coil zero. The current waveform has its greatest slope where it crosses the time axis. This causes the maximum voltage to be induced across the coil, either positively or negatively, according to whether the current is increasing or decreasing, respectively. See Fig. 20-1*b*.

It is not difficult to justify what waveform of voltage must appear across the coil, although it requires some mathematics to *prove* that the waveform is sinusoidal in shape. In fact, the voltage waveform is actually a cosine curve given by

$$v = V_m \cos \omega t$$

This is actually a sine curve that has been shifted to the left, or advanced by a quarter of a cycle or 90°. Since the voltage waveform reaches its peak value one-quarter of a cycle *before* the current waveform reaches its own peak, we say that the voltage *leads* the current by 90°.

Conversely, we can say that the current in the coil *lags* behind the applied coil voltage by 90° in a *purely* inductive circuit.

This angle is often referred to as a *phase angle,* because it shows by what angle the two waveforms of current and voltage are *out of phase.* Recall that in a purely resistive circuit the current and voltage are in-phase; the phase angle is 0°.

It is easy to remember that the current *lags* the voltage if we recall what happens in a dc circuit. When voltage is applied to a coil, the current builds up slowly to its final value (Fig. 19-2). Even in a dc circuit there is a delay between the application of voltage across the coil and the current flowing in the coil.

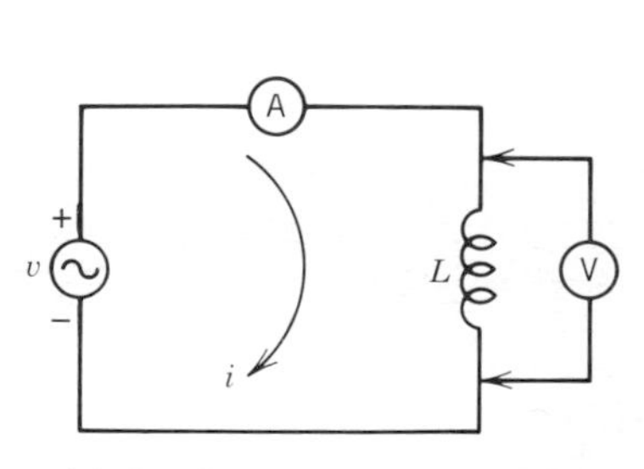

(*a*) Pure inductance in an ac circuit

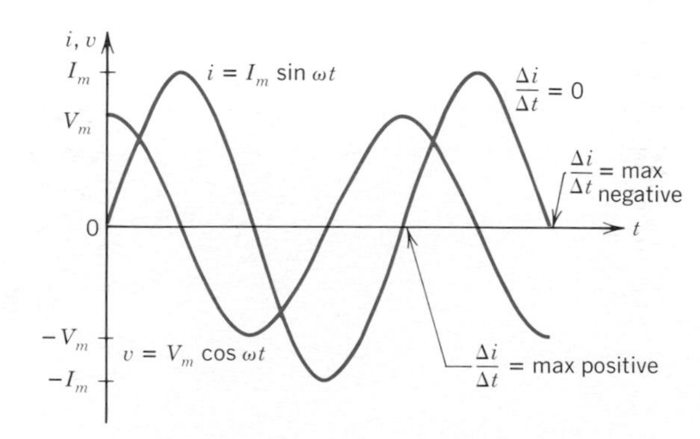

(*b*) Waveforms of current and voltage to show current lagging voltage by 90°

FIGURE 20-1
Effect of inductance in an ac circuit.

20-2 INDUCTIVE REACTANCE

Since both the current and voltage waveforms in Fig. 20-1 are sinusoidal, we can obtain their rms values:

$$V = \frac{V_m}{\sqrt{2}} \tag{15-17}$$

and

$$I = \frac{I_m}{\sqrt{2}} \tag{15-16}$$

These are the values of the voltmeter and ammeter readings in Fig. 20-1*a*. Now recall that the opposition to current in a resistive circuit (either dc or ac) is given by the ratio:

$$\frac{V_R}{I_R} = R \tag{15-15}$$

Similarly, the opposition to current provided by a *pure* inductance is given by the ratio V_L/I_L. Although this ratio of voltage to current must have units of ohms, we cannot call it resistance. (A pure inductance has *no* (zero) resistance.) The term used is *inductive reactance, X_L*. Thus, for a pure inductor,

$$\mathbf{X_L = \frac{V_L}{I_L} \quad \text{ohms } (\Omega)} \tag{20-1}$$

where: X_L is the inductive reactance in ohms, Ω

V_L is the voltage across the pure inductance in volts, V

I_L is the current through the inductance, in amperes, A

EXAMPLE 20-1

In the circuit of Fig. 20-2, a 28.28-V p-p sinusoidal voltage causes an ammeter reading of 10 mA. What is the opposition to current caused by the inductance?

SOLUTION

The rms value of the applied voltage is given by

$$V = \frac{V_{p-p}}{2\sqrt{2}} \tag{15-17}$$

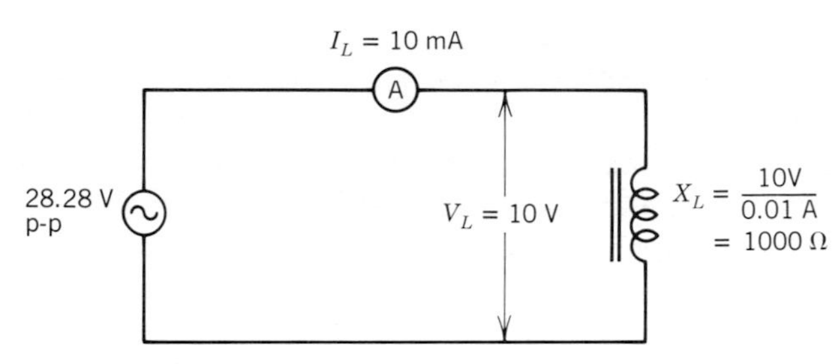

FIGURE 20-2
Circuit for Example 20-1, showing $X_L = V_L/I_L$.

$$= \frac{28.28\text{ V}}{2\sqrt{2}} = 10\text{ V}$$

$$X_L = \frac{V_L}{I_L} \tag{20-1}$$

$$= \frac{10\text{ V}}{10 \times 10^{-3}\text{ A}}$$

$$= 1000\ \Omega = \mathbf{1\ k\Omega}$$

EXAMPLE 20-2

How much current will flow through a coil of negligible resistance whose inductive reactance is 500 Ω when 120 V are applied?

SOLUTION

$$X_L = \frac{V_L}{I_L} \tag{20-1}$$

Therefore,

$$I_L = \frac{V_L}{X_L}$$

$$= \frac{120\text{ V}}{500\ \Omega} = \mathbf{0.24\ A}$$

EXAMPLE 20-3

What is the voltage drop across a coil of inductive reactance 1.5 kΩ and negligible resistance when a current of 70 mA flows through the coil?

SOLUTION

$$X_L = \frac{V_L}{I_L} \tag{20-1}$$

Therefore, $V_L = I_L X_L$

$$= 70 \times 10^{-3}\text{ A} \times 1.5 \times 10^3\ \Omega$$

$$= \mathbf{105\ V}$$

20-3 FACTORS AFFECTING INDUCTIVE REACTANCE

The inductive reactance or opposition to current is due to the induced emf in the inductance. Anything that increases this emf must also increase the inductive reactance.

We know that

$$v_L = L \frac{\Delta i}{\Delta t} \qquad (17\text{-}1)$$

Plainly, an increase in L must increase the inductive reactance. Further, $\frac{\Delta i}{\Delta t}$, the slope of the current waveform, must depend upon the frequency. A higher frequency means that the peak value of current is reached in a shorter time, so $\frac{\Delta i}{\Delta t}$ must be higher. Combining these two effects and using the angular frequency $\omega = 2\pi f$, we obtain

$$\mathbf{X_L = 2\pi f L = \omega L \quad \text{ohms } (\Omega)} \qquad (20\text{-}2)$$

where: X_L is the inductive reactance in ohms, Ω
f is the frequency in hertz, Hz
L is the inductance in henrys, H
ω is the angular frequency in radians per second, rad/s

Note that this equation is valid only for sinusoidal ac circuits. (A full derivation, using the principles we have covered, is given in Appendix A-10.)

Equation 20-2 states that there is a linear increase in the inductive reactance of a coil as the frequency increases. This is represented in the graph of Fig. 20-3 drawn for a constant inductance of $L = 1.6$ H. Note that when $f = 0$ Hz (direct current), the inductive reactance is zero. **This means that inductance has no steady-state effect or opposition to current in a direct current circuit.**

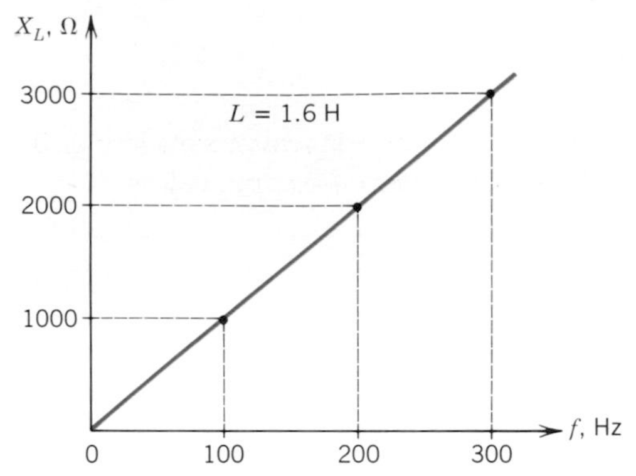

FIGURE 20-3
Linear increase of X_L with frequency, f, for a constant inductance of 1.6 H.

EXAMPLE 20-4

A 120-V, 60-Hz source is applied to a 5-H coil of negligible resistance.

a. How much current flows?

b. What is the equation for current?

SOLUTION

a. $X_L = 2\pi f L$ (20-2)

$= 2\pi\ 60 \text{ Hz} \times 5 \text{ H}$

$= 1885\ \Omega$

Therefore, $I_L = \frac{V_L}{X_L}$ (20-1)

$= \frac{120 \text{ V}}{1885\ \Omega} = 0.064 \text{ A} = \mathbf{64\ mA}$

b. $I_m = I_L \times \sqrt{2}$ (15-20)

$= 0.064 \text{ A} \times \sqrt{2}$

$= 0.091 \text{ A}$

$i = I_m \sin 2\pi f t$

$= 0.091 \sin 2\pi 60 t$

$= \mathbf{0.091 \sin 377t \text{ A}}$

EXAMPLE 20-5

A sine wave signal generator, set to 1-kHz and 10-V output, causes a current of 50 mA through a coil of negligible resistance. What is the inductance of the coil?

SOLUTION

$X_L = \frac{V_L}{I_L}$ (20-1)

$= \frac{10 \text{ V}}{50 \times 10^{-3} \text{ A}} = 200\ \Omega$

$X_L = 2\pi f L$ (20-2)

Therefore, $L = \dfrac{X_L}{2\pi f}$

$$= \frac{200\ \Omega}{2\pi \times 1 \times 10^3\ \text{Hz}}$$

$$= 0.032\ \text{H} = \mathbf{32\ mH}$$

The total opposition to current arising from the combined resistance and inductance of the coil (called impedance) is considered in Section 24-3.

20-4 APPLICATION OF INDUCTORS IN FILTER CIRCUITS

It should be clear from the previous discussion that a coil with negligible resistance provides a much higher opposition to alternating current than to direct current. This is illustrated in the two circuits shown in Fig. 20-4. With direct current, the inductive reactance of the coil is zero. The only opposition to current in the circuit is the 20 Ω of coil resistance and 100 Ω of lamp resistance. With an applied emf of 120 V dc this results in a current of 1 A dc. The lamp lights.

With alternating current, an inductance of 4 H at a frequency of 60 Hz causes an inductive reactance of 1500 Ω. The resistance of the circuit is negligible in comparison with X_L so that the current is determined by V_L/X_L. The current of 80 mA is insufficient to light the lamp.

The different opposition to alternating and direct currents provided by a coil is made use of in a dc power supply *filter* circuit. After the alternating current has been converted to direct current by diodes, an ac component remains and must be filtered out to leave pure direct current. This may be done using a coil, often called a *choke*, to allow the passage of direct current but block the alternating current.

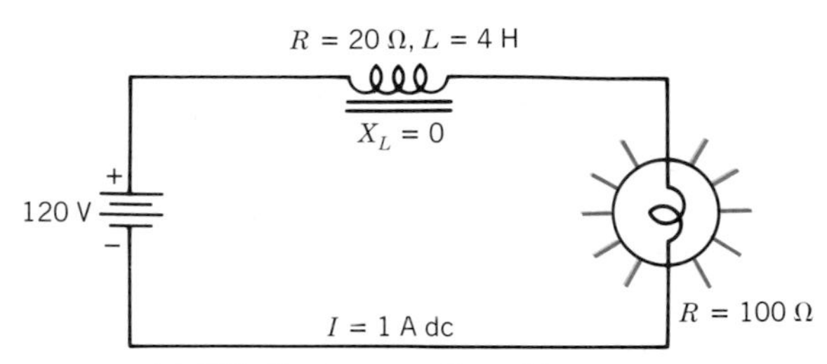

(*a*) With direct current, the coil has zero inductive reactance. The lamp lights.

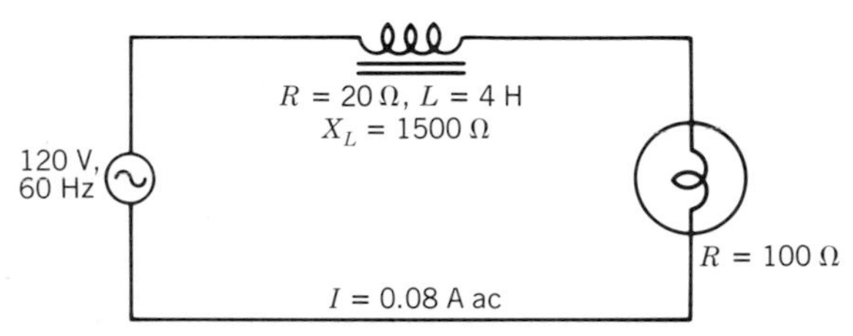

(*b*) With alternating current, the high inductive reactance limits the current to 80 mA. The lamp does not light.

FIGURE 20-4
Comparison of inductance in dc and ac circuits.

Consider the simple filter circuit in Fig. 20-5. To simplify matters, let us assume that the coil's resistance is negligible compared with R. Also assume that the inductive reactance of the coil, X_L, is much larger than R.

Consider a waveform applied to the input that contains both direct current and alternating current. The dc portion appears across R without being affected by the coil. But the ac portion undergoes a form of voltage division between X_L and R. With the assumptions made, the ac output voltage across R is given by

$$V_{\text{out(ac)}} = V_{\text{in(ac)}} \times \frac{R}{X_L} \tag{20-3}$$

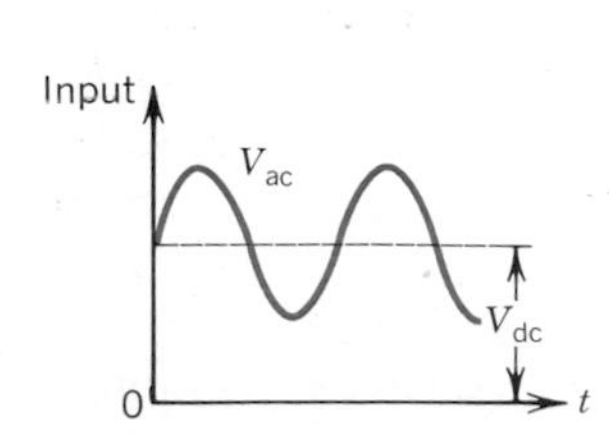

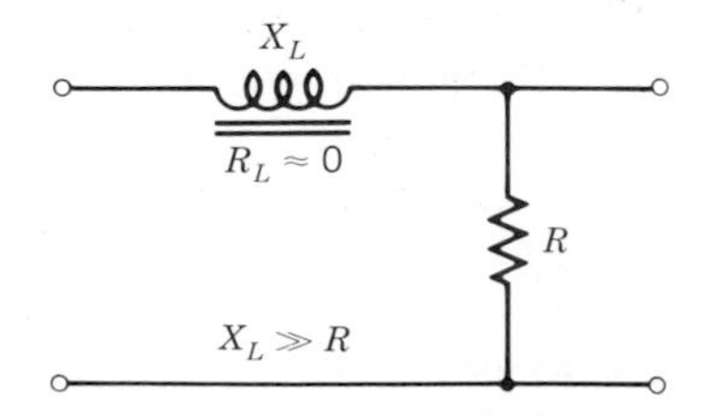

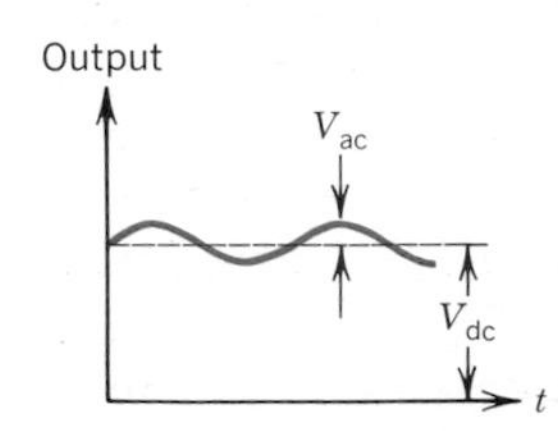

FIGURE 20-5
Inductor used in filter circuit to reduce ac ripple component.

EXAMPLE 20-6

A waveform containing 10 V dc and 10 V ac is applied to the filter of Fig. 20-5 with $L = 10$ H and $R = 100\ \Omega$. If the ac component has a frequency of 60 Hz, determine:

a. The dc output voltage.

b. The ac output voltage.

SOLUTION

a. The dc output = the dc input = 10 V
(assuming that the resistance of the coil is zero).

b. $$X_L = 2\pi fL \qquad (20\text{-}2)$$
$$= 2\pi \times 60\text{ Hz} \times 10\text{ H}$$
$$= 3770\ \Omega$$

$$V_{out(ac)} = V_{in(ac)} \times \frac{R}{X_L} \qquad (20\text{-}3)$$

$$= 10\text{ V} \times \frac{100\ \Omega}{3770\ \Omega} = \mathbf{0.27\ V}$$

Note that Example 20-6 shows how the ac *ripple* component is reduced by an inductor, as shown in Fig. 20-5.

20-5 APPLICATION OF INDUCTANCE IN A FLUORESCENT LAMP CIRCUIT

Figure 20-6*a* shows a preheat fluorescent lamp circuit. It is typical of many desk-type fluorescent lamps, but the principle of operation applies to larger lamps also.

The inductor (ballast) is used to induce a momentary high voltage to fire the lamp. The ballast then *limits* the current through the lamp, after the lamp is lit, because of the coil's inductive reactance. The operation of the lamp circuit is as follows.

The fluorescent lamp is a glass tube with a tungsten filament sealed in each end. The inner surface of the tube is coated with a phosphor material—the particular type determines the color of light produced. Most of the air is removed during manufacture, and a small amount of argon gas and mercury are admitted to the sealed tube.

When the momentary-contact ON switch is pushed CLOSED, and held closed for several seconds, a complete series path exists for current to flow through the two filaments and the ballast (inductor). This allows the filaments to become heated, emitting electrons. A dull glow is observed at each end of the tube. When the ON switch is released (OPEN), the current through the ballast is interrupted, causing a high voltage to be momentarily induced. This voltage, along with the 120-V input, is sufficient to cause the lamp to "fire." This means that the current is conducted through the ionized gas in the tube from one filament to the other. (This in itself, however, does *not* produce the visible light emitted by a fluorescent lamp.)

As the electrons move through the tube, they collide with the now-gaseous mercury ions. This causes the mercury valence electrons to be dislodged from their shell and raised to a higher-energy level. As these electrons fall back to their stable orbits, the energy they absorbed is released in the form of invisible ultraviolet "light." When the ultraviolet radiation strikes the inside of the tube, the phosphor coating *fluoresces,* emitting visible light.

The visible light produced by this process wastes much less energy in the form of heat compared to an incandescent lamp. As a result, a fluorescent lamp produces approximately three times as much light (measured in lumens) compared to an incandescent lamp of the same wattage. In addition, the life of a typical fluorescent tube is between 10,000 and 20,000 h, compared to only 1000 to 1500 h for most incandescent lamps.

It should be noted that the ballast gets its name from the second function it provides. After the lamp is lit, a typical 14-W lamp requires only 55 V to maintain the proper current through the lamp. The opposition to alternating current caused by the inductance, its inductive reactance, drops the applied 120 V to the required value across the lamp.

Fluorescent lamps that use a *single* on-off switch in the supply line for control purposes employ a *starter* in place of the ON switch, as shown in Fig. 20-6*b*.

The starter resembles a neon glow lamp, but it also has a bimetallic strip. Initially, the starter is an open circuit. When power is applied to the fluorescent lamp circuit, 120 V appears across the open starter electrodes. Ionization of the neon gas makes the starter glow, heating the bimetallic strip until it bends and makes contact with the other electrode. This completes a path for current through the filaments of the fluorescent lamp—similar to our closing the ON switch—so the lamp preheats. But a few seconds after the starter switch has closed, no further

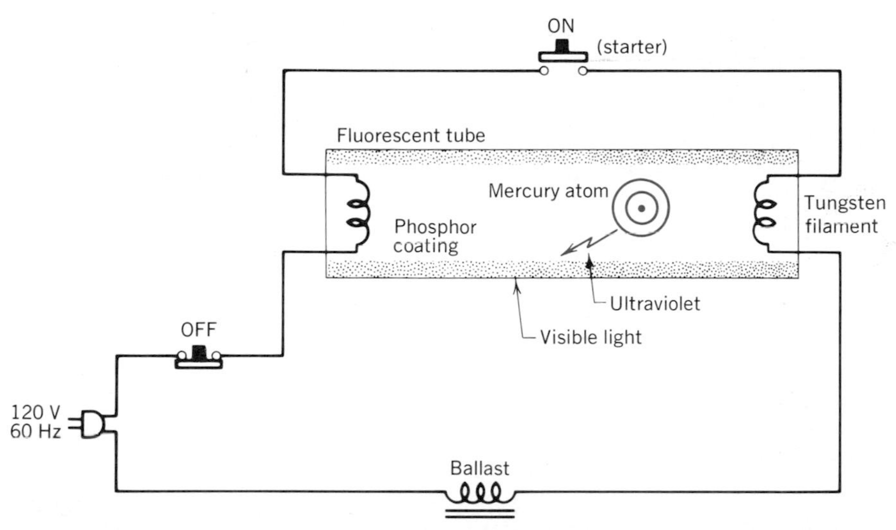

(*a*) Schematic diagram

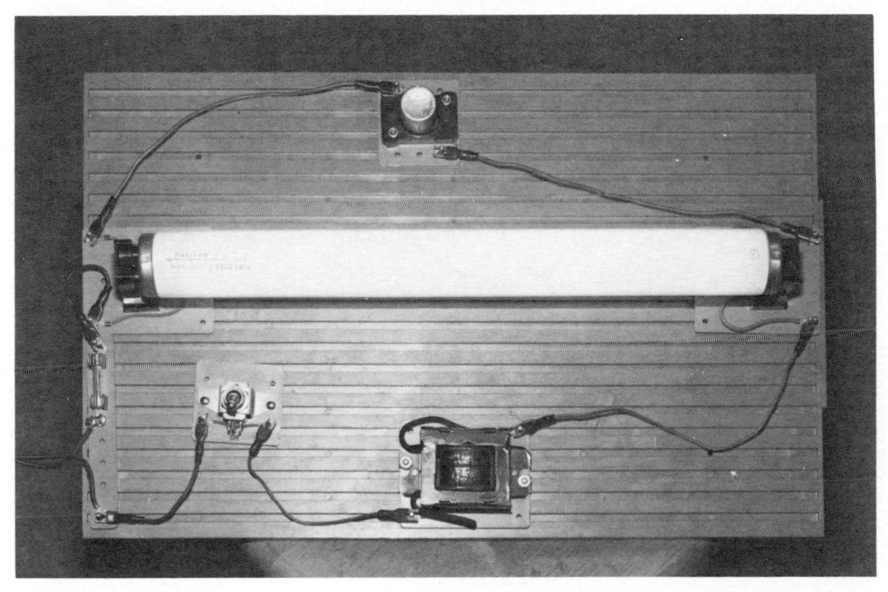

(*b*) Circuit that uses a starter

FIGURE 20-6
Use of an inductor in a preheat fluorescent lamp circuit.

heat is developed; therefore, the bimetal strip cools, bends, and opens the contacts, interrupting the current as required. Once the lamp fires, there is insufficient voltage across the lamp (and starter) to initiate ionization in the starter, so the starter contacts remain open.

A recent development is a fluorescent lamp that can be used in incandescent sockets. It consists of a replaceable fluorescent tube that is *double-folded* to occupy a space of $7\frac{1}{2}$ in. in length by $3\frac{1}{4}$ in. in diameter. A ballast and starter switch are located in the base of the lamp. Consuming only 27 W, the compact fluorescent lamp produces 1000 lumens (a little more than a 60-W incandescent lamp) but has an average life of 7500 h. See Fig. 20-7.

Other fluorescent lamps are the *rapid-start,* which does not need a starter, and the *instant-start,* which has a single-pin base and a step-up transformer to produce the high ionization voltage to fire the lamp.

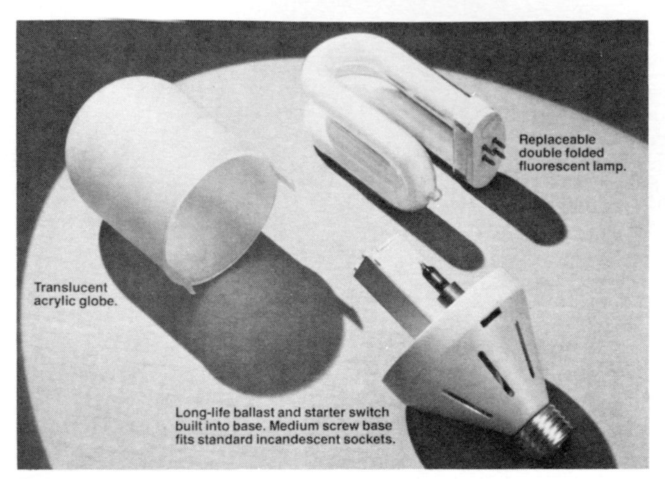

FIGURE 20-7
Compact fluorescent lamp designed to replace incandescent lamps. (Courtesy of Westinghouse Electric Corporation.)

20-6 SERIES INDUCTIVE REACTANCES

When two or more coils are connected in series, the total inductive reactance is given by

$$X_{L_T} = X_{L_1} + X_{L_2} + X_{L_3} + \cdots + X_{L_n} \quad (20\text{-}4)$$

This equation assumes that there is no mutual inductance between the coils. The current through the coils may be obtained from

$$I = \frac{V}{X_{L_T}} \quad (20\text{-}5)$$

and the voltage across each coil from $V_L = IX_L$.

EXAMPLE 20-7

Two coils of 8 and 4 H with negligible resistance are series-connected across a 120-V, 60-Hz source, as in Fig. 20-8. Calculate:

a. The total inductive reactance.
b. The reading of an ammeter in series with the coils.
c. The voltage drop across each coil.

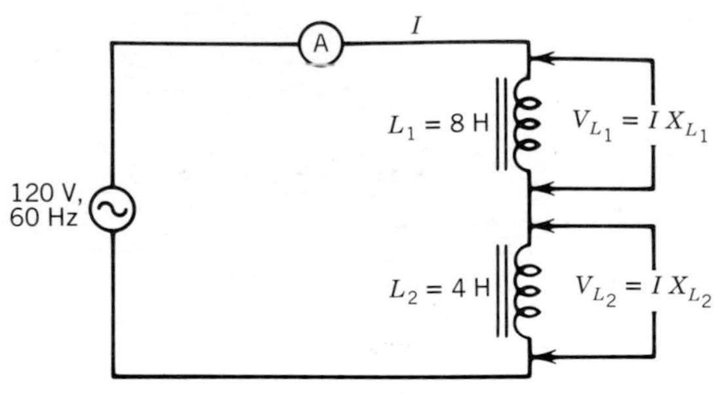

FIGURE 20-8
Circuit for Example 20-7 with series inductive reactances.

SOLUTION

a.
$$X_{L_1} = 2\pi f L_1 \quad (20\text{-}2)$$
$$= 2\pi \times 60 \text{ Hz} \times 8 \text{ H}$$
$$= 3016\ \Omega \approx 3\ \text{k}\Omega$$
$$X_{L_2} = 2\pi f L_2 \quad (20\text{-}2)$$
$$= 2\pi \times 60 \text{ Hz} \times 4 \text{ H}$$
$$= 1508\ \Omega \approx 1.5\ \text{k}\Omega$$
$$X_{L_T} = X_{L_1} + X_{L_2} \quad (20\text{-}4)$$
$$= 3\ \text{k}\Omega + 1.5\ \text{k}\Omega = \mathbf{4.5\ k\Omega}$$

b.
$$I = \frac{V}{X_{L_T}} \quad (20\text{-}5)$$
$$= \frac{120\ \text{V}}{4.5\ \text{k}\Omega} = \mathbf{26.67\ mA}$$

c.
$$V_{L_1} = IX_{L_1} \quad (20\text{-}1)$$
$$= 26.67 \times 10^{-3}\ \text{A} \times 3 \times 10^3\ \Omega$$
$$= \mathbf{80\ V}$$
$$V_{L_2} = IX_{L_2} \quad (20\text{-}1)$$
$$= 26.67 \times 10^{-3}\ \text{A} \times 1.5 \times 10^3\ \Omega$$
$$= \mathbf{40\ V}$$

Note that $V_{L_1} + V_{L_2} = 120$ V, the supply voltage. Also, note that $V_{L_1} = 2V_{L_2}$ because $L_1 = 2\,L_2$.

20-6.1 Series Inductive Reactances with Mutual Reactance

We saw in Chapter 17 that two series-connected coils that have mutual inductance between them have a total inductance given by

$$L_T = L_1 + L_2 \pm 2\,M \qquad (17\text{-}10)$$

When an alternating current flows through these coils, the total inductive reactance is

$$X_{L_T} = X_{L_1} + X_{L_2} \pm 2\,X_M \qquad \text{ohms } (\Omega) \quad (20\text{-}6)$$

where: X_{L_1} is the inductive reactance of L_1 in ohms, Ω
X_{L_2} is the inductive reactance of L_2 in ohms, Ω
$X_M = 2\pi fM$ is the mutual reactance in ohms, Ω
M is the mutual inductance between the coils in henrys, H

The mutual reactance is positive if the magnetic fields are aiding, negative if the fields are opposing.

As before, the current through the coils is given by $I = V/X_{L_T}$. The voltage across each coil is

$$V_L = I(X_L \pm X_M) \qquad (20\text{-}7)$$

EXAMPLE 20-8

Assume that the two coils in Example 20-7 now have a coefficient of coupling of 0.6. The coils are connected with their fields aiding, as shown by the dot notation in Fig. 20-9. Calculate:

a. The total inductive reactance.
b. The current through the coils.
c. The voltage across each coil.

SOLUTION

a. $M = k\sqrt{L_1 L_2}$ (17-12)

$= 0.6\sqrt{8\text{ H} \times 4\text{ H}}$

$= 3.394\text{ H} \approx 3.4\text{ H}$

$X_M = 2\pi fM$

$= 2\,\pi \times 60\text{ Hz} \times 3.4\text{ H}$

$= 1282\ \Omega \approx 1.3\text{ k}\Omega$

$X_{L_T} = X_{L_1} + X_{L_2} + 2\,X_M$ (20-6)

$= 3\text{ k}\Omega + 1.5\text{ k}\Omega + 2 \times 1.3\text{ k}\Omega$

$= \mathbf{7.1\ k\Omega}$

b. $I = \dfrac{V}{X_{L_T}}$ (20-5)

$= \dfrac{120\text{ V}}{7.1\text{ k}\Omega} = \mathbf{16.9\ mA}$

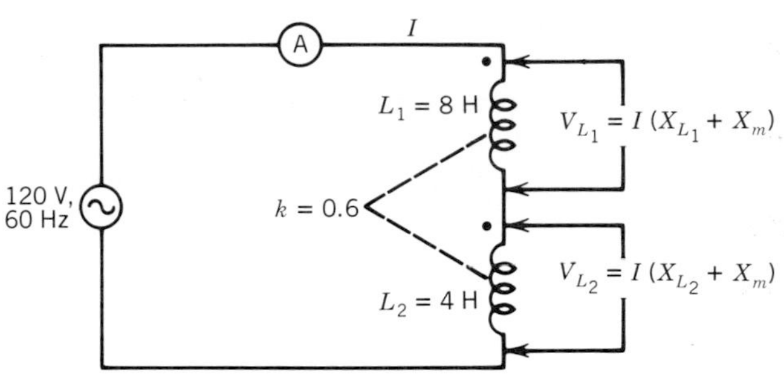

FIGURE 20-9
Circuit for Example 20-8 with series inductive reactances and positive mutual reactance.

c. $V_{L_1} = I\,(X_{L_1} + X_M)$ (20-7)

$= 16.9 \times 10^{-3}\text{ A}\,(3 \times 10^3\ \Omega + 1.3 \times 10^3\ \Omega)$

$= 16.9 \times 10^{-3}\text{ A} \times 4.3 \times 10^3\ \Omega$

$= \mathbf{72.7\ V}$

$V_{L_2} - I\,(X_{L_2} + X_M)$ (20-7)

$= 16.9 \times 10^{-3}\text{ A}\,(1.5 \times 10^3\ \Omega + 1.3 \times 10^3\ \Omega)$

$= 16.9 \times 10^{-3}\text{ A} \times 2.8 \times 10^3\ \Omega$

$= \mathbf{47.3\ V}$

Note once again that $V_{L_1} + V_{L_2} = 120$ V. However, V_{L_1} is *not* twice V_{L_2} because of the effect of the mutual reactance X_M.

20-7 PARALLEL INDUCTIVE REACTANCES

The parallel connection of two or more inductive reactances provides a reduction in the overall reactance compared with the branch having the lowest reactance. Thus, similar to resistances in parallel, the total inductive reactance is given by

$$\frac{1}{X_{L_T}} = \frac{1}{X_{L_1}} + \frac{1}{X_{L_2}} + \frac{1}{X_{L_3}} + \cdots + \frac{1}{X_{L_n}} \qquad (20\text{-}8)$$

EXAMPLE 20-9

Two coils of inductance 8 and 4 H, with negligible resistance, are connected in parallel across a 120-V, 60-Hz supply, as in Fig. 20-10. Determine the total current drawn from the supply.

SOLUTION

As earlier,
$$X_{L_1} = 3\ \text{k}\Omega$$
$$X_{L_2} = 1.5\ \text{k}\Omega$$

$$I_{L_1} = \frac{V}{X_{L_1}} \qquad (20\text{-}1)$$
$$= \frac{120\ \text{V}}{3\ \text{k}\Omega} = 40\ \text{mA}$$

$$I_{L_2} = \frac{V}{X_{L_2}} \qquad (20\text{-}1)$$
$$= \frac{120\ \text{V}}{1.5\ \text{k}\Omega} = 80\ \text{mA}$$

$$\text{Total current } I_T = I_{L_1} + I_{L_2} \qquad (6\text{-}1)$$
$$= 40\ \text{mA} + 80\ \text{mA} = \mathbf{120\ mA}$$

or
$$\frac{1}{X_{L_T}} = \frac{1}{X_{L_1}} + \frac{1}{X_{L_2}} \qquad (20\text{-}8)$$
$$X_{L_T} = \frac{X_{L_1} \times X_{L_2}}{X_{L_1} + X_{L_2}}$$
$$= \frac{3\ \text{k}\Omega \times 1.5\ \text{k}\Omega}{3\ \text{k}\Omega + 1.5\ \text{k}\Omega} = 1\ \text{k}\Omega$$

Therefore,
$$I = \frac{V}{X_{L_T}} \qquad (20\text{-}5)$$
$$= \frac{120\ \text{V}}{1\ \text{k}\Omega} = \mathbf{120\ mA}$$

or
$$L_T = \frac{L_1 \times L_2}{L_1 + L_2} \qquad (17\text{-}5)$$
$$= \frac{8\ \text{H} \times 4\ \text{H}}{8\ \text{H} + 4\ \text{H}}$$
$$= \frac{32}{12}\ \text{H} = 2\frac{2}{3}\ \text{H}$$
$$X_{L_T} = 2\pi f L_T \qquad (20\text{-}2)$$
$$= 2\ \pi \times 60\ \text{Hz} \times 2\frac{2}{3}\ \text{H}$$
$$= 1000\ \Omega$$
$$I = \frac{V}{X_{L_T}} \qquad (20\text{-}5)$$
$$= \frac{120\ \text{V}}{1\ \text{k}\Omega} = \mathbf{120\ mA}$$

20-8 QUALITY OF A COIL, Q

At high frequencies, the usefulness of a coil is judged not only by its inductance, but also by the *ratio* of its inductive reactance to the *ac* or *effective* resistance of the coil. See Fig. 20-11. This ratio is called the *quality* of the coil and is given the symbol Q:

$$Q = \frac{X_L}{R_{ac}} \qquad (20\text{-}9)$$

where: Q is the quality of the coil (a dimensionless number)

X_L is the reactance of the coil in ohms, Ω

R_{ac} is the ac resistance of the coil in ohms, Ω

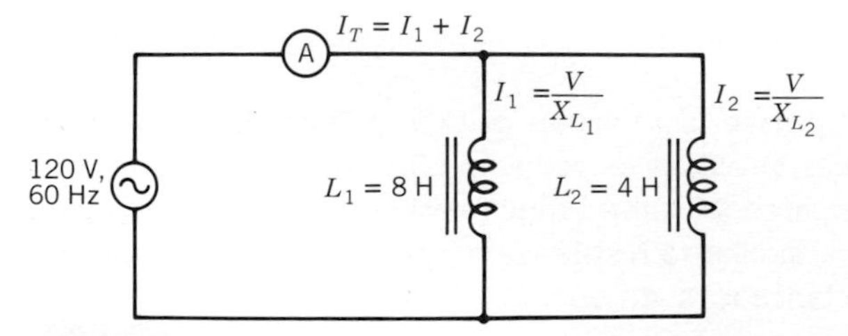

FIGURE 20-10
Circuit for Example 20-9 with parallel inductive reactances.

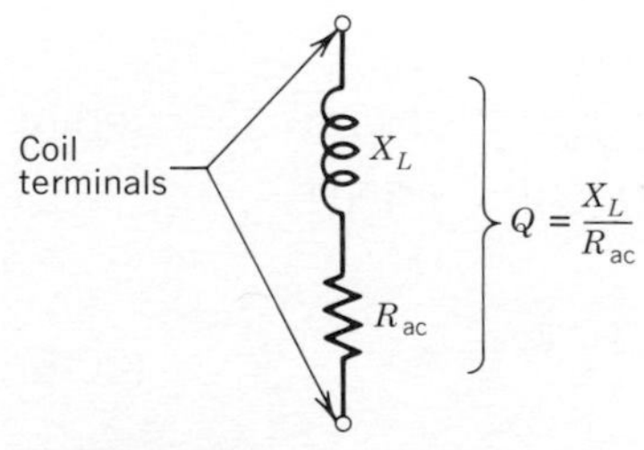

FIGURE 20-11
Equivalent circuit of a coil.

The Q of a coil is also known as the *storage factor,* since it is proportional to the energy storage capability of the coil.

EXAMPLE 20-10

A 200-μH coil, when operated at a frequency of 1.5 MHz, has a total ac resistance of 100 Ω. What is the quality of the coil?

SOLUTION

$$Q = \frac{X_L}{R_{ac}} = \frac{2\pi fL}{R_{ac}} \tag{20-9}$$

$$= \frac{2\pi \times 1.5 \times 10^6 \text{ Hz} \times 200 \times 10^{-6} \text{ H}}{100\ \Omega}$$

$$= \mathbf{18.8}$$

20-9 EFFECTIVE RESISTANCE, R_{ac}

Note the use of *effective* (ac) resistance in Eq. 20-9. This is different from, and higher than, the dc (ohmic) resistance measured with an ohmmeter, for the following reason.

Consider the changing magnetic field (set up by the alternating current) around, and in, the wire conductors that make up the coil. The field has the greatest rate of change, $\frac{\Delta\Phi}{\Delta t}$, at the *center* of the conductor. Thus the induced voltage, $v = \frac{\Delta\Phi}{\Delta t}$, and the opposition to the current, has the greatest value at the center of the conductor. As a result, at radio frequencies (hundreds of kHz and above), most of the current flows along the *surface* of the conductor and very little, if any, at the center. (The current becomes denser toward the surface.) This effectively decreases the cross-sectional area of the conductor and increases its resistance.

This is called the *skin effect.* At microwave frequencies, the effect is so pronounced that hollow conductors are used because current flows only on the surface. At radio frequencies, an additional loss occurs due to the *radiation* of energy. The combined higher resistance is called the ac resistance, R_{ac}.

The Q of a coil, at radio frequencies, is relatively constant with frequency. This is because R_{ac} increases with the frequency at approximately the same rate that X_L increases, maintaining an almost constant ratio. Thus, in Example 20-10, at a frequency of 3 MHz, both X_L and R_{ac} would approximately double so that Q remains approximately 18.

At power line frequencies of 60 Hz, the skin effect is negligible. And yet, a significant difference between dc and ac resistance can occur. If the coil is wrapped around a *magnetic core,* we have both hysteresis and eddy current losses, as discussed in Section 17-6.3. Since both of these losses convert electrical energy to heat, their effect is the dissipation of electrical power, as if an equivalent resistance is connected in the circuit. The total ac resistance of a circuit can be obtained from a wattmeter reading of total power. That is, $R_{ac} = P/I^2$, which includes the additional effects of hysteresis, eddy currents, skin effect, and so on, over and above the dc (ohmic) resistance.

EXAMPLE 20-11

An inductor of 0.5 H is connected in series with an ammeter, a wattmeter, and a 60-Hz supply, as in Fig. 20-12. If the ammeter indicates 250 mA and the wattmeter indicates 5 W, calculate:

a. The ac resistance of the coil.

b. The Q of the coil.

SOLUTION

a. $$R_{ac} = \frac{P}{I^2} \tag{15-12}$$

$$= \frac{5 \text{ W}}{(0.25 \text{ A})^2} = \mathbf{80\ \Omega}$$

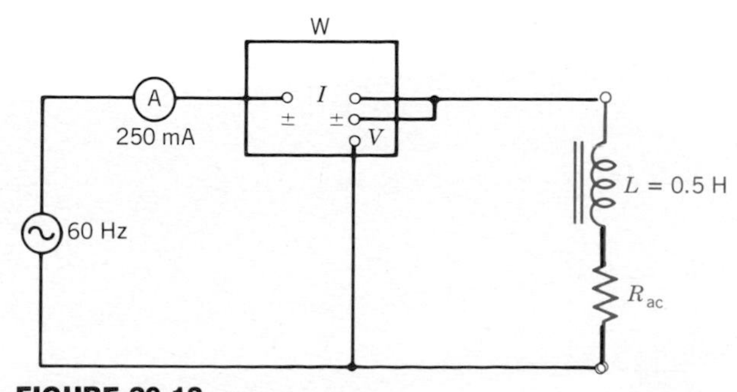

FIGURE 20-12
Circuit for Example 20-11.

b. $$Q = \frac{X_L}{R_{ac}} \qquad (20\text{-}9)$$

$$= \frac{2\pi \times 60 \text{ Hz} \times 0.5 \text{ H}}{80 \ \Omega} = \mathbf{2.4}$$

20-10 MEASURING INDUCTANCE

An instrument that operates on the principle of the Wheatstone bridge, introduced in Section 13-7, may be used to measure the inductance and quality (Q) of a coil. The impedance bridge, shown in Fig. 20-13, can also measure the dc and ac resistance of the coil. Both controls are adjusted, first one then the other, until minimum deflection of the galvanometer is obtained. The Q is read directly off the left dial, the inductance off the right using the appropriate multiplying factor. An internally generated 1-kHz signal is used for the measurement, but an external signal generator may be connected to measure the Q at any other desired frequency. Inductances from 1 μH to 1100 H may be measured with Q values from 0.02 to 1000. (A digital instrument that measures inductance from 1 μH to 10 H is shown in Fig. 21-9.)

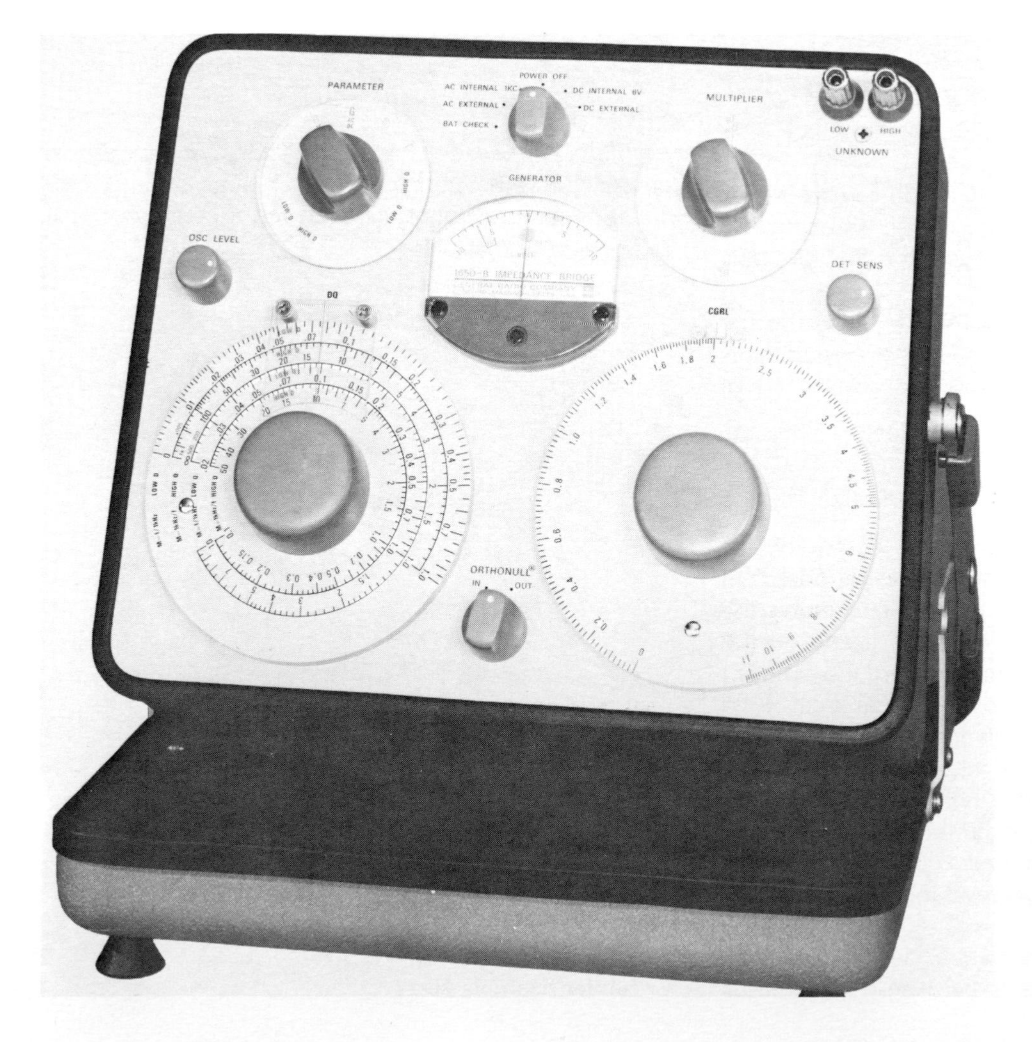

FIGURE 20-13
Impedance bridge suitable for measuring inductance and Q of a coil. (Courtesy of GenRad Inc., Concord, Mass.)

SUMMARY

1. A sinusoidal voltage across a pure inductance causes a sinusoidal current that lags behind the voltage by 90°.

2. The opposition to alternating current provided by a pure inductance is called inductive reactance and is given by $X_L = 2\pi fL$.

3. The current through an inductor that has negligible resistance is given by $I_L = V_L/X_L$.

4. A coil with low resistance provides a much higher opposition to alternating current than to direct current. This property is put to use in filter circuits to remove unwanted alternating current.

5. The ballast in a fluorescent lamp circuit provides two functions. It induces a momentary high voltage, when current through the coil is interrupted, to fire the lamp. The inductive reactance of the coil then limits the current through the lamp after the lamp is lit.

6. Invisible ultraviolet energy produces visible light from a fluorescent lamp by making a phosphor coating inside the lamp fluoresce. This makes the lamp approximately three times more efficient than an incandescent lamp.

7. The equations for combining inductive reactances connected in series and parallel are the same as those for resistances if no mutual inductance is involved.

8. The total inductive reactance of two series-connected coils that have mutual inductance is given by $X_{L_T} = X_{L_1} + X_{L_2} \pm 2X_M$.

9. The voltage across each of the series-connected coils that have mutual inductance is given by $V_L = I(X_L \pm X_M)$.

10. The quality of a coil is given by $Q = X_L/R_{ac}$.

11. The effective or ac resistance is higher than the dc resistance and includes the effects of skin effect, radiation loss, eddy current, and hysteresis losses.

12. Inductance can be measured directly on an impedance bridge. Both ac and dc resistance of the coil can be measured as well as the quality, Q.

SELF-EXAMINATION

Answer T or F, or, for multiple choice, a, b, c or d
(Answers at back of book)

20-1. In an ac circuit, inductance determines both the waveform of the current and the rms value of the current. ________

20-2. The phase angle between voltage across a pure inductance and the current through it is always 90° even if connected in series with a pure resistance. ________

20-3. In a purely inductive circuit it is possible for there to be zero current through the inductance at the instant when the voltage is not zero. ________

20-4. Inductive reactance may be thought of as an effective ac resistance to the flow of alternating current. ________

20-5. A coil of 2-H inductance and negligible resistance is used at a frequency of 400 Hz. Its inductive reactance is

a. 5000 Ω
b. 5.1 kΩ
c. 1600 Ω
d. 1600π Ω

20-6. If the coil in Question 20-5 is connected to a 240-V 400-Hz supply, the current is

a. 47.7 mA
b. 4.8 mA
c. 15 mA
d. 0.15 A

20-7. The use of an inductor to filter out unwanted ac voltage relies upon a high inductive reactance compared with its dc resistance. ________

20-8. One of the functions of a ballast in a fluorescent lamp circuit is the production of a momentary high voltage to fire the lamp. ________

20-9. A fluorescent lamp has a high efficiency because of the intense white heat produced by the phosphor coating. ________

20-10. The mercury in a fluorescent lamp is necessary for the production of ultraviolet energy. ________

20-11. The inductive reactance of the ballast in a fluorescent lamp circuit is used to limit current only at the instant the lamp is being turned on. ________

20-12. The same equations can be used to find total inductive reactance for series- and parallel-connected reactances as are used for resistances, with no restrictions. ________

20-13. The dot notation used to show when mutual reactance is additive is the same as that used for mutual inductance. ________

20-14. The quality of a coil is the comparison of the coil's resistance to the coil's inductive reactance. ________

20-15. At high frequencies the Q of a coil is relatively constant. ________

20-16. Alternating current resistance is the effective resistance responsible for ac power dissipation in a circuit and can be much larger than the dc resistance. ________

20-17. An impedance bridge works on the general principle of a Wheatstone bridge to measure inductance. ________

REVIEW QUESTIONS

1. a. Why does inductance in an ac circuit determine both *how* the current changes and how *much* current flows?

b. What is the phase relationship between current and voltage in a pure inductance?

c. How do you think this would be different in a coil that has as much resistance as inductive reactance?

2. a. Why is it not correct to characterize inductive reactance as the effective ac resistance due to a coil's inductance?

b. Why should an *increase* in inductance or frequency *reduce* the current in an inductive circuit if the voltage is unchanged?

3. a. In what way can direct current be thought of as zero frequency alternating current when applied to an inductor?

b. Does this mean that inductance has no effect at all when the frequency of the applied voltage is zero?

4. If an alternating voltage of 120 V rms is as effective in producing heat as 120 V dc, explain in your own words why the lamp lights in Fig. 20-4*a* but does not light in Fig. 20-4*b*.

5. a. Describe how an inductor can be used to filter out unwanted ac voltage from an input containing both alternating current and direct current.

b. Under what conditions can Eq. 20-3 be used?

6. Explain in detail how a fluorescent lamp circuit works, including the two functions of the ballast and the production of visible light.

7. Under what conditions can the equations for combining resistances in series and parallel be applied to inductive reactances in series and parallel?

8. Describe how you would use an ac source of known frequency, an ac ammeter, and an ac voltmeter to determine the mutual inductance between two series-connected coils whose interconnections can be interchanged. Assume that the coils have negligible resistance.

9. a. Explain why the ac resistance of a circuit can be considerably higher than the dc resistance measured with an ohmmeter.

b. What resistance value is used in determining the Q of a coil?

c. Why doesn't Q increase with frequency?

10. Describe two ways of measuring the inductance of a coil.

PROBLEMS

(Answers to odd numbers at back of book)

20-1. What inductive reactance does a coil have if an applied emf of 120 V causes a current of 50 mA? The resistance of the coil is negligible.

20-2. A coil with an inductive reactance of 2.5 kΩ and negligible resistance carries a current of 20 mA. What peak-to-peak voltage does an oscilloscope indicate when connected across the coil?

20-3. What inductance is necessary to cause a voltage drop of 15 V due to 5 mA of current having a frequency of 400 Hz?

20-4. What voltage drop appears across a 30-mH inductance with a 10-mA p-p current having a frequency of 1 kHz?

20-5. What frequency of current will cause a voltage drop of 12 V p-p across a 200-μH inductance when a current of 1 mA flows?

20-6. A 12-V p-p sinusoidal voltage is applied to an 8-H coil of negligible resistance. How much current flows at the following frequencies?

a. 100 Hz.

b. 10 kHz.

c. 1 MHz.

20-7. A 5-V, 10-kHz signal generator is applied to a 0.1-H coil of negligible resistance. Determine:

a. How much current flows.

b. The equation for current assuming a sinusoidal input.

20-8. The output of a full-wave rectifier has a 12-V dc component and an 8-V p-p ac component having a frequency of 120 Hz. It is applied to a filter circuit like that in Fig. 20-5, with $L = 4$ H and $R = 47\ \Omega$. Determine:

a. The dc output voltage.

b. The ac output voltage.

20-9. Repeat Problem 20-8 using $L = 8$ H and $R = 33\ \Omega$.

20-10. Assuming that the ballast in Fig. 20-6 has negligible resistance and a voltage drop of 75 V, calculate:

a. The inductive reactance of the ballast when the lamp is lit and drawing 0.3 A.

b. The inductance of the ballast.

20-11. Three coils of 4, 8, and 12 H are series-connected across a 60-V, 400-Hz supply. Assuming negligible resistance, calculate:

a. The current through the coils.

b. The voltage across each coil.

20-12. Two series-connected coils, one having twice the inductance of the other, draw a current of 200 mA when connected to a 120-V, 60-Hz source. Determine the inductance of each coil assuming negligible resistance and no mutual inductance.

20-13. Two coils of 30 and 50 mH are series-connected with a coefficient of coupling of 0.8. They are connected across a 10-V, 1-kHz sinusoidal source with their fields aiding. Assuming negligible resistance, calculate:

a. The total inductive reactance.

b. The current through the coils.

c. The voltage across each coil.

20-14. Repeat Problem 20-13 assuming opposing fields.

20-15. Repeat Problem 20-12 assuming a coefficient of coupling of 0.5 and aiding fields.

20-16. Repeat Problem 20-12 assuming a coefficient of coupling of 0.5 and opposing fields.

20-17. Two equal coils with negligible resistance are connected in series with a 20-V, 5-kHz source. With their fields aiding the current is 1 mA; with their fields opposing the current is 5 mA. Calculate:

a. The mutual inductance between the coils.

b. The self-inductance of the coils.

c. The coefficient of coupling between the coils.

20-18. Three inductors of negligible resistance are connected in parallel across a 400-Hz supply. They draw a total current of 500 mA. The first coil carries a current of 100 mA; the second coil has 40 V across it; and the third coil has an inductance of 50 mH. Calculate:

a. The total inductive reactance of the circuit.

b. The inductances of the first two coils.

20-19. The three coils in Problem 20-11 are connected in parallel across a 60-V, 400-Hz supply. Calculate:

a. The total inductive reactance of the circuit.

b. The total current drawn from the supply.

20-20. A 47-μH coil, when operated at a frequency of 10 MHz, has an ac resistance of 70 Ω.

a. What is the quality of the coil?

b. What would you expect the Q to be at 20 MHz?

20-21. A wattmeter in an ac circuit indicates 40 W when a series ac ammeter shows 0.15 A.

a. Determine the ac resistance of the circuit.

b. If an ohmmeter measurement shows 1200 Ω of resistance, what resistance is due to hysteresis and eddy current losses?

CHAPTER TWENTY-ONE
CAPACITANCE

We now consider the third property of electric circuits, *capacitance*. A capacitor consists of two metal plates separated by an *insulator* or *dielectric*. When connected to a voltage source, a momentary charging current deposits charge on the plates, establishing an *electric field*. Energy is stored in this field and may be returned by *discharging* the capacitor through a *load*. We examine in this chapter the factors that affect the capacitance of a capacitor and the various types of capacitors. We also consider the series and parallel connection of capacitors, and the maximum working voltage of each combination.

21-1 BASIC CAPACITOR ACTION

A capacitor consists of two conductors separated by an insulator or *dielectric.* In its simplest form, a capacitor may be thought of as two parallel conductive (metal) plates with air as the dielectric. In many practical capacitors, the "plates" may be aluminum foil with some dielectric (such as paper, mica, etc.) between the plates. This allows the capacitor to be rolled into a cylinder to conserve space. Many different types of dielectric and methods of fabrication are used to give characteristics for specific applications. These are covered in Section 21-5.

Consider a parallel-plate capacitor connected to a dc source through a resistor and switch as shown in Fig. 21-1*a*.

Upon closing the switch, there is a momentary current or flow of charge. Electrons are deposited upon one plate, and an equal number is removed from the other. The capacitor is now said to be *charged,* one plate positively, the other negatively. If the dc source is removed, a high-resistance voltmeter, connected across the charged capacitor, indicates a potential difference equal to the previously applied emf. This is shown in Fig. 21-1*b*.

Now if the capacitor is *discharged* across a load, as in Fig. 21-1*c*, electrons flow in the *opposite* direction to redistribute the original charge. Electrons flow from the plate having a surplus of negative charge to the plate having a deficiency. The similarity of this action to the two rods charged by friction in Section 2-4 should be noted. In both cases, current flows until the excess charge on one plate has been transferred to the other, at which point there is zero potential difference between the two.

The momentary nature of the charging and discharging currents may be observed on an oscilloscope connected across the resistor R. The shape of these current pulses is considered in Section 22-2.

21-2 CAPACITANCE

Let us assume that we can measure the charge deposited on either one of the plates of a capacitor. (This can be done with a suitable "measuring amplifier.") Figure 21-

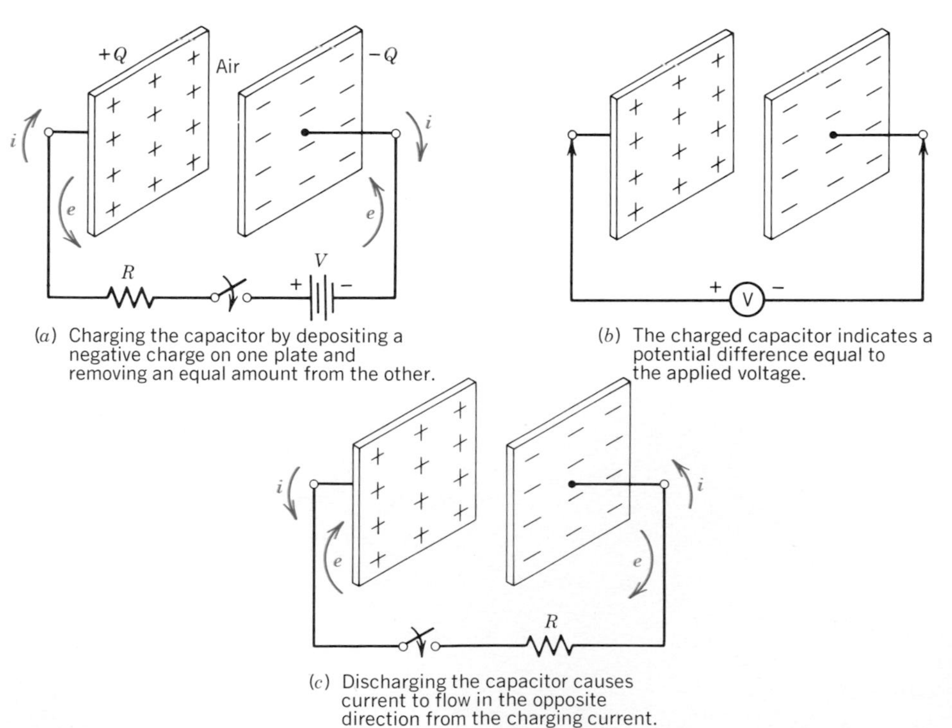

(*a*) Charging the capacitor by depositing a negative charge on one plate and removing an equal amount from the other.

(*b*) The charged capacitor indicates a potential difference equal to the applied voltage.

(*c*) Discharging the capacitor causes current to flow in the opposite direction from the charging current.

FIGURE 21-1
Basic capacitor charging and discharging action.

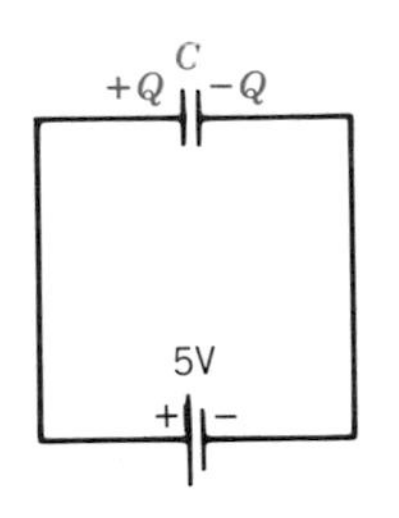

(a) 5 V deposits charge Q

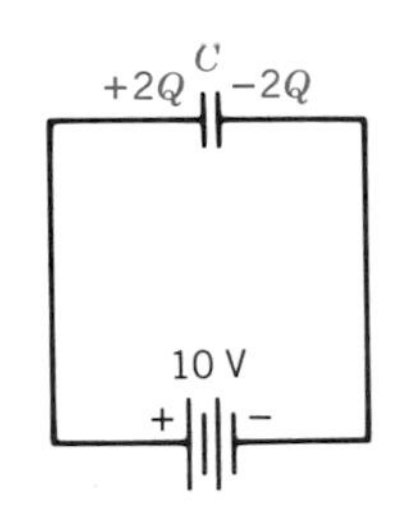

(b) 10 V deposits charge $2Q$

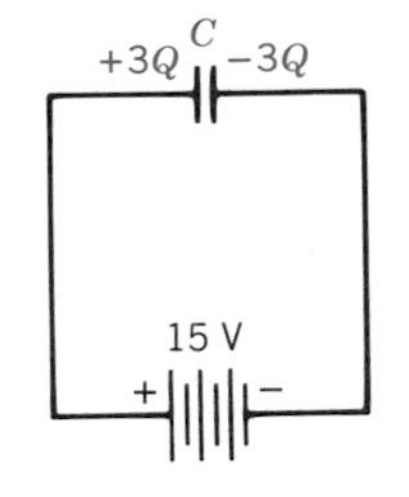

(c) 15 V deposits charge $3Q$

FIGURE 21-2
Relation between the voltage and charge for a given capacitor C.

2a shows that a 5-V source deposits a charge Q on each plate: one positive, $+Q$; the other negative, $-Q$. If the voltage source is increased to 10 V, the charge is doubled to $2Q$. If the voltage is increased to 15 V, the accumulated charge becomes $3Q$, where Q is the initial charge due to 5 V. **For any *given capacitor,* it is found that the ratio of the charge on one plate to the potential difference across the plates is a constant. This constant is a property of the capacitor called its *capacitance.***

Thus capacitance is defined as

$$C = \frac{Q}{V} \quad \text{farads (F)} \qquad (21\text{-}1)$$

where: C is the capacitance of the capacitor in farads, F
Q is the charge per plate in coulombs, C
V is the potential difference across the capacitor in volts, V

Thus 1 farad = 1 coulomb/volt or 1 F = 1 C/V.

A capacitor has a capacitance of 1 F if 1 C of charge must be deposited to raise the potential difference by 1 V across the capacitor.

A farad is a very large amount of capacitance. (An air capacitor with a plate separation of 1 mm would require plates approximately 10.5 km ($6\frac{1}{2}$ miles) on a side to produce a capacitance of 1 F.)

More practical units of capacitance are the microfarad and picofarad.

$$1 \text{ microfarad} = 1\ \mu\text{F} = 1 \times 10^{-6}\ \text{F}$$
$$1 \text{ picofarad} = 1\ \text{pF} = 1 \times 10^{-12}\ \text{F}$$

Practical capacitors are available from 1 pF to 500,000 μF.

EXAMPLE 21-1

A certain capacitor requires 50 μC of charge to be deposited on the plates to raise the potential difference to 2 V.

a. What is the capacitance of the capacitor?
b. How much charge is required to raise the voltage across the capacitor from 0 to 10 V?
c. If the charge on the plates is reduced to 10 μC, what voltage now exists across the capacitor?

SOLUTION

a. $$C = \frac{Q}{V} \qquad (21\text{-}1)$$
$$= \frac{50 \times 10^{-6}\ \text{C}}{2\ \text{V}}$$
$$= 25 \times 10^{-6}\ \text{F} = \mathbf{25\ \mu F}$$

b. $$Q = VC$$
$$= 10\ \text{V} \times 25 \times 10^{-6}\ \text{F}$$
$$= 250 \times 10^{-6}\ \text{C} = \mathbf{250\ \mu C}$$

c. $$V = \frac{Q}{C}$$
$$= \frac{10 \times 10^{-6}\ \text{C}}{25 \times 10^{-6}\ \text{F}} = \mathbf{0.4\ V}$$

21-3 FACTORS AFFECTING CAPACITANCE

It can be shown (see Appendix A-11) that capacitance is determined by the physical factors of the plate area, plate separation, and dielectric material as given by

$$C = K\frac{\epsilon_0 A}{d} \quad \text{farads (F)} \qquad (21\text{-}2)$$

where: C is the capacitance in farads, F

A is the area of one side of one plate in square meters, m^2

d is the separation of the two plates in meters, m

ϵ_0 is the permittivity of free space (air) = $8.85 \times 10^{-12} \frac{C^2}{Nm^2}$

K is the dielectric constant of the particular material used between the plates. (See Table 21-1.)

See Fig. 21-3 for some of the quantities involved and two alternate capacitor symbols.

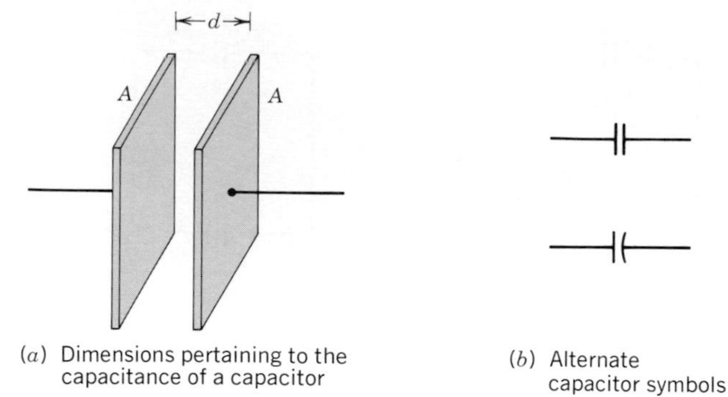

(a) Dimensions pertaining to the capacitance of a capacitor

(b) Alternate capacitor symbols

FIGURE 21-3
Physical factors that determine capacitance and alternate symbols.

EXAMPLE 21-2

Two metal plates, each 5 × 6 cm, are separated from each other by 0.5 mm. Calculate the capacitance for the following conditions:

a. The dielectric is air.

b. The dielectric is mica.

c. The dielectric is strontium titanate (ceramic material).

d. The separation is reduced to 0.25 mm with the same ceramic dielectric.

SOLUTION

a. $K_{air} = 1$

$$C = K \frac{\epsilon_0 A}{d} \qquad (21\text{-}2)$$

$$= 1 \times 8.85 \times 10^{-12} \frac{C^2}{Nm^2} \times \frac{5 \times 10^{-2} \text{ m} \times 6 \times 10^{-2} \text{ m}}{0.5 \times 10^{-3} \text{ m}}$$

$$= 53.1 \times 10^{-12} \frac{C^2}{Nm}$$

$$= 53.1 \times 10^{-12} \text{ F} = \mathbf{53.1\ pF}$$

b. $K_{mica} = 5$

$$C = K \frac{\epsilon_0 A}{d} \qquad (21\text{-}2)$$

$$= 5 \times 53.1 \text{ pF} = \mathbf{265.5\ pF}$$

c. $K_{ceramic} = 7500$

$$C = K \frac{\epsilon_0 A}{d} \qquad (21\text{-}2)$$

$$= 7500 \times 53.1 \text{ pF} = 398{,}250 \text{ pF} = \mathbf{0.4\ \mu F}$$

d. If the separation is cut in half, the capacitance is doubled.

$$C = 2 \times 0.4\ \mu\text{F} = \mathbf{0.8\ \mu F}$$

TABLE 21-1
Dielectric constant, K
(Average values at room temperature)

Material	K
Vacuum (and air)	1
Glass	5
Mica	5
Rubber	3
Neoprene	7
Bakelite (plastic)	6
Polyethylene	2
Vinylite	3
Teflon	2
Paper	3
Polyesters	4
Polystyrene	2.5
Ceramics—Porcelain	7
Titanium dioxide	14–110
Strontium titanate	7500

21-4 EFFECT OF A DIELECTRIC

In the previous section we saw that a larger plate area increases the capacitance. This seems reasonable because a larger area will allow a greater accumulation of charge. But why should a smaller separation also *increase* the capacitance? Also, why should a material other than air increase the capacitance? To answer these and other questions, we must consider the electric field between the plates of a charged capacitor and the effect on this field of inserting a dielectric.

21-4.1 Electric Field

First consider an atom of a dielectric material in an uncharged capacitor as in Fig. 21-4*a*. The electrons of the atom follow an approximately circular path as they orbit the atom's nucleus. After the capacitor is charged, however, this path becomes increasingly elliptical. See Fig. 21-4*b*. This is because the negatively charged electron tends to move closer toward the positive plate and away from the negative plate. It is this "stressing" of the electrons from a more circular to a more elliptical orbit that accounts for the *momentary* charging current when voltage is first applied. There is no continuous current because there are no free electrons in the dielectric material. The charging current is often referred to as a "displacement" current, resulting from the shifting of the atoms of the dielectric in their orbital path. The atoms are then said to be polarized. When the capacitor is discharged, the electrons return to their original, more circular path.

The force acting on the electrons can be represented by *electric lines of force* drawn between the two charged plates, as in Fig. 21-4*b*. It is conventional to use an arrow to indicate the direction of force on a *positive* charge. The force on an electron is then opposite to this arrow. The region between the two charged plates is called an *electric field*. The strength or *intensity* of this electric field determines the force on, and the resulting distortion of, the electrons in orbit around the nucleus of the dielectric's atoms. It is possible that the electric field intensity could be high enough so that the electrons are "torn" from their orbits. This results in current flowing continuously; then the dielectric is said to have broken down or *ruptured*. Evidently, a capacitor must be designed to withstand a high electric field intensity.

The intensity of the electric field is given by

$$E = \frac{V}{d} \qquad \text{volts/meter (V/m)} \qquad (21\text{-}3)$$

where: E is the electric field intensity in volts per meter, V/m

V is the potential difference across the plates in volts, V

d is the plate separation in meters, m

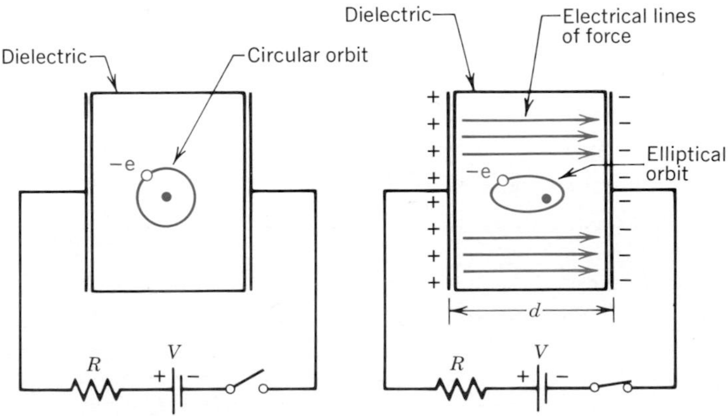

(*a*) Electrons in dielectric have a more circular orbit in an uncharged capacitor.

(*b*) Electric field causes the electron orbit to become more elliptical

FIGURE 21-4
Effect of electric field upon electron orbit in an atom of a dielectric in a capacitor.

EXAMPLE 21-3

The capacitor in Example 21-2 is charged to a potential difference of 20 V. With a ceramic dielectric determine the electric field intensity for the following separations:

a. 0.5 mm

b. 0.25 mm

SOLUTION

a. $E = \frac{V}{d}$ (21-3)

$= \frac{20 \text{ V}}{0.5 \times 10^{-3} \text{ m}}$

$= 40 \times 10^3 \text{ V/m} = \mathbf{40 \text{ kV/m}}$

b. $E = \frac{V}{d}$ (21-3)

$= \frac{20 \text{ V}}{0.25 \times 10^{-3} \text{ m}} = \mathbf{80 \text{ kV/m}}$

Note that for a given voltage, a smaller separation causes a higher electric field intensity. This means that more electric lines of force would have to be drawn to represent the field between the two plates. But electric lines of force originate on a positive charge and terminate on a negative charge. This implies that there must be a greater charge accumulation on the plates if the electric field intensity is higher. But a greater charge for the same voltage means greater capacitance. Hence a *smaller* separation of plates, all other factors being unchanged, causes an *increase* in capacitance.

21-4.2 Dielectric Strength

The field *intensities* calculated in Example 21-3 are independent of the type of dielectric. That is, it does not matter whether the dielectric is ceramic or paper. However, the dielectric *strength* of the material determines whether the capacitor can withstand the electric field intensity in any given situation. Thus the separation of the capacitor plates *and* the type of dielectric determine the permissible dc working voltage (WVDC) of a capacitor. Table 21-2 repeats some of the dielectric strengths shown in Section 4-3.

TABLE 21-2
Average Dielectric Strengths

Dielectric	Dielectric Strength, kV/mm
Air	4
Porcelain	8
Ceramics	10
Plastic	16
Paper	20
Glass	28
Teflon	60
Mica	100

EXAMPLE 21-4

A 600-WVDC, 0.1-μF paper capacitor is to be constructed in a tubular form. Calculate:

a. The minimum separation of the plates using a safety factor of 1.5.

b. The area of one of the metal foil plates.

c. The length of metal foil for one plate if the capacitor is to be only 4 cm long.

SOLUTION

a. Maximum field intensity $E_m = \frac{V}{d}$ (21-3)

Therefore, $d = \frac{V}{E_m}$

$= \frac{600 \text{ V}}{20 \times 10^3 \text{ V/mm}}$

$= 3 \times 10^{-2} \text{ mm}$

Applying a safety factor of 1.5, $d = 1.5 \times 3 \times 10^{-2}$ mm $=$ **$\mathbf{4.5 \times 10^{-2}}$ mm**

b. $C = K \frac{\epsilon_0 A}{d}$ (21-2)

Therefore, $A = \frac{Cd}{K\epsilon_0}$

$= \frac{0.1 \times 10^{-6} \text{ F} \times 4.5 \times 10^{-5} \text{ m}}{3 \times 8.85 \times 10^{-12} \text{ C}^2/\text{Nm}^2}$

$= 0.17 \frac{\text{F} \times \text{m}}{\text{C}^2/\text{Nm}^2}$

But $1F = 1\ C^2/Nm$; therefore,

$$A = 0.17\ \frac{C^2/Nm \times m}{C^2/Nm^2} = \mathbf{0.17\ m^2}$$

c. Area = length × width

Therefore, $\text{length} = \frac{\text{area}}{\text{width}}$

$$= \frac{0.17\ m^2}{4 \times 10^{-2}\ m} = \mathbf{4.25\ m}$$

NOTE When the capacitor is rolled into a tubular form, one plate will make contact with the other. A second sheet of paper is used for insulating purposes and has the effect of doubling the capacitance because of a parallel combination. (See Section 21-6.) Thus the required area in Example 21-4 need only be half as much, with a foil length of only **2.12 m.**

Note also that paper has five times the dielectric strength of air. This means that for a given voltage rating, the plates are five times as close as when air is used. This requires only one-fifth the area needed by an air dielectric and contributes to a smaller-sized capacitor.

21-4.3 Induced Electric Field

The effect of a dielectric (other than air) on a capacitor can now be examined. Refer to Fig. 21-5. Consider a pair of charged plates, separated by air, with a voltmeter connected as in Fig. 21-5*a*. An electric field exists as shown, E_{air}.

Now assume that a glass dielectric is inserted between the plates as in Fig. 21-5*b*. It is observed that the voltmeter reading drops to one-fifth its initial value ($K_{glass} = 5$). Why? Although there are no free electrons in the dielectric, an *induced* charge is set up where the dielectric comes in contact with the metal plates. This is due to the *polarization* of the atoms in the dielectric when placed between the plates. The induced dielectric charges set up an electric field, E_{diel}, in the dielectric that *opposes* the initial field, E_{air}. (This accounts for the term *di*-electric, "di" meaning *opposing*.)

The resulting electric field is now given by

$$E_{total} = E_{air} - E_{diel} \tag{21-4}$$

and E_{total} is evidently weaker than before. But we know that the electric field intensity is

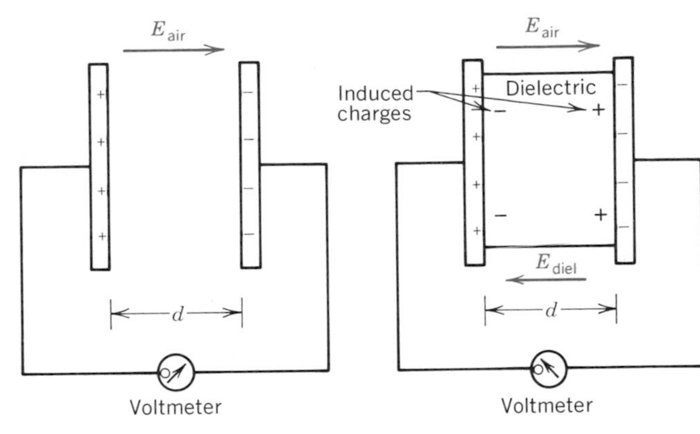

(*a*) Charged capacitor with plates separated by air.

(*b*) Insertion of a dielectric sets up an opposing electric field due to induced charges.

FIGURE 21-5
Effect of a dielectric upon the electric field of a capacitor.

$$E_{total} = \frac{V}{d} \tag{21-3}$$

But since d is unchanged, V must have decreased, proving the results of Fig. 21-5.

But how does this insertion of a dielectric cause an *increase* in capacitance? Consider that a dc source is connected across the plates. As the dielectric is being inserted, the voltage across the plates will try to drop. But this will allow the dc source to deposit more charge on the plates to restore the voltage back to the applied voltage. After fully inserting the dielectric, a new, *larger* charge will exist on the metal plates for the *same* potential difference across the plates. Since

$$C = \frac{Q}{V} \tag{21-1}$$

the result is an increase in capacitance.

The ratio of the capacitance *with* the dielectric to the capacitance with *air* is called the *relative* permittivity or dielectric constant, K.

$$K = \frac{C_{diel}}{C_{air}} \tag{21-5}$$

This is a measure of the ease with which the dielectric *permits* an opposing electric field to be set up in a given material. The *absolute* permittivity of a dielectric is $K\epsilon_0$. Thus the capacitance of a capacitor is

$$C = K\frac{\epsilon_0 A}{d} \tag{21.2}$$

We can now identify three functions of a dielectric in a capacitor:

1. It solves the mechanical problem of keeping two metal plates separated by a very small distance.
2. It increases the maximum voltage that can be applied, before causing breakdown, compared with air.
3. It increases the amount of capacitance, compared with air, for a given set of dimensions.

21-5 TYPES OF CAPACITORS

The type of dielectric determines the properties and the applications of a capacitor. Table 21-3 shows the major types of capacitors, their typical values, working voltage, and common applications.

As we have described, a capacitor consists of two metal plates separated by an insulating material. These "plates" may consist of metal films in a *ceramic disc,* silver coatings in a *tubular ceramic,* or aluminum foils in a *paper capacitor,* as shown in Fig. 21-6*a*, 21-6*b*, and 21-6*c*. A paper capacitor often has a black band to indicate which lead is connected to the outer foil. In some applications, it is desirable to connect this lead to ground to provide shielding by the outer foil. However, the capacitance is the same regardless of the way the capacitor is connected in the circuit.

To increase the effective area of the capacitor, plates may be interleaved as in Fig. 21-6*d*. This construction is typical of *mica* capacitors. This arrangement is also used with the *variable* type of *air* capacitors; one set of plates is fixed in position and the other set can be rotated in or out to increase or reduce capacitance, respectively. These are often used in low-capacitance tuning circuits in radios.

One type of capacitor that must be connected in a circuit with proper polarity is the *polarized electrolytic.* The most common type is the *aluminum* electrolytic, shown in Fig. 21-6*e*. A paste electrolyte is placed between two sheets of aluminum foil and a dc *forming* voltage is applied. A current flows, and, by a process of electrolysis, a molecular-thin layer of bubbles of alumi-

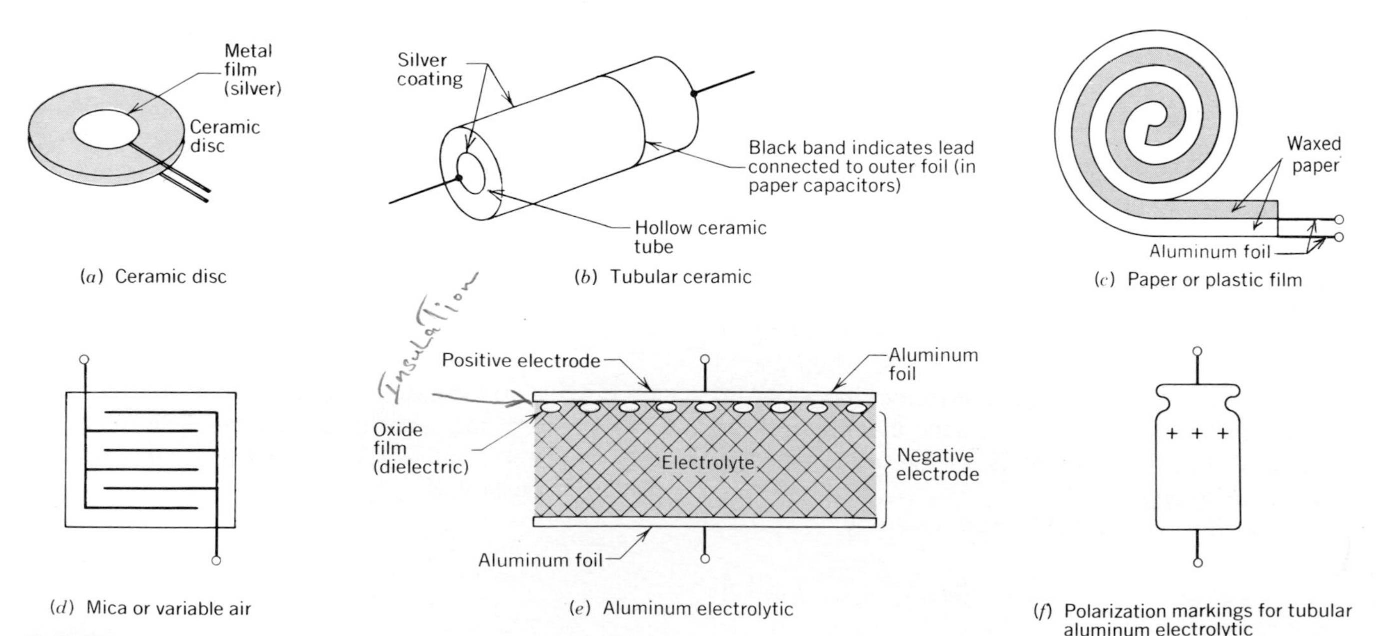

FIGURE 21-6
Construction of various types of capacitors.

TABLE 21-3
Major Types of Capacitors

Type	Capacitance	Voltage WVDC	Applications
Monolithic ceramics	1 pF–10 μF	50–200	UHF, RF coupling
Disc and tube ceramics	1 pF–1 μF	50–500	General, VHF
Paper	0.001–1 μF	200–1600	Motors, power supplies
Film—polypropylene	0.001–0.47 μF	400–1600	TV vertical circuits, RF
Polyester	0.001–1 μF	100–600	Entertainment electronics
Polystyrene	0.001–1 μF	100–200	General, high stability
Polycarbonate	0.01–18 μF	50–200	General
Metallized polypropylene	4–60 μF	400 VAC 60Hz	AC motors
Metallized polyester	0.01–10 μF	100–600	Coupling, RF filtering
Electrolytic-aluminum	1–500,000 μF	5–500	Power supplies, filters
Tantalum	0.1–1000 μF	3–125	Small space requirement High reliability low leakage
Nonpolarized (either Al or Ta)	0.47–220 μF	16–100	Loudspeaker crossovers
Mica	330 pF–0.05 μF	50–100	High frequency
Silver—mica	5–820 pF	50–500	High frequency
Variable—ceramic	1–5 to 16–100 pF	200	Radio, TV, communications
Film	0.8–5 to 1.2–30 pF	50	Oscillators, Antenna, RF circuits
Air	10–365 pF	50	Broadcast receiver
Teflon	0.25–1.5 pF	2000	VHF, UHF

num oxide builds up on the *positive* electrode. This serves as the dielectric. The rest of the electrolyte and the other aluminum foil make up the negative electrode. The very thin dielectric is responsible for very large values of capacitance (thousands of microfarads) in a relatively small space.

If an electrolytic capacitor is connected with reverse polarity, the insulating layer of bubbles is removed from the positive electrode, a high current flows, and the capacitor becomes hot and may *explode.* The polarization of the capacitor is clearly marked, as shown in Fig. 21-6*f*, to ensure proper connection. It is considered good practice to use an electrolytic capacitor at a voltage close to its design value. If, for example, a 20–μF 400-V capacitor is used in a circuit at only 10 V, there may not be sufficient voltage to maintain the oxide layer in good condition, and the capacitance value may change. In addition, the 400-V capacitor is bulkier and more expensive than the 10-V version.

21-5.1 Other Properties of Capacitors

One disadvantage of the aluminum electrolytic is its relatively large *leakage current.* This is a current that flows through the capacitor due to the imperfection of the dielectric or to surface paths from one plate to the other. The effect can be represented by an *insulation resistance,* R_p, connected in parallel with the capacitor as shown in Fig. 21-7.

Insulation resistances vary from 3000 MΩ for mica capacitors to over 100,000 MΩ for plastic film capacitors. For electrolytics, the leakage current is specified instead of the insulation resistance. For aluminum electrolytics, this current is in the region of 0.01 to 0.04 mA per μF of capacitance, for working voltages from 100 to 500 V. Any capacitor showing a leakage current in the range of 10 to 15 mA should normally be replaced because of the damaging heat produced internally in the capacitor.

In critical applications requiring low leakage currents, a *tantalum* capacitor can be used. It too is an electrolytic

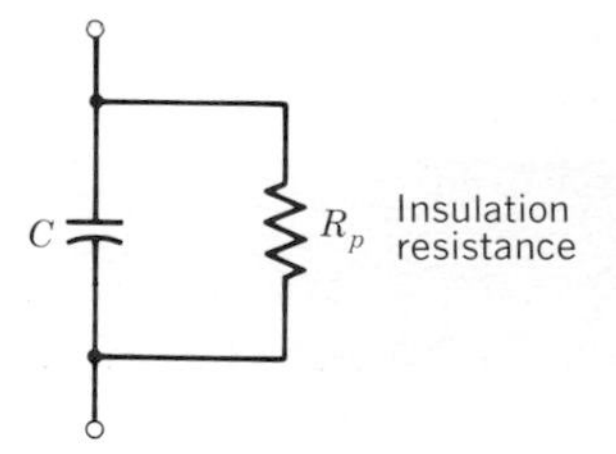

FIGURE 21-7
Equivalent circuit of a capacitor.

and must be connected with correct polarity, although some can withstand a 3-V reverse voltage. The tantalum capacitor has a lower maximum capacitance and voltage rating than the aluminum capacitor but occupies less space and has a high reliability. Typical capacitors are shown in Fig. 21-8.

Many capacitors have an *operating temperature* range from −55 to +125°C. A new capacitor designed for hostile environments operates from 25 to 300°C. It has a silicon nitride dielectric and uses tungsten for the metal plates.

Like resistors, capacitors have various *temperature coefficients* and *tolerances*. Tolerances may range from ±1% up to ±20%. Some capacitors (because the minimum value is important) may have a specified tolerance of −20%, +80%. Ceramic capacitors are available that have positive, negative, or virtually zero temperature coefficients. They can be used to compensate against temperature variations because a capacitance can be made to automatically increase or decrease with an increase in temperature. A typical temperature coefficient may be +750 parts per million (ppm) per degree Celsius.

Capacitors are marked in various ways to show their values. Some use a color-coded stripe method similar to resistors, others a series of colored dots. In very small capacitors, it is understood that the number, printed on the outside, refers to picofarads. In some cases, the marking of MF is used, which means microfarads (or millionths of a farad).

There are many instruments that can measure the capacitance and other properties of a capacitor. Figure 21-9 shows two modern instruments that, between them, can measure capacitance from 0.1 pF to 200,000 μF, with an accuracy of 0.1%.

Capacitors can also be checked for open or short circuits by means of an ohmmeter (covered in Section 22-3.1).

21-6 CAPACITORS IN PARALLEL

Figure 21-10*a* shows two capacitors connected in parallel with a dc voltage source, *V*. We wish to find the value of the single capacitor that is equivalent to the two parallel-connected capacitors.

The charge that accumulates on each capacitor is given by

$$Q_1 = VC_1 \tag{21-1}$$

and

$$Q_2 = VC_2$$

If one capacitor, C_T, is to take the place of C_1 and C_2, it must store a total charge

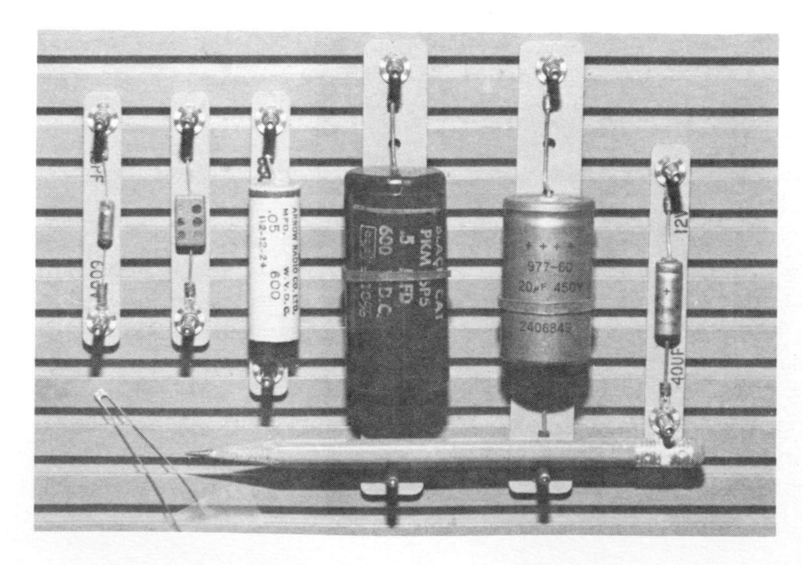

FIGURE 21-8
Typical capacitors. From left to right: 33-pF, 60-V; 15-pF, 600-V; 100-pF, mica; 0.05-μF, 600-V; 0.5-μF, 600-V, paper; 20-μF, 450-V; 40-μF, 12-V aluminum electrolytic.

(*a*) Digital capacitor-inductor meter can check capacitors from 1 pF to 200,000 μF, from 3 V to 600 V.

(*b*) Digital meter automatically checks capacitors from 0.1 pF to 1999 μF at 0.1% accuracy.

FIGURE 21-9

Capacitance meters. [Photograph in (*a*) courtesy of Sencore, Inc. in (*b*) courtesy of Data Precision Corp.]

$$Q_T = VC_T \qquad (21\text{-}1)$$

where $\quad Q_T = Q_1 + Q_2.$

Therefore,

$$VC_T = VC_1 + VC_2$$
$$= V(C_1 + C_2)$$

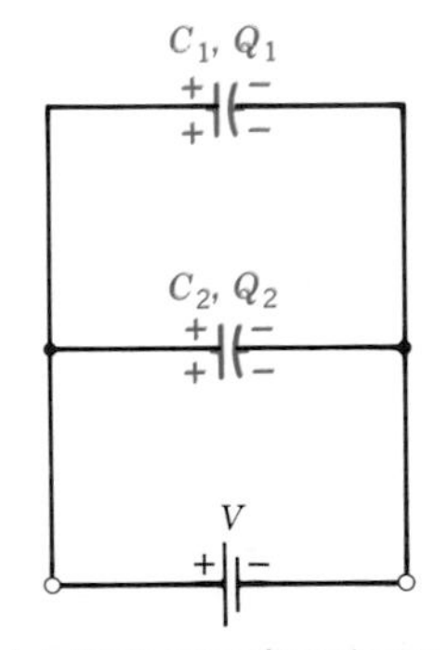

(*a*) Each capacitor charges to the same voltage due to separate charges Q_1 and Q_2.

$C_T, Q_T = Q_1 + Q_2$

V

(*b*) The single equivalent capacitor charges to a total charge of $Q_1 + Q_2$.

FIGURE 21-10

Equivalent capacitance of two parallel-connected capacitors.

or

$$C_T = C_1 + C_2$$

In general, $\quad C_T = C_1 + C_2 + C_3 + \cdots + C_N \quad$ **(21-6)**

This equation gives the total combined capacitance for any number of *parallel*-connected capacitors. Note how the equation for *parallel*-connected capacitors is similar to a *series* connection of resistors. That is, an increase in capacitance results due to an effective increase in plate area, A, in Eq. 21-2. It should be evident that capacitors are connected in parallel to increase the capacitance. However, if capacitors of different voltage rating are parallel-connected, the combination is limited to the working voltage of the lowest voltage-rating capacitor in the group, as shown in Example 21-5.

EXAMPLE 21-5

The following three capacitors are connected in parallel with each other: 5 μF, 25 V; 10 μF, 20 V; and 1 μF, 50 V. Determine:

a. The total capacitance.
b. The maximum working voltage of the group.
c. The maximum charge that the parallel-connected group can store without damage to any capacitor.

SOLUTION

a. $C_T = C_1 + C_2 + C_3$ (21-6)

$= 5 + 10 + 1\ \mu F$

$= \mathbf{16\ \mu F}$

b. Maximum working voltage = **20 V**

c. $Q_T = VC_T$ (21-1)

$= 20\ V \times 16\ \mu F$

$= \mathbf{320\ \mu C}$

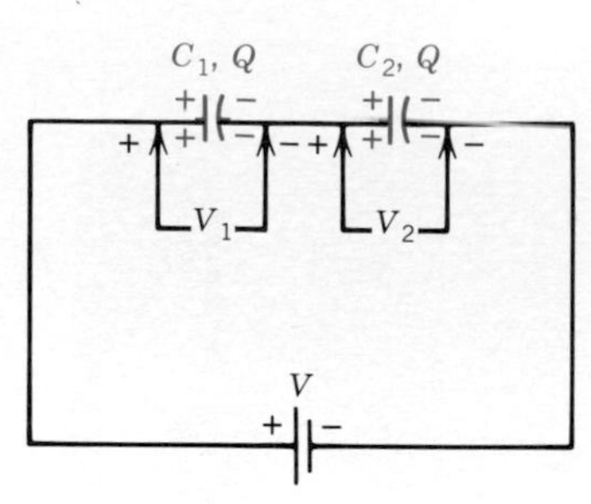

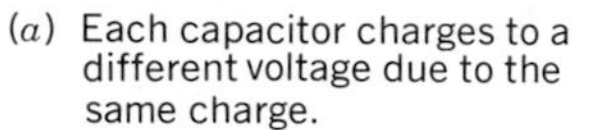

(a) Each capacitor charges to a different voltage due to the same charge.

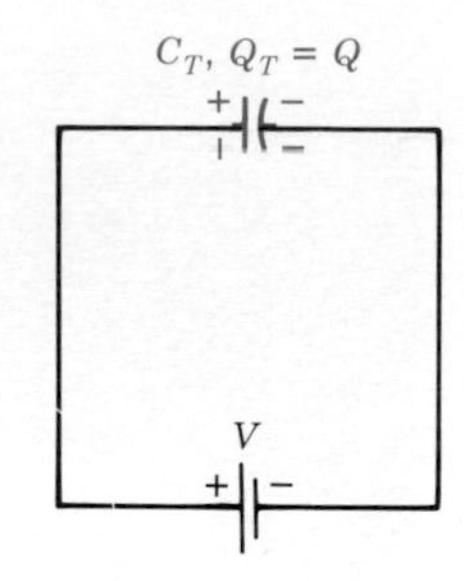

(b) The single equivalent capacitor charges to Q due to total voltage V.

FIGURE 21-11
Equivalent capacitance of two series-connected capacitors.

21-7 CAPACITORS IN SERIES

When capacitors are connected in series with a dc source, V, as in Fig. 21-11a, the momentary charging current is the same throughout the circuit. This means that the same charge, Q, is deposited on each capacitor.

The voltage to which each capacitor charges is then given by:

$$V_1 = \frac{Q}{C_1} \quad (21\text{-}1)$$

$$V_2 = \frac{Q}{C_2}$$

But Kirchhoff's voltage law requires that

$$V = V_1 + V_2 \quad (5\text{-}1)$$

Therefore,
$$V = \frac{Q}{C_1} + \frac{Q}{C_2}$$

and
$$V = Q\left(\frac{1}{C_1} + \frac{1}{C_2}\right)$$

For one equivalent capacitor, C_T, to store a charge Q due to a total voltage V it is required that

$$V = \frac{Q}{C_T} \quad (21\text{-}1)$$

Therefore,
$$\frac{Q}{C_T} = Q\left(\frac{1}{C_1} + \frac{1}{C_2}\right)$$

and
$$\frac{1}{C_T} = \frac{1}{C_1} + \frac{1}{C_2}$$

In general,
$$\frac{1}{C_T} = \frac{1}{C_1} + \frac{1}{C_2} + \frac{1}{C_3} + \cdots + \frac{1}{C_N} \quad \mathbf{(21\text{-}7)}$$

This equation gives the total combined capacitance for *any* number of *series*-connected capacitors. Note how the equation is similar to a *parallel* connection of resistors. This means a *decrease* in capacitance results with the total capacitance always less than the smallest of the individual capacitors. This may be thought of as resulting from an effective *increase* in the separation, d, of the capacitor plates in Eq. 21-2.

Other equations to find total capacitance for series-connected capacitors follow. For two series-connected capacitors only:

$$C_T = \frac{C_1 \times C_2}{C_1 + C_2} \quad (21\text{-}8)$$

For N equal series-connected capacitors each of capacitance C:

$$C_T = \frac{C}{N} \quad (21\text{-}9)$$

EXAMPLE 21-6

Two capacitors—5 μF, 20 V and 10 μF, 20 V—are connected in series across a 30-V dc supply. Calculate:

a. The total capacitance of the series combination.
b. The charge on each capacitor.
c. The voltage across each capacitor.

SOLUTION

a. $C_T = \frac{C_1 \times C_2}{C_1 + C_2}$ (21-8)

$= \frac{5\ \mu F \times 10\ \mu F}{5\ \mu F + 10\ \mu F}$

$= \frac{50}{15}\ \mu F = \mathbf{3\frac{1}{3}\ \mu F}$

b. $Q = VC_T$ (21-1)

$= 30\ V \times 3\frac{1}{3}\ \mu F$

$= \mathbf{100\ \mu C}$

c. $V_1 = \frac{Q}{C_1}$ (21-1)

$= \frac{100\ \mu C}{5\ \mu F} = \mathbf{20\ V}$

$V_2 = \frac{Q}{C_2}$

$= \frac{100\ \mu C}{10\ \mu F} = \mathbf{10\ V}$

21-7.1 Voltage Division Across Capacitors in Series

The main reason for connecting capacitors in series is to *increase* the working voltage. However, Example 21-6 shows an important restriction. The total working voltage of the combination is *not* simply the sum of the individual working voltages (unless all the capacitors are identical in capacitance and voltage rating). In the example above, the 5-μF capacitor is working at its limit of 20 V when only 30 V are applied to the series combination.

For two capacitors in series,

$$Q = C_1V_1 = C_2V_2$$

and

$$\frac{V_2}{V_1} = \frac{C_1}{C_2} \qquad (21\text{-}10)$$

This equation implies that the smaller capacitor has the *larger* voltage across it, and vice versa.

For the special case of only two capacitors in series, it can be shown that

$$V_1 = V \times \frac{C_2}{C_1 + C_2} \quad \text{volts (V)} \qquad (21\text{-}11a)$$

$$V_2 = V \times \frac{C_1}{C_1 + C_2} \quad \text{volts (V)} \qquad (21\text{-}11b)$$

where: V_1 is the voltage across capacitor C_1 in volts, V
V_2 is the voltage across capacitor C_2 in volts, V
V is the total voltage across the series connection of capacitors C_1 and C_2 in volts, V

These voltage-divider equations for capacitors are similar to the current division equations for two parallel-connected resistors.

For more than two capacitors in series, the voltage across each capacitor can be obtained using the method shown in Example 21-6; that is, by finding the charge on each capacitor and applying Eq. 21-1.

EXAMPLE 21-7

Two capacitors—$C_1 = 50\ \mu F$, 16 V and $C_2 = 40\ \mu F$, 10 V—are connected in series. Calculate:

a. The maximum working voltage of this combination.
b. The voltage across each capacitor when working at the maximum total voltage.
c. The total capacitance.

SOLUTION

a. Since C_2 is the smaller capacitor, it will have the higher voltage across it, which must be limited to 10 V.

$$V_2 = V \times \frac{C_1}{C_1 + C_2} \qquad (21\text{-}11b)$$

Maximum voltage $V = V_2 \times \frac{C_1 + C_2}{C_1}$

$= 10\ V \times \frac{50\ \mu F + 40\ \mu F}{50\ \mu F}$

$= 10\ V \times \frac{9}{5} = \mathbf{18\ V}$

b. Since C_2 is working at its limit of 10 V, C_1 must have a voltage across it of **8 V**.

Or
$$V_1 = V \times \frac{C_2}{C_1 + C_2} \quad \text{(21-11a)}$$
$$= 18\text{ V} \times \frac{40\ \mu\text{F}}{50\ \mu\text{F} + 40\ \mu\text{F}}$$
$$= 18\text{ V} \times \frac{4}{9} = \mathbf{8\ V}$$
$$V_2 = V \times \frac{C_1}{C_1 + C_2} \quad \text{(21-11b)}$$
$$= 18\text{ V} \times \frac{50\ \mu\text{F}}{50\ \mu\text{F} + 40\ \mu\text{F}}$$
$$= 18\text{ V} \times \frac{5}{9} = \mathbf{10\ V}$$

c.
$$C_T = \frac{C_1 \times C_2}{C_1 + C_2} \quad \text{(21-8)}$$
$$= \frac{50\ \mu\text{F} \times 40\ \mu\text{F}}{50\ \mu\text{F} + 40\ \mu\text{F}}$$
$$= \frac{2000}{90}\ \mu\text{F} = \mathbf{22.2\ \mu F}$$

21-8 ENERGY STORED IN A CAPACITOR

We have seen that after a capacitor has been charged and the charging voltage has been removed, a potential difference remains across the capacitor because of the charge on the plates. If a load is connected across the capacitor, such as a photographic flash bulb, there is a momentary flash of light. This shows that energy is stored in the capacitor and can be transferred upon discharge through a load. In fact, a capacitor is the only device that can store electrical energy apart from an electric voltaic cell. (In some applications, capacitors can store charge for months, being used to indicate measurements of pressure or temperature in remote locations. An inductor also stores energy but only as long as current is flowing through it.)

The energy can be thought of as being stored in the electric field of the capacitor, similar to the energy stored in the magnetic field of an inductor. The expression for stored energy in a capacitor is similar to that for an inductor ($\frac{1}{2} LI_f^2$). But an inductor stores energy only so long as the current is maintained. The energy stored in a capacitor is given by

$$W = \tfrac{1}{2} CV_f^2 \quad \text{joules (J)} \quad \text{(21-12)}$$

where: W is the energy stored in the capacitor in joules, J
C is the capacitance in farads, F
V_f is the final voltage across the capacitor in volts, V

EXAMPLE 21-8

A 40-μF, 450-V capacitor and a 20-μF, 500-V capacitor are connected in parallel across a 400-V dc supply. Calculate:

a. The energy stored by each capacitor.
b. The total energy stored.

SOLUTION

a. For the 40 μF,
$$W_1 = \tfrac{1}{2} CV_f^2 \quad \text{(21-12)}$$
$$= \tfrac{1}{2} \times 40 \times 10^{-6}\text{ F} \times (400\text{ V})^2$$
$$= \mathbf{3.2\ J}$$

For the 20 μF,
$$W_2 = \tfrac{1}{2} CV_f^2$$
$$= \tfrac{1}{2} \times 20 \times 10^{-6}\text{ F} \times (400\text{ V})^2$$
$$= \mathbf{1.6\ J}$$

b. Total energy $= W_1 + W_2$
$$= 3.2\text{ J} + 1.6\text{ J} = \mathbf{4.8\ J}$$

or
$$W_T = \tfrac{1}{2} C_T V_f^2 \quad \text{(21-12)}$$
$$= \tfrac{1}{2} \times 60 \times 10^{-6}\text{ F} \times (400\text{ V})^2$$
$$= \mathbf{4.8\ J}$$

It should be noted that capacitors charged to the energy levels in Example 21-8 are considered dangerous. To avoid shock and to prevent damage to measuring instruments (such as ohmmeters), always discharge capacitors, with the power removed, before working on them.

SUMMARY

1. A capacitor consists of two (metal) plates separated by an insulator called a dielectric.

2. When connected to a source of voltage, a capacitor charges to the applied emf, with a positive charge on one plate and an equal but negative charge on the other.

3. Capacitance C is defined as the ratio of the charge Q in coulombs per plate to the potential difference V between the plates: $C = \frac{Q}{V}$.

4. The basic unit of capacitance is the farad, F with 1 μF $= 1 \times 10^{-6}$ F and 1 pF $= 1 \times 10^{-12}$ F used for practical values of capacitors.

5. The capacitance of a capacitor increases with the area of plates, A, and with a decrease in the plate separation, d, as given by $C = K \frac{\epsilon_0 A}{d}$.

6. ϵ_0 is the permittivity of free space and K is the dielectric constant, given in Table 21-1.

7. The momentary charging current, or displacement current, is due to the stressing of the valence electrons in the atoms of the dielectric when an electric field is set up. The atoms are said to be polarized.

8. The dielectric strength in kV/mm is the electric field intensity that must not be exceeded to avoid a dielectric breaking down. It determines a capacitor's working voltage.

9. A dielectric in a capacitor has an induced electric field that opposes the electric field that arises from charges on the plates. This permits more charge to be deposited for a given voltage compared with air and increases the capacitance.

10. The type of capacitor is determined by the dielectric. Electrolytic capacitors rely upon a chemical process to establish a very thin dielectric that is polarity sensitive but produces a very large capacitance in a small volume.

11. When capacitors are parallel-connected, the total capacitance is the sum of the individual capacitances with a working voltage equal to the lowest individual voltage rating.

12. The capacitance decreases when capacitors are series-connected (similar to parallel-connected resistors). The smallest capacitor has the largest voltage across it.

13. The energy stored in a capacitor (in the electric field of the dielectric) is given by $W = \frac{1}{2} C V_f^2$.

SELF-EXAMINATION

Answer T or F or a, b, c, d, or e for multiple choice
(Answers at back of book)

21-1. Capacitance is the property of a capacitor to store electrical charge and energy. ______

21-2. The charge stored on one plate of a capacitor is the same as the charge on the other plate. ______

21-3. Capacitance is defined as the ratio of the potential difference across the capacitor to the charge stored per plate. ______

21-4. If a charge of 6 μC raises the potential difference of a capacitor from 0 to 3 V, the capacitance is

a. 2 pF
b. 18 μF
c. 2 μF
d. 0.5 μF
e. none of the above

21-5. A capacitance of 0.005 μF is the same as

a. 500 pF
b. 50 nF
c. 5×10^{-8} F
d. 5000 pF
e. none of the above

21-6. If the area of a given capacitor's plates is doubled and the plate separation is reduced by one-half, the new capacitance compared with the initial value is

a. not changed
b. quadrupled
c. doubled
d. eight times larger
e. one-eighth as large

21-7. If a dielectric with $K = 6$ is inserted between the plates of an air capacitor, the new capacitance compared with the initial value is

a. 6 times smaller
b. 36 times larger
c. 6 times larger
d. not changed
e. none of the above

21-8. The momentary charging current into a capacitor is due to the dielectric becoming polarized. ________

21-9. If a voltage of 500 V is applied to a capacitor whose plates are separated by 0.5 mm, the electric field intensity is

a. 250 kV/m
b. 1 kV/mm
c. 1000 kV/mm
d. 250 V/mm
e. either part (a) or (d)

21.10. The dielectric strength of paper is higher than that of air. ________

21-11. A smaller plate separation increases the capacitance because the electrical field intensity is higher. This means that there is a greater charge on the plates for the same applied voltage. ________

21-12. The insertion of a dielectric between the plates of a charged capacitor (with no voltage applied) causes an increase in the electric field intensity and potential difference. This accounts for the increase in capacitance with a dielectric other than air. ________

21-13. The dielectric constant of a material is the ratio of the capacitance with the dielectric to the capacitance with air. ________

21-14. Large values of capacitance are obtainable only in electrolytic types of capacitors. ________

21-15. A tantalum capacitor, although an electrolytic type, is not polarity-sensitive. ________

21-16. A high insulation resistance indicates a large leakage current in a capacitor. ________

21-17. The high capacitance of an electrolytic capacitor results from a molecular-thin oxide layer acting as the dielectric between the plates. ________

21-18. The equations for calculating the total capacitance of parallel-connected capacitors are similar to those for parallel-connected resistors. ________

21-19. The maximum working voltage of parallel-connected capacitors is determined by the lowest individual working voltage. ________

21-20. The working voltage of a series-connected combination of capacitors is the sum of the individual working voltages. ________

21-21. A parallel combination of 0.5 μF, 100 V and 1 μF, 50 V has an equivalent value of

a. 1.5 μF, 100 V
b. 3 μF, 50 V
c. 1.5 μF, 50 V
d. 1.5 μF, 150 V
e. none of the above

21-22. A series connection of the two capacitors in Question 21-21 has a total capacitance of

a. $\frac{1}{3}$ μF

b. 3 μF

c. 1.5 μF

d. 0.3 μF

e. none of the above

21-23. When capacitors are series-connected, the largest capacitor has the largest voltage across it. ______

21-24. Voltage division across series-connected capacitors can be calculated using the same equations as those used for voltage division across series-connected resistors. ______

21-25. The energy stored by a capacitor can be doubled by either doubling the capacitance or the voltage. ______

REVIEW QUESTIONS

1. a. What is the mathematical definition of *capacitance?*

b. What is 1 coulomb/volt called?

c. Why are microfarads and not farads used to identify a capacitor's capacitance?

2. a. What are the factors that determine the capacitance of a capacitor?

b. Which of these factors also determine the working voltage?

3. a. If a perfect insulator were used for the dielectric in a capacitor, would there still be a momentary charging current?

b. If so, how can you explain this?

4. What is the difference between the electric field intensity of a charged capacitor and the dielectric field strength?

5. Explain why a decrease in plate separation causes an increase in capacitance.

6. a. Why does the insertion of a dielectric (in place of air) cause a decrease in voltage across the plates of a charged capacitor?

b. What is the origin of the word *dielectric?*

c. Explain why a dielectric increases the capacitance compared with air.

7. What are three functions provided by a dielectric?

8. a. What is the basic construction difference between an electrolytic capacitor and the other types?

b. What accounts for the very large values of capacitance?

c. What restriction applies to an electrolytic?

9. a. What does a capacitor's insulation resistance represent?

b. What initial and final readings do you think an ohmmeter would indicate when connected across an uncharged capacitor?

c. How can the final reading of the ohmmeter differ if an electrolytic is being checked compared with a plastic film capacitor? Would the reading change if the ohmmeter connections were reversed?

10. a. What is the point of connecting capacitors in parallel?

b. How is the maximum working voltage determined?

c. If electrolytic capacitors are being parallel-connected, what precaution must be observed?

11. a. Why, when capacitors are series-connected, does the smallest capacitor have the highest voltage across it?

b. Under what condition is the maximum working voltage of series-connected capacitors equal to the sum of the individual capacitor working voltages?

12. How is the energy stored in a capacitor different from the energy stored by an inductor?

PROBLEMS

(Answers to odd numbers at back of book)

21-1. How much capacitance is required to store 100 μC of charge on each plate of a capacitor when connected to a 20-V dc supply?

21-2. A 0.015-μF capacitor is connected to a 6-V battery. How much charge is deposited on each plate?

21-3. What voltage must a 1500-pF capacitor be connected to for a charge of 0.03 μC per plate?

21-4. A charge of 2.5 μC raises the potential difference of a capacitor to 100 V.

a. What is the capacitance of the capacitor?

b. How much charge must be removed to drop the voltage to 20 V?

c. What is the potential difference across the capacitor when the charge is increased to 4 μC?

21-5. A sheet of paper is 0.008 cm thick. Calculate:

a. The capacitance of two metal plates each 21.6 × 28 cm (approximately $8\frac{1}{2}$ by 11 in.) when separated by one sheet of paper.

b. The maximum working voltage of the capacitor in part (a).

21-6. Repeat Problem 21-5 for the same thickness of mica.

21-7. A 1-μF capacitor is to be constructed with a strontium titanate (ceramic) dielectric that is 0.1 mm thick. Calculate:

a. The necessary plate area.

b. The maximum working voltage assuming a safety factor of three.

21-8. Repeat Problem 21-7 using a bakelite (plastic) dielectric.

21-9. A 0.0015-μF paper capacitor must have a maximum working voltage of 600 V, with a safety factor of three. Determine the necessary area of each plate.

21-10. Repeat Problem 21-9 using a bakelite (plastic) dielectric.

21-11. Determine the electric field intensity in the capacitor of Problem 21-9 when operating at its maximum working voltage.

21-12. Two metal plates of 20-cm diameter are separated by 1 cm of air. A charge of 0.003 μC is deposited on the plates. Calculate:

a. The capacitance of the two plates.

b. The potential difference across the plates.

c. The electric field intensity.

d. The new potential difference when a 1-cm-thick sheet of glass is inserted between the plates.

e. The new electric field intensity.

f. The new capacitance of the plates.

g. The *additional* charge necessary to restore the potential difference to the initial value in part (b).

21-13. The following capacitors are parallel-connected: 47 μF, 25 V; 25 μF, 35 V; and 50 μF, 16 V. Determine:

a. The total capacitance.

b. The maximum working voltage of the group.

c. The maximum charge that the group can store.

21-14. A 40-μF capacitor is connected across a 12-V dc supply. What size of capacitor must be connected in parallel with the first so that a total charge of 840 μC is stored?

21-15. Four 25-μF, 25-V capacitors are connected in series to a 50-V source. Calculate:

a. The total capacitance of the series combination.

b. The charge on each capacitor.

c. The voltage across each capacitor.

d. The maximum working voltage of the combination.

21-16. Repeat Problem 21-15 using two 25-μF, 25-V capacitors and two 10-μF, 30-V capacitors.

21-17. A 0.015-μF capacitor is connected in series with a 0.02-μF capacitor and a 200-V dc supply. Calculate:

a. The total capacitance.

b. The minimum voltage rating of each capacitor.

21-18. What is the largest value of capacitance that can safely be connected in series with a 500-μF, 12-V capacitor and a 15-V dc supply?

21-19. What size of capacitor must be connected in series with a 40-μF capacitor to give a total capacitance of 24 μF?

21-20. If the two capacitors in Problem 21-19 are each rated at 15 V, what is the maximum working voltage of the series combination?

21-21. Find the maximum energy that the capacitors in Problem 21-13 can store.

21-22. Find the maximum energy that the capacitors in Problem 21-15 can store.

21-23. What size of capacitor, charged to 250 V, is necessary to store an energy of 1 J?

21-24. A capacitor is to be charged with 0.01 C to store 2 J of energy. Calculate:

a. The necessary voltage.

b. The amount of capacitance.

21-25. Calculate the total capacitance of the series-parallel-connected capacitors in Fig. 21-12*a* and 21-12*b*.

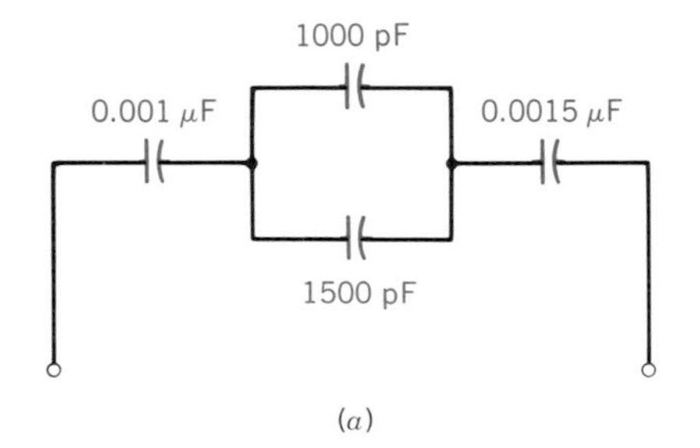

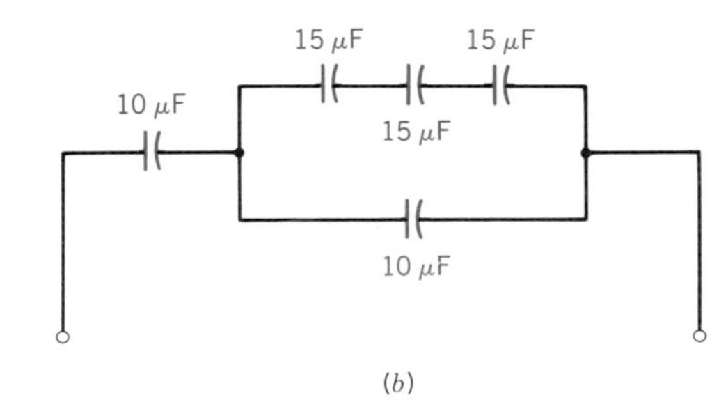

FIGURE 21-12
Circuits for Problem 21-25.

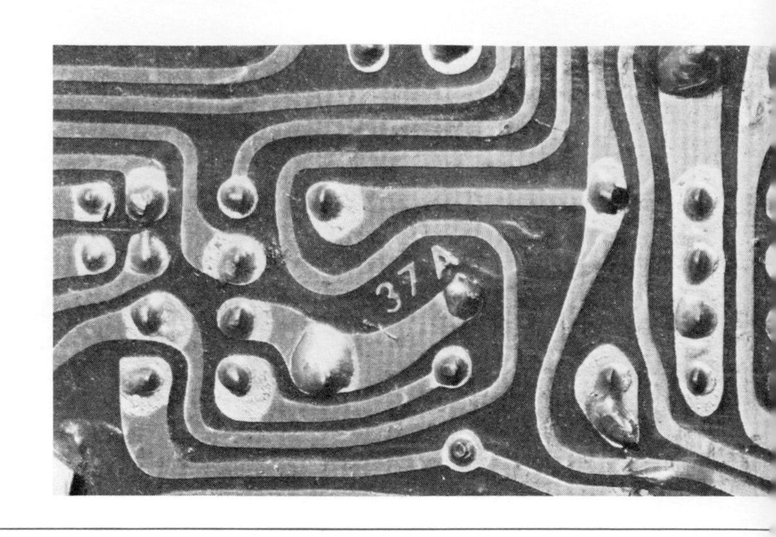

CHAPTER TWENTY-TWO

CAPACITANCE IN AC CIRCUITS

Although no *steady* current flows in a dc capacitive circuit, there is a *momentary* charging (or discharging) current whenever the voltage across the capacitor is *changing.* (This should be compared with the induced *voltage* across a coil, appearing only when the *current* through the coil is changing.) To determine how long it takes to charge (or discharge) a capacitor, we must determine the *time constant* (RC) of the circuit. The time constant allows us to examine the *transient responses* in the series RC circuit and to determine how *steady-state* conditions are reached. Similar to inductive circuits, a graphical method of universal time-constant curves can be used to solve for instantaneous capacitor voltage and current.

One of the applications of a capacitor in a dc circuit is an electronic *photoflash* unit. This application relies upon *slowly charging* a capacitor to limit the current drawn from the battery or dc supply, and then *suddenly discharging* all (or some) of the stored energy to provide a bright light for a short time. Other applications include a short time-constant-*differentiating* circuit, and long time-constant-*filtering* and *delay* circuits.

Finally, we consider how an ohmmeter can be used to check a capacitor and how an oscilloscope can be used to measure capacitance.

22-1 CURRENT IN A CAPACITIVE CIRCUIT

We know that when a dc voltage is applied to a capacitor, the charge stored is given by

$$Q = CV \qquad (21\text{-}1)$$

Let us assume that depositing an additional small charge, ΔQ, raises the potential difference across the capacitor by a small amount, ΔV. Therefore,

$$\Delta Q = C\,\Delta V$$

If these changes take place in a short time Δt, then:

$$\frac{\Delta Q}{\Delta t} = C\frac{\Delta V}{\Delta t}$$

But $\frac{\Delta Q}{\Delta t}$ is the average current i during this short interval, Δt. Thus

$$i = C\frac{\Delta V}{\Delta t}$$

To emphasize that this is the current into (or out of) the capacitor and we must consider the *changing* voltage across the capacitor, let us use the following:

$$i_C = C\left(\frac{\Delta v_C}{\Delta t}\right) \quad \textbf{amperes (A)} \quad \textbf{(22-1)}$$

where: i_C is the current into or out of the capacitor in amperes, A

C is the capacitance in farads, F

Δv_C is the change in voltage across the capacitor in volts, V

Δt is the time interval in seconds, s

$\frac{\Delta v_C}{\Delta t}$ is the rate of change of voltage in volts per second, V/s

Equation 22-1 states that current flows into (or out of) a capacitor *only* when the voltage across the capacitor is *changing*. This should be compared with the corresponding equation for an inductor:

$$v_L = L\left(\frac{\Delta i}{\Delta t}\right) \qquad (17\text{-}1)$$

where the *voltage* is induced only when the *current* is changing.

EXAMPLE 22-1

The voltage across a 4-μF capacitor is raised from 16 to 24 V in 2 ms. Calculate:

a. The average rate of change of voltage across the capacitor.
b. The average current into the capacitor during the 2-ms interval.
c. The increase in charge on the capacitor.

SOLUTION

a. $$\frac{\Delta v_C}{\Delta t} = \frac{24\text{ V} - 16\text{ V}}{2 \times 10^{-3}\text{ s}} = \frac{8\text{ V}}{2 \times 10^{-3}\text{ s}} = \mathbf{4000\ V/s}$$

b. $$i_C = C \times \left(\frac{\Delta v_C}{\Delta t}\right) \qquad (22\text{-}1)$$

$$= 4 \times 10^{-6}\text{ F} \times 4000\,\frac{\text{V}}{\text{s}} = 16 \times 10^{-3}\text{ A} = \mathbf{16\ mA}$$

c. $$\Delta Q = C\,\Delta V = 4 \times 10^{-6}\text{ F} \times 8\text{ V} = 32 \times 10^{-6}\text{ C} = \mathbf{32\ \mu C}$$

EXAMPLE 22-2

The voltage across a 50-μF capacitor varies as shown in Fig. 22-1a. Draw the waveform of the capacitor current.

SOLUTION

During the first 5 ms:

$$i_C = C \times \frac{\Delta v_C}{\Delta t} \qquad (22\text{-}1)$$

$$= 50 \times 10^{-6}\text{ F} \times \frac{14\text{ V} - 4\text{ V}}{5 \times 10^{-3}\text{ s}}$$

$$= 50 \times 10^{-6}\text{ F} \times 2 \times 10^{3}\,\frac{\text{V}}{\text{s}} = \mathbf{100\ mA}$$

Note that the rate of change of voltage, $\frac{\Delta v_C}{\Delta t}$, is constant at 2 kV/s during this interval, so the current into

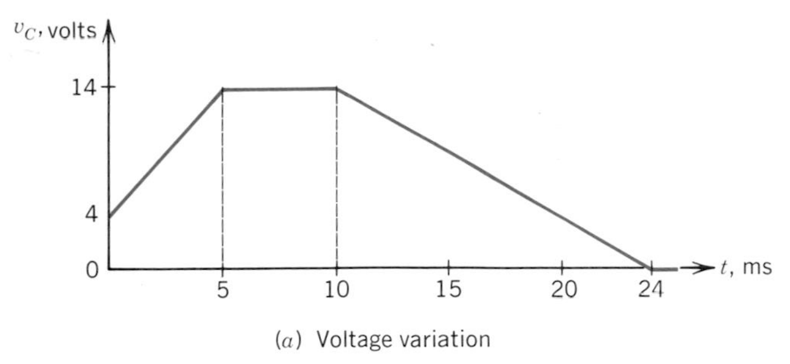

(a) Voltage variation

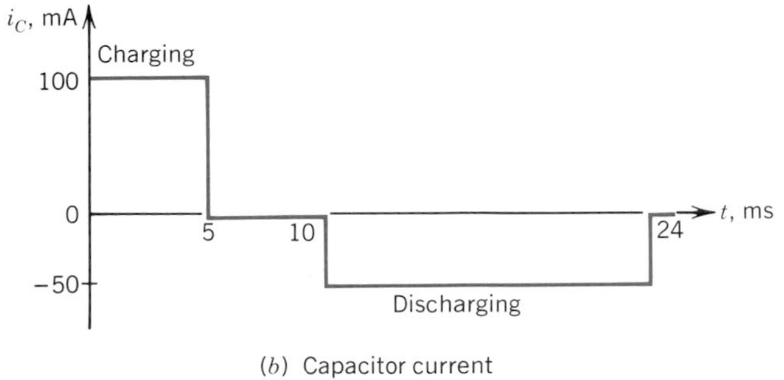

(b) Capacitor current

FIGURE 22-1
Waveforms for Example 22-2.

the capacitor (causing it to charge), is constant at 100 mA. See Fig. 22-1*b*.

From 5 to 10 ms there is no change in voltage. This means that $\frac{\Delta v_C}{\Delta t}$ is zero, and the capacitor current is also zero.

During the interval from 10 to 24 ms the voltage is *decreasing* at a constant rate. The capacitor current is given by

$$i_C = C \times \frac{\Delta v_C}{\Delta t} \tag{22-1}$$

$$= 50 \times 10^{-6}\,\text{F} \times \frac{0\ \text{V} - 14\ \text{V}}{24 - 10\ \text{ms}}$$

$$= 50 \times 10^{-6}\,\text{F} \times \left(-1 \times 10^{3}\,\frac{\text{V}}{\text{s}}\right) = \mathbf{-50\ mA}$$

The negative sign means that the current is flowing in the opposite direction, indicating that the capacitor is *discharging*. See Fig. 22-1*b*.

22-2 CHARGING A CAPACITOR

Example 22-2 suggests that if a capacitor is charged with a constant current, the capacitor voltage increases *linearly*. We now examine the rate at which a capacitor charges when connected suddenly across a dc voltage, with a series resistance, as in Fig. 22-2*a*.

Let us assume that the capacitor is initially discharged so that v_C is zero. This means that when the switch is first closed, all the applied voltage must appear across the resistor. (The capacitor appears, momentarily, as a short

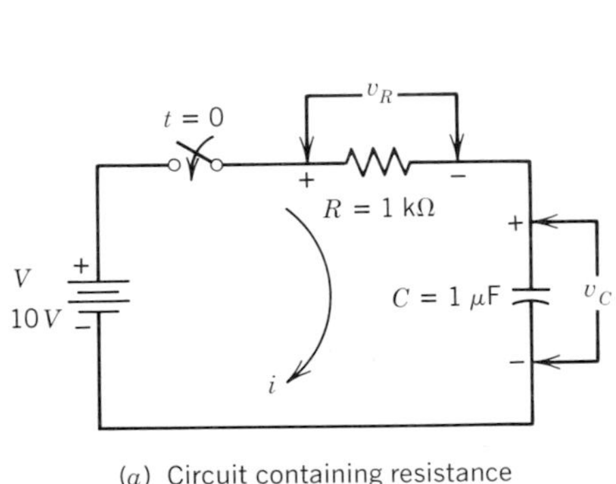

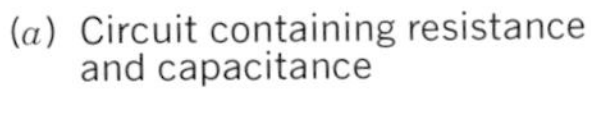
(a) Circuit containing resistance and capacitance

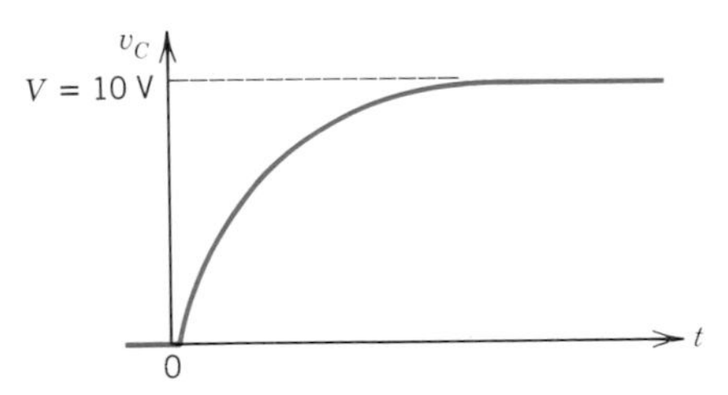

(b) Capacitor voltage variation with time

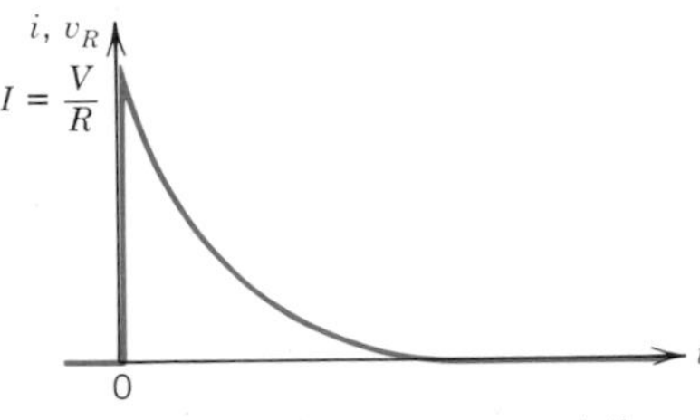

(c) Current and resistor voltage variation with time

FIGURE 22-2
Voltage and current waveforms for a charging capacitor.

circuit.) The *initial* charging current is thus determined only by the applied voltage and the resistance:

$$I_{\text{initial}} = \frac{V}{R} = \frac{10 \text{ V}}{1 \text{ k}\Omega} = 10 \text{ mA}$$

The current immediately increases from 0 to $\frac{V}{R}$, at the instant the switch is closed, as shown in Fig. 22-2*c*.

This current causes a charge to be deposited on the capacitor so that some short time later, the capacitor voltage has increased above zero, as shown in Fig. 22-2*b*.

But Kirchhoff's voltage law, as shown in Fig. 22-2*a*, requires that

$$V = v_R + v_C$$

This means that with some charge on the capacitor, the resistor voltage, v_R, is less than when the switch was first closed:

$$v_R = V - v_C$$

And since $v_R = iR$, the current must have some lower value now. But we also know that

$$i = C\frac{\Delta v_C}{\Delta t} \qquad (22\text{-}1)$$

If i is lower than its initial value, the rate of change of capacitor voltage, $\frac{\Delta v_C}{\Delta t}$, is also lower. That is, the capacitor voltage is still increasing, as shown in Fig. 22-2*b*, but at a *decreasing* rate. Similarly, the charging current continues, but also at a decreasing rate, as in Fig. 22-2*c*.

Eventually, the capacitor voltage reaches the applied voltage, V, there is no further change in voltage, and the charging current drops to zero (Fig. 22-2*c*). If the switch is now opened, the capacitor voltage remains at its *steady-state* value of V. The increasing portion of v_C (Fig. 22-2*b*) and the momentary decreasing charging current (Fig. 22-2*c*) are called *transient responses*. They are similar to the transients occurring in an inductive circuit with the roles of current and voltage interchanged.

To determine how long it takes for the capacitor to fully charge to its steady-state value, we must consider the *time constant*, τ, of the circuit.

22-3 TIME CONSTANT, τ

Since the largest charging current occurs at the instant of closing the switch, the highest rate of change of capacitor voltage must occur at this instant, that is,

$$i = I_{\text{initial}} = \frac{V}{R}$$

and

$$i = C\left(\frac{\Delta v_C}{\Delta t_0}\right) \qquad (22\text{-}1)$$

Thus

$$C\left(\frac{\Delta v_C}{\Delta t_0}\right) = \frac{V}{R}$$

and the *initial* rate of rise of the capacitor voltage is

$$\frac{\Delta v_C}{\Delta t_0} = \frac{V}{RC} \quad \text{volts per second (V/s)} \qquad (22\text{-}2)$$

where: $\frac{\Delta v_C}{\Delta t_0}$ is the *initial* rate of rise of the capacitor voltage in volts per second, V/s
V is the applied dc voltage in volts, V
R is the series resistance in ohms, Ω
C is the capacitance in farads, F

EXAMPLE 22-3

For the circuit and values given in Fig. 22-2*a*, determine:

a. The initial rate of rise of the capacitor voltage.
b. The initial charging current.
c. The final capacitor voltage.

SOLUTION

a. $$\frac{\Delta v_C}{\Delta t_0} = \frac{V}{RC} \qquad (22\text{-}2)$$

$$= \frac{10 \text{ V}}{1 \times 10^3 \ \Omega \times 1 \times 10^{-6} \text{ F}}$$

$$= \mathbf{10 \times 10^3 \ V/s}$$

b. Initial charging current $= \frac{V}{R}$

$$= \frac{10 \text{ V}}{1 \times 10^3 \ \Omega} = \mathbf{10 \ mA}$$

c. Final capacitor voltage $= V =$ **10 V**

Now let us consider how long it would take the capacitor voltage to reach its final value *if* it continued to increase at its initial rate of change of $\frac{V}{RC}$ volts per second. Let us call this length of time the time constant, τ (tau). See Fig. 22-3.

We know from Fig. 22-3 that the voltage does not increase at this rate. In fact, when $t = \tau$, the actual capacitor voltage is only 0.63 of its final value. This is the more usual definition of the time constant. (See Section 22-5.)

Let the dotted line that is tangent to the voltage curve reach the final value of the capacitor voltage in a time interval τ. This dotted line represents the initial rate of change of the capacitor voltage. It has a slope given by

$$\text{initial slope} = \frac{\text{rise}}{\text{run}} = \frac{\text{vertical intercept}}{\text{horizontal intercept}}$$

$$= \frac{V}{\tau} \quad \text{volts per second}$$

But the initial slope $= \dfrac{\Delta v_C}{\Delta t_0} = \dfrac{V}{RC}$

volts per second (V/s) (22-2).

Therefore,

$$\frac{V}{\tau} = \frac{V}{RC}$$

and $\quad \tau = RC \quad$ **seconds (s)** $\quad$ **(22-3)**

where: τ is the time constant of the circuit in seconds, s

R is the resistance of the circuit in ohms, Ω

C is the capacitance of the circuit in farads, F

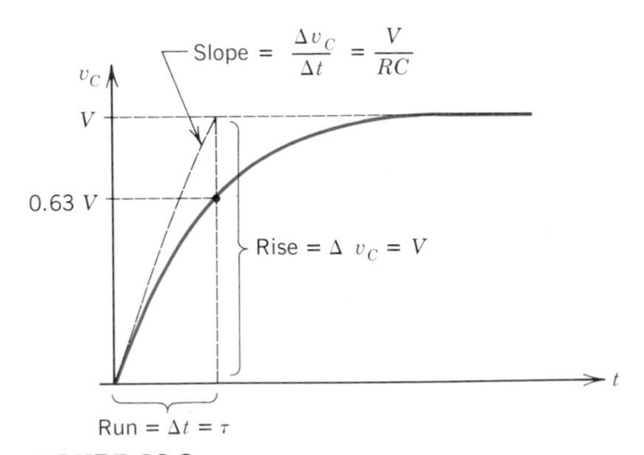

FIGURE 22-3
Determination of time constant, τ.

Equation 22-3 gives the time required for a capacitor to reach 63% of its final steady-state voltage. This time is equal to the product of the circuit capacitance and resistance.

It can also be shown (see Section 22-5) that it takes approximately five time constants for the capacitor to charge to its final steady-state voltage. This time is not dependent upon the voltage applied, as shown by the following examples.

EXAMPLE 22-4

A 5-μF capacitor and a 10-kΩ resistor are series-connected across a 12-V dc supply. Calculate:

a. The time constant of the circuit.
b. The initial rate of rise of the capacitor voltage.
c. The capacitor voltage after one time constant.
d. The time required for the capacitor to charge to 12 V.

SOLUTION

a. $\tau = RC$ (22-3)

$$= 10 \times 10^3\ \Omega \times 5 \times 10^{-6}\ \text{F}$$

$$= 50 \times 10^{-3}\ \text{s} = \mathbf{50\ ms}$$

b. $\dfrac{\Delta v_C}{\Delta t_0} = \dfrac{V}{RC}$ (22-2)

$$= \frac{12\ \text{V}}{50 \times 10^{-3}\ \text{s}} = \mathbf{240\ V/s}$$

c. After one time constant

$$v_C = 0.63 \times V$$

$$= 0.63 \times 12\ \text{V} = \mathbf{7.56\ V}$$

d. Time to reach 12 V = $5\,\tau$

$$= 5 \times 50\ \text{ms} = \mathbf{250\ ms}$$

EXAMPLE 22-5

The capacitance in the previous example is doubled to 10 μF; the resistance is cut in half to 5 kΩ; and the voltage is doubled to 24 V dc. Calculate:

a. The time constant of the circuit.
b. The initial rate of rise of the capacitor voltage.
c. The capacitor voltage after one time constant.
d. The time required for the capacitor to charge to 24 V.

SOLUTION

a. $\tau = RC$ (22-3)

$= 5 \times 10^3\ \Omega \times 10 \times 10^{-6}$ F

$= 50 \times 10^{-3}$ s = **50 ms**

b. $\frac{\Delta v_C}{\Delta t_0} = \frac{V}{RC}$ (22-2)

$= \frac{24 \text{ V}}{50 \times 10^{-3} \text{ s}} =$ **480 V/s**

c. After one time constant

$v_C = 0.63 \times V$

$= 0.63 \times 24$ V = **15.12 V**

d. Time to reach 24 V = 5 τ

= 5 × 50 ms = **250 ms**

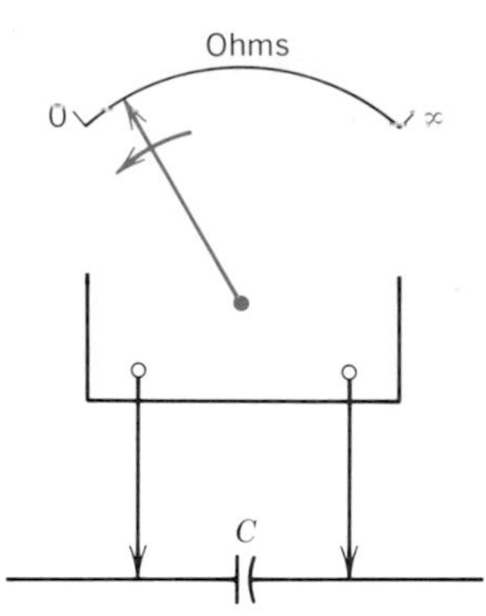

(*a*) Ohmmeter initially deflects toward zero when first connected to the capacitor.

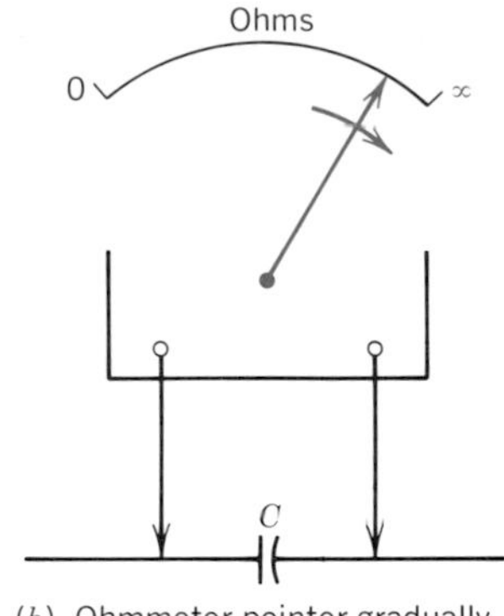

(*b*) Ohmmeter pointer gradually moves toward infinity as the capacitor charges.

FIGURE 22-5
Checking a capacitor with an ohmmeter.

Both capacitors reach 63% of their final steady-state value (7.56 V for the first capacitor, 15.12 V for the second) in the *same* time of 50 ms. Also, both charge to their *final* values of 12 and 24 V in the *same* time of 5RC or 250 ms. This is achieved by the rate of rise of voltage being twice as great in Example 22-5 as in Example 22-4. See Fig. 22-4.

22-3.1 Checking Capacitors with an Ohmmeter

When an ohmmeter is connected across an uncharged capacitor, there is an initial charging current determined by the resistance of the ohmmeter and its cell voltage. (Recall that an ohmmeter is essentially a series circuit of resistance, dry cell, and meter movement. The ohmmeter charges the capacitor, eventually, to the voltage of the internal cell or cells.)

The effect on the ohmmeter is a sudden deflection of the pointer from infinity toward zero ohms because of the relatively large initial charging current. The capacitor initially acts like a short circuit (very low resistance). See Fig. 22-5*a*.

As the capacitor charges to the ohmmeter cell voltage, the ohmmeter pointer moves gradually back toward infinity. The rate at which it moves is determined by the time constant of the capacitor and the ohmmeter resistance. See Fig. 22-5*b*. The final reading of the ohmmeter is an indication of the insulation resistance of the capacitor (Section 21-5.1). For most capacitors, this resistance is practically infinite (on the average ohmmeter). For electrolytic capacitors, however, the final ohmmeter reading will be noticeably less than infinite and may depend upon the polarity applied to the capacitor.

It should be evident that if the ohmmeter pointer goes to zero, and stays there, the capacitor is shorted. No initial deflection from infinity toward zero indicates an open capacitor. However, capacitors must be checked on a suitable ohmmeter range to give a visible charging deflection. On the very low resistance ranges, charging can be so instantaneous that no initial deflection toward zero is observable. This is especially true for very small capacitors of a few picofarads, where the time constant is very small. Using a higher resistance range, say $R \times 1$ MΩ, is necessary for these capacitors. However, the higher resistance limits the initial charging current, so not much, if any, deflection of the pointer from infinity will result on

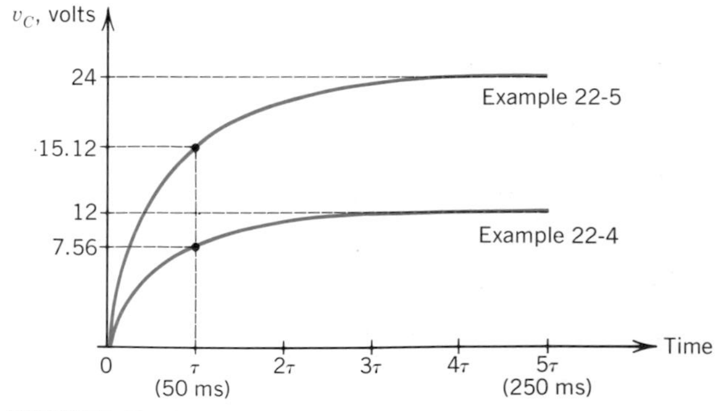

FIGURE 22-4
Capacitor charging curves for Examples 22-4 and 22-5.

any ohmmeter range for very small capacitors. Also, if the capacitor is connected in a complete circuit, the ohmmeter readings may be affected by other components.

Obviously, the use of an ohmmeter to check a capacitor must be done carefully. It provides only a simple check to see if the capacitor has any obvious problems.

22-4 DISCHARGING A CAPACITOR

We have seen how a charged capacitor stores energy ($W = \frac{1}{2} CV^2$). Let us now examine how a capacitor *discharges*, returning this energy to a load.

Consider the circuit in Fig. 22-6*a*. If the switch has remained in position 1 for five or more time constants, the capacitor is fully charged (to 10 V, in this example). When the single-pole, double-throw switch is thrown to position 2, the capacitor immediately begins to discharge. The initial discharge current is determined by the initial capacitor voltage and the resistance:

$$I_{\text{initial}} = \frac{V_{\text{initial}}}{R} = \frac{10 \text{ V}}{1 \times 10^3 \ \Omega} = 10 \text{ mA}$$

This occurs because, by Kirchhoff's voltage law, the initial capacitor voltage must be equal to the voltage across the resistor. Since the direction of the discharging current is opposite to the charging current, the waveform of the capacitor current in Fig. 22-6*c* is shown negative.

Since charge is being removed from the capacitor, the potential difference across the capacitor drops from its initial value, as shown in Fig. 22-6*b*. The initial rate of decrease of capacitor voltage is again given by

$$\frac{\Delta v_C}{\Delta t_0} = \frac{V_{\text{initial}}}{RC} \qquad (22\text{-}2)$$

After one time constant, $\tau = RC$, the capacitor voltage has *changed* 63%. This means that the voltage across the

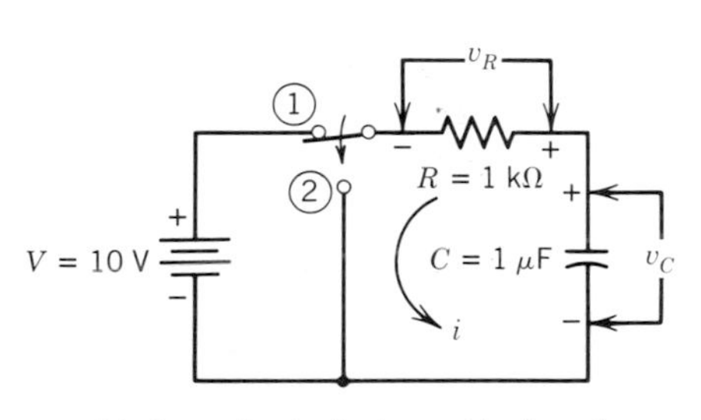

(*a*) Capacitor is discharged by throwing switch to position 2.

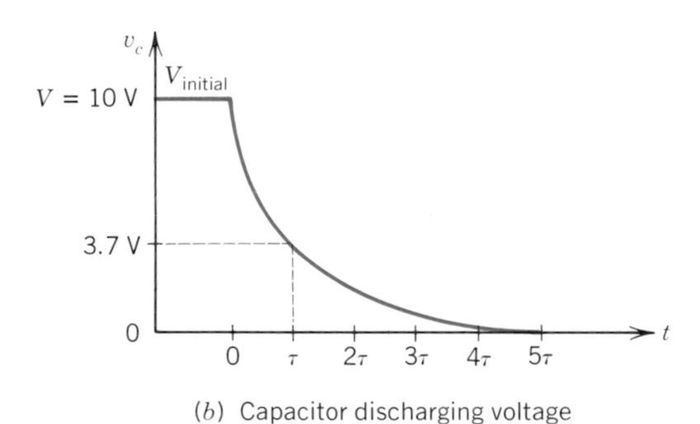

(*b*) Capacitor discharging voltage

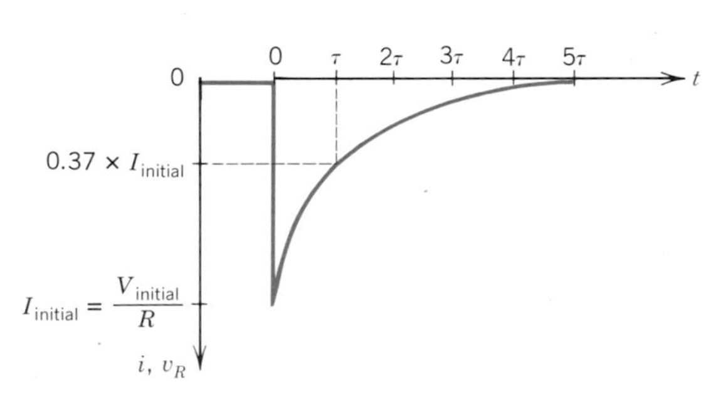

(*c*) Capacitor discharging current

FIGURE 22-6
Voltage and current waveforms for a discharging capacitor.

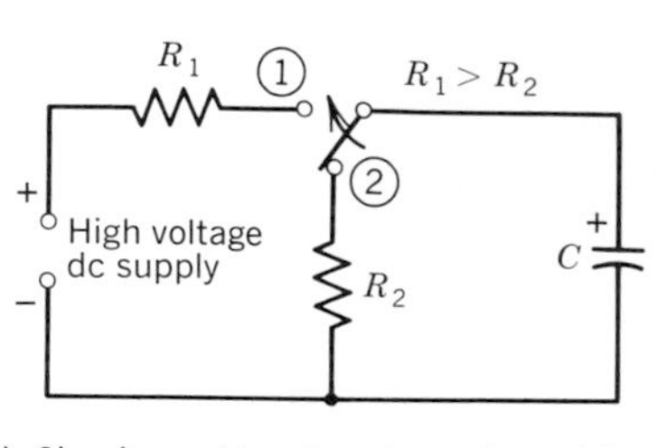

(a) Circuit provides slow charge in position 1 and fast discharge in position 2.

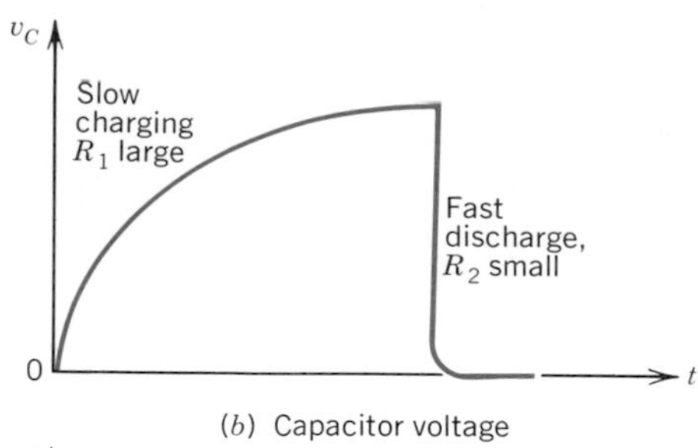

(b) Capacitor voltage

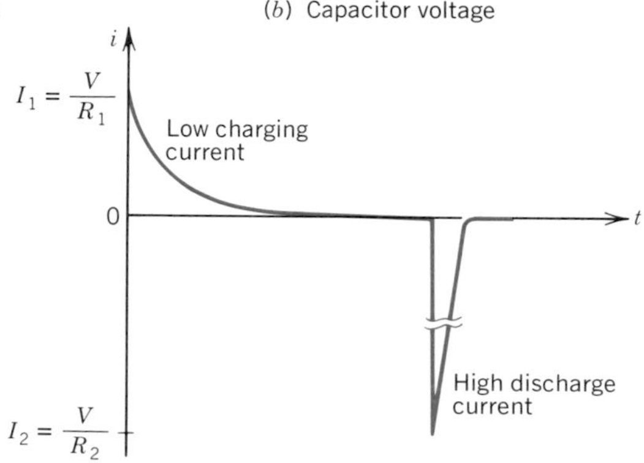

(c) Capacitor current

FIGURE 22-7
Typical RC circuit for short-duration, high-current pulses.

capacitor is now only 0.37 of its initial value (3.7 V). Similarly, the current has dropped to 0.37 of its initial value (3.7 mA) after one time constant.

After five time constants, $5\ \tau = 5\ RC$, both the decreasing capacitor voltage and the discharging current are approximately zero. The transient is ended and the capacitor is completely discharged. All of the energy stored in the capacitor has been dissipated in the resistor.

NOTE It takes the same time for the capacitor to discharge as it does to charge if the same series resistance is used. For Fig. 22-6a, $\tau = RC = 50$ ms for both charging and discharging. Thus it takes 250 ms to completely charge or discharge. Example 22-6 shows a circuit that has different charge and discharge time constants.

22-4.1 Application to a Capacitor Photoflash Unit

The circuit shown in Fig. 22-7 can produce short-duration, high-current pulses without drawing a large current from the supply. The large resistance R_1 limits the peak charging current I_1 and allows a gradual charging of the capacitor due to the large time constant $\tau_1 = R_1C$. When the switch is thrown to position 2, the low resistance of R_2 permits a high discharge current. The duration of this current is determined by the small time constant, $\tau_2 = R_2C$. This system is used to provide the high surge current required in electric spot welding, radar transmitter tubes, photoflash units, and so on. The amount of energy delivered by each discharge pulse can be controlled (for a given capacitance) by the voltage used to charge the capacitor.

EXAMPLE 22-6

A capacitor photoflash unit uses a circuit similar to that in Fig. 22-7 with a 300-V supply, a 500-μF capacitor, and a current limiting resistor of 4 kΩ. Assume that the resistance of the lamp, R_2, stays constant at 10 Ω when the shutter contact discharges the capacitor in position 2. Calculate:

a. The time required for the capacitor to fully charge.
b. The peak charging current.
c. The time required for the capacitor to fully discharge.
d. The peak discharging current.
e. The energy stored by the capacitor.
f. The average power produced by the lamp, in watts, during the discharge period.

SOLUTION

a. Charging time $= 5R_1C$

$$= 5 \times 4 \times 10^3\ \Omega \times 500 \times 10^{-6}\ \text{F}$$

$$= \mathbf{10\ s}$$

b. Peak charging current, $I_1 = \dfrac{V}{R_1}$

$$= \frac{300\ \text{V}}{4 \times 10^3\ \Omega} = \mathbf{75\ mA}$$

c. Discharging time $= 5R_2C$

$$= 5 \times 10\ \Omega \times 500 \times 10^{-6}\ \text{F}$$

$$= \mathbf{25\ ms}$$

d. Peak discharging current,

$$I_2 = \frac{V}{R_2}$$

$$= \frac{300\ \text{V}}{10\ \Omega} = \mathbf{30\ A}$$

e. Energy stored,

$$W = \tfrac{1}{2}CV^2 \qquad (21\text{-}12)$$

$$= \tfrac{1}{2} \times 500 \times 10^{-6}\ \text{F} \times (300\ \text{V})^2$$

$$= \mathbf{22.5\ J}$$

f. Power, $P = \dfrac{W}{t}$ (3-6)

$$= \frac{22.5\ \text{J}}{0.025\ \text{s}}$$

$$= 900\ \text{J/s} = \mathbf{900\ W}$$

22-5 UNIVERSAL TIME-CONSTANT CURVES

If we wish to determine capacitor voltage and current at times other than one and five time constants, we can use the universal time-constant curves. These give the instantaneous values as a percentage of the initial or steady-state values, with time given in time constants. See Fig. 22-8. (These are the same curves as those used in Fig. 19-4 for series *RL* circuits.) Note that at one time constant, τ, the capacitor has charged to 63% of its final steady-state voltage, and the charging current has dropped to only 37% of its initial value. In either case, a *change* of 63% occurs in one time constant.

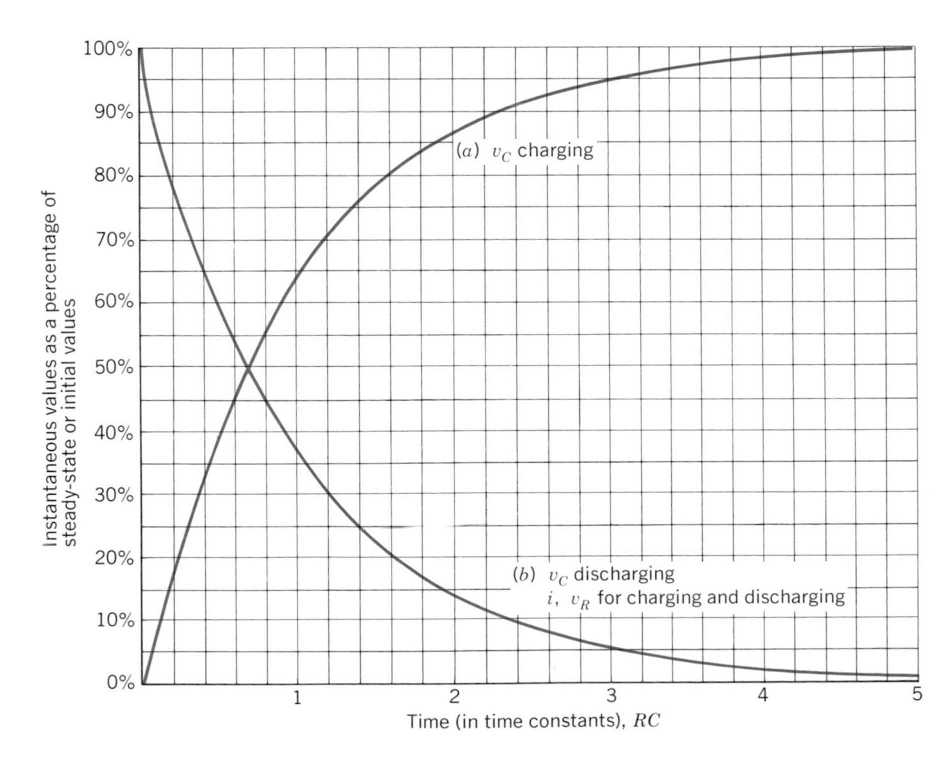

FIGURE 22-8
Universal time-constant curves for charge and discharge voltage and current in a capacitive circuit with direct current.

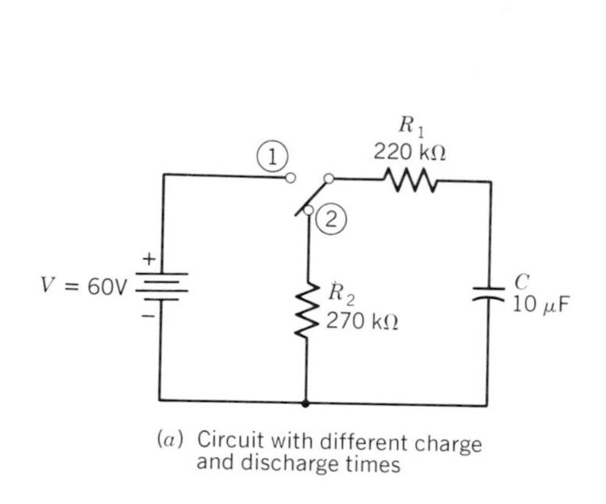

(a) Circuit with different charge and discharge times

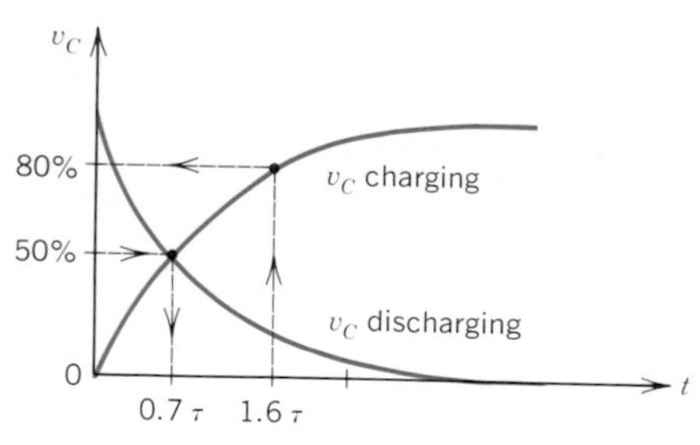

(b) Sketch of universal time-constant curves

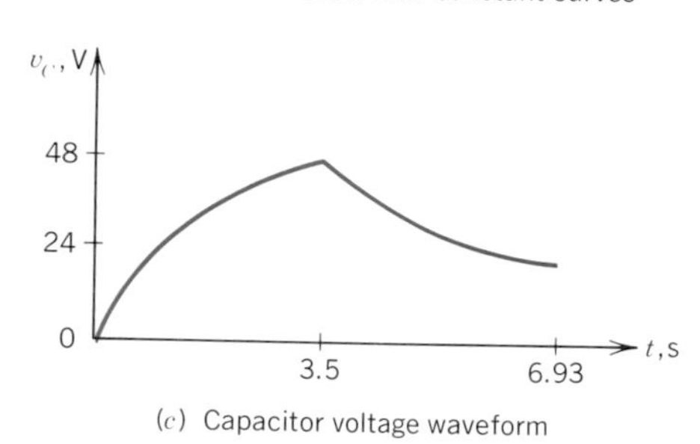

(c) Capacitor voltage waveform

FIGURE 22-9
Circuit and waveforms for Example 22-7.

These curves also confirm that the charging or discharging of a capacitor is essentially complete after five time constants.

EXAMPLE 22-7

Refer to the circuit of Fig. 22-9*a*. After the switch has remained in position 2 for a long time (until the capacitor is completely discharged), the switch is thrown to position 1. Use the universal time constant curves to determine:

a. The capacitor voltage 3.5 s later.
b. How long it takes for the capacitor voltage to discharge to 24 V if the switch is returned to position 2 after being in position 1 for 3.5 s.
c. The capacitor voltage waveform for parts (a) and (b).

SOLUTION

a. $\tau_1 = R_1 C$ (22-3)

$= 220 \times 10^3\ \Omega \times 10 \times 10^{-6}\ \text{F}$

$= 2.2\ \text{s}$

Charge time $t = 3.5$ s

$$= \frac{3.5\ \text{s}}{2.2\ \text{s}/\tau} = 1.59\ \tau \approx 1.6\ \tau$$

From Fig. 22-8*a*, when $t = 1.6\ \tau$, $v_C = 80\%$ of V (see Fig. 22-9*b*).

Therefore, $v_C = 80\%$ of V

$= 0.8 \times 60\ \text{V} =$ **48 V**

b. Initial capacitor voltage = 48 V

Capacitor voltage

$= 24$ V

$$= \frac{24\ \text{V}}{48\ \text{V}} \times 100\% \text{ of initial value}$$

$= 50\%$ of initial value

From Fig. 22-8*b*, when $v_C = 50\%$ of initial value,

$t = 0.7\ \tau$ (see Fig. 22-9*b*)

Discharge time constant

$\tau_2 = (R_1 + R_2)C$

$= (220 \times 10^3 + 270 \times 10^3)\ \Omega \times 10 \times 10^{-6}\ \text{F}$

$= 4.9$ s

Discharge time to reach 24 V $= 0.7\ \tau$

$= 0.7 \times 4.9$ s

$=$ **3.43 s**

c. See Fig. 22-9*c*.

Note that the *discharge* time constant depends upon the *total* series resistance, $R_1 + R_2$.

Instead of using a graphical approach, we can calculate the capacitor voltage and current at any instant, for charging or discharging, using exponential equations given in Appendix A-12. A problem similar to Example 22-7 is solved using these equations.

22-6 OSCILLOSCOPE MEASUREMENT OF CAPACITANCE

Although it is possible to observe the charging and discharging of a capacitor on an oscilloscope using a mechanical switching circuit, it is only practical for relatively large time constants. Where the time constant is small, repetitive switching at a high-enough frequency is only possible using the output of a square wave signal generator, connected to the capacitive circuit. The frequency is adjusted until the capacitor voltage waveform shows that a steady state has been reached, as shown in Fig. 22-10. That is, the half-period of the square wave output must be equal to or greater than five time constants, that is, $\frac{T}{2} \geq 5RC$.

With the oscilloscope connected across the capacitor, the time required to reach 63% of the final voltage is the time constant, τ. See Fig. 22-10*c*. If the total resistance of the circuit is known, the capacitance can be obtained. See Example 22-8.

If the oscilloscope is connected across the resistor (with R and C interchanged to avoid common ground problems), the time constant is given by the time required for v_R to fall to 37% of its initial value. See Fig. 22-10*d*. This method is particularly useful if the capacitance of the circuit is not clearly defined. That is, *stray* capacitance in the circuit (due to the method of wiring, etc.) may make it impossible to connect the oscilloscope across the capacitance itself. See Problem 22-20.

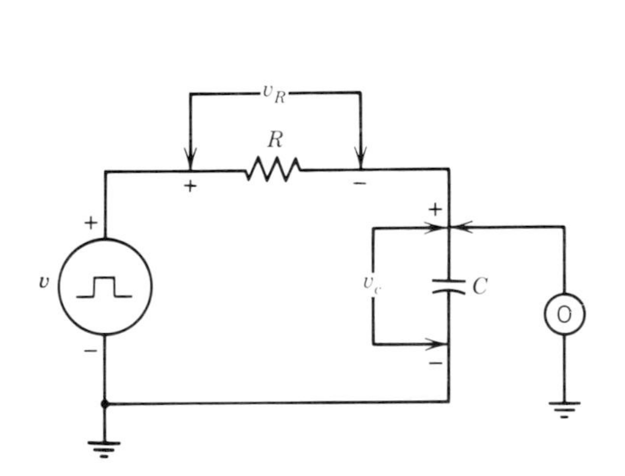

(*a*) Square wave generator connected to a capacitive circuit

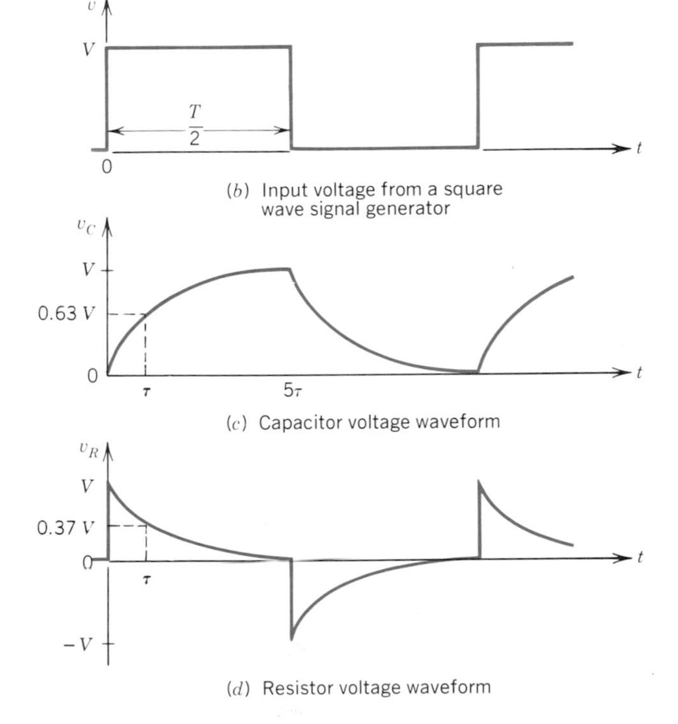

(*b*) Input voltage from a square wave signal generator

(*c*) Capacitor voltage waveform

(*d*) Resistor voltage waveform

FIGURE 22-10
Voltage waveforms in a capacitive circuit with a square wave signal generator.

EXAMPLE 22-8

An oscilloscope is connected across a capacitor as in Fig. 22-10*a*. The display reaches a final vertical value of 5 cm. It requires a horizontal distance of 1.5 cm for the capacitor voltage display to reach 3.15 cm. If the oscilloscope horizontal calibration is 2 ms/cm, determine:

a. The time constant of the circuit.
b. The capacitance if $R = 4.7\ \text{k}\Omega$.
c. The maximum frequency allowable for the signal generator.

SOLUTION

a. 63% of 5 cm = 0.63 × 5 cm

$$= 3.15 \text{ cm (vertical distance)}$$

The *horizontal* distance of 1.5 cm is thus proportional to the time constant:

$$\tau = 1.5 \text{ cm} \times 2 \text{ ms/cm}$$
$$= \mathbf{3\ ms}$$

b. $\tau = RC$ (22-3)

Therefore, $C = \dfrac{\tau}{R}$

$$= \frac{3 \times 10^{-3}\ \text{s}}{4.7 \times 10^{3}\ \Omega}$$
$$= 0.64 \times 10^{-6}\ \text{F} = \mathbf{0.64\ \mu F}$$

c. $\dfrac{T}{2} \geq 5RC$

Therefore, $T_{\text{min}} = 10\,RC$

$$= 10 \times 3 \text{ ms} = 30 \text{ ms}$$

$$f_{\text{max}} = \frac{1}{T} \qquad (14\text{-}9a)$$

$$= \frac{1}{30 \times 10^{-3}\ \text{s}} = \mathbf{33.3\ Hz}$$

22-7 SHORT TIME-CONSTANT DIFFERENTIATING CIRCUIT

Consider the circuit of Fig. 22-11*a* with a time constant very small compared with the period of the applied voltage (e.g., $5\ RC \leq T/10$). This means that the capacitor charges very rapidly, producing a waveform, as in Fig. 22-11*c*, that is very similar to the input voltage waveform.

Now we know that the current in the circuit is given by

$$i_C = C\frac{\Delta v_C}{\Delta t} \qquad (22\text{-}1)$$

and $v_R = iR = RC\dfrac{\Delta v_C}{\Delta t}$. But since $\Delta v_C \approx v$ (the input voltage),

$$v_R \approx RC\frac{\Delta v}{\Delta t} \qquad (22\text{-}4)$$

Thus the resistor voltage, v_R, is proportional to the *slope* of the input voltage, $\dfrac{\Delta v}{\Delta t}$.

The mathematical process of finding the slope of a curve is called *differentiation.* ($\dfrac{\Delta v}{\Delta t}$ is an approximation to the derivative, $\dfrac{dv}{dt}$, found by *differentiating* a mathematical equation, so v_R is said to be the output of a *differentiating circuit.*)

Thus the resistor voltage gives a sharp, narrow positive pulse, when the *leading* edge of the input voltage goes positive, because of the very steep slope of the input waveform at this instant.

The resistor voltage then drops to zero, corresponding to zero slope of the input voltage. A sharp negative pulse is then produced at the *trailing* edge of the input voltage, showing the negative slope of the input waveform. These fast rising and falling voltage pulses are useful in digital logic circuits for triggering purposes.

22-8 LONG TIME-CONSTANT CIRCUITS

The circuit of Fig. 22-12*a* shows an application in which it is desirable to have a very long time constant. With the switch open, the diode produces a half-wave-rectified voltage variation across the load (Fig. 22-12*b*). Although this output is direct current, it is not very smooth because it contains a large ac ripple voltage.

When the switch is closed, the capacitor is placed in parallel with the load. Now the capacitor can charge to the peak voltage V_m when the input voltage increases. But when the input voltage decreases, the capacitor

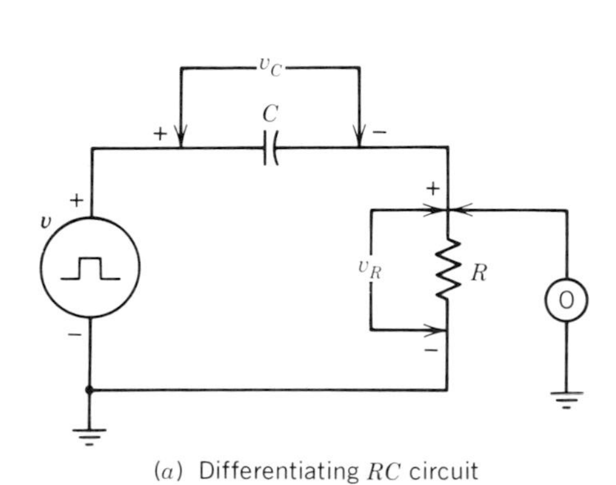

(a) Differentiating RC circuit

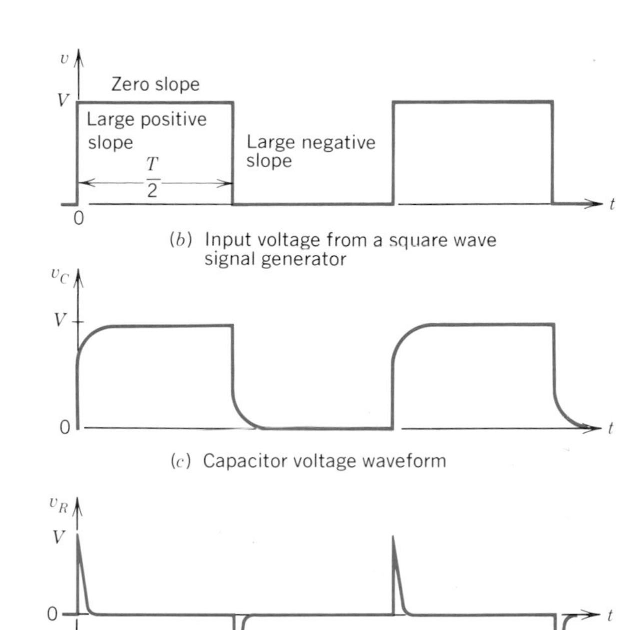

(b) Input voltage from a square wave signal generator

(c) Capacitor voltage waveform

(d) Resistor voltage waveform is proportional to the slope of the input voltage waveform.

FIGURE 22-11
Voltage waveforms in a differentiating circuit.

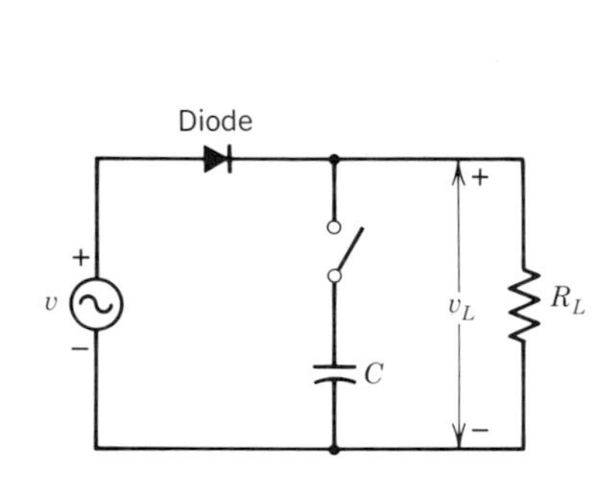

(a) Capacitor filters voltage fluctuations when the switch is closed.

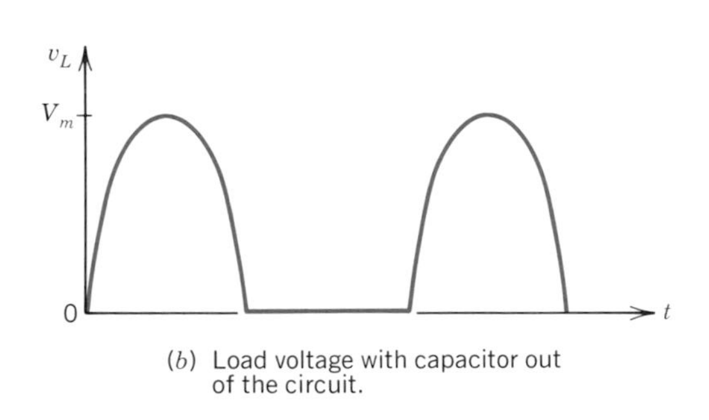

(b) Load voltage with capacitor out of the circuit.

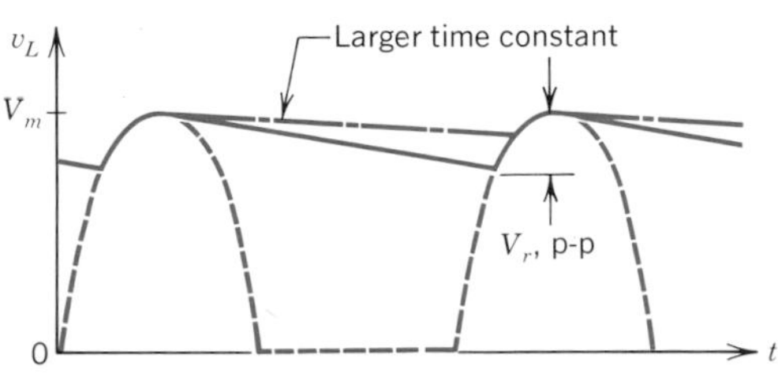

(c) Load voltage with capacitor filtering, for two different time constants.

FIGURE 22-12
Use of a capacitor to filter voltage fluctuations.

discharges through the load, attempting to keep the voltage near V_m. We are using the basic property of a capacitor to *oppose* any change of voltage, to smooth the load voltage. The larger the time constant R_LC, the higher will be the output load voltage, as shown in Fig. 22-12*c*. The energy delivered to the load during the discharge period is replaced during the very short charging period when the input voltage again nears V_m. Note that the ac ripple voltage content is very small due to the action of the capacitor.

EXAMPLE 22-9

A 1000-μF capacitor is used to provide filtering for a 100-Ω load, as in Fig. 22-12*a*, with a peak voltage input of 9 V and a frequency of 60 Hz. Assuming that the capacitor discharges for a full period of the input waveform, calculate:

a. The lowest value of load voltage.

b. The peak-to-peak ripple voltage across the load as indicated in Fig. 22-12*c*.

c. The effect on load and ripple voltage of using a 2000-μF capacitor.

SOLUTION

a.
$$\tau = RC \qquad (22\text{-}3)$$
$$= 100\ \Omega \times 1000 \times 10^{-6}\ \text{F}$$
$$= 0.1\ \text{s}$$
$$T = \frac{1}{f} \qquad (14\text{-}9)$$

Therefore, discharge time,

$$t = \frac{1}{60\ \text{Hz}}$$
$$= 16.6\ \text{ms}$$
$$= \frac{16.6\ \text{ms}}{0.1\ \text{s}/\tau} = 0.166\ \tau$$

From curve (b) in Fig. 22-8, when $t = 0.17\ \tau$,

$$v_C = 85\%\ \text{of}\ V_{\text{initial}}$$
$$= 0.85 \times 9\ \text{V}$$
$$= \mathbf{7.65\ V}$$

b. Peak-to-peak ripple voltage $= V_{r,\text{p-p}}$
$$= v_{C_{\max}} - v_{C_{\min}}$$
$$= 9\ \text{V} - 7.65\ \text{V}$$
$$= \mathbf{1.35\ V}$$

c. When $C = 2000\ \mu\text{F}$, $\tau = 0.2$ s and $t = 0.083\ \tau$
$$v_C = 92\%\ \text{of}\ V_{\text{initial}} = \mathbf{8.28\ V}$$
and
$$V_{r,\text{p-p}} = \mathbf{0.72\ V}$$

Example 22-9 shows that use of a larger capacitor reduces ripple voltage.

22-8.1 RC Delay Circuits

An RC circuit with a long time constant can be used to introduce a delay, as in the timing circuit of Fig. 22-13.

The neon lamp acts as an open circuit until a firing voltage between 60 and 80 V is reached. This means that the voltage across the capacitor charges toward 100 V at a rate determined by the time constant $(R_1 + R_2)C$. But when the firing voltage of the neon lamp is reached, the capacitor discharges through the neon lamp, lighting it briefly. Because of the low resistance of the conducting lamp, the capacitor voltage drops quickly and the lamp is extinguished. Since the lamp is once again an open circuit, the capacitor recharges, providing a controlled delay before the lamp again fires. The rate of flashing can be varied by adjusting R_2. (See Problem 22-22.)

The delay introduced by the capacitor in the circuit of Fig. 22-13 illustrates a useful principle. For example, if it is required to delay the operation of a dc relay following the application of voltage to the coil, a capacitor connected across the coil will achieve this. For longer delays, a resistor in series with the parallel-connected coil and capacitor can be used, although this will also affect the final voltage applied to the coil. (See Problem 22-24.)

Finally, the voltage across the capacitor in an *RC* circuit with a medium or long time constant, is proportional to the *integral* of the input voltage. The circuit is referred to as an *integrating* circuit because it "sums up" or "averages out" the input voltage. See Appendix A-13 for an application of an integrating circuit that displays a *B-H* hysteresis curve on an oscilloscope.

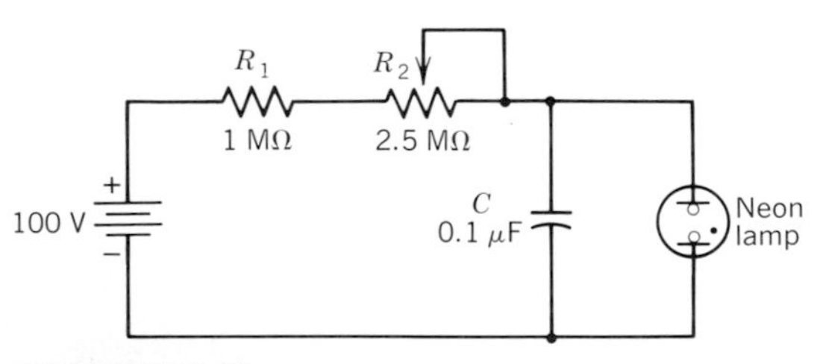

FIGURE 22-13
RC relaxation oscillator timing circuit.

SUMMARY

1. Capacitor current only flows when voltage across the capacitor changes, as given by the equation:

$$i_C = C\frac{\Delta v_C}{\Delta t}$$

2. The initial charging current of a capacitor is determined only by the applied voltage and series resistance, as given by:

$$I_{\text{initial}} = \frac{V}{R}$$

3. The gradual rise of capacitor voltage during charging is determined by the time constant:

$$\tau = RC.$$

4. The time constant is the time required for the capacitor voltage to rise to 63% of its final steady-state value or for the charging current to fall to 37% of its initial value.

5. After five time constants, the transient response is complete, and a capacitor is considered to be fully charged (or discharged).

6. The initial rate of change of capacitor voltage, during charging or discharging, is given by

$$\frac{\Delta v_C}{\Delta t} = \frac{V}{RC}$$

7. If a capacitor continued to charge (or discharge) at its initial rate of change, the capacitor voltage and current would reach their steady-state values in one time constant.

8. An ohmmeter can be used to make a quick check of a capacitor by watching for an initial deflection toward zero followed by a gradual charging effect toward infinity.

9. The high discharge current available from a capacitor can be used to advantage in an electronic photoflash unit to light a lamp very brightly for a very short time.

10. Universal time-constant curves can be used to graphically determine instantaneous capacitor voltage and current with time given in terms of time constants.

11. Exponential equations, used to determine algebraically the values of capacitor voltage and current at any instant, are given in Appendix A-12.

12. An oscilloscope and square wave signal generator can be used to determine the capacitance of a circuit by measuring the circuit's time constant.

13. If a series RC circuit has a short time constant $\left(5\,RC \leq \frac{T}{10}\right)$, the voltage waveform across the resistor is an approximation to the derivative or the slope of the input voltage, and the circuit is called a differentiating circuit.

14. Long time-constant circuits can be used to provide filtering in dc power supplies or to cause a delay in timing and relay circuits.

SELF-EXAMINATION

Answer T or F or a, b, c, d, or e for multiple choice
(Answers at back of book)

22-1. When a dc voltage is applied to a capacitor the voltage across the capacitor cannot change instantaneously. ______

22-2. Capacitor current can only flow if the voltage across the capacitor is changing. ________

22-3. Capacitor charging current has its greatest rate of change at one time constant. ________

22-4. A larger capacitor has a higher initial charging current when a dc voltage is applied than a smaller capacitor. ________

22-5. If a capacitor voltage increased from 25 to 50 V in 5 ms, the rate of change of voltage is

a. 5 V/s
b. 5 kV/s
c. 5 mV/s
d. 125 mV/s
e. none of the above

22-6. Given the rate of change in Question 22-5, what capacitor current will a capacitance of 10 μF cause?

a. 5 mA
b. 50 A
c. 50 mA
d. 1.25 A
e. 5 μA

22-7. If 10 V dc is applied to a series combination of a 2-kΩ resistor and a 5-μF capacitor, the initial charging current is

a. 20 mA
b. 0.2 mA
c. 10 mA
d. 5 μA
e. 5 mA

22-8. A series circuit of 2.2-MΩ resistance and 0.1-μF capacitance has a time constant of

a. 0.22 s
b. 0.22 ms
c. 0.22 μs
d. 22 s
e. 22 ms

22-9. Checking a capacitor with an ohmmeter determines only whether the capacitor is good or bad, not the actual capacitance. ________

22-10. A 0.1-μF capacitor, initially charged to 20 V, is discharged through a 10-kΩ resistor. After 1 ms the capacitor voltage is

a. 6.3 V
b. 3.7 V
c. 7.4 V
d. 12.6 V
e. none of the above

22-11. The charge and discharge times of a capacitor are always equal. ________

22-12. The universal time constant curves of Fig. 22-8 can be used either for the charging or discharging of a capacitor. ________

22-13. A capacitor's voltage after discharging for five time constants is exactly zero. ______

22-14. If a series RC circuit with $R = 20\ \text{k}\Omega$ shows a time constant on an oscilloscope of 50 ms, the circuit capacitance is

a. 20 μF
b. 40 pF
c. 20 pF
d. 2.5 μF
e. 2.5 pF

22-15. The output of a differentiating circuit is taken across the resistor in a series RC circuit. ______

22-16. When a capacitor is used in a filter circuit, a higher load contributes to a longer time constant and therefore a smoother output voltage. ______

22-17. A larger capacitor used across the coil of a relay will increase the delay in the operation of the relay. ______

REVIEW QUESTIONS

1. Which two equations can you cite to show that capacitors and inductors have opposite effects as far as voltage and current are concerned?

2. a. How would you interpret a negative $\dfrac{\Delta v_C}{\Delta t}$?

b. What effect does it have on the capacitor current?

3. Explain in your own words why the capacitor voltage increases *gradually* (and not instantaneously) in a series RC circuit with direct current applied.

4. Give two definitions of a time constant as applied to a series RC circuit.

5. Why is the rate of rise of the capacitor voltage highest when direct current is first applied to a series RC circuit?

6. a. Why doesn't it take longer (for a given RC circuit) to charge to a final steady-state value of 20 V than it does for a final value of 10 V?

b. What makes this possible?

7. Describe the effect on an ohmmeter of checking:

a. A shorted capacitor.
b. An open capacitor.
c. A very small capacitor on a very low range.
d. A very large capacitor on a medium range.

8. a. If the capacitor in a photoflash unit charges to the applied voltage, why not use the voltage source (battery) to light the flash lamp directly?

b. What other advantages does the use of a capacitor provide?

c. If you had the choice of increasing the lamp brightness by increasing the size of the capacitor or the applied voltage, which would be more effective and why?

9. Describe how you would measure the capacitance in a series RC circuit using a square wave generator and an oscilloscope connected across:

a. The capacitor.
b. The resistor.

10. If a sinusoidal voltage is applied to a differentiating circuit, draw the output waveform, in proper time relationship.

11. a. What is the result of a long time constant in a capacitor filtering circuit?

b. How would you expect the peak-to-peak ripple output voltage to change as the load resistance is increased?

12. Draw the capacitor voltage waveform for the neon lamp oscillating circuit in Fig. 22-13. Indicate when the lamp is on and off. What will happen to the flashing frequency if R_2 is increased?

PROBLEMS

(Answers to odd-numbered problems at back of book)

22-1. The voltage across a 0.2-μF capacitor is increased uniformly from 50 to 100 V in 10 ms. Calculate:

a. The rate of change of the capacitor voltage.

b. The capacitor current.

c. The increase in stored charge.

22-2. The voltage across a 1500-pF capacitor is decreased uniformly from 60 to 20 V in 100 μs. Calculate:

a. The rate of change of the capacitor voltage.

b. The capacitor current.

c. The decrease in stored charge.

22-3. The voltage across a 15-μF capacitor increases uniformly from 3 to 15 V in 4 ms, remains constant for 6 ms, and then decreases uniformly to zero in 6 ms. Draw the waveforms of the capacitor voltage and current showing all values.

22-4. The voltage across a 200-pF capacitor varies as shown in Fig. 22-14. Draw the waveform of capacitor current.

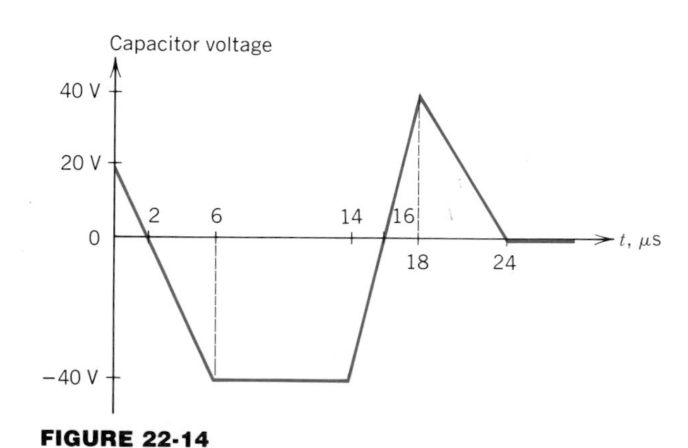

FIGURE 22-14
Capacitor voltage waveform for Problem 22-4.

22-5. A 6-V battery is connected in series with a 22-kΩ resistor and a 0.15-μF capacitor. Calculate:

a. The initial rate of increase of the capacitor voltage.

b. The initial charging current.

c. The final capacitor voltage.

22-6. When a 20-V dc supply is connected to a series RC circuit, the initial rate of rise of the capacitor voltage is determined to be 500 V/s. Calculate:

a. The time constant of the circuit.

b. The time for the capacitor to reach its steady-state value if it continues to charge at the initial rate of rise.

c. The approximate time required for the capacitor to charge to 20 V.

22-7. A 250,000-μF capacitor and a 100-Ω resistor are series-connected across a 5-V dc supply. Calculate:

a. The time constant of the circuit.

b. The initial rate of rise of the capacitor voltage.

c. The capacitor voltage after one time constant.

d. The approximate time required for the capacitor to charge to 5 V.

e. The initial charging current.

22-8. It is required that the initial charging current in a series RC circuit be limited to 50 mA when a 200-V dc supply is connected. It is also required that the charging transient be complete in 100 ms. Determine:

a. The resistance of the circuit.

b. The capacitance of the circuit.

c. The charging current after one time constant.

d. The capacitor voltage after one time constant.

22-9. After a 40-μF capacitor was fully charged to 450 V, a 2.2-MΩ resistor was connected across the capacitor. Calculate:

a. The initial discharge current.

b. The current after one time constant.

c. The capacitor voltage after one time constant.

d. The time to completely discharge the capacitor.

22.10. A 500,000-μF capacitor is charged to 6 V. Calculate:

a. The resistance load that will fully discharge the capacitor in one minute.

b. The initial discharge current.

c. The capacitor current after one time constant.

d. The capacitor voltage after one time constant.

e. The initial rate of decrease of capacitor voltage upon discharge.

22-11. Repeat Example 22-6 assuming that $C = 300\ \mu$F and $V = 400$ V.

22-12. Consider the circuit of Fig. 22-7. Determine:

a. The voltage necessary if a 1000-μF capacitor is to store 30 J of energy.

b. The value of R_1 if the initial charging current is to be limited to 100 mA.

c. The time required for the capacitor to fully charge.

d. The discharge time if $R_2 = 5\ \Omega$.

e. The peak discharging current.

f. The average power dissipated in R_2.

22-13. Repeat Example 22-7 with $R_1 = 150\ \text{k}\Omega$, $R_2 = 68\ \text{k}\Omega$, $C = 40\ \mu\text{F}$, and $V = 80$ V.

22-14. Refer to Fig. 22-9. After the switch has been left in position 1 for more than five time constants, it is thrown to position 2. Determine, using Fig. 22-8,

a. How long it takes for the capacitor voltage to drop to 30 V.

b. What time is required for the voltage to increase to 45 V if the switch is then returned to position 1.

22-15. Refer to the circuit of Fig. 22-9*a*. $R_1 = 4.7\ \text{k}\Omega$, $R_2 = 3.3\ \text{k}\Omega$, $C = 0.15\ \mu\text{F}$, and $V = 150$ V. The capacitor is initially uncharged. Calculate:

a. The capacitor voltage 1.5 ms after the switch is thrown to position 1.

b. The capacitor current at this instant.

c. The time required for the capacitor to discharge to 20 V if the switch is returned to position 2 after being in position 1 for 1.5 ms.

d. The capacitor current at this instant.

22-16. Repeat Problem 22-15 with $C = 0.3\ \mu\text{F}$.

22-17. An oscilloscope is connected across a capacitor as in Fig. 22-10*a*. The oscilloscope display reaches a final value of 6 cm. It requires a horizontal distance of 1.2 cm for the capacitor voltage display to reach 3.8 cm. If the oscilloscope time axis has a calibration of 100 μs/cm, determine:

a. The time constant of the circuit.

b. The capacitance if $R = 100\ \Omega$.

c. The maximum permitted frequency of the signal generator.

22-18. A series *RC* circuit of unknown resistance and capacitance has a square wave signal generator of 10-V amplitude applied to it. A 10-Ω resistor is connected in series with the generator, and an oscilloscope across the resistor displays a peak value of 50 mV. The waveform drops to 18.5 mV in a horizontal distance of 2.2 cm. If the time axis of the oscilloscope has a calibration of 5 μs/cm, calculate:

a. The peak capacitor current.

b. The resistance of the original circuit.

c. The time constant of the circuit.

d. The capacitance of the circuit.

22-19. A differentiating circuit consists of a 10-Ω resistor in series with a 0.15-μF capacitor. Determine:

a. The maximum frequency up to which this circuit should operate satisfactorily.

b. The approximate slope of the input square wave voltage if the peak resistor voltage is 0.5 V.

22-20. Refer to Fig. 22-13. Assuming that it requires one time constant for the capacitor to charge to the firing voltage of the neon lamp, and the discharge of the capacitor takes only one-tenth of this time, determine the maximum and minimum flashing rate possible with the given circuit.

22-21. Repeat Example 22-9 for a load resistance of 1 kΩ.

22-22. A 100-Ω, 12-V dc relay will operate satisfactorily at 10 V. It is required to delay the operation of the relay by 10 ms when the voltage is applied by connecting a 20-Ω resistor in series with the relay and a capacitor in parallel with the relay. Determine the necessary size of capacitor. (*Hint:* With the capacitor removed, obtain the Thévenin equivalent circuit (as seen by the capacitor) for the 12-V source in series with the 20-Ω resistor and 100-Ω relay coil).

The following problems are designed to be solved using exponential equations A12-1 through A12-4 found in Appendix A-12.

22-23. Solve Problem 22-13 using Eqs. A12-1 through A12-4.

22-24. Solve Problem 22-14 using Eqs. A12-1 through A12-4.

22-25. Solve Problem 22-15 using Eqs. A12-1 through A12-4.

22-26. Repeat Problem 22-15 with $C = 0.3\ \mu F$ using Eqs. A12-1 through A12-4.

CHAPTER TWENTY-THREE

CAPACITANCE IN AC CIRCUITS

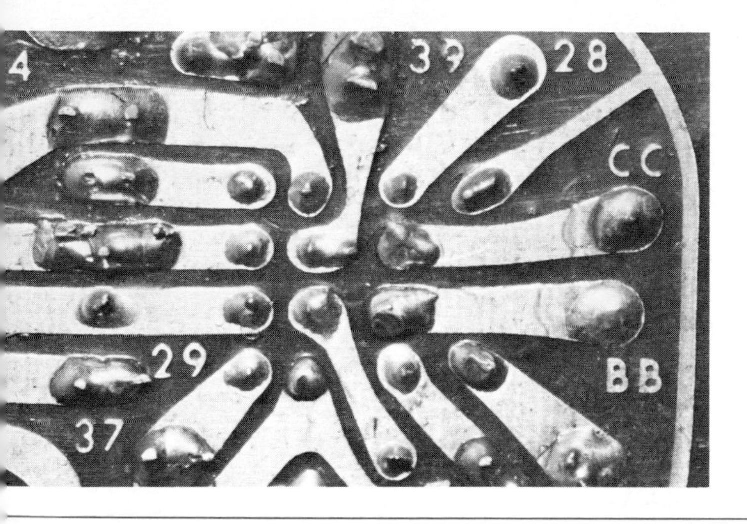

In this chapter we consider the current and voltage relationships of a capacitor in a sinusoidal ac circuit. Because the voltage is *continuously* changing, there is some rms capacitor current at all times. The ratio of capacitor voltage to current is called *capacitive reactance* (X_C) and is the opposition in ohms provided by the capacitor. The reactance is shown to be inversely proportional to both the frequency and the capacitance. Thus, for direct current, the capacitor is an open circuit, meaning that it can be used to "block" direct current but "pass" alternating current.

The series and parallel connection of capacitive reactances can be treated similarly to resistances and inductances. But as in dc circuits, it is found that a higher voltage occurs across the *smaller* capacitor in a series circuit.

Finally, the *dissipation factor, D,* of a capacitor is defined. This is an indication of the total losses that take place in a capacitor, at a given frequency, and includes the *dielectric hysteresis* effect produced by the alternating electric field.

23-1
EFFECT OF A CAPACITOR IN AN AC CIRCUIT

When a capacitor is connected across a sinusoidally varying voltage, there will be a continuous charging and discharging of the capacitor. Consequently, an ac ammeter connected in series with the capacitor will indicate some effective value of current. Although it may appear that current is flowing *through the capacitor,* the charge is being transferred from one plate to the other, back and forth, *through the ac supply.* We now determine the relation between the capacitor voltage and current, and the opposition to current provided by the capacitor.

23-1.1 Phase Angle Relationship Between Voltage and Current

Let us assume that a sinusoidal voltage is applied to a capacitor as in Fig. 23-1. The voltage waveform is given by

$$v = V_m \sin \omega t$$

We can deduce the waveform of the resulting current if we use the relationship:

$$i = C\left(\frac{\Delta v}{\Delta t}\right) \qquad (22\text{-}1)$$

Refer to Fig. 23-1*b* and note that at instants 2 and 4 of the voltage waveform the slope $\frac{\Delta v}{\Delta t}$ is zero. Hence the current at these instants must also be zero. At instants 1 and 5, the voltage curve has its maximum *positive* slope, $\left(\frac{\Delta v}{\Delta t}\right)$, so the current must be at a *positive* peak, I_m. Similarly, when the voltage is *decreasing* at its maximum rate at instant 3, the capacitor current must have its maximum *negative* value, $-I_m$.

The manner in which the current varies between these instants can be shown to follow a cosine curve given by

$$i = I_m \cos \omega t$$

The current waveform is of the same general shape (and frequency) as the sine curve of voltage but displaced one-quarter of a cycle to the left. Note that the current waveform reaches its peak value one-quarter of a cycle

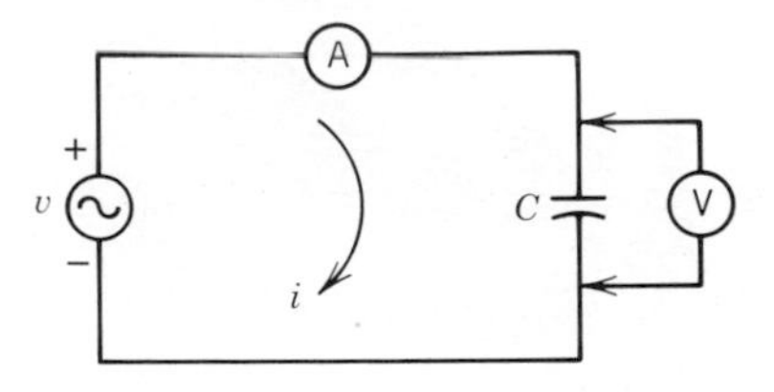

(*a*) Pure capacitance in an ac circuit

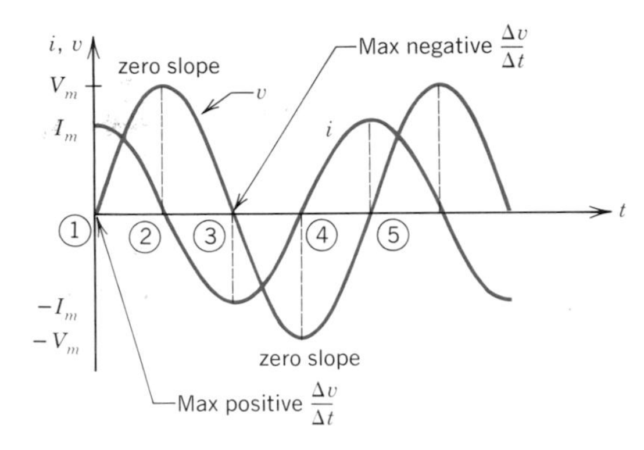

(*b*) Waveforms of current and voltage

FIGURE 23-1
Effect of a capacitor in an ac circuit.

before the voltage waveform. **Thus we say that the current in a capacitor *leads* the voltage across the capacitor by 90°. (This assumes that the capacitor itself has negligible series resistance, which is usually the case.)** This leading angle is referred to as the *phase angle* because it shows by what angle the two waveforms of current and voltage are *out of phase.*

It is easy to remember that the current *leads* the voltage if we recall what happens in a capacitive dc circuit. When the voltage is first applied, the maximum charging current *immediately* flows (and decreases gradually to zero) while the capacitor voltage builds up *gradually* to maximum after starting at zero.

23-2
CAPACITIVE REACTANCE

Since both the voltage and current waveforms in Fig. 23-1*b* are sinusoidal, we can obtain their rms values:

$$V = \frac{V_m}{\sqrt{2}} \qquad (15\text{-}17)$$

and
$$I = \frac{I_m}{\sqrt{2}} \qquad (15\text{-}16)$$

These are the values of the voltmeter and ammeter readings in Fig. 23-1*a*. The ratio of these two, $\frac{V}{I}$, we call the opposition to current, or *capacitive reactance*, X_C. That is:

$$X_C = \frac{V_C}{I_C} \quad \text{ohms}(\Omega) \qquad (23\text{-}1)$$

where: X_C is the capacitive reactance in ohms, Ω
V_C is the voltage across the pure capacitance in volts, V
I_C is the capacitor current in amperes, A

EXAMPLE 23-1

A sinusoidal voltage of peak-to-peak value 100 V is connected across a capacitor, as shown in Fig. 23-2. A series-connected ac milliammeter indicates 50 mA. What is the opposition to the current caused by the capacitor?

SOLUTION

$$V = \frac{V_{p-p}}{2\sqrt{2}} \qquad (15\text{-}19)$$
$$= \frac{100 \text{ V}}{2\sqrt{2}} \approx 35 \text{ V}$$
$$X_C = \frac{V_C}{I_C} \qquad (23\text{-}1)$$
$$\approx \frac{35 \text{ V}}{50 \times 10^{-3} \text{ A}} \approx \mathbf{700\ \Omega}$$

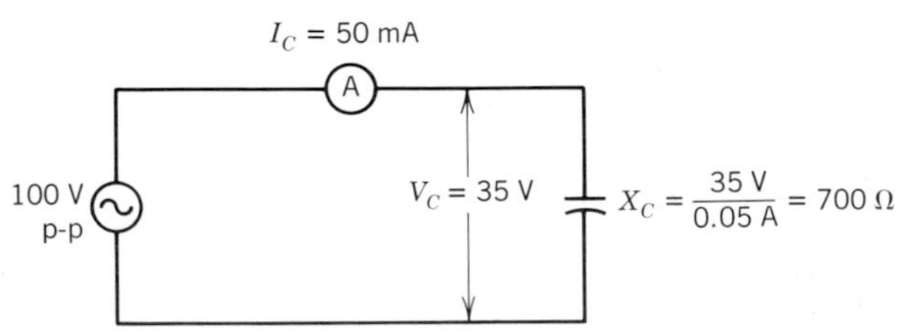

FIGURE 23-2
Circuit for Example 23-1, showing how $X_C = \frac{V_C}{I_C}$.

EXAMPLE 23-2

What is the capacitor current when a 120-V, 60-Hz source is applied to a capacitor whose reactance is 200 Ω?

SOLUTION

$$I_C = \frac{V_C}{X_C} \qquad (23\text{-}1)$$
$$= \frac{120 \text{ V}}{200\ \Omega} = \mathbf{0.6\ A}$$

EXAMPLE 23-3

What peak voltage occurs across a capacitor of reactance 2 kΩ when the capacitor current is 25 mA?

SOLUTION

$$V_C = I_C X_C \qquad (23\text{-}1)$$
$$= 25 \times 10^{-3} \text{ A} \times 2 \times 10^3\ \Omega$$
$$= 50 \text{ V}$$

Peak capacitor voltage,
$$V_m = \sqrt{2} \times V_C \qquad (15\text{-}18)$$
$$= \sqrt{2} \times 50 \text{ V}$$
$$= \mathbf{70.7\ V}$$

23-3 FACTORS AFFECTING CAPACITIVE REACTANCE

The capacitor current depends upon the capacitance and the change in capacitor voltage as given by

$$i_C = C\left(\frac{\Delta v_C}{\Delta t}\right) \qquad (22\text{-}1)$$

If, for a given supply voltage and frequency, the capacitance C is *increased*, a larger current must flow. This means a *decrease* in capacitive reactance, so X_C is *inversely proportional* to C.

Now consider an increase in the frequency f of the capacitor supply voltage. This means that the voltage across the capacitor must reach its maximum value in a shorter time (reduced period T). This leads to an *increase* in the maximum $\frac{\Delta v_C}{\Delta t}$ and an *increase* in the peak current I_m. Once again, a reduction in X_C results, so X_C is *inversely proportional* to the frequency, f, as well as C.

Combining these two effects, and using angular frequency $\omega = 2\pi f$, we obtain:

$$X_C = \frac{1}{2\pi fC} = \frac{1}{\omega C} \quad \text{ohms } (\Omega) \qquad (23\text{-}2)$$

where: X_C is the capacitive reactance in ohms, Ω
f is the frequency in hertz, Hz
C is the capacitance in farads, F
ω is the angular frequency in radians per second, rad/s

NOTE This equation is valid only for sinusoidal ac circuits. (A full derivation of X_C is given in Appendix A-14.)

EXAMPLE 23-4

What capacitive reactance will a 0.5-μF capacitor have at the following frequencies?

a. 60 Hz.
b. 1 kHz.

SOLUTION

a. $$X_C = \frac{1}{2\pi fC} \qquad (23\text{-}2)$$

$$= \frac{1}{2\pi \times 60 \text{ Hz} \times 0.5 \times 10^{-6} \text{ F}}$$

$$= \mathbf{5305\ \Omega}$$

b. $$X_C = \frac{1}{2\pi fC} \qquad (23\text{-}2)$$

$$= \frac{1}{2\pi \times 1 \times 10^3 \text{ Hz} \times 0.5 \times 10^{-6} \text{ F}}$$

$$= \mathbf{318\ \Omega}$$

EXAMPLE 23-5

A capacitor draws 1.5 A from a 120-V, 60-Hz supply. Calculate the capacitance.

SOLUTION

$$X_C = \frac{V}{I} \qquad (23\text{-}1)$$

$$= \frac{120 \text{ V}}{1.5 \text{ A}} = 80\ \Omega$$

$$X_C = \frac{1}{2\pi fC} \qquad (23\text{-}2)$$

$$C = \frac{1}{2\pi f X_C}$$

$$= \frac{1}{2\pi \times 60 \text{ Hz} \times 80\ \Omega} = \mathbf{33\ \mu F}$$

23-3.1 Comparison of Capacitance in Dc and Ac Circuits

It should be clear from Eq. 23-2 and Example 23-4 that the capacitive reactance decreases as the frequency increases.

The graph in Fig. 23-3 shows that a capacitor approaches a *short circuit* (very low reactance) at *high* frequencies; at *low* frequencies the capacitor approaches an *open* circuit. Put another way, when $f = 0$, the capacitive reactance is infinite. This means that in a dc circuit, ($f = 0$) no current flows (except a momentary transient current when voltage is first applied.) **Thus a capacitor is said to *block* direct current but allows the flow of alternating current.** This is shown in the circuits of Fig. 23-4.

In Fig. 23-4*a*, the reactance of the 50-μF capacitor at 60 Hz is only 53 Ω. This is negligible compared with the 576-Ω resistance of the lamp; therefore, the rated lamp current of 0.21 A flows and the lamp lights brightly. If a capacitor smaller than 50 μF is used, the reactance will be higher and the current in the circuit will be smaller. The lamp will be less bright. (The combined opposition to

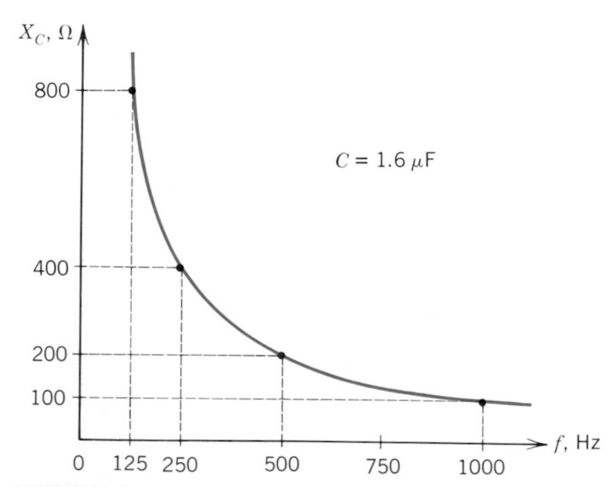

FIGURE 23-3
Decrease of X_C with increase in frequency f for a constant capacitance of 1.6 μF.

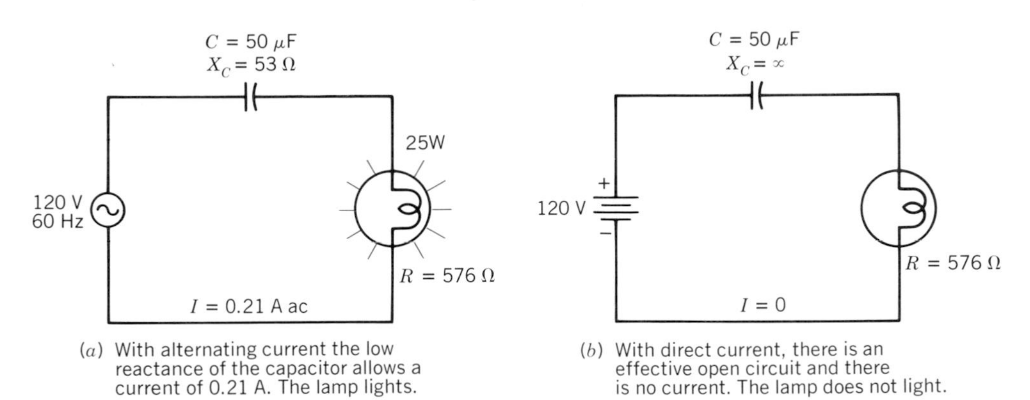

(*a*) With alternating current the low reactance of the capacitor allows a current of 0.21 A. The lamp lights.

(*b*) With direct current, there is an effective open circuit and there is no current. The lamp does not light.

FIGURE 23-4
Comparison of capacitance in ac and dc circuits.

alternating current due to capacitance and resistance is called impedance. This is covered in Section 24-6.)

When a dc voltage is applied, as in Fig. 23-4*b*, no steady current flows and the lamp is OFF, regardless of the size of capacitance.

The ability of a capacitor to block direct current but "pass" alternating current is made use of in ac amplifiers. When an amplified voltage is coupled from one stage of amplification to the next, a coupling capacitor is used to block the direct current from the biasing of the previous stage but allows the ac component of signal voltage to be applied to the next stage. See Fig. 23-5.

If the coupling capacitor is large, so that its reactance is small compared with R, practically all of the ac input voltage appears across R for further amplification.

23-4 SERIES CAPACITIVE REACTANCES

When capacitors are *series*-connected, the total reactance is

$$X_{C_T} = X_{C_1} + X_{C_2} + X_{C_3} + \cdots + X_{C_n} \qquad \textbf{(23-3)}$$

The current for a given applied voltage can be obtained from

$$I = \frac{V}{X_{C_T}} \qquad (23\text{-}4)$$

and the voltage across each capacitor from $V_C = IX_C$.

NOTE When capacitors are connected in series with each other, the overall capacitance drops, but this causes an overall increase in the reactance compared with one capacitor. This explains why the total reactance is additive as in Eq. 23-3.

EXAMPLE 23-6

Two capacitors of 8 and 4 μF are series-connected across a 120-V, 60-Hz supply as in Fig. 23-6. Calculate:

a. The total capacitive reactance.
b. The current drawn by the capacitors.
c. The voltage across each capacitor.

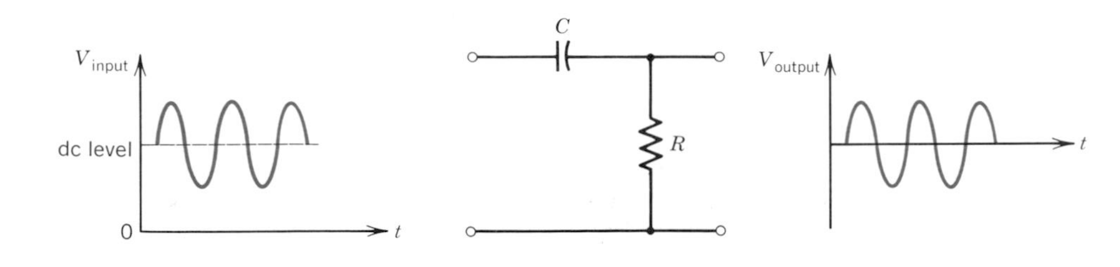

FIGURE 23-5
Blocking action of the capacitor couples the ac component of the input voltage to R but not the dc component.

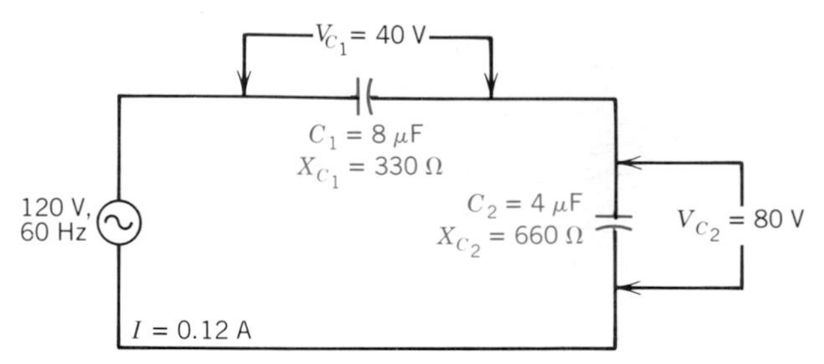

FIGURE 23-6
Circuit for Example 23-6.

SOLUTION

a. $X_{C_1} = \dfrac{1}{2\pi f C_1}$ (23-2)

$= \dfrac{1}{2\pi \times 60\ \text{Hz} \times 8 \times 10^{-6}\ \text{F}}$

$\approx 330\ \Omega$

$X_{C_2} = \dfrac{1}{2\pi f C_2}$ (23-2)

$= \dfrac{1}{2\pi \times 60\ \text{Hz} \times 4 \times 10^{-6}\ \text{F}}$

$\approx 660\ \Omega$

$X_{C_T} = X_{C_1} + X_{C_2}$ (23-3)

$\approx 330\ \Omega + 660\ \Omega \approx \mathbf{990\ \Omega}$

b. $I = \dfrac{V}{X_{C_T}}$ (23-4)

$= \dfrac{120\ \text{V}}{990\ \Omega} \approx \mathbf{0.12\ A}$

c. $V_{C_1} = IX_{C_1}$ (23-1)

$= 0.12\ \text{A} \times 330\ \Omega \approx \mathbf{40\ V}$

$V_{C_2} = IX_{C_2}$ (23-1)

$= 0.12\ \text{A} \times 660\ \Omega \approx \mathbf{80\ V}$

Note that the voltage divides in such a way that the highest voltage, 80 V, appears across the smallest capacitor, 4 μF, as is the case with dc voltages.

23-5 PARALLEL CAPACITIVE REACTANCES

When capacitors are parallel-connected, the overall increase in capacitance causes a decrease in reactance.

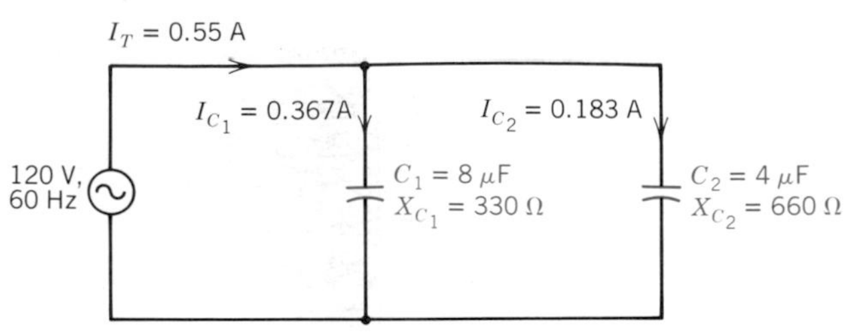

FIGURE 23-7
Circuit for Example 23-7.

Thus, similar to resistances in parallel, the total capacitive reactance is given by

$$\frac{1}{X_{C_T}} = \frac{1}{X_{C_1}} + \frac{1}{X_{C_2}} + \frac{1}{X_{C_3}} + \cdots + \frac{1}{X_{C_n}} \quad \textbf{(23-5)}$$

EXAMPLE 23-7

Two capacitors of 8 and 4 μF are connected in parallel across a 120-V, 60-Hz supply as in Fig. 23-7. Calculate:

a. The total capacitive reactance.
b. The total current drawn from the supply.

SOLUTION

a. From Example 23-6, $X_{C_1} = 330\ \Omega$

$X_{C_2} = 660\ \Omega$

$\dfrac{1}{X_{C_T}} = \dfrac{1}{X_{C_1}} + \dfrac{1}{X_{C_2}}$ (23-5)

Therefore, $X_{C_T} = \dfrac{X_{C_1} \times X_{C_2}}{X_{C_1} + X_{C_2}}$

$= \dfrac{330\ \Omega \times 660\ \Omega}{330\ \Omega + 660\ \Omega} \approx \mathbf{220\ \Omega}$

b. $I_T = \dfrac{V}{X_{C_T}}$ (23-4)

$= \dfrac{120\ \text{V}}{220\ \Omega} = \mathbf{0.55\ A}$

23-6 DISSIPATION FACTOR OF A CAPACITOR

Whenever a capacitor is charged by a dc source, we know that the atoms of the dielectric become polarized in one direction. If alternating current is applied, the electrons of the dielectric are stressed, first in one direction, then in the other, due to the alternating electric field. This causes an energy loss, in the dielectric, in the form of heat. Since it is similar to the hysteresis loss in a magnetic core, the effect is called *dielectric hysteresis*.

The effect can be represented in the equivalent circuit of a capacitor by a resistor (an energy-dissipating element) connected in parallel with the capacitance. See Fig. 23-8.

Recall, however, that we had already considered a parallel resistance to represent the *insulation resistance* of a capacitor, in Section 21-5.1. The two resistance effects can be combined into the single resistance, R_p, as shown in Fig. 23-8. The dielectric hysteresis effect can be thought of as an *added* ac phenomenon, similar to the ac resistance of a coil arising from magnetic hysteresis and skin effects, and so on.

Similar to the way in which we considered the *quality*, Q, or *storage factor* of a coil, we can obtain the *dissipation factor, D,* of a capacitor:

$$D = \frac{X_C}{R_p} = \frac{1}{2\pi f C R_p} \qquad (23\text{-}6)$$

where: D is the dissipation factor of the capacitor (dimensionless)

X_C is the capacitive reactance of the capacitor in ohms, Ω

R_p is the total parallel ac resistance of the capacitor in ohms, Ω

As the name suggests, D is an indication of how much power is dissipated in a capacitor at a given frequency. A perfect, lossless capacitor has D equal to zero. Of course, if the capacitor is used with only direct current applied, power dissipation is due only to the relatively high insulation resistance (leakage current). Also, the additional power dissipated with alternating current applied depends upon the actual ac voltage across the capacitor.

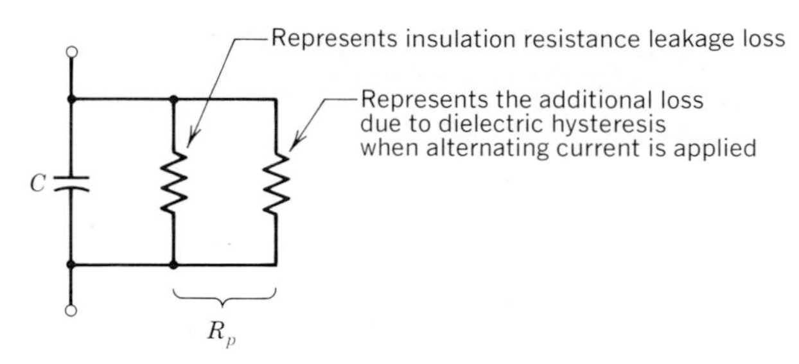

FIGURE 23-8
Ac equivalent circuit for a capacitor taking into account insulation resistance and dielectric hysteresis losses.

The D factors for nonelectrolytic capacitors are in the range of 0.0005 for polystyrene and 0.02 for ceramics at 1 kHz. Electrolytics, however, may have D factors from 0.15 at 475 WVDC up to 0.75 at 6 WVDC (specified for an ac component having a frequency of 60 Hz). It is especially important, when checking electrolytics, that the D factor be within allowable limits, since overheating can occur if there is any appreciable ac component of voltage across the capacitor. In fact, electrolytics used in ac motor starting applications should be replaced if the dissipation factor exceeds 0.15 or 15%.

Generally, unless problems develop, losses in capacitors (unlike those in inductors), are so small that they are neglected except in special cases.

Some capacitor checkers show the *power factor* of the capacitor instead of the dissipation factor. (The two factors are approximately equal for low loss capacitors). In this case, corresponding acceptable values of the power factor range from 0.15 to 0.6 for the same working voltages mentioned above. (See Section 25-6 for a discussion of the power factor.)

Finally, it should be mentioned that the reciprocal of D is the quality, Q, of the capacitor. Although this term is used primarily with coils, a high Q means a low-loss capacitor.

SUMMARY

1. A sinusoidal voltage across a capacitor causes a sinusoidal current that leads the voltage by 90°

2. The opposition to alternating current offered by a capacitor is called the capacitive reactance; for a sinusoidal waveform, it is given by

$$X_C = \frac{1}{2\pi fC}$$

3. The voltage across a capacitor in an ac circuit is $V_C = I_C X_C$.

4. A capacitor blocks direct current but passes alternating current, permitting the use of a capacitor for coupling purposes in an ac amplifier.

5. A series connection of capacitors gives an increase in the overall capacitive reactance similar to series-connected resistors, with the largest voltage occurring across the smallest capacitor.

6. The equations for parallel-connected capacitive reactances are similar to those for parallel-connected resistors, with the largest current flowing in the branch with the highest capacitance.

7. When a capacitor is used in an ac circuit, there is an additional loss called dielectric hysteresis, caused by the repeated reversal of the alternating electric field.

SELF-EXAMINATION

Answer T or F or a, b, c, d, e for multiple choice
(Answers at back of book)

23-1. In a sinusoidal ac circuit, the capacitor current is a maximum when the capacitor voltage passes through zero. ______

23-2. The phase angle between the capacitor voltage and the current is always 90°, no matter what else is connected in series with the capacitor, when the applied voltage is sinusoidal. ______

23-3. The only time the capacitor current is zero in a sinusoidal ac circuit is when the capacitor voltage is at its maximum values. ______

23-4. Capacitive reactance is the ratio of the rms capacitor current to the rms capacitor voltage. ______

23-5. A 1-μF capacitor, at a frequency of 500 Hz, has a capacitive reactance of:

a. 1000 Ω
b. 1000π Ω
c. $1000/\pi$ Ω
d. 300 Ω
e. any of the above

23-6. If the capacitor in Question 23-5 is connected to a 10-V, 500-Hz source, the current is approximately:

a. 10 mA
b. 31.4 mA
c. 3.14 mA
d. 0.314 A
e. 50 mA

23-7. If the frequency of the voltage applied to a capacitor is doubled, the current is reduced 50%. ______

23-8. If a very large capacitor is used in series with a lamp, the lamp will light no matter whether the applied voltage is alternating current or direct current. ______

23-9. The use of a capacitor to couple alternating current from one stage to another in an amplifier depends upon the capacitor's ability to block alternating current and pass direct current. ______

23-10. When two equal capacitors are series-connected across a 120-V ac source, the voltage across each capacitor is 60 V. ______

23-11. When two unequal capacitors are parallel-connected across a 120-V, 60-Hz source, the smaller capacitor will have the smaller current. ______

23-12. The same type of equation is used to combine series and parallel capacitive reactances as used for combining series and parallel capacitances. ______

23-13. Dielectric hysteresis is an effect that is considered only in ac applications. ______

23-14. The dissipation factor of a capacitor is a constant—independent of the frequency of operation. ______

23-15. If the dissipation factor of a capacitor is excessive, overheating of the capacitor can occur if used with a significant alternating voltage. ______

REVIEW QUESTIONS

1. If a capacitor consists of an insulator between two metal plates, how is it possible for a series-connected ac ammeter to show a steady ac current?

2. How does the phase angle between the capacitor current and voltage in an ac circuit differ from those for a purely inductive or purely resistive circuit? Why the difference?

3. a. Why is it not correct to characterize capacitive reactance as the effective ac resistance of a capacitor?

b. Why should an *increase* in capacitance or frequency *increase* current in a capacitive circuit if the voltage is unchanged?

4. a. In what way can direct current be thought of as zero frequency alternating current as applied to a capacitor?

b. How much opposition does a capacitor offer to direct current?

c. What are at least two practical applications of this result?

5. Describe how you could use a voltmeter, ammeter, and ac voltage source of known frequency to measure the capacitance of a capacitor.

6. Why is the total capacitive reactance of capacitors in series given by the *sum* of the individual reactances?

7. Why is there no mutual reactance between capacitors as there may be with inductors?

8. Justify why series-connected capacitors cause the largest voltage to occur across the smallest capacitor.

9. a. What do you understand by the term *dielectric hysteresis?*

b. When is it a factor to take into account?

c. What are typical dissipation factors for electrolytic capacitors?

d. Why do you think electrolytics have higher D values than a paper capacitor?

PROBLEMS

(Answers to odd-numbered problems at back of book)

23-1. What capacitive reactance does a capacitor have if an applied emf of 230 V causes a current of 0.4 A?

23-2. A capacitor draws 20 mA when connected to a 10-V sinusoidal supply. What is its reactance?

23-3. What is the voltage across a 0.1-μF capacitor when the current is 5 mA at 400 Hz?

23-4. If a capacitor's current is 15 mA p-p at a frequency of 1.5 kHz, what does a voltmeter indicate across the 0.015-μF capacitor?

23-5. A 20-V p-p sinusoidal voltage at 50 Hz is connected across a capacitor. If a series-connected ammeter indicates 4 mA, calculate the capacitance.

23-6. A capacitor and a 2.2-kΩ resistor are series-connected across a 60-Hz supply. A voltmeter across the capacitor indicates 3.5 V when an oscilloscope across the resistor shows a peak value of 8 V. How much capacitance does the circuit have?

23-7. A 6.3-V sinusoidal voltage is applied to a 0.5-μF capacitor. How much current flows at the following frequencies?

a. 60 Hz

b. 6 kHz

c. 600 kHz

23-8. The following voltage is applied to a 1500-pF capacitor:

$$v = 50 \sin 10{,}000\, \pi t \text{ volts}$$

Determine the equation of the current waveform.

23-9. What is the maximum frequency that can be applied to a 4-μF capacitor if the capacitor current must not exceed 50 mA with an applied voltage of 24 V?

23-10. What is the minimum frequency that can be applied with a 10-μA constant-current, sine wave signal generator to a 0.002-μF capacitor to limit the peak capacitor voltage to 5 V?

23-11. Three capacitors—0.1 μF, 0.15 μF, and 0.2 μF—are series-connected across a 24-V, 400-Hz supply. Calculate:

a. The total capacitive reactance.

b. The current drawn by the capacitors.

c. The voltage across each capacitor.

d. The total capacitance of the circuit.

23-12. Three capacitors are connected in series across a 120-V, 60-Hz supply. The voltage across the first capacitor is 35 V; the capacitance of the second is 1.5 μF; and the current through the third is 15 mA. Calculate the capacitance of the first and third capacitors.

23-13. Three capacitors—0.1 μF, 0.15 μF, and 0.2 μF—are parallel-connected across a 48-V, 60-Hz supply. Calculate:

a. The total capacitive reactance.

b. The current drawn by each capacitor.

c. The total current drawn by the capacitors.

d. The total capacitance of the circuit.

23-14. Three capacitors are connected in parallel across a 240-V, 50-Hz supply, and draw a total current of 4 A. If the first capacitor is 20 μF, and the second capacitor has a current of 1.5 A, determine the capacitance of the second and third capacitors.

23-15. A 40-μF electrolytic capacitor dissipates a power of 1 W when used in a filter circuit with an ac voltage of 10 V at a frequency of 120 Hz. Calculate:

a. The effective parallel resistance R_p that determines the capacitor's losses.

b. The dissipation factor of this capacitor at this frequency.

CHAPTER TWENTY-FOUR

ALTERNATING CURRENT CIRCUITS

In this chapter we introduce a method of representing voltage and current waveforms by phasor notation. Two phasors, drawn at right angles to each other, can represent the phase angle relations between current and voltage in pure L or C circuits.

When resistance and inductance are connected in series, their individual voltages must be added taking into account their phase angles. This leads us to the expression for the total opposition to current, called *impedance* (Z), where R and X_L must be added using the square root of the sum of the squares of the individual oppositions. A similar expression may be used for a series RC circuit except that the current *leads* the applied voltage in a capacitive circuit. The current *lags* the applied voltage in an inductive circuit.

In a *series RLC* circuit, a canceling effect occurs between inductive and capacitive reactances. This causes an overall reduction of the total impedance. If $X_L = X_C$, the circuit displays a purely resistive property, a condition called *resonance.*

The *parallel* connection of R, L, and C is analyzed by obtaining the total current and obtaining the impedance by an application of "Ohm's law for an ac circuit," $Z = V/I$.

Finally, methods of measuring a circuit's phase angle between current and applied voltage are examined. One method involves a *Lissajous* pattern and is applicable to a single-beam oscilloscope; the other requires a dual-trace (or dual-beam) oscilloscope.

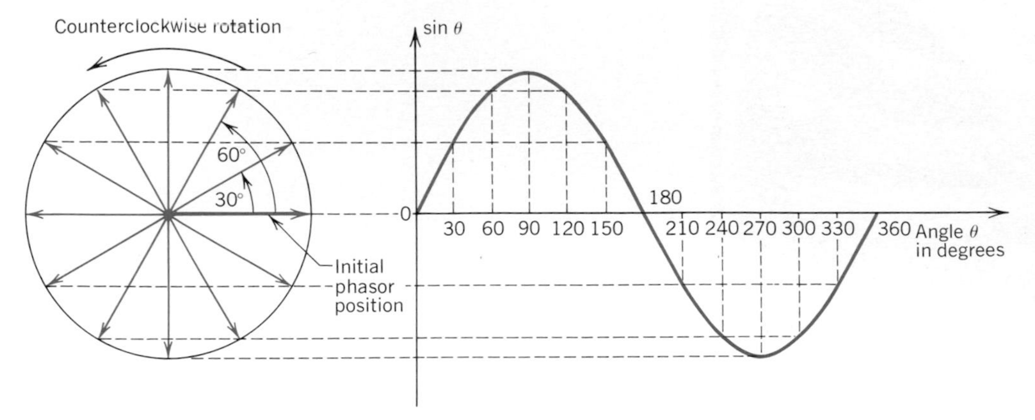

FIGURE 24-1
Showing how a sine wave can be constructed by using horizontal projections from a rotating radius vector.

24-1 REPRESENTATION OF A SINE WAVE BY A PHASOR

To show the phase angle relationship between the current and voltage in an ac circuit, it is convenient to use a single line, with an arrow, called a *vector* or *phasor,** to represent each waveform. The length of the line represents the magnitude and the direction of the line represents the phase angle with respect to some reference.

Consider a radius of a circle, initially in a horizontal position, rotated counterclockwise through the 360° of a complete circle. Let us consider the radius at various angles, say every 30°, and plot a graph of the vertical distance from the end of this line to the horizontal axis against the angle. An easy way of doing this is to project lines horizontally, for each angle, using some suitable scale for the x-axis in degrees. See Fig. 24-1.

It should be clear from right-angled trigonometry that the vertical distances plotted are proportional to the *sine* of each angle involved. (Recall that $\sin\theta$ is the ratio of the opposite side to the hypotenuse in a right triangle. Since the hypotenuse in this case is always the same length—the radius of a circle—the opposite side is directly proportional to the sine of the angle θ.) Thus the waveform produced by connecting each of the plotted points is a *sine wave,* with an amplitude or peak value equal to the length of the phasor. (If very small angles are used for the horizontal projections, say every 5°, a very accurate sine wave can be drawn.)

* A vector has both magnitude and direction. It is used to represent physical quantities such as a force or velocity. A phasor also has magnitude and direction, but it represents a quantity that also *varies* in magnitude with *time* (such as an alternating voltage or current).

Therefore, we can use a horizontal line (phasor) to represent a sinusoidal waveform. The phasor is always assumed to rotate counterclockwise, at a speed equal to the frequency of the waveform. Although, strictly speaking, the length of the phasor should be the *peak* value of the waveform, we can choose the length to represent the *rms* value of any given quantity, since this is generally what is measured and used in ac circuits.

Now consider the waveform represented by a phasor drawn vertically upward, and also assumed to rotate counterclockwise through a complete circle. Since the original reference phasor is considered to start at 0°, this second waveform starts at its maximum vertical height. As the phasor rotates through 90°, the horizontal projections show a *decreasing* vertical height, reaching zero in one-quarter of a cycle. As shown by Fig. 24-2, the waveform now goes negative for 180°, and finally positive for the last quarter-cycle.

The resulting waveform is a *cosine* curve. This, simply, is a sine curve moved *ahead* by 90°, or *leading* a sine curve by a quarter-cycle. Note, then, that if two phasors A and B are drawn in a single diagram, with A ahead of B by 90°, we have represented two waveforms that are 90° out of phase. See Fig. 24-3a.

Observe, since the phasors are *always* assumed to rotate counterclockwise, how obvious it is that *A leads B* by 90°, or *B lags A* by 90°. This *phase angle* relationship is shown in Fig. 24-3b, which also shows the waveforms represented by the phasors. Note the difference in magnitudes and the phase relations between the two waves. Both types of representations are used. The phasor

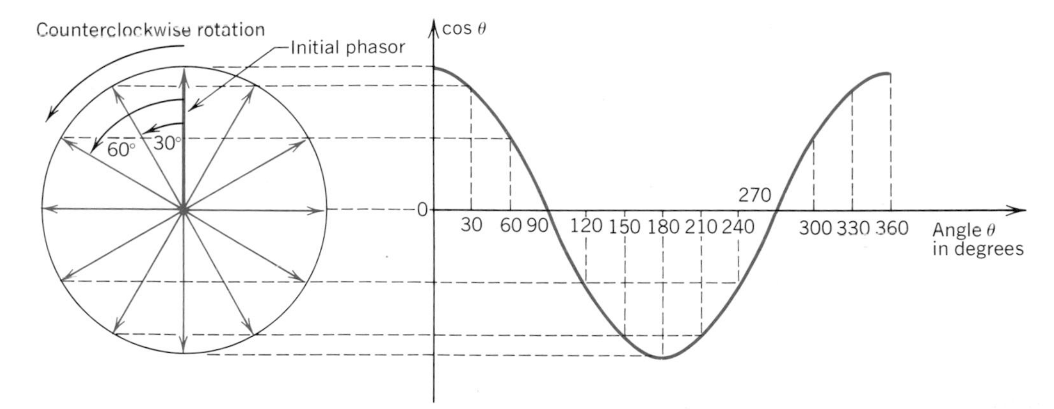

FIGURE 24-2
Showing how a vertical phasor represents a cosine curve.

diagram is used for *sinusoids* at the *same* frequency. Waveforms are used to represent sinusoidal *and* nonsinusoidal waves at *any* frequency. However, phasor diagrams are easier to draw and interpret than are the waveforms they represent.

24-2 VOLTAGE AND CURRENT PHASORS IN PURE *RLC* CIRCUITS

Let us see how phasors can be used to represent the voltage and current waveforms in sinusoidal ac circuits with pure resistance, inductance, or capacitance at a given frequency. Refer to Fig. 24-4.

Since the same voltage is applied to each circuit, the voltage phasor has been drawn in the horizontal reference position for each. For pure resistance, the current and voltage are in phase, so the phasors are drawn "in line" with each other, indicating a 0° phase angle. See Fig. 24-4*a*.

We know that current *lags* voltage by 90° in a purely inductive circuit. The phasor diagram of Fig. 24-4*b* clearly shows this. Similarly, it is very clear from the phasor diagram of Fig. 24-4*c* that the capacitive current *leads* the voltage by 90°.

Note that the current phasor has been drawn arbitrarily shorter than the voltage phasor. Except where phasors are drawn to scale in some graphical solutions, the lengths are not important. As noted, they are usually indicated with rms values, with capital letters as shown.

24-3 SERIES *RL* CIRCUIT

Now let us apply the principle of phasor representation to a series circuit containing pure resistance and pure inductance. See Fig. 24-5.

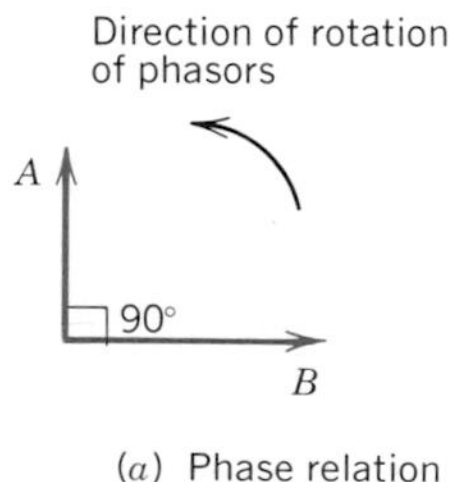

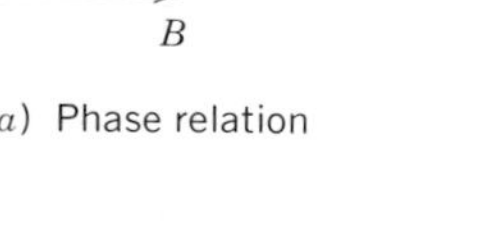

(*a*) Phase relation

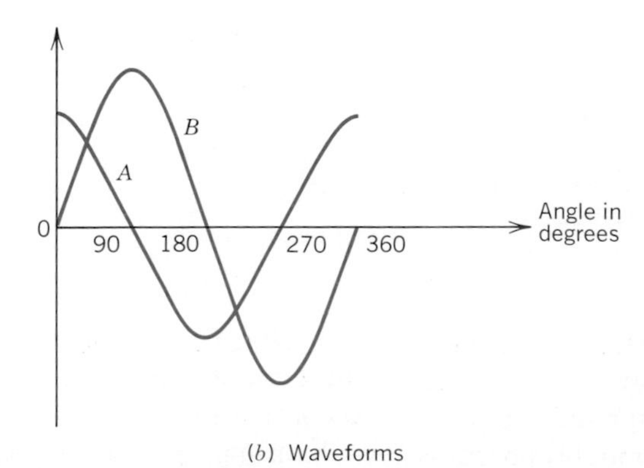

(*b*) Waveforms

FIGURE 24-3
Showing how two phasors *A* and *B*, with *A* leading *B* by 90°, represent two waveforms out of phase by 90°.

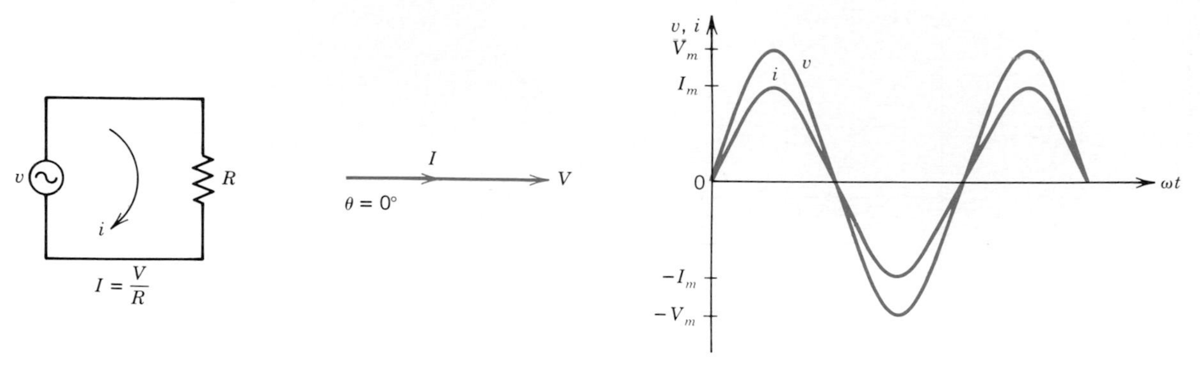

(a) Purely resistive circuit, I and V are in phase.

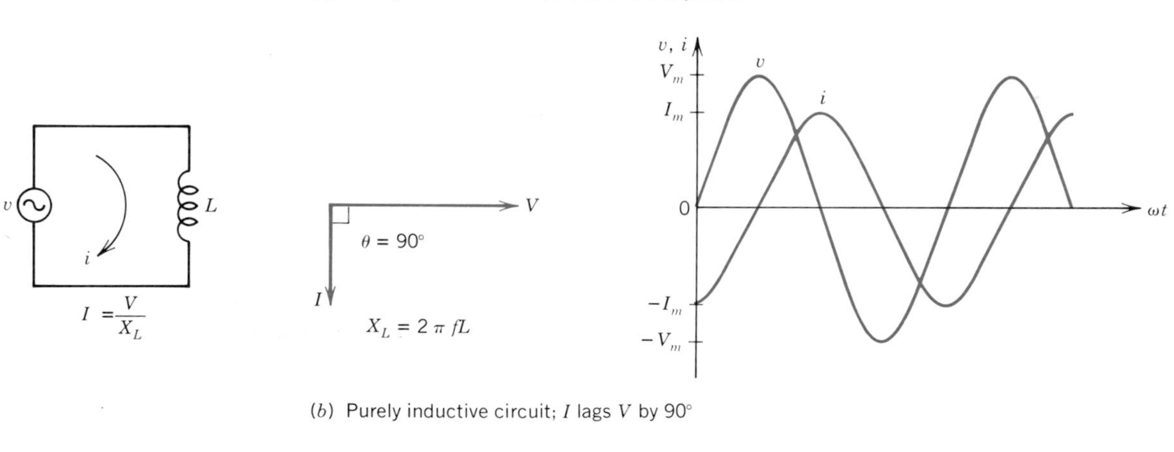

(b) Purely inductive circuit; I lags V by 90°

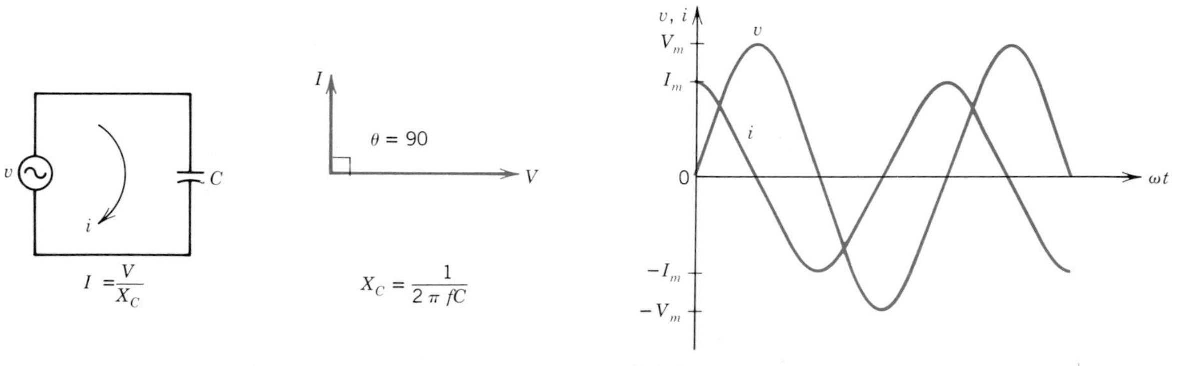

(c) Purely capacitive circuit, I leads V by 90°

FIGURE 24-4
Phasor representations and waveforms of V and I in pure resistance, inductance, and capacitance circuits.

Since we are considering a series circuit, it is convenient if we draw the current phasor in the horizontal reference position because it is "common" to both resistor and inductor. Superimposed upon this phasor is the voltage phasor across the resistor, V_R. This is because the current and voltage are always in phase with each other in a pure resistor.

Similarly, the voltage phasor across the inductor, V_L, is drawn 90° ahead of, or leading, the current phasor. This is because we know the current always lags the inductor voltage by 90° in a pure inductance.

Let us assume, for ease of discussion, that the resistance R is equal to the inductive reactance X_L of the coil at the frequency of the applied voltage. In Fig. 24-5*a*, $R = X_L = 50\ \Omega$. Then, since both components have the same current through them (let us say $I = 2$ A in our example), they must have the same magnitude (size) of voltage across them, that is,

$$V_R = I_R R \qquad (15\text{-}15b)$$
$$= 2\text{ A} \times 50\ \Omega = 100\text{ V}$$

$$V_L = I_L X_L \qquad (20\text{-}1)$$
$$= 2\text{ A} \times 50\ \Omega = 100\text{ V}$$

However, these two voltages are 90° out of phase with each other. This means that the total voltage across the series combination cannot be obtained simply by adding V_R to V_L algebraically.* We must take into account the angle between them. That is, Kirchhoff's voltage law applies, as it does in any circuit. The applied voltage V is the (phasor) sum of V_R and V_L with the phase angle also

* We can, however, add the instantaneous values of v_R and v_L algebraically at *each instant of time,* to obtain v in Fig. 24-5*c*.

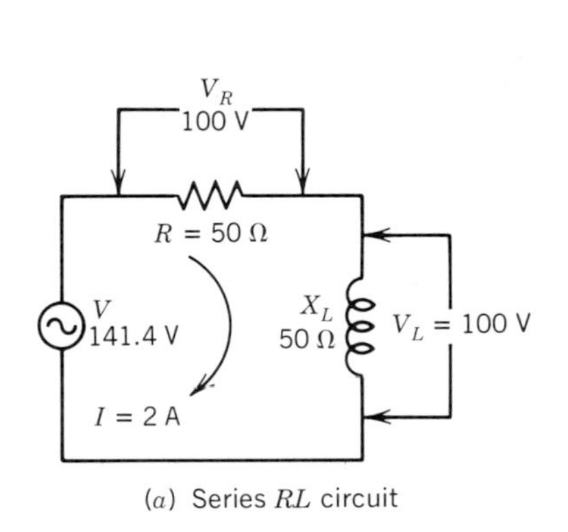

(*a*) Series *RL* circuit

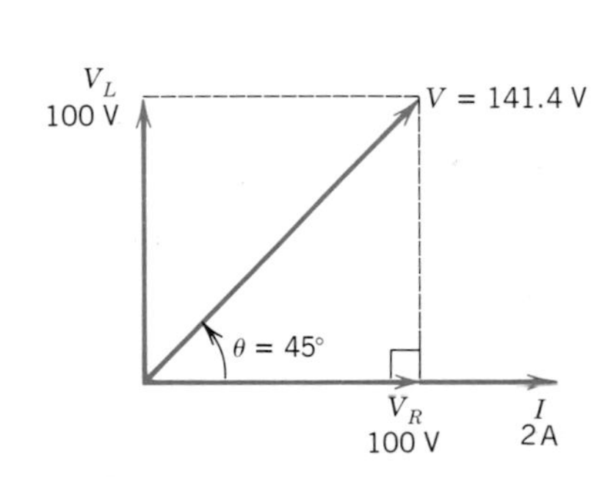

(*b*) Phasor diagram

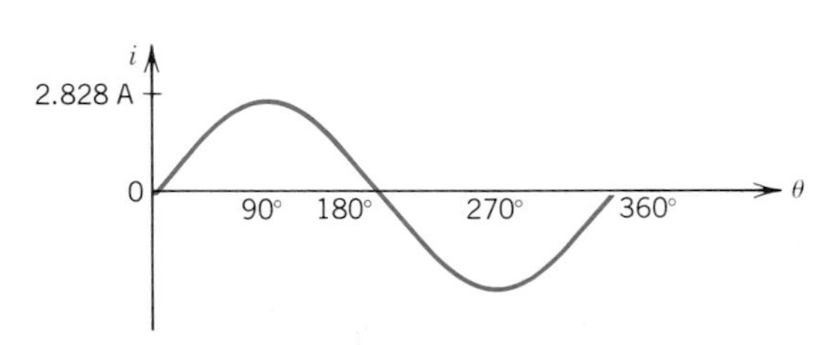

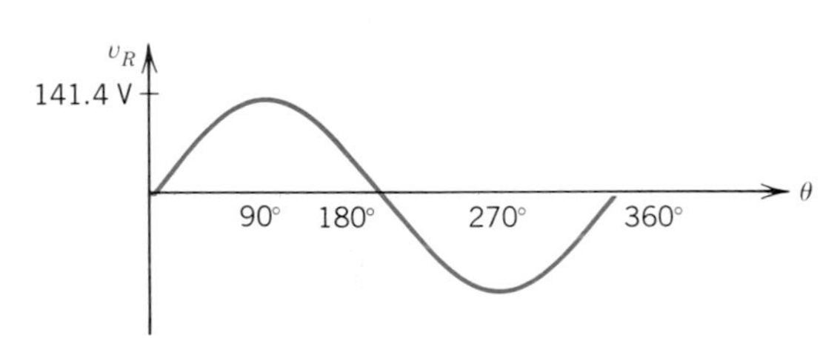

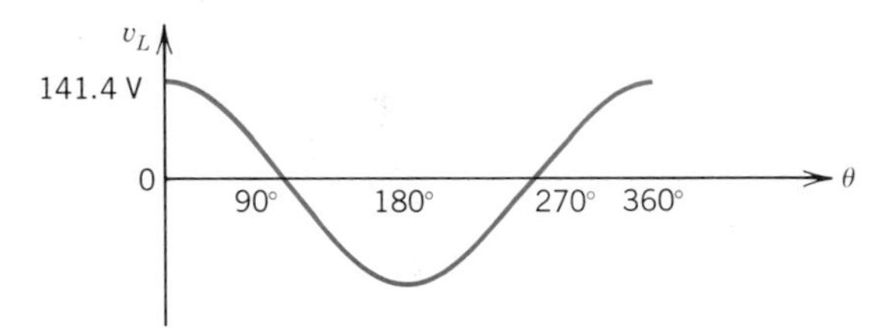

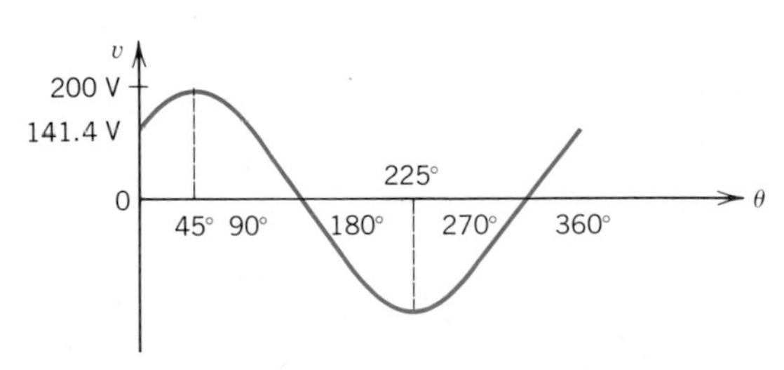

(*c*) Current and voltage waveforms

FIGURE 24-5
Series *RL* circuit with phasor diagram and waveforms.

involved in the addition. This we call *vector* or *phasor addition*.

This phasor addition can be carried out simply by constructing a parallelogram (a square in this case) and drawing the diagonal. This is shown in Fig. 24-5*b*. Clearly, the *phasor* sum V is less than the *algebraic* sum of V_L and V_R. Also, because V is the hypotenuse of a right-angled triangle, it is given by

$$\mathbf{V = \sqrt{V_R^2 + V_L^2} \qquad \text{volts (V)}} \qquad \mathbf{(24\text{-}1)}$$

where symbols are as above.

For our numerical values,

$$\begin{aligned} V &= \sqrt{(100\ \text{V})^2 + (100\ \text{V})^2} \\ &= \sqrt{10{,}000 + 10{,}000}\ \text{V} \\ &= \sqrt{20{,}000}\ \text{V} = \mathbf{141.4\ V} \end{aligned}$$

Thus the rms value of the applied voltage is 141.4 V, with a peak value of 200 V. (The waveforms in Fig. 24-5*c* show the peak values and how they would be displayed on an oscilloscope.)

The phase angle θ by which the current lags the applied voltage can now be obtained from the right triangle in Fig. 24-5*b*:

$$\tan\theta = \frac{V_L}{V_R} \qquad (24\text{-}2)$$

or

$$\mathbf{\theta = \tan^{-1}\frac{V_L}{V_R} \qquad \text{or} \qquad \text{arc tan}\,\frac{V_L}{V_R}} \qquad \mathbf{(24\text{-}2a)}$$

where symbols are defined as above.

The symbols $\tan^{-1}$ or arc tan V_L/V_R mean that θ is the angle whose tangent is V_L/V_R.* It may be obtained from a scientific calculator using the INV and TAN buttons as follows:

For our numerical values,

$$\begin{aligned} \theta &= \tan^{-1}\frac{100\ \text{V}}{100\ \text{V}} \\ &= \tan^{-1} 1 \end{aligned}$$

Enter 1 into the calculator, press INV, press TAN. The result is

$$\theta = \mathbf{45°}$$

Note that in Fig. 24-5*c* the applied voltage V reaches its peak value 45° before the current I does. This means that V leads I by 45° or I lags V by 45°.

This result should be no surprise. If a circuit is purely resistive, the phase angle is 0°; if a circuit is purely inductive, the phase angle is 90°; if the circuit is made up of equal values of resistance and inductive reactance, we should expect the angle to be midway between 0° and 90° or **45°**.

EXAMPLE 24-1

A coil of negligible resistance and a 100-Ω resistor are series-connected across a 120-V, 60-Hz supply. A voltmeter connected across the resistor indicates 60 V. Calculate:

a. The reading of a voltmeter connected across the coil.
b. The phase angle between the applied voltage and the current.
c. The current in the circuit.
d. The inductance of the coil.

SOLUTION

See Fig. 24-6.

a. $V = \sqrt{V_R^2 + V_L^2}$ (24-1)

$V^2 = V_R^2 + V_L^2$

Therefore,

$$\begin{aligned} V_L &= \sqrt{V^2 - V_R^2} \\ &= \sqrt{(120\ \text{V})^2 - (60\ \text{V})^2} \\ &= \sqrt{14{,}400 - 3600}\ \text{V} \\ &= \sqrt{10{,}800}\ \text{V} = \mathbf{104\ V} \end{aligned}$$

b. $\theta = \tan^{-1}\dfrac{V_L}{V_R}$ (24-2a)

$$\begin{aligned} &= \tan^{-1}\frac{104\ \text{V}}{60\ \text{V}} \\ &= \tan^{-1} 1.73 = \mathbf{60°} \end{aligned}$$

c. $I = \dfrac{V_R}{R}$ (15-15a)

$$= \frac{60\ \text{V}}{100\ \Omega} = \mathbf{0.6\ A}$$

* See Appendix A-6, inverse trigonometric functions.

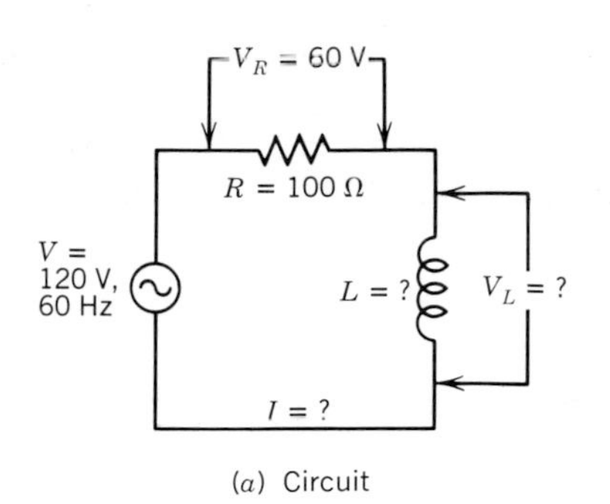

(a) Circuit

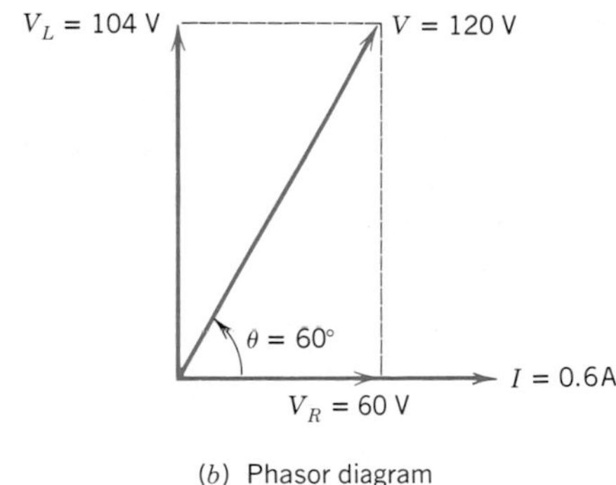

(b) Phasor diagram

FIGURE 24-6
Circuit and phasor diagram for Example 24-1.

d. $X_L = \frac{V_L}{I}$ (20-1)

$= \frac{104\text{ V}}{0.6\text{ A}} = 173\ \Omega$

$= 2\pi fL$ (20-2)

Therefore,

$$L = \frac{X_L}{2\pi f}$$

$$= \frac{173\ \Omega}{2\ \pi \times 60\text{ Hz}} = \mathbf{0.46\ H}$$

24-4 IMPEDANCE OF A SERIES RL CIRCUIT

The total opposition to current in a series *RL* circuit is called the impedance, *Z*. It is the ratio of the total applied voltage V to the current I. Impedance is measured in ohms as are resistance and inductive reactance. But, as shown by the following, impedance is the *vector* sum of resistance and reactance.

Consider the "voltage triangle" for a series RL circuit as in Fig. 24-7*a*. This is similar to the phasor diagram in Fig. 24-5*b* with V_L transferred to make a closed triangle.

Given $\qquad V = \sqrt{V_R^2 + V_L^2} \qquad$ (24-1)

and $\qquad V_R = IR,\ V_L = IX_L$

then
$$V = \sqrt{(IR)^2 + (IX_L)^2}$$
$$= \sqrt{I^2R^2 + I^2X_L^2}$$
$$= \sqrt{I^2(R^2 + X_L^2)}$$
$$= I\sqrt{R^2 + X_L^2}$$

and
$$\frac{V}{I} = \sqrt{R^2 + X_L^2}$$

But $\frac{V}{I}$ is the impedance, Z.

Therefore, $\qquad \mathbf{Z = \sqrt{R^2 + X_L^2}}$ **ohms (Ω)** $\qquad$ **(24-3)**

where: Z is the impedance in ohms, Ω
R is the resistance in ohms, Ω
X_L is the inductive reactance in ohms, Ω

and $\qquad \mathbf{I = \frac{V}{Z}}$ **amperes (A)** $\qquad$ **(24-4)**

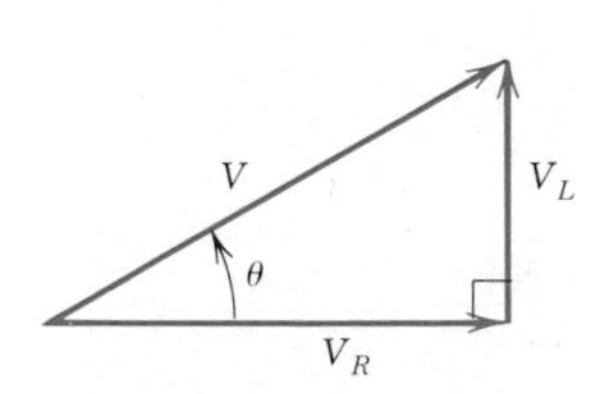

(a) Voltage triangle

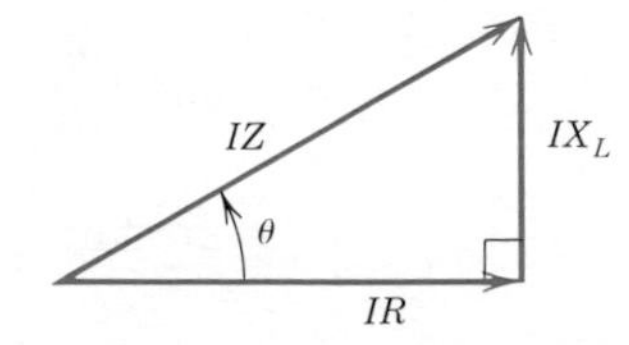

(b) Equivalent voltage triangle

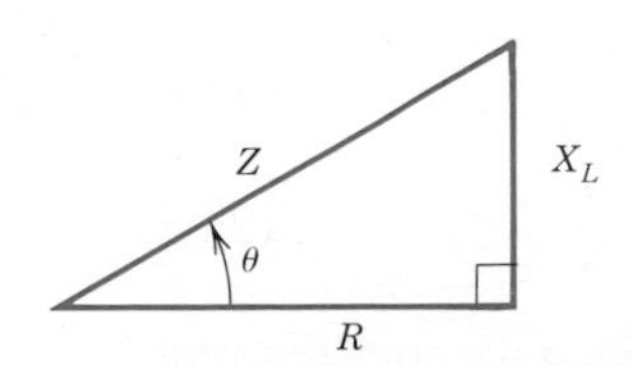

(c) Impedance triangle

FIGURE 24-7
Voltage and impedance triangles for a series *RL* circuit.

where: I is the circuit current in amperes, A
V is the applied voltage in volts, V
Z is the circuit impedance in ohms, Ω

Equation 24-4 is often referred to as Ohm's law for an ac circuit, since it is the Ohm's law equivalent of $I = V/R$ for a dc circuit. We shall soon see that Z represents the total opposition to current for any combination of RLC whether in series or parallel. However, the expression used to obtain Z will vary according to the circuit.

Note that the impedance diagram in Fig. 24-7c is drawn *without* arrows and is not referred to as a phasor diagram. This is because X_L, R, and Z do not vary with time, as do V_L, V_R, and V. Note also that the phase angle θ between the current (V_R) and voltage in Fig. 24-7a can now be obtained from the impedance triangle in Fig. 24-7c. This is because the two triangles are *similar,* and θ is the same.

$$\tan \theta = \frac{X_L}{R} \tag{24-5}$$

and

$$\theta = \tan^{-1}\frac{X_L}{R} \tag{24-5a}$$

where: θ is the phase angle between the current and the applied voltage in a series RL circuit, in degrees or radians
X_L is the inductive reactance in ohms, Ω
R is the total circuit resistance in ohms, Ω

NOTE It is the inductive *reactance* compared with the resistance that determines the phase angle, not inductance alone. This means that the phase angle can change if the frequency of the applied voltage changes, even if the inductance and resistance remain constant.

We can also see, from Fig. 24-7c, that if the impedance and phase angle are known, we can obtain the resistance and inductive reactance:

$$R = Z \cos \theta \tag{24-6}$$

$$X_L = Z \sin \theta \tag{24-7}$$

These equations will be useful when we consider practical coils, in which the coil resistance is not negligible.

EXAMPLE 24-2

A 30-mH coil of negligible resistance and a 200-Ω resistor are connected in series across a 10-V, 1-kHz sinusoidal supply. Determine:

a. The impedance of the circuit.
b. The current in the circuit.
c. The phase angle between the applied voltage and current.

SOLUTION

See Fig. 24-8.

a.
$$X_L = 2\pi f L \tag{20-2}$$
$$= 2\pi \times 1 \times 10^3 \text{ Hz} \times 30 \times 10^{-3} \text{ H}$$
$$= 188.5\ \Omega$$
$$Z = \sqrt{R^2 + X_L^2} \tag{24-3}$$
$$= \sqrt{200^2 + 188.5^2}\ \Omega$$
$$= \mathbf{274.8\ \Omega}$$

b.
$$I = \frac{V}{Z} \tag{24-4}$$
$$= \frac{10 \text{ V}}{274.8\ \Omega} = \mathbf{36.4\ mA}$$

c.
$$\theta = \tan^{-1}\frac{X_L}{R} \tag{24-5a}$$
$$= \arctan \frac{188.5\ \Omega}{200\ \Omega} = \mathbf{43.3°}$$

Even though a circuit may contain both resistance and reactance, the relative value of the two determines the nature of the impedance. Thus a series circuit is said to be predominantly inductive if the inductive reactance is in the order of 10 times as large as the resistance. For example, if $R = 1\ \Omega$ and $X_L = 10\ \Omega$,

$$Z = \sqrt{R^2 + X_L^2}$$
$$= \sqrt{1^2 + 10^2}\ \Omega$$
$$= \mathbf{10.05\ \Omega}$$

and

$$\theta = \tan^{-1}\frac{X_L}{R}$$
$$= \arctan \frac{10}{1} = \mathbf{84°}$$

Similarly, if R is 10 times X_L, the circuit is considered to be predominantly resistive.

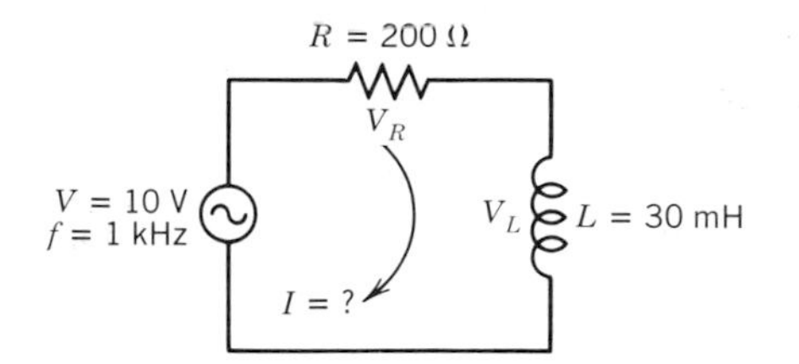

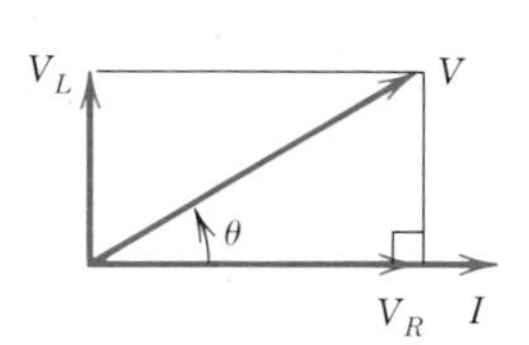

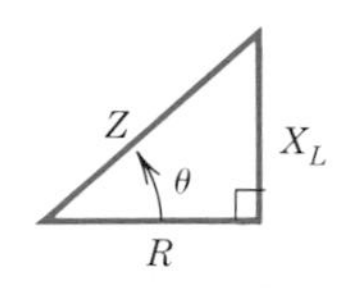

FIGURE 24-8
Circuit diagram, phasor diagram, and impedance triangle for Example 24-2.

24-4.1 Measuring Inductance of a Practical Coil

We have assumed, so far, that a coil's resistance is negligible compared with X_L. Although this may be true at higher frequencies (since X_L increases with frequency), the coil's resistance may *not* be negligible at power line (60 Hz) frequencies (where X_L is small). In this case, the voltage across the coil is *not* 90° out of phase with the current through the coil. Further, it is impossible to make measurements across either the resistive portion or the inductive portion of the coil *alone*. It then becomes necessary to insert a small resistance in series with the coil to develop a voltage waveform proportional to the current. The resistor also allows a determination of the phase angle using a dual-beam (or dual-trace) oscilloscope, as shown in Fig. 24-9.

If the dual-beam oscilloscope is connected as shown in Fig. 24-9*a*, channel 1 displays the applied voltage waveform with a peak value, V_m. This is approximately equal to the coil voltage because $v_{R'}$ is negligible since R' is only a few ohms. Channel 2 displays a voltage waveform

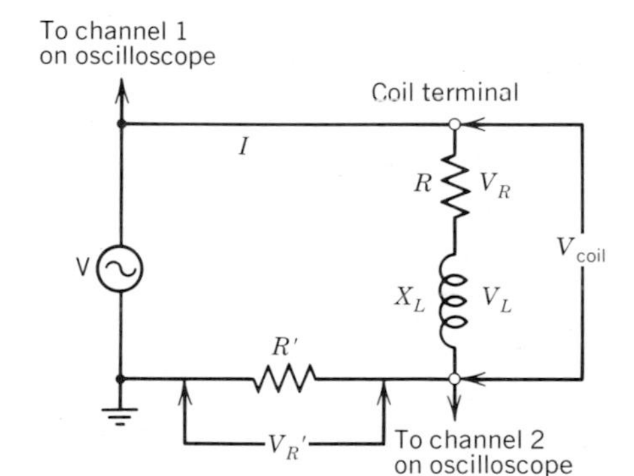

(*a*) Circuit of coil with small series resistor R'

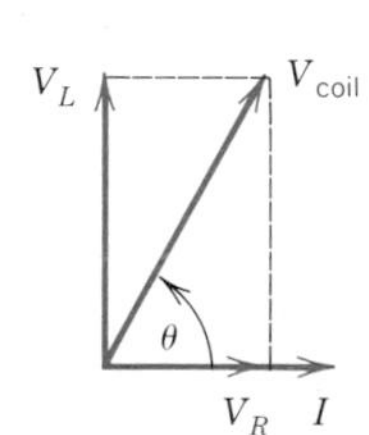

(*b*) Phasor diagram

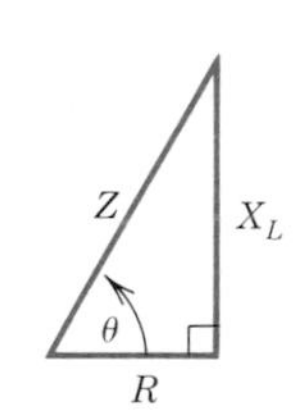

(*c*) Impedance diagram

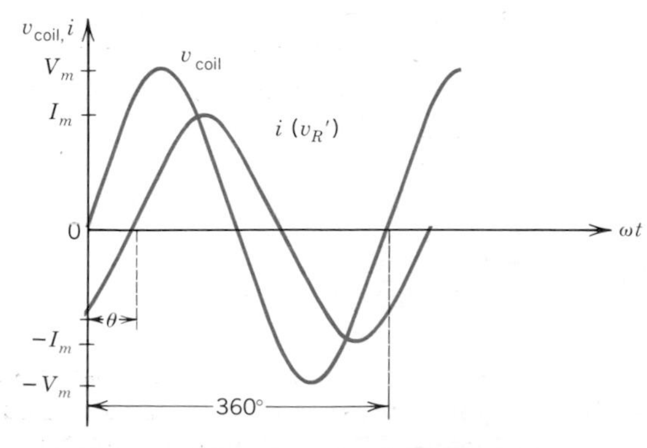

(*d*) Oscilloscope display of voltage waveforms across coil and R′

FIGURE 24-9
Voltage and current relations in a practical coil.

across R' that is proportional to the current in the circuit. That is,

$$I_m = \frac{V_{R'\,max}}{R'}$$

The impedance of the whole circuit, which is approximately equal to the impedance of the coil, can now be obtained:

$$Z = \frac{V_m}{I_m}$$

The phase angle θ can be determined from the oscilloscope by the direct ratio of the two horizontal distances representing θ and 360° in Fig. 24-9*d*. (This requires that the oscilloscope horizontal sweep be triggered by the same signal for both traces.) The resistive and inductive portions of the coil can now be obtained:

$$R = Z \cos \theta \qquad (24\text{-}6)$$

$$X_L = Z \sin \theta \qquad (24\text{-}7)$$

Note that R is actually the ac resistance of the coil, R_{ac}, so that the coil's Q can now be determined:

$$Q = \frac{X_L}{R_{ac}} \qquad (20\text{-}9)$$

and

$$L = \frac{X_L}{2\pi f} \qquad (20\text{-}2)$$

A method to find the phase angle using a *single-beam* oscilloscope is given in Section 24-8.

EXAMPLE 24-3

A coil and a 10-Ω resistor are series-connected across a 60-Hz supply. A dual-beam oscilloscope, connected as in Fig. 24-9*a*, displays a peak coil voltage of 18 V and a peak resistor voltage of 120 mV. A full cycle of the coil voltage occupies 5 cm and the beginning of the resistor voltage waveform is delayed 1 cm. Calculate:

a. The impedance of the circuit.
b. The phase angle between the applied voltage and current.
c. The ac resistance of the coil.
d. The inductive reactance of the coil.
e. The Q of the coil.
f. The inductance of the coil.

SOLUTION

a. $I_m = \frac{V_{R'\,max}}{R'}$

$$= \frac{120 \times 10^{-3}\ \text{V}}{10\ \Omega} = 12\ \text{mA}$$

$$Z = \frac{V_m}{I_m}$$

$$= \frac{18\ \text{V}}{12 \times 10^{-3}\ \text{A}} = \mathbf{1500\ \Omega}$$

b. By direct ratio:

$$\frac{\theta}{360°} = \frac{1\ \text{cm}}{5\ \text{cm}}$$

therefore, $\theta = 360° \times \frac{1}{5} = \mathbf{72°}$

c. $R_{ac} = Z \cos \theta \qquad (24\text{-}6)$

$$= 1500\ \Omega \times \cos 72° = \mathbf{464\ \Omega}$$

d. $X_L = Z \sin \theta \qquad (24\text{-}7)$

$$= 1500\ \Omega \times \sin 72° = \mathbf{1427\ \Omega}$$

e. $Q = \frac{X_L}{R_{ac}} \qquad (20\text{-}9)$

$$= \frac{1427\ \Omega}{464\ \Omega} = \mathbf{3.1}$$

f. $L = \frac{X_L}{2\pi f} \qquad (20\text{-}2)$

$$= \frac{1427\ \Omega}{2\pi \times 60\ \text{Hz}} = \mathbf{3.8\ H}$$

NOTE $Z = 1500\ \Omega$ is the impedance of the whole circuit so that $R_{ac} = 464\ \Omega$ actually includes the 10-Ω resistance of R'. (A more accurate value for the coil's ac resistance is therefore 454 Ω.)

24-5 SERIES RC CIRCUIT

Similar to the series RL circuit, we can draw the phasor diagram representing the voltage and current waveforms in a series RC circuit. See Fig. 24-10.

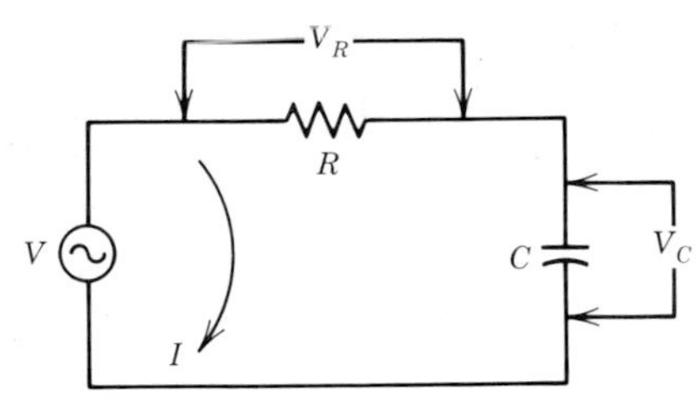

(a) Series RC circuit

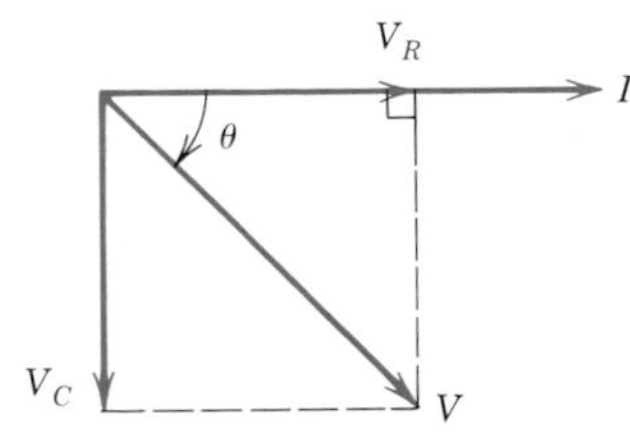

(b) Phasor diagram

FIGURE 24-10
Series RC circuit with phasor diagram.

Once again the current phasor has been drawn in the horizontal reference position, with the resistor voltage phasor "in-line" with it. Lagging I by 90° is the capacitor voltage phasor, and V is the phasor sum:

$$V = \sqrt{V_R^2 + V_C^2} \quad \text{volts (V)} \qquad (24\text{-}8)$$

and

$$\theta = -\text{arc tan}\,\frac{V_C}{V_R} \qquad (24\text{-}9)$$

where: V is the total applied voltage in volts, V

V_R is the voltage across the resistance in volts, V

V_C is the voltage across the pure capacitance in volts, V

θ is the phase angle between the applied voltage and the circuit current

Note that the phase angle θ is negative in comparison with an inductive circuit. That is, the current *leads* the applied voltage in a capacitive circuit but *lags* in an inductive circuit.

EXAMPLE 24-4

A capacitor and a 2.2-kΩ resistor are connected in series across a 60-Hz, sinusoidally alternating voltage. A series-connected ac ammeter indicates 15 mA, and an oscilloscope across the capacitor indicates 60 V peak-to-peak. Calculate:

a. The reading of a voltmeter connected across the resistor.
b. The rms voltage across the capacitor.
c. The applied voltage.
d. The phase angle between the current and applied voltage.
e. The amount of capacitance.

SOLUTION

See Fig. 24-11.

a. $V_R = IR$ (15-15b)

$= 15 \times 10^{-3}\text{ A} \times 2.2 \times 10^3\ \Omega = \mathbf{33\ V}$

b. $V_C = \dfrac{V_{p\text{-}p}}{2\sqrt{2}}$ (15-17)

$= \dfrac{60\text{ V}}{2\sqrt{2}} = \mathbf{21.2\ V}$

c. $V = \sqrt{V_R^2 + V_C^2}$ (24-8)

$= \sqrt{33^2 + 21.2^2}\text{ V} = \mathbf{39.2\ V}$

d. $\theta = -\tan^{-1}\dfrac{V_C}{V_R}$

$= -\text{arc tan}\,\dfrac{21.2\text{ V}}{33\text{ V}} = \mathbf{-32.7°}$

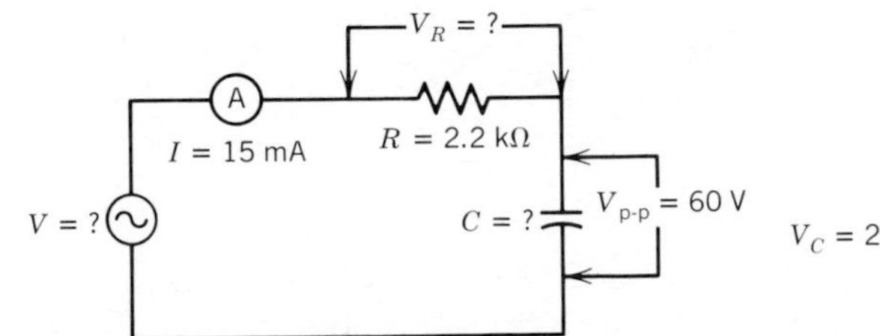

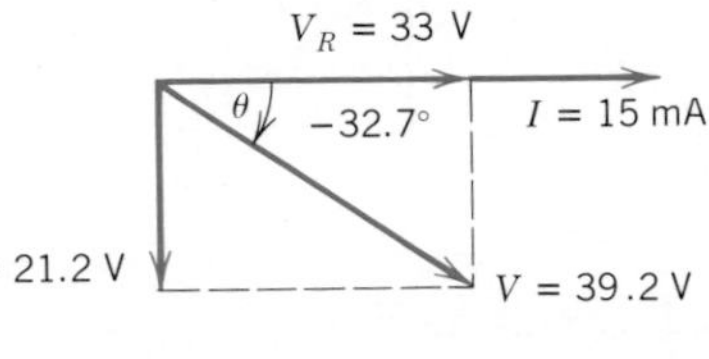

FIGURE 24-11
Circuit and phasor diagram for Example 24-4.

e. $X_C = \frac{V_C}{I}$ (23-1)

$= \frac{21.2 \text{ V}}{15 \times 10^{-3} \text{ A}} = 1.4 \text{ k}\Omega$

$C = \frac{1}{2\pi f X_C}$ (23-2)

$= \frac{1}{2\pi \times 60 \text{ Hz} \times 1.4 \times 10^3 \ \Omega}$

$= \mathbf{1.9 \ \mu F}$

The phasor diagram is shown in Fig. 24-11.

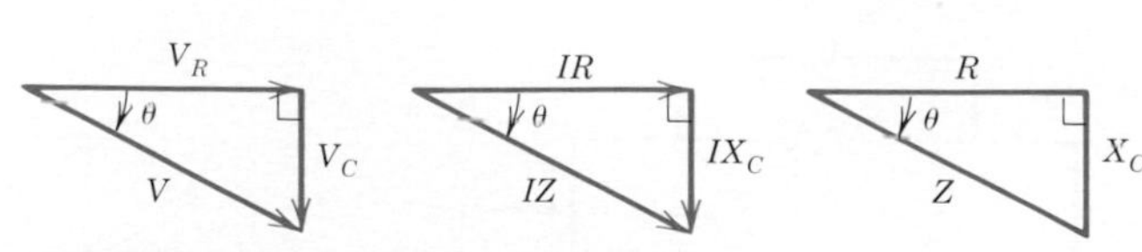

FIGURE 24-12
Voltage and impedance triangles for a series RC circuit.

24-6 IMPEDANCE OF A SERIES *RC* CIRCUIT

It can be shown, similar to an inductor, that the total opposition to current in a series RC circuit is given by the impedance Z. See Fig. 24-12.

$$Z = \sqrt{R^2 + X_C^2} \quad \text{ohms } (\Omega) \qquad (24\text{-}10)$$

$$I = \frac{V}{Z} \quad \text{amperes (A)} \qquad (24\text{-}4)$$

and

$$\theta = -\tan^{-1}\frac{X_C}{R} \qquad (24\text{-}11)$$

where: Z is the impedance in ohms, Ω
R is the resistance in ohms, Ω
X_C is the capacitive reactance in ohms, Ω
θ is the phase angle between the current I and the applied voltage V

EXAMPLE 24-5

a. How much resistance must be added in series with a 0.1-μF capacitor to limit the current to 5 mA when the series combination is connected to a 10-V, 1-kHz sine wave?

b. What is the phase angle between the applied voltage and current?

SOLUTION

See Fig. 24-13.

a. $X_C = \frac{1}{2\pi f C}$ (23-2)

$= \frac{1}{2\pi \times 1 \times 10^3 \text{ Hz} \times 0.1 \times 10^{-6} \text{ F}}$

$= 1.6 \text{ k}\Omega$

$Z = \frac{V}{I}$ (24-4)

$= \frac{10 \text{ V}}{5 \times 10^{-3} \text{ A}} = 2 \text{ k}\Omega$

$Z = \sqrt{R^2 + X_C^2}$ (24-10)

Therefore, $R = \sqrt{Z^2 - X_C^2}$

$= \sqrt{(2 \times 10^3)^2 - (1.6 \times 10^3)^2} \ \Omega$

$= \mathbf{1.2 \ k\Omega}$

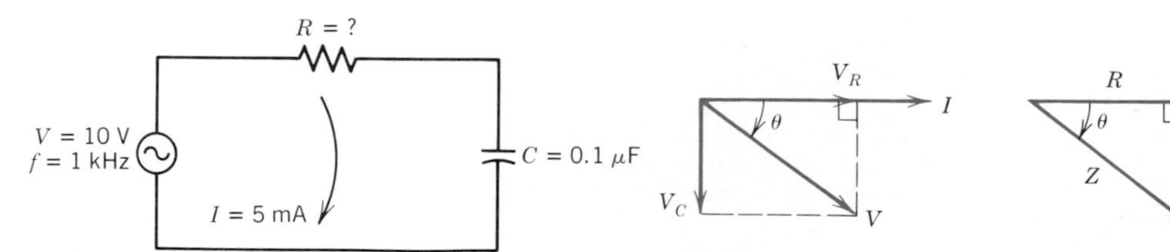

FIGURE 24-13
Circuit, phasor diagram, and impedance triangle for Example 24-5.

b. $\theta = -\tan^{-1}\dfrac{X_C}{R}$

$= -\text{arc tan}\ \dfrac{1.6\ \text{k}\Omega}{1.2\ \text{k}\Omega} = \mathbf{-53.1°}$

The phasor diagram is shown in Fig. 24-13.

24-7 SERIES *RLC* CIRCUIT

We have seen how the capacitance makes the current *lead* the applied voltage in a series *RC* circuit, and how the inductance makes the current *lag* in a series *RL* circuit. When *both* the capacitance and inductance are in series, we should expect a canceling effect. This is evident when we consider the phasor and impedance diagrams of Fig. 24-14.

The inductive voltage, V_L, is 180° out of phase with the capacitive voltage, V_C, in Fig. 24-14*b*. If V_L is larger than V_C, the difference, $V_L - V_C$, would be indicated on a voltmeter connected between *A* and *B* in Fig. 24-14*a* (assuming that the resistance of the coil is negligible). When $V_L - V_C$ is added to V_R vectorially, the result is the applied voltage, *V*. Thus

$$V = \sqrt{V_R{}^2 + (V_L - V_C)^2} \quad \text{volts (V)} \quad (24\text{-}12)$$

and

$$\theta = \tan^{-1}\frac{V_L - V_C}{V_R} \quad (24\text{-}13)$$

where: *V* is the applied voltage in volts, V

V_R is the voltage across the total circuit resistance in volts, V (includes the resistance of the coil if not negligible)

V_L is the voltage across the inductance in volts, V

V_C is the voltage across the capacitance in volts, V

$V_L - V_C$ is the net reactive voltage in volts, V

θ is the phase angle between the current and applied voltage in degrees or radians.

The impedance of a series *RLC* circuit can be obtained from the impedance diagram in Fig. 24-14*c*. The net reactance of the circuit is $X_L - X_C$ so that *Z* is given by

$$Z = \sqrt{R^2 + (X_L - X_C)^2} \quad \text{ohms } (\Omega) \quad (24\text{-}14)$$

and

$$\theta = \tan^{-1}\frac{X_L - X_C}{R} \quad (24\text{-}15)$$

and

$$I = \frac{V}{Z} \quad \text{amperes (A)} \quad (24\text{-}4)$$

where: *Z* is the circuit impedance in ohms, Ω

R is the circuit resistance in ohms, Ω

X_L is the circuit inductive reactance in ohms, Ω

$X_L - X_C$ is the net circuit reactance in ohms, Ω

θ is the phase angle between the current and applied voltage in degrees or radians

For the phasor diagram shown in Fig. 24-14*b*, the current *lags* the applied voltage *V* because the overall

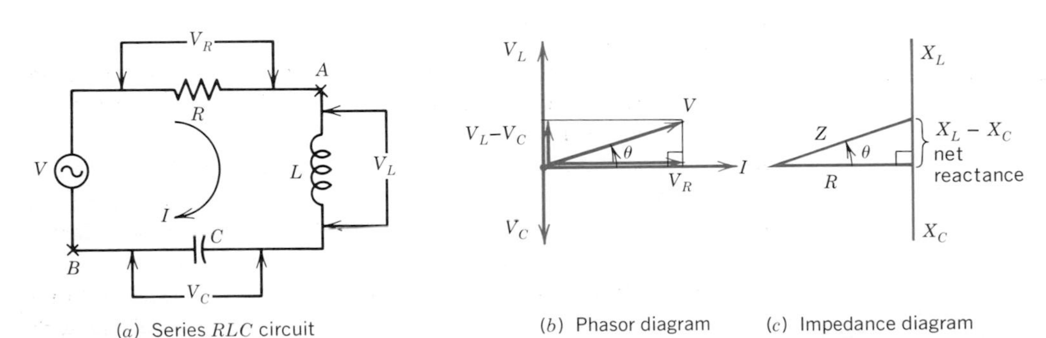

(*a*) Series *RLC* circuit (*b*) Phasor diagram (*c*) Impedance diagram

FIGURE 24-14
Series *RLC* circuit, phasor diagram, and impedance diagram.

circuit is inductive. That is, X_L is greater than X_C, making V_L larger than V_C. Clearly, if V_C is larger than V_L (X_C greater than X_L), then the current *leads* the applied voltage, and the overall circuit is said to be capacitive. Further, if $X_L = X_C$, then $V_L = V_C$, the phase angle is zero, and the circuit is purely resistive. This condition is called series *resonance*; it occurs when the inductive and capacitive reactances cancel each other. Resonance is a very important effect and is considered in detail in Chapter 26.

EXAMPLE 24-6

A 30-mH coil (of negligible resistance), a 0.005-μF capacitor, and a 1-kΩ resistor are series-connected across a 2-V, 10-kHz sinusoidal supply. Calculate:

a. The circuit impedance.

b. The circuit current.

c. The voltage across each component.

d. The phase angle between the current and applied voltage.

Draw the circuit, the phasor and the impedance diagrams indicating all values.

SOLUTION

a. $$X_L = 2\pi fL \qquad (20\text{-}2)$$
$$= 2\pi \times 10 \times 10^3 \text{ Hz} \times 30 \times 10^{-3} \text{ H}$$
$$= 1885\ \Omega$$

$$X_C = \frac{1}{2\pi fC} \qquad (23\text{-}2)$$
$$= \frac{1}{2\pi \times 10 \times 10^3 \text{ Hz} \times 0.005 \times 10^{-6} \text{ F}}$$
$$= 3183\ \Omega$$

$$Z = \sqrt{R^2 + (X_L - X_C)^2}\ \Omega \qquad (24\text{-}14)$$
$$= \sqrt{1000^2 + (1885 - 3183)^2}\ \Omega$$
$$= \mathbf{1639\ \Omega}$$

b. $$I = \frac{V}{Z} \qquad (24\text{-}4)$$
$$= \frac{2 \text{ V}}{1639\ \Omega} = \mathbf{1.2\ mA}$$

c. $$V_R = IR \qquad (15\text{-}15b)$$
$$= 1.2 \times 10^{-3} \text{ A} \times 1000\ \Omega = \mathbf{1.20\ V}$$
$$V_L = IX_L \qquad (20\text{-}1)$$
$$= 1.2 \times 10^{-3} \text{ A} \times 1885\ \Omega = \mathbf{2.26\ V}$$
$$V_C = IX_C \qquad (23\text{-}1)$$
$$= 1.2 \times 10^{-3} \text{ A} \times 3183\ \Omega = \mathbf{3.82\ V}$$

d. $$\theta = \tan^{-1}\frac{X_L - X_C}{R} \qquad (24\text{-}15)$$
$$= \tan^{-1}\frac{1885 - 3183\ \Omega}{1000\ \Omega}$$
$$= \text{arc tan} - 1.3 = \mathbf{-52.4°}$$

The diagrams are shown in Fig. 24-15.

NOTE The voltages across the capacitor and inductor in Example 24-6 are each larger than the input voltage. This is a common phenomenon in series *RLC* circuits when the inductive and capacitive reactances are of the same order of magnitude. This is because the overall impedance is *lower* than if the capacitance or the inductance alone is in the circuit with the resistance. This means that a relatively high current flows, producing a relatively high voltage across the reactances. Note too that because X_C is greater than X_L, the overall circuit is capacitive. Also, V_C is greater than V_L and the current *leads* the applied voltage by 52.4°.

24-8 OSCILLOSCOPE MEASUREMENT OF PHASE ANGLES

Although the phase angle between the two voltages can be observed on a dual-beam (or dual-trace) oscilloscope, there is a method of displaying this information on a single-beam oscilloscope that is very useful. It involves applying one of the voltages to the vertical (or Y input), the other to the horizontal (or X input). Since both waveforms are at the same frequency, the phase relation between them is always constant. The smaller voltage of the two is usually connected to the vertical input, since this input is the more sensitive. In order for the other voltage to be applied to the horizontal deflection plates, it is usually necessary to switch the time-base selector switch to "external" (EXT).

An example of how the phase angle is measured in a series *RLC* circuit is shown in Fig. 24-16, along with typical *Lissajous figures*.

Since the frequency of the two voltages is the same, a stationary pattern results, as in Fig. 24-16*b*. In general,

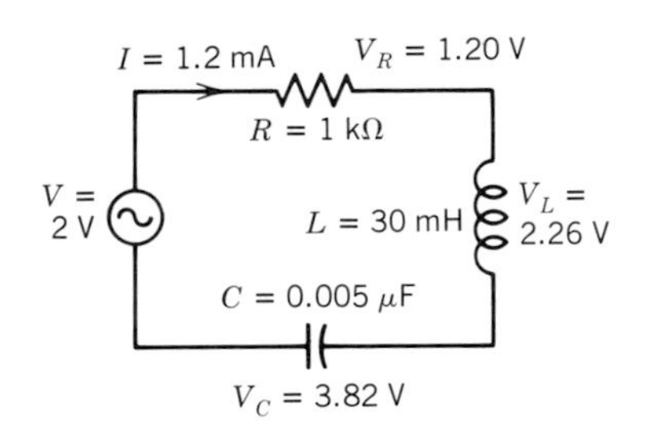

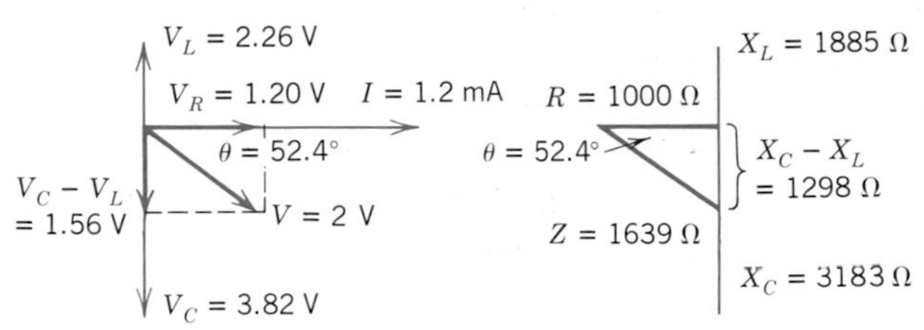

FIGURE 24-15
Series *RLC* circuit, phasor diagram, and impedance diagram for Example 24-6.

an ellipse is produced with an overall height BB' and intercepts on the y-axis at AA'. It can be shown that the sine of the phase angle θ between the two voltages applied to the oscilloscope is given by

$$\sin\theta = \frac{OA}{OB} = \frac{AA'}{BB'} \tag{24-16}$$

and

$$\theta = \arcsin\frac{OA}{OB} = \arcsin\frac{AA'}{BB'} \tag{24-16a}$$

Note that AA' and BB' are *distances* taken from the oscilloscope screen. It is *not* necessary that the oscilloscope be calibrated vertically or horizontally, but the Lissajous pattern must be *centered* horizontally.

In the case of a *circle*, AA' and BB' are equal, so $\theta = \arcsin 1 = 90°$. For a straight line, $AA' = 0$, so $\theta = \arcsin 0 = 0°$. If $\theta = 45°$, $\frac{OA}{OB} = \sin 45° = 0.7$. Whether this is a leading or lagging phase angle cannot be determined from the Lissajous display alone.*

In the circuit of Fig. 24-16*a*, the phase angle determined is between V_R and V. But since V_R is proportional

* But a dual trace or dual-beam presentation clearly shows which waveform is leading or lagging. The advantage of the Lissajous method is that the phase angle is determined quite accurately.

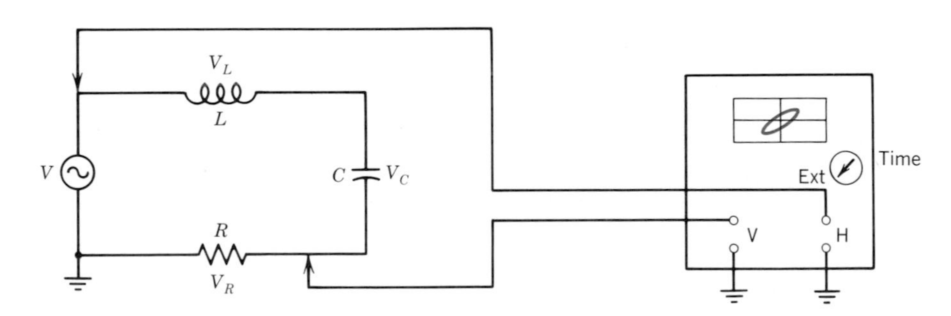

(*a*) Circuit connections to measure the phase angle

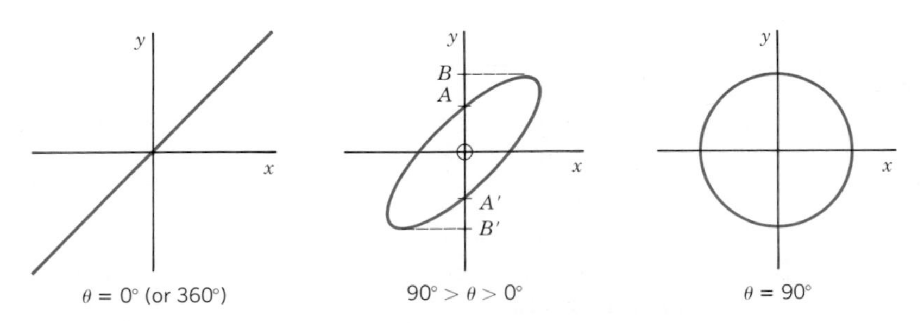

(*b*) Lissajous figures for various phase angles

FIGURE 24-16
Showing how to make phase angle measurements using Lissajous figures. (*Note:* the oscilloscope must be adjusted for equal vertical and horizontal deflections to obtain above figures, and in some oscilloscopes, the figures may slope the opposite way.)

to the current I (and in phase with I), the measured phase angle is between the current and the applied voltage. However, not all phase angle measurements necessarily give the phase angle between the current and applied voltage. For example, what meaning would the phase angle determined from the oscilloscope have if R and C were interchanged in Fig. 24-16*a*? Recall that in many cases one side of the supply is grounded, and so is the low side on the oscilloscope. Care must be taken to determine what voltages are being compared. (See Problem 24-24.)

EXAMPLE 24-7

An oscilloscope is used to make a phase angle measurement between the current and applied voltage in a series *RLC* circuit, as in Fig. 24-16*a*. The resulting ellipse is as shown in Fig. 24-16*b*, with $OA = 2$ cm and $OB = 3$ cm. Voltmeters across L and C indicate 8 and 5 V, respectively. Calculate:

a. The phase angle between the current and applied voltage.
b. The resistor voltage.
c. The applied voltage.
d. The reading of a voltmeter connected across the series LC combination.

Draw the phasor diagram.

SOLUTION

a. $$\theta = \sin^{-1}\frac{OA}{OB} \qquad (24\text{-}16a)$$

$$= \text{arc sin}\,\frac{2\text{ cm}}{3\text{ cm}} = \mathbf{41.8^\circ}$$

b. $$\tan\theta = \frac{V_L - V_C}{V_R} \qquad (24\text{-}13)$$

Therefore, $$V_R = \frac{V_L - V_C}{\tan\theta}$$

$$= \frac{8\text{ V} - 5\text{ V}}{\tan 41.8^\circ} = \mathbf{3.36\ V}$$

c. $$V = \sqrt{V_R^2 + (V_L - V_C)^2} \qquad (24\text{-}12)$$

$$= \sqrt{3.36^2 + (8-5)^2}\text{ V} = \mathbf{4.5\ V}$$

d. Voltage across LC combination $= V_L - V_C$
$= 8 - 5\text{ V} = \mathbf{3\ V}$

The phasor diagram is shown in Fig. 24-17.

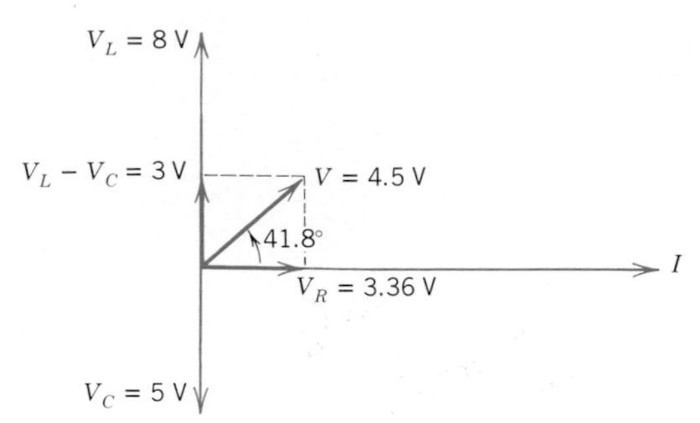

FIGURE 24-17
Phasor diagram for Example 24-7.

24-9 PARALLEL *RL* CIRCUIT

When a resistor and an inductor are connected in parallel across a sinusoidal supply, the total current drawn, by Kirchhoff's current law, is the phasor sum of each branch current. Since voltage is common to both branch elements, we use V as the horizontal reference phasor, as in Fig. 24-18*b*.

The current drawn by each branch is given by

$$I_R = \frac{V}{R} \qquad (15\text{-}15a)$$

and

$$I_L = \frac{V}{X_L} \qquad (20\text{-}1)$$

Therefore, the total current I_T is

$$I_T = \sqrt{I_R^2 + I_L^2} \quad \textbf{amperes (A)} \qquad \textbf{(24-17)}$$

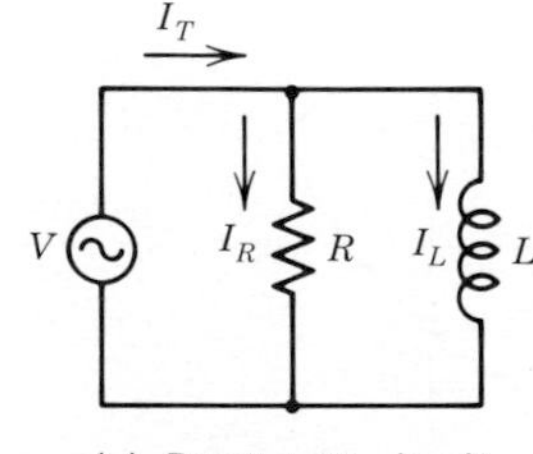

(*a*) Parallel *RL* circuit

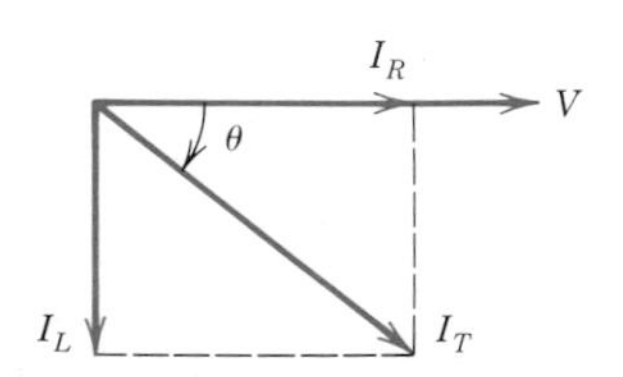

(*b*) Phasor diagram

FIGURE 24-18
Circuit and phasor diagram for a parallel *RL* circuit.

$$Z = \frac{V}{I_T} \qquad (24\text{-}4)$$

and

$$\theta = -\tan^{-1}\frac{I_L}{I_R} \qquad (24\text{-}18)$$

In this case, the negative phase angle shows that the total circuit current *lags* the applied voltage, as it does in a series *RL* circuit.

EXAMPLE 24-8

A 200-Ω resistor, and a coil of negligible resistance and 400-Ω reactance, are connected in parallel across a 40-V sinusoidal source. Calculate:

a. The total circuit current.
b. The impedance of the circuit.
c. The phase angle between the circuit current and applied voltage.

SOLUTION

a. $I_R = \frac{V}{R} \qquad (15\text{-}15a)$

$= \frac{40\ \text{V}}{200\ \Omega} = 0.2\ \text{A}$

$I_L = \frac{V}{X_L} \qquad (20\text{-}1)$

$= \frac{40\ \text{V}}{400\ \Omega} = 0.1\ \text{A}$

$I_T = \sqrt{I_R^2 + I_L^2} \qquad (24\text{-}17)$

$= \sqrt{0.2^2 + 0.1^2}\ \text{A} = \mathbf{0.224\ A}$

b. $Z = \frac{V}{I_T} \qquad (24\text{-}4)$

$= \frac{40\ \text{V}}{0.224\ \text{A}} = \mathbf{179\ \Omega}$

c. $\theta = -\tan^{-1}\frac{I_L}{I_R}$

$= -\text{arc tan}\,\frac{0.1\ \text{A}}{0.2\ \text{A}} = \mathbf{-26.6°}$

Note that the overall circuit is considered to be more resistive than inductive because the resistive current is larger than the inductive current.

(*a*) Parallel *RC* circuit

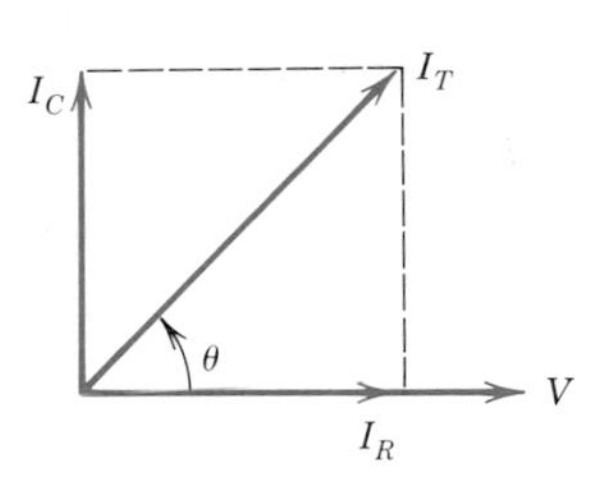

(*b*) Phasor diagram

FIGURE 24-19
Parallel *RC* circuit and phasor diagram.

24-10 PARALLEL *RC* CIRCUIT

The phasor diagram for the current and voltage in a parallel *RC* circuit is shown in Fig. 24-19. It should be obvious that the total circuit current is given by

$$I_T = \sqrt{I_R^2 + I_C^2} \quad \text{amperes (A)} \qquad (24\text{-}19)$$

$$Z = \frac{V}{I_T} \qquad (24\text{-}4)$$

and

$$\theta = \tan^{-1}\frac{I_C}{I_R} \qquad (24\text{-}20)$$

EXAMPLE 24-9

A 200-Ω resistor and a 100-Ω capacitive reactance are connected in parallel with a 40-V sinusoidal supply. Calculate:

a. The total circuit current.
b. The impedance of the circuit.
c. The phase angle between the circuit current and applied voltage.

SOLUTION

a. $I_R = \frac{V}{R} \qquad (15\text{-}15a)$

$= \frac{40\ \text{V}}{200\ \Omega} = 0.2\ \text{A}$

$I_C = \frac{V}{X_C} \qquad (23\text{-}1)$

$$= \frac{40\ \text{V}}{100\ \Omega} = 0.4\ \text{A}$$

$$I_T = \sqrt{I_R^2 + I_C^2} \quad (24\text{-}19)$$

$$= \sqrt{0.2^2 + 0.4^2}\ \text{A} = \mathbf{0.447\ A}$$

b. $$Z = \frac{V}{I_T} \quad (24\text{-}4)$$

$$= \frac{40\ \text{V}}{0.447\ \text{A}} = \mathbf{89.5\ \Omega}$$

c. $$\theta = \tan^{-1}\frac{I_C}{I_R} \quad (24\text{-}20)$$

$$= \text{arc tan}\,\frac{0.4\ \text{A}}{0.2\ \text{A}} = \mathbf{63.4°}$$

In this case, the circuit is more capacitive than resistive, (I_T leads V), because I_C is larger than I_R. This is a result of X_C being *smaller* than R. (Note how this is opposite to a series RC circuit.)

24-11 PARALLEL *RLC* CIRCUIT

Similar to a series RLC circuit, we should expect some canceling effect when inductance and capacitance are both connected in parallel across an ac voltage source. In the circuit of Fig. 24-20, it is assumed that X_C is *less* than X_L so that I_C is *greater* than I_L. The resulting total current leads the applied voltage, and the circuit is said to be overall capacitive.

Conversely, if X_L is less than X_C, then I_L is greater than I_C and the circuit is considered inductive. But if $X_L = X_C$, the total current is in phase with the applied voltage (θ is zero), and the circuit is purely resistive. This condition is called *parallel resonance* and is discussed further in Section 26-6.

The total current of the parallel RLC circuit is

$$I_T = \sqrt{I_R^2 + (I_C - I_L)^2} \quad \text{amperes (A)} \quad (24\text{-}21)$$

$$Z = \frac{V}{I_T} \quad (24\text{-}4)$$

and

$$\theta = \tan^{-1}\frac{I_C - I_L}{I_R} \quad (24\text{-}22)$$

EXAMPLE 24-10

The resistor, inductor, and capacitor from the previous two examples ($R = 200\ \Omega$, $X_L = 400\ \Omega$, and $X_C = 100\ \Omega$) are connected in parallel with the same voltage source of $V = 40$ V. Calculate:

a. The total circuit current.
b. The circuit impedance.
c. The phase angle between the circuit current and applied voltage.

SOLUTION

a. As before,

$$I_R = 0.2\ \text{A}$$
$$I_L = 0.1\ \text{A}$$
$$I_C = 0.4\ \text{A}$$
$$I_T = \sqrt{I_R^2 + (I_C - I_L)^2} \quad (24\text{-}21)$$
$$= \sqrt{0.2^2 + (0.4 - 0.1)^2}\ \text{A} = \mathbf{0.36\ A}$$

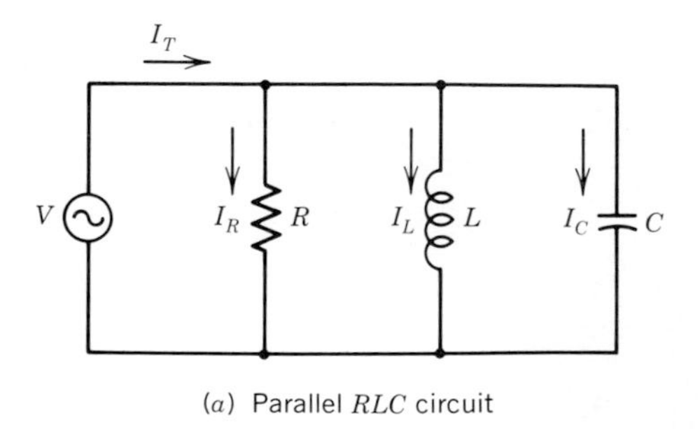

(a) Parallel RLC circuit

(b) Phasor diagram

FIGURE 24-20
Parallel RLC circuit and phasor diagram.

b. $Z = \dfrac{V}{I_T}$ (24-4)

$= \dfrac{40\text{ V}}{0.36\text{ A}} = \mathbf{111\ \Omega}$

c. $\theta = \tan^{-1}\dfrac{I_C - I_L}{I_R}$ (24-22)

$= \text{arc tan}\dfrac{0.4 - 0.1\text{ A}}{0.2\text{ A}} = \mathbf{56.3°}$

There are some important points to note in this example. First, observe that the *total* circuit current (0.36 A) is *less* than the capacitor branch current (0.4 A). It is not unusual in some parallel *RLC* circuits, where X_L and X_C are equal or close in value, for a very large current to circulate in the *tank* circuit consisting of *L* and *C*.

Second, note that the total impedance of 111 Ω is *not* less than the smallest reactance of $X_C = 100\ \Omega$. That is, in a parallel *RL or RC* circuit, the total impedance *Z is less* than the smallest branch resistance or reactance. But the combined effect of $X_L = 400\ \Omega$ and $X_C = 100\ \Omega$ actually produce, between them, a combined reactance of 133 Ω (40 V/0.3 A). It is obvious then, that we cannot combine parallel resistances and reactances in the same way that we treat a purely resistive parallel circuit.

We shall consider other combinations, such as series-parallel reactances, when we are better prepared to use complex numbers in Chapter 27.

SUMMARY

1. A sine wave may be represented by a horizontal line called a phasor, which is assumed to rotate counterclockwise at the frequency of the supply.

2. A cosine curve is represented by a vertical phasor; it leads a sine curve by 90°.

3. Phasors at right angles can be used to represent the rms values of the current and voltage in pure *R*, *L*, and *C* circuits.

4. A series *RL* circuit has a total applied voltage given by

$$V = \sqrt{V_R^2 + V_L^2}$$

and a phase angle between current and applied voltage given by

$$\theta = \tan^{-1}\frac{V_L}{V_R} = \tan^{-1}\frac{X_L}{R}$$

5. The total opposition to the current in a series *RL* circuit is called the impedance and is given by

$$Z = \sqrt{R^2 + X_L^2}$$

6. The current in an ac circuit is given by the equation: $I = \dfrac{V}{Z}$.

7. Practical coils that have a resistance that is not negligible have a phase angle between the current and coil voltage of less than 90°.

8. A series *RC* circuit has a total applied voltage given by

$$V = \sqrt{V_R^2 + V_C^2}$$

and a phase angle between the current and applied voltage given by

$$\theta = -\tan^{-1}\frac{V_C}{V_R} = -\tan^{-1}\frac{X_C}{R}$$

9. The impedance of a series RC circuit is given by

$$Z = \sqrt{R^2 + X_C^2}$$

10. The impedance of a series RLC circuit is given by:

$$Z = \sqrt{R^2 + (X_L - X_C)^2}$$

and

$$\theta = \tan^{-1}\frac{X_L - X_C}{R} = \tan^{-1}\frac{V_L - V_C}{V_R}$$

11. An oscilloscope display of two voltages in a stationary Lissajous figure can be used to obtain the phase angle between the two voltages using:

$$\theta = \sin^{-1}\frac{OA}{OB}$$

12. A parallel RLC circuit has a total line current:

$$I_T = \sqrt{I_R^2 + (I_C - I_L)^2}$$

$$Z = \frac{V}{I_T} \quad \text{and} \quad \theta = \tan^{-1}\frac{I_C - I_L}{I_R}$$

SELF-EXAMINATION

Answer T or F or a, b, c, d, e for multiple choice
(Answers at back of book)

24-1. A phasor is a line that represents a quantity that varies with time. ________

24-2. Phasors are assumed to rotate clockwise at the same frequency as the voltage or current they represent. ________

24-3. Two phasors drawn at right angles can represent the voltage and current waveforms in a purely capacitive or purely inductive circuit. ________

24-4. The phasor sum of the resistive and inductive voltages in a series RL circuit is *always* less than the algebraic sum. ________

24-5. If the resistive and inductive voltages in a series RL circuit are each 5 V, the total applied voltage and phase angle are

a. 7.07 V, 90°
b. 10 V, 45°
c. 7.07 V, $\pi/4$ rad
d. 10 V, 90°
e. none of the above

24-6. If the inductive and resistive voltages in a series *RL* circuit are equal, then the inductive reactance is equal to the resistance. ________

24-7. A series *RL* circuit has a resistance of 30 Ω and an inductive reactance of 40 Ω. The impedance is

a. 50 Ω
b. 70 Ω
c. 10 Ω
d. 35 Ω
e. none of the above

24-8. A series *RC* circuit has a resistance of 50 Ω and a capacitive reactance of 120 Ω and is connected to a 130-V supply. The current drawn is

a. 2.2 A
b. 1.1 A
c. 0.1 A
d. 1 A
e. none of the above

24-9. The voltage across a coil is always 90° out of phase with the current through the coil. ________

24-10. If a series *RC* circuit has a leading phase angle of 30°, the circuit is more capacitive than resistive. ________

24-11. If a series *RC* circuit has $X_C = 2R$, the phase angle between the current and applied voltage is

a. 30°
b. 63.4°
c. 26.6°
d. 60°
e. 45°

24-12. The phase angle of a series *RLC* circuit is determined only by the values of *R*, *L*, and *C*. ________

24-13. If a capacitor is added in series with a series *RL* circuit, the result is always a decrease in the impedance. ________

24-14. If a capacitor of the correct reactance is connected in series with a series *RL* circuit, the result may be a purely resistive circuit. ________

24-15. The determination of a circuit's phase angle requires an oscilloscope with accurately calibrated vertical and horizontal inputs. ________

24-16. If a parallel *RL* circuit contains a resistance of 100 Ω and an inductive reactance of 1000 Ω, the circuit is considered predominantly inductive. ________

24-17. The total current leads the applied voltage in a parallel *RL* circuit whereas it lags in a series *RL* circuit. ________

24-18. The total line current in a parallel *RLC* circuit must, because it obeys Kirchhoff's current law, always be larger than any one of the branch currents. ________

24-19. The total impedance of a parallel *RLC* circuit can be *larger* than the smallest branch resistance or reactance.

REVIEW QUESTIONS

1. Distinguish between a vector and a phasor, giving an example of each.

2. In what way would the waveform represented by a horizontal phasor be different if the phasor were assumed to rotate clockwise instead of counterclockwise?

3. Why, strictly speaking, should phasors be drawn or labeled with *peak* values and not *rms* values?

4. How would the waveforms of current and voltage for pure R, L, or C circuits be different (in Fig. 24-4) if the *current* phasor is drawn in the horizontal reference position?

5. What simple proof can you give, using voltage and impedance triangles, to show that the phase angle between the current and voltage in a series RL or RC circuit is determined by the ratio of the reactance to the resistance?

6. What is the difference between ac resistance, reactance, and impedance?

7. How would Eq. 24-3 be modified to obtain the total impedance of a series RL circuit in which the inductor's resistance R_L is not negligible?

8. Why is it not proper to refer to an impedance diagram as a phasor diagram?

9. Describe a method by which you could use a dual-beam oscilloscope to determine the resistance, inductance, and Q of a coil.

10. How would the impedance of a series RC circuit change, compared with an RL circuit, if the frequency of the applied voltage were increased?

11. How would you expect that the impedance and phase angle would change if you varied the frequency from direct current to some high value in a series RLC circuit? Sketch graphs of each.

12. How is it possible to develop larger voltages across the capacitor or inductor than the input voltage in a series RLC circuit? When is this effect most noticeable?

13. Sketch the circuit connections to a single-trace oscilloscope if you wanted to measure the phase angle between the current and applied voltage in a series RC circuit if one side of the supply and the low side of the oscilloscope are both grounded.

14. Give two ways in which a parallel RLC circuit differs from a parallel resistive circuit.

15. Show why $\tan \theta = -R/X_L$ for a *parallel* RL circuit, but $\tan \theta = -X_L/R$ for a *series* RL circuit.

PROBLEMS

(Answers to odd-numbered problems at back of book)

24-1. Construct accurately a graph of a sine wave using the method of a rotating line as in Fig. 24-1. Use a 1-in. radius circle and angles every 15°. Make the horizontal axis of the graph approximately 2 in. long.

24-2. Repeat Problem 24-1 to construct a cosine curve as in Fig. 24-2.

24-3. Draw two vectors, A and B, at right angles to each other as in Fig. 24-3. Make them 1 in. long and construct the two waveforms they represent (as in Problems 24-1 and 24-2).

Find the *phasor* sum, $A + B$, by constructing the diagonal as in Fig. 24-6b. Now use this line to construct the *waveform* represented by this phasor. Does it coincide with the *instantaneous* sums of the A and B *waveforms* taken every 30°?

24-4. Repeat Problem 24-3 but make A 1 in. long, and B 2 in. long.

24-5. A series RL circuit is connected to an ac supply. Voltmeters across the resistor and inductor (of negligible resistance) indicate 40 and 80 V, respectively. Calculate:

a. The applied voltage.

b. The phase angle between the current and applied voltage.

c. The relative value of X_L compared with R.

24-6. A 24-V sinusoidal emf is applied to a series RL circuit. If a voltmeter across the resistor indicates 12 V, calculate:

a. The voltage across the inductance.

b. The phase angle between the applied voltage and current.

c. The relative value of X_L compared with R.

24-7. A coil of negligible resistance and a 2.2-kΩ resistor are series-connected across a 24-V, 400-Hz supply. If a voltmeter connected across the coil indicates 10 V, calculate:

a. The reading of a voltmeter connected across the resistor.

b. The phase angle between the applied voltage and current.

c. The current in the circuit.

d. The inductance of the coil.

24-8. The phase angle between current and applied voltage in a series RL circuit is known to be 30° If the voltage across the 4.7-kΩ resistor is 15 V, calculate:

a. The inductor voltage.

b. The applied voltage.

c. The current in the circuit.

d. The inductive reactance of the coil.

24-9. A series RL circuit draws a current of 20 mA from a 24-V, 60-Hz source. If the inductive reactance is known to be twice as large as the resistance, calculate:

a. The circuit impedance.

b. The circuit resistance.

c. The circuit's inductive reactance.

d. The resistor voltage.

e. The inductor voltage.

f. The inductance of the coil.

24-10. Repeat Problem 24-9 with $R = 2X_L$.

24-11. A 200-μH coil of negligible resistance and a 47-Ω resistor are series-connected across a 20-V peak-to-peak, 10-kHz, sinusoidal signal generator. Calculate:

a. The circuit impedance.

b. The current in the circuit.

c. The phase angle between the current and applied voltage.

24-12. A 100-mH coil of unknown resistance is connected in series with a 220-Ω resistor and an 18-V peak-to-peak sine wave at 500 Hz. If a series-connected milliammeter indicates 12 mA, calculate:

a. The circuit impedance.

b. The resistance of the coil.

c. The voltage across the resistor.

d. The impedance of the coil.

e. The voltage across the coil.

24-13. A coil of unknown inductance and resistance is connected in series with a 4.7-Ω resistor and a 6.3-V, 60-Hz sinusoidal supply. A dual-beam oscilloscope, connected as in Fig. 24-9*a*, indicates a peak-to-peak resistor voltage of 80 mV. A full cycle of the resistor voltage occupies 10 cm and is delayed 2.25 cm compared with the coil voltage. Calculate:

a. The circuit impedance.
b. The phase angle between the current and applied voltage.
c. The ac resistance of the coil.
d. The inductive reactance of the coil.
e. The Q of the coil.
f. The inductance of the coil.

24-14. A coil with a Q of 2 is connected in series with a resistor and a 60-Hz supply. If the coil voltage is 100 V and the resistor voltage is 60 V, calculate the supply voltage. Draw the phasor diagram.

24-15. A capacitor and a 33-Ω resistor are connected in series across a 60-Hz sinusoidal supply. A series-connected ac ammeter indicates 1.65 A, and an oscilloscope across the capacitor indicates 150 V peak. Calculate:

a. The applied voltage.
b. The phase angle between the current and applied voltage.
c. The amount of capacitance.

24-16. The current in a series RC circuit is 40 mA. If the phase angle between the current and applied voltage is −50° and the resistor is 330 Ω, calculate:

a. The applied voltage.
b. The impedance of the circuit.
c. The size of the capacitor if the frequency is 60 Hz.

24-17. a. How much resistance must be added in series with a 200-pF capacitor to limit the current to 80 μA when the series combination is connected to a 150-mV, 1-MHz sine wave?

b. What is the circuit's phase angle between the applied voltage and current?

24-18. a. How much capacitance must be added in series with a 50-Ω resistor to limit the current to 1.5 A when the series combination is connected to a 340-V peak-to-peak, 60-Hz sine wave?

b. What is the circuit's phase angle between the applied voltage and current?

24-19. An 8-H coil (of negligible resistance), a 0.0035-μF capacitor, and a 2.2-kΩ resistor are series-connected across a 3.5-V, variable-frequency sinusoidal signal generator. For a frequency setting of 600 Hz, calculate:

a. The circuit impedance.
b. The circuit current.
c. The voltage across each component.
d. The phase angle between the current and applied voltage.

Draw the circuit, phasor, and impedance diagrams indicating all values.

24-20. Repeat Problem 24-19 using a frequency of 951 Hz. Compare the results with Problem 24-19.

24-21. Repeat Problem 24-19 using a frequency of 1200 Hz. Compare the results with Problems 24-19 and 24-20.

24-22. A relay coil has a resistance of 300 Ω and requires a current of 150 mA for its operation. When operated from a 60-Hz source, the required voltage is 180 V. Calculate the value of the capacitance in series with the relay coil that will allow its operation from a 120-V, 60-Hz source.

24-23. An oscilloscope is used to make a phase angle measurement between the current and applied voltage in a series *RLC* circuit. The resulting ellipse has an overall height of 6 cm and intercepts on the *y*-axis 5 cm apart. If a voltmeter connected across the series combination of *L* and *C* indicates 30 V, calculate:

a. The phase angle between the current and applied voltage.
b. The resistor voltage.
c. The applied voltage.

Draw a phasor diagram given that the circuit is capacitive.

24-24. Determine the oscilloscope display when it is connected as in the circuit of Fig. 24-21 to show a Lissajous figure. Justify your answer.

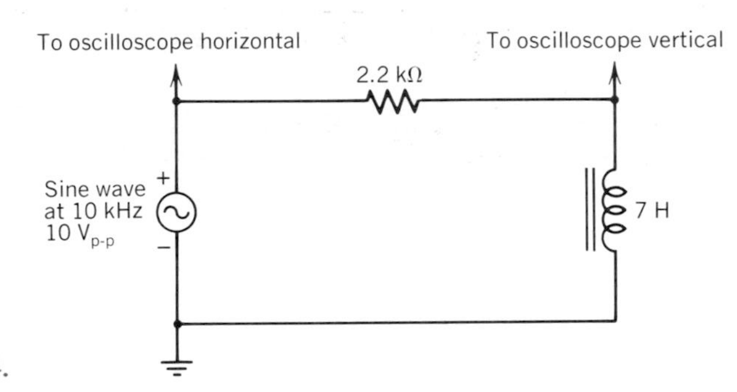

FIGURE 24-21
Circuit for Problem 24-24.

24-25. A 0.5-H inductor of negligible resistance and a 1.2-kΩ resistor are connected in parallel across a 48-V, 400-Hz source. Calculate:

a. The total current drawn from the source.
b. The impedance of the circuit.
c. The phase angle between the total current and applied voltage.

24-26. Repeat Problem 24-25 with the inductor replaced by a 0.2-μF capacitor.

24-27. A parallel *RLC* circuit consists of a 0.5-H coil (of negligible resistance), a 2.2-kΩ resistor, and a 0.03-μF capacitor connected across a 5-V, 1-kHz source. Calculate:

a. The total circuit current.
b. The circuit impedance.
c. The phase angle between the circuit current and applied voltage.

Draw a phasor diagram showing all values.

24-28. a. Calculate the size of the capacitor to be added in parallel with the circuit of Problem 24-27 to make the phase angle between the current and applied voltage equal zero.

b. Determine the new line current and impedance after adding the capacitor. Compare the results with Problem 24-27.

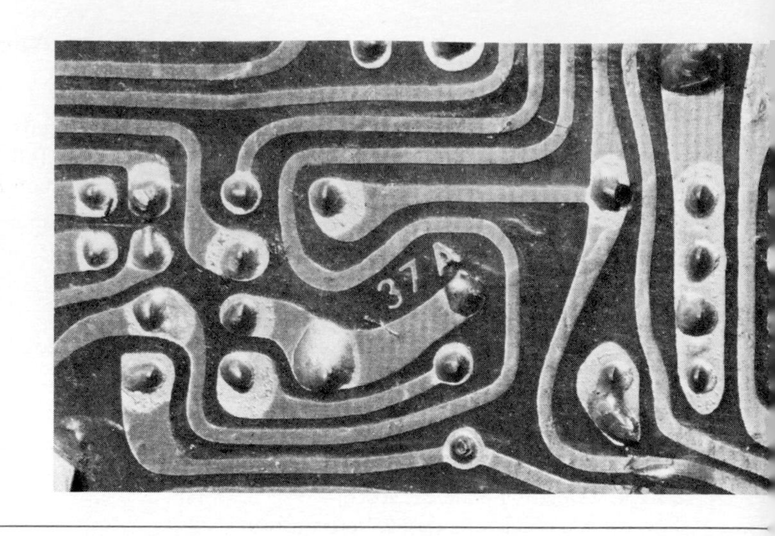

CHAPTER TWENTY-FIVE
POWER IN AC CIRCUITS

There are three types of power in an ac circuit. One is *true power,* measured in watts, which is the power dissipated in a resistance when electrical energy is converted to a different form. Capacitors and inductors, on the other hand, alternately store energy and return it to the source. Such transferred power is called *reactive power,* measured in volt-amperes reactive or vars. Unlike true power, reactive power can do *no* useful work. The last type of power is called *apparent power.* It is simply the product of the total applied voltage and the total circuit current, and has units of volt-amperes, VA.

The three powers are related and can be represented in a right-angled power triangle, from which the *power factor* can be obtained as the ratio of true power to apparent power. For inductive loads, the power factor is called *lagging* to distinguish it from the *leading* power factor in a capacitive load.

A circuit's power factor determines how much current is necessary from the source to deliver a given true power. Since a *low* power factor requires a higher current than a unity power factor circuit, methods of correcting the power factor are examined. This is usually accomplished by connecting a large capacitance across the line to offset the naturally inductive loads of motors and fluorescent lamps.

Finally, we show how to measure and calculate power in a three-phase, three-wire system, and how to maximize power transfer to a load from an ac source with a given internal impedance.

25-1 POWER IN PURE RESISTANCE

We have seen (Section 15-3) that current and voltage are in phase with each other in a purely resistive circuit, or in any *portion* of an ac circuit where only pure resistance is considered. This in-phase relationship can be represented by the phasor diagram in Fig. 25-1*b* or the waveforms in Fig. 25-1*c*. The instantaneous power dissipated in the resistance is given by the product of v and i as shown in Fig. 25-1*c*.

Note that the power varies at twice the source frequency, as was shown in Section 15-3.1, and has an average value given by

$$P = I_R^2 R = V_R^2/R = V_R I_R \quad \text{watts} \quad (15\text{-}12 \text{ to } 15\text{-}14)$$

To distinguish power *dissipated* in a resistance in the form of heat from other types of power, we use the term *real power* or *true power*. The symbol P (with no subscript), and the unit of watt (W) are used exclusively for real or true power. True power describes the conversion of electrical energy into another form such as heat, light, rotating mechanical power, and so forth. The true power in watts used by a dc or ac circuit is indicated by a properly connected wattmeter. Only if the circuit contains some power dissipating element will the wattmeter indicate any power.

EXAMPLE 25-1

A wattmeter, connected in an ac circuit, indicates 50 W. An ammeter, connected in series with the only resistance in the whole circuit, reads 1.5 A. Determine the resistance of the ac circuit.

SOLUTION

$$P = I_R^2 R \quad (15\text{-}12)$$

$$R = \frac{P}{I_R^2} = \frac{50\text{ W}}{(1.5\text{ A})^2} = \mathbf{22.2\ \Omega}$$

25-2 POWER IN PURE INDUCTANCE

If an ac circuit contains only inductance, the voltage and current are 90° out of phase, as shown by the phasor diagram in Fig. 25-2*b*. The result of multiplying the v and i waveforms is a power curve that again has a frequency twice that of the source, as shown in Fig. 25-2*c*. However, over a complete cycle of input voltage, the power curve has an average value of zero. That is, the power curve shows an equal alternation of positive and negative power above and below the zero time axis.

What is the meaning of *positive* and *negative power*? Refer to Fig. 25-2*c*. The shaded portion of the power curve *above* the zero axis represents energy being delivered *to* the inductor (or load) *from* the source. This positive power actually represents a storage of energy in the magnetic field of the inductance. The shaded portion of the power curve *below* the zero axis represents energy returned *to* the source *from* the inductor. This negative power indicates that a flow of energy is taking place in the

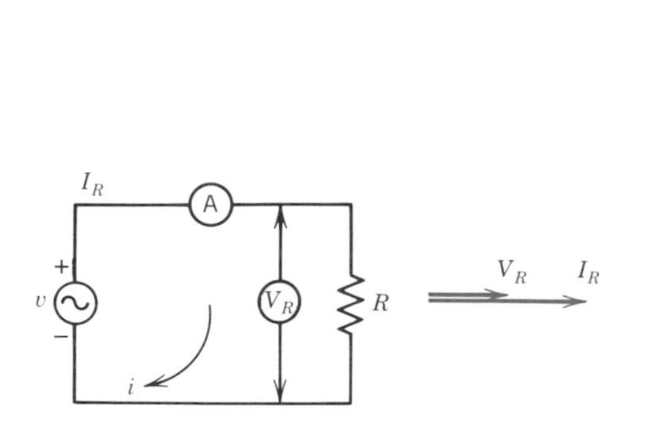

(*a*) Purely resistive circuit (*b*) Phasor diagram

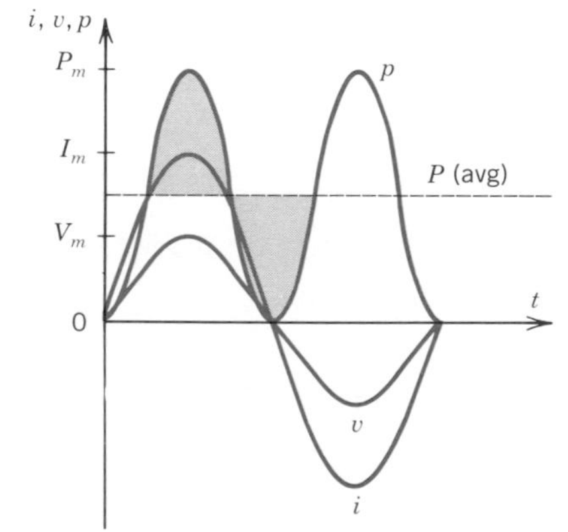

(*c*) Current, voltage, and power waveforms

FIGURE 25-1
Power in a pure resistance.

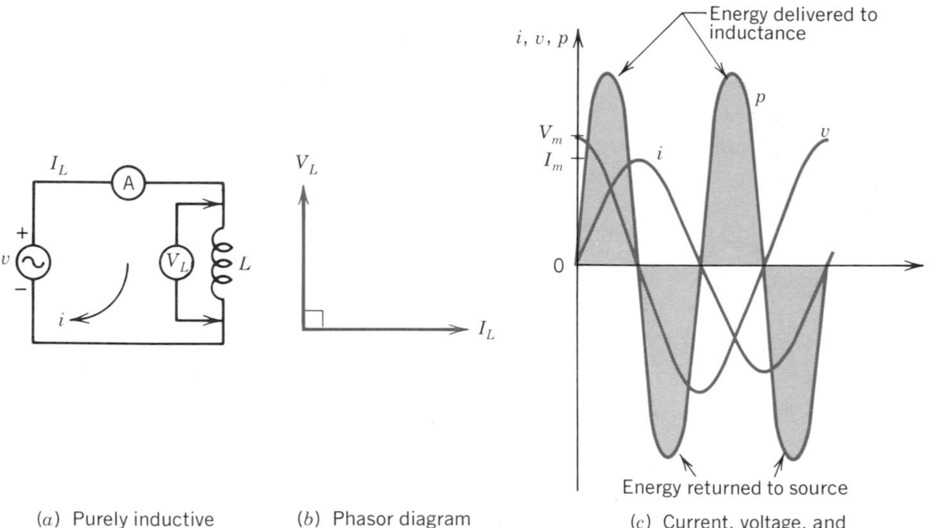

(a) Purely inductive circuit (b) Phasor diagram (c) Current, voltage, and power waveforms

FIGURE 25-2
Power in a pure inductance.

opposite direction (from load to source) when the coil's magnetic field collapses.

In a pure inductor, with zero resistance, the power *returned* during one-quarter cycle is the same as was *delivered* during the previous quarter-cycle. **It is clear then that *the average true power, P, is zero, in a pure inductance.*** This simply means that if a wattmeter is connected in the circuit of Fig. 25-2*a*, it reads zero. Further, the load coil in Fig. 25-2*a* would never become warm because no power is dissipated in the coil (i.e., converted into another form or "used up").

However, the source must be capable of delivering power for a quarter of a cycle, even though this power will be returned during the next quarter-cycle. This stored or transferred power is called reactive power, P_q.

In the case of a purely inductive circuit, the reactive power is given by

$$P_q = V_L I_L \qquad \text{volt-amperes reactive (vars)} \qquad (25\text{-}1)$$

where: P_q is the reactive (or quadrature) power* in volt-amperes reactive, vars

V_L is the voltage across the inductance in volts, V

I_L is the current through the inductance in amperes, A

* The subscript q, in P_q, represents quadrature power. Some texts use the letter Q instead of P_q to represent reactive power. However, Q is also used for the quality of a coil. We shall avoid confusion by using P_q for reactive power.

Since
$$V_L = I_L X_L \qquad (20\text{-}1)$$

then,
$$P_q = I_L^2 X_L \quad \text{vars} \qquad (25\text{-}2)$$

and,
$$P_q = \frac{V_L^2}{X_L} \quad \text{vars} \qquad (25\text{-}3)$$

where X_L is the inductive reactance in ohms.

Note how the equations for reactive power are similar to those for true power with X_L used in place of R. But we must remember to use vars for the unit of reactive power, not watts.

EXAMPLE 25-2

Calculate the reactive power of a circuit that has an inductance of 4 H when it draws 1.4 A from a 60-Hz supply.

SOLUTION

$$X_L = 2\pi f L \qquad (20\text{-}2)$$
$$= 2\pi \times 60\ \text{Hz} \times 4\ \text{H}$$
$$= 1508\ \Omega$$
$$P_q = I_L^2 X_L \qquad (25\text{-}2)$$
$$= (1.4\ \text{A})^2 \times 1508\ \Omega$$
$$= 2956\ \text{vars} = \mathbf{2.96\ kvar}$$

Note that 1 kvar = 1 kilovar = 1000 vars.

The importance of reactive power will be seen in a later section but note the basic difference between true and reactive power: True power, in watts, is the product of the voltage and current that are *in phase* with each other; reactive power, in vars, is the product of the voltage and current that are *90° out of phase* with each other. Also, because the current *lags* the voltage in an inductive circuit, inductive reactive power is often called *lagging* reactive power.

25-3 POWER IN PURE CAPACITANCE

For pure capacitance, the voltage and current are once again 90° out of phase with each other, the current leading as shown by the phasor diagram in Fig. 25-3*b*.

The product of v and i gives a power curve as shown in Fig. 25-3*c*. Again we see that the energy *delivered* to the capacitor and stored in the electric field is represented as a *positive* quantity. A quarter of a cycle later, *all* of this energy is *returned* to the source, as the capacitor discharges. **Thus *the average true power, P, is zero in a pure capacitance.*** This means that no power is dissipated in a pure capacitor and no heat develops.

However, as with a coil, reactive power P_q is drawn by the capacitor, and the source must be able to supply this power.

For a purely capacitive circuit, the reactive power is given by

$$P_q = V_C I_C \qquad \textbf{volt-amperes reactive (vars)} \qquad \textbf{(25-4)}$$

where: P_q is the reactive power in volt-amperes reactive, vars

V_C is the voltage across the capacitance in volts, V

I_C is the current through the capacitance in amperes, A

Since $$V_C = I_C X_C \qquad (23\text{-}1)$$

then $$P_q = I_C^2 X_C \qquad \text{vars} \qquad (25\text{-}5)$$

and $$P_q = \frac{V_C^2}{X_C} \qquad \text{vars} \qquad (25\text{-}6)$$

where X_C is the capacitive reactance in ohms.

Again we see the similarity in the equations for reactive and true power, with X_C used in place of R. We must use vars, not watts, for the reactive power.

EXAMPLE 25-3

A reactive power of 100 vars is drawn by a 10-μF capacitor due to a current of 0.87 A. Calculate the frequency.

SOLUTION

$$P_q = I_C^2 X_C \qquad (25\text{-}5)$$

$$X_C = \frac{P_q}{I_C^2}$$

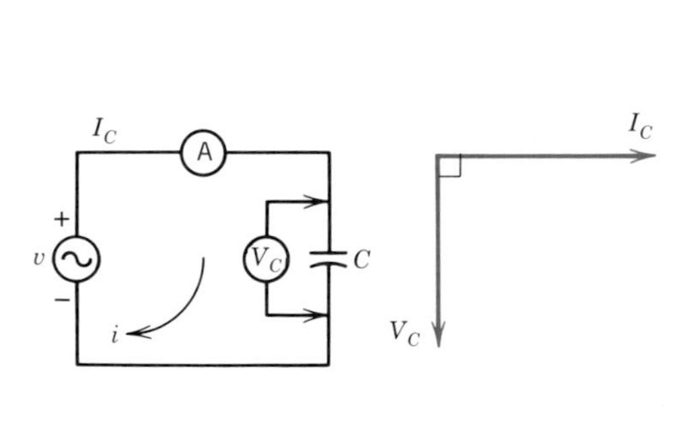

(*a*) Purely capacitive circuit

(*b*) Phasor diagram

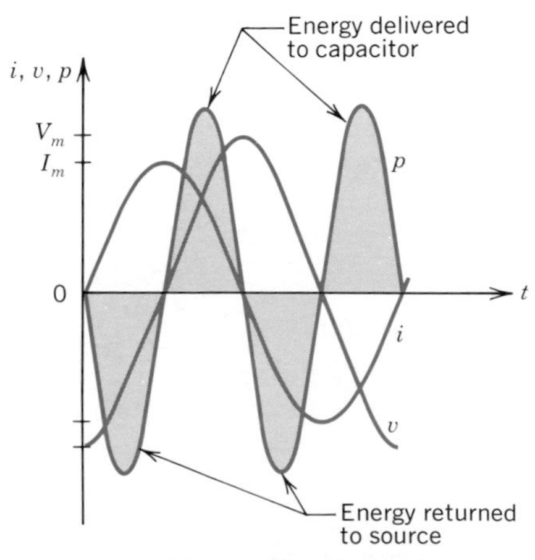

(*c*) Current, voltage and power waveforms

FIGURE 25-3
Power in a pure capacitance.

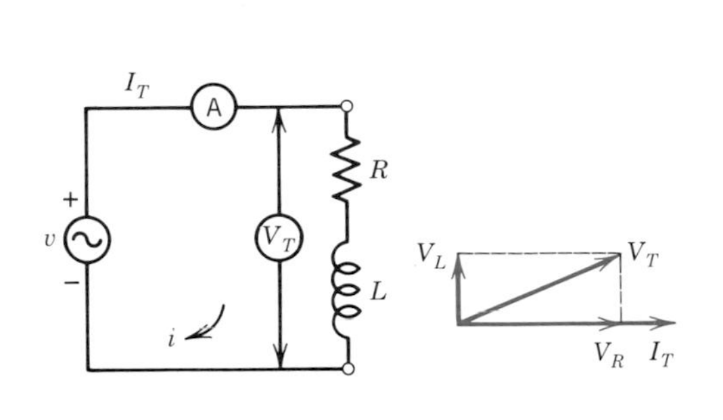

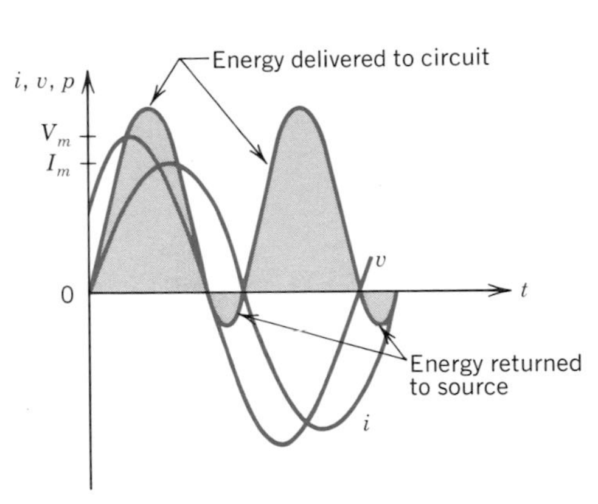

(*a*) Series *RL* circuit (*b*) Phasor diagram (*c*) Current, voltage, and power waveforms

FIGURE 25-4
Power in a series *RL* circuit.

$$= \frac{100 \text{ vars}}{(0.87 \text{ A})^2} = 132\ \Omega$$

$$X_C = \frac{1}{2\pi fC} \tag{23-2}$$

$$f = \frac{1}{2\pi X_C C}$$

$$= \frac{1}{2\pi \times 132\ \Omega \times 10 \times 10^{-6}\text{ F}} = \mathbf{120\ Hz}$$

It was noted that inductive reactive power is referred to as lagging reactive power. Similarly, because the current *leads* the voltage in a capacitive circuit, capacitive reactive power is known as *leading* reactive power. It is interesting to note that reactive power can be measured by using a *varmeter.* This is actually a wattmeter, used in conjunction with a phase-shifting circuit to provide a voltage 90° out of phase from the true load voltage. The product of the meter's current and voltage now gives an indication of the volt-amperes reactive. However, whether the vars are leading or lagging cannot be determined directly from the varmeter. A method of doing this is considered in Section 25-6.

25-4 POWER IN A SERIES *RL* CIRCUIT

We have seen that inductance is always accompanied by resistance. Thus coils in motors, generators, relay coils, and so on, contain both resistance and inductance. When an ac voltage is applied, the current, *I,* is neither *in* phase nor 90° *out* of phase with the applied voltage *V,* as shown in Fig. 25-4*b*.

This means, unlike pure resistance and pure reactance, that the product of the voltmeter and ammeter readings in Fig. 25-4*a* is a combination of the true and (quadrature) reactive power. **We call the product of total *V* and total *I apparent power,* P_s. Since it is neither true power in watts nor reactive power in vars, we use a new unit—the *volt-ampere,* VA—to measure the apparent power.**

$$P_s = V_T I_T \qquad \textbf{volt-amperes (VA)} \tag{25-7}$$

where: P_s is the apparent power* in volt-amperes, VA
V_T is the total applied voltage in volts, V
I_T is the total circuit current in amperes, A

Note the use of the subscript *T* to emphasize that the apparent power is the product of the *total* applied voltage and current. That is, multiplying the voltmeter reading in volts by the ammeter reading in amperes gives the total apparent circuit power in volt-amperes. It will later be shown that P_s is the vector sum of P_q and *P*. (See Section 25-5.)

But $$I_T = \frac{V_T}{Z} \tag{24-4}$$

Thus $$P_s = \frac{V_T^2}{Z} \tag{25-8}$$

* Some texts use the letter *S* to describe apparent power. We shall use the older but more descriptive notation P_s.

and

$$P_s = I_T^2 Z \qquad (25\text{-}9)$$

where: Z is the circuit impedance in ohms.

Once again we recognize a similarity in the equations for apparent power and true power, with Z used in place of R, and volt-amperes used in place of watts.

EXAMPLE 25-4

A current of 2.6 A flows through a series circuit of 300-Ω resistance and 400-Ω inductive reactance. Calculate the apparent power drawn by the circuit.

SOLUTION

$$Z = \sqrt{R^2 + X_L^2} \qquad (24\text{-}3)$$

$$= \sqrt{(300\ \Omega)^2 + (400\ \Omega)^2} = 500\ \Omega$$

$$P_s = I_T^2 Z$$

$$= (2.6\ \text{A})^2 \times 500\ \Omega$$

$$= 3380\ \text{VA} = \mathbf{3.38\ kVA}$$

Note that 1000 VA = 1 kilovolt-ampere = 1 kVA.

25-4.1 Measuring the Inductance of a Coil

Figure 25-4*c* shows a power curve that is more positive than negative. This means that there is some net true power delivered to the circuit and dissipated in the resistance. This true power must be indicated on a wattmeter connected as shown in Fig. 25-5.

If a voltmeter and ammeter are also connected as shown, it is possible from the three meter readings to determine the coil's resistance and inductance, if the frequency is known, as in Example 25-5.

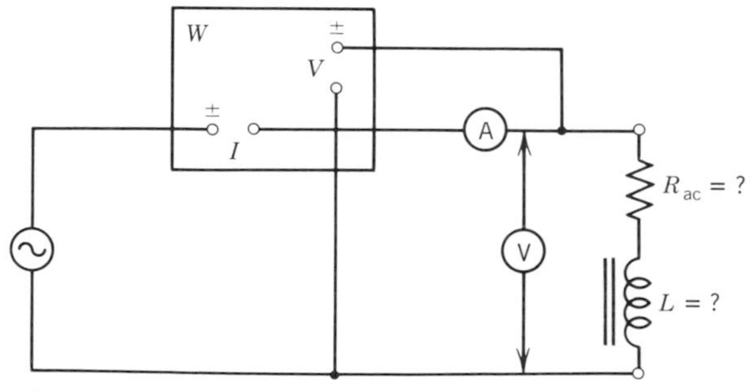

FIGURE 25-5
Measuring the inductance of a coil.

EXAMPLE 25-5

A coil is connected to the 120-V, 60-Hz line as in Fig. 25-5. The wattmeter indicates 525 W and the ammeter indicates 5 A. Calculate the inductance of the coil.

SOLUTION

$$Z = \frac{V}{I} \qquad (24\text{-}4)$$

$$= \frac{120\ \text{V}}{5\ \text{A}} = 24\ \Omega$$

$$R = \frac{P}{I_R^2} \qquad (15\text{-}12)$$

$$= \frac{525\ \text{W}}{(5\ \text{A})^2} = 21\ \Omega$$

$$Z = \sqrt{R^2 + X_L^2} \qquad (24\text{-}3)$$

$$X_L = \sqrt{Z^2 - R^2}$$

$$= \sqrt{(24\ \Omega)^2 - (21\ \Omega)^2} = 11.6\ \Omega$$

$$X_L = 2\pi f L \qquad (20\text{-}2)$$

$$L = \frac{X_L}{2\pi f}$$

$$= \frac{11.6\ \Omega}{2\ \pi \times 60\ \text{Hz}} = \mathbf{30.8\ mH}$$

NOTE The calculated resistance (21 Ω) is the total *ac* resistance of the coil, including hysteresis and eddy current losses if a magnetic core is present.

Note also that this method of measuring the coil's inductance (and resistance) is limited to power line frequencies because of the frequency limitation of most wattmeters.

25-5 THE POWER TRIANGLE

We have now identified three different types of power in an ac circuit: true power in watts, reactive (quadrature) power in vars, and apparent power in volt-amperes. Let us see if there is a simple relation among the three. Refer to Fig. 25-6.

We know (see Fig. 25-6*b*) that

$$V_T = \sqrt{V_R^2 + V_L^2} \qquad (24\text{-}1)$$

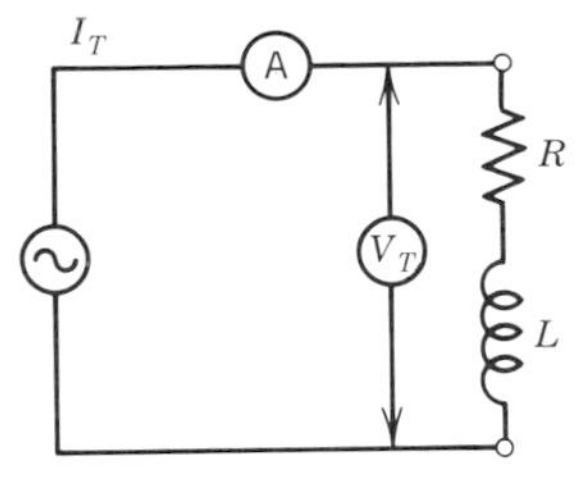

(a) Series RL circuit

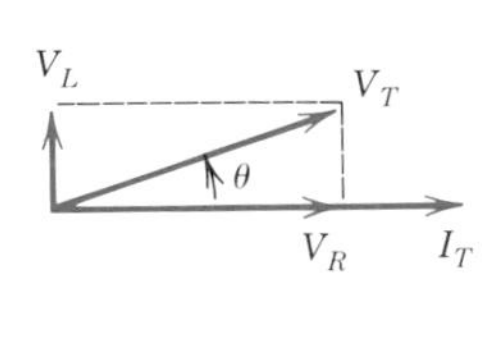

(b) Phasor diagram

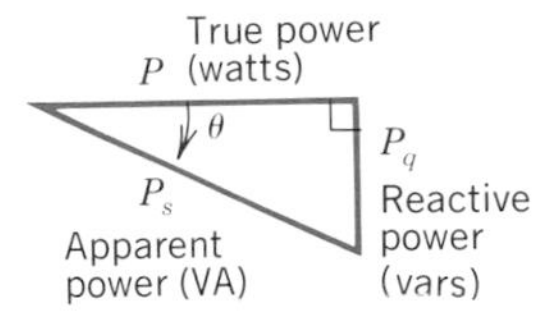

(c) Power triangle

FIGURE 25-6
Power triangle relations for a series *RL* circuit.

Therefore, $V_T I_T = \sqrt{(V_R I_T)^2 + (V_L I_T)^2}$

But

$$V_T I_T = P_s \quad \text{(VA)} \qquad (25\text{-}7)$$

$$V_R I_T = P \quad \text{(watts)} \qquad (15\text{-}14)$$

$$V_L I_T = P_q \quad \text{(vars)} \qquad (25\text{-}1)$$

Therefore,

$$P_s = \sqrt{P^2 + P_q^2} \quad \text{volt-amperes (VA)} \qquad (25\text{-}10)$$

where: P_s is the apparent power in volt-amperes, VA
P is the true power in watts, W
P_q is the reactive power in volt-amperes reactive, vars

This relation can be represented in a *power triangle*, as in Fig. 25-6*c*. The apparent power is represented by the hypotenuse of the right triangle. The true power is the product of the current and voltage in phase with each other and is drawn horizontally. The out-of-phase product of V_L and I_T gives the reactive power and is drawn vertically *downward*. This is a convention used to show a *lagging* inductive reactive power, corresponding to a lagging current. (A capacitive reactive power is drawn vertically *upward*, corresponding to a *leading* current.)

Note that no arrows are shown on the power triangle because the power varies at *twice* the frequency of the voltage and current in the circuit. The triangle is used to show the Pythagorean relationship of Eq. 25-10.

EXAMPLE 25-6

A series circuit of 300-Ω resistance and 400-Ω inductive reactance draws a current of 2.6 A. Calculate:

a. The true power.
b. The inductive reactive power.
c. The apparent power.

SOLUTION

a. $P = I_R^2 R$ (15-12)
$= (2.6\text{ A})^2 \times 300\ \Omega =$ **2028 W**

b. $P_q = I_L^2 X_L$ (25-2)
$= (2.6\text{ A})^2 \times 400\ \Omega =$ **2074 vars**

c. $P_s = \sqrt{P^2 + P_q^2}$ (25-10)
$= \sqrt{(2028\text{ W})^2 + (2074\text{ vars})^2}$
$= 3380\text{ VA} =$ **3.38 kVA**

EXAMPLE 25-7

A coil connected across the 120-V, 60-Hz line draws a current of 5 A. A series-connected wattmeter indicates 525 W. Calcultate:

a. The apparent power.
b. The reactive power.

SOLUTION

a. $P_s = V_T I_T$ (25-7)
$= 120\text{ V} \times 5\text{ A} =$ **600 VA**

b. $P_s = \sqrt{P^2 + P_q^2}$ (25-10)
$P_q = \sqrt{P_s^2 - P^2}$
$= \sqrt{(600\text{ VA})^2 - (525\text{ W})^2} =$ **290 vars**

25-5.1 Power Triangle for a Parallel *RL* Circuit

When a pure resistance and a pure inductance are parallel-connected across an ac supply, as in Fig. 25-7*a*, the total current is (see Fig. 25-7*b*):

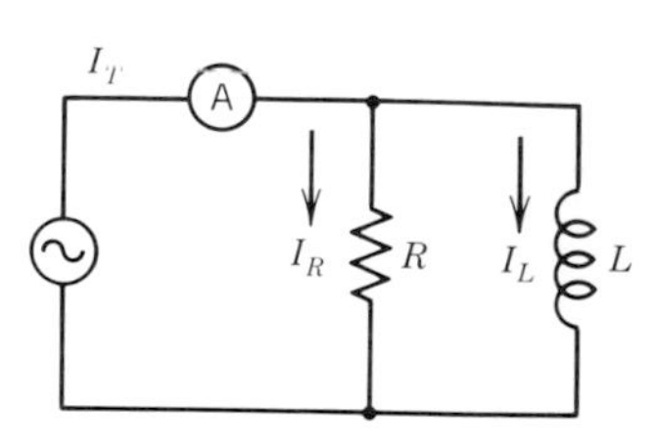

(a) Parallel RL circuit

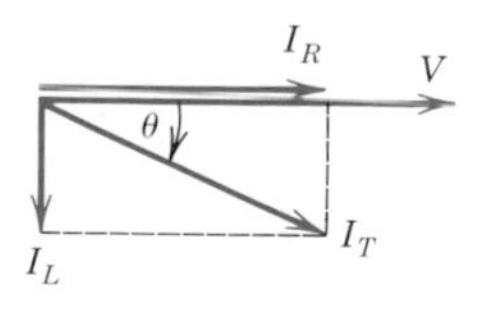

(b) Phasor diagram

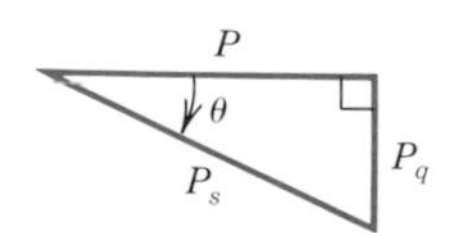

(c) Power triangle

FIGURE 25-7
Power triangle relations for a parallel RL circuit.

$$I_T = \sqrt{I_R^2 + I_L^2} \tag{24-17}$$

Therefore, $V_T I_T = \sqrt{(V_T I_R)^2 + (V_T I_L)^2}$

and
$$P_s = \sqrt{P^2 + P_q^2} \tag{25-10}$$

where all terms are as already defined.

Again we can draw a power triangle, as in Fig. 25-7*c*. Evidently, we can use Eq. 25-10 for either a series or parallel RL combination.

EXAMPLE 25-8

A 300-Ω resistor and an inductive reactance of 400 Ω are parallel-connected across a 120-V, 60-Hz supply. Calculate:

a. The true power.

b. The reactive power.

c. The apparent power.

SOLUTION

a. $P = \dfrac{V_R^2}{R}$ (15-13)

$= \dfrac{(120\text{ V})^2}{300\ \Omega} = \mathbf{48\ W}$

b. $P_q = \dfrac{V_L^2}{X_L}$ (25-3)

$= \dfrac{(120\text{ V})^2}{400\ \Omega} = \mathbf{36\ vars}$

c. $P_s = \sqrt{P^2 + P_q^2}$

$= \sqrt{(48\text{ W})^2 + (36\text{ vars})^2} = \mathbf{60\ VA}$

25-5.2 Power Triangle for a Series or Parallel RC Circuit

Using the relationship

$$V_T = \sqrt{V_R^2 + V_C^2} \tag{24-8}$$

for a series RC circuit as in Fig. 25-8*a* and

$$I_T = \sqrt{I_R^2 + I_C^2} \tag{24-19}$$

for a parallel RC circuit as in Fig. 25-8*b*, we can show that

$$P_s = \sqrt{P^2 + P_q^2} \tag{25-10}$$

where all terms are previously listed.

Note that the power triangle in Fig. 25-8*c* has the leading capacitive reactive power drawn vertically *upward*. This corresponds to a current that *leads* the applied voltage, for both series and parallel RC circuits.

In summary, we can say that the Pythagorean power equation, Eq. 25-10, can be used for all circuit combinations, series or parallel, RL or RC.

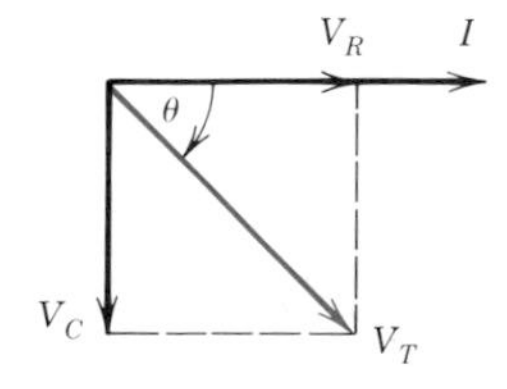

(a) Phasor diagram for a series RC circuit

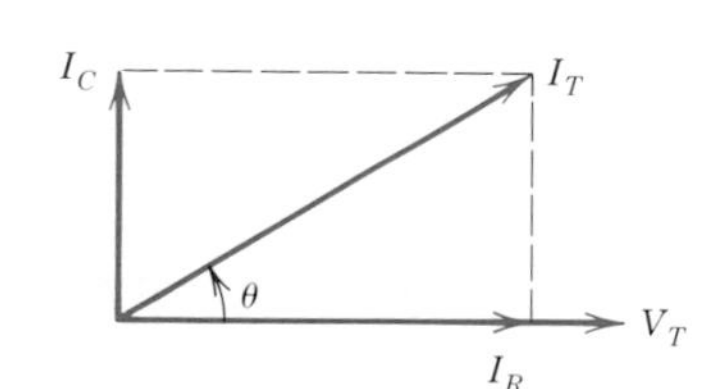

(b) Phasor diagram for a parallel RC circuit

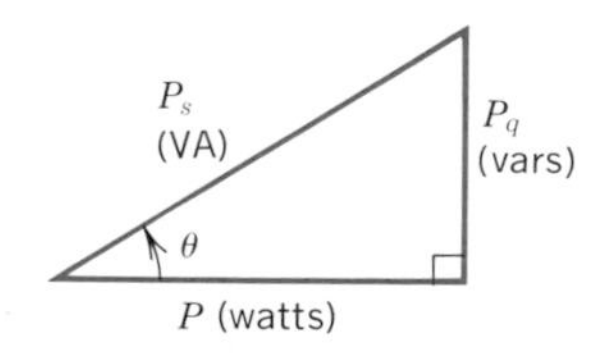

(c) Power triangle for either a series or parallel RC circuit

FIGURE 25-8
Phasor diagrams and power triangle for series and parallel RC circuits.

25-5.3 Power Triangle for a Series or Parallel *RLC* Circuit

When a circuit contains both inductance and capacitance, there is a transfer of reactive power back and forth between them. This can be seen by examining Fig. 25-2*c* and Fig. 25-3*c*. During the first quarter of a cycle, energy is being *delivered* to the inductance (positive power) at the very same time that the capacitor is *returning* energy to the source (negative power). If both the capacitor and the inductor are in the same circuit, in series or parallel, this exchange of reactive power results in an effective reduction in reactive power that must be supplied by the source. See the power triangle in Fig. 25-9.

The power triangle is now represented by the following equation:

$$P_s = \sqrt{P^2 + (P_{qC} - P_{qL})^2} \quad \text{volt-amperes, (VA)} \qquad (25\text{-}11)$$

where: P_s is the apparent power in volt-amperes, VA
P is the true power in watts, W
P_{qC} is the capacitive reactive power in volt-amperes reactive, vars
P_{qL} is the inductive reactive power in volt-amperes reactive, vars

Figure 25-9 has been drawn arbitrarily with $P_{qC} > P_{qL}$. This results in a net reactive power that is capacitive. If $P_{qC} < P_{qL}$, the circuit draws a net inductive reactive power. Also, if $P_{qC} = P_{qL}$, *no* reactive power is supplied by the source. This is an important condition and is used in power factor correction in Section 25-8.

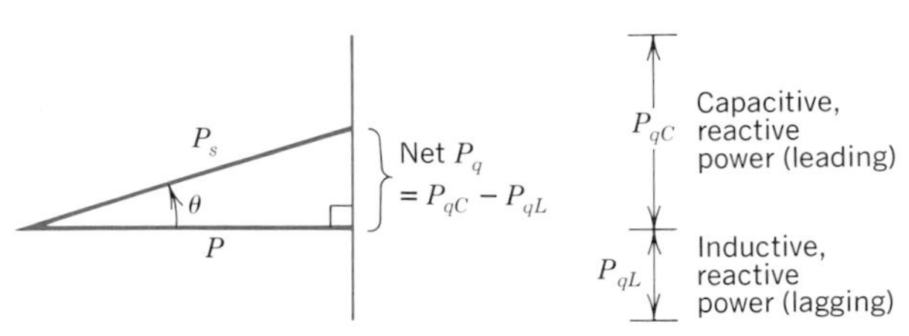

FIGURE 25-9
Power triangle for a series or parallel *RLC* circuit.

EXAMPLE 25-9

A coil and a 12-μF capacitor are connected in parallel across a 240-V, 60-Hz line. The coil has an inductance of 0.25 H and a resistance of 75 Ω. Calculate:

a. The true power.
b. The net reactive power.
c. The apparent power.
d. The total line current.

Draw the circuit diagram and the power triangle.

SOLUTION

a. The coil's inductive reactance

$$X_L = 2\pi f L \qquad (20\text{-}2)$$
$$= 2\pi \times 60 \text{ Hz} \times 0.25 \text{ H}$$
$$= 94 \ \Omega$$

The coil's impedance

$$Z = \sqrt{R^2 + X_L^2} \qquad (24\text{-}3)$$
$$= \sqrt{(75 \ \Omega)^2 + (94 \ \Omega)^2} = 120 \ \Omega$$

The current through the coil

$$I = \frac{V}{Z} \qquad (24\text{-}4)$$
$$= \frac{240 \text{ V}}{120 \ \Omega} = 2 \text{ A}$$

The true power

$$P = I_R^2 R \qquad (15\text{-}12)$$
$$= (2 \text{ A})^2 \times 75 \ \Omega$$
$$= \mathbf{300 \ W}$$

b. The capacitor's reactance

$$X_C = \frac{1}{2\pi f C} \qquad (23\text{-}2)$$
$$= \frac{1}{2\pi \times 60 \text{ Hz} \times 12 \times 10^{-6} \text{ F}}$$
$$= 221 \ \Omega$$

Capacitive reactive power

$$P_{qC} = \frac{V_C^2}{X_C} \qquad (25\text{-}6)$$
$$= \frac{(240 \text{ V})^2}{221 \ \Omega} = 261 \text{ vars}$$

Inductive reactive power

$$P_{qL} = I_L^2 X_L \qquad (25\text{-}2)$$
$$= (2 \text{ A})^2 \times 94 \ \Omega$$
$$= 376 \text{ vars}$$

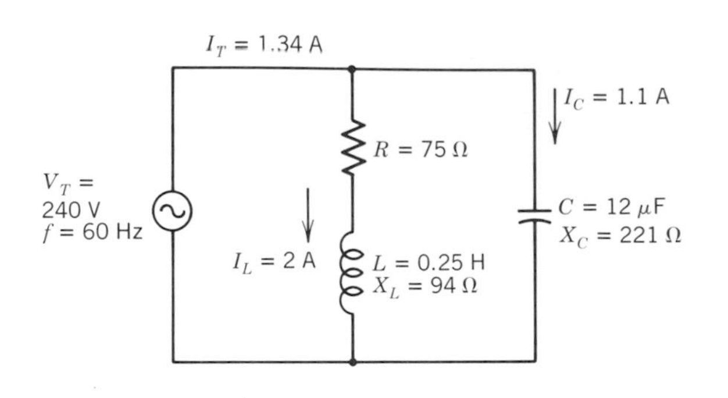

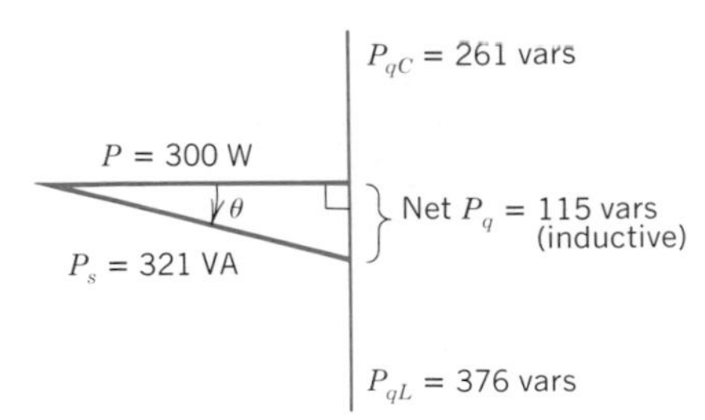

FIGURE 25-10
Circuit and power triangle for Example 25-9.

$$\text{Net reactive power} = P_{qL} - P_{qC}$$
$$= 376 - 261 \text{ vars}$$
$$= \mathbf{115\ vars} \text{ (inductive)}$$

c. Apparent power

$$P_s = \sqrt{P^2 + (P_{qC} - P_{qL})^2} \tag{25-11}$$
$$= \sqrt{(300\text{ W})^2 + (261\text{ vars} - 376\text{ vars})^2}$$
$$= \mathbf{321\ VA}$$

d. $P_s = V_T I_T$ (25-7)

$$\text{Therefore,}\quad I_T = \frac{P_s}{V_T}$$
$$= \frac{321\text{ VA}}{240\text{ V}} = \mathbf{1.34\ A}$$

The circuit and power triangle are shown in Fig. 25-10.

NOTE Example 25-9 provides a method of finding the total line current *without* a phasor addition of the coil and capacitor currents. This phasor addition is not simple because the coil current is neither 90 nor 180° out of phase with the capacitor current.

25-6 POWER FACTOR

The ratio of the true power delivered to an ac circuit compared to the apparent power that the source must supply is called the *power factor* of the load.

If we examine *any* power triangle, as in Fig. 25-11, we see that the ratio of the true power to the apparent power is the *cosine* of the angle θ:

$$\text{power factor} = \frac{P}{P_s} = \cos\theta \tag{25-12}$$

where: $\cos\theta$ is the symbol for power factor (dimensionless)
P is the true power dissipated in the load in watts, W
P_s is the apparent power drawn by the load in volt-amperes, VA

EXAMPLE 25-10

A wattmeter connected to an ac circuit indicates 1 kW. If a voltmeter across the supply indicates 250 V when a series-connected ammeter reads 5 A, determine the power factor of the circuit.

SOLUTION

$$\text{Apparent power } P_s = V_T I_T \tag{25-7}$$
$$= 250\text{ V} \times 5\text{ A} = 1250\text{ VA}$$

$$\text{Power factor} = \frac{P}{P_s} \tag{25-12}$$
$$= \frac{1000\text{ W}}{1250\text{ VA}} = \mathbf{0.8}$$

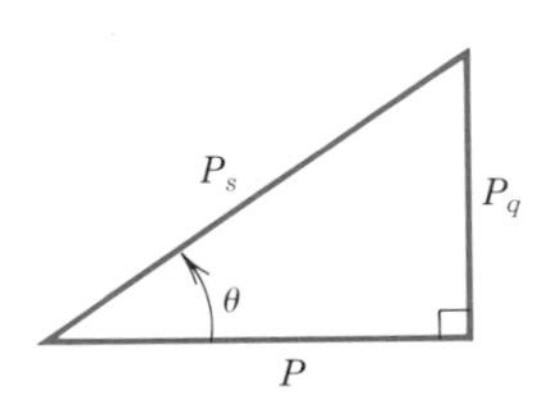

(*a*) Power triangle for a series or parallel *RC* circuit

(*b*) Power triangle for a series or parallel *RL* circuit

FIGURE 25-11
Power triangles showing that $P/P_s = \cos\theta$.

If we rearrange Eq. 25-12, we obtain

$$P = P_s \times \cos\theta$$

or $$P = V_T I_T \cos\theta \qquad \text{watts (W)} \qquad (25\text{-}12a)$$

Equation 25-12a is very important. It shows that the power factor ($\cos\theta$) is simply the factor by which the apparent power ($V_T I_T$) must be multiplied to obtain the true power in watts.

This statement can be verified by considering the phasor diagrams in Fig. 25-12.

In Fig. 25-12*a*, the phasor diagram for a series *RL* circuit shows that

$$\frac{V_R}{V_T} = \cos\theta$$

Therefore, $$V_R = V_T \cos\theta$$

But $$P = V_T I_T \cos\theta \qquad (25\text{-}12a)$$

Rearranging, we obtain $P = I_T(V_T \cos\theta)$

and therefore, $$P = I_T V_R \qquad (15\text{-}14)$$

This equation gives the true power because it is the product of the current and the *component* of the applied voltage that is *in-phase* with the current.

Similarly, for a parallel circuit, as in Fig. 25-12*b*,

$$\frac{I_R}{I_T} = \cos\theta$$

Therefore: $$I_R = I_T \cos\theta$$

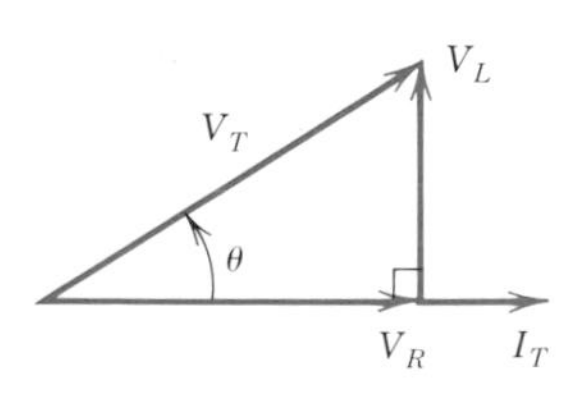

(*a*) Phasor diagram for a series *RL* circuit

(*b*) Phasor diagram for a parallel *RL* circuit

FIGURE 25-12
Phasor diagrams for series and parallel *RL* circuits.

But $$P = V_T(I_T \cos\theta) \qquad (25\text{-}12a)$$

and therefore, $$P = V_T I_R$$

This equation gives true power because it is the product of applied voltage and the *component* of the total current that is *in-phase* with the voltage.

To summarize, the equation $P = V_T I_T \cos\theta$ is an equation that gives the true power in *any* single-phase ac circuit. It gives the product of the *voltage and current components that are in phase,* in watts. This is what a wattmeter is designed to measure: the component of voltage that is in phase with current, or vice versa.

In a similar manner, we can show that the equation

$$P_q = V_T I_T \sin\theta \qquad \text{(vars)} \qquad (25\text{-}13)$$

gives the product of the 90° *out-of-phase* components of voltage and current, resulting in a circuit's *reactive* power. $\sin\theta = P_q/P_s$ is called the *reactive factor.*

It should be clear from the preceding discussion that the power factor angle, θ, between P and P_s, is the *same* as the phase angle, θ, between the current and voltage. Since θ is determined by the circuit elements, the power factor is determined by the nature of the load.

Thus, if the circuit is purely resistive, $\theta = 0°$, the power factor $\cos\theta$ is 1 and Eq. 25-12a reduces to $P = V_T I_T$. The apparent power is equal to the true power. Also, $\sin\theta = 0$ and Eq. 25-13 gives $P_q = 0$; there is no reactive power.

If the circuit is purely reactive, $\theta = 90°$, $\cos\theta$ is zero, and Eq. 25-12a gives $P = 0$. Now $\sin\theta = 1$, and Eq. 25-13 gives $P_q = V_T I_T$; all the apparent power is reactive power.

EXAMPLE 25-11

A 240-V, 1-hp single-phase ac motor is running at full load and drawing a current of 6 A. An oscilloscope measurement determines a 41° phase angle between the current and applied voltage. Calculate:

a. The power factor of the motor.
b. The apparent power drawn by the motor.
c. The true power delivered to the motor.
d. The reactive power drawn by the motor.
e. The efficiency of the motor.

SOLUTION

a. Power factor $= \cos\theta$ (25-12)

$= \cos 41° = \mathbf{0.75}$

b. Apparent power

$P_s = V_T I_T$ (25-7)

$= 240\text{ V} \times 6\text{ A} = \mathbf{1440\ VA}$

c. True power $P = V_T I_T \cos\theta$ (25-12a)

$= 1440\text{ VA} \times 0.75 = \mathbf{1080\ W}$

d. Reactive power

$P_q = V_T I_T \sin\theta$ (25-13)

$= 1440\text{ VA} \times \sin 41° = \mathbf{945\ vars}$

e. Efficiency $= \dfrac{P_{\text{out}}}{P_{\text{in}}}$

$$= \frac{1\text{ hp} \times 746\text{ W/hp}}{1080\text{ W}} \times 100\%$$

$= \mathbf{69\%}$

Note that the determination of efficiency compares useful output power to *true* input power in watts.

Although the power factor is a *positive* dimensionless quantity, it is useful to distinguish between a capacitive load and an inductive load. Because inductive loads have current *lagging* applied voltage, we say that an inductive circuit has a *lagging power factor*. Similarly, a capacitive load has a *leading power factor*. In each case, the power factor can be anywhere from 0 to 1, depending on the phase angle between the current and voltage.

Power factor *meters* are available (see Fig. 25-13) that indicate what the power factor is and whether it is leading or lagging. Most industrial installations have a lagging power factor because of the large number of ac induction motors that are inherently inductive.

EXAMPLE 25-12

Using the results of Example 25-9, find:

a. The power factor of the parallel combination of the coil and capacitor.
b. The phase angle between the applied voltage and current.

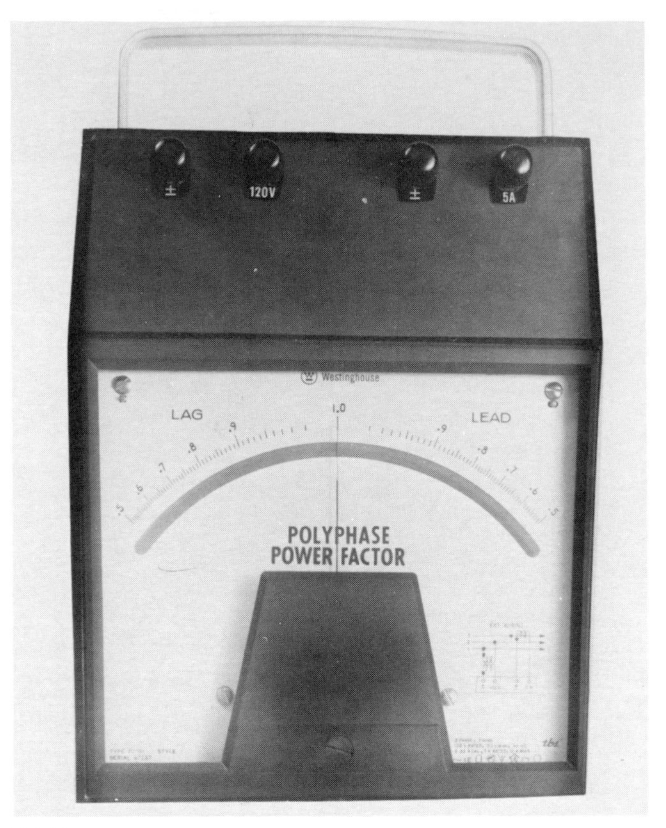

FIGURE 25-13
One type of power factor meter. This meter incorporates a solid-state power factor transducer and a direct-current measuring mechanism. The instrument is suitable for three-phase, three-wire operation using a potential and a current transformer as shown in the external wiring diagram. (Courtesy of Westinghouse Electric Corp.)

SOLUTION

a. True power $P = 300$ W

Net reactive power $P_q = 115$ vars (inductive)

Apparent power $P_s = 321$ VA

Power factor $= \dfrac{P}{P_s}$ (25-12)

$= \dfrac{300\text{ W}}{321\text{ VA}}$

$= \mathbf{0.93\ lagging}$

b. $\cos\theta = 0.93$

$\theta = \arccos 0.93$

$= \mathbf{21.6°}$

Note how an almost unity power factor implies a voltage and current almost in phase with each other.

25-6.1 Detrimental Effect of a Low Power Factor

To show the important effect of the power factor, let us consider a 120-V, 60-Hz, 1-hp motor. Let us assume that it is 100% efficient so that it draws a true power of 746 W. Such a motor has a typical power factor of 0.75, lagging.

To deliver 746 W from 120 V at a power factor of 0.75 requires a current of

$$I = \frac{P}{V \times \cos\theta} \qquad (25\text{-}12a)$$

$$= \frac{746 \text{ W}}{120 \text{ V} \times 0.75} = \mathbf{8.3\ A}$$

Now let us assume that we can modify the motor in some way to make the power factory unity (1). The current now required is

$$I = \frac{P}{V \times \cos\theta} \qquad (25\text{-}12a)$$

$$= \frac{746 \text{ W}}{120 \text{ V} \times 1} = \mathbf{6.2A}$$

Evidently, it requires a higher current to deliver a given quantity of true power if the power factor of the load is less than unity. This higher current means that more energy is wasted in the feeder wires serving the motor. In fact, if an industrial installation has a power factor less than 85% (0.85) over all, a "power factor penalty" is assessed by the electric utility company. It is for this reason that *power factor correction* is necessary in large installations.

25-7 POWER FACTOR CORRECTION

The reason that load current is higher when the power factor is less than unity is that some of the current is required to supply the reactive power. Figure 25-14 shows the power triangle and phasor diagrams before power factor correction, for the 1-hp motor example, in Section 25-6.1.

In Fig. 25-14*b*, which has been drawn with the voltage phasor as reference, the 8.3-A motor current has been resolved into two components. Component $I_x = 6.2$ A is in phase with the applied voltage and is responsible for delivering the *useful* true power, P, equal to 746 W. Component $I_y = 5.4$ A supplies the inductive reactive power, P_q, equal to 653 vars, which serves *no* useful purpose.

Now consider adding a capacitor in parallel with the motor as in Fig. 25-15*a*.* Further, let us choose a capacitor whose reactance will draw a leading current of 5.4 A, as in Fig. 25-15*b*.

* Although it is possible to achieve unity power factor by connecting a capacitor in *series* with the load, power factor correction is *not* carried out this way for two reasons:

1 A reduction of the overall impedance (due to the canceling of the reactances) leads to an *increase* in line current, not a decrease.

2 The increased line current increases the voltage across the inductive load to a level *higher* than the normal operating value.

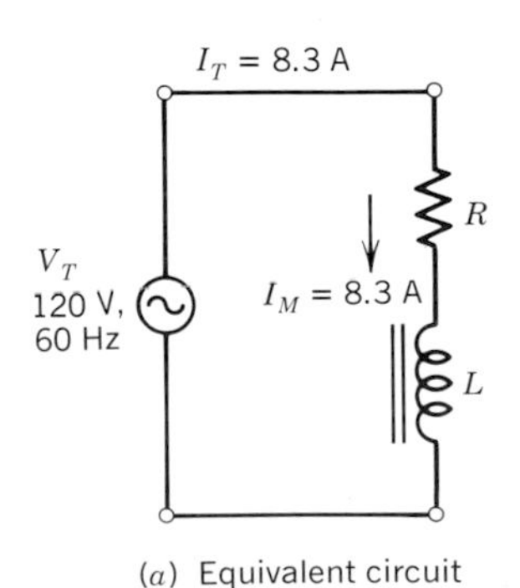

(*a*) Equivalent circuit of motor

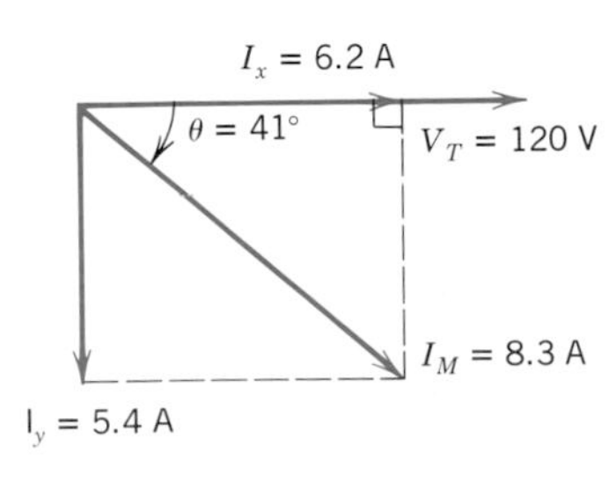

(*b*) Phasor diagram with voltage used as reference

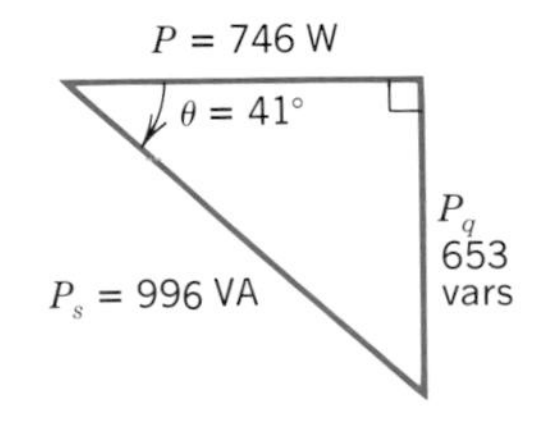

(*c*) Power triangle

FIGURE 25-14
Phasor diagram and power triangle for a 1-hp motor before power factor correction.

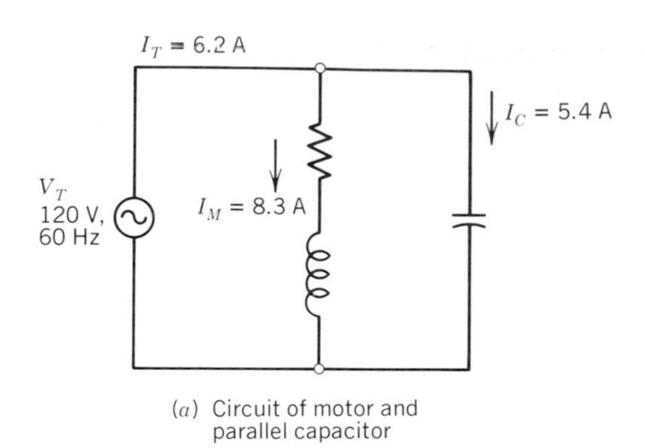

(a) Circuit of motor and parallel capacitor

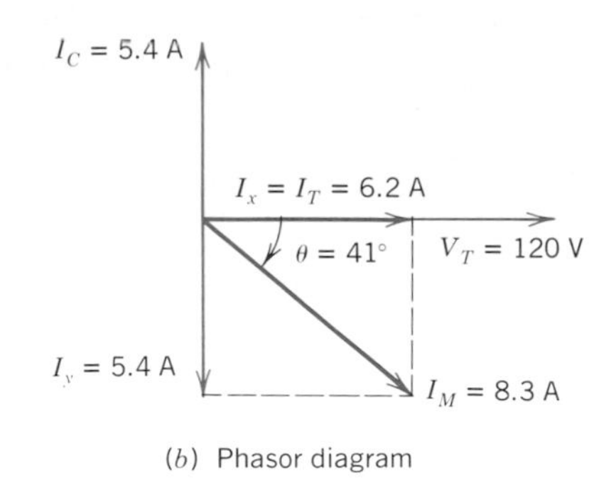

(b) Phasor diagram

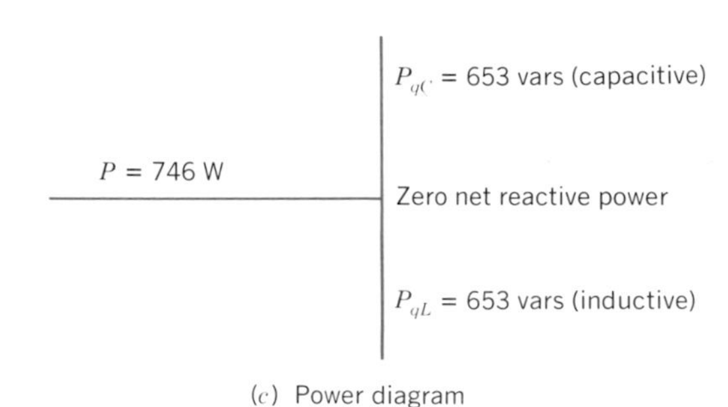

(c) Power diagram

FIGURE 25-15
Phasor and power diagrams for 1-hp motor after power factor correction.

The total line current (supplied by the source) is now given by $I_T = I_x = 6.2$ A, since $I_C + I_y = 0$ (equal and opposite currents).

$$X_C = \frac{V_C}{I_C} \qquad (23\text{-}1)$$

$$= \frac{120 \text{ V}}{5.4 \text{ A}} = 22\ \Omega$$

But
$$X_C = \frac{1}{2\pi f C} \qquad (23\text{-}2)$$

Therefore,
$$C = \frac{1}{2\pi f X_C}$$

$$= \frac{1}{2\pi \times 60 \text{ Hz} \times 22\ \Omega} = \mathbf{120\ \mu F}$$

The function of the capacitor is to provide the inductive portion of the motor with the reactive power it needs instead of drawing this power from the source. In other words, the reactive power is transferred back and forth between the capacitor and inductance. In fact, for unity power factor ($\cos\theta = 1$), we can choose the capacitor to draw the *same* reactive power as the inductance:

Inductive reactive power

$$P_{qL} = V_T I_T \sin\theta \qquad (25\text{-}13)$$

$$= 120 \text{ V} \times 8.3 \text{ A} \times \sin 41° = 653 \text{ vars}$$

Capacitive reactive power
$$P_{qC} = \frac{V_C{}^2}{X_C} \qquad (25\text{-}6)$$

Therefore,
$$X_C = \frac{V_C{}^2}{P_{qC}}$$

For unity power factor correction,

$$P_{qC} = P_{qL} = 653 \text{ vars}$$

Therefore,
$$X_C = \frac{(120 \text{ V})^2}{653 \text{ vars}} = 22\ \Omega$$

and
$$C = \mathbf{120\ \mu F}$$

It should be noted that if a wattmeter is connected to measure the power delivered to the motor, before power factor correction, it would indicate 746 W. This is because the wattmeter's operation takes into account the phase angle of 41° between the voltage and current. After power factor correction, the wattmeter still reads 746 W.* A series-connected ammeter, however, will decrease from 8.3 to 6.2 A following power factor correction.

This illustrates an important precaution to observe when using a wattmeter in highly reactive circuits. If the power factor is very close to zero, the wattmeter deflection will be very small. And yet, a *large* current may be passing through the wattmeter. It is therefore standard practice to use a series-connected ammeter to ensure that the wattmeter's current rating is *not* unknowingly exceeded.

Similarly, a transformer's rating is given in kilovolt-amperes (or kVA) because if an inductive (or capacitive) load is connected, an appreciable current can flow even though very little *true* power is delivered. Thus a 1-kVA transformer rated at 250 V can deliver a maximum of 4 A. If the load power factor is 0.7, the true power is only 700 W, but the transformer would be working at its limit of 1000 VA. The transformer must not be asked to deliver 1000 W unless the load is purely resistive (power factor of unity).

Although it is possible to correct the power factor of each inductive load (see Fig. 25-16), large industries often use banks of parallel capacitors connected across the line where the service enters the building. Also, it is not usually economical to correct to unity power factor. Therefore, just sufficient capacitance is added to improve the power factor so that it is higher than the penalty level. It is also possible to improve the power factor using large *synchronous* motors that can be adjusted to run at a leading current to offset the inductive load. These motors can then be used to provide some useful work as well as help to correct the power factor.

* It should be noted that the power factor of a motor varies considerably with the variation in mechanical load connected to the motor. When the motor is idling, or supplying a very light load, the power factor is very low. Only if the motor is running at rated load does the power factor rise to approximately 0.7 or 0.8. Capacitors for power factor correction are usually chosen on the basis of full load. However, there are now available *power factor controllers* that monitor the load current and automatically reduce the applied voltage when the load is light, to help improve the power factor and reduce losses.

FIGURE 25-16
The rectangular enclosure in the foreground contains power factor correction capacitors for the 3-phase 480-V motor in the background. The unit contains six capacitor cells to provide a terminal-to-terminal capacitance of 144 μF rated at 25 kVAR. (Courtesy of Aerovox® Inc.)

EXAMPLE 25-13

A 15-W desk-type fluorescent lamp has an effective resistance of 200 Ω when operating. It is in series with a ballast that has a resistance of 80 Ω and an inductance of 0.9 H. The lamp and ballast are operated at 120 V, 60 Hz. Calculate:

a. The current drawn by the lamp.
b. The apparent power.
c. The power in the fluorescent lamp.
d. The power dissipated in the ballast.
e. The overall power factor.
f. The reactive power.
g. The size of the capacitor to provide unity power factor.
h. The current drawn after power factor correction.

SOLUTION

a. $Z = \sqrt{R_T^2 + X_L^2}$ (24-3)

$R_T = R_{\text{lamp}} + R_{\text{ballast}} = 200\ \Omega + 80\ \Omega = 280\ \Omega$

$X_L = 2\pi fL$ (20-2)

$= 2\pi \times 60\ \text{Hz} \times 0.9\ \text{H} = 339\ \Omega$

Therefore, $Z = \sqrt{(280\ \Omega)^2 + (339\ \Omega)^2} = 440\ \Omega$

$I = \frac{V}{Z}$ (24-4)

$= \frac{120\ \text{V}}{440\ \Omega} = \mathbf{0.27\ A}$

b. $P_s = V_T I_T$ (25-7)

$= 120\ \text{V} \times 0.27\ \text{A} = \mathbf{32.4\ VA}$

c. $P_{\text{lamp}} = I_R^2 R_{\text{lamp}}$ (15-12)

$= (0.27\ \text{A})^2 \times 200\ \Omega = \mathbf{14.6\ W}$

d. $P_{\text{ballast}} = I_R^2 R_{\text{ballast}}$ (15-12)

$= (0.27\ \text{A})^2 \times 80\ \Omega = \mathbf{5.8\ W}$

e. Power factor $= \frac{P}{P_s}$ (25-12)

$= \frac{20.4\ \text{W}}{32.4\ \text{VA}} = \mathbf{0.63}$

f. $P_q = V_T I_T \sin\theta$ (25-13)

$\cos\theta = 0.63$

$\theta = \cos^{-1} 0.63 = 51°$

Therefore, $P_{qL} = 32.4\ \text{VA} \times \sin 51° = \mathbf{25.2\ vars}$

g. $P_{qC} = P_{qL} = 25.2$ vars

$= \frac{V_C^2}{X_C}$ (25-6)

so that $X_C = \frac{V_C^2}{P_{qC}}$

$= \frac{(120\ \text{V})^2}{25.2\ \text{vars}} = 571\ \Omega$

$X_C = \frac{1}{2\pi fC}$ (23-2)

Therefore, $C = \frac{1}{2\pi f X_C}$

$= \frac{1}{2\pi \times 60\ \text{Hz} \times 571} = \mathbf{4.6\ \mu F}$

h. $P = V_T I_T \cos\theta$ (25-12a)

The total power has not changed,

$P = 14.6\ \text{W} + 5.8\ \text{W}$

$= 20.4\ \text{W}$

$\cos\theta = 1$

Therefore, $I_T = \frac{P}{V_T \cos\theta}$

$= \frac{20.4\ \text{W}}{120\ \text{V} \times 1} = \mathbf{0.17\ A}$

25-8 POWER IN THREE-PHASE CIRCUITS

The power delivered to a load (balanced or unbalanced, i.e., equal or unequal line currents) in a three-phase, three-wire system can be measured using two wattmeters connected as in Fig. 25-17.

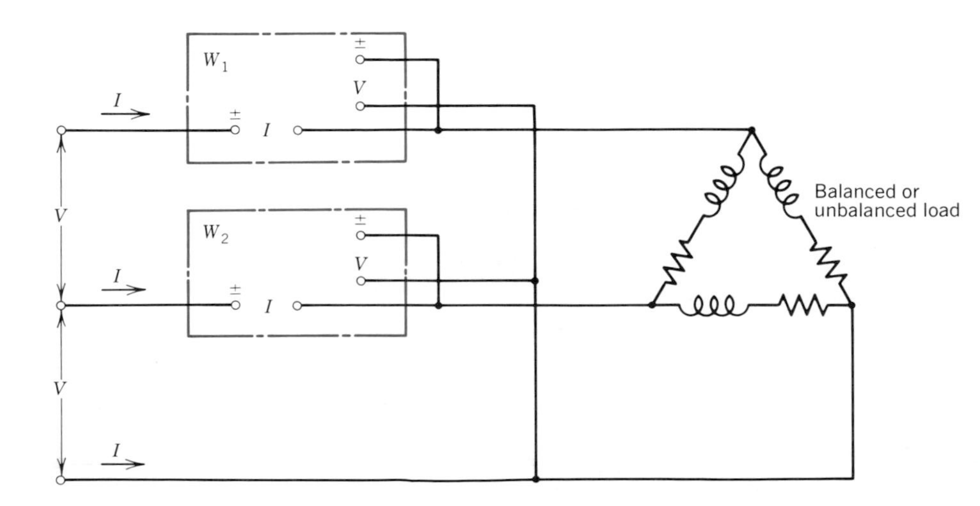

FIGURE 25-17
Measuring total power in a three-phase three-wire system using two wattmeters.

The total power is given by the sum of the two wattmeter readings

$$W_T = W_1 + W_2 \qquad (25\text{-}14)$$

If one of the wattmeters reads backward, when they are connected symmetrically with respect to the marked terminals, the current coil connections should be reversed on this wattmeter and the resulting reading treated as a negative quantity in Eq. 25-14.

If the three-phase load is balanced—that is, equal line currents and voltages—the total true power can be calculated using:

$$P = \sqrt{3}\, VI \cos\theta \qquad \text{watts (W)} \qquad (25\text{-}15)$$

where: P is the total true power in the load in watts, W

V is the line-to-line voltage in volts, V

I is the line current in amperes, A

θ is the phase angle between V and I

EXAMPLE 25-14

Two wattmeters are connected as in Fig. 25-17 to measure the power delivered to a three-phase, delta-connected motor. One wattmeter reads 432 W and the other reads 1092 W. If the motor draws 2.5 A on each line at 440 V, calculate the power factor of the motor.

SOLUTION

$$P = W_T = W_1 + W_2 \qquad (25\text{-}14)$$

$$= 432\text{ W} + 1092\text{ W} = 1524\text{ W}$$

$$P = \sqrt{3}\, VI \cos\theta \qquad (25\text{-}15)$$

Therefore, $\cos\theta = \dfrac{P}{\sqrt{3}\, VI}$

$$= \frac{1524\text{ W}}{\sqrt{3} \times 440\text{ V} \times 2.5\text{ A}} \approx \mathbf{0.8}$$

25-9 MAXIMUM POWER TRANSFER

In Chapter 9 we saw that maximum power is transferred to a load from a dc source when the load resistance, R, equals the internal resistance of the source, r.

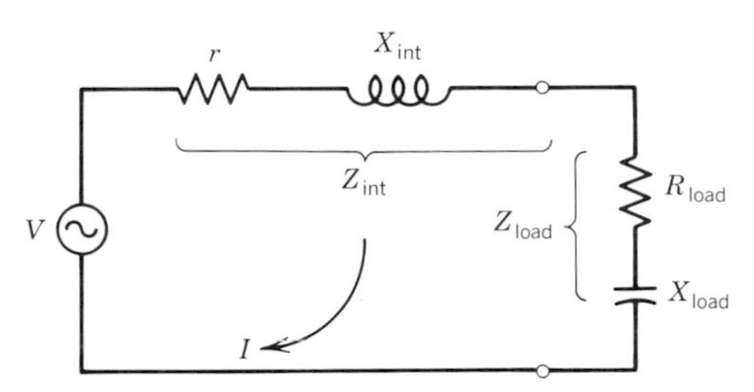

FIGURE 25-18
Load impedance must be the conjugate of the internal impedance for maximum power transfer.

For an *ac* source, we must consider the internal *impedance* of the source to determine what load receives maximum power. Refer to Fig. 25-18.

The total impedance of the circuit is a minimum if the *capacitive* reactance of the load equals the *inductive* reactance of the source (or vice versa if the internal impedance of the source happens to be capacitive). Maximum current flows, and, if $R_{\text{load}} = r$, we have maximum power transfer, as in a dc circuit. **A load that has the same resistance as the source and the opposite (but equal) reactance, is said to be the *conjugate* of the source impedance. Thus, for maximum power transfer in an ac circuit, the load impedance must be the conjugate of the source impedance.**

If a transformer is used to match a load to a source (as in Section 18-1.4), Eq. 18-10 must be modified to the more general form:

$$\mathbf{Z_p = a^2 Z_L \qquad \text{ohms } (\Omega) \qquad (25\text{-}16)}$$

where: Z_p is the impedance reflected into the primary in ohms, Ω

Z_L is the impedance of the load on the secondary in ohms, Ω

a is the turns ratio of the transformer, $\frac{N_p}{N_s}$, dimensionless

Thus, if a load has a resistance of 5 Ω in series with an inductive reactance of 2 Ω, and if the transformer turns ratio is 4:1, the impedance reflected into the primary consists of a resistance of $4^2 \times 5\ \Omega = 80\ \Omega$ in series with an inductive reactance of $4^2 \times 2\ \Omega = 32\ \Omega$. The necessary source impedance to provide maximum power transfer would consist of 80 Ω of resistance in series with 32 Ω of capacitive reactance. (This neglects the resistance and reactance of the transformer itself.) See Problem 25-28.

EXAMPLE 25-15

A 10-V, 60-Hz supply has an internal resistance of 10 Ω and an inductance of 10 mH. A relay coil is to be connected whose inductance is known to be 50 mH. Determine:

a. The resistance of the coil for maximum power transfer.
b. The size of capacitor to be connected in series with the coil to provide maximum power transfer.
c. The power delivered to the coil with the capacitor.
d. The power delivered to the coil without the capacitor.

SOLUTION

a. Coil resistance $R_{\text{load}} = r = \mathbf{10\ \Omega}$

b. The capacitive reactance of the capacitor must equal the *total* inductive reactance of the source and load to minimize the impedance.

$$X_L = 2\pi f L_T \qquad (20\text{-}2)$$
$$= 2\pi \times 60\ \text{Hz} \times (10 + 50) \times 10^{-3}\ \text{H}$$
$$= 22.6\ \Omega$$

$$X_C = \frac{1}{2\pi f C} \qquad (23\text{-}2)$$

Therefore, $C = \dfrac{1}{2\pi f X_C}$

$$= \frac{1}{2\pi \times 60\ \text{Hz} \times 22.6\ \Omega}$$
$$= \mathbf{117\ \mu F}$$

c. Since the reactances of the circuit have canceled each other, the impedance of the circuit is now purely resistive and equal to 20 Ω.

$$I = \frac{V}{Z} \qquad (24\text{-}4)$$
$$= \frac{10\ \text{V}}{20\ \Omega} = 0.5\ \text{A}$$
$$P_{\text{coil}} = I_R^2 R \qquad (15\text{-}12)$$
$$= (0.5\ \text{A})^2 \times 10\ \Omega = \mathbf{2.5\ W}$$

d. $Z = \sqrt{R_T^2 + X_L^2} \qquad (24\text{-}3)$

$$= \sqrt{(20\ \Omega)^2 + (22.6\ \Omega)^2} = 30.2\ \Omega$$
$$I = \frac{V}{Z} \qquad (24\text{-}4)$$
$$= \frac{10\ \text{V}}{30.2\ \Omega} = 0.33\ \text{A}$$
$$P_R = I_R^2 R \qquad (15\text{-}12)$$
$$= (0.33\ \text{A})^2 \times 10\ \Omega = \mathbf{1.1\ W}$$

Example 25-15 illustrates an important effect called *series resonance*. When the capacitive and inductive reactance cancel, the circuit becomes purely resistive and maximum current flows. This condition exists at only one frequency, so the circuit is said to be *frequency-selective*. Such a circuit can be used to "tune in" a desired frequency, as explained in the next chapter.

SUMMARY

1. Power dissipated in resistance is called true power, measured in watts. It can be calculated using the equations:

$$P = I_R^2 R = \frac{V_R^2}{R} = V_R I_R$$

2. Power delivered to a pure inductance is called reactive power, measured in vars. Reactive power is returned to the source and is not dissipated in the inductance. Inductive reactive power can be calculated using the equations:

$$P_q = I_L^2 X_L = \frac{V_L^2}{X_L} = V_L I_L$$

3. Reactive power delivered to a capacitor during one quarter-cycle is returned to the source during the next quarter-cycle. Capacitive reactive power, in vars, can be calculated using

$$P_q = I_C^2 X_C = \frac{V_C^2}{X_C} = V_C I_C$$

4. The apparent power delivered to a series or parallel *RL* or *RC* circuit is the product of the applied voltage and total circuit current, measured in volt-amperes, VA. Apparent power can be calculated using

$$P_s = I_T^2 Z = \frac{V_T^2}{Z} = V_T I_T$$

5. True power, reactive power, and apparent power can be represented in a power triangle with the inductive reactive power drawn vertically downward and the capacitive reactive power drawn vertically upward. The Pythagorean power equation, $P_s^2 = P^2 + P_q^2$ applies to *any* ac circuit combination.

6. If an ac circuit contains both inductance and capacitance, there is an exchange of reactive power between the two so that the source must supply only the net reactive power

$$P_{qC} - P_{qL}$$

7. The power factor of a circuit is defined as the ratio of the true power in the circuit to the apparent power and is given by

$$\cos\theta = P/P_s$$

8. The true power in a single-phase ac circuit can always be obtained using

$$P = V_T I_T \cos\theta \qquad \text{watts}$$

and the reactive power can be obtained using

$$P_q = V_T I_T \sin\theta \qquad \text{vars}$$

9. Inductive loads have a lagging power factor; capacitive loads have a leading power factor.

10. A low power factor means that a higher current is required to deliver a given amount of true power than if the power factor is near unity.

11. Power factor correction requires the parallel connection of a capacitor to draw the same amount of leading capacitive reactive power as the initial lagging inductive reactive power of the load.

12. The total power delivered to a three-phase, three-wire system can be measured using two wattmeters and finding the algebraic sum of their readings.

13. In a balanced three-phase load, the total true power can be calculated using

$$P = \sqrt{3}\, VI \cos\theta$$

14. Maximum power transfer occurs in an ac circuit when the load impedance is the conjugate of the internal source impedance.

SELF-EXAMINATION

Answer true or false or a, b, c, d for multiple choice
(Answers at back of book)

25-1. Only resistance causes true power to be dissipated. ________

25-2. A wattmeter always shows the true power dissipated in an ac circuit, even if the circuit contains inductance and capacitance. ________

25-3. A purely resistive ac circuit only involves positive power whereas a purely inductive or capacitive circuit shows equal positive and negative power alternations. ________

25-4. The apparent power delivered to a purely inductive circuit is zero. ________

25-5. If a purely inductive circuit has a reactance of 120 Ω, the reactive power drawn by the circuit from a 120-V, 60-Hz supply is

a. 14,400 vars
b. 14.4 kvars
c. 120 vars
d. either part (a) or (b)

25-6. If a capacitor of 240-Ω reactance draws a current of 2 A, the reactive power supplied the capacitor is

a. 960 vars
b. 480 vars
c. 60 vars
d. 240 vars

25-7. If a circuit has an impedance of 800 Ω and draws a current of 10 A, the apparent power is

a. 8000 VA
b. 80 kVA
c. 8 kVA
d. either part (a) or (c)

25-8. If a wattmeter indicates 300 W and a varmeter indicates 400 vars, the apparent circuit power is

a. 2500 VA
b. 25,000 VA
c. 500 VA
d. 700 VA

25-9. The Pythagorean power equation only applies to series and parallel *RL* circuits. ________

25-10. When an ac circuit contains both inductance and capacitance, in series or parallel, the net reactive power is always given by $P_q = P_{qC} - P_{qL}$. ________

25-11. If a circuit has more inductive reactive power than capacitive reactive power, the circuit has a lagging power factor. ________

25-12. The power factor is the factor by which the reactive power must be multiplied to obtain the true power. ________

25-13. True power is the product of the in-phase components of voltage and current while reactive power is the product of the 90° out-of-phase components of voltage and current. ________

25-14. If a wattmeter indicates 400 W, a series-connected ammeter indicates 2 A, and a parallel-connected voltmeter indicates 250 V, the power factor is

a. 0.2
b. 0.25
c. 0.4
d. 0.8

25-15. The major problem with low power factor circuits is the relatively high current necessary to deliver a given true power. ________

25-16. Power factor correction (to unity power factor) requires the addition of a capacitor in parallel with the circuit to make the applied voltage and circuit current in phase with each other. ________

25-17. When an ac motor has had its power factor corrected, the motor is more efficient. ________

25-18. If, in trying to correct a load's power factor, too large a capacitor is connected, there may be no reduction in the line current and in fact the current may increase. ________

25-19. Two wattmeters can be used to measure the total power in a three-phase, three-wire system only if the three line currents are equal. ________

25-20. If an ac source has an impedance consisting of 100 Ω resistance and a capacitive reactance of 50 Ω, the required load impedance for maximum power transfer is

a. $R = 100\ \Omega, X_L = 100\ \Omega$
b. $R = 100\ \Omega, X_C = 50\ \Omega$
c. $R = 50\ \Omega, X_C = 50\ \Omega$
d. $R = 100\ \Omega, X_L = 50\ \Omega$

REVIEW QUESTIONS

1. Distinguish, in your own words, between true power and reactive power.

2. Explain why a wattmeter has zero deflection when connected in a purely inductive or capacitive circuit.

3. What do you understand by "positive" power and "negative" power?

4. Why can't we use watts to describe reactive and apparent power?

5. Under what conditions is the apparent power equal to

a. True power?
b. Reactive power?

6. a. Describe a method to measure the inductance of a coil that requires a wattmeter.
b. Why is this method limited to power line frequencies?
c. If you measured the resistance of the coil with an ohmmeter, would you expect the reading to be higher or lower than the resistance obtained from the wattmeter and ammeter readings? Why?

7. a. How does a power triangle drawn for an inductive load differ from one drawn for a capacitive load?
b. Why are no arrows shown on power triangles?
c. Does a power triangle drawn for a series *RL* circuit differ from one drawn for a parallel *RL* circuit? Why or why not?
d. Under what conditions does a power diagram for a series or parallel *RLC* circuit *not* result in a triangle?

8. a. Describe in your own words what is meant by "power factor."
b. Describe two methods of measuring a circuit's power factor.
c. What is the range of possible values for the power factor of a circuit?

9. How must the apparent power quantity be modified to obtain:
a. True power?
b. Reactive power?

10. Explain how the equation $P = V_T I_T \cos \theta$ gives the product of the *in-phase* components of current and voltage for true power in any single-phase ac circuit (series or parallel).

11. What is meant by leading and lagging power factors?

12. a. Why is a low power factor considered detrimental when supplying ac power from a source?
b. Give two reasons why a low power factor can be costly both to the consumer and the supplier of ac power.

13. a. What is the general principle used to select a capacitor to provide unity power factor correction?
b. What is the primary effect of correcting the power factor?
c. How can a motor *and* its supply system, including feeder wires, be considered more *efficient* following power factor correction if the true power delivered to the motor is unchanged?

14. a. Draw a circuit diagram showing how to connect two wattmeters to measure the total power delivered to a load in a three-phase, three-wire system.
b. What would you do to obtain the total power if one of the wattmeters read backward?

15. a. What is the maximum power transfer theorem as applied to an ac circuit?
b. What is meant by the term *conjugate?*

PROBLEMS

(Answers to odd-numbered problems at back of book)

25-1. A wattmeter, connected in an ac circuit, indicates 240 W. An ammeter, connected in series with the wattmeter, reads 4.5 A. Determine the resistance of the ac circuit.

25-2. An oscilloscope indicates 340 V p-p across the only resistor in a series ac circuit that draws a current of 600 mA. Calculate:

a. The resistance of the circuit.

b. The reading of a wattmeter connected to the circuit.

25-3. A 2-H coil of negligible resistance is connected across the 120-V, 60-Hz line. Calculate:

a. The coil's reactive power.

b. The coil's apparent power.

25-4. A coil of 100-Ω resistance and 0.8-H inductance draws a current of 400 mA from a 60-Hz supply. Determine the reading of a varmeter connected to the circuit.

25-5. A bank of capacitors used for power factor correction draws a reactive power of 20 kvars from a 550-V, 60-Hz, single-phase supply. Determine the amount of capacitance.

25-6. How much reactive power is supplied to a 50-μF motor-starting capacitor operating at 220 V, 60 Hz?

25-7. What is the apparent power supplied to a circuit that draws a current of 300 mA at 230 V?

25-8. What maximum current can be supplied by a 7.5-kVA transformer with a 240-V secondary?

25-9. A series circuit of 250-Ω resistance and 150-Ω inductive reactance draws a current of 1.8 A. What is the apparent power drawn by the circuit?

25-10. A coil is connected to the 220-V, 60-Hz line in series with an ammeter and wattmeter. If the ammeter indicates 1.25 A and the wattmeter indicates 80 W, calculate:

a. The inductance of the coil.

b. The quality of the coil.

25-11. A series circuit of 220-Ω resistance and 0.4-H inductance draws a current of 700 mA from a 60-Hz supply. Calculate:

a. The true power.

b. The reactive power.

c. The apparent power.

25-12. A relay coil connected across a 24-V, 50-Hz supply draws a current of 200 mA. If a series-connected wattmeter indicates 3 W, calculate:

a. The apparent power.

b. The reactive power.

25-13. A 470-Ω resistor and a 150-mH coil of negligible resistance are connected in parallel across a 100-V, 400-Hz supply. Calculate:

a. The true power.

b. The reactive power.

c. The apparent power.

d. The circuit's power factor.

25-14. Repeat Problem 25-13 assuming that the coil has a resistance of 500 Ω.

25-15. A 1-μF capacitor and a 100-Ω resistor are series-connected across a 20-V, 1-kHz source. Calculate:

a. The true power.

b. The reactive power.

c. The apparent power.

d. The circuit's power factor.

25-16. Repeat Problem 25-15 with the two components connected in *parallel* across the source.

25-17. A 50-μF capacitor, a 47-Ω resistor and a 100-mH coil of negligible resistance are connected in parallel across the 120-V, 60-Hz line. Calculate:

a. The true power.
b. The net reactive power.
c. The apparent power.
d. The total line current.
e. The power factor of the circuit.

25-18. Repeat Problem 25-17 assuming that the coil has a resistance of 40 Ω. Draw a power triangle.

25-19. A circuit drawing a current of 8.5 A from a 220-V, 60-Hz supply is known to have a lagging power factor of 0.6. Calculate:

a. The phase angle between the applied voltage and current.
b. The apparent power.
c. The true power.
d. The reactive power.

25-20. A 220-V, $1\frac{1}{2}$-hp, single-phase ac motor is running at full load and drawing a current of 9 A. The phase angle between the applied voltage and the current is measured to be 35°. Calculate:

a. The power factor of the motor.
b. The apparent power.
c. The true power.
d. The reactive power.
e. The efficiency of the motor.

25-21. How much current is drawn by a 120-V, 60-Hz, 75% efficient, 1-hp, single-phase motor when operating at the following power factors:

a. 0.7 lagging.
b. 0.9 lagging.
c. 1.0.
d. 0.9 leading?

25-22. Repeat Problem 25-21 assuming the use of a three-phase motor.

25-23. For the circuit of Problem 25-19 calculate:

a. The capacitance required to provide unity power factor.
b. The line current following power factor correction.

25-24. For the circuit of Problem 25-20 calculate:

a. The capacitance required to provide unity power factor at 60 Hz.
b. The line current following power factor correction.

25-25. Repeat Example 25-13 assuming that the ballast is replaced by one that has a resistance of 50 Ω and an inductance of 1.1 H. What visual effect would be noticed with the new ballast?

25-26. What capacitance must be connected across a 120-V, 60-Hz, 75% efficient, 1-hp, single-phase motor operating at a 0.7 lagging power factor to produce an overall 0.9 leading power factor?

25-27. Two wattmeters, connected to measure the total power delivered to a 5-hp, three-phase, delta-connected motor, indicate 2860 and 1440 W, respectively. If the motor draws 5.2 A on each line at 550 V, calculate:

a. The motor's power factor.

b. The motor's efficiency.

25-28. A 40-V p-p sinusoidal source at 1 kHz has an internal impedance consisting of 100-Ω resistance in series with a 4-μF capacitor. A transformer is to be selected to maximize power transfer to a 4-Ω resistive load. Calculate:

a. The necessary turns ratio of the transformer.

b. The amount of inductance that must be in series with the 4-Ω resistance load to provide maximum power transfer.

c. The power delivered to the load with the transformer and inductance, assuming that the transformer is 100% efficient.

d. The power delivered to the load without the inductance.

e. The power delivered to the load if connected directly across the source.

25-29. A transmitter operating at 2 MHz delivers maximum power to an antenna when the antenna is "tuned" to represent a load of 150-pF capacitance in series with a resistance of 150 Ω. Determine the internal impedance of the transmitter.

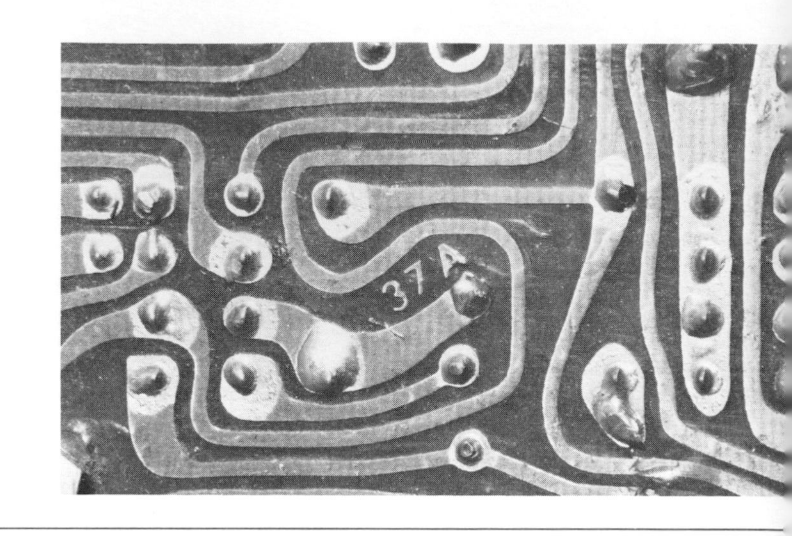

CHAPTER TWENTY-SIX
RESONANCE

The antenna of a radio or TV receiver has voltages induced in it by the radio frequency signals from many different stations and channels. How it is possible to select a desired *band* of frequencies and reject the others is discussed in this chapter.

Both series and parallel *RLC* circuits can be tuned to *resonate* at some desired frequency, equal to that assigned to the particular broadcasting station. The resonant circuit, as well as being able to receive a single *center* or *carrier* frequency, is *sensitive* to some *band* of frequencies emitted by the transmitter. By the proper selection of the components, we can make the circuit's *bandwidth* receive all of the transmitted information but reject the frequencies of adjacent stations. This requires that stations operate at specific frequencies assigned by the FCC and adjust their bandwidths so that there is no bandwidth overlap.

In a *series RLC* circuit, we find that there is a *resonant rise* of voltage across the reactive components which (if the *Q* of the circuit is large) is many times *larger* than the *input* voltage. Similarly, a resonant rise of the *tank* current occurs in a parallel *RLC* circuit, which can be much *larger* than the *line* current. Since both of these effects are frequency *selective,* we can *discriminate* against all other undesired frequencies when we *tune-in* a desired station.

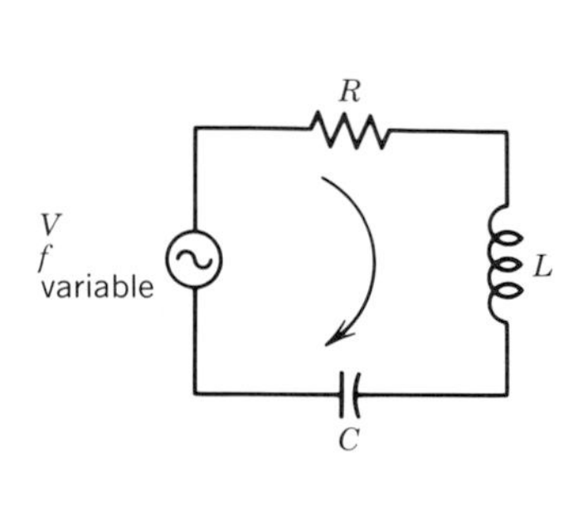

(a) Series RLC circuit

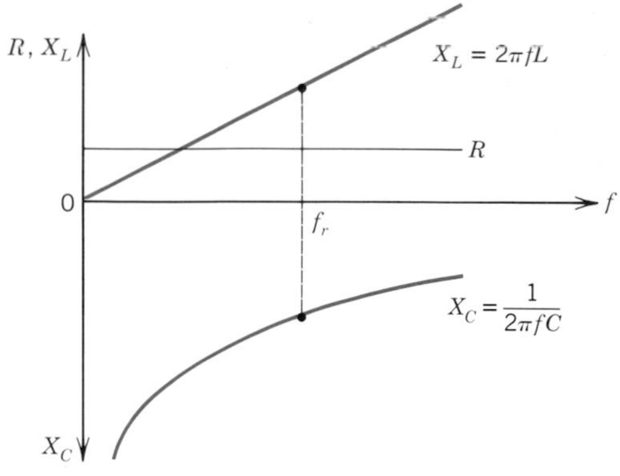

(b) Variation in X_L and X_C with f

FIGURE 26-1
Effect of a variable frequency on a series *RLC* circuit.

26-1 SERIES RESONANCE

In Section 24-7, we examined the basic relation among the current, voltage, and impedance in a series *RLC* circuit. We now examine how such a circuit reacts to a *variable frequency* source. See Fig. 26-1.

We know that

$$Z = \sqrt{R^2 + (X_L - X_C)^2} \qquad (24\text{-}14)$$

$$I = \frac{V}{Z} \qquad (24\text{-}4)$$

and

$$\theta = \arctan \frac{X_L - X_C}{R} \qquad (24\text{-}15)$$

The way in which X_L and X_C vary as the frequency increases is shown in Fig. 26-1*b*. The inductive reactance, X_L, is a *positive increasing* quantity; the capacitive reactance, X_C, is represented as a *negative decreasing* quantity. Clearly, there is some frequency, f_r, where the algebraic sum of X_L and X_C ($X_L - X_C$) is zero. That is, the net reactance of the circuit is zero.

Thus, if the frequency of the source is adjusted to f_r, the reactances cancel and the impedance of the circuit is a *minimum* and equal to R (see Eq. 24-14). Also, from Eq. 24-4, when the impedance Z is a *minimum*, the current I is a *maximum* and equal to V/R. From Eq. 24-15, the phase angle θ is zero. That is, the current and applied voltage are in phase. All of these conditions occur at what is called the *resonant frequency*, f_r, and the series circuit is said to be in a condition of *series resonance*.

The conditions in the circuit at frequencies below, at, and above the resonant frequency are illustrated in Example 26-1.

EXAMPLE 26-1

A series *RLC* circuit consists of $R = 10\ \Omega$, $L = 200\ \mu\text{H}$, and $C = 50$ pF. A 100-mV variable frequency supply is connected to the circuit. Determine the following for frequencies of (1) 1 MHz, (2) 1.59 MHz, and (3) 2.2 MHz and present in tabular form. Also draw a phasor diagram for each frequency.

a. The impedance.
b. The phase angle.
c. The current.
d. V_R, V_L, and V_C.

SOLUTION

1. a. At $f = 1 \times 10^6$ Hz,

$$X_L = 2\pi fL \qquad (18\text{-}2)$$
$$= 2\pi \times 1 \times 10^6\ \text{Hz} \times 200 \times 10^{-6}\ \text{H}$$
$$= 1257\ \Omega$$

$$X_C = \frac{1}{2\pi fC} \qquad (23\text{-}2)$$
$$= \frac{1}{2\pi \times 1 \times 10^6\ \text{Hz} \times 50 \times 10^{-12}\ \text{F}}$$
$$= 3183\ \Omega$$

$$Z = \sqrt{R^2 + (X_L - X_C)^2} \qquad (24\text{-}14)$$
$$= \sqrt{(10\ \Omega)^2 + (1257\ \Omega - 3183\ \Omega)^2}$$
$$= \mathbf{1926\ \Omega} \quad \text{(capacitive)}$$

b. $\theta = \arctan \dfrac{X_L - X_C}{R}$ (24-15)

$= \arctan \dfrac{1257\ \Omega - 3183\ \Omega}{10\ \Omega} = \mathbf{-89.7°}$

c. $I = \dfrac{V}{Z}$ (24-4)

$= \dfrac{100\ \text{mV}}{1.926\ \text{k}\Omega} = \mathbf{51.9\ \mu A}$

d. $V_R = IR$ (13-15b)

$= 51.9 \times 10^{-6}\ \text{A} \times 10\ \Omega$

$= \mathbf{0.5\ mV}$

$V_L = IX_L$ (20-1)

$= 51.9 \times 10^{-6}\ \text{A} \times 1257\ \Omega$

$= \mathbf{65.2\ mV}$

$V_C = IX_C$ (23-1)

$= 51.9 \times 10^{-6}\ \text{A} \times 3183\ \Omega$

$= \mathbf{165.2\ mV}$

2. a. At $f = 1.59 \times 10^6$ Hz,

$X_L = 2\pi fL$ (18-2)

$= 2\pi \times 1.59 \times 10^6\ \text{Hz} \times 200 \times 10^{-6}\ \text{H}$

$\approx 2\ \text{k}\Omega$

$X_C = \dfrac{1}{2\pi fC}$ (23-2)

$= \dfrac{1}{2\pi \times 1.59 \times 10^6\ \text{Hz} \times 50 \times 10^{-12}\ \text{F}}$

$\approx 2\ \text{k}\Omega$

$Z = \sqrt{R^2 + (X_L - X_C)^2}$ (24-14)

$= \sqrt{(10\ \Omega)^2 + (2\ \text{k}\Omega - 2\ \text{k}\Omega)^2}$

$= \mathbf{10\ \Omega}$ (resistive)

b. $\theta = \arctan \dfrac{X_L - X_C}{R}$ (24-15)

$= \arctan \dfrac{2\ \text{k}\Omega - 2\ \text{k}\Omega}{10\ \Omega} = \mathbf{0°}$

c. $I = \dfrac{V}{Z}$ (24-4)

$= \dfrac{100\ \text{mV}}{10\ \Omega} = \mathbf{10\ mA}$

d. $V_R = IR$ (13-15b)

$= 10 \times 10^{-3}\ \text{A} \times 10\ \Omega = \mathbf{100\ mV}$

$V_L = IX_L$ (20-1)

$= 10 \times 10^{-3}\ \text{A} \times 2 \times 10^3\ \Omega = \mathbf{20\ V}$

$V_C = IX_C$ (23-1)

$= 10 \times 10^{-3}\ \text{A} \times 2 \times 10^3\ \Omega = \mathbf{20\ V}$

3. Similarly, at $f = 2.2$ MHz,

a. $X_L = 2764\ \Omega$, $X_C = 1447\ \Omega$,

$Z = \mathbf{1317\ \Omega}$ (inductive)

b. $\theta = \mathbf{89.6°}$

c. $I = \mathbf{75.9\ \mu A}$

d. $V_R = \mathbf{0.76\ mV}$, $V_L = \mathbf{209.8\ mV}$, $V_C = \mathbf{109.8\ mV}$

See Table 26-1.

The Phasor diagrams for the three frequencies are shown in Fig. 26-2.

In Example 26-1, the inductive and capacitive reactances are equal when the frequency is 1.59 MHz, the circuit is purely resistive, and maximum current flows. Thus the resonant frequency of the circuit must be 1.59 MHz.

Table 26-1
Circuit Conditions in a Variable Frequency Circuit

Frequency	X_L, Ω	X_C, Ω	$X_L - X_C$, Ω	Z, Ω	θ	I, mA	V_R	V_L	V_C
1 MHz	1257	3183	−1926	1926	−89.7° (Capacitive)	0.05	0.5 mV	65 mV	165 mV
1.59 MHz	2000	2000	0	10	0° (Resistive)	10.0	100 mV	20 V	20 V
2.2 MHz	2764	1447	1317	1317	89.6° (Inductive)	0.08	0.8 mV	210 mV	110 mV

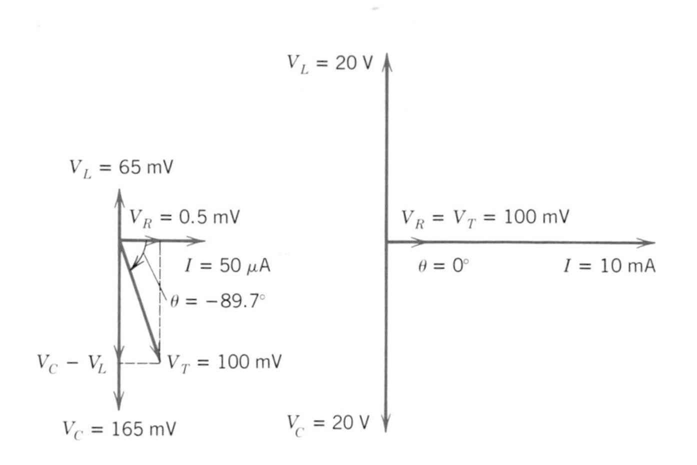

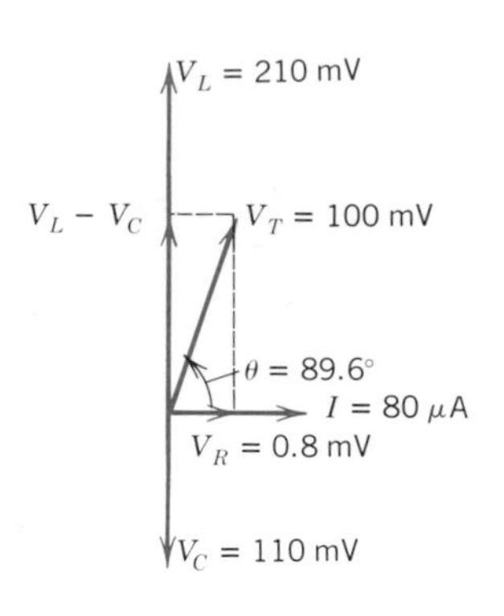

FIGURE 26-2
Phasor diagrams for series *RLC* circuit of Example 26-1 (not drawn to scale).

26.2 RESONANT FREQUENCY

We can find the resonant frequency of any series circuit by noting that resonance occurs when $X_L = X_C$:

$$X_L = X_C$$

$$2\pi f_r L = \frac{1}{2\pi f_r C}$$

$$f_r^2 = \frac{1}{4\pi^2 LC}$$

Therefore, $$f_r = \frac{1}{2\pi\sqrt{LC}} \quad \text{hertz (Hz)} \qquad (26\text{-}1)$$

where: f_r is the resonant frequency of a series circuit in hertz, Hz
L is the inductance of the circuit in henrys, H
C is the capacitance of the circuit in farads, F

Note that the resonant frequency of a series circuit is independent of the circuit's resistance.

EXAMPLE 26-2

Calculate the resonant frequency of the circuit in Example 26-1, where $R = 10\ \Omega$, $L = 200\ \mu\text{H}$, and $C = 50$ pF.

SOLUTION

$$f_r = \frac{1}{2\pi\sqrt{LC}} \qquad (26\text{-}1)$$

$$= \frac{1}{2\pi\sqrt{200 \times 10^{-6}\ \text{H} \times 50 \times 10^{-12}\ \text{F}}}$$

$$= 1.59 \times 10^6\ \text{Hz}$$

$$= \mathbf{1.59\ MHz}$$

EXAMPLE 26-3

What value of capacitance is necessary to provide series resonance with a 30-mH coil at 1 kHz?

SOLUTION

$$f_r = \frac{1}{2\pi\sqrt{LC}} \qquad (26\text{-}1)$$

Therefore, $$4\pi^2 f_r^2 LC = 1$$

and $$C = \frac{1}{4\pi^2 f_r^2 L}$$

$$= \frac{1}{4\pi^2 \times (1 \times 10^3\ \text{Hz})^2 \times 30 \times 10^{-3}\ \text{H}}$$

$$= \mathbf{0.84\ \mu F}$$

Example 26-1 showed the characteristics of a series circuit at three specific frequencies. If calculations for Z, I, and θ are made at other frequencies and plotted in a graph, the effect above and below the resonant frequency is as shown in Fig. 26-3.

We note that at the resonant frequency, f_r:

1. **The circuit *impedance* is a *minimum*, equal to the resistance R.**
2. **The circuit *current* is a *maximum*, equal to V/R.**
3. **The circuit reactance is zero, since $X_C = X_L$.**
4. **The circuit current is in phase with the applied voltage; θ is zero.**

26-3 SELECTIVITY OF A SERIES CIRCUIT

The graph of I versus f in Fig. 26-3*b* shows that the circuit is *selective* in its *response* to a certain *band* of frequencies around the resonant frequency. That is, at frequencies that are much higher or lower than the resonant frequency, f_r, the current in the circuit is very low. We say that the circuit has a low *response* or is not very *sensitive* to these frequencies. The frequency *selectivity* of a resonant circuit is its most important characteristic. In order to compare the selectivity of one circuit with another, we refer to the *bandwidth* of the resonant circuit. See Fig. 26-4.

Consider the two frequencies f_1 and f_2 in Fig. 26-4. These have been selected where the circuit current is 0.707 times the maximum current at the resonant frequency, f_r. The bandwidth, Δf, is defined to be the band of frequencies between f_1 and f_2 or:

$$\Delta f = f_2 - f_1 \qquad \text{hertz (Hz)} \qquad (26\text{-}2)$$

where: Δf is the bandwidth of a resonant circuit in hertz, Hz

f_2 is the upper half-power frequency in hertz, Hz

f_1 is the lower half-power frequency in hertz, Hz

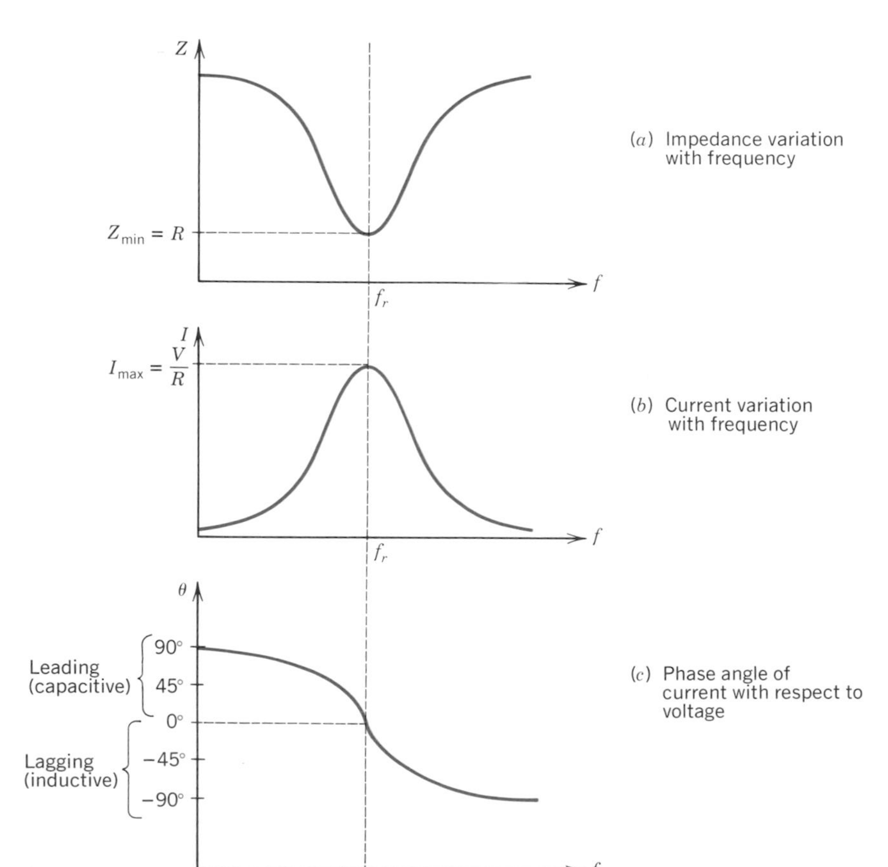

FIGURE 26-3
Effect of varying frequency on impedance, current, and phase angle in a series *RLC* circuit.

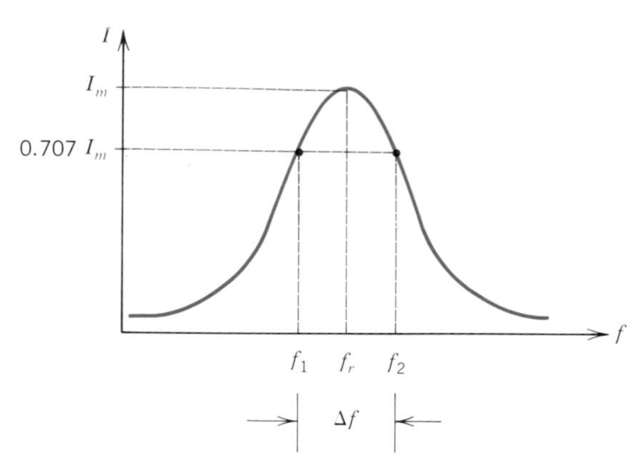

FIGURE 26-4
Bandwidth of a resonant circuit.

Note that f_1 and f_2 have been referred to as half-power frequencies.* This is because at these frequencies the power (true power) delivered to the circuit is one-half the power delivered at the resonant frequency, f_r. That is,

At $\quad f = f_r, \quad P = I_m{}^2R$

At $\quad f = f_1, \quad P = (0.707\,I_m)^2R$

$$= 0.5\,I_m{}^2R$$

$$= \tfrac{1}{2} \times \text{power delivered at } f_r$$

Since a 50% change in the power level is not considered large (a 3-dB change in the power output from an audio amplifier is barely noticeable), the response of the circuit is considered relatively *flat* between f_1 and f_2. Beyond f_1 and f_2, the power drops off rapidly, and the circuit is considered as *rejecting* these frequencies. Thus, by the proper selection of the L and C components, a series resonant circuit can be made to "tune-in" a given band of frequencies, centered at the circuit's resonant frequency, and reject (but not completely) the adjacent frequencies. In most high-frequency communication circuits, f_1 and f_2 are symmetrically placed above and below f_r. Thus the "edge frequencies," f_1 and f_2, can be obtained as follows:

*f_1 and f_2 are also referred to as the lower and upper 3-dB frequencies. The notation dB is short for decibels, which are used to show a comparison between the two power levels P_2 and P_1 according to the equation dB = 10 log P_2/P_1. If $P_2 = 2\,P_1$, dB = 10 log 2 = 10 × 0.3 = 3 dB. Thus a 3-dB ratio refers to a 2:1 power level, such as occurs between f_1 and f_r or f_2 and f_r.

$$f_1 = f_r - \frac{\Delta f}{2} \qquad (26\text{-}2a)$$

$$f_2 = f_r + \frac{\Delta f}{2} \qquad (26\text{-}2b)$$

where terms are as above.

EXAMPLE 26-4

A series circuit has a resonant frequency of 400 kHz and a bandwidth of 12 kHz. Determine the "edge" or upper and lower half-power frequencies.

SOLUTION

$$f_1 = f_r - \frac{\Delta f}{2} \qquad (26\text{-}2a)$$

$$= 400\text{ kHz} - \frac{12\text{ kHz}}{2} = \mathbf{394\ kHz}$$

$$f_2 = f_r + \frac{\Delta f}{2} \qquad (26\text{-}2b)$$

$$= 400\text{ kHz} + \frac{12\text{ kHz}}{2} = \mathbf{406\ kHz}$$

It can be shown that the bandwidth and the resonant frequency are related by the equation:

$$\Delta f = \frac{f_r}{Q} \quad \textbf{hertz (Hz)} \qquad \textbf{(26-3)}$$

where: Δf is the bandwidth of the circuit in hertz, Hz
f_r is the resonant frequency of the circuit in hertz, Hz
Q is the quality of the circuit, dimensionless

Recall from Section 20-8 that the quality of a coil was defined as

$$Q = \frac{X_L}{R_{ac}} \qquad (20\text{-}9)$$

where R_{ac} is the total circuit resistance for a series RLC circuit. Thus, although resistance does not determine a series circuit's resonant frequency, it does affect its bandwidth.

If Q is very large, the bandwidth Δf is very narrow, and we have a highly selective circuit. If Δf is *too* narrow, the circuit will not respond to all the incoming frequencies, as from a radio transmitter, so that some information is lost. Conversely, if Q is very small, the bandwidth is very broad, and interference from adjacent "channels" or frequencies can occur because the circuit is not sufficiently selective. Each communication system has its own resonant frequency and bandwidth requirements.

For example, the intermediate frequency amplifiers in an AM (amplitude modulation) broadcast band receiver operate at an intermediate frequency of 455 kHz and require a bandwidth of approximately 10 kHz; an FM (frequency modulation) receiver operates at an intermediate frequency of 10.7 MHz and requires a 200-kHz bandwidth.

EXAMPLE 26-5

Determine the necessary Q values of circuits used in AM and FM intermediate amplifiers, using the data above.

SOLUTION

$$\Delta f = \frac{f_r}{Q} \quad (26\text{-}3)$$

Therefore, $$Q = \frac{f_r}{\Delta f}$$

For AM: $$Q = \frac{455 \text{ kHz}}{10 \text{ kHz}} = \mathbf{45.5}$$

For FM: $$Q = \frac{10.7 \text{ MHz}}{0.2 \text{ MHz}} = \mathbf{53.5}$$

The Q values in Example 26-5 are quite easy to attain commercially. In fact, it is because these Q values *are* practical that the commercially used intermediate frequencies of 455 kHz and 10.7 MHz were chosen.

EXAMPLE 26-6

Find the bandwidth of the circuit in Example 26-1 with $L = 200\ \mu\text{H}$, $R = 10\ \Omega$, and $C = 50$ pF.

SOLUTION

$$f_r = \frac{1}{2\pi\sqrt{LC}} \quad (26\text{-}1)$$

$$= \frac{1}{2\pi\sqrt{200 \times 10^{-6}\text{ H} \times 50 \times 10^{-12}\text{ F}}}$$

$$= 1.59 \text{ MHz}$$

$$Q = \frac{X_L}{R_{ac}} \quad (20\text{-}9)$$

$$= \frac{2\pi \times 1.59 \times 10^6 \text{ Hz} \times 200 \times 10^{-6}\text{ H}}{10\ \Omega}$$

$$= 200$$

$$\Delta f = \frac{f_r}{Q} \quad (26\text{-}3)$$

$$= \frac{1.59 \times 10^6 \text{ Hz}}{200} \approx \mathbf{8\ kHz}$$

26-4 THE Q OF A SERIES-RESONANT CIRCUIT

We can determine another expression for the Q of a circuit to show why it determines the bandwidth.

For *any* resonant circuit, the Q can be defined as the ratio of the coil or capacitor's reactive power to the true power.

$$Q = \frac{P_q}{P} \quad (26\text{-}4)$$

$$= \frac{I^2 X_L}{I^2 R_{ac}}$$

For a series circuit, the same current flows through the coil and resistor.

Therefore: $$Q = \frac{X_L}{R_{ac}} = \frac{2\pi f_r L}{R_{ac}} = 2\pi f_r \frac{L}{R_{ac}} \quad (20\text{-}9)$$

But at resonance:

$$f_r = \frac{1}{2\pi\sqrt{LC}} \quad (26\text{-}1)$$

Therefore, $$2\pi f_r = \frac{1}{\sqrt{LC}}$$

Thus $$Q = \frac{1}{\sqrt{LC}}\frac{L}{R_{ac}}$$

and $$Q = \frac{1}{R_{ac}}\sqrt{\frac{L}{C}} \quad (26\text{-}5)$$

where: Q is the quality of a series-resonant circuit, dimensionless

R_{ac} is the total ac resistance of the circuit in ohms, Ω

L is the inductance in henrys, H

C is the capacitance in farads, F

EXAMPLE 26-7

Determine the Q of a series-resonant circuit in which $R = 10\ \Omega$, $L = 200\ \mu$H, and $C = 50$ pF.

SOLUTION

$$Q = \frac{1}{R_{ac}}\sqrt{\frac{L}{C}} \tag{26-5}$$

$$= \frac{1}{10\ \Omega}\sqrt{\frac{200 \times 10^{-6}\ \text{H}}{50 \times 10^{-12}\ \text{F}}}$$

$$= \frac{1}{10\ \Omega}\sqrt{4 \times 10^{6}\ \Omega^2} = \mathbf{200}$$

Let us examine Eq. 26-5 to see the reason for its effect on the bandwidth. We note that if R_{ac} alone is reduced, Q is increased and the bandwidth is decreased, without affecting the resonant frequency. Thus, if in Example 26-7, R is reduced from 10 to 5 Ω, Q is doubled to 400. The reason for the reduction in bandwidth can be seen in Fig. 26-5*a*. The current at resonance must now be double its initial value, causing a *steeper* rise in current around the resonant frequency, f_r.

Similarly, we see that an increase in the L/C ratio in Eq. 26-5 increases Q. Thus, when $L = 200\ \mu$H and $C = 50$ pF, $L/C = 4 \times 10^6$; if L is increased to 400 μH and C is reduced to 25 pF, the L/C ratio equals 16×10^6 so that Q is now double its initial value. (Note that the resonant frequency for this example is unchanged, since the product LC is still the same.)

The steeper rise in the I versus f curve in Fig. 26-5*b* shows that a narrower bandwidth results with a higher L/C ratio. The reason for this is found in the graphs of X_L and X_C versus f in Fig. 26-1. A larger L/C ratio implies a larger value of L and a smaller value of C. This leads to *steeper* slopes of the X_L and X_C graphs. This means that as the frequency increases, the point of resonance, where X_L equals X_C, is reached more *rapidly*. If L were small and C were large, the two graphs would be relatively "flat," and a large change in f would be necessary to make X_L equal X_C for resonance.

Thus Q is the one term that combines the effect of all three components in determining the shape of the resonance curve of a series RLC circuit.

26-5 RESONANT RISE OF VOLTAGE IN A SERIES *RLC* CIRCUIT

One of the most useful features of a series RLC circuit is the *resonant rise of voltage* across the capacitor and inductor as resonance is approached. This is seen in

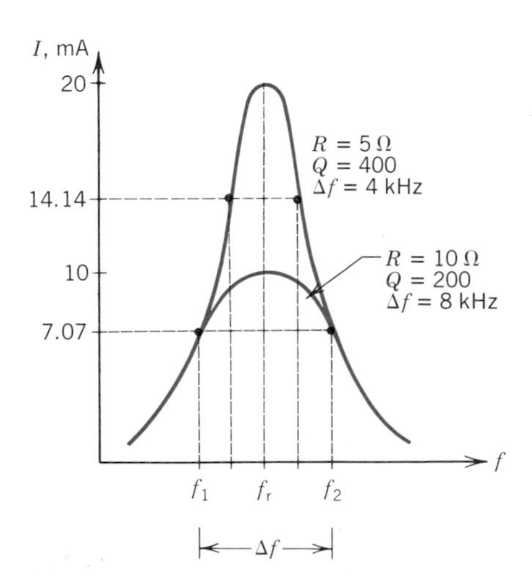

(*a*) A reduction in R increases Q and decreases the bandwidth.

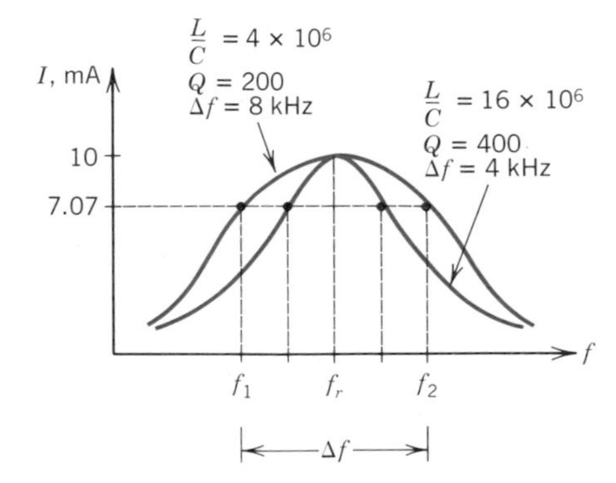

(*b*) An increase in the L/C ratio increases Q and decreases the bandwidth.

FIGURE 26-5 Effect of changes in R and L/C on the Q and bandwidth of a series resonant circuit.

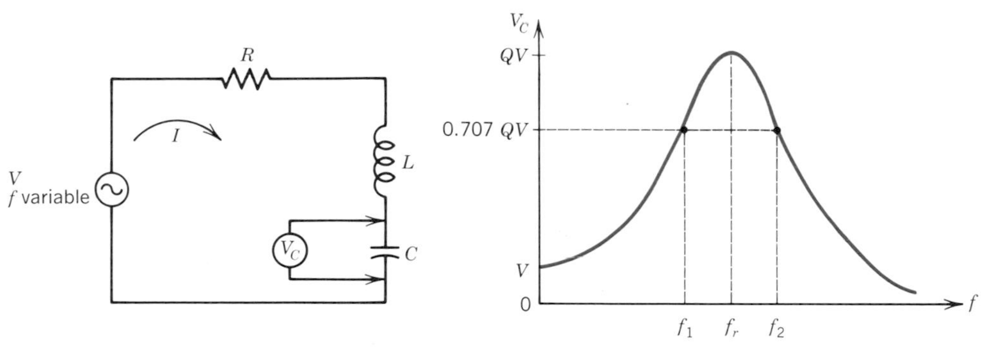

(a) Series RLC circuit (b) Resonant rise in capacitor voltage

FIGURE 26-6
Effect on the capacitor voltage of varying the frequency of a series resonant *RLC* circuit.

Table 26-1 where, at the resonant frequency of 1.59 MHz, both the capacitor and inductor voltages are 20 V, with a supply voltage of only 100 mV.

Although the net reactance in the circuit is zero, allowing a large current to circulate, the individual reactances are quite large. The product of a large current (10 mA) and a large reactance (2 kΩ) produces a high voltage (20 V). Actually, because the capacitive reactance decreases with increasing frequency, the product IX_C reaches a maximum at a frequency slightly *lower* than the resonant frequency, f_r. Similarly, the product IX_L reaches a maximum at a frequency slightly *higher* than the resonant frequency. For high Q circuits, however, this difference is negligible and we can think of the capacitor (and inductor) as having a frequency response curve that peaks at f_r and has a bandwidth Δf given by Eq. 26-3. See Fig. 26-6*b*.

Note that the curve in Fig. 26-6*b* is not symmetrical. At very low frequencies, the capacitor voltage approaches the supply voltage, V (an open circuit). At very high frequencies, V_C approaches zero because X_C and I also approach zero. However, for high Q, high-frequency circuits, f_1 and f_2 can be considered to be symmetrically located below and above f_r.

We can obtain an expression for the capacitor (and inductor) voltage at resonance as follows:

$$V_C = IX_C \qquad (23\text{-}1)$$

But, at resonance:

$$I = \frac{V}{R} \qquad (15\text{-}15a)$$

and

$$X_C = X_L$$

Therefore,

$$V_C = \frac{V}{R}X_L$$

$$= \frac{X_L}{R}V$$

and

$$V_C = QV \qquad \text{volts (V)} \qquad (26\text{-}6)$$

where: V_C is the capacitor voltage at resonance in volts, V
Q is the total circuit quality, dimensionless
V is the voltage applied to the circuit at resonance, in volts, V

Because Q is usually much larger than unity, we see a significant resonant rise in the capacitor voltage at resonance. A similar voltage rise occurs (but 180° out of phase) across the inductor. However, because of its resistance, the coil's voltage is not given simply by Eq. 26-6, although the two are close in value for high Q circuits.

EXAMPLE 26-8

A coil of 8-H inductance and 400-Ω resistance is series-connected with a variable capacitor and an ammeter across the 120-V, 60-Hz line. Calculate:

a. The capacitor voltage when the capacitor is adjusted for maximum circuit current.
b. The necessary capacitor for the condition in part (a).

SOLUTION

a. At resonance, the circuit current is a maximum, so $f_r = 60$ Hz.

$$Q = \frac{X_L}{R} \quad (20\text{-}9)$$

$$= \frac{2\pi \times 60 \text{ Hz} \times 8 \text{ H}}{400\ \Omega} = 7.5$$

$$V_C = QV \quad (26\text{-}6)$$

$$= 7.5 \times 120 \text{ V} = \mathbf{900\ V}$$

b. $$f_r = \frac{1}{2\pi\sqrt{LC}} \quad (26\text{-}1)$$

$$C = \frac{1}{4\pi^2 f^2 L}$$

$$= \frac{1}{4\pi^2 \times (60 \text{ Hz})^2 \times 8 \text{ H}} = \mathbf{0.88\ \mu F}$$

Note, in Example 26-8, how it is possible for dangerously high voltages to be produced at resonance in a series *RLC* circuit.

Figure 26-7 shows how a capacitor's resonant rise of voltage is used to advantage in a frequency selective circuit, such as in a radio receiver or TV receiver.

The antenna, in Fig. 26-7*a*, "picks up" many different frequencies from the electromagnetic waves radiated by neighboring radio transmitters. These signals, in the order of several millivolts for an AM receiver, and only a few microvolts for an FM receiver, must be "tuned-in" by the resonant circuit. Assume that the circuit is tuned to a resonant frequency of 1400 kHz (e.g., the carrier frequency of a local AM station), by adjusting *C*, as in Fig. 26-7*b*. Only the information contained in the band of frequencies associated with *that* resonant frequency will appear across the capacitor for amplification and further processing. If the circuit's bandwidth is correct, all other adjacent signals will be rejected, or "tuned-out." The adjustment of *C* to a different value makes the circuit *sensitive* to a different band of frequencies, and another station may be tuned in.

EXAMPLE 26-9

What is the necessary range of a variable capacitor to use with a 200-μH coil in a series circuit to tune in the standard radio broadcast band from 535 to 1605 kHz.

SOLUTION

$$f_r = \frac{1}{2\pi\sqrt{LC}} \quad (26\text{-}1)$$

$$C = \frac{1}{4\pi^2 f_r^2 L}$$

For $f_r = 535$ kHz,

$$C = \frac{1}{4\pi^2 \times (535 \times 10^3 \text{ Hz})^2 \times 200 \times 10^{-6} \text{ H}}$$

$$= \mathbf{442\ pF}$$

For $f_r = 1605$ kHz,

$$C = \frac{1}{4\pi^2 \times (1605 \times 10^3 \text{ Hz})^2 \times 200 \times 10^{-6} \text{ H}}$$

$$= \mathbf{49\ pF}$$

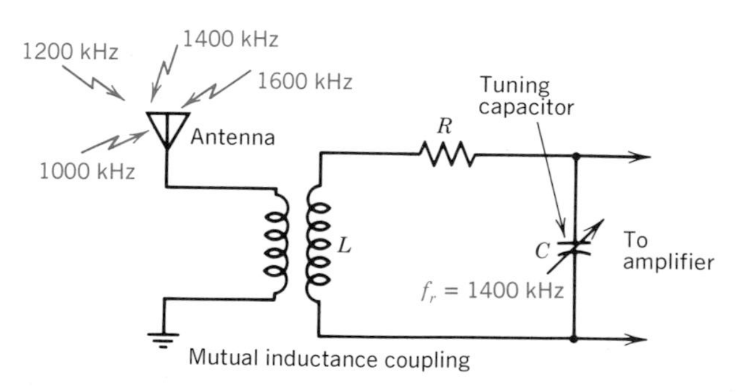

(*a*) Series *RLC* circuit tuned to 1400 kHz by *C* receives only one band of frequencies centered at 1400 kHz.

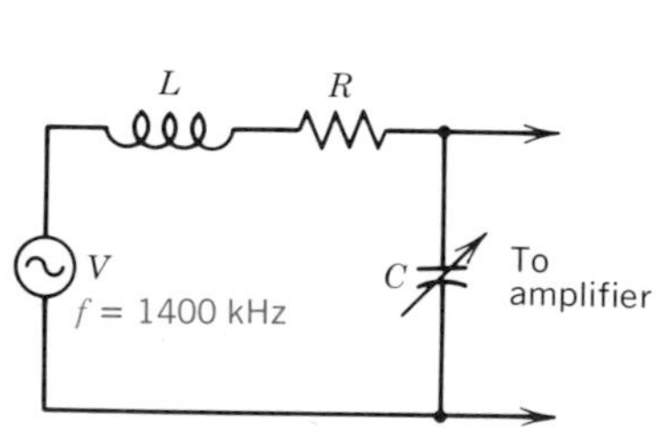

(*b*) Equivalent circuit.

FIGURE 26-7
Resonant rise of capacitor voltage in a series *RLC* circuit is used to select a desired band of frequencies.

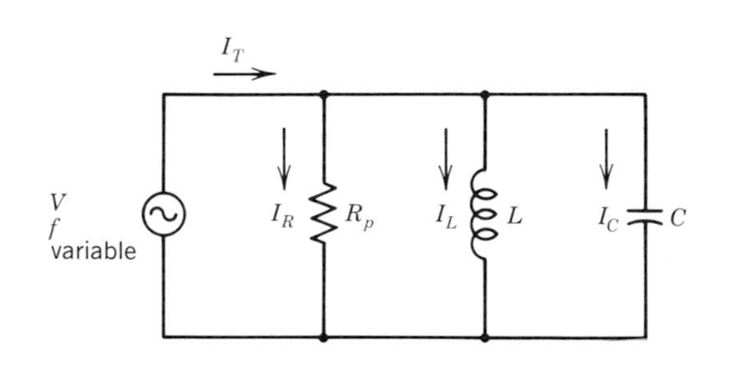

(*a*) Theoretical parallel resonant circuit. (Inductor has no resistance.)

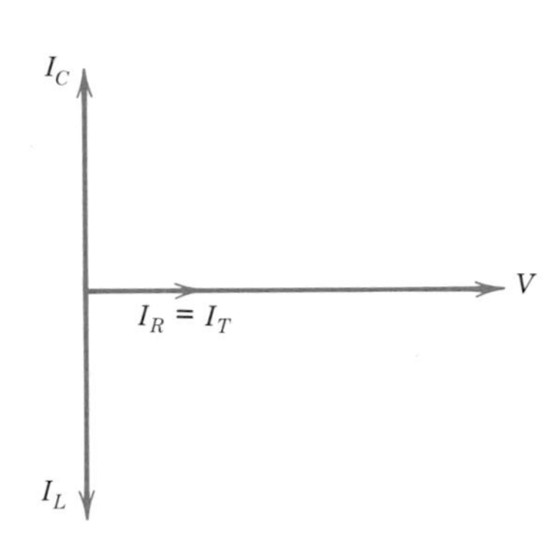

(*b*) Phasor diagram.

FIGURE 26-8
Resonance in a theoretical parallel resonant circuit.

26-6 RESONANCE IN A THEORETICAL, PARALLEL *RLC* CIRCUIT

Consider now the effect of varying frequency in a *parallel* *RLC* circuit, as in Fig. 26-8*a*.

The circuit shown is theoretical because we have assumed an ideal inductor having zero resistance. (Later we shall take into account the effect of the coil's resistance in Section 26-8.)

We know that at any frequency,

$$I_T = \sqrt{I_R^2 + (I_C - I_L)^2} \qquad (24\text{-}21)$$

$$I_T = \frac{V}{Z} \qquad (24\text{-}4)$$

and

$$\theta = \arctan \frac{I_C - I_L}{I_R} \qquad (24\text{-}22)$$

Also,

$$I_C = \frac{V}{X_C} \qquad (23\text{-}1)$$

and

$$I_L = \frac{V}{X_L} \qquad (20\text{-}1)$$

If the source frequency is adjusted until $X_L = X_C$, then $I_L = I_C$. This means that $(I_C - I_L) = 0$; also, the total circuit current I_T is a *minimum* and equal to I_R. But if I_T is a *minimum,* the impedance Z must be a *maximum.* (See Eq. 24-4.)

$$Z = \frac{V}{I_T} = \frac{V}{I_R} = R_p$$

Thus the circuit is purely resistive, $\theta = 0°$, and the circuit current and applied voltage are in phase with each other. (The L and C components apparently act as an open circuit, with infinite impedance, as far as the source is concerned.)

This condition is called *parallel resonance.* It is also sometimes referred to as "antiresonance" because the effect on circuit current and impedance is opposite to the series resonant effect, as shown in Fig. 26-9.

Since parallel resonance occurs in *this* circuit when $X_L = X_C$, as it does in a series *RLC* circuit, the resonant frequency is given by the same equation:

$$f_r = \frac{1}{2\pi\sqrt{LC}} \qquad (26\text{-}1)$$

EXAMPLE 26-10

A 100-μH coil of negligible resistance, a 50-pF capacitor, and a 100-kΩ resistor are connected in parallel across a variable frequency 50-mV signal generator. Calculate:

a. The resonant frequency of the circuit.

b. Each branch current when the signal generator is adjusted to the resonant frequency.

Show the currents in a circuit diagram.

SOLUTION

a. $$f_r = \frac{1}{2\pi\sqrt{LC}} \qquad (26\text{-}1)$$

$$= \frac{1}{2\pi\sqrt{100 \times 10^{-6}\ \text{H} \times 50 \times 10^{-12}\ \text{F}}}$$

$$= 2.25 \times 10^6\ \text{Hz} = \mathbf{2.25\ MHz}$$

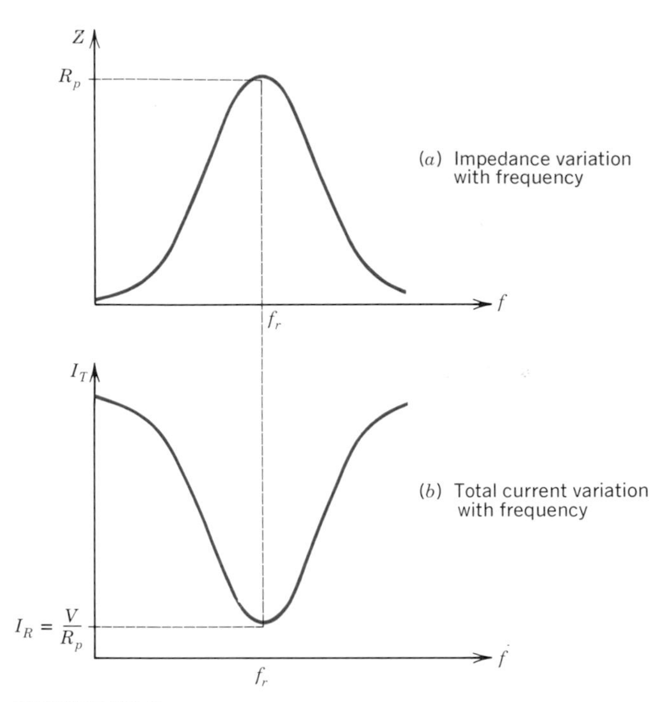

FIGURE 26-9
Variation of impedance and total current in a parallel *RLC* circuit with a variable frequency source.

b. $I_R = \frac{V}{R}$ (15-15a)

$$= \frac{50 \text{ mV}}{100 \text{ k}\Omega} = \mathbf{0.5\ \mu A}$$

$$X_L = 2\pi fL \qquad (20\text{-}2)$$

$$= 2\pi \times 2.25 \times 10^6 \text{ Hz} \times 100 \times 10^{-6} \text{ H}$$

$$= 1414\ \Omega$$

$$I_L = \frac{V}{X_L} \qquad (20\text{-}1)$$

$$= \frac{50 \text{ mV}}{1.4 \text{ k}\Omega} = \mathbf{35.7\ \mu A}$$

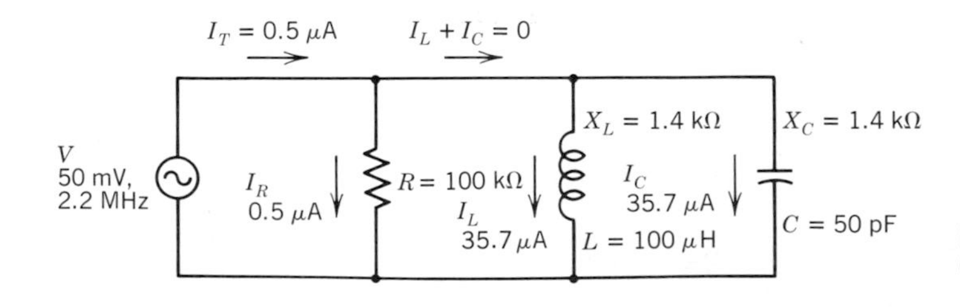

FIGURE 26-10
Circuit currents for Example 26-10.

$$X_C = \frac{1}{2\pi fC} \qquad (23\text{-}2)$$

$$= \frac{1}{2\pi \times 2.25 \times 10^6 \text{ Hz} \times 50 \times 10^{-12} \text{ F}}$$

$$= 1414\ \Omega$$

$$I_C = \frac{V}{X_C} \qquad (23\text{-}1)$$

$$= \frac{50 \text{ mV}}{1.4 \text{ k}\Omega} = \mathbf{35.7\ \mu A}$$

See Fig. 26-10.

Note that the inductor and capacitor currents are equal in magnitude and 180° out of phase with each other. Thus $I_L + I_C$ in Fig. 26-10 is zero, even though the individual currents are much larger than the total line current. This shows how the theoretical parallel LC circuit is an *open* circuit, since it draws no current from the source.

The type of signal generator referred to in Example 26-10 is called a radio frequency (RF) generator. An RF generator produces signals from approximately 100 kHz to several hundred megahertz. The output voltage is much smaller than from an audio generator, sometimes as small as 0.1 V.

A typical RF generator is shown in Fig. 26-11. In addition to producing a sine wave output, some produce an *amplitude-modulated* output, where the radio-frequency signal (carrier) has its amplitude varied (modulated) at an *audio* rate, such as 400 Hz. This type of signal is useful for checking the operation of communications receivers.

26-6.1 The Q of a Theoretical, Parallel Resonant Circuit

We have seen (Section 26-4) that the Q of a resonant circuit is given by the following equation:

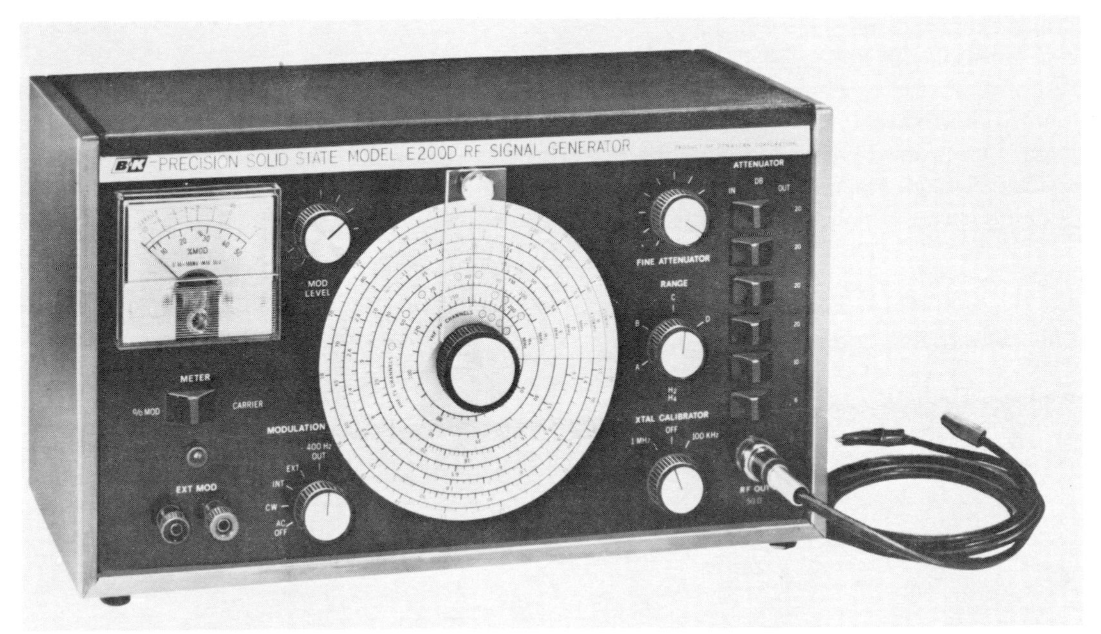

FIGURE 26-11
A radio-frequency signal generator. (Courtesy of Dynascan Corporation, B & K-Precision Product Group.)

$$Q = \frac{P_q}{P} \tag{26-4}$$

For the theoretical parallel *RLC* resonant circuit, with no resistance in the inductor,

$$P_q = \frac{V^2}{X_L} \tag{25-3}$$

and

$$P = \frac{V^2}{R_p} \tag{15-13}$$

Thus

$$Q = \frac{V^2/X_L}{V^2/R_p}$$

and

$$\mathbf{Q = \frac{R_p}{X_L}} \tag{26-7}$$

where: Q is the quality of a parallel *RLC* circuit, dimensionless

R_p is the resistance of the circuit in ohms, Ω

X_L is the inductive reactance of the *zero resistance* inductor in ohms, Ω

We have seen (Example 26-10) that R_p is usually much larger than X_L, so $Q > 1$ as in a series *RLC* circuit.

Note that the impedance of the parallel circuit at resonance can now be expressed in terms of Q:

$$Z = R_p = QX_L \tag{26-7a}$$

where terms are as previously given. Thus a high Q circuit has a high impedance at resonance.

We can also express the currents in terms of Q:

$$V = I_L X_L = I_R R_p$$

Therefore,

$$I_L = \frac{R_p}{X_L} \times I_R$$

But

$$\frac{R_p}{X_L} = Q \tag{26-7}$$

and

$$I_R = \text{line current} = I_T$$

Thus

$$I_L = I_C = Q \times I_T \tag{26-8}$$

where symbols are as already stated.

Equation 26-8 states that there is a *resonant rise of current* in the *tank circuit* consisting of the parallel LC combination. This corresponds to the resonant rise of *voltage* in a *series* circuit. Very large currents can circulate in the tank circuit as energy is transferred back and forth between the coil and capacitor, even if the line current is very low, because the Q of the circuit is high (Eq. 26-8).

EXAMPLE 26-11

For the circuit in Example 26-10, determine:

a. Q.

b. The inductor and capacitor currents at resonance.

c. The impedance at resonance.

SOLUTION

a. $$Q = \frac{R_p}{X_L} \qquad (26\text{-}7)$$

$$= \frac{100\ \text{k}\Omega}{1.4\ \text{k}\Omega} = \mathbf{71.4}$$

b. $$I_L = I_C = Q \times \text{line current} \qquad (26\text{-}8)$$

$$= 71.4 \times 0.5\ \mu\text{A} = \mathbf{35.7\ \mu A}$$

c. $$Z = QX_L \qquad (26\text{-}7a)$$

$$= 71.4 \times 1.4\ \text{k}\Omega = \mathbf{100\ k\Omega}$$

26-7 SELECTIVITY OF A PARALLEL RESONANT CIRCUIT

The bandwidth of a parallel resonant circuit (theoretical or practical) is the band of frequencies between f_1 and f_2 on the resonance curve, where the total impedance drops to 0.707 of the parallel resistance R_p. See Fig. 26-12.

Again it can be shown that

$$\Delta f = \frac{f_r}{Q} \qquad (26\text{-}3)$$

so that a high Q results in a very selective circuit, as in the case for a series resonant circuit.

EXAMPLE 26-12

Determine the bandwidth of the circuit in Example 26-10.

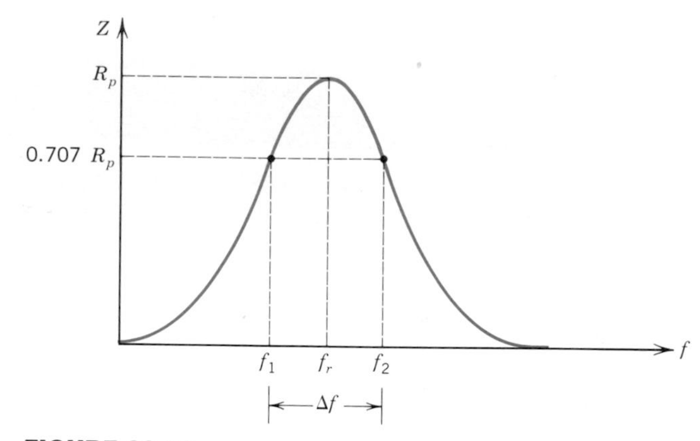

FIGURE 26-12
Bandwidth of a parallel resonant circuit.

SOLUTION

$$\Delta f = \frac{f_r}{Q} \qquad (26\text{-}3)$$

$$= \frac{2.25 \times 10^6\ \text{Hz}}{71.4} = \mathbf{31.5\ kHz}$$

To see the usefulness of a parallel circuit to select certain frequencies, we must now include the internal resistance of the source. This internal resistance is purposely made high for reasons described below. See Fig. 26-13.

If we represent the impedance of the parallel RLC circuit at *any* frequency by Z, as in Fig. 26-13b, we obtain a series voltage-divider circuit, where

$$V_{\text{out}} = V \times \frac{Z}{Z + R_{\text{int}}} \qquad (5\text{-}6)$$

At resonance:

$$V_{\text{out}} = V \times \frac{R_p}{R_p + R_{\text{int}}} \qquad (5\text{-}6a)$$

That is, at resonance, when Z is a maximum and equal to R_p, the maximum voltage is developed across the tank circuit. At frequencies above or below resonance, the impedance Z decreases, as in Fig. 26-12, so the output voltage also decreases. If R_{int} were zero or small, the voltage across the tank circuit would be the same as the source voltage, at all frequencies, and no selectivity would result. For this reason, the R_{int} is deliberately

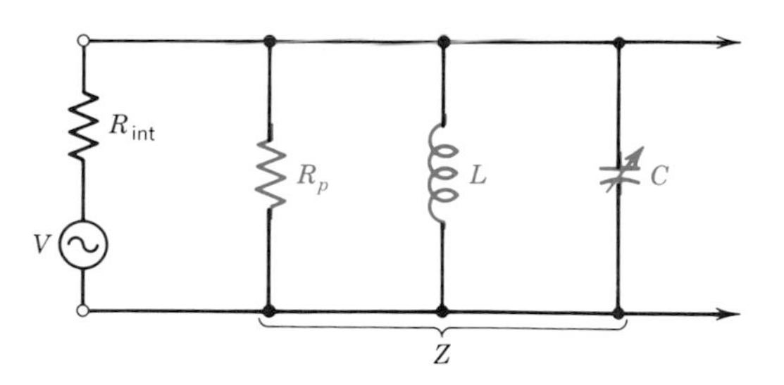

a. Parallel resonant circuit with high internal resistance source

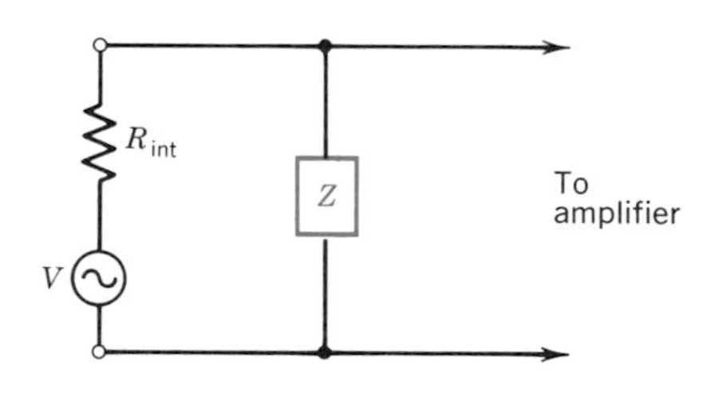

b. Equivalent circuit. At resonance, $Z = R_p$

FIGURE 26-13
Selectivity of a parallel resonant circuit is provided by a high internal resistance in the source.

increased so that high selectivity is produced in the range of the resonant frequency.

EXAMPLE 26-13

A parallel *RLC* circuit has $R_p = 100\ \text{k}\Omega$. The circuit is connected to a 10-μV source with an internal resistance of 50 kΩ. Calculate:

a. The output voltage across the tank circuit when the capacitor is adjusted to provide resonance.
b. The output voltage for a frequency far from resonance where the tank circuit's impedance is 10 kΩ.

SOLUTION

a. $$V_{out} = V \times \frac{R_p}{R_p + R_{int}} \qquad (5\text{-}6a)$$

$$= 10\ \mu\text{V} \times \frac{100\ \text{k}\Omega}{100\ \text{k}\Omega + 50\ \text{k}\Omega}$$

$$= \mathbf{6.7\ \mu V}$$

b. $$V_{out} = V \times \frac{Z}{Z + R_{int}} \qquad (5\text{-}6)$$

$$= 10\ \mu\text{V} \times \frac{10\ \text{k}\Omega}{10\ \text{k}\Omega + 50\ \text{k}\Omega} = \mathbf{1.7\ \mu V}$$

NOTE The addition of Z and R_{int} in part (b) above should actually be done *vectorially*, since, off resonance, the impedance is no longer pure resistance. But the error is small, since $R_{int} > Z$ and the total circuit is primarily resistive.

The selectivity of a parallel resonant circuit is used to produce sine wave voltages from *nonsinusoidal* pulses. These repetitive pulses have some *fundamental* sinusoidal frequency and many *harmonics*. (A harmonic is some multiple of the fundamental.) Thus, if a circuit is resonant to only one frequency, it will reject the others and produce a pure sine wave. This is made use of in Class C radio frequency amplifiers that conduct for less than half a cycle of a sine wave and produce pulses at some basic frequency. (This mode of operation produces a highly efficient amplifier for large power outputs.) A parallel resonant circuit can also be used as a frequency doubler or *multiplier* if the circuit is tuned to the second harmonic rather than the fundamental. Finally, the output of an RF signal generator may not be a good sine wave until it is used with some resonant circuit, series or parallel, to reject the harmonics.

26-8 RESONANCE IN A PRACTICAL, TWO-BRANCH PARALLEL *RLC* CIRCUIT

We now consider the effect of the coil's resistance on the resonant frequency of a two-branch parallel *RLC* circuit. See Fig. 26-14.

Assume that the frequency of the source has been adjusted until $X_L = X_C$. Because of the (series) resistance of the coil, R_s, there is *less* current through the coil than

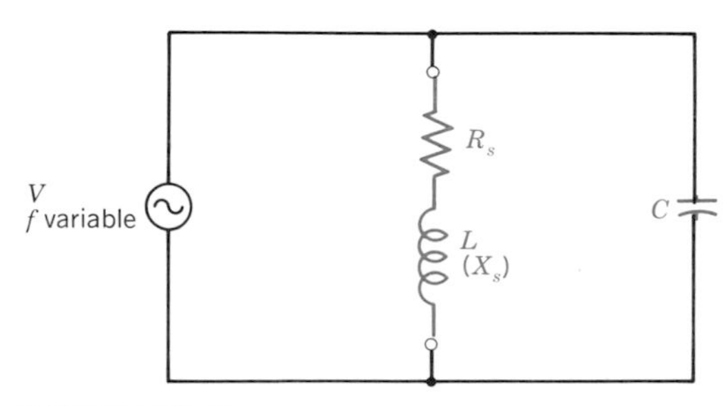

FIGURE 26-14
Practical parallel *RLC* circuit.

through the capacitor. This means that the reactive power in L and C are *not* equal, so we do *not* have resonance. That is, for resonance, in series or parallel, we require the following essentially identical conditions:

1 $P_{qL} = P_{qC}$.

2 $\cos\theta = 1$.

3 $\theta = 0°$.

4 $Z = R$.

To make $P_{qL} = P_{qC}$, we must *lower* the frequency slightly, reducing X_L so that I_L *increases* until $I_L^2X_L = I_C^2X_C$. When this occurs, it can be shown that the resonant frequency is given by

$$f_r = \frac{1}{2\pi\sqrt{LC}} \times \sqrt{\frac{Q^2}{1+Q^2}} \quad \text{hertz (Hz)} \quad (26\text{-}9)$$

where: f_r is the resonant frequency of a practical parallel circuit in hertz, Hz

L is the inductance of the coil in henrys, H

C is the capacitance of the circuit in farads, F

Q is the quality of the coil, dimensionless

Note that

$$f_r < \frac{1}{2\pi\sqrt{LC}}$$

as we expected, so the coil's resistance *does* determine the circuit's resonant frequency. However, if $Q \geq 10$, the resonant frequency, f_r, is given by $\frac{1}{2\pi\sqrt{LC}}$ with a maximum error of only 0.5%. Thus Eq. 26-1 gives the resonant frequency of series *RLC* circuits *and* high Q, parallel *RLC* circuits.

Note also, that the Q in Eq. 26-9 is specified as the Q of the coil itself. The reason for this can be seen in the equivalent circuit of Fig. 26-15*b*.

By the use of network analysis techniques, the practical two-branch parallel circuit in Fig. 26-15*a* can be converted to the equivalent three-branch theoretical circuit in Fig. 26-15*b*, where

$$R_p = \frac{R_s^2 + X_s^2}{R_s} = \frac{Z_s^2}{R_s} \quad (26\text{-}10)$$

and

$$X_p = \frac{R_s^2 + X_s^2}{X_s} = \frac{Z_s^2}{X_s} \quad (26\text{-}11)$$

where symbols are as previously stated.

But we know that for the theoretical circuit:

$$Q = \frac{R_p}{X_p} \quad (26\text{-}7)$$

$$= \frac{R_s^2 + X_s^2}{R_s} \div \frac{R_s^2 + X_s^2}{X_s}$$

Therefore,

$$Q = \frac{X_s}{R_s} = \frac{2\pi fL}{R_s} \quad (26\text{-}12)$$

where: Q is the quality of the coil (and the whole parallel circuit if there is no additional parallel resistance), dimensionless

L is the inductance of the coil in henrys, H

R_s is the series resistance of the coil itself in ohms, Ω

We can now express Eq. 26-9 in the following form:

$$f_r = \frac{1}{2\pi\sqrt{LC}} \times \sqrt{1 - \frac{CR_s^2}{L}} \quad \text{hertz (Hz)} \quad (26\text{-}13)$$

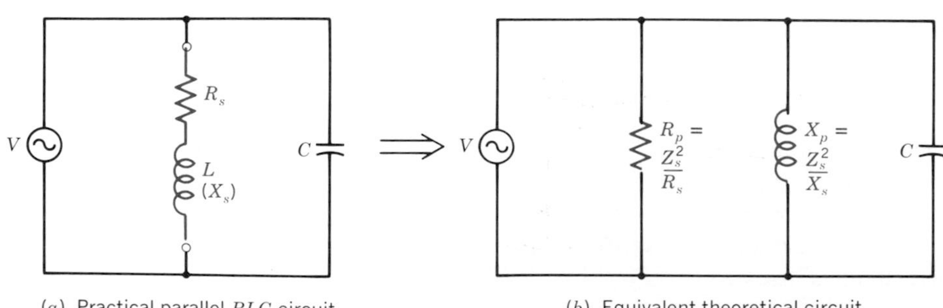

(*a*) Practical parallel *RLC* circuit

(*b*) Equivalent theoretical circuit

FIGURE 26-15
A practical parallel resonant circuit may be converted to an equivalent theoretical parallel circuit.

where symbols are as already stated. Note that if $R_s = 0$, Eq. 26-13 reduces to Eq. 26-1,

$$f_r = \frac{1}{2\pi\sqrt{LC}}$$

EXAMPLE 26-14

A 10-μH coil having an ac resistance of 5 Ω is connected in parallel with a 0.01-μF capacitor across a radio frequency signal generator. Calculate:

a. The resonant frequency of the two-branch parallel circuit.

b. The Q of the circuit.

c. The bandwidth of the circuit.

d. The impedance of the circuit when the generator is tuned to the resonant frequency.

SOLUTION

a. $$f_r = \frac{1}{2\pi\sqrt{LC}} \times \sqrt{1 - \frac{CR_s^2}{L}} \qquad (26\text{-}13)$$

$$= \frac{1}{2\pi\sqrt{10 \times 10^{-6}\ \text{H} \times 0.01 \times 10^{-6}\ \text{F}}} \times \sqrt{1 - \frac{0.01 \times 10^{-6} \times 5^2}{10 \times 10^{-6}}}$$

$$= 503\ \text{kHz} \times 0.987 = \mathbf{496\ kHz}$$

b. $$Q = \frac{X_s}{R_s} = \frac{2\pi fL}{R_s} \qquad (26\text{-}12)$$

$$= \frac{2\pi \times 496 \times 10^3\ \text{Hz} \times 10 \times 10^{-6}\ \text{H}}{5\ \Omega}$$

$$= \frac{31\ \Omega}{5\ \Omega} = \mathbf{6.2}$$

c. $$\Delta f = \frac{f_r}{Q} \qquad (26\text{-}3)$$

$$= \frac{496\ \text{kHz}}{6.2} = \mathbf{80\ kHz}$$

d. $$Z = R_p = \frac{R_s^2 + X_s^2}{R_s} \qquad (26\text{-}10)$$

$$= \frac{(5\ \Omega)^2 + (31\ \Omega)^2}{5\ \Omega}$$

$$= \mathbf{197\ \Omega} \quad \text{(pure resistance)}$$

Note that since $Q = 6.2$ is less than 10, the resonant frequency is a little lower than that given by Eq. 26-1.

Also note the low impedance (resistance) of the circuit at resonance. This is because the circuit's Q is very low, that is,

$$Z = R_p = \frac{R_s^2 + X_s^2}{R_s}$$

$$= R_s + \frac{X_s^2}{R_s}$$

$$= R_s\left(1 + \frac{X_s^2}{R_s^2}\right)$$

$$= R_s(1 + Q^2) \quad \text{ohms } (\Omega) \qquad (26\text{-}14)$$

$$= 5\ \Omega\ (1 + 6.2^2) = \mathbf{197\ \Omega}$$

26-8.1 Damping a Parallel Resonant Circuit

Purposely reducing the Q of a tuned parallel circuit by introducing resistance is called *damping*. This can be achieved for the practical two-branch circuit in two ways.

1. The series resistance of the coil, R_s, in Fig. 26-14 can be *increased*, as seen in Eq. 26-12, reducing the Q. This also has an interesting effect on the resonant frequency as shown by Eq. 26-13. If R_s is increased until $\frac{CR_s^2}{L} \geq 1$, there is *no* resonant frequency. This is called *critical* damping and is used in high frequency circuits (as in TV receivers) to keep the coils and their distributed capacitances from *ringing* (oscillating at some undesired frequency).

2. Resistance can be added in *parallel* with the circuit. This *reduces* the effective value of R_p in Eq. 26-7, so the Q is also reduced. This is called *shunt* damping and is the preferred method to maintain a symmetrical resonant response, although both series and parallel damping combined is also used.

Damping is sometimes used in communications receivers (see Fig. 26-16) to broaden the bandwidth when looking for a weak signal. That is, a weak signal can easily be "tuned through" without detection by a circuit that has a very narrow bandwidth. However, once the weak signal is found when "tuned" in, a narrow bandwidth would be an asset to block out competing, adjacent stronger signals.

EXAMPLE 26-15

In the parallel two-branch circuit of Example 26-14, with the 10-μH coil having a resistance of 5 Ω, and the 0.01-μF capacitor, determine:

FIGURE 26-16
A communications receiver that uses parallel resonant circuits. (Courtesy of Trio-Kenwood Communications, Inc.)

a. The minimum resistance to be *added* in series with the coil to make sure the circuit has no resonant frequency.
b. The resistance to be added in *series* with the coil to broaden the bandwidth to 100 kHz and the new resonant frequency.
c. The shunting resistance to be connected in *parallel* with the coil to broaden the bandwidth to 100 kHz.

SOLUTION

a. To avoid resonance: $\frac{CR_s^2}{L} \geq 1$

Therefore, minimum resistance

$$R_s = \sqrt{\frac{L}{C}}$$

$$= \sqrt{\frac{10 \times 10^{-6}\ \text{H}}{0.01 \times 10^{-6}\ \text{F}}} = 31.6\ \Omega$$

Since the coil's resistance is 5 Ω, the *minimum* resistance to be added = 31.6 − 5 Ω = **26.6 Ω**.

b. For a bandwidth of 100 kHz:

$$Q = \frac{f_r}{\Delta f} \tag{26-3}$$

$$= \frac{496\ \text{kHz}}{100\ \text{kHz}} = 4.96$$

$$f_r = \frac{1}{2\pi\sqrt{LC}} \times \sqrt{\frac{Q^2}{1 + Q^2}} \tag{26-9}$$

$$= 503\ \text{kHz} \times \sqrt{\frac{4.96^2}{1 + 4.96^2}}$$

$$= 503\ \text{kHz} \times 0.98 = \mathbf{493\ kHz}$$

Thus the new Q must be 4.93 to produce a bandwidth of 100 kHz at f_r = 493 kHz.

$$Q = \frac{X_s}{R_s} = \frac{2\pi fL}{R_s} \tag{26-12}$$

$$R_s = \frac{2\pi f_r L}{Q}$$

$$= \frac{2\pi \times 493 \times 10^3\ \text{Hz} \times 10 \times 10^{-6}\ \text{H}}{4.93}$$

$$= 6.3\ \Omega$$

The resistance to be added in series with the coil = 6.3 − 5 Ω = **1.3 Ω**.

c. For the equivalent parallel circuit with R_p:

$$Q = \frac{R_p}{X_p} \tag{26-7}$$

Initially, with Δf = 80 kHz, the *equivalent* R_p = 197 Ω. For a Δf of 100 kHz, the *effective* R_s = 6.3 Ω:

$$R_p = \frac{R_s^2 + X_s^2}{R_s} \tag{26-10}$$

$$= \frac{(6.3\ \Omega)^2 + (31\ \Omega)^2}{(6.3\ \Omega)} = 159\ \Omega$$

We need a resistance in parallel with 197 Ω to produce 159 Ω:

$$R_T = \frac{R_1 \times R_2}{R_1 + R_2} \tag{6-10}$$

$$159 = \frac{197 \times R_2}{197 + R_2}$$

$$R_2 = \mathbf{824\ \Omega}$$

Example 26-15 shows how the resonant frequency of a parallel *RLC* circuit can be changed (3 kHz here) by adding resistance in parallel with the overall circuit, or in series with the coil itself.

26-9 OPERATING PRINCIPLE OF A SUPERHETERODYNE AM RECEIVER

Series and parallel resonant circuits are widely used in AM, FM, and TV receivers. Figure 26-17 shows a block diagram of an AM radio receiver. (The principle of operation also applies to an FM radio receiver but with different frequencies.)

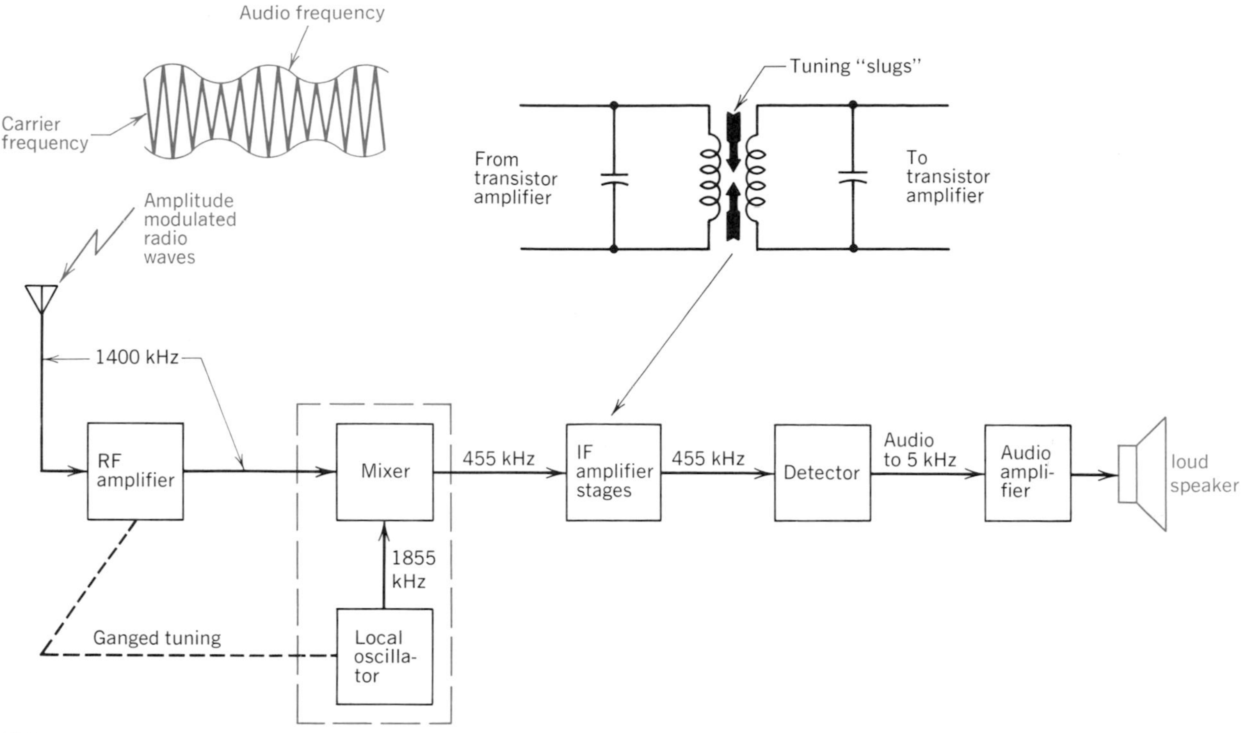

FIGURE 26-17
Block diagram of a superheterodyne AM radio receiver.

We have already seen (in Fig. 26-7), how a series resonant circuit can be tuned to select just one of the many bands of frequencies arriving at the antenna. This antenna signal is fed to a radio frequency amplifier, which resonantly tunes to the incoming signal whenever we turn the tuning dial. The incoming signal is very weak (a few microvolts) and must be amplified thousands of times before it is audible. This amplification is carried out in stages.

In the early days of radio, each stage had to be tuned to the signal being received so that each stage had to be tunable over the full band, from 535 to 1605 kHz. This was called a tuned radio frequency receiver (TRF); it had many disadvantages.

Modern receivers use *frequency conversion* or *heterodyne* circuits to produce a signal whose *radio* frequency is always the same (455 kHz) but which contains the *audio* frequencies of whichever station is tuned in. In this way, all further stages of radio frequency amplification take place at the one *fixed* frequency called the *intermediate* frequency (IF).

Assume that we wish to listen to an AM radio station broadcasting at 1400 kHz. When we turn the dial to this frequency, the radio frequency amplifier is tuned to 1400 kHz. Also, "ganged" together, a *local oscillator* produces an RF sine wave equal to 1400 kHz + 455 kHz, or 1855 kHz. The 1400-kHz signal (containing the "intelligence" or desired audio) and the 1855 kHz signal (unmodulated) are applied to the *mixer*. As the name suggests, the mixer or *converter* produces an output signal, which, among other things, contains frequencies of the *sum* and *difference* of the input signals. That is, frequencies of 1855 kHz + 1400 kHz = 3255 kHz and 1855 kHz − 1400 kHz = 455 kHz are available at the output. Each of these two frequencies contains the same audio as the initial signal. However, for practical considerations, only the *difference* frequency of 455 kHz is used.

Thus all further intermediate frequency amplifier stages are tuned to this frequency, since this is the only frequency produced, no matter which station is tuned in. These stages use parallel *LC* circuits with inductive coupling from one stage to the next, as shown in Fig. 26-

17. The coupling can be adjusted by ferrite slugs to *align* the receiver at the proper frequency and provide the correct bandwidth of approximately 10 kHz. This bandwidth is necessary to prevent overlap between stations. As a result, the highest audio frequency permitted is 5 kHz. When the carrier (1400 kHz, in our example) is amplitude-modulated by a frequency of 5 kHz, *sideband* frequencies are produced, 5 kHz *above* and *below* the carrier. All of these frequencies must be retrieved, except in a single side band receiver, to reproduce the original sound. This process of *demodulation* is carried out in the detector stage and is followed by audio amplification to drive a loudspeaker.

SUMMARY

1. When a series circuit is in resonance, the circuit's impedance is a minimum (equal to R), the current is a maximum (equal to V/R), and the applied voltage and current are in phase with each other.

2. The resonant frequency of a series RLC circuit is given by

$$f_r = \frac{1}{2\pi\sqrt{LC}}$$

3. The bandwidth of a resonant circuit is the band of frequencies, Δf, between the two frequencies, f_1 and f_2, at which the power delivered to the circuit is one-half the power delivered at the resonant frequency.

$$\Delta f = \frac{f_r}{Q}$$

4. The Q of a series resonant circuit can be obtained using:

$$Q = \frac{P_q}{P} = \frac{X_L}{R_{ac}} = \frac{1}{R_{ac}}\sqrt{\frac{L}{C}}$$

5. At resonance in a series RLC circuit, there is a resonant rise of voltage across the capacitor and inductor given by

$$V_L = V_C = QV$$

6. At resonance in a theoretical *parallel* RLC circuit, the circuit *impedance* is a *maximum* (equal to R_p), the line *current* is a *minimum* (equal to V/R_p), and the circuit is purely resistive with the applied voltage and current in phase with each other.

7. The resonant frequency of a theoretical parallel RLC circuit is given by

$$f_r = \frac{1}{2\pi\sqrt{LC}}$$

8. The Q of a theoretical parallel RLC circuit is given by

$$Q = \frac{R_p}{X_L}$$

9. There is a resonant rise of current in the tank circuit of a parallel RLC circuit given by

$$I_L = I_C = Q \times I_R$$

10. The bandwidth of a parallel resonant circuit is the band of frequencies between f_1 and f_2, where the circuit impedance drops to 0.707 of the parallel resistance R_p, and is given by

$$\Delta f = \frac{f_r}{Q}$$

11. The selectivity of a parallel resonant circuit depends upon the source having a high internal resistance to cause a voltage-divider effect that maximizes the output voltage at resonance.

12. Because of a coil's resistance, the resonant frequency of a practical RLC circuit is slightly less than the theoretical unless $Q \geq 10$.

13. A parallel circuit can be damped to reduce Q and increase the bandwidth by adding series resistance to the coil or parallel resistance to the whole circuit.

14. A superheterodyne receiver uses a frequency mixer to produce a difference or intermediate frequency so that amplification can take place at a fixed frequency.

SELF-EXAMINATION

Answer true for false or a, b, c, d for multiple choice
(Answers at back of book)

26-1. Series resonance always occurs where $X_L = X_C$. ________

26-2. At a series circuit's resonant frequency, the impedance is a maximum and the current is a minimum. ________

26-3. A series circuit behaves, as far as the source is concerned, as if it is a pure resistance when tuned to resonance. ________

26-4. The capacitor voltage can never be larger than the applied voltage in a series RLC circuit or Kirchhoff's voltage law would be violated. ________

26-5. The resonant frequency of a series RLC circuit is not determined by the resistance at all. ________

26-6. If a series circuit has $R = 2\ \Omega$, $L = 20$ mH, and $C = 2\ \mu$F, the approximate value of the circuit's resonant frequency is

a. 8 kHz
c. 8 MHz
b. 800 Hz
d. 80 kHz

26-7. The approximate value of Q in Question 26-6 is

a. 8
c. 32
b. 16
d. 2

26-8. The bandwidth of the circuit in Question 26-6 is approximately:

a. 50 Hz
c. 250 Hz
b. 4 kHz
d. 100 Hz

26-9. The Q of a series circuit can be increased only by reducing the circuit's resistance. ________

26-10. If a 10-V_{p-p}, sine wave signal generator is connected to the series circuit in Question 26-6 and adjusted for resonance, the peak-to-peak voltage across the capacitor is

a. 16 V
c. 160 V
b. 80 V
d. 20 V

26-11. At resonance, the same voltage occurs across the inductor as across the capacitor for all conditions. ________

26-12. The resonant frequency of a theoretical parallel circuit is calculated using the same equation as for a series *RLC* circuit. ________

26-13. At parallel resonance the circuit has its maximum impedance and is purely reactive. ________

26-14. A 10-μH coil of negligible resistance, a 10-pF capacitor, and a 100-kΩ resistor are connected in parallel. The circuit's approximate resonant frequency is

a. 1.6 MHz
b. 160 kHz
c. 16 MHz
d. 160 MHz

26-15. The *Q* and bandwidth of the circuit in Question 26-14 are

a. 50, 800 kHz
b. 100, 160 kHz
c. 10, 16 kHz
d. 20, 80 kHz

26-16. The principal difference between the practical and theoretical parallel circuits is the resistance of the inductor. ________

26-17. A practical parallel circuit has a resonant frequency slightly higher than that given by the equation for the theoretical circuit. ________

26-18. A parallel resonant circuit can be damped either by adding resistance in series with the coil or in parallel with the coil. ________

26-19. A coil's series resistance can be increased to a point where the parallel *RLC* circuit will not resonate. ________

26-20. The advantage of a superheterodyne receiver is that most of the radio frequency amplification takes place at the fixed intermediate frequency. ________

REVIEW QUESTIONS

1. What are the conditions that characterize a series resonant circuit?

2. Why is the resonant frequency of a practical series circuit not affected by resistance whereas a practical parallel circuit's resistance does affect the resonant frequency?

3. a. What do you understand by the term *selectivity of a circuit*?
b. How is selectivity measured?
c. What is the origin of the term *half-power frequencies*?
d. Why are the half-power frequencies sometimes referred to as 3-dB frequencies?
e. What is meant by "edge frequencies"?

4. a. What is the *general* definition of the *Q* of a circuit?
b. What equation shows that *Q* is not determined only by *L* and *R*?
c. Justify why a reduction in *R* or an increase in *L*/*C* increases *Q*.

5. a. How is it possible for the voltage across the capacitor in a series resonant circuit to be larger than the input voltage?
b. How is this fact used to select only certain frequencies from an antenna?

6. a. Why is parallel resonance sometimes called antiresonance?
b. What is theoretical about a pure *RLC* circuit?
c. If it were possible to make a coil having zero resistance, what would be the impedance of a parallel coil and capacitor at resonance?
d. What would be the line current and tank current for this condition?

7. Explain how it is necessary for the source, connected to a parallel resonant circuit, to have a high resistance to produce a frequency selective function.

8. Justify why the resonant frequency of a practical parallel circuit is *less* than that for a theoretical circuit having the same L and C values.

9. Why do you think the addition of sufficient resistance in series with a coil used in a parallel *RLC* circuit could prevent the circuit from having a resonant frequency?

10. a. What do you understand by the term *damping*?

b. How can damping be achieved?

11. a. What range of frequencies must the local oscillator in Fig. 26-17 be able to produce to tune in the whole AM broadcast band?

b. If an FM tuner uses an intermediate frequency of 10.7 MHz, what range of frequencies must be generated in the local oscillator to receive incoming frequencies from 88-108 MHz?

c. What is the major difference between a superheterodyne receiver and a tuned radio frequency receiver?

d. What does it mean to *align* a receiver?

e. Why does an AM receiver not require a wider bandwidth than 10 kHz?

PROBLEMS

(Answers to odd-numbered problems at back of book)

26-1. A series *RLC* circuit consists of a 150-μH coil having a resistance of 12 Ω, and a 75-pF capacitor. A 20-mV radio frequency signal generator is connected to the circuit. For frequencies of (1) 1.485 MHz, (2) 1.500 MHz, and (3) 1.515 MHz, determine:

a. The impedance.

b. The phase angle.

c. The current.

d. V_R, V_L, and V_C.

26-2. Repeat Problem 26-1 using a 300-μH coil (12-Ω resistance) and a 37.5-pF capacitor.

26-3. Calculate the resonant frequency of the circuit in Problem 26-1.

26-4. Calculate the resonant frequency of the circuit in Problem 26-2.

26-5. What size of capacitance is required to provide series resonance with a 1-mH coil at 10-kHz?

26-6. What value of inductance will cause series resonance with a 0.0015-μF capacitor at 455 kHz?

26-7. A series circuit has a resonant frequency of 250 kHz and a bandwidth of 20 kHz. If 6 mW of power is delivered to the circuit at 250 kHz, calculate:

a. The edge frequencies.

b. The power delivered to the circuit when operating at the edge frequencies.

c. The Q of the circuit.

26-8. The power delivered to a series *RLC* circuit is monitored at different frequencies. The power is noted to be a maximum, and then two frequencies are found, at 800 and 900 Hz, where the power drops to one-half of maximum. Determine:

a. The bandwidth of the circuit.

b. The resonant frequency.

c. The Q of the circuit.

26-9. For the circuit in Problem 26-1, calculate:

a. Q.

b. The circuit's bandwidth.

26-10. For the circuit in Problem 26-2, calculate:

a. Q.

b. The circuit's bandwidth.

26-11. Determine the Q of a series resonant circuit in which $R = 15\ \Omega$, $L = 50\ \mu\text{H}$, and $C = 75$ pF.

26-12. A circuit must have a resonant frequency of 10.7 MHz and a bandwidth of 150 kHz. If an 8-Ω resistance coil is to be used, what is the required L/C ratio.

26-13. A series RLC circuit is resonant at 5 kHz when a 50-mH coil having a resistance of 150 Ω is used. If a 10-V p-p signal is applied to the circuit calculate:

a. The capacitor voltage at resonance.

b. The circuit capacitance.

c. The circuit's bandwidth.

d. The power delivered to the circuit at resonance.

26-14. An oscilloscope is connected across a capacitor in a series RLC circuit. The oscilloscope shows a maximum peak-to-peak voltage of 150 V when the input voltage of 24 V is adjusted in frequency to 400 Hz. If a series-connected wattmeter indicates 20 W, calculate:

a. The Q of the circuit.

b. The circuit resistance.

c. The inductance.

d. The capacitance.

26-15. The local oscillator frequency in an AM receiver must always be 455 kHz above the incoming frequency that ranges from 535 to 1605 kHz. If a 100-μH coil is used in the oscillator, calculate the range of a variable capacitor to tune the incoming frequencies.

26-16. If the coil in Problem 26-15 has a resistance of 15 Ω, calculate:

a. The Q of the oscillator circuit at each end of the tuning range.

b. The bandwidth of the oscillator circuit at each end of the tuning range.

26-17. A 0.5-mH coil of negligible resistance, a 0.0025-μF capacitor, and a 50-kΩ resistor are connected in parallel across a 10-mV, radio frequency signal generator. Calculate:

a. The circuit's resonant frequency.

b. Each branch current at resonance.

c. The total line current at resonance.

26-18. Repeat Problem 26-17 using a 50-μH coil.

26-19. For the circuit of Problem 26-17, calculate:

a. The Q of the circuit at resonance.

b. The impedance of the circuit at resonance.

c. The tank current at resonance.

d. The circuit's bandwidth.

26-20. For the circuit of Problem 26-18, calculate:

a. The Q of the circuit at resonance.

b. The impedance of the circuit at resonance.

c. The tank current at resonance.

d. The circuit's bandwidth.

26-21. Assume now that the signal generator of Problem 26-17 has an internal resistance of 40 kΩ. Calculate the output voltage across the tank circuit at

a. Resonance.

b. A frequency where the impedance of the circuit has dropped to 5 kΩ (assume resistive).

26-22. A 10-mV signal generator of unknown internal resistance is connected in parallel with a 20-kΩ resistor, a capacitor, and a coil of negligible resistance. The frequency is adjusted until, at 100 kHz, the maximum power transfer occurs *and* maximum circulating tank current occurs with a value of 20 μA. Calculate:

a. The internal resistance of the source.

b. The Q of the parallel RLC circuit.

c. The inductance of the coil.

d. The amount of capacitance.

26-23. A practical parallel resonant circuit is known to have a Q of 4 and a theoretical resonant frequency given by

$$f_r = \frac{1}{2\pi\sqrt{LC}}$$

of 150 kHz. What is the circuit's true resonant frequency?

26-24. A practical parallel RLC circuit is *measured* to have a resonant frequency of 200 kHz but has a *calculated* resonant frequency of 210 kHz when the equation

$$f_r = \frac{1}{2\pi\sqrt{LC}}$$

is used. Determine the Q of the circuit.

26-25. A 25-μH coil having an ac resistance of 4 Ω is connected in parallel with a 0.05-μF capacitor across a radio frequency signal generator. Calculate:

a. The resonant frequency of the circuit.

b. The Q of the circuit.

c. The circuit's bandwidth.

26-26. Repeat Problem 26-25 using 1 Ω for the coil's ac resistance.

26-27. For the parallel circuit of Problem 26-25, calculate:

a. The minimum resistance to be added in series with the coil to ensure that the circuit has no resonant frequency.

b. The resistance to be added in *series* with the coil to increase the bandwidth to 40 kHz, and the new resonant frequency.

26-28. For Problem 26-27, find the resistance to be connected in *parallel* with the coil to increase the bandwidth to 50 kHz.

26-29. The local oscillator frequency in an FM receiver must always be 10.7 MHz above the incoming frequency that ranges from 88– 108 MHz. If a 1-μH coil is used in a parallel RLC circuit in the oscillator, calculate the range of a variable capacitor to tune-in the incoming frequencies.

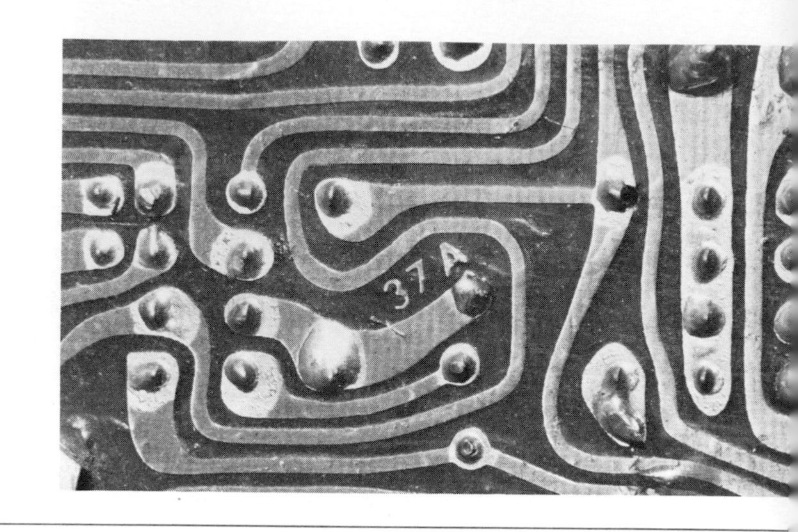

CHAPTER TWENTY-SEVEN
COMPLEX NUMBERS

We have seen how ac circuits, because of reactive components, have phase angles associated with current, voltage, and impedance. We now examine two *coordinate systems* that include the angular information as well as the magnitude.

The *polar system* directly indicates the *magnitude* and *phase angle.* The magnitude is always a positive quantity that corresponds to the reading of a meter, such a voltmeter or ammeter. The phase angle, which may be positive or negative, is measured from a (positive) horizontal zero degree reference position. The polar system is particularly useful for multiplying and dividing phasor quantities, yielding answers that include both magnitude and phase angle information.

The *rectangular coordinate* system consists of a *real* axis and an *imaginary j* axis. A *j* "operator" is used to distinguish between one phasor and another phasor that is 90° out of phase with the first, producing a *complex number* in *rectangular form.* Complex numbers in rectangular form are especially suitable for addition and subtraction although multiplication and division are also possible if the laws of algebra are observed.

A complex number in rectangular form can be used to represent a series circuit containing resistance and reactance. Then, after a rectangular-to-polar *conversion*, the equation $I = V/Z$ can be used to obtain current in a circuit. Similarly, parallel circuits may be solved using the concept of *admittance* Y, and *susceptance* B, which are the reciprocals of impedance and reactance respectively. Finally, we apply complex number theory to a series parallel circuit, and show how Thévenin's theorem may be used to solve ac circuits.

27-1 POLAR COORDINATES

We saw in Section 24-3 that the total voltage applied to a series RL circuit is the *phasor* sum of V_R and V_L, as shown in Fig. 27-1. That is, the addition of two or more voltages that are not in phase with each other must take into account the phase angle between them. For two voltages 90° out of phase, as in Fig. 27-1*b*, the *magnitude* of the total voltage is

$$|V_T| = \sqrt{V_R^2 + V_L^2} \quad (24\text{-}1)$$

and the phase angle for V_T with respect to I is

$$\theta = \arctan \frac{V_L}{V_R} \quad (24\text{-}2a)$$

These two pieces of information for V_T, the *magnitude* and *phase angle,* may be combined into a *single notation,* called the *polar form*:

$$\mathbf{V_T = \sqrt{V_R^2 + V_L^2} \,\angle \arctan \frac{V_L}{V_R}} \quad \textbf{volts (V)} \quad \textbf{(27-1)}$$

where the symbols are as above.

EXAMPLE 27-1

Given $V_R = 30$ V, $V_L = 40$ V, find the polar form of the total voltage applied to the series RL circuit:

SOLUTION

The magnitude of V_T is given by

$$V_T = \sqrt{V_R^2 + V_L^2} \quad (24\text{-}1)$$
$$= \sqrt{(30\text{ V})^2 + (40\text{ V})^2} = \mathbf{50\ V}$$

The phase angle is

$$\theta = \arctan \frac{V_L}{V_R} \quad (24\text{-}2a)$$
$$= \arctan \frac{40\text{ V}}{30\text{ V}} = \mathbf{53.1°}$$

The polar form of V_T is

$$V_T = \mathbf{50\ V \angle +53.1°}$$

The answer to Example 27-1 is read as: V_T equals 50 V *at an angle of* plus 53.1°.

The positive sign on the angle is sometimes used to distinguish the phasor from circuits in which the phase angle is negative. That is, a polar coordinate specifies a quantity (such as a voltage, impedance, current, etc.) in terms of a *positive* magnitude and either a positive or negative angle.

A positive angle is measured *counterclockwise* from the 0° horizontal reference line; a *negative* angle is measured *clockwise* from this reference line.

As a further example, the impedance of the series RL circuit, in polar form (see Fig. 27-1*c*) is given by

$$\mathbf{Z = \sqrt{R^2 + X_L^2} \,\angle \arctan \frac{X_L}{R}} \quad \textbf{ohms }(\Omega) \quad \textbf{(27-2)}$$

This is because the magnitude of the impedance is

$$|Z| = \sqrt{R^2 + X_L^2} \quad (24\text{-}3)$$

and the phase angle is

$$\theta = \arctan \frac{X_L}{R} \quad (24\text{-}5a)$$

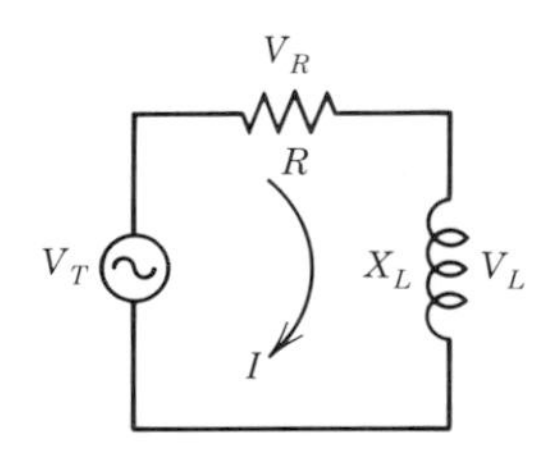

(*a*) Circuit diagram

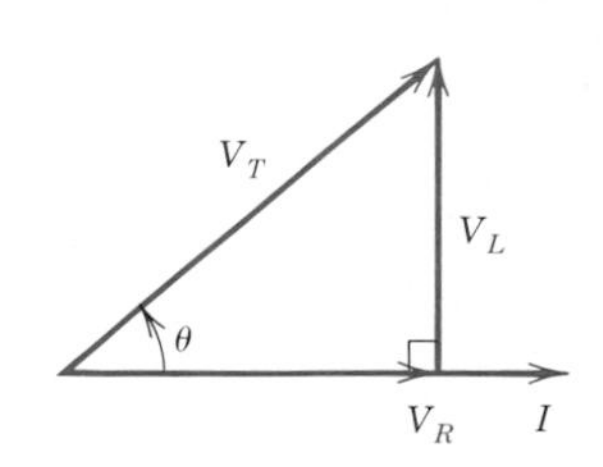

(*b*) Voltage triangle

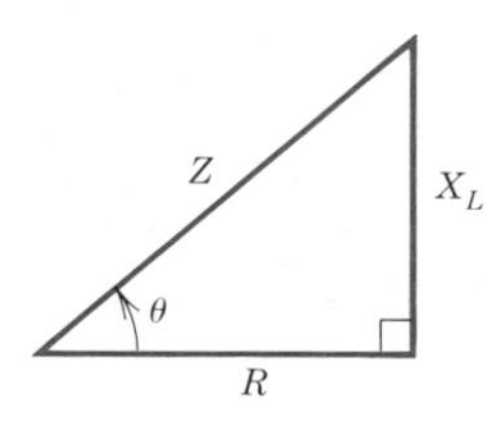

(*c*) Impedance triangle

FIGURE 27-1
Voltage and impedance triangles for a series *RL* circuit to show the use of polar coordinates.

EXAMPLE 27-2

A series RL circuit has a resistance of 4 Ω and an inductive reactance of 3 Ω. Find the circuit's impedance in polar form.

SOLUTION

$$|Z| = \sqrt{R^2 + X_L^2} \qquad (24\text{-}3)$$

$$= \sqrt{(4\ \Omega)^2 + (3\ \Omega)^2} = \mathbf{5\ \Omega}$$

$$\theta = \text{arc tan}\,\frac{X_L}{R} \qquad (24\text{-}5a)$$

$$= \text{arc tan}\,\frac{3\ \Omega}{4\ \Omega} = \mathbf{36.87°}$$

The polar form of Z is

$$Z = \mathbf{5\ \Omega\ \angle 36.9°}$$

Therefore, Z equals 5 Ω at an angle of plus 36.9°.

Similarly, for a series RC circuit, the total voltage and impedance in polar form are

$$V_T = \sqrt{V_R^2 + V_C^2}\ \angle -\text{arc tan}\,\frac{V_C}{V_R} \quad \text{volts (V)} \qquad (27\text{-}3)$$

$$Z = \sqrt{R^2 + X_C^2}\ \angle -\text{arc tan}\,\frac{X_C}{R} \quad \text{ohms }(\Omega) \qquad (27\text{-}4)$$

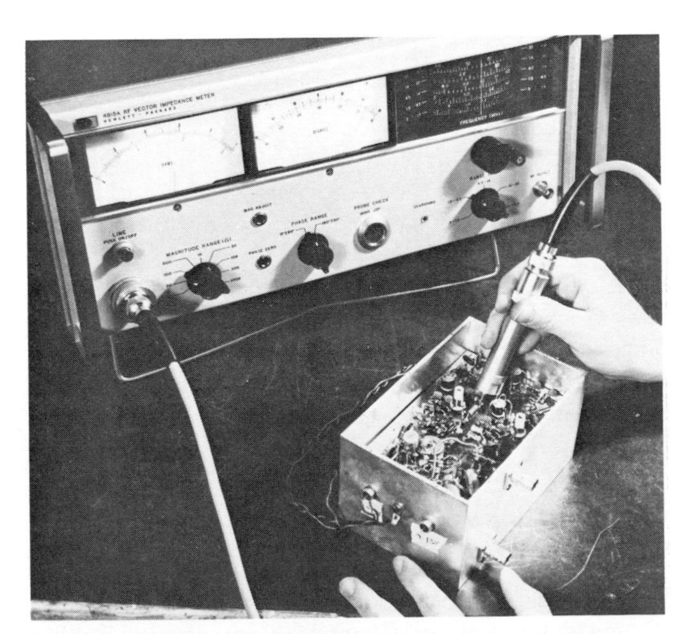

(*a*) Vector impedance meter measures the magnitude and phase angle of circuit impedance using a single probe

Thus, as an example, $V_T = 47\ \text{V} \angle -41°$, $Z = 15.6\ \Omega \angle -64°$.

In a similar way, the *current* in a circuit can be expressed in polar form, such as $I_1 = 3.5\ \text{A} \angle -72°$ or $I_T = 56\ \text{mA} \angle +12°$.

At this time, you may wonder what is the point of expressing an electrical quantity in polar coordinates. Is it merely a convenient notation?

The answer is that polar coordinates provide mathematical methods of *multiplying* and *dividing* electrical quantities that are very easy to perform. The resulting answers give magnitudes *and* phase angles. Also, the polar form is the reading you obtain from a meter. An ammeter or voltmeter reading is the magnitude of the polar form of the quantity.

Some instruments provide *both* magnitude *and* phase angle, such as the vector impedance meter and vector voltmeter shown in Fig. 27-2.

27-2 MULTIPLICATION AND DIVISION USING POLAR COORDINATES

In general, the polar form of any quantity can be expressed as

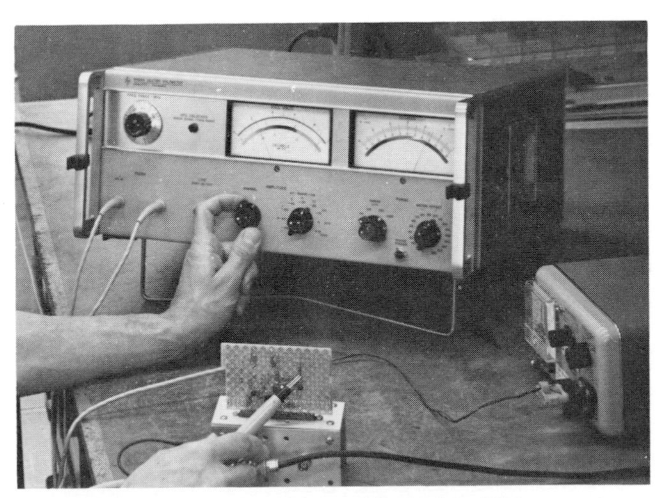

(*b*) Vector voltmeter measures the magnitude of, and the phase difference between, two voltage vectors from 1-1000 MHz

FIGURE 27-2

Meters to measure polar impedance and voltage. (Courtesy Hewlett-Packard Company.)

$$M \angle \theta$$

where: M = the magnitude
θ = the phase angle.

We know that M may have units of volts, ohms, and so on, and θ may be positive or negative. Let us assume that we have two phasors given in polar form:

$$M_1 \angle \theta_1, \qquad M_2 \angle \theta_2$$

See Fig. 27-3.

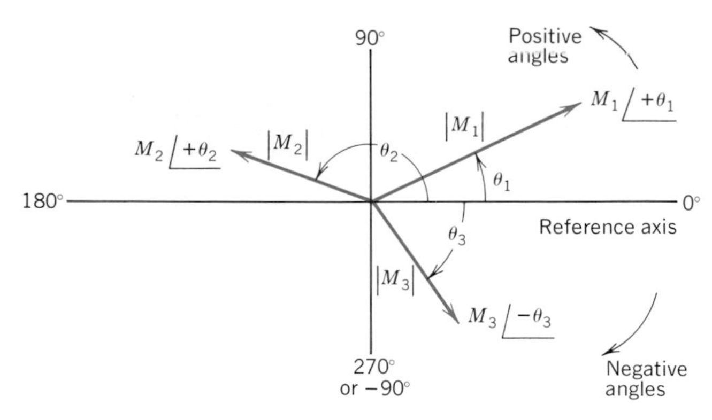

FIGURE 27-3
Three phasors given in polar form.

It can be shown that the *product* of two phasors given in polar form is found as follows:

$$\mathbf{M_1 \angle \theta_1 \times M_2 \angle \theta_2 = M_1 M_2 \angle \theta_1 + \theta_2} \qquad \textbf{(27-5)}$$

That is, to multiply two phasors given in polar form, multiply their magnitudes and algebraically *add* their phase angles.

Similarly, for division,

$$\mathbf{\frac{M_1 \angle \theta_1}{M_2 \angle \theta_2} = \frac{M_1}{M_2} \angle \theta_1 - \theta_2} \qquad \textbf{(27-6)}$$

That is, to divide two phasors given in polar form, divide their magnitudes and algebraically *subtract* their phase angles.

EXAMPLE 27-3

Given three phasors: $A = 7.5 \angle 53°$, $B = 3 \angle 12°$, and $C = 5 \angle -60°$, find:

a. $A \times B$
b. $A \div B$
c. $A \times C$
d. $A \div C$
e. $C \div B$

SOLUTION

a. $A \times B = 7.5 \angle 53° \times 3 \angle 12°$ (27-5)
$= 7.5 \times 3 \angle 53° + 12°$
$= \mathbf{22.5 \angle 65°}$

b. $A \div B = \dfrac{7.5 \angle 53°}{3 \angle 12°}$ (27-6)
$= \dfrac{7.5}{3} \angle 53° - 12°$
$= \mathbf{2.5 \angle 41°}$

c. $A \times C = 7.5 \angle 53° \times 5 \angle -60°$ (27-5)
$= 7.5 \times 5 \angle 53° + (-60°)$
$= \mathbf{37.5 \angle -7°}$

d. $A \div C = \dfrac{7.5 \angle 53°}{5 \angle -60°}$
$= \dfrac{7.5}{5} \angle 53° - (-60°)$ (27-6)
$= \mathbf{1.5 \angle 113°}$

e. $C \div B = \dfrac{5 \angle -60°}{3 \angle 12°}$
$= \dfrac{5}{3} \angle -60° - 12°$ (27-6)
$= \mathbf{1.67 \angle -72°}$

NOTE The result of multiplying (or dividing) two phasors has no simple *geometric* meaning with respect to the original phasors. In other words, the result cannot be found *graphically*. However, the result has a simple interpretation when applied to a circuit, as shown by Example 27-4.

27-2.1 Application of Polar Coordinates to Series Circuits

We can apply the polar coordinate system to the alternating current theory of Chapter 24 as shown by the following examples.

EXAMPLE 27-4

A circuit consists of a 50-Ω resistor in series with a capacitive reactance of 120 Ω. A 120-V, 60-Hz supply is connected across the circuit. Determine the current and the phase angle it makes with the applied voltage.

SOLUTION

$$Z = \sqrt{R^2 + X_C{}^2}\ \angle -\text{arc tan}\frac{X_C}{R} \qquad (27\text{-}4)$$

$$= \sqrt{50^2 + 120^2}\ \Omega\ \angle -\text{arc tan}\frac{120}{50}$$

$$= 130\ \Omega\ \angle -67.4^\circ$$

$$I = \frac{V}{Z} \qquad (24\text{-}4)$$

$$= \frac{120\text{ V}\ \angle 0^\circ}{130\ \Omega\ \angle -67.4^\circ} = \mathbf{0.92\ A\ \angle +67.4^\circ}$$

This states that the circuit current is 0.92 A and it *leads* the applied voltage by 67.4°. See Fig. 27-4.

Note how the polar form of 120 V is 120 V $\angle 0^\circ$, since the applied voltage is made the reference phasor in this problem for convenience.

EXAMPLE 27-5

Use polar coordinates to determine the magnitude and phase angle of the voltage in Example 27-4 across:

a. The resistor.

b. The capacitor.

SOLUTION

a. $V_R = IR \qquad (15\text{-}15b)$

$$= 0.92\text{ A}\ \angle +67.4^\circ \times 50\ \Omega\ \angle 0^\circ$$

$$= \mathbf{46\ V\ \angle 67.4^\circ}$$

b. $V_C = IX_C \qquad (23\text{-}1)$

$$= 0.92\text{ A}\ \angle +67.4^\circ \times 120\ \Omega\ \angle -90^\circ$$

$$= \mathbf{110\ V\ \angle -22.6^\circ}$$

Note how the resistance has a polar form of 50 Ω $\angle 0^\circ$ (along the reference axis) and X_C has a polar form of

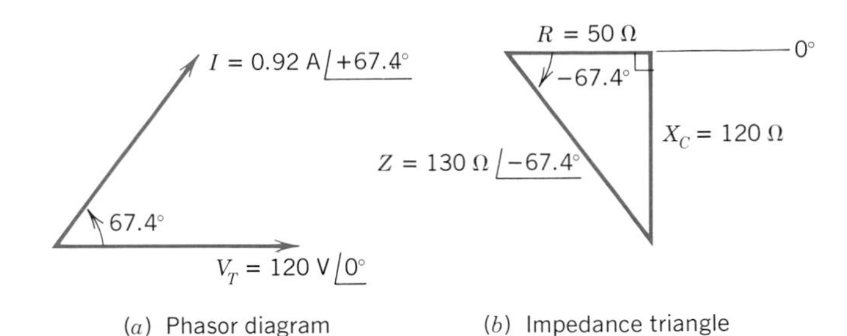

(*a*) Phasor diagram (*b*) Impedance triangle

FIGURE 27-4
Diagrams for Example 27-4.

120 Ω $\angle -90^\circ$. The phasor diagram in Fig. 27-5*b* shows how the V_R and V_C phasors *are* 90° out of phase, as required.

Note, however, that the phasor diagram is drawn rotated counterclockwise (67.4°) from the diagram in Fig. 24-11*b*. This is because the diagram of Fig. 27-5*b* uses the applied *voltage* in the reference position of 0°, whereas Fig. 24-11*b* uses *current* as the reference. Current is generally used as a reference in *series* circuit because it is the *same* in all components.

EXAMPLE 27-6

A series circuit with an applied emf of 48 V has a current of 6.4 A that lags the applied voltage by 40°. Determine:

a. The type of series circuit.

b. The amount of impedance in the circuit, using polar coordinate notation.

c. The resistance and reactance of the circuit.

d. The voltages throughout the circuit.

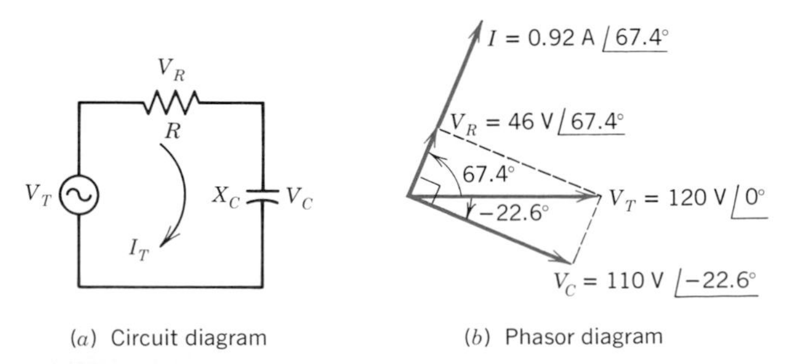

(*a*) Circuit diagram (*b*) Phasor diagram

FIGURE 27-5
Diagrams for Example 27-5.

SOLUTION

a. Since the current *lags* the voltage, the circuit must be **inductive.**

b. $Z = \frac{V}{I}$ (24-4)

$$= \frac{48\ \text{V}\ \angle 0^\circ}{6.4\ \text{A}\ \angle -40^\circ} = \mathbf{7.5\ \Omega\ \angle 40^\circ}$$

c. $R = Z \cos \theta$ (24-6)

$= 7.5\ \Omega \times \cos 40^\circ = \mathbf{5.75\ \Omega}$

$X_L = Z \sin \theta$ (24-7)

$= 7.5\ \Omega \times \sin 40^\circ = \mathbf{4.8\ \Omega}$

d. $V_R = IR$ (15-15b)

$= 6.4\ \text{A} \angle -40^\circ \times 5.75\ \Omega \angle 0^\circ$

$= \mathbf{36.8\ V \angle -40^\circ}$

$V_L = IX_L$ (20-1)

$= 6.4\ \text{A} \angle -40^\circ \times 4.8\ \Omega \angle 90^\circ$

$= \mathbf{30.7\ V \angle 50^\circ}$

See the phasor diagram and impedance triangle in Fig. 27-6.

Note how the inductive reactance has a polar form of $4.8\ \Omega \angle +90^\circ$, and how V_R, in phase with I, is 90° out of phase with V_L.

27-3 THE *j* OPERATOR

The polar coordinate system is ideal for performing multiplication and division. However, addition (and subtraction) is not directly possible in the polar form. Conversion to another coordinate system is necessary before this can be readily accomplished.

Consider a polar coordinate given by $1 \angle 90^\circ$. From Eq. 27-5 we know that

$$1 \angle 90^\circ \times 1 \angle 90^\circ = 1 \angle 180^\circ$$

But $\quad 1 \angle 180^\circ = -1$

and $\quad 1 \angle 90^\circ \times 1 \angle 90^\circ = (1 \angle 90^\circ)^2$

Therefore, $\quad (1 \angle 90^\circ)^2 = -1$

Taking the square root of both sides:

$$1 \angle 90^\circ = \sqrt{-1}$$

We know from mathematics that $\sqrt{-1}$ is an *imaginary* number (there is no *real* number that, when squared, is equal to -1). Mathematicians use the letter i to represent this imaginary number, $\sqrt{-1}$. However, to avoid confusion with the symbol for current, we use the letter symbol j to represent $\sqrt{-1}$. But more important, we can now see that

$$j = \sqrt{-1} = \mathbf{1 \angle 90^\circ} \quad \text{or} \quad j1 = j \times 1 = 1 \angle 90^\circ$$

This means that if we write $j2$ (or $j \times 2$), we represent $2 \angle 90^\circ$. That is, *the effect of placing a j in front of a real number is to cause that number to be rotated through an angle of +90°.* (Counterclockwise.)

It is for this reason that *j* is referred to as *an operator*. It *operates on* or modifies a number by *rotating it* through a positive angle of 90°.

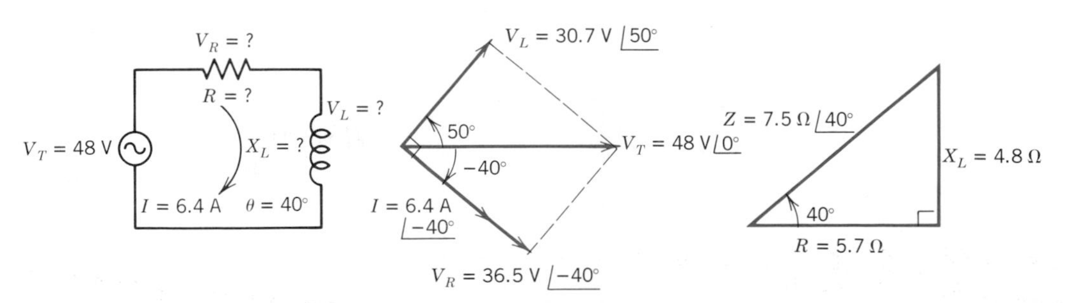

FIGURE 27-6
Diagrams for Example 27-6.

Thus we can think of an *imaginary axis* with divisions of $j1, j2, j3$, and so on, drawn vertically upward from the *real axis*, as in Fig. 27-7. This is the *positive* imaginary axis.

Now consider the effect of multiplying j by j.

$$j^2 = j \times j = \sqrt{-1} \times \sqrt{-1} = -1$$

Similarly, $j^2 \times 2 = -2$, and so on.

Thus, multiplying a positive imaginary number by j, operates on the number to rotate it a further 90° to the negative real axis.

If we now multiply -1 by j, we obtain

$$-1 \times j = 1\,\underline{/180^\circ} \times 1\,\underline{/90^\circ} = 1\,\underline{/270^\circ} = -j$$

This establishes a *negative* imaginary axis due to the additional 90° rotation. Note that

$$-1 \times j = j^2 \times j = j^3 = -j$$

Finally, $\quad -j \times j = -j^2 = -(-1) = 1$

or $\quad j^3 \times j = j^4 = 1$

See Fig. 27-8.

27-4 RECTANGULAR COORDINATE SYSTEM

We now have a method of distinguishing between a phasor that is horizontal and another phasor that is 90° out of phase. Simply putting a j in front of the quantity indicates that a phase angle of $+90°$ is associated with that number. Similarly, a $-j$ in front of a quantity indicates that a phase angle of $-90°$ is associated with that number. For example, if a voltage is the *sum* of two voltages, one 90° out of phase with the other, we can show this sum with an addition sign if we use the j term.

Thus, for a series *RL* circuit, where V_L leads the resistor voltage V_R by 90°, the total applied voltage is given, in *rectangular form*, by

$$V_T = V_R + jV_L \tag{27-7}$$

Similarly, $\quad Z = R + jX_L$

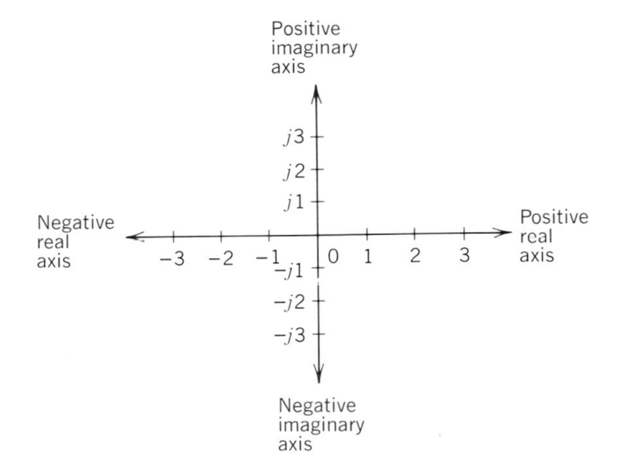

FIGURE 27-7
Real and imaginary axes make up the complex number plane or the rectangular coordinate system.

See Fig. 27-9.

It is conventional practice to write the *real* number *first*, followed by the imaginary number. Such a number, because it is the combination of a real and an imaginary number, is called a *complex number*.

For a series *RC* circuit, it follows that the *rectangular coordinate* system gives us

$$V_T = V_R - jV_C \tag{27-9}$$

and

$$Z = R - jX_C \tag{27-10}$$

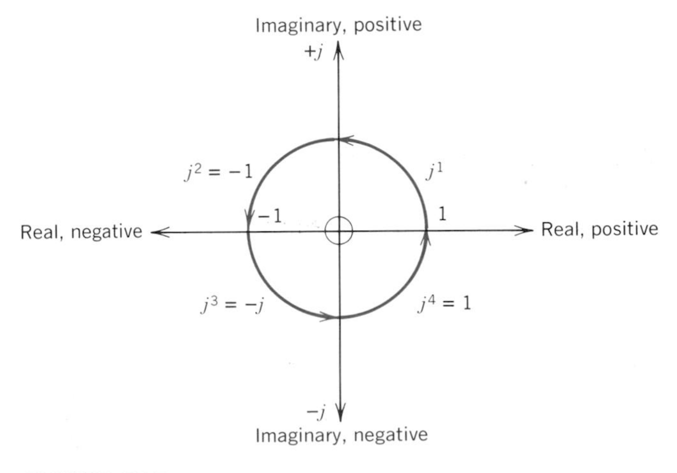

FIGURE 27-8
Showing how *j* operates on *any* number, real or imaginary, to rotate that number through an angle of +90°.

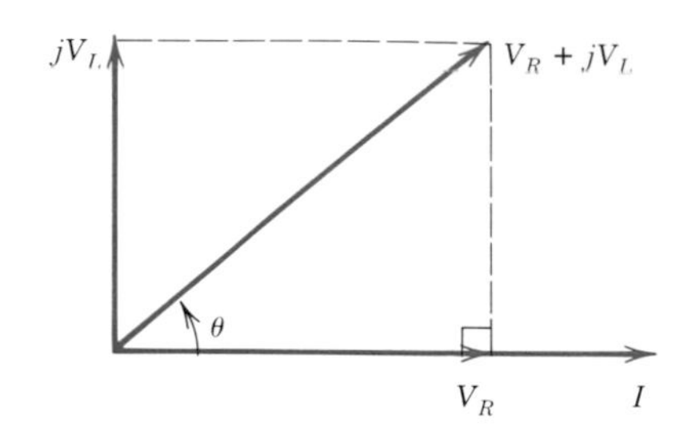

(a) Phasor diagram

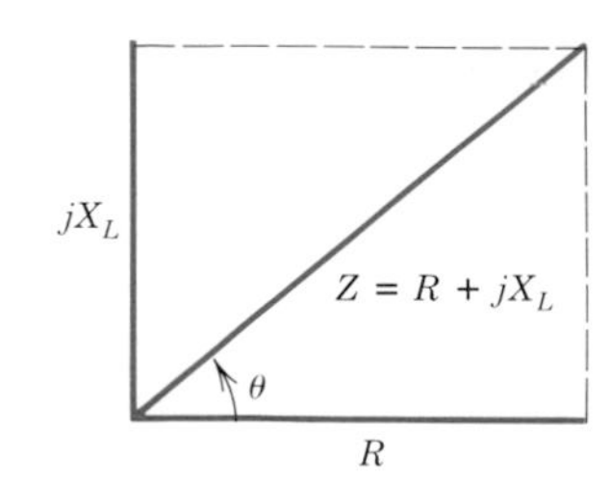

(b) Impedance diagram

FIGURE 27-9
Rectangular coordinate representations for a series *RL* circuit.

Thus the impedance of a series *RC* circuit may appear as $Z = 20 - j50\ \Omega$. This simply means that the circuit consists of a 20-Ω resistance in series with a capacitor whose reactance is 50 Ω.

Also, an impedance of $Z = 4.7 + j2$ kΩ represents a series circuit with a 4.7-kΩ resistance and an inductive reactance of 2 kΩ.

For a series *RLC* circuit,

$$Z = R + j(X_L - X_C) \qquad (27\text{-}11)$$

$$V_T = V_R + j(V_L - V_C) \qquad (27\text{-}12)$$

See Fig. 27-10.

Again, it may seem that all that we have gained with rectangular coordinates is a compact system of combining components that are 90 or 180° out of phase. The real value, however, is the mathematical manipulation that this system permits for solving complex circuit problems.

27-5 ADDITION AND SUBTRACTION USING RECTANGULAR COORDINATES

Addition and subtraction of phasors is especially simple when the phasors are given in rectangular form. (Recall that polar quantities do *not* lend themselves to addition and subtraction.)

Assume two phasors given in rectangular coordinates:

$$a + jb, \qquad c + jd$$

The sum of the two phasors is found by algebraically adding together first the real parts, then the imaginary parts, respectively,

$$(a + jb) + (c + jd) = (a + c) + j(b + d) \qquad (27\text{-}13)$$

The geometric result of the addition of two phasors is shown in Fig. 27-11*a*. This applies whether the phasors are given in polar or rectangular coordinates. Similarly, the subtraction of two phasors requires the algebraic subtraction of their real and imaginary parts:

$$(a + jb) - (c + jd) = (a - c) + j(b - d) \qquad (27\text{-}14)$$

Figure 27-11*b* shows that subtraction can be thought of as adding the negative of the phasor to be subtracted. That is, if $V_1 = a + jb$, then $-V_1 = -a - jb$. Adding $-V_1$ to V_2 results in $V_2 - V_1$.

EXAMPLE 27-7

Given three complex phasors: $A = 3 + j7$, $B = 4 - j5$, and $C = -6 + j8$. Find:

a. $A + B$.
b. $A - B$.

$R = 20\ \Omega$
$X_C = 50\ \Omega$
$Z = 20 - j50\ \Omega$

(a) Series *RC* circuit

$R = 4.7$ kΩ
$X_L = 2$ kΩ
$Z = 4.7 + j2$ kΩ

(b) Series *RL* circuit

$R = 40\ \Omega$
$X_L = 80\ \Omega$
$X_C = 30\ \Omega$
$Z = 40 + j(80 - 30)\ \Omega = 40 + j50\ \Omega$

(c) Series *RLC* circuit

FIGURE 27-10
Impedance of series circuits in rectangular coordinates.

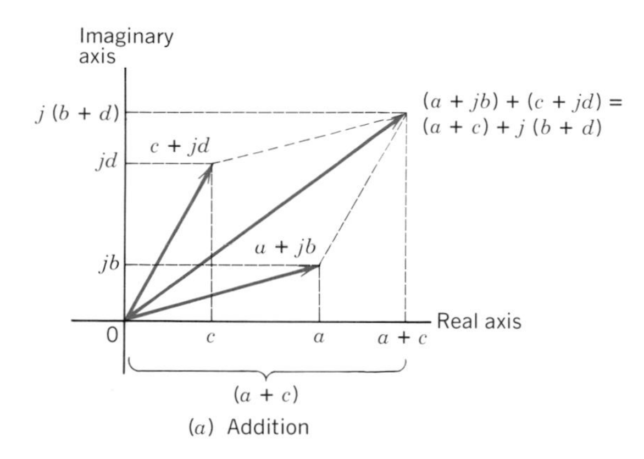

(*a*) Addition

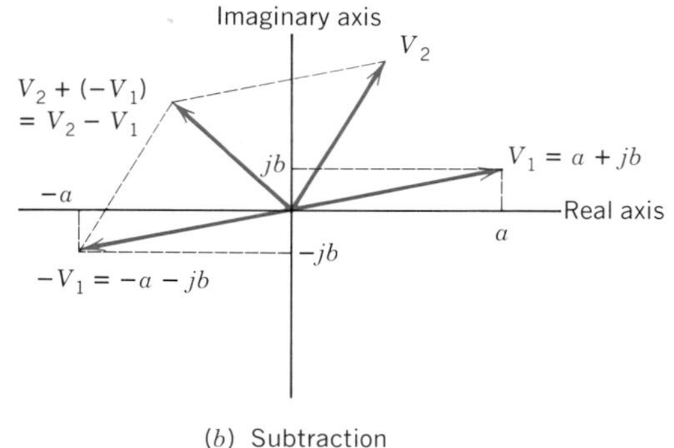

(*b*) Subtraction

FIGURE 27-11
Addition and subtraction of complex numbers in rectangular coordinates.

c. $B + C$.
d. $B - C$.

SOLUTION

a. $A + B = 3 + j7 + 4 - j5$
$= (3 + 4) + j(7 - 5)$ (27-13)
$= \mathbf{7 + j2}$

b. $A - B = (3 + j7) - (4 - j5)$
$= (3 - 4) + j(7 - (-5))$ (27-14)
$= \mathbf{-1 + j12}$

or $A - B = A + (-B)$
$= 3 + j7 + (-4 + j5)$
$= (3 - 4) + j(7 + 5)$
$= \mathbf{-1 + j12}$

c. $B + C = (4 - j5) + (-6 + j8)$
$= (4 - 6) + j(-5 + 8)$ (27-13)
$= \mathbf{-2 + j3}$

d. $B - C = (4 - j5) - (-6 + j8)$
$= (4 + 6) + j(-5 - 8)$ (27-14)
$= \mathbf{10 - j13}$

27-5.1 Application of Rectangular Coordinates to Series Circuits

The total impedance of a series circuit is the sum of the impedances throughout the circuit.

$$Z_T = Z_1 + Z_2 + Z_3 + \cdots Z_n \qquad (27\text{-}15)$$

Although Eq. 27-15 applies to polar *and* rectangular coordinates, it is the rectangular form that is most useful, as shown by Example 27-8.

EXAMPLE 27-8

Two coils and a capacitor are connected in series. The first coil has a resistance of 20 Ω and a reactance of 50 Ω; the second coil has a resistance of 15 Ω and a reactance of 35 Ω; the capacitor has a reactance of 10 Ω. Find the total series impedance.

SOLUTION

$Z_T = Z_1 + Z_2 + Z_3$ (27-15)
$= (20 + j50) + (15 + j35) - j10$
$= (20 + 15) + j(50 + 35 - 10)$
$= \mathbf{35 + j75}$

See Fig. 27-12.
Note that the original circuit is *equivalent* to a series circuit of 35-Ω *resistance* and 75-Ω *inductive* reactance.

$X_{L1} = 50\ \Omega$ $X_{L2} = 35\ \Omega$
$R_1 = 20\ \Omega$ $R_2 = 15\ \Omega$ $X_C = 10\ \Omega$
$(20 + 15) + j\,(50 + 35 - 10) = 35 + j\,75\ \Omega$
$R_T = 35\ \Omega$ $X_{LT} = 75\ \Omega$

FIGURE 27-12
Addition of impedances in rectangular coordinates for Example 27-8.

We know that Kirchhoff's voltage law must be obeyed in *any* series circuit. Thus

$$V_T = V_1 + V_2 + V_3 + \cdots + V_n \qquad (27\text{-}16)$$

as shown by Example 27-9.

EXAMPLE 27-9

Three loads are series-connected across the 120-V line. The first load has a voltage given by $40 + j30$ V; the second load has a voltage given by $25 - j90$ V. Find the rectangular form of the voltage across the third load.

SOLUTION

$$V_T = V_1 + V_2 + V_3 \qquad (27\text{-}16)$$

$$120 + j0 = (40 + j30) + (25 - j90) + V_3$$

Therefore,

$$V_3 = (120 + j0) - (40 + j30) - (25 - j90)\text{ V}$$
$$= (120 - 40 - 25) + j(0 - 30 - (-90))\text{ V}$$
$$= \mathbf{55 + j60\ V}$$

Note that the answer given for V_3 in Example 27-9 is *not* what a voltmeter would read. A voltmeter indicates the *polar* form, not the rectangular form. We shall see how to *convert* from rectangular to polar in Section 27-7.

27-6 MULTIPLICATION AND DIVISION USING RECTANGULAR COORDINATES

Although the process of multiplication (and division) is *most easily* carried out using polar coordinates, the process is not difficult using the rectangular form.

Given two complex numbers

$$a + jb, \qquad c + jd$$

the product is given by

$$(a + jb)(c + jd) = ac + jad + jbc + j^2bd$$
$$= ac + j^2bd + j(ad + bc)$$

But $\qquad j^2 = -1$

Therefore,

$$(a + jb)(c + jd) = (ac - bd) + j(ad + bc)$$

This is not to be used as a formula. It shows that multiplication in rectangular form follows the *distributive law* of algebra. The result is a complex number with a real and an imaginary part.

Similarly, the division of two complex numbers given in rectangular form also follows the laws of algebra, requiring a *rationalization* of the denominator.

Given two complex numbers

$$a + jb, \qquad c + jd$$

their quotient is given by

$$\frac{a + jb}{c + jd} = \frac{a + jb}{c + jd} \times \frac{c - jd}{c - jd}$$
$$= \frac{ac - jad + jbc - j^2bd}{c^2 - jcd + jcd - j^2d^2}$$
$$= \frac{(ac + bd) + j(bc - ad)}{c^2 + d^2}$$
$$= \frac{ac + bd}{c^2 + d^2} + j\frac{bc - ad}{c^2 + d^2}$$

Note that $c - jd$ is the *conjugate* of $c + jd$. Multiplying the numerator and denominator by $c - jd$ does not change the original fraction, but it does provide a real number in the final denominator because the imaginary components cancel. See Example 27-10.

EXAMPLE 27-10

Given two complex numbers, $A = 5 + j8$, $B = 7 - j4$, find:

a. $A \times B$.

b. A/B.

SOLUTION

a. $A \times B = (5 + j8)(7 - j4)$
$$= 35 - j20 + j56 - j^232$$
$$= (35 + 32) + j(56 - 20)$$
$$= \mathbf{67 + j36}$$

b.
$$A/B = \frac{5 + j8}{7 - j4}$$
$$= \frac{5 + j8}{7 - j4} \times \frac{7 + j4}{7 + j4}$$
$$= \frac{35 + j20 + j56 + j^2 32}{49 + j28 - j28 - j^2 16}$$
$$= \frac{(35 - 32) + j(20 + 56)}{(49 + 16) + j0}$$
$$= \frac{3 + j76}{65}$$
$$= \frac{3}{65} + j\frac{76}{65}$$
$$\approx \mathbf{0.05 + j1.17}$$

We shall see in Section 27-8 that an alternative to the procedures shown in Example 27-10 requires the conversion of the rectangular form *into* polar coordinates. However, if the result is required in rectangular form, another conversion must be made *from* the polar form, following the multiplication or division process. In this case, the procedures shown in Example 27-10 may be easier.

27-6.1 Application of Rectangular Coordinates to Parallel Circuits

If two impedances Z_1 and Z_2 are connected in parallel, their total impedance is

$$Z_T = \frac{Z_1 \times Z_2}{Z_1 + Z_2} \qquad (27\text{-}17)$$

where all impedances are given in complex numbers.

EXAMPLE 27-11

A coil of 5-Ω resistance and 8-Ω reactance is connected in parallel with a second coil having a resistance of 4 Ω and a reactance of 6 Ω. Find the total impedance.

SOLUTION

$$Z_T = \frac{Z_1 \times Z_2}{Z_1 + Z_2} \qquad (27\text{-}17)$$
$$= \frac{(5 + j8)(4 + j6)}{(5 + j8) + (4 + j6)}\ \Omega$$

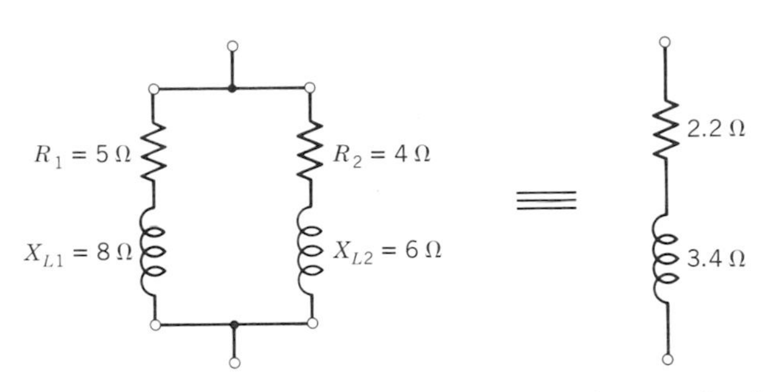

(*a*) Original parallel circuit (*b*) Equivalent series circuit

FIGURE 27-13
Circuits for Example 27-11.

$$= \frac{20 + j30 + j32 + j^2 48}{(5 + 4) + j(8 + 6)}\ \Omega$$
$$= \frac{(20 - 48) + j(30 + 32)}{9 + j14}\ \Omega$$
$$= \frac{-28 + j62}{9 + j14}\ \Omega$$
$$= \frac{-28 + j62}{9 + j14} \times \frac{9 - j14}{9 - j14}\ \Omega$$
$$= \frac{-252 + j392 + j558 - j^2 868}{81 + 196}\ \Omega$$
$$= \frac{(868 - 252) + j(392 + 558)}{81 + 196}\ \Omega$$
$$= \frac{616 + j950}{277}\ \Omega$$
$$= \frac{616}{277} + j\frac{950}{277}\ \Omega$$
$$\approx \mathbf{2.2 + j3.4\ \Omega}$$

See Fig. 27-13.

NOTE The total parallel impedance is equivalent to a *series* combination of a 2.2-Ω resistance and a 3.4-Ω inductive reactance. A complex number in rectangular coordinates is *always* interpreted as a *series* impedance.

EXAMPLE 27-12

A coil having a resistance of 4 Ω and an inductive reactance of 2 Ω is connected in parallel with a capacitor whose reactance is 3 Ω. Find the total impedance.

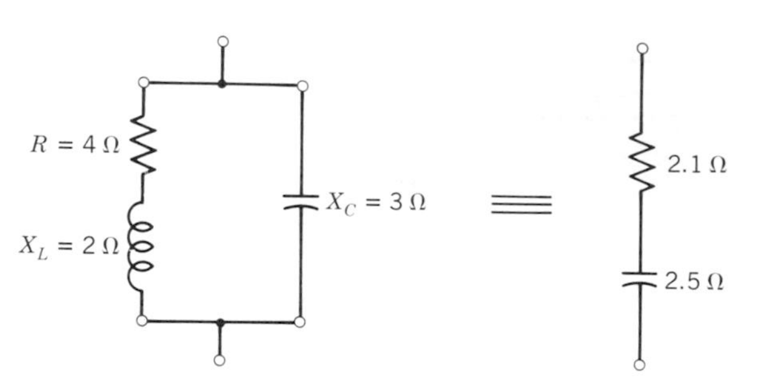

FIGURE 27-14
Circuits for Example 27-12.

SOLUTION

$$Z_T = \frac{Z_1 \times Z_2}{Z_1 + Z_2} \qquad (27\text{-}17)$$

$$= \frac{(4 + j2)(-j3)}{(4 + j2) + (0 - j3)}\ \Omega$$

$$= \frac{-j12 - j^2 6}{4 - j1}\ \Omega$$

$$= \frac{6 - j12}{4 - j1}\ \Omega$$

$$= \frac{6 - j12}{4 - j1} \times \frac{4 + j1}{4 + j1}\ \Omega$$

$$= \frac{24 + j6 - j48 - j^2 12}{16 - j^2 1}\ \Omega$$

$$= \frac{36 - j42}{17}\ \Omega$$

$$= \frac{36}{17} - j\frac{42}{17}\ \Omega$$

$$\approx \mathbf{2.1 - j2.5\ \Omega}$$

See Fig. 27-14.

Note that in this case the equivalent circuit is capacitive. If more than two impedances are parallel-connected, they can be combined two at a time, as we did for parallel resistances. (See Section 6-4.)

Finally, note that we now have a method of obtaining the impedance of a parallel circuit *directly*. We do not have to assume a voltage, find the individual currents and their sum, and finally obtain Z_T from V/I_T, as we did in Chapter 24.

27-7 CONVERSION FROM RECTANGULAR TO POLAR FORM

It may be more convenient to work with a complex number if it is in polar form, especially for multiplication and division. A phasor can be converted from rectangular to polar coordinates using the Pythagorean relationship, as shown in Fig. 27-15.

Given a complex number $a + jb$, it can be converted to the form $M \angle \theta$ as follows:

$$a + jb = \sqrt{a^2 + b^2} \angle \arctan \frac{b}{a} \qquad (27\text{-}18)$$

Care must be taken if the real number is negative as shown by Example 27-13.

EXAMPLE 27-13

Convert the following to polar form:

a. $3 + j4$.
b. $5 - j12$.
c. $-4 + j6$.
d. $-4 - j6$.

SOLUTION

a. $3 + j4 = \sqrt{3^2 + 4^2} \angle \arctan \frac{4}{3} \qquad (27\text{-}18)$

$= \mathbf{5 \angle 53.1°}$

b. $5 - j12 = \sqrt{5^2 + 12^2} \angle \arctan -\frac{12}{5} \qquad (27\text{-}18)$

$= \mathbf{13 \angle -67.4°}$

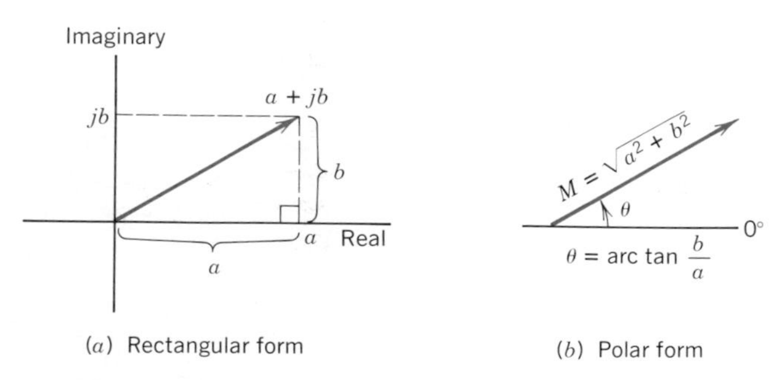

FIGURE 27-15
Conversion from rectangular to polar form.

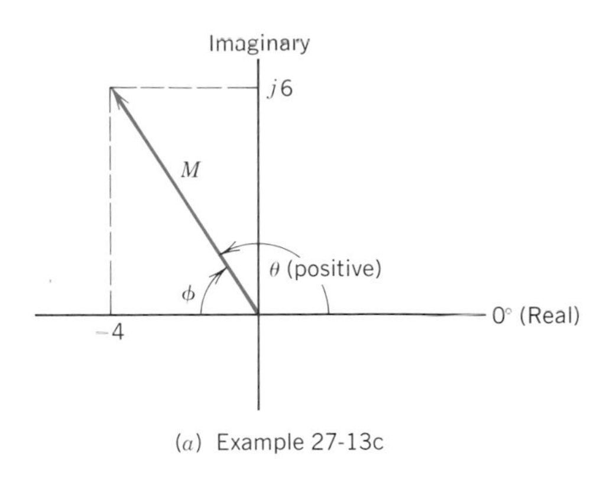

(a) Example 27-13c

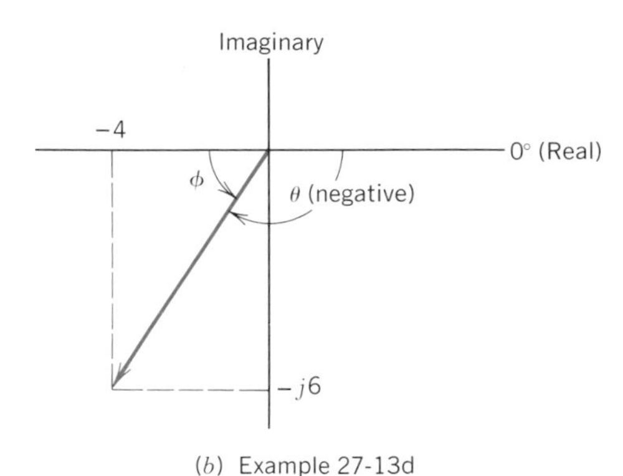

(b) Example 27-13d

FIGURE 27-16
Sketches for Example 27-13.

c. Make a sketch of the complex number as in Fig. 27-16*a*.

$$M = \sqrt{4^2 + 6^2} = 7.2$$
$$\phi = \text{arc tan } \tfrac{6}{4} = 56.3^\circ$$
$$\theta = 180^\circ - 56.3^\circ = 123.7^\circ$$

Therefore: $-4 + j6 = \mathbf{7.2 \,\underline{/123.7^\circ}}$.

d. Make a sketch as in Fig. 27-16*b*.

$M = 7.2$

$\theta = -(180^\circ - 56.3^\circ) = -123.7^\circ$

Therefore, $-4 - j6 = \mathbf{7.2 \,\underline{/-123.7^\circ}}$

or $-4 - j6 = 7.2 \,\underline{/180^\circ + \phi}$

$= 7.2 \,\underline{/180^\circ + 56.3^\circ}$

$= \mathbf{7.2 \,\underline{/+236.3^\circ}}$

Note that the polar coordinate *angle* can be given as a positive or negative number, but the *magnitude* is always positive. Generally, if the positive angle is more than 180°, it is conventional to express the coordinates with a negative angle.

Most scientific calculators have a coded program for rectangular/polar conversions that avoid the difficulties arising from numbers being in the second and third quadrants. These programs vary, but the following one applies for one of the Texas Instruments type of calculator.*

Convert $-4 - j6$ to polar coordinates.

Enter	Press	Display	Comments
−4	x :y	0	Enters real part
−6	INV 2nd 18	−123.7°	Displays θ
	x :y	7.2	Displays *M*

Therefore, $-4 - j6 = \mathbf{7.2 \,\underline{/-123.7^\circ}}$

EXAMPLE 27-14

A coil having an ac resistance of 20 Ω and an inductive reactance of 50 Ω is connected in parallel with a capacitor that has an ac resistance of 10 Ω and a capacitive reactance of 20 Ω. Calculate the total impedance.

SOLUTION

$$Z_T = \frac{Z_1 \times Z_2}{Z_1 + Z_2} \qquad (27\text{-}17)$$

$$= \frac{(20 + j50)(10 - j20)}{20 + j50 + 10 - j20}\ \Omega$$

$$= \frac{53.9 \,\underline{/68.2^\circ} \times 22.4 \,\underline{/-63.4^\circ}}{30 + j30}\ \Omega$$

$$= \frac{1207.4 \,\underline{/4.8^\circ}}{42.4 \,\underline{/45^\circ}}$$

$$= \mathbf{28.5\ \Omega \,\underline{/-40.2^\circ}}$$

---

* In some calculators (such as the TI-55 in Fig. 1-1), instead of pressing INV 2nd "18" (the code for rectangular-to-polar conversion) press INV 2nd $P \rightarrow R$ or the button marked $\rightarrow P$.

27-8 CONVERSION FROM POLAR TO RECTANGULAR FORM

The answer to Example 27-14 suggests a total impedance that is capacitive. However, to determine how much is capacitive and how much is resistive requires a *polar-to-rectangular* conversion. That is, we can only identify an impedance in terms of series resistance and reactance when the impedance is given in rectangular coordinates.

The conversion from polar to rectangular can be done using right-angled trigonometry, as shown in Fig. 27-17.

Given a complex number in polar form $M \angle \theta$, it can be converted to the form $a + jb$ as follows:

$$\boldsymbol{M \angle \theta = M \cos \theta + jM \sin \theta} \qquad \textbf{(27-19)}$$

This equation applies to all angles, positive or negative, with no restrictions.

EXAMPLE 27-15

Determine the components of the equivalent series impedance in Example 27-14, where $Z_T = 28.5\ \Omega \angle -40.2°$.

SOLUTION

$$\begin{aligned} Z_T &= 28.5\ \Omega \angle -40.2° \\ &= 28.5 \cos(-40.2°) + j28.5 \sin(-40.2°) \\ &= 28.5 \times 0.76 + j28.5 \times (-0.65) \\ &= \mathbf{21.7 - j18.5\ \Omega} \end{aligned}$$

Thus the total impedance is equivalent to a resistance of 21.7 Ω in series with a capacitor having a reactance of 18.5 Ω.

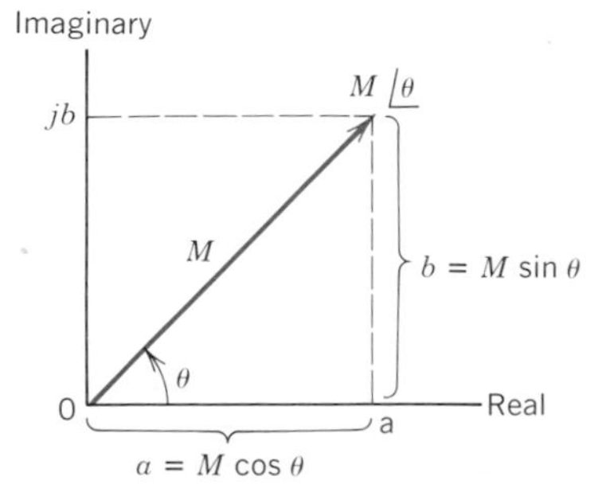

FIGURE 27-17
Conversion from polar to rectangular form.

It is wise, when making conversions, to make a sketch of the initial coordinates, whether polar or rectangular, to check that the converted number falls in the correct quadrant.

EXAMPLE 27-16

Two coils and a capacitor are connected in series across a sinusoidal supply. The voltages across the coils are $50\ \text{V} \angle 20°$ and $30\ \text{V} \angle 50°$. The capacitor's voltage is $75\ \text{V} \angle -80°$. Calculate the supply voltage.

SOLUTION

By Kirchhoff's voltage law,

$$V_T = V_1 + V_2 + V_3 \qquad (27\text{-}16)$$

$$\begin{aligned} V_1 &= 50\ \text{V} \angle 20° = 50 \cos 20° \\ &\quad + j50 \sin 20° = 47 + j17.1\ \text{V} \\ V_2 &= 30\ \text{V} \angle 50° = 30 \cos 50° \\ &\quad + j30 \sin 50° = 19.3 + j23\ \text{V} \\ V_3 &= 75\ \text{V} \angle -80° = 75 \cos(-80°) \\ &\quad + j75 \sin(-80°) = 13 - j73.9\ \text{V} \\ V_T &= (47 + 19.3 + 13) \\ &\quad + j(17.1 + 23 - 73.9)\ \text{V} \end{aligned}$$

Therefore, $V_T = 79.3 - j33.8\ \text{V} = \mathbf{86.2\ V \angle -23.1°}$

Note that it is necessary to convert to the polar form to obtain the total voltage that would be indicated by a voltmeter.

Similar to Section 27-7, a polar to rectangular conversion can be made on some scientific calculators as follows:*

Convert $5 \angle 30°$ to rectangular coordinates.

Enter	Press	Display	Comments
5	x :y	0	Enters magnitude
30°	2nd 18	2.5	Value of imaginary
	x :y	4.3	Value of real

* In some calculators (such as the TI-55 in Fig. 1-1), instead of pressing "18" (the code for polar-to-rectangular conversion) press the button marked $P \rightarrow R$ or $\rightarrow R$.

Therefore, $5\,\underline{/30^\circ} = 4.3 + j2.5$

Note that for the TI calculator, the imaginary part of the complex number is displayed first. A rough sketch of $5\,\underline{/30^\circ}$ would make it clear that the smaller number must be the imaginary component.

EXAMPLE 27-17

A parallel circuit has two branches and draws a total current of 58 mA $\underline{/35^\circ}$ from a 120-V, 60-Hz source. If the current in one branch is 35 mA $\underline{/-20^\circ}$, calculate:

a. The current in the other branch, as indicated by an ammeter.
b. The total impedance of the parallel circuit.
c. The total true power dissipated in the whole circuit.
d. The total reactive power in the whole circuit.
e. The power factor of the whole circuit.

SOLUTION

a. By Kirchhoff's current law,

$$I_T = I_1 + I_2 \tag{6-1}$$

Therefore,

$$\begin{aligned} I_2 &= I_T - I_1 \\ &= 58\,\underline{/35^\circ} - 35\,\underline{/-20^\circ}\text{ mA} \\ &= (47.5 + j33.3) - (32.9 - j12)\text{ mA} \\ &= 14.6 + j45.3\text{ mA} \\ &= \mathbf{47.6\text{ mA}\,\underline{/72.1^\circ}} \end{aligned}$$

b. $$Z_T = \frac{V}{I_T} \tag{24-4}$$

$$= \frac{120\text{ V}\,\underline{/0^\circ}}{58\text{ mA}\,\underline{/35^\circ}}$$

$$= \mathbf{2.1\text{ k}\Omega\,\underline{/-35^\circ}}$$

c. $$\begin{aligned} Z_T &= 2.1\text{ k}\Omega\,\underline{/-35^\circ} \\ &= \mathbf{1.7 - j1.2\text{ k}\Omega} \end{aligned}$$

This means that the circuit is equivalent to a current of 58 mA flowing through a resistance of 1.7 kΩ as far as total true power is concerned.

$$\begin{aligned} P_T &= I_R{}^2R \qquad (15\text{-}12) \\ &= (58 \times 10^{-3}\text{ A})^2 \times 1.7 \times 10^3\ \Omega \\ &= \mathbf{5.7\text{ W}} \end{aligned}$$

d. $$\begin{aligned} P_q &= I_C{}^2X_C \qquad (25\text{-}5) \\ &= (58 \times 10^{-3}\text{ A})^2 \times 1.2 \times 10^3\ \Omega \\ &= \mathbf{4\text{ vars}} \end{aligned}$$

e. $$\begin{aligned} P_s &= I^2Z \qquad (25\text{-}9) \\ &= (58 \times 10^{-3}\text{ A})^2 \times 2.1 \times 10^3\ \Omega \\ &= \mathbf{7.1\text{ VA}} \end{aligned}$$

f. $$\text{Power factor} = \cos\theta = \frac{P_T}{P_s} \tag{25-12}$$

$$= \frac{5.7\text{ W}}{7.1\text{ VA}} = \mathbf{0.8}$$

See Fig. 27-18.

27-9 APPLICATION OF COMPLEX NUMBERS TO SERIES-PARALLEL CIRCUITS

In Section 6-7, we solved for currents and voltage drops in a series-parallel *resistance* circuit. We can now apply complex number theory to any ac circuit using a similar approach to that used in Section 6-7, except that we must include the phase angles. That is, voltage-divider and current-divider rules apply equally well to dc resistance

I_T = 58 mA $\underline{/35^\circ}$
120 V, 60 Hz
I_1 = 35 mA $\underline{/-20^\circ}$
I_2 = 47.6 mA $\underline{/72.1^\circ}$

(*a*) Circuit diagram

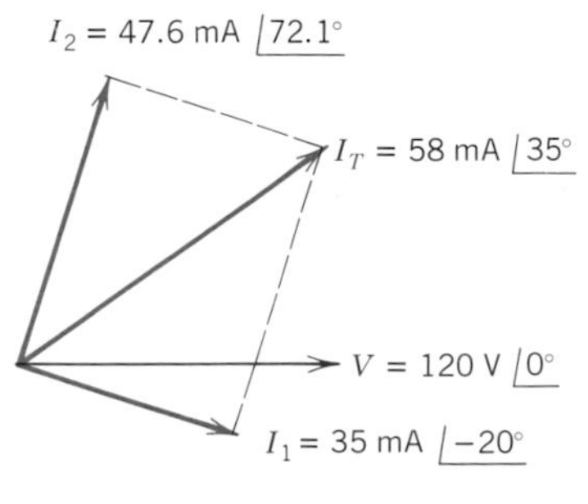

(*b*) Phasor diagram

FIGURE 27-18 Circuit and phasor diagrams for Example 27-17.

circuits and ac circuits when complex numbers are used, as shown in Example 27-18.

EXAMPLE 27-18

Given the circuit in Fig. 27-19, calculate:

a. The total impedance of the whole circuit.
b. The total supply current.
c. The reading of a voltmeter connected across Z_1.
d. The current through the capacitor.

SOLUTION

a. $X_{L1} = 2\pi f L_1$ (20-2)

$$= 2\pi \times 400 \text{ Hz} \times 50 \times 10^{-3} \text{ H}$$

$$= 125.7\ \Omega$$

$$Z_1 = R_1 + jX_{L1} = 100 + j125.7\ \Omega$$

$$X_{L2} = 2\pi f L_2 = 251.4\ \Omega$$

$$Z_2 = R_2 + jX_{L2} = 220 + j251.4\ \Omega$$

$$X_C = \frac{1}{2\pi f C} \qquad (23\text{-}2)$$

$$= \frac{1}{2\pi \times 400 \text{ Hz} \times 2 \times 10^{-6} \text{ F}}$$

$$= 198.9\ \Omega$$

$$Z_3 = R_3 - jX_C = 150 - j198.9\ \Omega$$

$$Z_T = Z_1 + Z_4$$

$$= Z_1 + Z_2 \| Z_3$$

$$= Z_1 + \frac{Z_2 \times Z_3}{Z_2 + Z_3}$$

$$= 100 + j125.7 + \frac{(220 + j251.4) \times (150 - j198.9)}{(220 + j251.4) + (150 - j198.9)}\ \Omega$$

$$= 100 + j125.7 + \frac{334.1 \angle 48.8^\circ \times 249.1 \angle -53^\circ}{370 + j52.5}\ \Omega$$

$$= 100 + j125.7 + \frac{83{,}224 \angle -4.2^\circ}{373.7 \angle 8.1^\circ}\ \Omega$$

$$= 100 + j125.7 + 222.7 \angle -12.3^\circ\ \Omega$$

$$= 100 + j125.7 + 217.6 - j47.4\ \Omega$$

$$= 317.6 + j78.3\ \Omega$$

$$= \mathbf{327.1\ \Omega \angle 13.8^\circ}$$

b. $I_T = \dfrac{V}{Z}$ (24-4)

$$= \frac{24 \text{ V} \angle 0^\circ}{327.1\ \Omega \angle 13.8^\circ} = 0.0734 \text{ A} \angle -13.8^\circ$$

$$= \mathbf{73.4 \text{ mA} \angle -13.8^\circ}$$

c. $V_1 = I_T Z_1$ (24-4)

$$= 0.0734 \text{ A} \angle -13.8^\circ \times (100 + j125.7\ \Omega)$$

$$= 0.0734 \text{ A} \angle -13.8^\circ \times 160.6\ \Omega \angle 51.5^\circ$$

$$= \mathbf{11.8 \text{ V} \angle 37.7^\circ}$$

d. $I_C = I_{Z3} = I_T \times \dfrac{Z_2}{Z_2 + Z_3}$ (6-12)

$$= 73.4 \text{ mA} \angle -13.8^\circ \times \frac{220 + j251.4}{(220 + j251.4) + (150 - j198.9)}$$

$$= 73.4 \text{ mA} \angle -13.8^\circ \times \frac{334.1\ \Omega \angle 48.8^\circ}{373.7\ \Omega \angle 8.1^\circ}$$

$$= 73.4 \text{ mA} \angle -13.8^\circ \times 0.894 \angle 40.7^\circ$$

$$= \mathbf{65.6 \text{ mA} \angle 26.9^\circ}$$

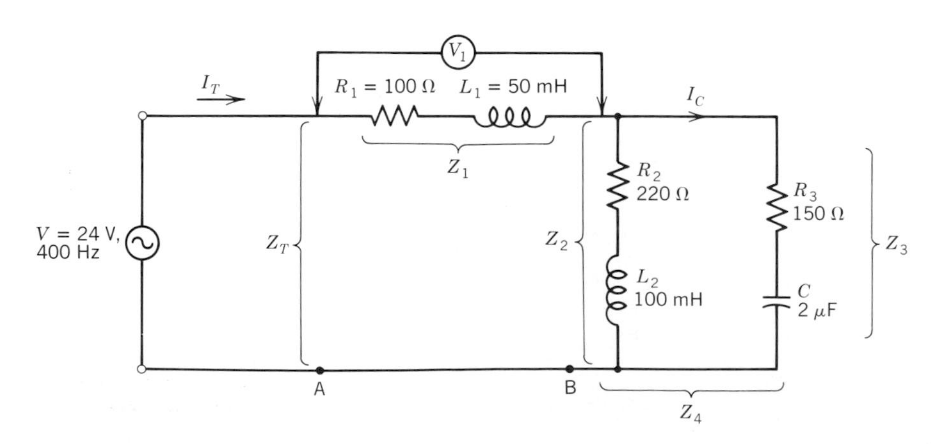

FIGURE 27-19
Circuit for Example 27-18.

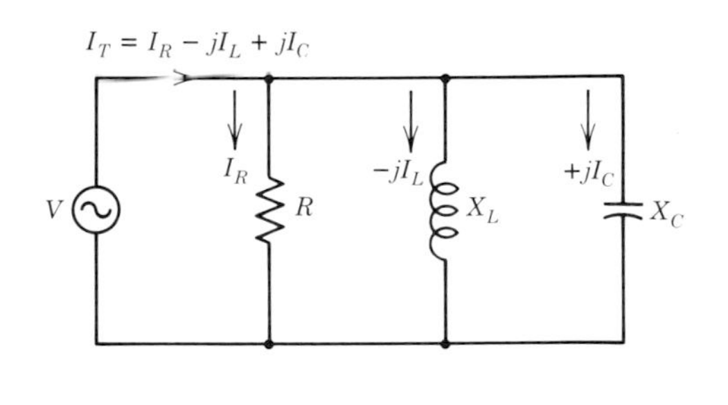

(a) Circuit diagram

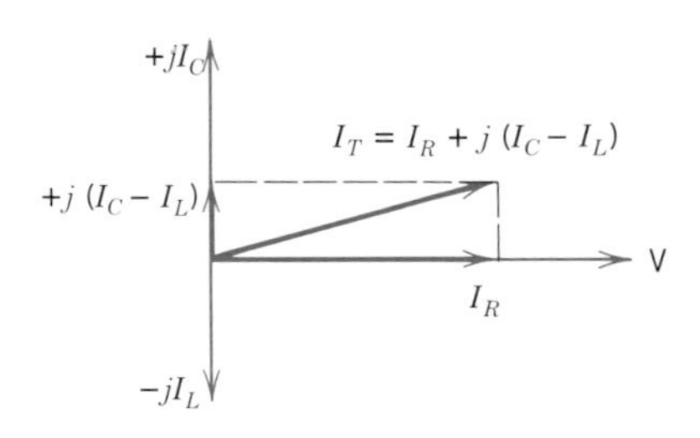

(b) Phasor diagram

FIGURE 27-20
Complex numbers applied to a parallel circuit.

27-10 CONDUCTANCE, SUSCEPTANCE, AND ADMITTANCE

There is an alternate method for solving parallel ac circuits. It involves the reciprocal of resistance, reactance, and impedance; these reciprocals are called conductance, susceptance, and admittance, respectively. Refer to Fig. 27-20.

We can express the total current in a pure parallel circuit in rectangular form:

$$I_T = I_R - jI_L + jI_C$$

or

$$\frac{V}{Z_T} = \frac{V}{R} - j\frac{V}{X_L} + j\frac{V}{X_C}$$

Therefore,

$$\frac{1}{Z_T} = \frac{1}{R} - j\frac{1}{X_L} + j\frac{1}{X_C}$$

From Eq. 2-5 we defined $1/R$ as *conductance G*, measured in siemens, S. Conductance is the ability of a pure resistance to pass electric current.

We now define $1/X_L$ as the inductive *susceptance* B_L, and $1/X_C$ as the capacitive *susceptance* B_C, measured in siemens, S. Susceptance is the ability of a pure inductance or pure capacitance to pass alternating current.

Similarly, we define $1/Z$ as *admittance* Y, measured in siemens, S. Admittance is the *overall* ability of a circuit to pass alternating current. Note that G, B, and Y are all measured in siemens, S:

$$B_L = \frac{1}{X_L} \quad \text{siemens (S)} \qquad (27\text{-}20)$$

$$B_C = \frac{1}{X_C} \quad \text{siemens (S)} \qquad (27\text{-}21)$$

$$Y = \frac{1}{Z} \quad \text{siemens (S)} \qquad (27\text{-}22)$$

where all terms are as previously defined.

We can now write, for a parallel RLC circuit,

$$Y = G - jB_L + jB_C \quad \text{siemens (S)} \quad (27\text{-}23)$$

where: Y is the total admittance of a pure parallel circuit in siemens, S
G is the conductance in siemens, S
B_L is the inductive susceptance in siemens, S
B_C is the capacitive susceptance in siemens, S

Note that in contrast to a series impedance expression ($Z = R + jX_L - jX_C$), inductive *susceptance* is preceded by a *negative j* and capacitive susceptance is preceded by a *positive j*.

Once the total admittance of a parallel circuit has been obtained, the following information is available:

1 The impedance can be found from

$$Z_T = \frac{1}{Y_T} \quad \text{ohms } (\Omega) \qquad (27\text{-}22)$$

2 The components of an *equivalent parallel* circuit can be identified if the admittance is given in rectangular form. (This corresponds to the equivalent *series* components for *impedance* given in rectangular form.)

For example, if the admittance of a complex ac circuit, series or parallel, is reduced to

$$Y_T = 0.01 + j0.02 \text{ S}$$

we can interpret this as an equivalent parallel circuit with

$G = 0.01$ S and $B_C = 0.02$ S. That is, the equivalent circuit would consist of a resistance

$$R = \frac{1}{G} = \frac{1}{0.01 \text{ S}} = 100 \ \Omega$$

in parallel with a capacitor having a capacitive reactance of:

$$X_C = \frac{1}{B_C} = \frac{1}{0.02 \text{ S}} = 50 \ \Omega$$

See Fig. 27-21*a*.

We can also find the polar form of an admittance and then find the equivalent impedance. For example,

$$Y_T = 0.01 + j0.02 \text{ S}$$
$$= 0.022 \text{ S} \angle 63.4^\circ$$

$$Z_T = \frac{1}{Y_T} \qquad (27\text{-}22)$$
$$= \frac{1}{0.022 \text{ S} \angle 63.4^\circ} = 45.5 \ \Omega \angle -63.4^\circ$$
$$= 20.4 - j40.7 \ \Omega$$

This represents a *series* circuit of 20.4-Ω resistance and 40.7-Ω capacitive reactance. See Fig. 27-21*b*. The two circuits in Fig. 27-21 are equivalent to each other in all *ac* applications. This is the method by which a practical resonant circuit was converted to a theoretical parallel resonant circuit in Section 26-8. (See Fig. 26-15 and Eqs. 26-10 and 26-11.)

It should be noted that in any parallel circuit:

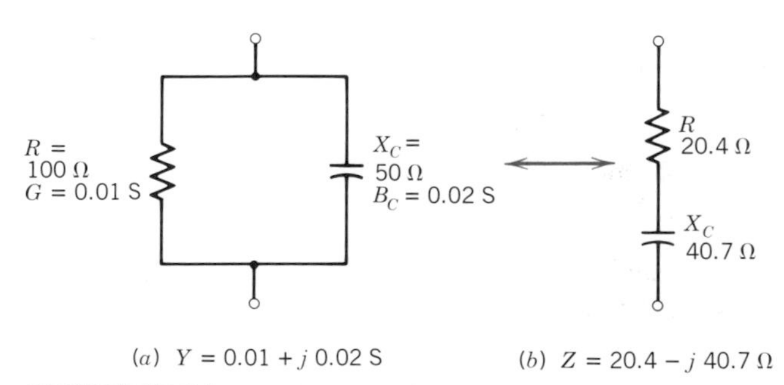

(*a*) $Y = 0.01 + j\,0.02$ S (*b*) $Z = 20.4 - j\,40.7\ \Omega$

FIGURE 27-21
Equivalent parallel and series circuits.

$$Y_T = Y_1 + Y_2 + Y_3 + \cdots + Y_n \quad \text{siemens (S)} \qquad (27\text{-}24)$$

as shown in Example 27-19.

EXAMPLE 27-19

For the parallel circuit given in Fig. 27-22*a*, calculate the following component values using the method of admittances:

a. An equivalent parallel circuit.
b. An equivalent series circuit.

SOLUTION

a. $Z_1 = 20 + j50 \ \Omega = 53.9 \ \Omega \angle 68.2^\circ$

$$Y_1 = \frac{1}{Z_1} \qquad (27\text{-}22)$$
$$= \frac{1}{53.9 \ \Omega \angle 68.2^\circ} = 0.0185 \text{ S} \angle -68.2^\circ$$
$$= 0.0069 - j0.0172 \text{ S}$$

$Z_2 = 25 + j40 \ \Omega = 47.2 \ \Omega \angle 58^\circ$

$$Y_2 = \frac{1}{Z_2} \qquad (27\text{-}22)$$
$$= \frac{1}{47.2 \ \Omega \angle 58^\circ} = 0.0212 \text{ S} \angle -58^\circ$$
$$= 0.0112 - j0.0180 \text{ S}$$

$Z_3 = 40 - j40 \ \Omega = 56.6 \ \Omega \angle -45^\circ$

$$Y_3 = \frac{1}{Z_3} \qquad (27\text{-}22)$$
$$= \frac{1}{56.6 \ \Omega \angle -45^\circ} = 0.0177 \text{ S} \angle 45^\circ$$
$$= 0.0125 + j0.0125 \text{ S}$$

$$Y_T = Y_1 + Y_2 + Y_3 \qquad (27\text{-}24)$$
$$= (0.0069 - j0.0172) + (0.0112 - j0.0180) + (0.0125 + j0.0125) \text{ S}$$
$$= 0.0306 - j0.0227 \text{ S} = 0.0381 \text{ S} \angle -36.6^\circ$$

For an equivalent parallel circuit:

$$Y_T = G - jB_L$$
$$R = \frac{1}{G} \qquad (2\text{-}5)$$
$$= \frac{1}{0.0306 \text{ S}} = \mathbf{32.7 \ \Omega}$$

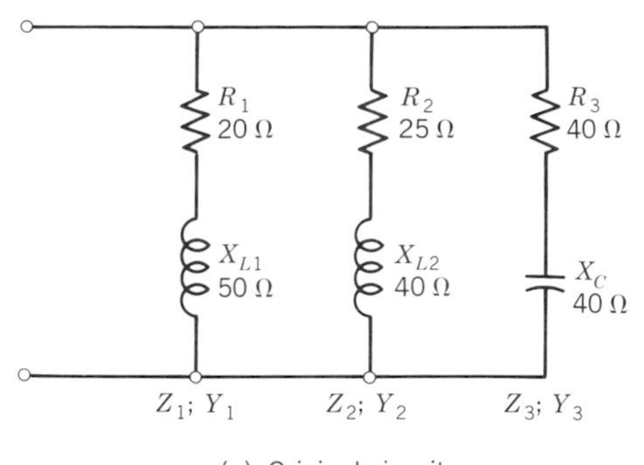

(*a*) Original circuit

R 32.7 Ω X_L 44.1 Ω

$Y_T = 0.0306 - j\,0.0227$ S

(*b*) Equivalent parallel circuit

R 21 Ω X_L 15.6 Ω

$Z_T = 21 + j\,15.6$ Ω

(*c*) Equivalent series circuit

FIGURE 27-22
Circuits for Example 27-19.

$$X_L = \frac{1}{B_L} \tag{27-20}$$

$$= \frac{1}{0.0227\text{ S}} = \mathbf{44.1\ \Omega}$$

See Fig. 27-22*b*.

b. For an equivalent series circuit:

$$Z_T = \frac{1}{Y_T} \tag{27-22}$$

$$= \frac{1}{0.0381\text{ S}\ \underline{/-36.6^\circ}} = 26.2\ \Omega\ \underline{/36.6^\circ}$$

$$= 21 + j15.6\ \Omega$$

$$R = \mathbf{21\ \Omega}, \qquad X_L = \mathbf{15.6\ \Omega}$$

See Fig. 27-22*c*.

27-11 APPLICATION OF COMPLEX NUMBERS TO THÉVENIN'S THEOREM

In Section 10-4 we saw how Thévenin's theorem could be used to convert a complicated circuit containing electromotive forces and resistances into a simple equivalent series circuit of one electromotive force and one resistance.

We can now restate Thévenin's theorem as it applies to an ac circuit containing emfs and impedances:

Any linear two-terminal network consisting of fixed impedances and sources of emf, can be replaced by a single source of emf, V_{Th}, in series with a single impedance, Z_{Th}, whose values are given by the following:

1. **V_{Th} is the open circuit voltage at the terminals of the original circuit.**
2. **Z_{Th} is the impedance looking back into the original network from the two terminals with all sources of emf shorted out and replaced by their internal impedance.**

EXAMPLE 27-20

Given the circuit in Fig. 27-23, use Thévenin's theorem to find the current through the capacitor. (Note that this circuit is the same as that in Fig. 27-19, with Z_1 the internal impedance of the source.)

SOLUTION

First remove the load, Z_3, to find the *open-circuit* voltage between A and B, as in Fig. 27-24.

$$V_{\text{Th}} = V \times \frac{Z_2}{Z_1 + Z_2} \tag{5-5}$$

$$= 24\text{ V}\ \underline{/0^\circ}$$

$$\times \frac{220 + j251.4\ \Omega}{(100 + j125.7) + (220 + j251.4)\ \Omega}$$

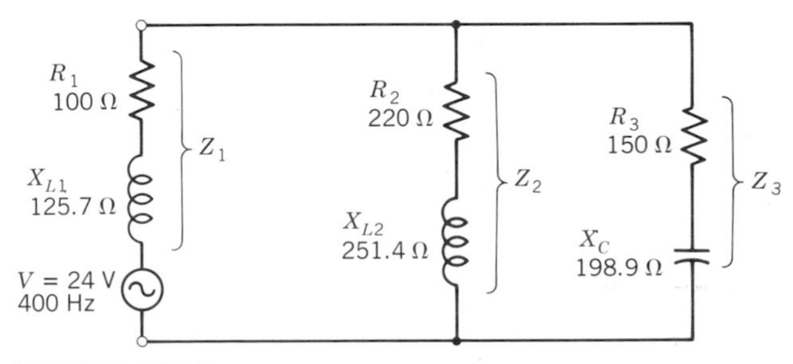

FIGURE 27-23
Circuit for Example 27-20.

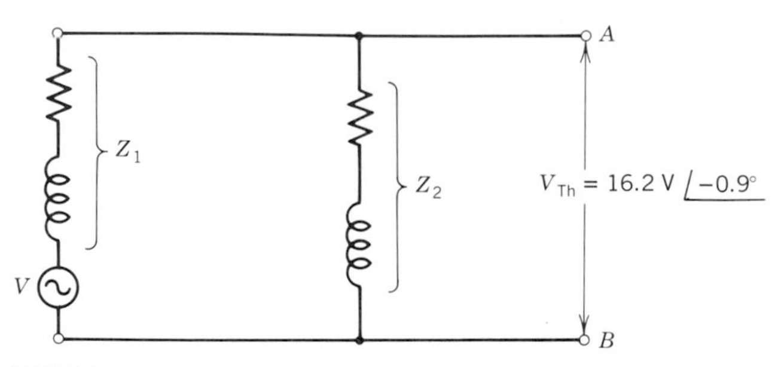

FIGURE 27-24
Obtaining the equivalent Thévenin voltage for Example 27-20.

$$= 24\text{ V}\,\angle 0^\circ \times \frac{334.1\,\angle 48.8^\circ\ \Omega}{320 + j377.1\ \Omega}$$

$$= 24\text{ V}\,\angle 0^\circ \times \frac{334.1\ \Omega\,\angle 48.8^\circ}{494.6\ \Omega\,\angle 49.7^\circ}$$

$$= 24\text{ V}\,\angle 0^\circ \times 0.675\,\angle -0.9^\circ$$

$$= \mathbf{16.2\,V\,\angle -0.9^\circ}$$

To find the Thévenin impedance, short the voltage source, V, and replace it with its internal impedance, Z_1, as in Fig. 27-25.

This places Z_1 and Z_2 in parallel with each other.

$$Z_{Th} = Z_1 \| Z_2$$

$$= \frac{Z_1 \times Z_2}{Z_1 + Z_2} \qquad (27\text{-}17)$$

$$= \frac{(100 + j125.7) \times (220 + j251.4)}{(100 + j125.7) + (220 + j251.4)}\ \Omega$$

$$= \frac{160.6\,\angle 51.5^\circ \times 334.1\,\angle 48.8^\circ}{320 + j377.1}\ \Omega$$

$$= \frac{53{,}656\,\angle 100.3^\circ}{494.6\,\angle 49.7^\circ}\ \Omega$$

$$= \mathbf{108.5\ \Omega\,\angle 50.6^\circ = 68.9 + 83.8\ \Omega}$$

The Thévenin equivalent circuit is shown in Fig. 27-26.

$$I = \frac{V_{Th}}{Z_{Th} + Z_3} \qquad (24\text{-}4)$$

$$= \frac{16.2\text{ V}\,\angle -0.9^\circ}{68.9 + j83.8 + 150 - j198.9}$$

$$= \frac{16.2\text{ V}\,\angle -0.9^\circ}{218.9 - j115.1\ \Omega}$$

$$= \frac{16.2\text{ V}\,\angle -0.9^\circ}{247.3\ \Omega\,\angle -27.7^\circ}$$

$$= 0.0655\text{ A}\,\angle 26.8^\circ = \mathbf{65.5\ mA\,\angle 26.8^\circ}$$

The solution for this problem should be compared with that for Example 27-18. Although the amount of work is similar, the Thévenin approach is preferred if the current is to be calculated for a number of different loads. Then only the last calculation for I must be repeated.

It should be noted that all of the network analysis methods covered in Chapter 10—branch currents, loop or mesh currents, superposition, and Norton's theorem—can be applied to ac circuits (as we did with Thévenin's), provided that complex numbers are used to represent the voltage, current, and impedance. However, the calculations are usually long and tedious, whereas the Thévenin approach is relatively simple. The reader is referred to circuit analysis texts for details on other methods of solving complex ac circuits.

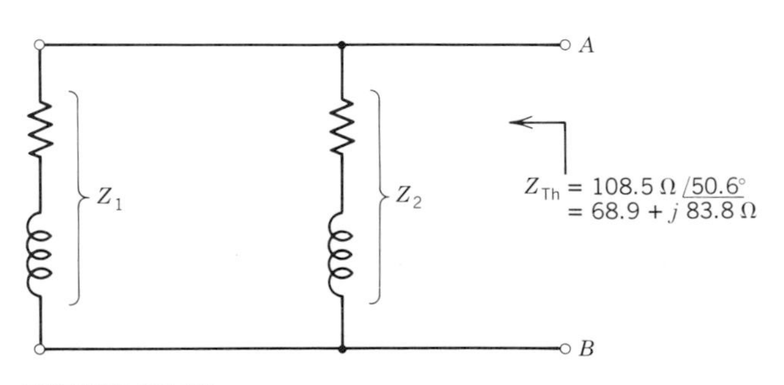

FIGURE 27-25
Obtaining the equivalent Thévenin impedance for Example 27-20.

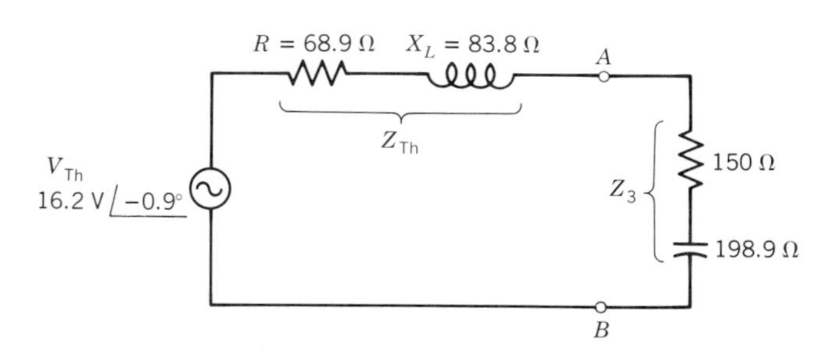

FIGURE 27-26
Thévenin equivalent circuit for Example 27-20.

SUMMARY

1. The polar form of a complex number consists of a positive magnitude and a phase angle that may be positive or negative, such as $5\ \text{V}\angle{-15^\circ}$.

2. To multiply two complex numbers given in polar form, multiply their magnitudes and algebraically *add* their phase angles.

3. To divide two complex numbers given in polar form, divide their magnitudes and algebraically *subtract* their phase angles.

4. When the voltage and impedance are given in polar form, the equation $I = V/Z$ gives the magnitude and phase angle of the current with respect to the voltage.

5. The polar form of the current or voltage gives the reading of an ammeter or voltmeter, respectively.

6. The j operator, equal to $\sqrt{-1}$ or $1\angle 90^\circ$, rotates a number through an angle of $+90^\circ$.

7. The rectangular coordinate system consists of a *real* axis and an *imaginary* axis.

8. A complex number in rectangular form consists of a real part and an imaginary part, such as $Z = R + jX$.

9. When an impedance is given in rectangular form, the real part is the resistance portion and the imaginary part is the series reactive portion (inductive if positive, capacitive if negative).

10. The sum (difference) of two complex numbers given in rectangular form is found by algebraically adding (subtracting) the real parts, then the imaginary parts, respectively.

11. The multiplication of complex numbers in rectangular form follows the distributive law of algebra, whereas division requires a rationalization of the denominator. Both involve the fact that $j^2 = -1$.

12. The total impedance of a series circuit is given by $Z_T = Z_1 + Z_2 + Z_3$, and so on, and of a parallel circuit by

$$Z_T = \frac{Z_1 \times Z_2}{Z_1 + Z_2}$$

for two branches, all impedances given in complex number form.

13. A complex number in rectangular form can be converted to polar form using:

$$a + jb = \sqrt{a^2 + b^2}\ \angle \arctan\frac{b}{a}$$

14. A complex number in polar form can be converted to rectangular form using:

$$M\angle\theta = M\cos\theta + jM\sin\theta$$

15. Admittance, Y, is the reciprocal of impedance; susceptance, B, is the reciprocal of reactance; and conductance, G, is the reciprocal of resistance. All are measured in siemens (S).

16. If the admittance of a circuit is given in rectangular form, $Y = G + j(B_C - B_L)$, the resistance and reactive components of an equivalent parallel circuit can be identified, as well as the circuit's impedance, $Z = \dfrac{1}{Y}$.

17. The total admittance of a parallel circuit is given by $Y_T = Y_1 + Y_2 + Y_3$, and so on.

18. Thévenin's theorem can be applied to an ac circuit if complex numbers are used to represent the impedances and voltage.

SELF-EXAMINATION

Match the expressions in the right-hand column with those in the left-hand column. (Answers at back of book)

27-1.	$30 - j40$ V =	a.	0.25 S
27-2.	$R = 10\ \Omega$, $X_L = 20\ \Omega$; $Z =$	b.	$30\text{ V}\ \underline{/5^\circ}$
27-3.	$V_R = 60$ V, $V_L = 80$ V; $V_T =$	c.	$10 + j20\ \Omega$
27-4.	$I_R = 5$ A, $I_L = 6$ A, $I_C = 7$ A; $I_T =$	d.	$2\text{ A}\ \underline{/60^\circ}$
27-5.	$I = 3\text{ A}\ \underline{/-40^\circ}$, $Z = 10\ \Omega\ \underline{/45^\circ}$; $V =$	e.	$72\ \underline{/-45^\circ}$
27-6.	$V = 120\text{ V}\ \underline{/10^\circ}$, $Z = 60\ \Omega\ \underline{/-50^\circ}$; $I =$	f.	$50\text{ V}\ \underline{/-53.1^\circ}$
27-7.	$24\ \underline{/-20^\circ} \times 3\ \underline{/-25^\circ}$	g.	$7 - j8$
27-8.	$3\ \underline{/-25^\circ} \div 24\ \underline{/-40^\circ}$	h.	$5 + j1$ A
27-9.	$5\ \underline{/43^\circ} \times 3.4\ \underline{/47^\circ}$	i.	-4
27-10.	$6\ \underline{/95^\circ} \div 1.5\ \underline{/-85^\circ}$	j.	$-7 - j1$
27-11.	$(3 + j7) + (4 - j15)$	k.	$0.125\ \underline{/15^\circ}$
27-12.	$(-14 - j6) - (-7 - j5)$	l.	$8 + j9$
27-13.	$(6 - j4) \times (6 + j4)$	m.	$j17$
27-14.	$(3 - j5)^2$	n.	52
27-15.	$(43 - j6) \div (2 - j3)$	o.	$50 + j50$
27-16.	$Z = 50\ \Omega\ \underline{/-53.1^\circ}$; $Y =$	p.	$-16 - j30$
27-17.	$70.7\ \underline{/45^\circ} =$	q.	$0.02\text{ S}\ \underline{/53.1^\circ}$
27-18.	$8 + j9$: conjugate =	r.	4 S
27-19.	$X_C = 0.25\ \Omega$; $B_C =$	s.	$100\text{ V}\ \underline{/53.1^\circ}$
27-20.	$X_L = 4\ \Omega$; $B_L =$	t.	$8 - j9$

REVIEW QUESTIONS

1. a. When an impedance is given in polar form, how can you tell whether it is inductive or capacitive?

b. How can you determine if there is more resistance than reactance?

2. If an impedance is given in rectangular form, how do you determine the resistive and reactive components?

3. What are two advantages of the polar notation?

4. For what mathematical operations is the rectangular notation especially suited?

5. What are the polar forms for j^2 and j^3?

6. The voltage across a pure capacitor in rectangular form is given by $-j150$ V. Is there anything *imaginary* about this voltage? Explain.

7. Although it is possible to multiply and divide phasors given in rectangular form (without conversion), how can addition and subtraction be performed directly (without conversion) in polar form?

8. a. What do you understand by the term *conjugate*?

b. Give two applications in which the conjugate is used.

9. A circuit contains both resistance and inductance. How could you express the apparent power in terms of the true and reactive power using the rectangular form?

10. a. Given the admittance of a circuit in rectangular form, how do you interpret the real and imaginary parts?

b. How would you determine the equivalent *series* circuit components?

11. How is Thévenin's theorem different when applied to an ac circuit compared with a dc circuit?

12. An ac bridge consists of a Wheatstone bridge circuit with an ac voltage source across one set of diagonally opposite corners and an ac null indicator across the other set of corners. If the resistance symbols are replaced by impedances, with Z_1 and Z_4 opposite each other, and Z_2 and Z_3 opposite each other, show that balance of the ac bridge occurs when the general bridge equation is satisfied:

$$Z_1 \times Z_4 = Z_2 \times Z_3$$

PROBLEMS

27-1. A series *RL* circuit has 80 V across the resistance and 40 V across the inductor. Determine the total applied voltage in rectangular and polar form.

27-2. A series *RC* circuit has 30 V across the resistance and 90 V across the capacitor. Determine the total applied voltage in rectangular and polar form.

27-3. A series *RC* circuit has 25 Ω of resistance and 40 Ω of reactance. Determine the impedance in rectangular and polar form.

27-4. A series *RL* circuit has 470 Ω of resistance and 300 Ω of reactance. Determine the impedance in rectangular and polar form.

27-5. A 68-Ω resistor and a 30-mH inductor of negligible resistance are series-connected across a 400-Hz sinusoidal supply. Determine the circuit impedance in rectangular and polar form.

27-6. A 2.2-kΩ resistor and a 0.1-μF capacitor are series-connected across a 1-kHz sinusoidal supply. Determine the impedance in rectangular and polar form.

27-7. Given $A = 50 \angle 60^\circ$ and $B = 4 \angle -25^\circ$, find:

a. $A \times B$

b. $A \div B$

c. $B \div A$

27-8. Given $C = 2 \times 10^{-3} \angle -51.2^\circ$ and $D = 4 \times 10^{-6} \angle -30.8^\circ$, find:

a. $C \times D$

b. $C \div D$

c. $D \div C$

27-9. For the circuit of Problem 27-3, assume an applied voltage of 120 V and find, in magnitude and phase:

a. The current.

b. The resistor voltage.

c. The capacitor voltage.

Draw a phasor diagram.

27-10. For the circuit of Problem 27-4, assume an applied emf of 230 V and find, in magnitude and phase:

a. The current.

b. The resistor voltage.

c. The inductor voltage.

Draw a phasor diagram.

27-11. For the circuit of Problem 27-5, assume an applied emf of 48 V and find, in magnitude and phase:

a. The current.

b. The resistor voltage.

c. The inductor voltage.

Draw a phasor diagram.

27-12. For the circuit of Problem 27-6, assume an applied emf of 20 V peak-to-peak and find, in magnitude and phase, rms values of:

a. The current.

b. The resistor voltage.

c. The capacitor voltage.

Draw a phasor diagram.

27-13. A series circuit with an applied emf of 60 V has a current of 250 mA that leads the applied voltage by 30°. Determine:

a. The type of series circuit.

b. The circuit impedance in polar form.

c. The resistance and reactance of the circuit.

27-14. A current of 40 mA flows when an 80-V dc source is applied to a series circuit. When a 120-V, 60-Hz source is applied, a current of 30 mA flows. Determine:

a. The circuit impedance in rectangular form.

b. The circuit impedance in polar form.

c. The inductance of the circuit.

27-15. Given $A = 3 - j7$ and $B = 8 + j11$, find:

a. $A + B$

b. $A - B$

c. $B - A$

27-16. Given $C = -4 - j6$, $D = 4 - j9$, and $E = 11 + j10$, find:

a. $C + D + E$

b. $D - E - C$

c. $2C + 3(D + E)$

27-17. A coil having a resistance of 50 Ω and a reactance of 150 Ω is connected in series with a capacitor whose reactance is 50 Ω. Find the total impedance in rectangular and polar form.

27-18. Three loads are series-connected across the 230-V line. The first load has a voltage given by $50 + j70$ V; the second load has a voltage given by $100 - j150$ V. Find the voltage across the third load in rectangular and polar form.

27-19. Given $A = 2 - j6$ and $B = 7 + j9$, find:

a. $A \times B$

b. $A \div B$

c. $B \div A$, in rectangular form

27-20. Given $C = -4 - j5$, $D = 3 - j10$ and $E = 12 + j2$, find:

a. $C \times D \times E$

b. $C \div D$

c. C^2, in rectangular form.

27-21. Two coils, one of 10-Ω resistance and 25-Ω reactance, the other of 15-Ω resistance and 30-Ω reactance, are connected in parallel across a 24-V, 60-Hz source. Calculate:

a. The total circuit impedance in rectangular and polar form.

b. The total current.

c. Each branch current.

27-22. Repeat Problem 27-21 with a 100-μF capacitor connected in parallel with the coils.

27-23. A coil having a resistance of 400 Ω and a reactance of 1 kΩ is connected in parallel with a capacitor whose reactance is 1.2 kΩ. When connected to a 100-mV source, calculate:

a. The total circuit impedance.

b. The total current.

c. Each branch current.

27-24. Repeat Problem 27-23 assuming that the capacitor has a resistance of 1 kΩ connected in series with it.

27-25. Convert the following to polar form:

a. $6 + j10$

b. $3 - j7$

c. $-8 - j6$.

27-26. Convert the following to polar form:

a. $10j$

b. -5

c. $3j^3$

27-27. Find the total impedance of the two parallel-connected coils in Problem 27-21 using rectangular to polar conversion.

27-28. What rectangular impedance must be connected in parallel with $5 + j4$ Ω to produce a total impedance of $2 - j1$ Ω? (*Hint:* Let the impedance be $a + jb$ and equate $1/(2 - j1)$ to $1/(a + jb) + 1/(5 + j4)$).

27-29. Convert the following to rectangular form:

a. $65 \angle 125^\circ$.

b. $12 \angle -72^\circ$.

c. $20 \angle 200^\circ$.

27-30. A coil having an ac resistance of 8 Ω and an inductance of 200 μH is connected in parallel with a 1-μF capacitor that has an ac resistance of 4 Ω. Determine the components of the equivalent *series* impedance at a frequency of 10 kHz.

27-31. Repeat Problem 27-30 for a frequency of 5 kHz.

27-32. Two capacitors, a coil, and a resistor are connected in series. The voltages across the capacitors are $12\ \text{V} \angle -85^\circ$ and $15\ \text{V} \angle -90^\circ$; the coil voltage is $20\ \text{V} \angle 45^\circ$; and the resistor voltage is $10\ \text{V} \angle 0^\circ$. Calculate the supply voltage.

27-33. A parallel circuit has two branches and draws a total current of $120\ \text{mA} \angle -50^\circ$ from a 48-V, 400-Hz source. If the current in one branch is $30\ \text{mA} \angle 30^\circ$, calculate:

a. The reading of an ammeter connected in the other branch.

b. The total impedance of the circuit.

c. The total power dissipated in the whole circuit.

27-34. A coil and a pure capacitance, connected in parallel across a $120\text{-V} \angle 0^\circ$, 60-Hz supply, draw a total current of $4\ \text{A} \angle 0^\circ$. If the capacitor draws a current of $2\ \text{A} \angle 90^\circ$, calculate:

a. The reading of a wattmeter connected to measure the total power in the circuit.

b. The current through the coil.

c. The coil's resistance.

d. The coil's inductance.

e. The capacitance of the capacitor.

27-35. Repeat Example 27-18 using a frequency of 1 kHz.

27-36. Repeat Example 27-18 with an impedance equal to Z_1 connected between A and B.

27-37. For the parallel circuit given in Fig. 27-27, calculate the values of

a. An equivalent parallel circuit.

b. An equivalent series circuit, using the method of admittances, for a frequency of 1 kHz.

27-38. Repeat Problem 27-37 using a frequency of 100 Hz.

27-39. A 50-mV, 1-kHz sinusoidal source with an internal impedance of 12-kΩ resistance and 1-H inductance is connected to the parallel circuit of Fig. 27-27. Use Thévenin's theorem to calculate the current through the 1.5-H coil.

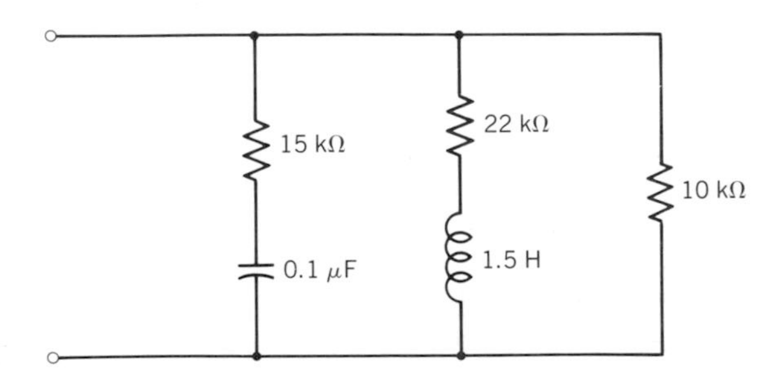

FIGURE 27-27
Circuit for Problems 27-37 and 27-38.

27-40. Repeat Problem 27-39 but find the current through the 10-kΩ resistor.

27-41. The ac bridge circuit of Fig. 27-28 balances when Z_1 is a 2.2-kΩ resistor, Z_2 is a 0.1-μF capacitor in series with a 100-kΩ resistor, Z_3 is a 5.1-kΩ resistor, and f = 1 kHz. Given that balance occurs when $Z_1Z_4 = Z_2Z_3$, determine the series components that make up Z_4.

27-42. The ac bridge circuit of Fig. 27-28 balances when Z_2 is a 15-kΩ resistor, Z_3 is a 6.8-kΩ resistor, Z_4 consists of a 0.015-μF capacitor in parallel with a 2.2-MΩ resistor, and f = 1 kHz. Determine:

a. The series components that make up Z_1.

b. The parallel components that make up Z_1.

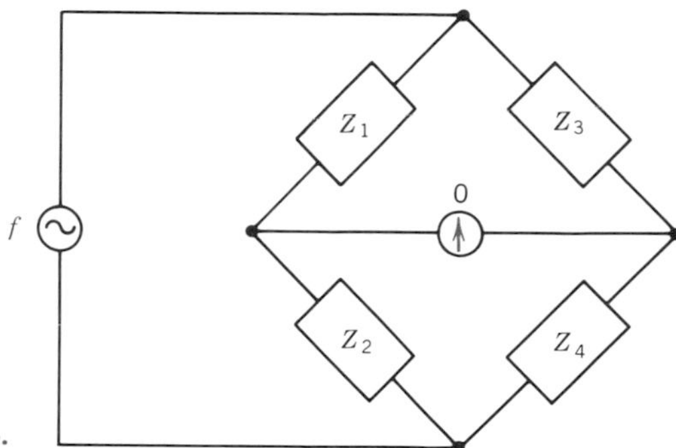

FIGURE 27-28
Circuit for Problems 27-41, 27-42 and 27-43.

27-43. An ac bridge used to determine accurately the inductance of a coil is called a Maxwell bridge. It has the form shown in Fig. 27-28, where Z_1 consists of a variable resistor R_1 in parallel with a capacitor C_s, Z_2 is a variable resistor R_2, Z_3 is a fixed resistor R_3, and Z_4 is the unknown inductance L_x in series with an unknown resistance R_x. Given that at balance $Z_1Z_4 = Z_2Z_3$, derive the equations for R_x and L_x in terms of the known components.

HINT: Use $Z_4 = Y_1Z_2Z_3$, then equate real components to each other and imaginary components to each other.

PART THREE

BASIC ELECTRONICS

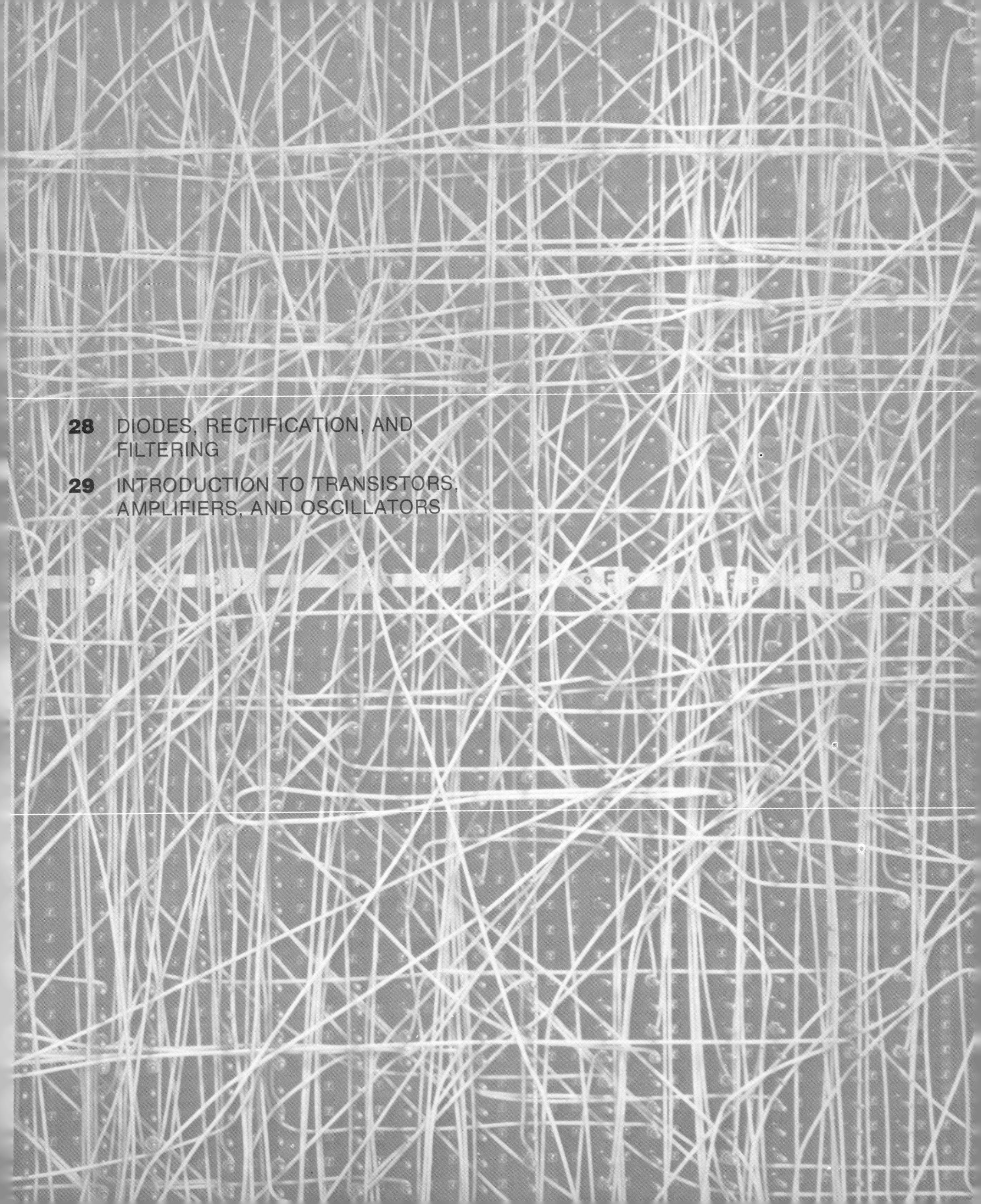

28 DIODES, RECTIFICATION, AND FILTERING

29 INTRODUCTION TO TRANSISTORS, AMPLIFIERS, AND OSCILLATORS

CHAPTER TWENTY-EIGHT
DIODES, RECTIFICATION, AND FILTERING

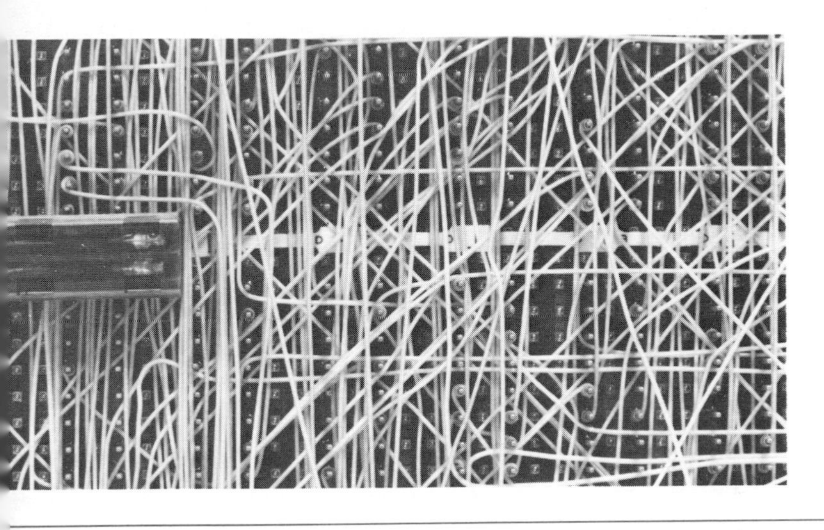

A diode is a device that permits current to flow through it in only one direction, depending upon the way the diode is connected with respect to the polarity of the applied voltage. Diodes are used extensively in power supplies to convert alternating current into direct current. In this application, they are often referred to as *rectifiers*. However, diodes can be used in a number of nonrectifying circuits, in radio and TV, computers, and industrial electronics. Light-emitting diodes (LEDs) even emit light when current passes through them.

Most modern solid-state diodes are constructed from the semiconductor material silicon. When an extremely pure wafer of silicon is modified by the addition of "impurity" materials, a *PN junction* is formed. When the *PN* junction is *forward-biased* by an external voltage (positive to *P*, negative to *N*) current easily flows through the diode. The diode is said to be *forward-biased* and *conducting*. Under conditions of *reverse-bias* (positive to *N*, negative to *P*), practically no current flows.

The conversion of alternating current into direct current can be accomplished in *half-wave* or *full-wave*, diode-rectifying circuits. But since this leaves an appreciable *ripple* in the output voltage, *filtering* is usually added. This may simply consist of a large capacitor connected in parallel with the load to smooth the output voltage; or some combination of inductance and capacitance may be employed to produce an even smoother dc output voltage.

28-1 PURE SILICON

As shown in Fig. 28-1 an atom of silicon has 14 electrons in orbit around the positive nucleus. The outer or valence shell has four electrons. If the valence shell could obtain four more electrons, this outermost shell would be complete and produce strong valence electron bonding. This is how a crystal of silicon is formed, by *covalent bonding,* as shown in Fig. 28-1*b*.

The central silicon atom is seen to be *sharing* four electrons from four neighboring atoms. But each of these atoms in turn is also sharing for bonding purposes one electron from the central atom. Thus no atom *exclusively* possesses eight valence electrons. However, each atom *covalently* bonds with another atom on this basis, forming a crystal of silicon.

28-1.1 Pure Silicon at Room Temperature

If a silicon crystal were at absolute zero (−273°C), all the valence electrons in the outer orbits would remain in the valence shell. None would have enough thermal energy to escape and become *free electrons,* so the material would be a perfect insulator. At room temperature, however, some of the valence electrons gain enough thermal energy to escape and break a covalent or shared bond. They are now *free* electrons and available to provide conduction, much as free electrons in a copper wire. However, the number of free electrons in pure silicon is very small, only about one for every 10^{12} atoms, giving pure silicon a very high resistivity of 2300 Ω-m at room temperature. (The ρ for copper, a good conductor, is 1.72×10^{-8} Ω-m at the same temperature.)

When an electron breaks a bond, it leaves a *vacancy* behind. Such an incomplete (unshared and broken) covalent bond is called a *hole.* This is shown in Fig. 28-2, where the hole is given a positive sign, for the *absence* of an electron. The important point is that holes can contribute to current much as the free electrons can. For every electron liberated by heat, there is a hole produced. We can refer to this combination as a *thermally generated electron-hole pair.* As the temperature increases, so does the number of electron-hole pairs.

28-1.2 Current in Pure Silicon

When a dc voltage is connected across a piece of pure silicon, a force acts on *both* the free electrons and the holes, as shown in Fig. 28-2*b*.

It is quite easy to visualize the flow of free electrons in the direction shown. But how can a hole flow in the opposite direction?

It is relatively easy for a valence electron in a neighboring atom to leave its covalent bond and fill this hole. But a hole is now created where the electron moved from and the hole effectively has moved in the *opposite* direction to the electron. This process continues, the holes moving to the right due to the valence electrons moving to the left. Note that this mechanism does *not* involve the motion of *free* electrons, so we represent it by the movement of positive charge or holes *only*. (This is supported by the Hall effect mentioned in Section 11-11.) This motion of charge is in *addition* to that due to *free* electrons. Consequently, the *total* current flowing in pure silicon is the *sum* of the free electron flow *and* the hole flow.

External to the silicon itself, only electrons can flow in the copper conducting wires. They must then provide as

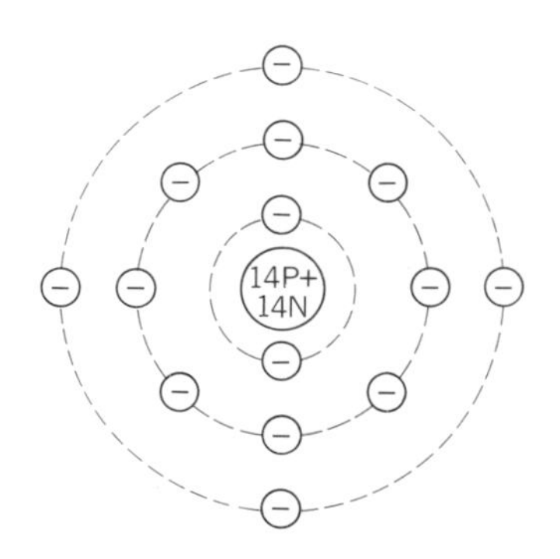

(*a*) Bohr concept of an atom of silicon.

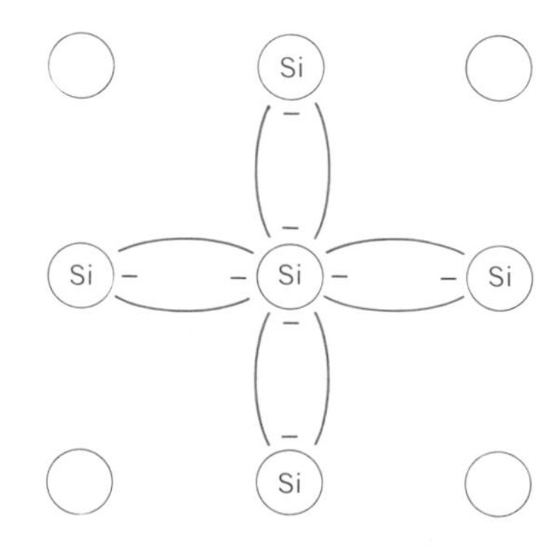

(*b*) Covalent bonding of silicon atoms form part of a crystal of pure silicon.

FIGURE 28-1
Diagrams representing atomic and crystal structure of silicon.

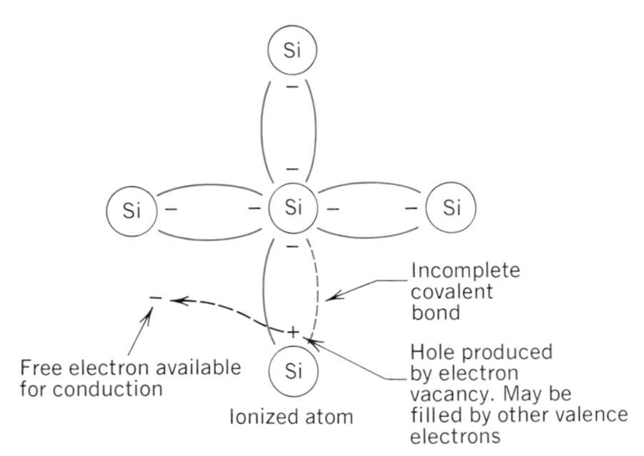

(*a*) Production of electron-hole pairs in pure silicon at room temperature

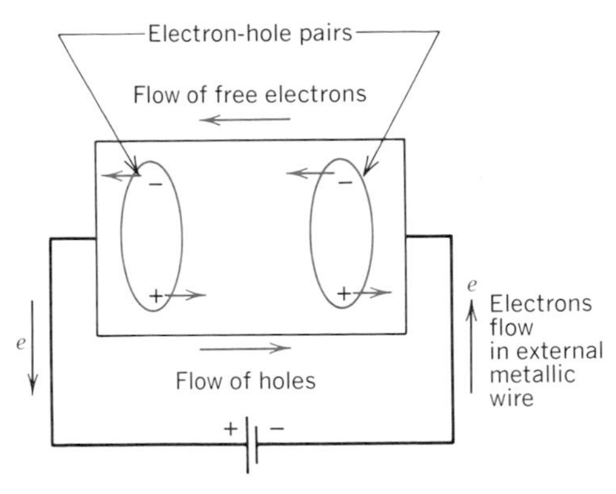

(*b*) Total current in pure silicon is due to the sum of electron and hole flow arising from thermally generated electron-hole pairs and applied voltage.

FIGURE 28-2
Pure silicon at room temperature.

much charge flow as the sum of the free electrons and holes combined. Also, the current that does flow through pure silicon is very temperature-dependent, increasing with a rise in temperature.

28-2 *N*-TYPE SILICON

The conductivity of pure (intrinsic) silicon (practically an insulator) can be significantly increased (to that of a semiconductor) by the addition of certain (impurity) materials. The silicon is then called extrinsic or a *doped* or *impure* semiconductor. These impurities are either *pentavalent* or *trivalent* atoms.

Phosphorus is a *pentavalent* material, meaning that it has five electrons in its outer valence shell. When a small percentage of phosphorus is added to silicon, the phosphorus atoms displace some of the silicon atoms, to form the arrangement shown in Fig. 28-3*a*. Since only four of the five electrons are necessary for covalent bonding purposes, one electron is *donated* to the crystalline structure. That is, at room temperature, this fifth unbonded electron is easily detached from its parent atom

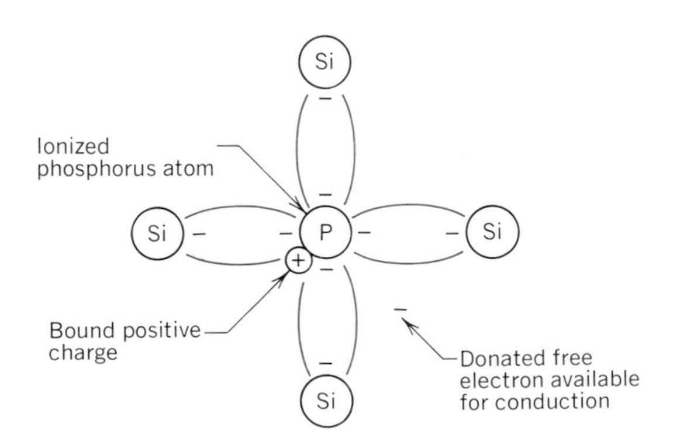

(*a*) Portion of a silicon crystal with a phosphorus atom substituted for a silicon atom to donate free electrons.

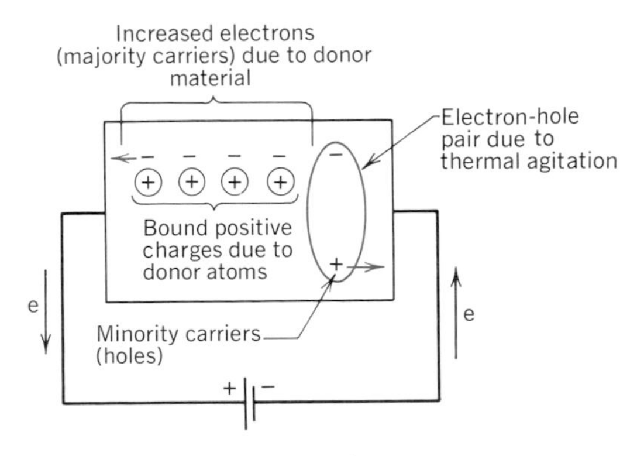

(*b*) Current in *N*-type silicon is due to electrons (majority carriers) and holes (minority carriers).

FIGURE 28-3
N-type silicon resulting from doping with pentavalent phosphorus impurity atoms.

to become a free electron. The phosphorus impurity is known as a *donor N-type impurity* material. The atom itself is ionized, acquiring a bound positive charge that is not free to move.

The doped silicon is called *N*-type (or negative type), since it now has many more free electrons than the pure silicon had. Also, *N*-type silicon has many more free electrons than holes. Its resistivity is much lower and it is capable of generating many more free electrons at room temperature than pure silicon.

28-2.1 Current in *N*-Type Silicon

Figure 28-3*b* represents the large increase in the number of free electrons now available (as a result of *N*-type doping) for conduction purposes. As in pure silicon, thermal agitation still causes electron-hole pairs. But because of the abundance of the donor electrons, the recombination of holes and electrons occurs rapidly so that there are even fewer holes in *N*-type silicon than in pure silicon. Because the free electron density is predominant (over holes), the silicon is called *N*-type (*N* for negative), even though the crystal as a whole is still electrically neutral. (Recall that every atom is electrically neutral. The pure silicon and the pure phosphorus are *each* electrically neutral.)

When an electromotive force is connected across the *N*-type silicon, both electrons and holes flow as before. But this time, the current is due primarily to the free donor electrons so that they are called the *majority current carriers.* The holes are termed the *minority current carriers,* since hole flow (in the opposite direction) due to valence electron motion is very small, in comparison.

28-3 *P*-TYPE SILICON

Now consider the addition of a *trivalent* impurity, aluminum, to silicon. Since there are only three electrons in the aluminum valence shell, the substitution of a silicon atom by an aluminum atom leaves an *incomplete* bond. This bond is usually completed by a neighboring silicon atom giving up an electron for this purpose. The aluminum atom thus acquires a bound negative charge, and a hole is created where this electron came from. This is shown in Fig. 28-4*a*.

Since the aluminum atom has created holes that can accept electrons, aluminum is known as an *acceptor* impurity. **The aluminum impurity has caused the *P*-type silicon to have many more holes than free electrons.**

28-3.1 Current in *P*-Type Silicon

Since the holes predominate in *P*-type silicon, holes are now the majority current carriers. The few, free valence

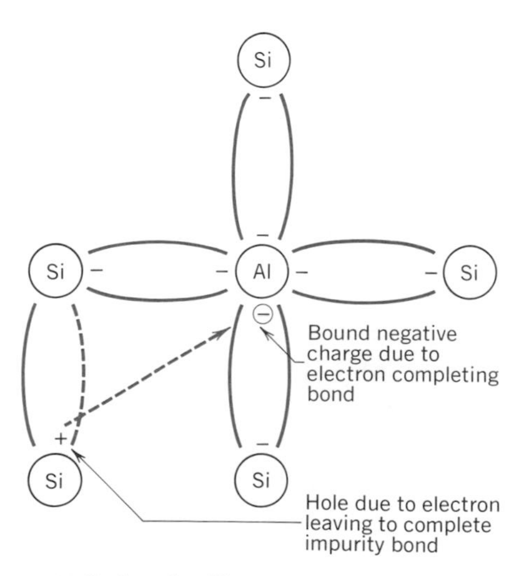

(*a*) Portion of a silicon crystal with an aluminum atom substituted for a silicon atom to provide holes

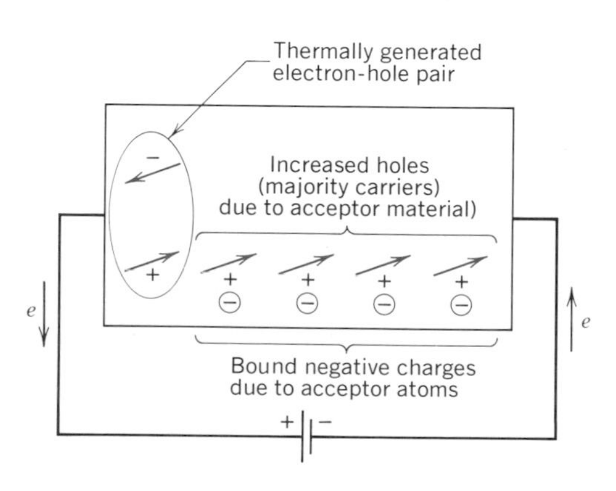

(*b*) Current in p-type silicon is due to holes (majority carriers) and electrons (minority carriers)

FIGURE 28-4
***P*-type silicon resulting from doping with trivalent aluminum impurity atoms.**

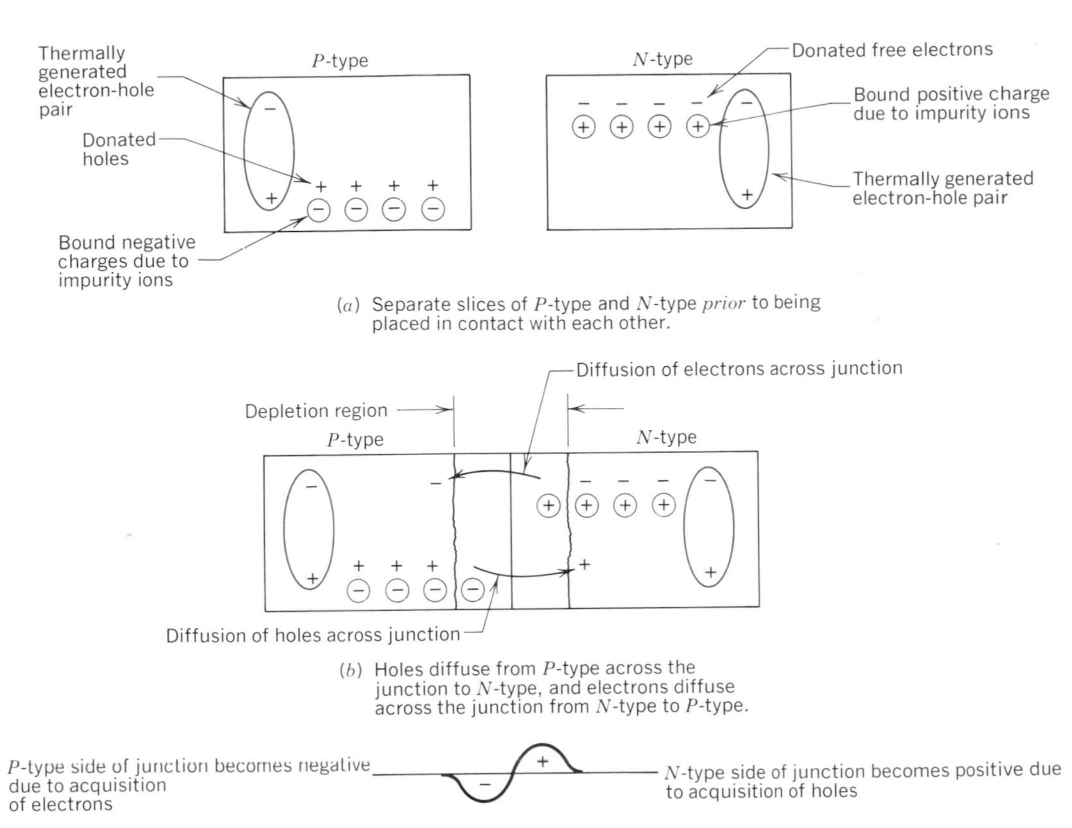

(*a*) Separate slices of *P*-type and *N*-type *prior* to being placed in contact with each other.

(*b*) Holes diffuse from *P*-type across the junction to *N*-type, and electrons diffuse across the junction from *N*-type to *P*-type.

(*c*) Potential barrier of approximately 0.7 V established across a *PN* junction due to diffusion of carriers.

FIGURE 28-5
Development of a potential barrier when a *PN* junction is formed.

electrons, thermally generated, are the minority current carriers. The holes being positive account for the name *P*-type. Figure 28-4*b* shows the motion of these carriers when an emf is applied. The total current is the sum of the majority hole flow and minority electron flow. Like *N*-type silicon, *P*-type silicon is also electrically neutral.

28-4 *PN* JUNCTIONS

Most modern *PN* junction diodes are a result of a diffusion process. This involves placing a wafer of *N*-type silicon and an oxide of the impurity material, such as aluminum, together in a furnace. When raised to several hundred degrees Celsius, a gas of the impurity atoms *diffuses* into the *N*-type semiconductor. This forms a thin *P*-type layer resulting in a *PN* junction. Metal ohmic contacts are made to the *P*- and *N*-type areas, and some form of encapsulation is provided to protect the brittle silicon chip.

Thus the *PN* junction is *not* formed by joining two pieces of *P*- and *N*-type materials. However, to understand what takes place at a *PN* junction's formation, let us consider placing an *N*-type and a *P*-type semiconductor side by side to "make" a *PN* junction. Remember that a *PN* junction is electrically neutral because the *P* and *N*-type materials are each electrically neutral. A *PN* junction is shown in Fig. 28-5.

Note that each free electron in the *N*-type region neutralizes the bound positive charge of the impurity donor ions; and in the *P*-type region, holes neutralize the bound negative charges.

At the moment of "joining together," some of the free electrons in the N-type region diffuse across the junction to the P-type region and combine with holes. Similarly, holes diffuse from the P-type to the N-type region and combine with electrons. This process establishes *unneutralized* bound negative and positive ions on opposite sides of the *junction*. These charges, referred to as *uncovered* charges, mean that the P-type region has acquired a net negative charge; the N-type region an equal net positive charge. This is shown in Fig. 28-5*c*.

Because of the diffusion of carriers across the junction, a potential barrier is established of about 0.7 V. Equilibrium is established because the N-type region, having become positive, will prevent any further diffusion of holes from the P-type region. Similarly, the P-type region, having a negative charge, will repel any further free electron diffusion. A very narrow region (about 0.5 μm thick) around the *PN* junction that is depleted of mobile charges now exists. This is referred to as the *depletion* or *space charge* region. The *PN* junction, as we shall see, is actually a *diode* that permits easy conduction when forward-biased and practically no conduction when reverse-biased.

28-5 FORWARD-BIASED *PN* JUNCTION

Now consider the application of an emf across the *PN* junction or *diode* with the positive side to the P-type, the negative side to the N-type. This is called forward bias. Refer to Fig. 28-6. Forward bias tends to lower the potential on the N side and raise it on the P side, thus *lowering* the potential barrier. Consequently, holes can now *drift* across the junction once more from P to N. Also, free electrons *drift* in the reverse direction across the junction from N to P.

The current across the junction is due primarily to these majority carriers, holes in the P-type and electrons in the N-type. The total current is due to the sum of the two. There is an additional (very small) current due to the *minority* carriers. These are electrons in the P-type and holes in the N-type. This minority current is small because of the small number of minority carriers. The depletion region also narrows somewhat under forward bias. (This is an important effect if the capacitance of the junction is considered.)

The diode is considered to have a *low* resistance under *forward-bias* conditions. If the forward voltage is increased, the potential barrier is further lowered, and the depletion region narrows further, allowing even more current to flow. The *conventional* direction of forward current (plus to minus) is shown in Fig. 28-6*b* to be the *same* as the arrow in the symbol for the diode.

28-6 REVERSE-BIASED *PN* JUNCTION

When the opposite polarity of external voltage is applied to the diode (positive to N-type, negative to P-type) a condition of *reverse bias* is said to exist. Figure 28-7 clearly shows how the majority carriers in each side are pulled away from the junction. Since this uncovers more

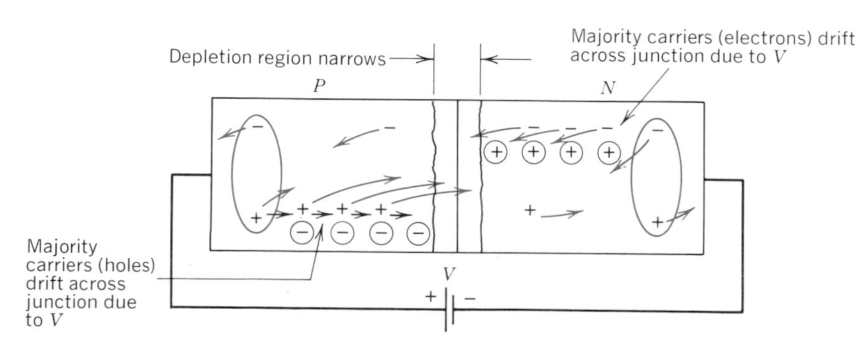

(*a*) Due to V lowering the potential barrier, the majority carriers can drift across the junction.

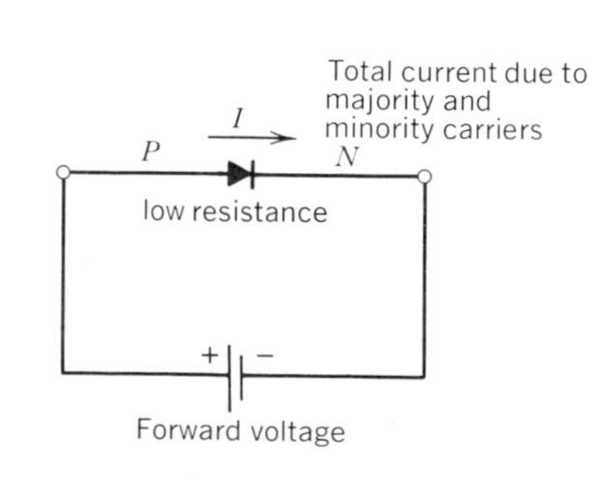

(*b*) Direction of conventional current for a forward-biased junction showing the symbol used for a diode

FIGURE 28-6
Forward-biased *PN* junction diode.

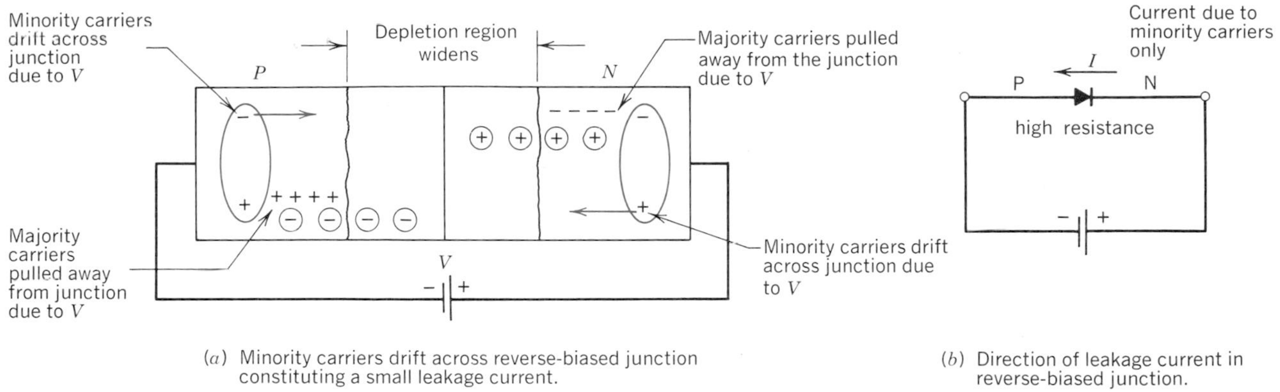

(*a*) Minority carriers drift across reverse-biased junction constituting a small leakage current.

(*b*) Direction of leakage current in reverse-biased junction.

FIGURE 28-7
Reverse-biased *PN* junction diode.

bound positive and negative ions, there is an *increase* in the potential barrier and the depletion region *widens.* *

At first it would seem that reverse bias will allow *no* current to flow at all. Certainly no *majority* carriers can cross the junction. But the *minority* carriers are unaffected by the potential barrier. In fact, as far as the *minority* carriers are concerned, the *PN* junction is *forward* biased! Consequently, a small reverse current does flow.

The reverse-biased minority current is strongly temperature dependent (because the minority carriers are due to thermally-generated electron-hole pairs.) This reverse or leakage current is in the order of only a few microamperes for low-current diodes. **Thus the diode is said to have a *high* resistance when *reverse* biased. The diode effectively *blocks* current compared with forward bias.**

* The charges stored on opposite sides of the depletion region cause a *capacitive* effect. This *junction capacitance* varies with the reverse voltage connected across the *PN* junction and is used to advantage in *varicap* or *varactor* diodes in communications tuning circuits.

28-7
DIODE *VI* CHARACTERISTICS

The forward- and reverse-bias properties of a diode are best shown in a plot of their current-voltage characteristics. A typical curve for a silicon diode is shown in Fig. 28-8.

It is seen that a very low, forward-bias voltage of approximately 0.6 V is necessary in the forward direction before the diode starts to conduct. This is often referred to as the *cut-in* or knee or threshold voltage. It is the voltage required to reduce the potential barrier so that majority carriers can start migrating across the junction.

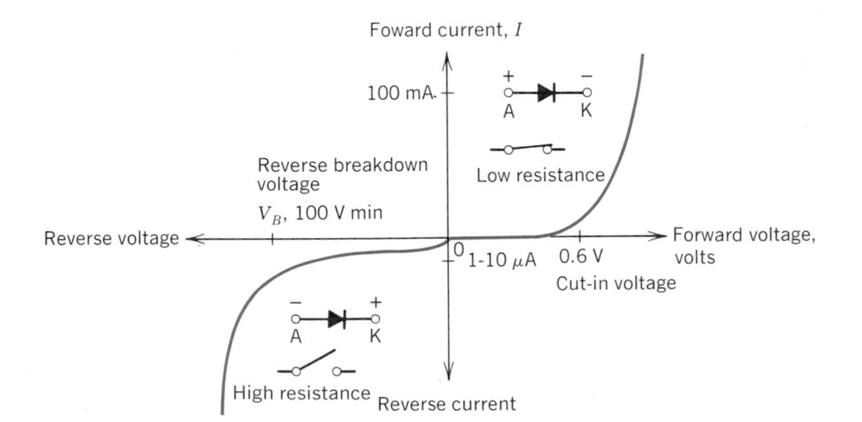

FIGURE 28-8
Forward and reverse *VI* characteristics of a silicon diode.

Beyond the cut-in voltage, the current increases sharply or exponentially even with very small voltage increases. At high-current levels with relatively low voltages, the curve becomes a straight line, approximating the voltage drop across a resistor. In this region, above the cutin voltage, the forward-biased diode can be thought of as a closed switch, having a low resistance, with approximately a 0.7 V drop across it.

When reverse-biased, even with high reverse voltages, the diode is effectively an open switch, with a high resistance, and a leakage current of only a few microamperes. In most rectifier diodes this current tends to increase *slightly* with a large increase in the reverse voltage. This is due to surface currents flowing across the *PN* junction.

However, there is some value of high reverse voltage, in the region of which there is a significant increase in reverse current. This is called the *reverse breakdown* voltage, V_B. If the diode is operated beyond this point, it is said to have broken down, since it is no longer effectively blocking the current.

One explanation of breakdown is found in the *avalanche* effect. Minority carriers, which constitute the reverse current, gain enough kinetic energy to knock bound electrons out of the covalent bonds. These electrons, in turn, collide with other atoms and increase the number of carriers available for reverse conduction.

This reverse voltage capability of a diode is often referred to as the peak inverse voltage (PIV) of the diode. Depending upon the type of diode, the peak inverse voltage may range from a minimum of about 100 V to several thousand volts.

In rectifier applications, the peak inverse voltage of the diode must not be exceeded or permanent damage can be sustained by the diode due to the overheating of the *PN* junction. If no damage has occurred, reducing the reverse bias below the breakdown level brings the diode back into normal blocking operation. If the reverse current is limited to a nondestructive value, the breakdown voltage can be used to advantage in *Zener* diodes.

28-7.1 Zener Diodes

In some diodes, breakdown can be made to occur very abruptly at accurately known values ranging from 2 to 200 V. See Fig. 28-9*a*. These are called Zener diodes. Since their reverse breakdown voltage (or Zener voltage) is maintained over a wide range of current, they can be used in voltage regulator circuits, as in Fig. 28-9*b*.

Here the reverse Zener diode voltage, V_Z, is also the voltage across the load R_L. If either the input voltage or the load current changes a different amount of current, I_Z, will flow through the Zener diode. But since V_Z changes little when I_Z changes, a nearly constant load voltage will result. The product of I_Z and V_Z must be limited to the power dissipation capability of the particular Zener diode. Also, I_Z must not be allowed to drop below some minimum value, I_{ZK}, in order to maintain voltage regulation.

28-7.2 Checking a Diode with an Ohmmeter

The terminals of a diode are often referred to as the anode and cathode. These terms are carryovers from the

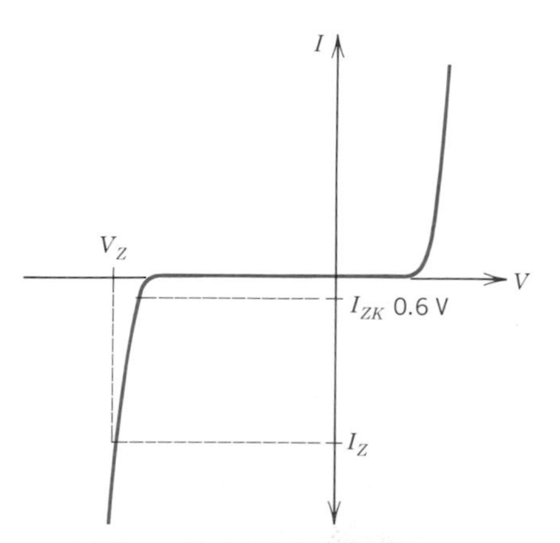

(*a*) Zener diode *VI* characteristic

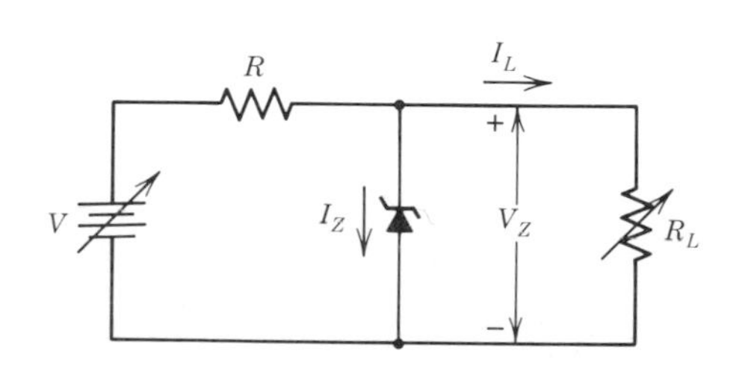

(*b*) Use of a Zener diode in a voltage-regulator circuit

FIGURE 28-9
Zener diode characteristics and application as a voltage regulator.

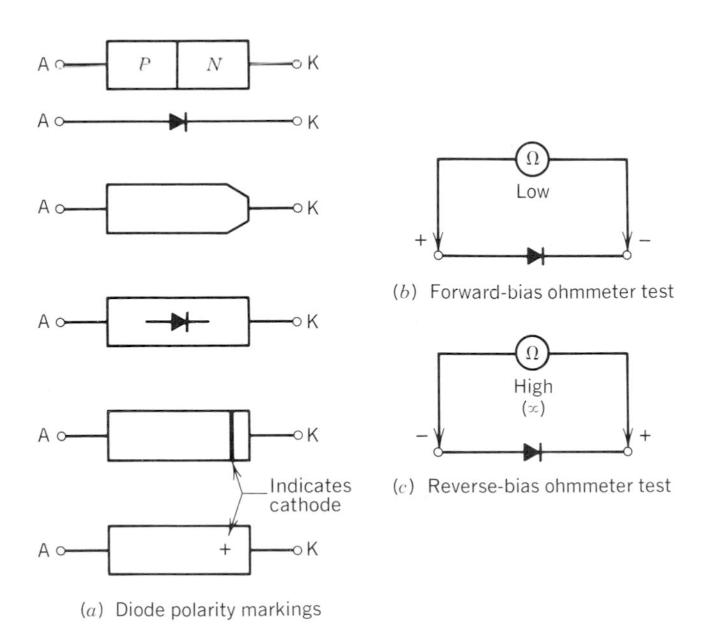

FIGURE 28-10
Showing how to determine the anode and cathode from the physical appearance of four types of diodes, and how an ohmmeter can be used to check a good diode.

electron tube diode and refer to the P-side and N-side, respectively. Figure 28-10*a* shows *four* types of diode encapsulation or marking to determine diode polarity. In all cases, the P-type side is the anode and the N-type side is the cathode.

An ohmmeter is used to determine a diode's polarity as well as check if the diode is good, open or shorted. The ohmmeter is essentially a voltage source in series with a resistor and an ammeter. A low resistance is indicated on the ohmmeter when forward-biasing the diode (Fig. 28-10*b*). A reading of close to infinity should result when the ohmmeter leads produce reverse bias (Fig. 28-10*c*). If *both* readings are low or both are high, however, the diode is either shorted or open, respectively.

Two precautions are in order. If you are using an ohmmeter to determine the polarity of the diode, you must know the polarity of the ohmmeter leads. Many VOMs used as an ohmmeter use the COMMON as the minus terminal and an OHMS terminal, which is positive. But the way that the batteries are connected may make the OHMS terminal negative and the COMMON terminal positive. The best check is to determine the polarity of the leads with a voltmeter.

The second precaution involves the use of some digital electronic multimeters in the ohms function. Quite often, on the X1 ohms range, a "good" diode indicates an open circuit regardless of how the leads are connected across the diode. This is because the cut-in voltage has not been exceeded; you should change to a higher ohms range (the middle ranges) for a proper check.

Diodes are available in a wide range of current and voltage ratings. Figure 28-11 shows a selection of silicon diodes that cover the range from 1A to the several-hundred-ampere, stud-mounted and hockey puck types.

28-8 THE GERMANIUM DIODE

Although most diodes used for rectification are made of silicon, some applications (still) require germanium diodes. Germanium is also a semiconductor but with a lower resistivity than silicon. It has four electrons in its valence shell so that doping is achieved similarly to

FIGURE 28-11
Selection of silicon diodes showing low current, stud and hockey puck types. (Courtesy of International Rectifier.)

silicon; pentavalent materials such as arsenic and antimony give *N*-type germanium whereas gallium and indium provide *P*-type germanium.

The *VI* characteristics for a germanium diode have the same general shape as for silicon. There are two important differences, however. Germanium, having a potential barrier of 0.2–0.4 V, has a cut-in voltage of 0.2 V compared with 0.6 V for silicon. Germanium also has a much higher (a minimum of 10 times higher) reverse current than a silicon diode. Both are affected by temperature in the same manner, but silicon diodes can be operated with their junctions as high as 175°C compared with a maximum of 85 to 100°C for germanium.

28-9 HALF-WAVE RECTIFICATION

Most electronic equipment must operate from a source of dc power. Although batteries can be used in portable equipment, an electronic *power supply* is often included allowing operation from the 120-V, 60-Hz supply. The purpose of the power supply is to convert the alternating current into direct current. Since a diode is sensitive to polarity, it can be used to rectify the alternating voltage waveform into a unidirectional waveform. The average value of this waveform is the dc voltage value.

The simplest rectifier involves a single diode, shown in Fig. 28-12*a*. Since many semiconductor (transistor) circuits operate at relatively low voltages of 6–20 V, a *transformer* is used to step down the ac voltage. In our example, let us assume that the secondary voltage is 6.3 V rms. Here R_L represents the resistance of the load to which we want to deliver dc power, such as a radio, tape recorder, amplifier, and so on. The input voltage to the diode is represented by a sine wave of peak value given by $V_m = 6.3\sqrt{2}\text{ V} \approx 9\text{ V}$, shown in Fig. 28-12*b*.

On the positive half-cycle of v_s, diode D_1 is forward-biased and conducts, allowing the current to flow through the load R_L. As a simplification, we neglect the cut-in voltage of the diode and its resistance. In this case, the voltage across the load will be a half sine wave of peak value 9 V, as shown in Fig. 28-12*c*.

When the input voltage *reverses* in polarity (shown by the circled polarity signs), the diode is *reverse-biased*. This blocking action means that no current flows and the voltage across the load is zero for half a cycle. Again we neglect the small reverse current that flows. Not until the input voltage again reverses will the load voltage increase toward the 9-V peak.

The half-wave, rectified output voltage across the load is sometimes called a *unidirectional pulsating* voltage (Fig. 28-12*c*). Although not smooth, the waveform contains direct current and is quite satisfactory (for some

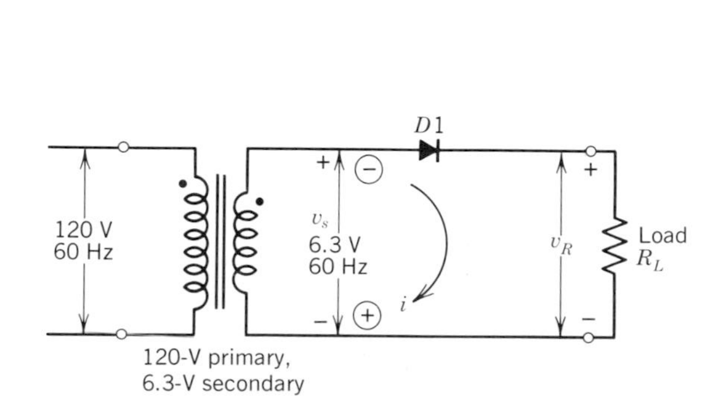

(*a*) Half-wave rectifier circuit. PIV = V_m.

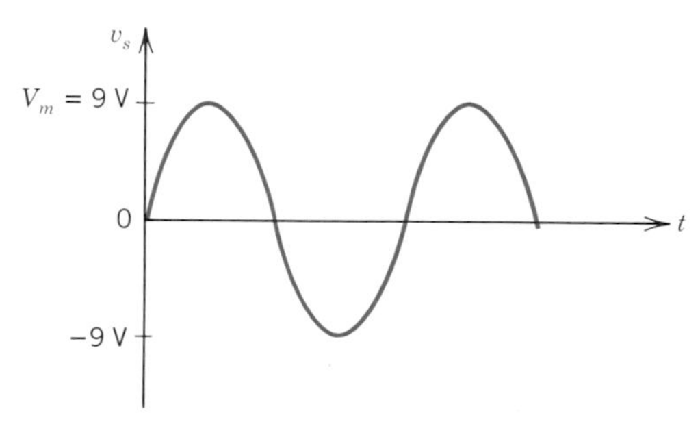

(*b*) Ac input to rectifying circuit.

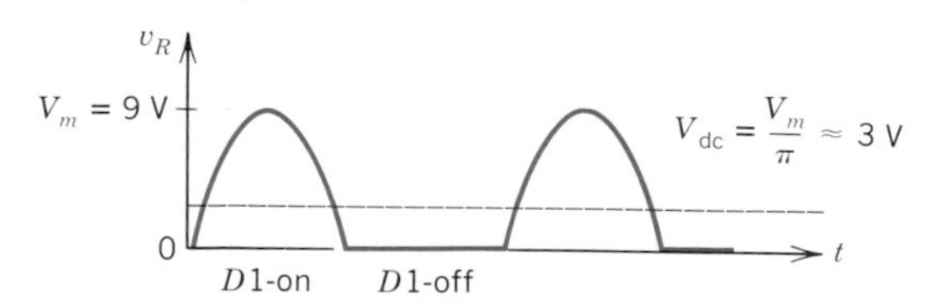

(*c*) Half-wave rectified output across load.

FIGURE 28-12
Half-wave rectification.

applications such as battery chargers or some dc motors.)

The average value of the half-wave, rectified waveform (Fig. 28-12*c*) can be shown to be

$$V_{dc} = \frac{V_m}{\pi} \quad \text{volts (V)} \qquad \textbf{(28-1)}$$

where: V_{dc} is the average or dc voltage in volts, V
V_m is the peak of the input voltage in volts, V

Note that V_{dc} is the reading of a dc voltmeter connected across the load, R_L. If a dc ammeter is connected in series with the load, it will indicate the average value of the current waveform that has the same wave shape as the voltage. This current is given by

$$I_{dc} = \frac{I_m}{\pi} \quad \text{amperes (A)} \qquad (28\text{-}2)$$

or by

$$I_{dc} = \frac{V_{dc}}{R_L} \quad \text{amperes (A)} \qquad (28\text{-}3)$$

where: I_{dc} is the average or dc current in amperes, A
I_m is the peak of the load current in amperes, A
R_L is the load resistance in ohms, Ω

EXAMPLE 28-1

A half-wave rectifier has an input voltage of 6.3 V ac, and a load of 220 Ω. Assuming an ideal diode (neglecting diode losses), calculate:

a. The dc voltage across the load.
b. The peak current through the load.
c. The reading of a dc ammeter in series with the load.
d. The dc power delivered to the load.

SOLUTION

a. $V_m = 6.3\ \text{V} \times \sqrt{2} = 8.9\ \text{V}$ (15-18)

$$V_{dc} = \frac{V_m}{\pi} \qquad (28\text{-}1)$$

$$= \frac{8.9\ \text{V}}{\pi} = \mathbf{2.8\ V}$$

b. $$I_m = \frac{V_m}{R_L} \qquad (15\text{-}17)$$

$$= \frac{8.9\ \text{V}}{220\ \Omega} = \mathbf{40.5\ mA}$$

c. $$I_{dc} = \frac{I_m}{\pi} \qquad (28\text{-}2)$$

$$= \frac{40.5\ \text{mA}}{\pi} = \mathbf{12.9\ mA}$$

d. $$P_{dc} = V_{dc} I_{dc} \qquad (3\text{-}7)$$

$$= 2.8\ \text{V} \times 0.0129\ \text{A} = \mathbf{0.036\ W}$$

$$= \mathbf{36\ mW}$$

Note that when the diode is reverse-biased, it is subjected to a peak inverse voltage of V_m, or 8.9 V in Example 28-1a. In high-voltage power supplies this is an important item to check when selecting the diode.

28-10 FULL-WAVE RECTIFICATION

Figure 28-13*a* shows how a second diode can be combined with the first to provide *full-wave* rectification. This circuit also requires a transformer with a *center-tapped* secondary for a common return path. The line-to-line secondary voltage in Fig. 28-13*a* is 12.6 V so that the line to center-tap voltage is 6.3 V, as shown. Thus this circuit is a combination of two of the half-wave rectifier circuits, with one diode rectifying each half-cycle of the input voltage.

On the *positive* half-cycle of v_s, $D1$ conducts and $D2$ is reverse-biased and open. On the *negative* half-cycle (polarity signs shown encircled) $D1$ is open and $D2$ conducts. The voltage waveform across the load is shown in Fig. 28-13*c* where once again we assume ideal diodes.

It should be evident that the *average* full-wave output voltage is now doubled compared to half-wave rectification:

$$V_{dc} = \frac{2V_m}{\pi} \quad \text{volts (V)} \qquad \textbf{(28-4)}$$

$$I_{dc} = \frac{2I_m}{\pi} \quad \text{amperes (A)} \qquad (28\text{-}5)$$

where: V_{dc} is the average or dc voltage in volts, V
V_m is the peak of the input voltage in volts, V
I_{dc} is the dc current in the load in amperes, A
I_m is the peak of the load current in amperes, A

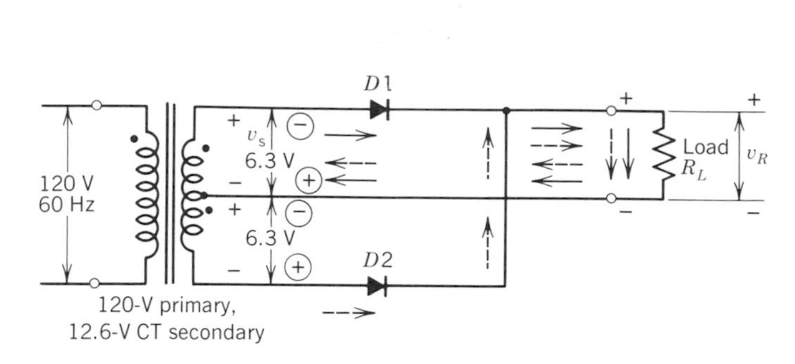

(a) Full-wave rectifier circuit. PIV = 2 V_m.

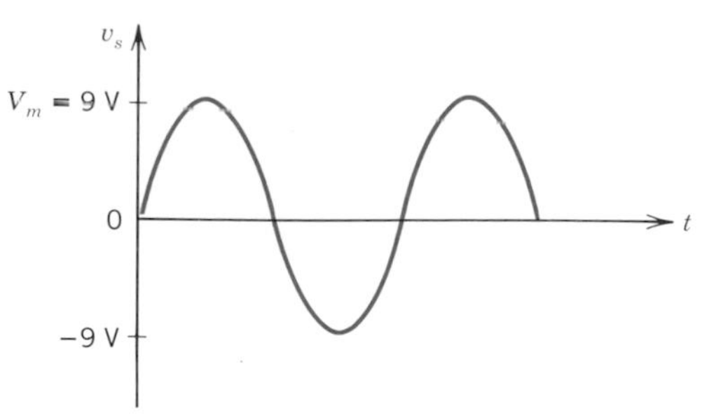

(b) Ac input to rectifying circuit.

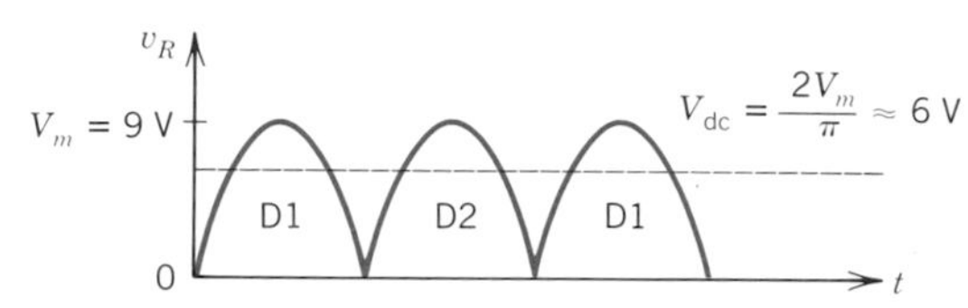

(c) Full-wave rectified output across load.

FIGURE 28-13
Full-wave rectification.

EXAMPLE 28-2

A 120-V primary, 12.6-V center-tapped secondary transformer is used in a two-diode, full-wave rectifier with a 220-Ω load. Neglecting diode losses (assume ideal diodes), calculate:

a. The dc (average) voltage across the load.
b. The peak current through the load.
c. The reading of a dc ammeter in series with the load.
d. The dc power delivered to the load.

SOLUTION

a. $V_m = 6.3 \text{ V} \times \sqrt{2} = 8.9 \text{ V}$ (15-18)

$$V_{dc} = \frac{2V_m}{\pi} \quad (28\text{-}4)$$

$$= \frac{2 \times 8.9 \text{ V}}{\pi} = \mathbf{5.7\ V}$$

b. $$I_m = \frac{V_m}{R_L} \quad (15\text{-}7)$$

$$= \frac{8.9 \text{ V}}{220\ \Omega} = \mathbf{40.5\ mA}$$

c. $$I_{dc} = \frac{2I_m}{\pi} \quad (28\text{-}5)$$

$$= \frac{2 \times 40.5 \text{ mA}}{\pi} = \mathbf{25.8\ mA}$$

d. $P_{dc} = V_{dc}I_{dc}$ (3-7)

$$= 5.7 \text{ V} \times 0.0258 \text{ A} = \mathbf{0.147\ W}$$

These answers should be compared to those for Example 28-1, which uses the same value of R_L and ac input voltage.

The peak inverse voltage of each diode in the full-wave rectifier is $2V_m$ or 17.8 V in Example 28-2. This is because the peak voltage of V_m across the load occurs at the same time that the input voltage to the nonconducting diode is a peak. The sum of these two polarity-aiding voltages is the diode's peak reverse voltage.

28-10.1 Bridge Rectifier

If *four* diodes are used in a *bridge* configuration, there is no need for a *center-tapped* (CT) transformer. This arrangement is shown in Fig. 28-14a and produces the same waveform as for the full-wave two-diode CT transformer rectifier (Fig. 28-13c). Note that the line-to-line transformer secondary voltage is only 6.3 V. Conduction alternates between diodes $D1$ and $D3$ for one half-cycle of the input and diodes $D2$ and $D4$ for the other half-cycle. This causes a PIV for each diode of only V_m.

For convenience, commercial bridge rectifiers contain *four* diodes encapsulated in the same package with two terminals provided for connecting the ac input and two terminals for the dc output. See Fig. 28-15.

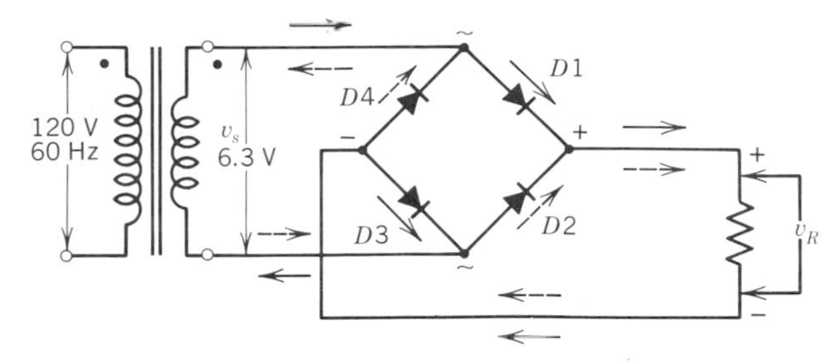

(a) Full-wave bridge rectifier circuit. PIV = V_m.

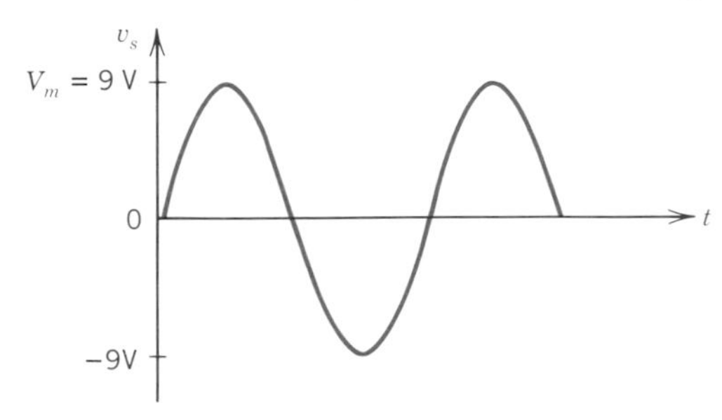

(b) Ac input voltage to rectifying circuit.

(c) Full-wave rectified output across load.

FIGURE 28-14
Full-wave rectification with a bridge rectifier.

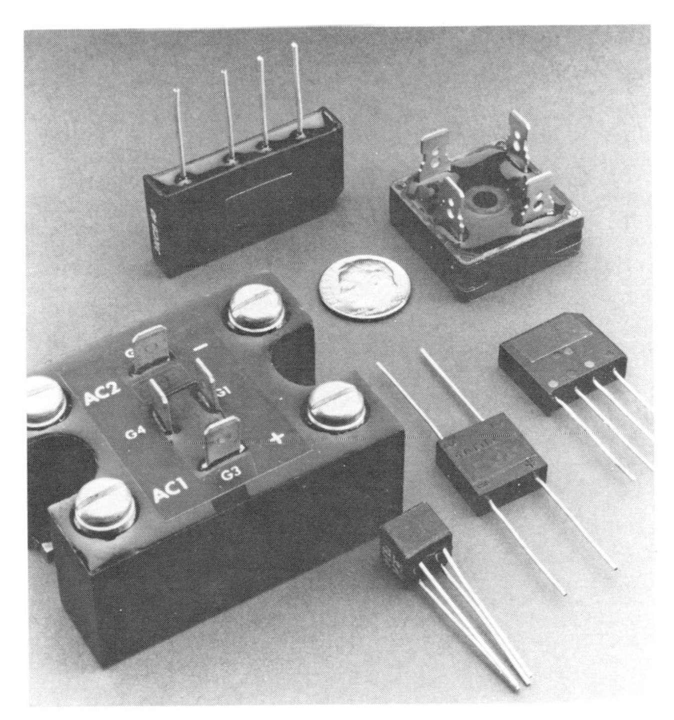

FIGURE 28-15
Encapsulated bridge rectifier packages. (Courtesy of International Rectifier.)

One advantage of the bridge rectifier is greater use of the transformer secondary winding in delivering the current to the load compared to either of the other two rectifiers. Therefore, for a given voltampere (apparent power) rating of the transformer, this circuit can deliver *more* dc power to the load than either of the two previous circuits. That is, a 1-kVA transformer can deliver 287 W of dc power in a *half-wave* rectifying circuit with no filtering, 693 W in a two-diode *full-wave* rectifying circuit, and 812 W in a *bridge* rectifier circuit.

EXAMPLE 28-3

A 120-V primary, 6.3-V secondary transformer is used in a full-wave bridge rectifier with a 220-Ω load. Neglecting diode losses, calculate:

a. The dc voltage across the load.
b. The dc current through the load.
c. The dc power delivered to the load.
d. The PIV of each diode.

SOLUTION

a. $V_m = 6.3 \times \sqrt{2} = 8.9 \text{ V}$ (15-18)

$$V_{dc} = \frac{2V_m}{\pi} \qquad (28\text{-}4)$$

$$= \frac{2 \times 8.9 \text{ V}}{\pi} = \mathbf{5.7 \text{ V}}$$

b. $$I_{dc} = \frac{V_{dc}}{R_L} \qquad (3\text{-}1a)$$

$$= \frac{5.7 \text{ V}}{220 \ \Omega} = \mathbf{25.9 \text{ mA}}$$

c. $$P_{dc} = V_{dc} I_{dc} \qquad (3\text{-}7)$$

$$= 5.7 \text{ V} \times 0.0259 \text{ A} = \mathbf{0.147 \text{ W}}$$

d. PIV = V_m = **8.9 V**

These answers should be compared to those from Example 28-2. (Numerical differences are due to rounding-off.)

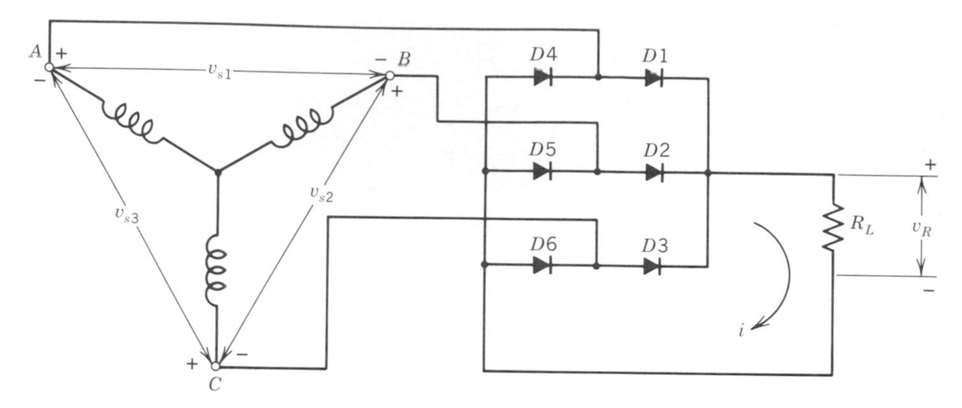

(*a*) Circuit of a three-phase, full-wave rectifier showing the six diodes.

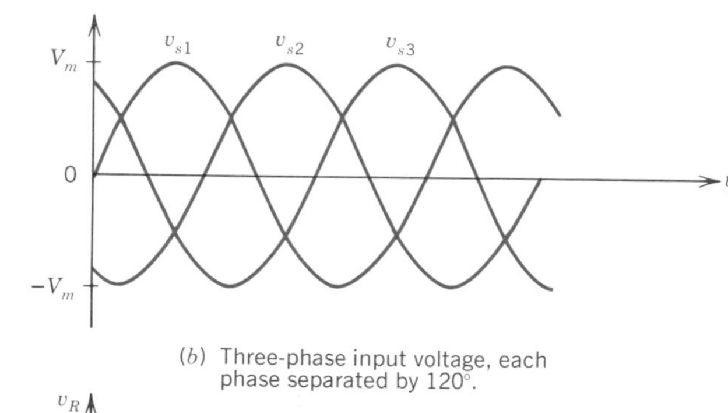

(*b*) Three-phase input voltage, each phase separated by 120°.

(*c*) Load voltage waveform showing dc output.

FIGURE 28-16
Circuit and waveforms for a three-phase full-wave rectifier.

28-10.2 Three-Phase, Full-Wave Rectifier

An important application of the rectifier is in the modern automobile electrical system. Since a three-phase alternator is used (to supply the electrical needs) an ac-to-dc conversion circuit is required. Also, because there are three phases, full-wave rectification requires six diodes, as shown in Fig. 28-16*a*. The diodes can be seen usually mounted in the alternator frame, which functions as a heat sink to cool the diodes.

Consider the interval from X to Y in Fig. 28-16*c*. Here v_{s1} is positive, which means that current will flow from A, through $D1$ and the load, and back through $D5$ into B (v_{s2} is negative). From Y *to* Z, v_{s3} is near its maximum negative value, which means that A is positive with respect to C. Consequently, the current again flows out of A, through $D1$ and the load, and back through $D6$ into C. Due to these conduction periods of only 60°, the output voltage never drops below $\frac{\sqrt{3}\, V_m}{2}$.

It is apparent from Fig. 28-16*c* that the output voltage has a *higher* average value than any of the previous circuits and can be shown to be

$$V_{dc} = \frac{3V_m}{\pi} \tag{28-6}$$

and

$$I_{dc} = \frac{3I_m}{\pi} \tag{28-7}$$

Also, the PIV of each diode is only V_m as in the bridge rectifier.

NOTE The three-phase voltages could be obtained from the secondary windings of three single-phase transformers whose primary and secondary windings would be connected in a wye as shown or a delta.

EXAMPLE 28-4

A three-phase, full-wave rectifier must develop a dc output voltage of 14.5 V (typical in an automobile).

a. Determine the necessary rms output voltage from the alternator.

b. What is the peak current through the diodes when the alternator is supplying (after rectification) 30 A dc to the load?

c. Determine the PIV of the diodes.

SOLUTION

a. $$V_{dc} = \frac{3V_m}{\pi} \tag{28-6}$$

Therefore, $V_m = \frac{\pi}{3} V_{dc}$

$= \frac{\pi \times 14.5}{3}\text{ V} = 15.2\text{ V}$

$$V_{rms} = V_m \times 0.707 \tag{15-17}$$

$= 15.2\text{ V} \times 0.707 = \mathbf{10.7\ V}$

b. $$I_{dc} = \frac{3I_m}{\pi} \tag{28-7}$$

Therefore, $I_m = \frac{\pi}{3} I_{dc}$

$= \frac{\pi}{3} \times 30\text{ A} = \mathbf{31.4\ A}$

c. $\text{PIV} = V_m = \mathbf{15.2\ V}$

Characteristics of the four rectifier circuits are compared in Table 28-1.

28-11 CAPACITOR FILTERING

The dc output voltage resulting from single-phase (and even three-phase) rectification contains too much ac ripple for most electronic applications. In an audio system this would show up as a 120-Hz "hum" from a full-wave rectifier circuit, this being the ripple frequency in the output for a 60-Hz input. A half-wave rectifier circuit has

TABLE 28-1
Comparison of Rectifier Circuits. (No filter connected.)

	Half-Wave	Full-Wave	Bridge	Three-Phase
Secondary voltage line-to-line ($V_m = \sqrt{2}V$)	V	$2V$	V	V
Number of diodes	1	2	4	6
PIV	V_m	$2V_m$	V_m	V_m
V_{dc}	$\frac{V_m}{\pi}$	$\frac{2V_m}{\pi}$	$\frac{2V_m}{\pi}$	$\frac{3V_m}{\pi}$
Output waveform	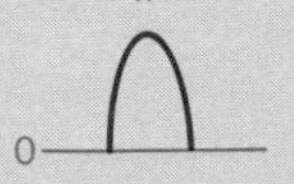			
Ripple frequency	f	$2f$	$2f$	$6f$

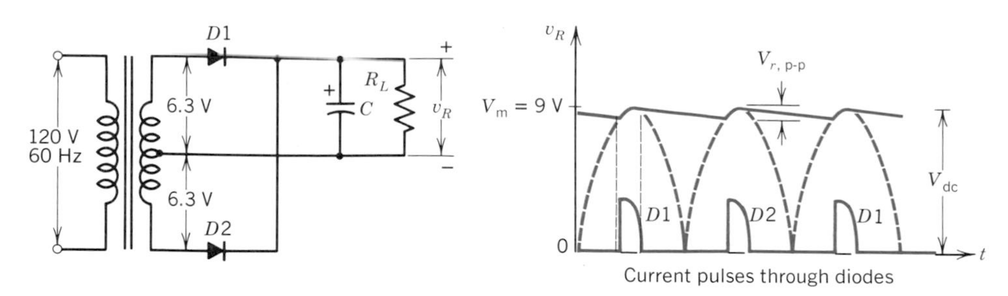

(a) Full-wave rectifier with capacitor filter

(b) Filtered output voltage showing peak-to-peak ripple voltage

FIGURE 28-17
Circuit and waveforms for simple capacitor filter.

a 60-Hz ripple frequency, but this circuit is seldom used where a very "smooth" output voltage is required.

A simple way to filter out much of the unwanted alternating current is to use a large (electrolytic) capacitor in parallel with the load. Suitable only for light loads, this method is shown in Fig. 28-17.

The capacitor charges up to the *peak* value of the input voltage and tries to maintain this value, V_m. As the full-wave input drops to zero, the capacitor discharges through R_L until the input voltage again increases to a value greater than the capacitor voltage. At this point a diode will again be forward-biased and there will be a pulse of current through the diode, recharging the capacitor.

It is apparent from the output waveform in Fig. 28-17*b* that there is a "ripple voltage," which is said to be "riding on a dc level." For light loads (large R_L), the dc output voltage will remain near V_m. However, as the load increases (a decrease in R_L), the discharge of C will be greater, resulting in more ripple and a lower dc output voltage.

28-11.1 Ripple Factor

The success of a filter is measured by the ripple factor, r, defined as

$$r = \frac{V_{r,\text{rms}}}{V_{\text{dc}}} \qquad (28\text{-}8)$$

where: r is the ripple factor (dimensionless)
$V_{r,\text{rms}}$ is the rms value of ripple voltage in volts, V
V_{dc} is the dc output voltage in volts, V

For a capacitor filter the ripple is approximately triangular with an rms value of

$$V_{r,\text{rms}} = \frac{V_{r,\text{p-p}}}{2\sqrt{3}} \quad \text{volts (V)} \qquad (28\text{-}9)$$

where: $V_{r,\text{rms}}$ is the rms value of ripple voltage in volts, V
$V_{r,\text{p-p}}$ is the peak-to-peak value of the ripple voltage in volts, V.

EXAMPLE 28-5

A full-wave rectifier supplies a load of 100 Ω with a filter capacitor of 1000 μF. The output is observed on an oscilloscope and found to have a peak of 9 V and a peak-to-peak ripple of 0.8 V. Calculate:

a. The rms value of the ripple voltage.
b. The dc output voltage.
c. The ripple factor.

SOLUTION

a. $$V_{r,\text{rms}} = \frac{V_{r,\text{p-p}}}{2\sqrt{3}} \qquad (28\text{-}9)$$
$$= \frac{0.8\text{ V}}{2\sqrt{3}} = \mathbf{0.23\ V}$$

b. $$V_{\text{dc}} = V_m - \left(\frac{V_{r,\text{p-p}}}{2}\right)$$
$$= 9\text{ V} - \left(\frac{0.8\text{ V}}{2}\right) = \mathbf{8.6\ V}$$

c. $$r = \frac{V_{r,\text{rms}}}{V_{\text{dc}}} \qquad (28\text{-}8)$$
$$= \frac{0.23\text{ V}}{8.6\text{ V}} = \mathbf{0.027}$$

or expressed as a percent,

$$r = \mathbf{2.7\%}$$

If the capacitor is recharged in a short time compared with half of the period, the theoretical value of the ripple factor is given by

$$r = \frac{1}{4\sqrt{3}fR_LC} = \frac{2410}{CR_L} \qquad (28\text{-}10)$$

where: r is the theoretical ripple factor (full-wave), dimensionless
C is the capacitance in μF
R_L is the load resistance in ohms
f is the supply frequency, 60 Hz.

EXAMPLE 28-6

Calculate the theoretical ripple factor from the information given in Example 28-5 and compare it with the measured value.

SOLUTION

$$r = \frac{2410}{CR_L} \qquad (28\text{-}10)$$

$$= \frac{2410}{1000\ \mu\text{F} \times 100\ \Omega} = 0.024 \text{ or } \mathbf{2.4\%}$$

This compares favorably with the measured value of 2.7%.

Note that Eq. 28-10 involves the *time constant* R_LC of the circuit in the denominator so that an increase in C or R_L (or both) reduces the ripple.

Two precautions should be observed in a capacitor filter. First, the large amount of capacitance required means that a polarized capacitor (such as an electrolytic) must be used. It must be connected with the proper polarity.

Second, any attempt to decrease the ripple simply by increasing the capacitance must be done carefully. A larger capacitor stores more energy, so it needs a larger pulse of current to recharge it. Also, this current must flow during an even shorter interval if there is less ripple. Both of these factors mean that the diodes must be able to handle the larger *peak* charging currents and must be selected accordingly.

28-12 INDUCTOR FILTERING

An alternative to using a *capacitor* in *parallel* with the load is to use an *inductor* in *series* with the load. Since the characteristic of an inductor is to oppose any change in current, the coil will have a *smoothing effect,* as shown in Fig. 28-18.

If we disregard the resistance of the inductor, the average or dc output voltage will be $\frac{2V_m}{\pi}$ as before. If the resistance is taken into account, the voltage will be reduced using the voltage-divider equation.

The ripple factor can be shown to be

$$r = \frac{1}{6\pi\sqrt{2}f(L/R_L)} = \frac{R_L}{1600L} \qquad (28\text{-}11)$$

where: r is the theoretical ripple factor (full-wave)
R_L is the load resistance in ohms, Ω
L is the inductance in henrys, H
f is the supply frequency, 60 Hz.

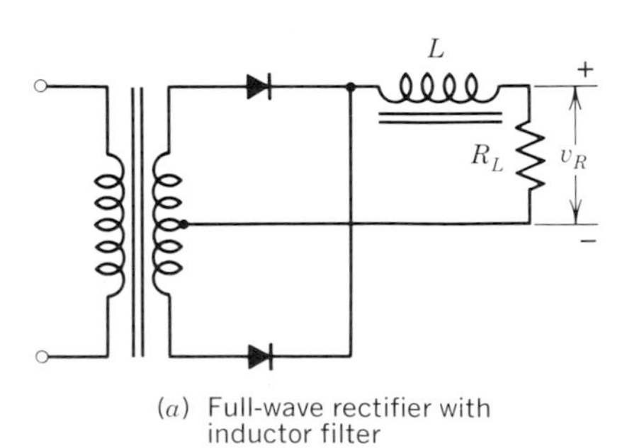

(*a*) Full-wave rectifier with inductor filter

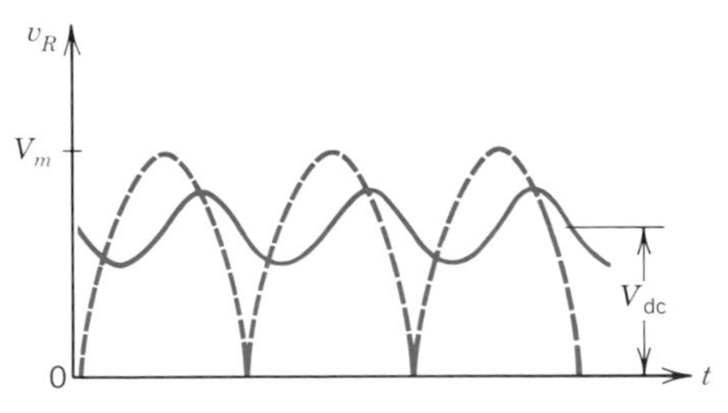

(*b*) Filtered output voltage showing average dc voltage

FIGURE 28-18
Circuit and waveforms for simple inductor filter.

EXAMPLE 28-7

A full-wave rectifier feeds a 100-Ω load. If a 6-H choke is used for filtering, calculate:

a. The theoretical ripple factor.
b. The dc output voltage if $V_m = 9$ V and choke resistance is neglected.
c. The dc output voltage if $V_m = 9$ V and the choke has a resistance of 25 Ω.

SOLUTION

a. $$r = \frac{R_L}{1600L} \qquad (28\text{-}11)$$

$$= \frac{100}{1600 \times 6} = \mathbf{0.01} \text{ or } \mathbf{1\%}$$

b. $$V_{dc} = \frac{2V_m}{\pi} \qquad (28\text{-}4)$$

$$= \frac{2 \times 9\text{ V}}{\pi} = \mathbf{5.7\ V}$$

c. $$V_{dc} = 5.7\text{ V} \times \frac{R_L}{R + R_L} \qquad (5\text{-}5)$$

$$= 5.7\text{ V} \times \frac{100\ \Omega}{25 + 100\ \Omega} = \mathbf{4.6\ V}$$

Note once again that the time constant L/R_L is in the denominator of Eq. 28-11, so a larger time constant (or small value of R_L) is still desirable for effective filtering.

28-13 CHOKE INPUT OR LC FILTER

The amount of ripple in the previous two filter circuits was determined in part by the load resistor; as the load changes, the amount of ripple also changes. If both a capacitor and an inductor are combined in one filter, the ripple is independent of the load. This is shown in Fig. 28-19. Neglecting the choke's resistance, the dc output voltage is $\frac{2V_m}{\pi}$. The ripple factor can be shown to be

$$r = \frac{0.83}{LC} \qquad (28\text{-}12)$$

where: r is the theoretical ripple factor (full-wave), dimensionless, for $f = 60$ Hz
L is the inductance in henrys, H
C is the capacitance in microfarads, μF

EXAMPLE 28-8

A full-wave rectifier uses an LC filter with $L = 6$ H and $C = 1000$ μF to feed a 100-Ω load. Calculate, neglecting the choke's resistance:

a. The theoretical ripple factor.
b. The dc output voltage if $V_m = 9$ V.

SOLUTION

a. $$r = \frac{0.83}{LC} \qquad (28\text{-}12)$$

$$= \frac{0.83}{6\text{ H} \times 1000\ \mu\text{F}} = \mathbf{0.00014} \text{ or } \mathbf{0.014\%}$$

b. $$V_{dc} = \frac{2V_m}{\pi} \qquad (28\text{-}4)$$

$$= \frac{2 \times 9\text{ V}}{\pi} = \mathbf{5.7\ V}$$

The answers for this example should be compared with those from Examples 28-6 and 28-7. In all three cases the load has been the same, but the LC filter has produced

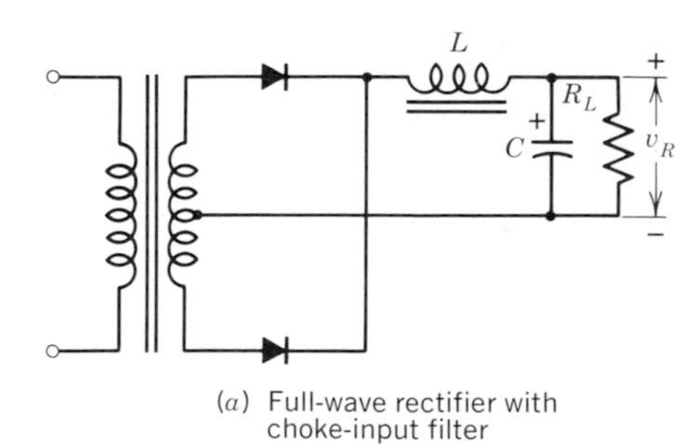

(a) Full-wave rectifier with choke-input filter

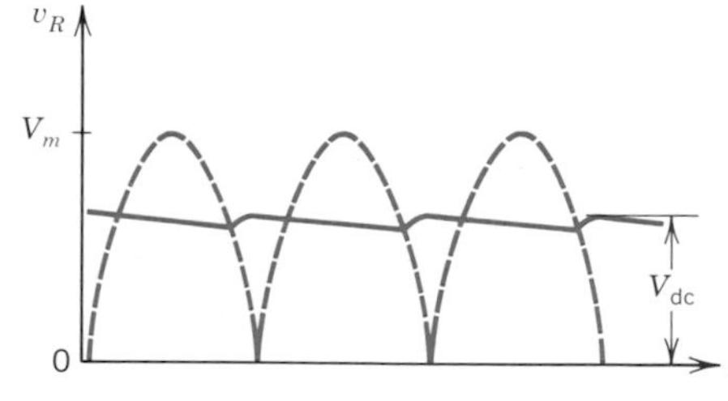

(b) Filtered output voltage showing average dc voltage

FIGURE 28-19 Circuit and waveforms for choke-input or LC filter.

TABLE 28-2
Summary of Filter Circuits

Type of Filter	Capacitor	Inductor	LC
Component values (R_L = 100 Ω)	C = 1000 μF	L = 6 H	L = 6 H, C = 1000 μF
Ripple factor, r (C in μF, L in H, R_L in Ω, f = 60 Hz)	$\frac{2410}{CR_L}$ (2.7%)	$\frac{R_L}{1600\,L}$ (1%)	$\frac{0.83}{LC}$ (0.014%)
Dc output voltage (neglecting choke resistance)	$V_m - \frac{V_{r,p-p}}{2}$ (8.6 V)	$\frac{2V_m}{\pi}$ (5.7 V)	$\frac{2V_m}{\pi}$ (5.7 V)

the lowest ripple of all. The capacitor filter provided the highest dc output voltage. This is summarized in Table 28-2. See Figure 28-20 for a 5-V dc power supply that can provide an output current of 6 A at 40°C. Note the large 5000-μF electrolytic capacitor used for filtering. This power supply also incorporates a voltage-regulating circuit.

It should be noted that in some very light load applications the inductor in the LC filter can be replaced by a resistor. The value of the resistor should be equal to the inductive reactance of the coil calculated at the ripple frequency of 120 Hz.

Finally, it should be noted that the ripple factor in a three-phase full-wave rectified output, as in Fig. 28-16c, is only 5.5%, *without* any filtering. This is comparable to the output of dc generators previously used in automobiles. In some industrial applications, where the large amounts of dc power required would make filtering impractical, transformers can be interconnected to give 6-phase, 9-phase or ever 12-phase operation to give even less ripple when rectified.

FIGURE 28-20
A 5-V dc power supply with an output of 6 A at 40°C. Note the large electrolytic capacitor used for filtering. (Courtesy of Lambda Electronics, Division of Veeco Instruments, Inc.)

SUMMARY

1. Pure silicon is a semiconductor having four valence electrons. A semiconductor crystal is formed by the covalent bonding of atoms.

2. At room temperature, pure silicon has a small number of thermally generated electron-hole pairs and a relatively high resistivity.

3. A hole is an incomplete bond due to an electron vacancy. A hole can contribute to the current by being filled with a valence electron. Holes move in the opposite direction to the electrons.

4. The current in pure silicon is the sum of the free electron flow and hole flow, and it is very temperature-dependent.

5. An *N*-type semiconductor is formed by adding an impurity material such as phosphorus that is pentavalent (donor), causing an increase in free electrons and a reduction in resistivity.

6. A *P*-type semiconductor is formed by adding an impurity material, such as aluminum, which is trivalent (acceptor), causing an increase in holes and a reduction in resistivity.

7. The majority current carriers in an *N*-type semiconductor are electrons, and the minority carriers are holes. The reverse is true for a *P*-type semiconductor.

8. When a *PN* junction is formed, carriers diffuse across the junction to establish a potential barrier of about 0.7 V for silicon.

9. A forward-biased *PN* junction diode has a low resistance. This is obtained by connecting the positive terminal of a dc source to the *P*-type side and the negative to the *N*-type side. The reverse is true for a reverse-biased diode.

10. The current through a forward-biased *PN* junction consists primarily of majority carriers. The small reverse current (when reverse-biased) is due entirely to thermally generated minority carriers.

11. The diode VI characteristics show that a cut-in voltage of more than 0.6 V is needed for the forward current to flow through a silicon diode.

12. Every diode has a peak inverse voltage above which the diode breaks down with an increase in the reverse current.

13. An ohmmeter with known polarity can be used to determine the anode and cathode terminals of a diode.

14. A simple dc power supply can involve half-wave or full-wave rectification with single-phase or three-phase supplies, as summarized in Table 28-1.

15. A filter circuit is used to smooth the output from a rectifier circuit and produce a "purer" direct current containing less ripple. This is done using a capacitor, inductor, or *LC* circuit.

16. The ripple factor is a measure of how well a filter circuit has removed the ac component from the output of a dc power supply. It is the ratio of the output ac ripple voltage to the dc output voltage.

17. The ripple in a capacitor filter increases as the load resistance decreases while the opposite is true in an inductor filter. An *LC* filter has a ripple independent of load. These factors are summarized in Table 28-2.

SELF-EXAMINATION

Answer true or false
(Answers at back of book)

28-1. Both silicon and germanium are semiconductors with a valence of four. ________

28-2. In pure silicon there are just as many holes as there are free electrons. ________

28-3. If the temperature of pure silicon increases, the number of free electrons increase and the number of holes decrease. ________

28-4. A hole may be thought of as a positive charge carrier. ________

28-5. A hole is the result of a broken or incomplete bond. ________

28-6. When current flows in pure silicon, there are two components: one due to free electron flow, the other to hole flow. ________

28-7. Because free electrons and holes flow in the same direction, the total current is the sum of the two. ________

28-8. The current in pure silicon goes up as the temperature decreases due to a reduction in resistance. ________

28-9. A donor impurity must be pentavalent to form an N-type semiconductor. ________

28-10. An N-type semiconductor is negatively charged. ________

28-11. If the majority carriers in a semiconductor are electrons, the material must be N-type. ________

28-12. In a P-type material the minority carriers are holes. ________

28-13. An acceptor impurity is so called because, being trivalent, it has an incomplete bond that can accept a valence electron and create a hole. ________

28-14. A P-type material, although it has a large number of positive holes, is still neutral. ________

28-15. A potential barrier is established across a PN junction because of the diffusion of carriers across the junction uncovering bound ion charges. ________

28-16. The potential barrier makes the P-side positive and the N-side negative. ________

28-17. The application of a forward bias reduces the potential barrier while a reverse bias increases the barrier. ________

28-18. Forward bias involves connecting the positive to the P-type and the negative to the N-type. ________

28-19. The resistance of a forward-biased diode is low because majority carriers can easily cross the PN junction. ________

28-20. Absolutely no current flows through a reverse-biased diode because the barrier is made so large. ________

28-21. Because minority carriers are thermally generated, there is a high temperature dependence of the reverse current in diodes. ________

28-22. The cut-in voltage of 0.6 V for a silicon diode is a direct result of the potential barrier across a PN junction. ________

28-23. The VI characteristics can be approximated by an open switch when forward-biased and a closed switch when reverse-biased. ________

28-24. The PIV is the maximum reverse voltage a diode can withstand before an appreciable current starts to flow. ________

28-25. The reverse breakdown voltage of a Zener diode can be used to advantage in voltage regulation because of the small change in voltage across the diode when the current through it changes. ________

28-26. If an ohmmeter indicates infinity when the positive side is connected to a diode's anode and the negative side is connected to the cathode, the diode is good. ________

28-27. A half-wave rectifier produces a positive half-cycle of voltage in the output for each cycle of ac input. ________

28-28. A full-wave rectifier circuit uses a minimum of two diodes and can use as many as four or six. ________

28-29. If a two-diode, full-wave, rectifier circuit uses a transformer with the same line-to-line secondary voltage as a half-wave rectifier, the output voltage will be twice that for the half-wave.

28-30. In a bridge rectifier one side of the dc output voltage is common with one side of the transformer secondary. ________

28-31. The amount of dc power delivered to a load in a full-wave bridge rectifier is four times as large as the dc power in a half-wave rectifier utilizing the same ac input voltage on the transformer secondary. ________

28-32. The ripple frequency in a three-phase rectifier is six times the supply frequency and accounts for the very low ripple factor. ______

28-33. A capacitor filter relies upon the energy stored in a capacitor to be discharged at a slow rate for good filtering. ______

28-34. A larger capacitor will reduce ripple without having any effect on the diodes. ______

28-35. Capacitor filtering is best suited for light loads whereas an inductor filter is best with heavy loads. ______

28-36. For a given amount of ripple, a choke-input filter requires a smaller capacitor than a pure capacitor filter. ______

28-37. Since the property of an inductor is to oppose ac current due to inductive reactance, the dc resistance of the choke has no effect on the dc output voltage in an inductor or *LC* filter. ______

REVIEW QUESTIONS

1. What do you understand by the term *covalent bonding?*

2. What is an electron-hole pair?

3. Explain why an increase in temperature increases the conductivity in a semiconductor but decreases it in most metallic conductors.

4. Explain why the total current flowing in pure silicon is the sum of the free electron flow and the hole flow.

5. When a pentavalent phosphorus atom replaces a silicon atom at room temperature, what is the effect on the following:

a. The number of free electrons in the overall silicon.

b. The number of electron-hole pairs.

6. For the condition in Question 5, why doesn't the liberation of the fifth electron from the phosphorus atom create as many holes as it does free electrons?

7. a. What are the majority carriers in *N*-type silicon?

b. What constitutes the minority carriers?

c. Is the material still neutral from a charge standpoint?

d. What is meant by "bound positive charges?"

8. a. What acceptor impurity is used with silicon for *P*-type doping?

b. How does this affect the electron-hole pairs?

c. What are the majority and minority carriers?

9. What prevents the migration of all the electrons from the *N*-type material across the *PN* junction to the *P*-type side at the moment a *PN* junction is formed?

10. a. Which *PN* junction, silicon or germanium, has the higher potential barrier to the flow of majority carriers?

b. How does this show up in the VI characteristics of the two diodes?

11. Explain how there can be *any* current flowing across a reverse-biased *PN* junction and why this current is very temperature-dependent.

12. A germanium atom has 32 electrons in orbit around its nucleus, silicon only 14. Can you suggest why the reverse current of a *PN* junction diode is larger when constructed from germanium instead of silicon?

13. Explain in your own words why a *PN* junction diode has a peak inverse voltage.

14. Explain in your own words how you would determine the anode and cathode of a diode, given an ohmmeter. What must you be sure of?

15. Explain the indications of an ohmmeter when checking diodes that are

a. Shorted.

b. Open.

c. Good.

16. If a diode is checked with an ohmmeter in the forward-biased connection, explain why the readings are not constant as you change from a high range to lower ranges. Will the resistance readings get larger or smaller? Why?

17. If a diode is forward-biased in a circuit, explain how, with a dc voltmeter, you could determine if it is germanium or silicon.

18. Consider the full-wave rectifier circuit of Fig. 28-13*a*. Describe the effects of:

a. Reversing both diode directions.

b. Reversing the direction of one diode.

19. Consider the bridge rectifier circuit of Fig. 28-14*a*. Describe the effects of:

a. Reversing all the diodes.

b. Reversing diode D1.

c. Reversing diodes D1 and D2.

20. Explain why the peak inverse voltage of a diode in a half-wave rectifier is $2V_m$ when a capacitor filter is used.

21. Explain why an increase in the time constant in a capacitor filter circuit decreases ripple.

22. Explain why repetitive peak diode current is increased as ripple is decreased in a capacitor filter circuit.

23. Describe how you would use an oscilloscope to experimentally determine the ripple factor of a dc power supply.

24. Suggest a reason why an inductor filter is more effective with a heavier load.

25. How does the ripple factor vary with load for the following filters:

a. Inductor.

b. Capacitor.

c. Choke input.

26. What advantage does an *LC* filter have over a capacitor filter? What disadvantage does it have?

PROBLEMS

(Answers to odd-numbered problems at back of book)

28-1. A 120-V primary, 24-V secondary transformer is used to supply a half-wave rectifier and a 470-Ω load. Neglecting diode losses, calculate:

a. The dc voltage across the load.

b. The peak diode current.

c. The dc load current.

d. The dc power delivered to the load.

e. The PIV across the diode.

Sketch the waveforms of the load voltage and current.

28-2. Repeat Problem 28-1, using a 48-V secondary.

28-3. If a load requires a dc voltage of 9 V from a half-wave rectifier, determine:

a. The rms input voltage to the rectifier.

b. The load resistance if the peak load current through the diode must not exceed 500 mA.

28-4. Repeat Problem 28-3 using a dc voltage of 12 V.

28-5. A 120-V, 60-Hz voltage is applied to the primary of a 1:2 step-up transformer whose secondary is center-tapped, allowing a load of 1 kΩ to be connected to a full-wave rectifier utilizing two diodes. Neglecting diode losses, determine:

a. The dc voltage across the load.

b. The dc current through the load.

c. The dc power delivered to the load.

d. The PIV across each diode.

e. The ripple frequency of the output.

28-6. Repeat Problem 28-5 using a 1:4 step-up transformer.

28-7. a. Specify what secondary transformer voltage is necessary to produce 18 V dc across a 50-Ω load connected to a two-diode, full-wave rectifier.

b. What will be the peak current through each diode?

28-8. Repeat Problem 28-7 for 30 V dc across a 100-Ω load.

28-9. Repeat Problem 28-5 using a bridge rectifier connected across the *full* secondary of the transformer.

28-10. a. What secondary voltage is necessary to be connected to a bridge rectifier to obtain a dc voltage of 100 V?

b. What will be each diode's PIV for this situation?

28-11. A 208-V, 60-Hz, three-phase system is applied to a full wave rectifier circuit that supplies dc power to a 50-Ω load. Determine:

a. The dc output voltage across the load.

b. The dc power in the load.

c. Each diode's PIV.

d. The ripple frequency of the output.

28-12. Repeat Problem 28-11 using a 277-V, 400-Hz, three-phase system.

28-13. a. A filtered dc power supply is known to have a 10% ripple factor. If the dc output voltage is 12 V, what is the rms value of ripple voltage in the output?

b. Assuming that this ripple is sinusoidal in nature, what is its peak-to-peak voltage?

c. Sketch the waveform that a dc-coupled oscilloscope would indicate when connected across this power supply.

28-14. Repeat Problem 28-13 assuming that the ripple voltage is triangular in nature.

28-15. The output of a filtered dc power supply is observed on an oscilloscope and found to have an approximately sinusoidal ripple of 4 V p-p "riding" on a dc level of 16 V.

a. Sketch this voltage waveform.

b. Determine the percent of ripple.

28-16. Repeat Problem 28-15 assuming a triangular ripple voltage.

28-17. A 12.6-V, center-tapped transformer operating at 60 Hz is used in a two-diode, full-wave rectifier with a 100-μF capacitor to provide filtering for a 1-kΩ resistive load. Calculate:

a. The theoretical percent ripple factor.

b. The dc output voltage.

28-18. Repeat Problem 28-17 using a 200-μF capacitor.

28-19. What minimum size of filter capacitor is necessary with a full-wave rectifier operating at 60 Hz to make sure that the ripple factor does not exceed 1% with a load of 200 Ω?

28-20. Repeat Problem 28-19 using a frequency of 400 Hz.

28-21. What value of inductance should be used in an inductor filter connected to a full-wave rectifier operating at 60 Hz if the ripple factor is not to exceed 3% for a 120-Ω load?

28-22. Calculate the dc output voltage for Problem 28-21 if a bridge circuit is used with a line-to-line voltage of 50 V, assuming that

a. The choke has negligible resistance.

b. The choke's resistance is 30 Ω.

28-23. An *LC* filter is to be used to provide a dc output with 0.5% ripple when operating from a full-wave rectifier operating at 60 Hz. To conserve the size of choke, a ratio for $L/C = 0.01$ is recommended (*L* in henrys, *C* in microfarads). Determine the required values of *L* and *C*.

28-24. The filter in Problem 28-23 is used with a 10-kΩ load and $V_m = 20$ V. Determine:

a. The dc output voltage neglecting the coil's resistance.

b. What value of resistance could be used to replace the inductor and still provide the same ripple.

c. The approximate dc output voltage with the resistor as in part (b).

CHAPTER TWENTY-NINE

INTRODUCTION TO TRANSISTORS, AMPLIFIERS, AND OSCILLATORS

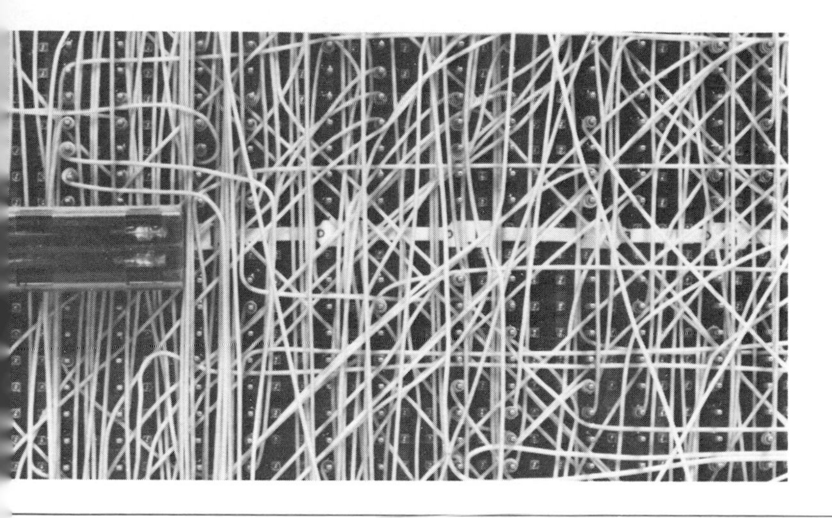

Discovered in 1947 by a team of Bell Telephone Laboratory scientists, the junction transistor has revolutionized electronics. It has replaced the bulky, slower, and heat-producing electron tube in almost every type of electronic equipment. Present-day, solid-state TV sets consume much less power (only a few tens of watts) compared with hundreds of watts in earlier tube-operated receivers.

A transistor consists essentially of two *PN* junction diodes back-to-back or front-to-front. Because both electrons *and* holes are involved in the operation of a junction transistor, they are often referred to as *bipolar junction transistors* (BJTs). This distinguishes them from the more recent *field effect transistors* (FETs), in which only electrons *or* holes flow, depending upon the type. Both BJTs and FETs are transistors.

Transistors can be used to *amplify* signals. That is, they increase the magnitude of the input signal current or voltage (or both) to a higher-voltage current or power level by controlling a power supply voltage. This can be done using one of three possible configurations: common emitter (CE), common base (CB), or common collector (CC). These names are derived from whichever of the transistor's three terminals (emitter, base, and collector) is common to both the input and output of the amplifier.

Almost all commercial amplifiers use some form of *negative feedback*. This involves feeding some of the amplifier's output back to the input to reduce the input signal. Although the result is a loss of gain, many beneficial effects occur, such as increased stability, increased bandwidth, and a reduction in the distortion of the amplifier's output. If *positive* feedback is supplied to the amplifier, the voltage gain increases and a condition of oscillation occurs. The amplifier now becomes an *oscillator* with a sinusoidal output, whose frequency is dependent upon the elements in the feedback circuit.

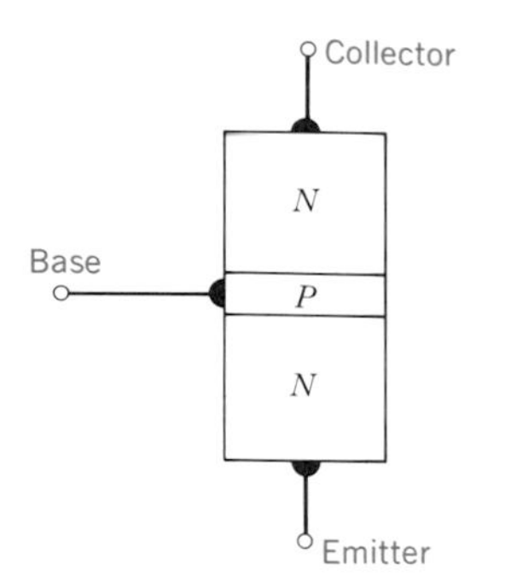

(*a*) Diagrammatic representation of an NPN sandwich

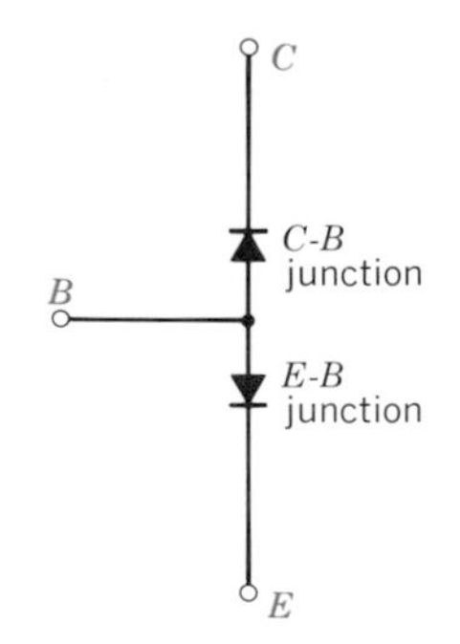

(*b*) Two-diode equivalent circuit of NPN transistor

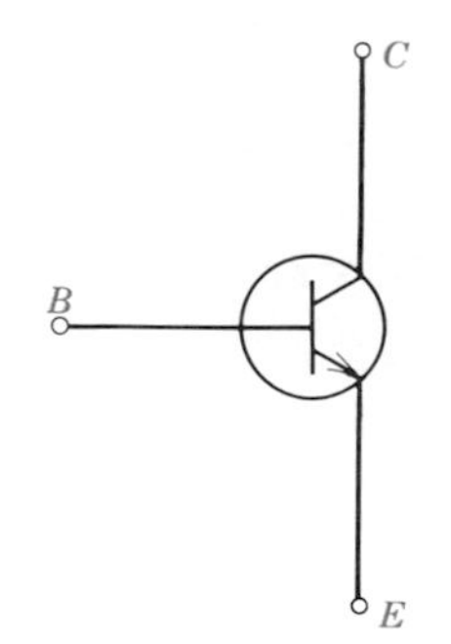

(*c*) Symbol used for NPN transistor

FIGURE 29-1
Different representations of an *NPN* transistor.

29-1 PHYSICAL CHARACTERISTICS OF TRANSISTORS

29-1.1 Types

There are two types of bipolar junction transistors (BJTs). They are the *NPN* and *PNP*. Figure 29-1 shows how an *NPN* transistor consists of a *P*-type material, the base, sandwiched between two *N*-type regions called the emitter and collector.

The base is very thin, only about 1–2% of the total length of the transistor. However, it is sufficient to produce two distinct *PN* junctions, represented in Fig. 29-1*b* as two diodes shown pointing "outward." The symbol in Fig. 29-1*c* uses an arrow, also pointing outward, to distinguish the emitter from the collector. The origin for these names will be clear when we consider how the transistor works.

If the transistor is constructed with an *N*-type base and *P*-type regions for the emitter and collector, a PNP type of transistor results. Its symbol, along with a simple two-diode equivalent circuit, is shown in Fig. 29-2. Note that the arrow on the emitter points "in."

29-1.2 Ohmmeter Checking of a Transistor

Since a transistor is normally *soldered* into a circuit, it is not readily removable for testing purposes. (Plug-in sockets like electron tubes are *not* generally used because oxidation of the contacts would have a large effect on the small voltages—only 0.6 V in some cases—found in transistor circuits). A method enabling in-circuit checking uses an ohmmeter as shown in Fig. 29-2*b* and 29-2*c*. Depending upon whether a junction is forward- or reverse-biased by the ohmmeter (no power is applied to

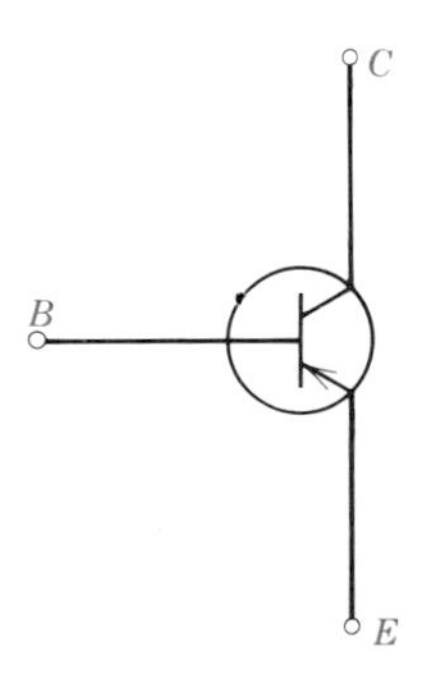

(*a*) Symbol for a *PNP* transistor

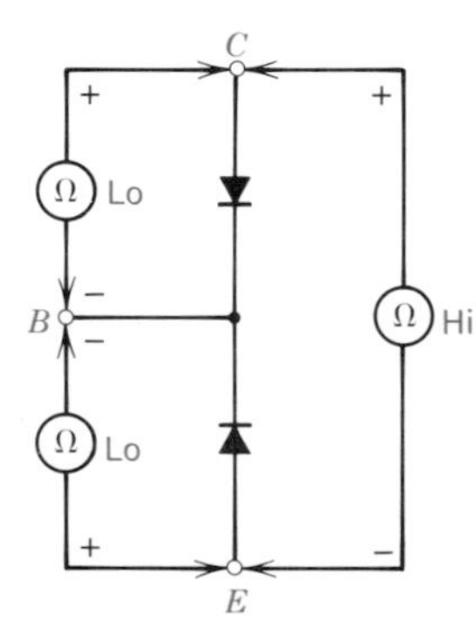

(*b*) Ohmmeter readings for given polarity

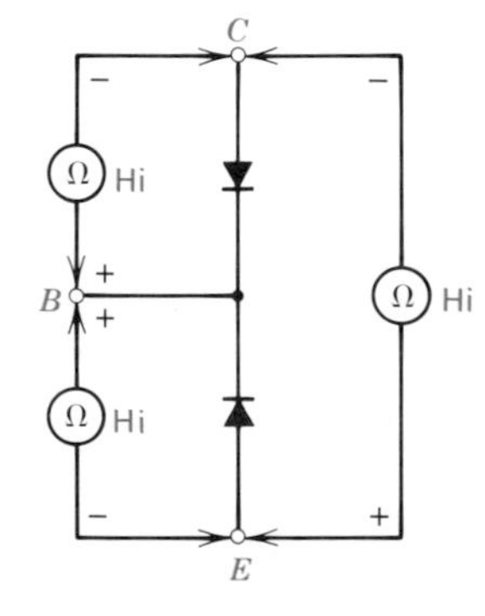

(*c*) Ohmmeter readings for reversed polarity

FIGURE 29-2
Checking a *PNP* transistor with an ohmmeter. (It is assumed that only one ohmmeter is connected to the transistor at any time.)

the circuit otherwise, of course) a low or high reading should result. A high reading is to be expected in *both* directions between C and E, except when punch-through has taken place. Care must be taken to include the effects of any other components connected across the transistor terminals before deciding that a transistor is bad. Also, using some digital multimeters on the *low-ohms range* may show a good junction to be open in both directions.

Furthermore, because low ranges of the ohmmeter can produce excessive current and high ranges can produce excessive voltage, care should be taken to use middle ohmmeter ranges.

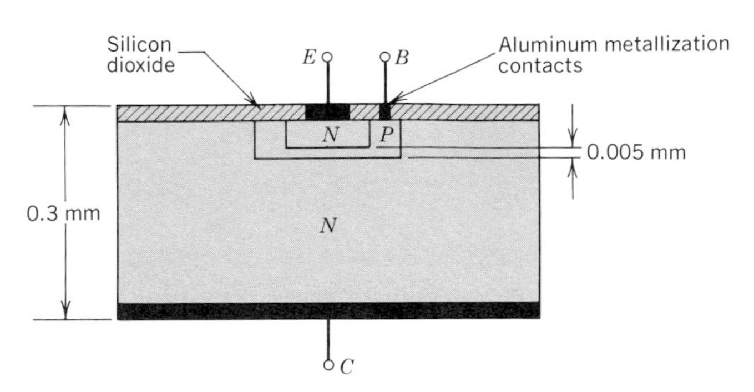

FIGURE 29-3
Typical construction of an *NPN* silicon planar transistor manufactured using the gaseous diffusion technique.

29-1.3 Construction

Although the earliest transistors were germanium, practically all transistors are now manufactured from silicon. Many methods of transistor construction were used. But most transistors now are made by a gaseous diffusion technique, closely related to integrated circuit (IC) construction.

Figure 29-3 shows a simplified cross-sectional view of an NPN silicon *planar* transistor. Starting with a block of N-type silicon, covered with a layer of silicon dioxide, the base-collector junction area is photoetched by photolithographic techniques. This simply exposes an area of the N-type material. The wafer of N-type material is then placed in a vessel (or "boat") with an oxide of the appropriate doping impurity such as boron, indium, or aluminum. The boat is passed slowly through a furnace with accurately known and controlled temperatures between 800 and 1200°C depending upon the type of junction desired. A gas of the impurity atoms diffuses into the exposed N-type material forming a thin layer of P-type material. This becomes the base region of the transistor.

The process is repeated but this time using an N-type doping impurity to diffuse the emitter into the base. A final layer of silicon dioxide is used to cover the structure, and aluminum metallization contacts are made through this layer with the NPN layers to provide the emitter, base, collector terminals.

It should be clear from Fig. 29-3 that the collector is a large region and the base is a very thin region between the emitter and collector. In other words, the transistor is not symmetrical, which may have been suggested by the representations in Fig. 29-1. Also, the base is very lightly doped in comparison with the emitter and collector. These are very important characteristics that will help explain the operation of a transistor.

29-1.4 Packaging and Lead Identification

The silicon wafer, with the transistor constructed in it, is extremely brittle and must be protected in some kind of encapsulation. Figure 29-4 shows four typical packages used. In the medium-power and high power types of transistors, the collector is often connected to the metal backing or housing. This is to help dissipate the heat from the transistor to the metal chassis or heat sink on which the transistor is mounted. Where the chassis is used as a common electrical connection (such as ground), it is sometimes necessary to use a mica washer to electrically insulate the collector from the chassis. In such cases a silicone grease is used to improve the heat flow between the transistor and the chassis.

Power ratings for the small-signal type of transistors range from 75–300 mW, and from 5–150 W for power transistors at ambient temperatures of 25°C.

Designations for transistors registered with the Electronic Industries Association (EIA) involve a 2N prefix such as 2N3391 or a five-digit number such as 40340. Whether a transistor is *NPN* or *PNP* can only be determined (without testing) from a transistor catalogue. (This also provides information concerning lead identification.) However, many foreign transistor types use a prefix of 2SA or 2SB for *PNP* (such as a 2SA 497) and 2SC or 2SD for *NPN* (such as 2SC 497). Often, because of lack of space, the 2S can be omitted from the number.

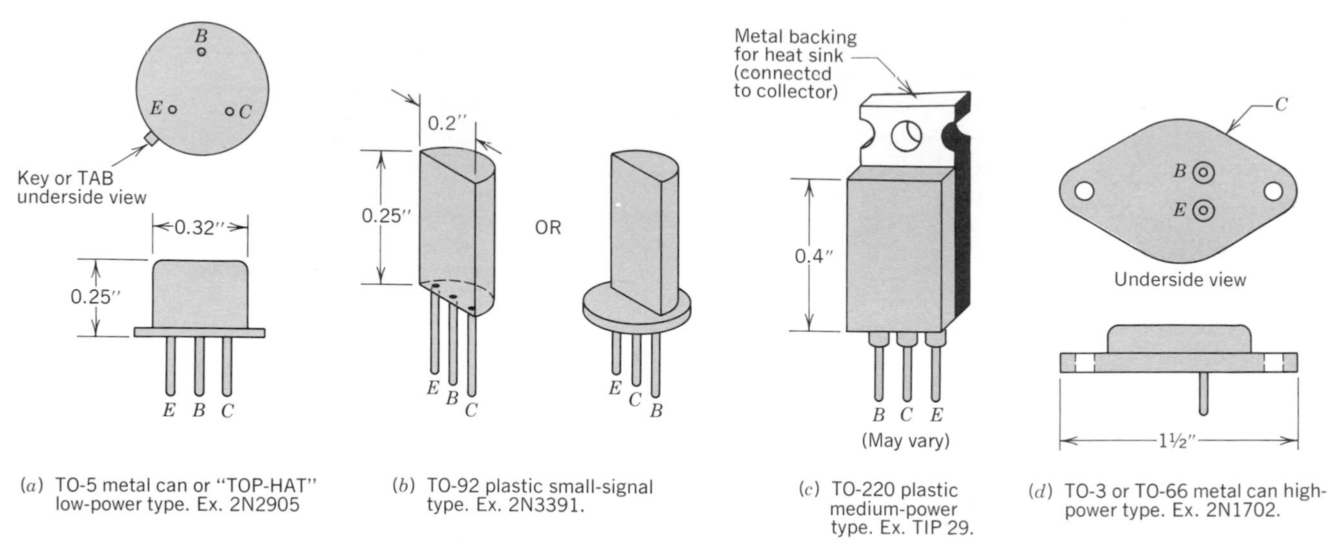

(a) TO-5 metal can or "TOP-HAT" low-power type. Ex. 2N2905

(b) TO-92 plastic small-signal type. Ex. 2N3391.

(c) TO-220 plastic medium-power type. Ex. TIP 29.

(d) TO-3 or TO-66 metal can high-power type. Ex. 2N1702.

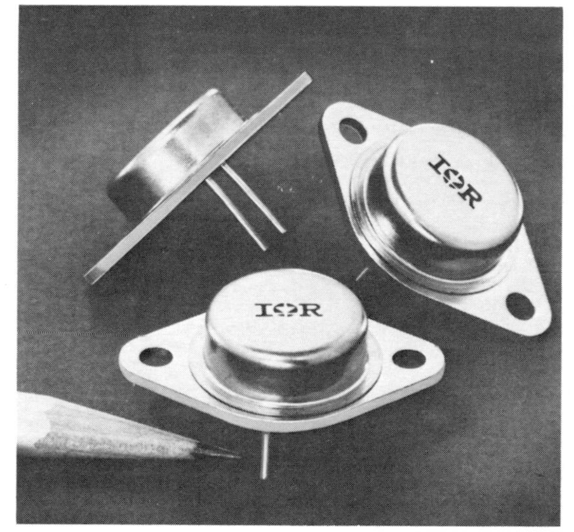

(e) Medium and high-power types

FIGURE 29-4
Typical transistor packages and lead identification. [Photographs in (e) courtesy of International Rectifier.]

29-2 BASIC COMMON-EMITTER AMPLIFYING ACTION

As already mentioned, a transistor amplifies an alternating, input voltage signal by controlling the voltage of the power supply to which it is connected. To understand how it does this, refer to Fig. 29-5.

If the emitter-base (EB) junction is forward-biased by V_{BB}, a current will flow as in any *PN* junction diode. In a silicon diode, the voltage from base to emitter, V_{BE}, is between 0.5 and 0.6 V for a base current, I_B, of 0.2 mA (see Fig. 29-5*a*).

Now consider that a dc supply voltage, $V_{CC} = 12$ V, is connected from collector to emitter with the base open. Clearly, this *reverse* biases the collector-base junction so that only a (minority) leakage current flows. This is referred to as I_{CEO} (the current from collector to emitter with the base open). Because this minority current (holes in the *N*-type material) is due to the thermally generated *minority* carriers, this current is very temperature-dependent (see Fig. 29-5*b*). Although I_{CEO} is very small in a silicon transistor (around 1 μA at 25°C), it causes instability at higher temperatures. (One method to reduce this effect will be examined in Section 29-6.)

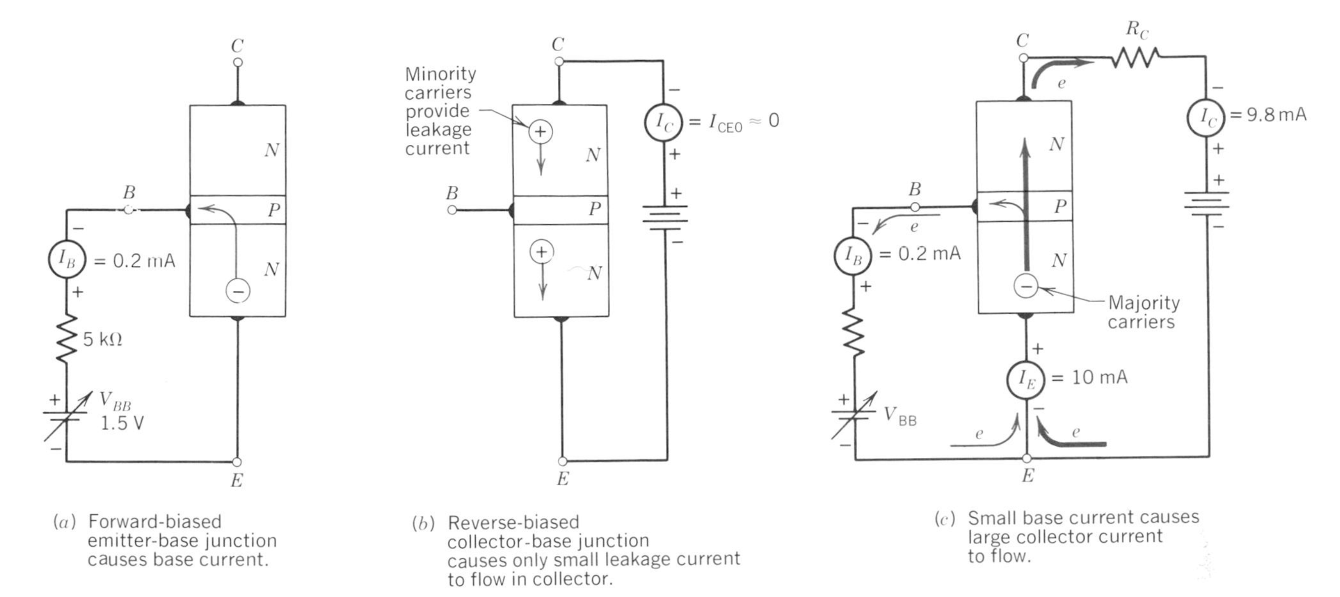

(*a*) Forward-biased emitter-base junction causes base current.

(*b*) Reverse-biased collector-base junction causes only small leakage current to flow in collector.

(*c*) Small base current causes large collector current to flow.

FIGURE 29-5
Current paths in an *NPN* transistor under different conditions of forward and reverse bias.

Now consider the effect of simultaneously applying forward bias to the emitter-base (EB) junction and reverse bias to the collector-base (CB) junction, as shown in Fig. 29-5c. The result is a *large* current flowing in the collector circuit, typically 50 to 200 times as large as the base current in small-signal transistors. (Collector resistor R_C is used to limit the collector current to a safe value.) How is this possible if the collector-base junction is reverse-biased?

The answer lies in the following three facts:

1 The base is very thin.
2 The base is very lightly doped compared with the emitter and collector.
3 The emitter-base (EB) junction is forward-biased.

That is, the forward biasing of the emitter-base junction allows electrons (the *majority* carriers) in the *N*-type emitter to crossover into the base. Once these electrons are in the base, they come under the influence of the *N*-type collector region, which is connected to the positive side of V_{CC}. The majority of the electrons (98% and over) emitted by the emitter are swept across the base and collected by the positive collector. Only a small base current flows because not many electrons (only 2% or less) recombine with holes in the base region. This is because the base is thin, so the electrons do not spend much time in this region on their way to the collector. Also, the low availability of holes in the lightly doped base do not encourage many recombinations.

Thus a small base current (used to forward-bias the emitter-base junction) can *control* a much larger collector current drawn from the supply, V_{CC}. And if the forward bias is increased, the collector current also increases. That is, the base current acts as a control valve or gate. How much collector current actually *will* flow depends upon how well the emitter-base junction has been *turned on*. When suitable bias currents are used, the control of the base current over the collector current is quite linear until a saturation effect occurs. This is shown in Fig. 29-6 using conventional current directions.

It has been noted that the base current does *not* generate or create collector current. The collector current is furnished by the collector supply, V_{CC}. The base current simply controls the conductivity between the collector and emitter (CE) terminals. Also, if the base current is zero, the transistor conductivity from the collector to emitter essentially drops to zero. From this standpoint it is not difficult to see how a transistor can be thought of either as a *valve* or as a *switch*. The latter is a very important application in digital logic devices.

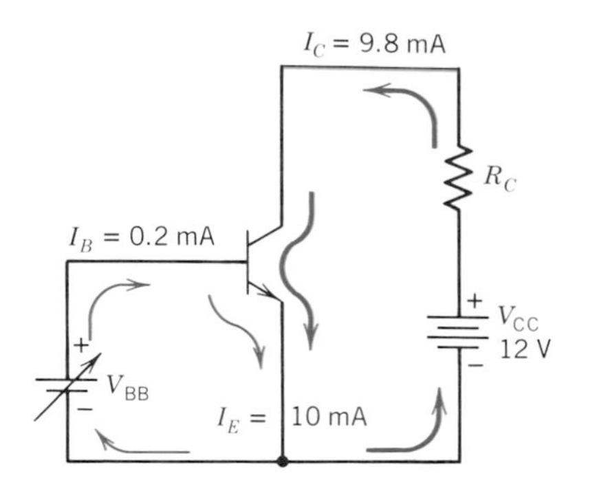

(a) A base current of 0.2 mA causes a collector current of 9.8 mA.

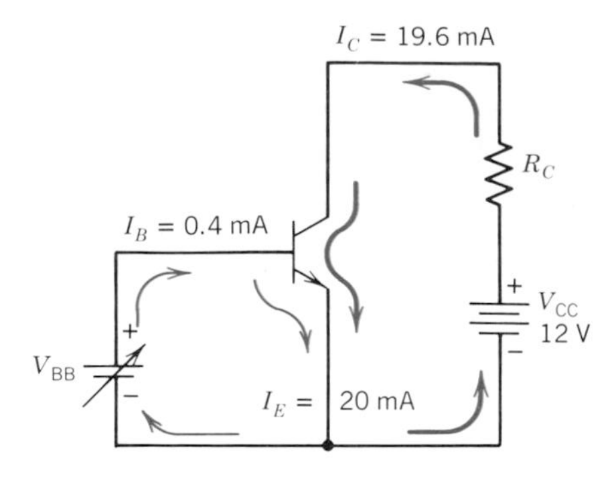

(b) A base current of 0.4 mA causes a collector current of 19.6 mA.

FIGURE 29-6
Conventional current directions in an *NPN* transistor showing how a base current determines a collector current.

29-3
CE COLLECTOR CHARACTERISTIC CURVES

The way in which base current controls the collector current can also be shown in a *graphical* form, called the *collector characteristic curves.* Figure 29-7*a* shows the circuit that can be used to obtain these curves experimentally, point by point. Alternatively, a *curve-tracer* will display this whole set of curves on a CRT screen, which can then be photographed (see Fig. 29-8). The characteristic curves in Fig. 29-7*b* show the influence of the collector-to-emitter voltage, V_{CE}. But, for a given collector voltage, the family of curves shows essentially equal separations for equal increments in base current. This effect shows the *linear* nature of the transistor over *most* of its characteristics when used as an ac amplifier.

The collector curves are very useful in transistor amplifier design and in comparing one transistor against another of the same type if matched characteristics are required in certain amplifier applications. This is especially important in complementary symmetry amplifiers, where a matched pair of NPN and PNP transistors is used together.

29-3.1 Forward-Current Transfer Ratio, β

We can also use the collector characteristic curves to calculate the "current gain" of the transistor, β, also known as the *forward-current transfer ratio.* Beta can also be defined in terms of dc current or small signal ac current. It is an indication of the control that the base current has over the collector current.

For dc:
$$\beta_{dc} = h_{FE} = \frac{I_C}{I_B} \qquad (29\text{-}1)$$

Evaluating β_{dc} (also known as h_{FE} in transistor manuals) at points R and S in Fig. 29-7*b*, we obtain

$$\text{At } R, \quad \beta_{dc} = \frac{I_C}{I_B} = \frac{5 \text{ mA}}{0.1 \text{ mA}} = \mathbf{50} \qquad (V_{CE} = 10 \text{ V})$$

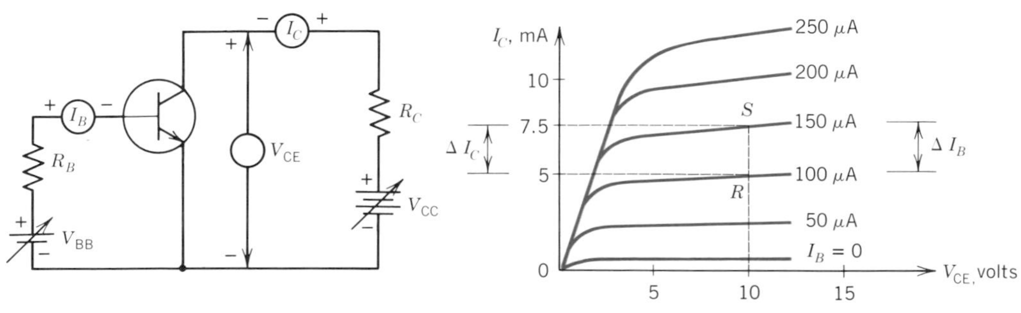

(a) Circuit to obtain collector characteristic curves for an *NPN* transistor in CE connection

(b) Typical collector characteristic curves for a small-signal type of transistor in the CE connection

FIGURE 29-7
Test circuit and CE collector characteristics for an *NPN* transistor.

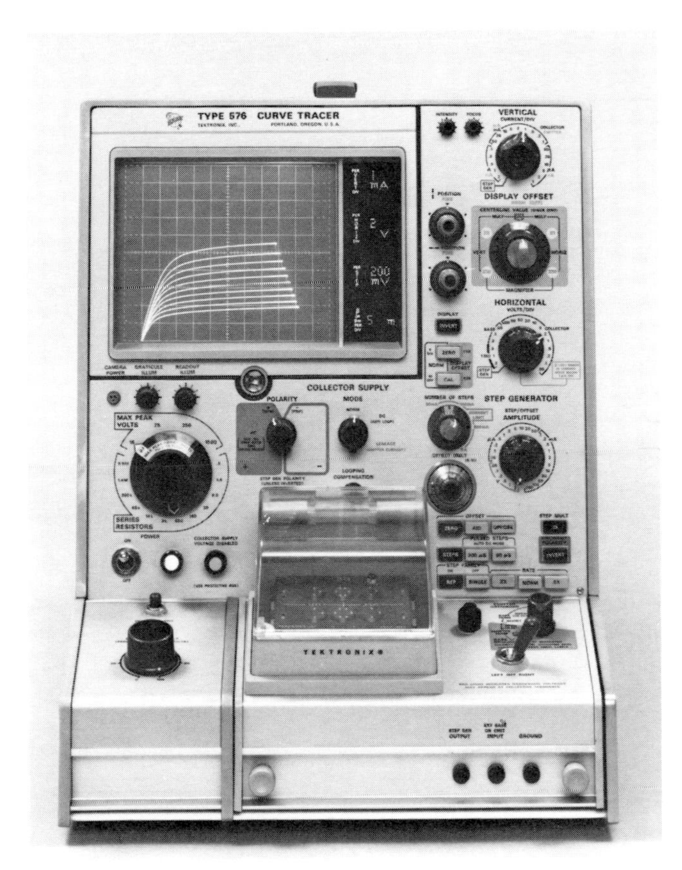

FIGURE 29-8
Type 576 curve tracer. (Courtesy of Tektronix, Inc.)

$$\text{At } S, \quad \beta_{\text{dc}} = \frac{I_C}{I_B} = \frac{7.5\text{ mA}}{0.15\text{ mA}} = \mathbf{50} \qquad (V_{\text{CE}} = 10\text{ V})$$

Instead of evaluating β at specific points on the collector characteristic, we can evaluate it as the ratio of a *small* change in the collector current for a given change in base current, ΔI_B, or

$$\boldsymbol{\beta_{\text{ac}} = h_{\text{fe}} = \frac{\Delta I_C}{\Delta I_B}}, \qquad \boldsymbol{V_{\text{CE}}} \textbf{ constant} \qquad \textbf{(29-2)}$$

Evaluating β_{ac} between points R and S in Fig. 29-7*b*, we find that

$$\beta_{\text{ac}} = \frac{\Delta I_C}{\Delta I_B} = \frac{7.5\text{ mA} - 5\text{ mA}}{0.15\text{ mA} - 0.1\text{ mA}} = \frac{2.5\text{ mA}}{0.05\text{ mA}} = \mathbf{50}$$

Over a large region of the characteristic curves, it is not unusual to find $\beta_{\text{dc}} = \beta_{\text{ac}}$ and fairly constant. As a result, the dc and ac subscripts are often dropped and we merely refer to the β (beta) of a transistor.

Values for β vary from 50–500 for *small-signal* transistors but may be only 10–50 for high-power transistors. However, even among transistors of the *same type*, it is not unusual for a 3–1 spread to exist between the maximum and minimum β values. Furthermore, the β of a given transistor is extremely sensitive to, and varies with, the temperature.

Note that we can now obtain the collector current for a transistor if we know its β, from the following equation:

$$I_C = \beta I_B + I_{\text{CEO}} \qquad (29\text{-}3)$$

where: I_{CEO} is the leakage (minority carrier) current
I_B is the base current
I_C is the collector current
β is the transistor's forward-current transfer ratio

EXAMPLE 29-1

A silicon small-signal transistor has a leakage current of 1 μA and $\beta = 200$. Determine the transistor's collector current when the base current is

a. 0.
b. 50 μA.
c. 100 μA.

SOLUTION

a. $I_C = \beta I_B + I_{\text{CEO}}$ (29-3)
$= 200 \times 0 + 1\ \mu\text{A}$
$= \mathbf{1\ \mu A}$

b. $I_C = \beta I_B + I_{\text{CEO}}$ (29-3)
$= 200 \times 50\ \mu\text{A} + 1\ \mu\text{A}$
$= 10\text{ mA} + 1\ \mu\text{A}$
$\approx \mathbf{10\text{ mA}}$

c. $I_C = \beta I_B + I_{\text{CEO}}$ (29-3)
$= 200 \times 100\ \mu\text{A} + 1\ \mu\text{A}$
$= 20\text{ mA} + 1\ \mu\text{A}$
$\approx \mathbf{20\text{ mA}}$

Example 29-1 shows that the reverse leakage current can be ignored except at extremely small values of base current.

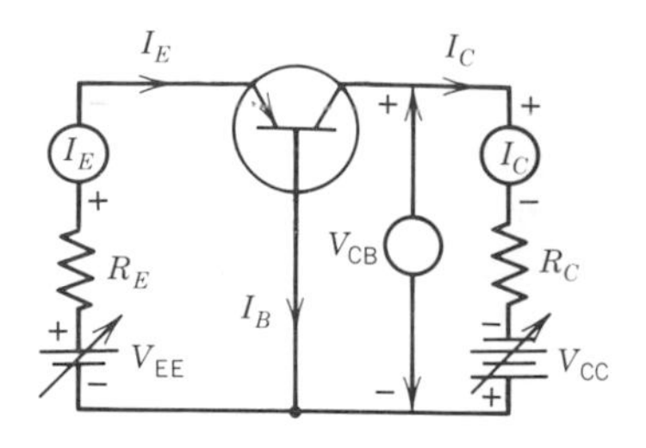

(*a*) Circuit to obtain collector characteristic curves for *PNP* transistor in CB connection

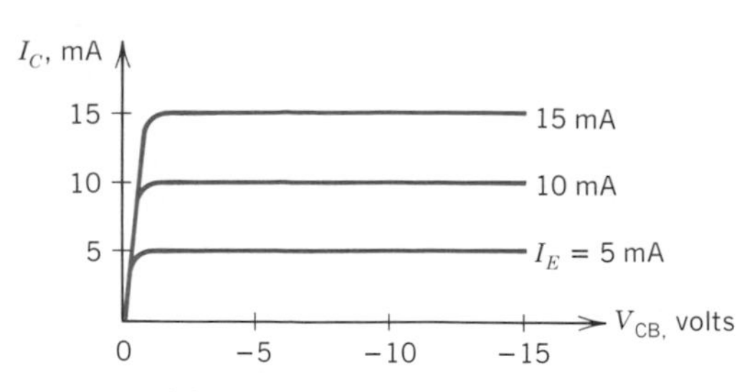

(*b*) Typical collector characteristic curves for a small-signal type of *PNP* transistor in the CB connection

FIGURE 29-9
Test circuit and CB collector characteristics for a *PNP* transistor.

Note that β has no units, since it is the ratio of two currents having the same units.

29-4 CB COLLECTOR CHARACTERISTIC CURVES

If the transistor is connected with the base terminal common to both input and output, it is said to be operating in the *common base* mode. In this case, the *emitter* is the *input* terminal. Refer to Fig. 29-9*a*, where, for circuit variation, a *PNP* transistor is used in the common base connection.

The voltage V_{EE} forward-biases the emitter-base (EB) junction so that holes in the *P*-type emitter flow into the base region. Once again, approximately 98% of these charge carriers are swept through the narrow, lightly doped base into the *P*-type collector region. So the base current is small as in the common emitter. Note that V_{EE} is necessary to forward-bias the emitter-base junction, but it is the "series" action of V_{EE} and V_{CC} that provide the emitter and collector current. If V_{CC} acts alone, with $I_E = 0$, only a leakage collector current flows, called I_{CBO}. This is the current from *c*ollector to *b*ase with the emitter *o*pen, and it is much smaller than I_{CEO} in a CE configuration.

Using the circuit of Fig. 29-9*a*, we can obtain the collector characteristic curves shown in Fig. 29-9*b*. The lines for each emitter current are practically horizontal, showing that the collector-base voltage, V_{CB}, has little effect on the collector current.

29-4.1 Forward-Current Transfer Ratio, α

This is a measure of how much collector current flows compared with the input emitter current. If we apply Kirchhoff's current law to the transistor, we obtain

$$I_E = I_C + I_B \tag{29-4}$$

It is clear then that the collector current is less than the emitter current by some small amount equal to the base current. Similarly to the CE connection, we can define the ratio of collector current to emitter current in terms of DC or AC quantities.

$$\alpha_{dc} = h_{FB} = \frac{I_C}{I_E} \tag{29-5}$$

$$\alpha_{ac} = h_{fb} = \frac{\Delta I_C}{\Delta I_E}, \; V_{CB} \text{ constant} \tag{29-6}$$

These would be difficult to evaluate from the characteristic curves because I_C and I_E differ by very small amounts. However, the dc and ac values are so close to each other that we can simply refer to the transistor's alpha.

EXAMPLE 29-2

Dc current measurements for a transistor in the CB connection show an emitter current of 6 mA and a base current of 120 μA. Determine the transistor's alpha.

SOLUTION

$$I_C = I_E - I_B \tag{29-4}$$
$$= 6 \text{ mA} - 0.12 \text{ mA}$$
$$= 5.88 \text{ mA}$$

$$\alpha = \frac{I_C}{I_E} \tag{29-5}$$
$$= \frac{5.88 \text{ mA}}{6 \text{ mA}} = \mathbf{0.98}$$

Note that the current gain in the CB connection is always slightly less than one (unity). That is, the output current is slightly less than the input current. How then can a CB connection be useful in an amplifier if it does not amplify the current?

If the output current is made to flow through a collector resistor R_C that is significantly larger than the input resistance, R_{in}, it is possible to have a significant voltage gain. That is, the output voltage is larger than the input voltage. In fact, the word *transistor* is derived from the words *trans*fer-res*istor* because of the ability of the device to transfer essentially the same current from a low-resistance input to a high-resistance output.

29-4.2 Relation Between α and β

It was shown above that the determination of α could not be done accurately because it involved small differences in currents. However, β can be determined much more accurately because of the large differences in collector and base currents. If a transistor's β is known, it can be shown that its α can be obtained quite accurately from

$$\alpha = \frac{\beta}{1 + \beta} \qquad (29\text{-}7)$$

Solving Eq. 29-7 for β yields:

$$\beta = \frac{\alpha}{1 - \alpha} \qquad (29\text{-}8)$$

Equation 29-8 shows that for a high β, a transistor must have α as close to unity as possible. This implies that high-gain transistors must have a very thin, lightly doped base to reduce the base current to a very low value. Under these conditions the collector current is practically equal to the emitter current and α is practically unity.

EXAMPLE 29-3

Use the information in Example 29-2 to obtain:

a. The transistor's β.

b. The transistor's α.

(Use $I_E = 6$ mA, $I_B = 120\ \mu$A.)

SOLUTION

a. $$I_C = I_E - I_B \qquad (29\text{-}4)$$

$$= 6 \text{ mA} - 0.12 \text{ mA}$$

$$= 5.88 \text{ mA}$$

$$\beta = \frac{I_C}{I_B} \qquad (29\text{-}1)$$

$$= \frac{5.88 \text{ mA}}{0.12 \text{ mA}} = \mathbf{49}$$

b. $$\alpha = \frac{\beta}{1 + \beta} \qquad (29\text{-}7)$$

$$= \frac{49}{1 + 49} = \mathbf{0.98}$$

Note that the collector current for a CB transistor connection can also be obtained from

$$I_C = \alpha I_E + I_{CBO} \qquad (29\text{-}9)$$

But since I_{CBO}, which is a thermally dependent current, is typically only 0.025 μA at 25°C, the collector current in a CB configuration is much less affected by temperature than is a CE connection.

29-5 BASIC COMMON-EMITTER AC AMPLIFIER

We have seen how a small dc base current can cause a larger dc collector current to flow through the transistor. If the base current is varied, the collector current will also vary. And, if this collector current is made to flow through a load resistor, an output voltage will be developed.

Figure 29-10*a* shows V_{BB} (the base supply voltage) causing a dc-base bias current I_B of 100 μA. This in turn causes a dc collector current, I_C, of 5 mA. (We are assuming that $\beta = 50$). A sinusoidal signal voltage v_s with its internal resistance R_s is connected from base to emitter through a coupling capacitor C_i. This capacitor simply prevents V_{BB} from sending dc current through the applied signal. However, capacitor C_i allows the ac signal current, i_s, to be superimposed on the dc bias current I_B to produce a total base input current of i_B. These three current waveforms are shown in Fig. 29-10*b*, 29-10*c*, and 29-10*d*, where it is assumed that the ac signal current is 100μA peak to peak.

This base current variation is amplified 50 times causing a 5 mA peak-to-peak variation in the collector current superimposed on a dc level of 5 mA, referred to as i_C. See Fig. 29-11*a*.

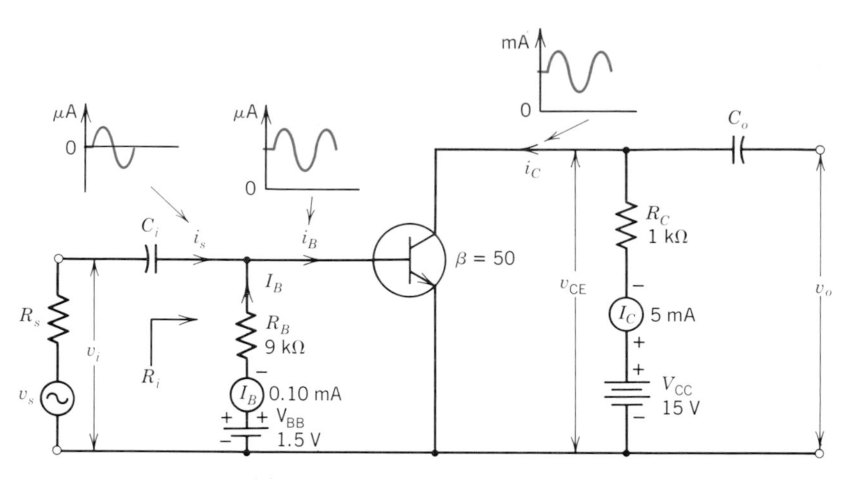

(*a*) *CE* ac amplifier circuit with two batteries

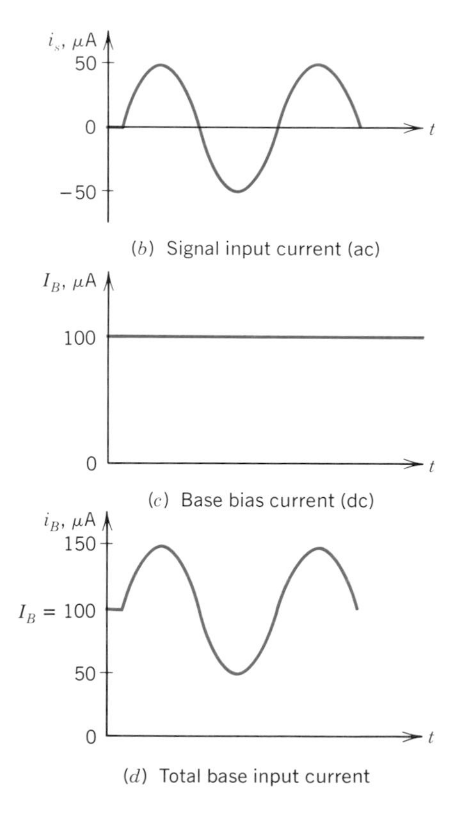

(*b*) Signal input current (ac)

(*c*) Base bias current (dc)

(*d*) Total base input current

FIGURE 29-10
Common emitter ac amplifier circuit and input current waveforms.

When this varying collector current flows through the 1-kΩ collector resistor R_C, the collector-to-emitter voltage varies between 7.5 and 12.5 V. See Fig. 29-11*b*. That is, as the collector current reaches its *peak* of 7.5 mA, v_{CE} *drops* to 7.5 V; and when i_C is at its *minimum* of 2.5 mA, v_{CE} *rises* to 12.5 V. This is because v_{CE} is equal to V_{CC} *minus* the voltage drop across R_C and is given by

$$v_{CE} = V_{CC} - i_C R_C \tag{29-10}$$

Capacitor C_o blocks the 10-V dc level in v_{CE} producing an output voltage, v_o, of 5 V peak to peak. See Fig. 29-11*c*. Since this voltage is an inverted sine wave compared with the input voltage, v_{BE}, we say the CE amplifier causes *phase inversion*. See Fig. 29-11*c* and 29-11*d*.

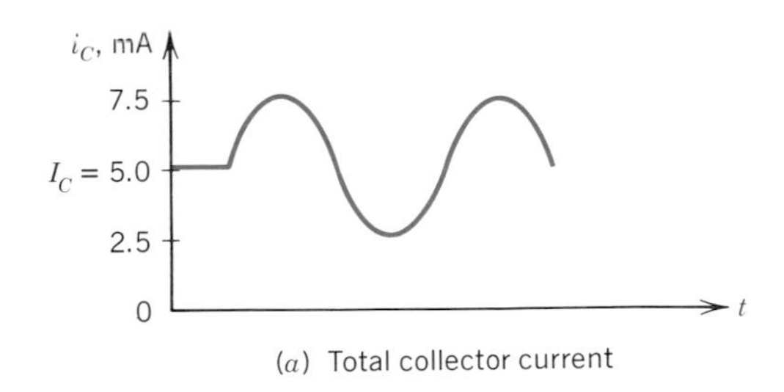

(a) Total collector current

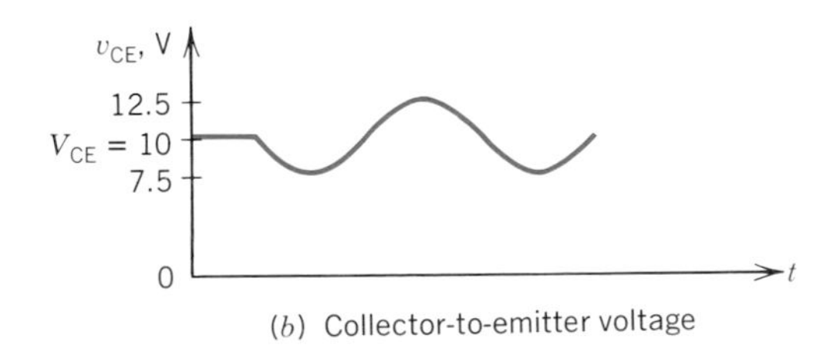

(b) Collector-to-emitter voltage

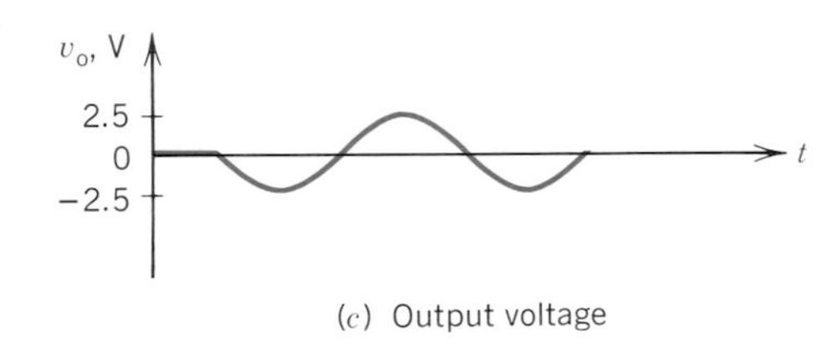

(c) Output voltage

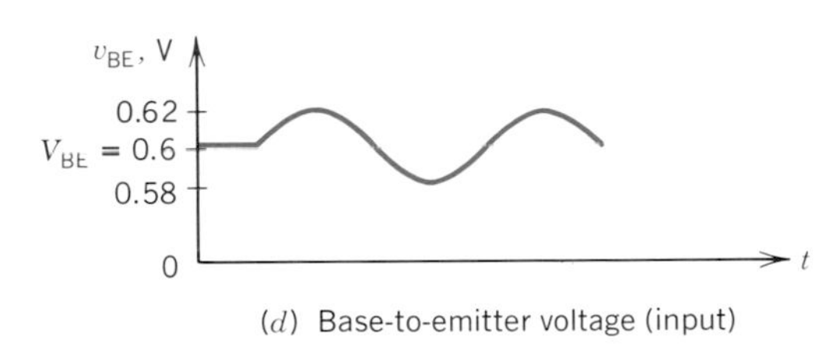

(d) Base-to-emitter voltage (input)

FIGURE 29-11
Current and voltage waveforms for a CE amplifier circuit.

29-5.1 Waveforms

From Fig. 29-10*d* we can see that a minimum value of dc bias current, I_B, into the base, is necessary to prevent the base current from dropping too low when an ac signal is applied. This prevents the transistor from being cutoff, which would introduce distortion into the output current. Similarly, the dc bias voltage of 0.6 V from base to emitter (Fig. 29-11*d*) is maintained by V_{BB}. It requires a change of only 0.04 V peak to peak (in our example) from the input signal voltage to vary the base and collector currents as shown.

If there is no distortion, the variations in current and voltage will be sinusoidal. This means that the dc readings of I_B, I_C, and V_{CE} will remain at 0.10 mA, 5 mA, and 10 V, respectively, whether a signal is applied or not.

29-5.2 Current, Voltage, and Power Gains

The ac current, voltage, and power amplifications can be defined as follows:

$$\text{Current gain, } A_i = \frac{i_c}{i_s} \qquad (29\text{-}11)$$

where: A_i is the current gain, dimensionless
i_c is the ac component of the output collector current, i_C
i_s is the ac component of the input signal current

$$\text{Voltage gain, } A_v = \frac{v_o}{v_i} \qquad (29\text{-}12)$$

where: A_v is the voltage gain, dimensionless
v_o is the ac component of the output collector voltage, v_{CE}
v_i is the ac component of the input signal voltage from base-to-emitter, v_{BE}

(If v_s is used instead of v_i, the *open circuit* voltage gain results, A_{vs}.)

Alternatively,
$$A_v = A_i \frac{R_o}{R_i} \approx -\beta \frac{R_o}{R_i} \qquad (29\text{-}13)$$

where: A_v is the voltage gain, dimensionless
A_i is the current gain, dimensionless, approximately equal to $-\beta$ for a CE single-stage amplifier
R_o is the total ac load resistance on the amplifier (R_C in this case), in ohms
R_i is the ac input resistance to the amplifier, in ohms, given by $\frac{v_i}{i_s}$

$$\text{Power gain, } A_p = \frac{P_o}{P_i} = A_v \times A_i \qquad (29\text{-}14)$$

where: A_p is the amplifier power gain, dimensionless
P_o is the ac output power
P_i is the ac input power

Sometimes, power gain A_p is given in terms of *decibels,* dB, defined as

$$\text{power gain in dB} = 10 \log \frac{P_o}{P_i} = 10 \log A_p \quad (29\text{-}15)$$

EXAMPLE 29-4

Using the data in Figs. 29-10 and 29-11, calculate the following ac characteristics of the amplifier.

a. Current gain.
b. Voltage gain.
c. Input resistance.
d. Power gain.

SOLUTION

a. $$A_i = \frac{i_c}{i_s} \quad (29\text{-}11)$$

$$= \frac{5 \text{ mA p-p}}{0.1 \text{ mA p-p}} = \mathbf{50}$$

b. $$A_v = \frac{v_o}{v_i} \quad (29\text{-}12)$$

$$= \frac{5 \text{ V p-p}}{0.04 \text{ V p-p}} = \mathbf{125}$$

c. $$R_i = \frac{v_i}{i_s}$$

$$= \frac{0.04 \text{ V p-p}}{0.1 \times 10^{-3} \text{ A p-p}} = \mathbf{400\ \Omega}$$

$$A_v = A_i \frac{R_o}{R_i} \quad (29\text{-}13)$$

$$= 50 \times \frac{1000\ \Omega}{400\ \Omega} = \mathbf{125}$$

d. $$A_p = A_v \times A_i \quad (29\text{-}14)$$

$$= 125 \times 50 = \mathbf{6250}$$

$$\text{power gain in dB} = 10 \log A_p$$
$$= 10 \log 6250$$
$$= 10 \times 3.8 = \mathbf{38\ dB}$$

Decibels are used to *linearize* power level *changes* to show, for example, what sound levels the human ear will notice. If an amplifier's output is doubled from 4–8 W, this corresponds to an increase of 10 log 2 = 3 dB. If the output is now doubled again from 8–16 W, the increase is again $10 \log \frac{16 \text{ W}}{8 \text{ W}} = 3$ dB.

This doubling of power will sound the same in either case (even though in one case there is an increase of 8 W and in the other there is an increase of only 4 W). Thus the mathematical description of 3 dB in each case has indicated the same change in hearing response regardless of the sound levels. This is now a linear response. Incidentally, a change of 3 dB can be heard, but it is *not* an appreciable change in power to the human ear.

29-5.3 CE Amplifier with Single Supply Voltage

The common emitter has two valuable features. One is the ability to produce a high *power* gain because it develops both current and voltage gains. (The CB and CC amplifiers only amplify voltage or current, respectively, but *not both*.) The second feature of a CE amplifier is the need for *only one* power supply. Figure 29-12*a* shows how the 15-V collector supply can be used to provide base bias if R_B is increased from 9 kΩ with V_{BB} = 1.5 V to 90 kΩ with V_{CC} = 15 V. This is possible because both supplies have their negative sides tied to the common emitter. (Note that this is not directly possible to do in a CB amplifier. See Fig. 29-9*a*).

Figure 29-12*b* indicates the preferred way of drawing the CE amplifier circuit. This is especially convenient when a number of CE amplifiers, all operating from the same supply voltage, are connected in cascade. This means that the output of one amplifier is connected, through a coupling capacitor, to the input of the next. In this case, the overall voltage gain is given by the *product* of the individual voltage gains of each amplifier. (If the individual stage gains are given in dB, the total power gain is the *sum* of the individual gains.)

29-6 PRACTICAL CE AC AMPLIFIER WITH BIAS STABILIZATION

We have seen that the collector current of a transistor in the CE connection is given by

$$I_C = \beta I_B + I_{CEO} \quad (29\text{-}3)$$

Both β and the minority leakage current I_{CEO} are temperature-dependent. Also, as the temperature *increases,* the input resistance *decreases,* causing an increase in I_B. Therefore, the collector current goes up with

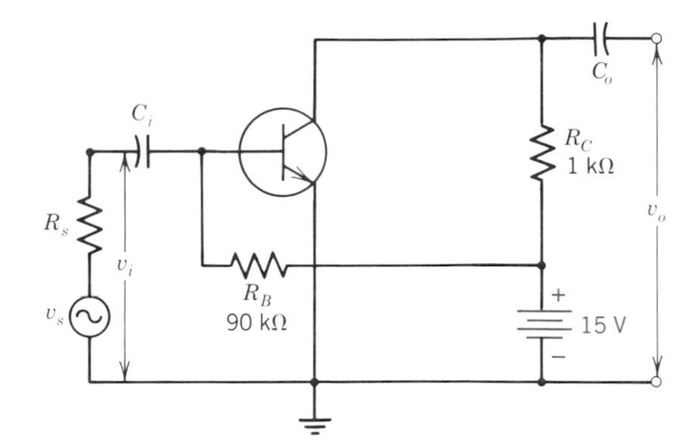

(*a*) *CE* ac amplifier operating from a single supply

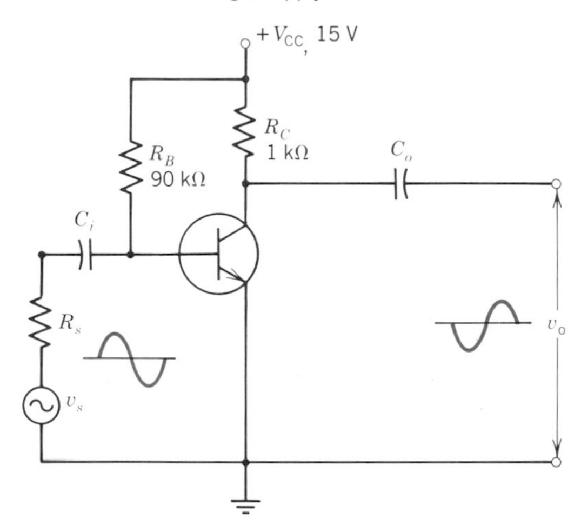

(*b*) *CE* ac amplifier circuit redrawn

FIGURE 29-12
CE amplifier operates with fixed bias from a single supply voltage.

the temperature. This increase in the collector current causes more heat to be dissipated within the transistor, raising its temperature further and causing still another increase in collector current. Thus any rise in the ambient temperature (in the medium surrounding the transistor) can lead to *thermal runaway*. This means that the amplifier's operating point changes, causing distortion, and, in large power transistors, could even lead to transistor damage. The solution lies in arranging the circuit so that it provides temperature stability. This is called *bias stabilization*.

One method of providing bias stabilization is to make any *increase* in the average value of the collector current, I_C, to cause a *decrease* in the base current, I_B. A simple way of doing this is to use a resistor in the emitter lead as shown in Fig. 29-13.

The voltage now responsible for forward-biasing the base-emitter junction is given by Kirchhoff's voltage law:

$$\begin{aligned} V_{\mathrm{BE}} &= V_{R_2} - V_{R_e} \qquad (29\text{-}16) \\ &= V_{R_2} - I_E R_e \\ &\approx V_{R_2} - I_C R_e \end{aligned}$$

The voltage across R_2, (i.e., V_{R_2}), resulting from the voltage divider of R_1 and R_2, tends to *forward-bias* the base-emitter junction. But the dc portion of the emitter current (practically equal to the collector current) develops a voltage across the emitter resistor, R_e, that tends to *reverse bias* the base-emitter junction. Furthermore, this reverse bias increases whenever the collector current increases. This produces a reduction in forward bias,

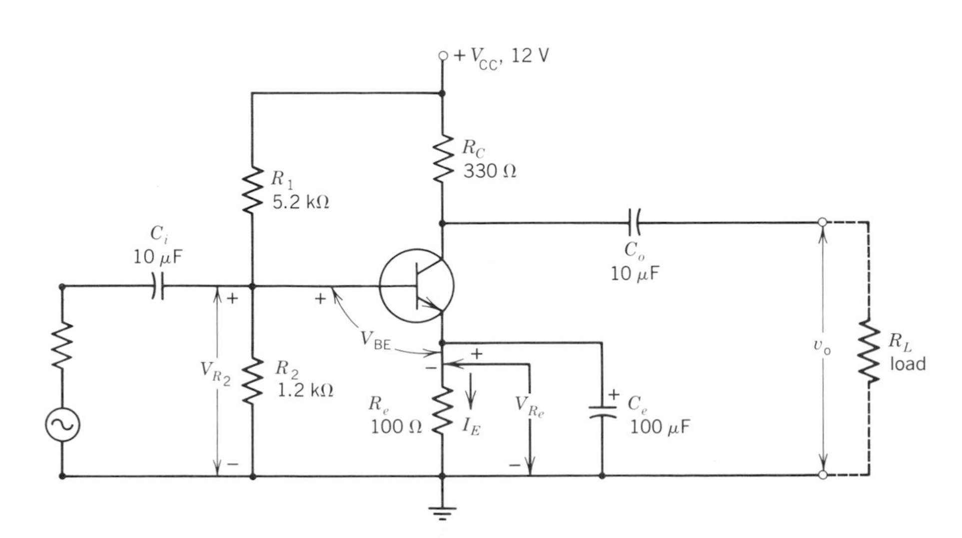

FIGURE 29-13
Typical CE ac amplifier circuit with emitter and voltage-divider bias to provide bias stabilization.

causing I_B to drop and keeping I_C nearly constant, whenever I_C tries to increase due to the temperature (Eq. 29-3).

The resistor R_e is bypassed by C_e so that the ac component of current will not develop a reduction in the ac input signal, which would decrease the ac voltage gain of the amplifier.

It should be noted that this bias stabilization is also useful if a transistor is replaced by one of the same type but having a very different value of β. The circuit of Fig. 29-13 makes the gain virtually independent of β and more dependent on the values of bias resistors selected.

Finally, a CB amplifier is the most stable configuration and has no need of bias stabilization. But since the CE amplifier does have the advantage of high-power gain, the emitter feedback circuit of Fig. 29-13 is almost universally used.

29-7 FREQUENCY RESPONSE OF A CE AMPLIFIER

An important characteristic of an amplifier is how constant its voltage (and power) gain is over the band of frequencies it is designed to handle. To show this information a frequency response curve can be plotted as in Fig. 29-14, either in terms of the voltage gain or power gain variation in dB.

The frequencies at which the voltage gain drops to 70.7% of the gain at some midfrequency, such as 400 Hz or 1 kHz, are called the lower and upper half-power frequencies, f_1 and f_2. Also, Δf, the band of frequencies between f_1 and f_2, is called the half-power or 3-dB bandwidth. Since a 3-dB change is not very noticeable to the human ear, the response is considered essentially "flat" over this bandwidth.

For acceptable audio (high-fidelity) amplifiers, f_1 is usually around 20 Hz and f_2 around 20 kHz to encompass the hearing range of most people. In some amplifiers that require a transformer in the output (to match to a low impedance speaker), the response may show a resonant rise due to distributed capacitance and inductance in the transformer.

The reason for the drop-off in response (Fig. 29-14) at *low* frequencies is the comparatively high capacitive reactance of C_i, C_o, and sometimes C_e in the circuit of Fig. 29-13. This causes a decrease in the base-emitter input voltage because of the voltage drop across C_i, and a reduction in the output voltage across R_L due to the voltage drop across C_o.

The reduction in gain at *high* frequencies is due to the capacitive effects of the reverse-biased *PN* junctions in the transistor itself. These junction reactances (essentially X_C) become very low at high frequencies and tend to "short" the input signal to ground.

A transistor's high-frequency capability is indicated in the manufacturers' specifications by its small-signal, cutoff frequency f_T. This is the frequency at which the short-circuit CE current gain drops to unity. Audio transistors have typical values of $f_T = 1$ MHz while RF (radio frequency) transistors have f_T values around 1 GHz.

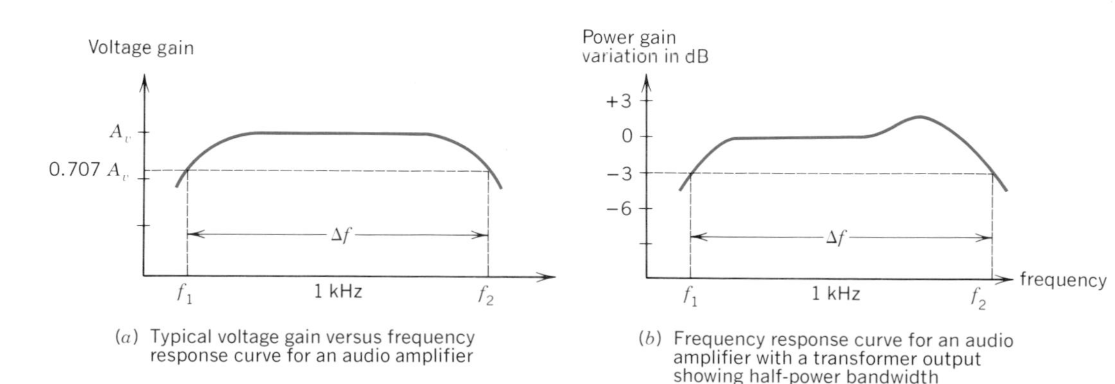

(*a*) Typical voltage gain versus frequency response curve for an audio amplifier

(*b*) Frequency response curve for an audio amplifier with a transformer output showing half-power bandwidth

FIGURE 29-14
Amplifier frequency response curves showing half-power bandwidths in terms of voltage gain and power gain in decibels.

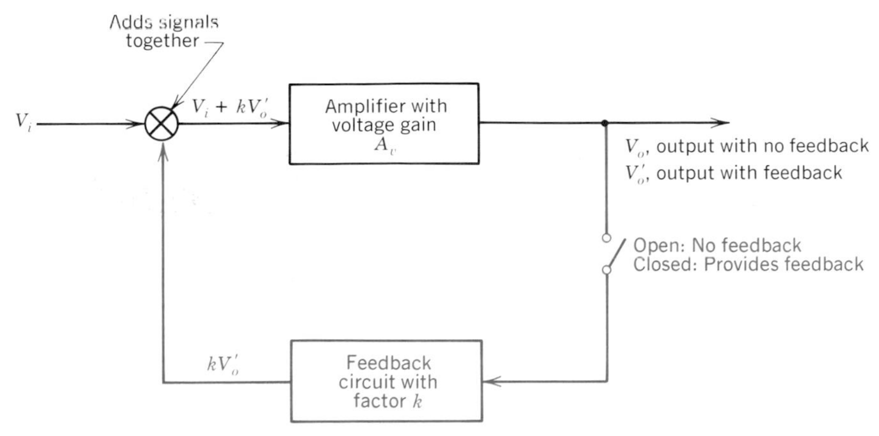

FIGURE 29-15
Block diagram showing how feedback can be introduced to an amplifier. Negative feedback results if kA_v is negative; positive feedback results if kA_v is positive.

The type of transistor configuration connection also determines the value of f_2. A CB amplifier is much better than a common-emitter at high frequencies and is often found in the front end of an FM receiver operating at frequencies around 100 MHz.

29-8 NEGATIVE FEEDBACK

The circuit of Fig. 29-13 operated on the principle of emitter feedback to stabilize the transistor in the event of a temperature rise. This is a form of negative feedback.

The principle of negative feedback is a very important one and is used widely in amplifiers, communications circuits, and control systems. Basically, it involves feeding a (voltage or current) signal back from the output of an amplifier or system to the input of the system. The general concept is shown in Fig. 29-15.

If the switch in the feedback path is open, there is *no* feedback and the voltage gain of the amplifier is

$$A_v = \frac{V_o}{V_i} \tag{29-12}$$

With the switch in Fig. 29-15 closed, a portion of the *new* output, equal to kV'_o, is added to the input. This changes the *new* input to the amplifier to $v_i + kV'_o$. Whether this increases or decreases the input depends upon the phase relationships or the signs (positive or negative) of k and A_v. In *negative* feedback, the input is deliberately decreased.

It can be shown that the overall gain of the amplifier *with* feedback is now given by

$$A_{vf} = \frac{V'_o}{V_i} = \frac{A_v}{1 - kA_v} \tag{29-17}$$

where: A_{vf} is the voltage gain with feedback
A_v is the voltage gain without feedback
k is the feedback circuit factor, usually between 0 and 1

Note that kA_v is the *loop gain*. For negative feedback, kA_v is negative, the denominator *increases* and A_{vf} decreases. But if kA_v is positive, this is called positive feedback, the denominator of Eq. 29-17 decreases, and A_{vf} increases. (Such increases can convert the amplifier to an oscillator.)

This is clearly shown in Example 29-5.

EXAMPLE 29-5

An amplifier has a voltage gain of −40. Calculate the overall voltage gain when feedback is provided as follows:

a. 10% *negative* feedback.
b. 20% *negative* feedback.
c. 1% *positive* feedback.

SOLUTION

a. When $k = 0.1$ (10%); $kA_v = 0.1 \times -40 = \quad 4$

$$A_{vf} = \frac{A_v}{1 - kA_v} \tag{29-17}$$

$$= \frac{-40}{1 - 0.1 \times (-40)} = \frac{-40}{1 - (-4)}$$

$$= \frac{-40}{1 + 4} = \mathbf{-8}$$

b. When $k = 0.2$ (20%); $kA_v = 0.2 \times -40 = -8$

$$A_{vf} = \frac{A_v}{1 - kA_v} \tag{29-17}$$

$$= \frac{-40}{1 - 0.2 \times (-40)} = \frac{-40}{1 - (-8)}$$

$$= \frac{-40}{1 + 8} = \mathbf{-4.4}$$

Note that the gain has decreased because negative feedback has increased from 10 to 20%. (The denominator has increased.)

c. $k = -0.01$ (1%); $kA_v = -0.01 \times -40 = +0.4$

$$A_{vf} = \frac{A_v}{1 - kA_v} \tag{29-17}$$

$$= \frac{-40}{1 - (-0.01)(-40)} = \frac{-40}{1 - 0.4} = \mathbf{-66.7}$$

Note that in the case of *positive* feedback, the denominator of Eq. 29-17 decreases, causing the overall gain (with feedback) to increase.

29-8.1 Reasons for Using Negative Feedback

It can be seen from the above example that *negative* feedback *reduces* the voltage gain while *positive* feedback *increases* the voltage gain. Why then should we want to use negative feedback if it is going to decrease our voltage gain? **The reduction in gain is the price that must be paid to obtain the following benefits with negative feedback:**

1. **A *reduction* in *distortion* of an amplifier's output.**
2. **A *widening* of an amplifier's *bandwidth*.**
3. **Often an *increase* in *input resistance*.**
4. **Often a *decrease* in *output resistance*.**
5. **Sometimes a *decrease* in amplifier *noise*.**
6. ***Stabilized* voltage gain (effects of changes reduced in the active device itself).**

FIGURE 29-16
Commercial audio amplifier. (Courtesy of McIntosh Laboratory, Inc.)

Almost all commercial amplifiers (see Fig. 29-16) use some form of negative feedback to achieve these benefits. These characteristics are *improved* by the same factor, $1 - kA_v$ (recall that the term kA_v is negative) as the voltage gain is *reduced*. That is,

$$R_{if} = R_i(1 - kA_v) \tag{29-18}$$

where: R_{if} is the input resistance *with* feedback
R_i is the input resistance *without* feedback

$$BW_f = BW(1 - kA_v) \tag{29-19}$$

BW_f is the bandwidth *with* feedback
BW is the bandwidth *without* feedback

$$D_f = \frac{D}{1 - kA_v} \tag{29-20}$$

D_f is the distortion *with* feedback
D is the distortion *without* feedback

We can show that the *gain-bandwidth product* of an amplifier is a *constant* (for a given transistor) by multiplying Eqs. 29-17 and 29-19:

$$A_{vf} = \frac{A_v}{1 - kA_v} \tag{29-17}$$

$$BW_f = BW(1 - kA_v) \tag{29-19}$$

$$A_{vf} \times BW_f = \frac{A_v}{(1 - kA_v)} \times BW(1 - kA_v)$$

or

$$A_{vf} \times BW_f = A_v \times BW \qquad (29\text{-}21)$$

where: $A_{vf} \times BW_f$ is the gain-bandwidth product *with* feedback

$A_v \times BW$ is the gain-bandwidth product *without* feedback

This important equation shows that the bandwidth can only be *increased* to a higher value, BW_f, at the expense of a corresponding *reduction* in gain to a lower value, A_{vf}.

EXAMPLE 29-6

A single-stage CE amplifier has an output of 6 V p-p when a 100-mV p-p input is applied. The amplifier's input resistance is 2 kΩ; its bandwidth is 10 kHz; and its distortion is 6%. After 15% negative feedback is introduced to the amplifier, calculate:

a. The voltage gain.
b. The input resistance.
c. The bandwidth.
d. The distortion.
e. The gain-bandwidth product. Compare this with the gain-bandwidth product before feedback.

SOLUTION

a. $$A_v = \frac{v_o}{v_i} \qquad (29\text{-}12)$$

$$= \frac{-6 \text{ V p-p}}{0.1 \text{ V p-p}} = -60$$

(The negative sign indicates the inherent phase inversion of a single-stage CE amplifier.)

$$A_{vf} = \frac{A_v}{1 - kA_v} \qquad (29\text{-}17)$$

$$= \frac{-60}{1 - 0.15 \times (-60)} = \frac{-60}{1 + 9} = \mathbf{-6}$$

b. $$R_{if} = R_i(1 - kA_v) \qquad (29\text{-}18)$$

$$= 2 \text{ k}\Omega(1 - 0.15 \times (-60))$$

$$= 2 \text{ k}\Omega(1 + 9) = \mathbf{20 \text{ k}\Omega}$$

c. $$BW_f = BW(1 - kA_v) \qquad (29\text{-}19)$$

$$= 10 \text{ kHz } (1 - 0.15 \times (-60))$$

$$= 10 \text{ kHz } (1 + 9) = \mathbf{100 \text{ kHz}}$$

d. $$D_f = \frac{D}{1 - kA_v} \qquad (29\text{-}20)$$

$$= \frac{6\%}{1 - 0.15(-60)} = \frac{6\%}{1 + 9} = \mathbf{0.6\%}$$

e. Gain-bandwidth product $= A_{vf} \times BW_f$ (29-21)

$$= (-6) \times 100 \text{ kHz}$$

$$= \mathbf{-600 \text{ kHz}}$$

Before feedback, gain-bandwidth product $= A_v Bw$

$$= (-60) \times 10 \text{ kHz}$$

$$= \mathbf{-600 \text{ kHz}}$$

29-8.2 Introduction of Negative Feedback in a CE Amplifier

Figure 29-17 shows how ac negative feedback can be introduced in a CE amplifier by not bypassing all of the emitter resistor. Note that this circuit is essentially the same as that of Fig. 29-12 with only a portion of the emitter resistor shunted by the capacitor. The unshunted portion of R_e is R_{e1} in Fig. 29-17.

Such shunting causes an ac voltage to be developed across R_{e1} instead of the entire emitter resistor R_e. (The circuit acts exactly as Fig. 29-13 using Eq. 29-16). The voltage across R_{e1} directly subtracts from the input voltage, v_i, reducing the base-emitter input voltage and thus the output v_o. That is,

$$v_{be} = v_i - v_{R_{e1}} \qquad (29\text{-}22)$$

Note the similarity of Eqs. 29-22 and 29-16.

It can be shown that the amount of voltage fed back is given by

$$k = \frac{R_{e1}}{R_o} \qquad (29\text{-}23)$$

where: k is the feedback factor (dimensionless)

R_{e1} is the unbypassed emitter resistor in ohms, Ω

R_o is the total ac load resistance equal to $R_C \| R_L$ in ohms, Ω

Since the voltage gain, A_v, of a CE amplifier is negative (phase inversion) the product kA_v is negative, producing negative feedback in this case.

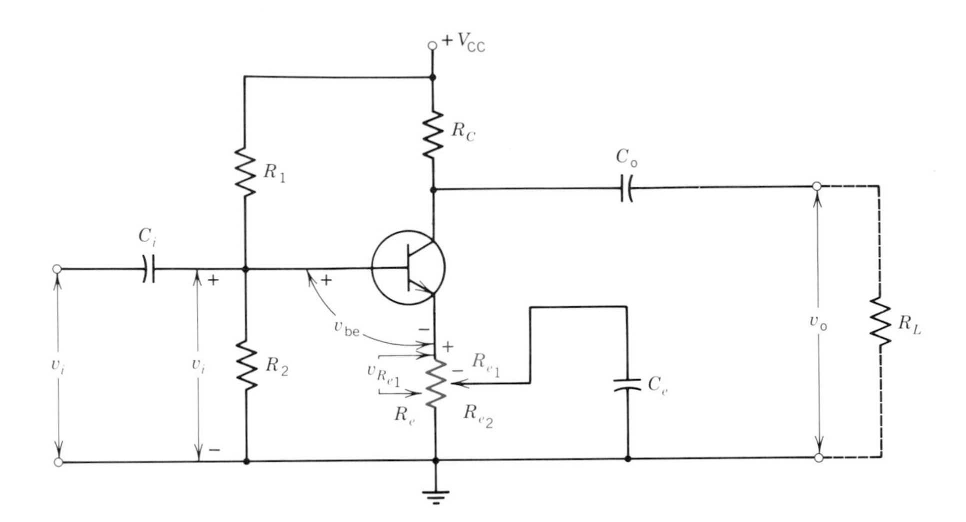

FIGURE 29-17
Showing how negative feedback can be introduced in a CE amplifier due to the unbypassed portion of the emitter resistor, R_{e1}.

EXAMPLE 29-7

A CE amplifier with a 1.5-kΩ collector resistor has an 82-Ω resistor in the emitter circuit. When R_e is completely bypassed, the amplifier has an input resistance of 1.2 kΩ. Given that the transistor's β is 100, calculate:

a. The amplifier's voltage gain (without feedback).

b. The voltage gain when the emitter bypass capacitor is not connected (with feedback).

c. The new voltage gain, with feedback, when a 1-kΩ load, R_L, is added.

SOLUTION

a. $$A_v = A_i \frac{R_o}{R_i} \approx -\beta \frac{R_o}{R_i} \qquad (29\text{-}13)$$

$$R_o = R_C,\ A_v \approx -100 \times \frac{1500\ \Omega}{1200\ \Omega} = \mathbf{-125}$$

b. $$k = \frac{R_e}{R_o} = \frac{R_e}{R_C} \qquad (29\text{-}23)$$

$$= \frac{82\ \Omega}{1500\ \Omega} = 0.055 \qquad (5.5\%)$$

$$A_{vf} = \frac{A_v}{1 - kA_v} \qquad (29\text{-}17)$$

$$= \frac{-125}{1 - 0.055 \times (-125)} = \frac{-125}{1 + 6.88}$$

$$= \mathbf{-15.9}$$

c. With $R_L = 1$ kΩ added, the voltage gain without feedback is changed because R_o is now

$$R_o = R_C \| R_L = \frac{R_C \times R_L}{R_C + R_L} \qquad (6\text{-}10)$$

$$= \frac{1500\ \Omega \times 1000\ \Omega}{1500\ \Omega + 1000\ \Omega} = 600\ \Omega$$

The new voltage gain *without* feedback is given by

$$A_v = A_i \frac{R_o}{R_i} \approx -\beta \frac{R_o}{R_i} \qquad (29\text{-}13)$$

$$\approx -100 \times \frac{600\ \Omega}{1200\ \Omega} = -50$$

The new feedback factor is

$$k = \frac{R_e}{R_o} \qquad (29\text{-}23)$$

$$= \frac{82\ \Omega}{600\ \Omega} = 0.137\ (13.7\%)$$

The new voltage gain *with* feedback is

$$A_{vf} = \frac{A_v}{1 - kA_v} \qquad (29\text{-}17)$$

$$= \frac{-50}{1 - 0.137 \times (-50)} = \frac{-50}{1 + 6.85} = \mathbf{-6.4}$$

29-9 POSITIVE FEEDBACK

Let us see the effect of increasing the amount of positive feedback in an amplifier until the loop gain, kA_v, equals one.

EXAMPLE 29-8

An amplifier has a voltage gain of 40 without feedback. Determine the voltage gains when positive feedback of the following amounts is applied:

a. $k = 0.01$.

b. $k = 0.02$.

c. $k = 0.025$.

SOLUTION

a. $$A_{vf} = \frac{A_v}{1 - kA_v} \qquad (29\text{-}17)$$

$$= \frac{40}{1 - 0.01 \times 40} = \frac{40}{0.6} = \mathbf{66.7}$$

b. $$A_{vf} = \frac{A_v}{1 - kA_v} \qquad (29\text{-}17)$$

$$= \frac{40}{1 - 0.02 \times 40} = \frac{40}{0.2} = \mathbf{200}$$

c. $$A_{vf} = \frac{A_v}{1 - kA_v} \qquad (29\text{-}17)$$

$$= \frac{40}{1 - 0.025 \times 40} = \frac{40}{0} = \mathbf{?}$$

Equation 29-17 suggests that the gain will be infinite when the loop gain reaches the critical value of $kA_v = +1$. Of course, the output voltage cannot be infinite and the amplifier will stop amplifying and start oscillating. That is, Eq. 29-17 no longer holds and an output voltage will develop without the need for any separate input. If the feedback path contains a frequency selective network, the requirement that $kA_v = 1$ is met at only one frequency. Sinusoidal oscillations result and we have a sine-wave oscillator.

29-9.1 Wien Bridge Oscillator Circuit

A circuit that can produce sine wave oscillations with a distortion of less than 1% is shown in Fig. 29-18.

The cascaded two-stage CE amplifier provides a 360° phase shift so that the output V_o is of the correct phase to feed back to the input. However, to ensure stable operation at a single frequency, the series-parallel RC network is used in the feedback path. It can be shown that V_2 is in phase with V_o only when $R = X_C$ at which point $V_2 = \frac{1}{3} V_o$. The amount of unbypassed emitter resistance in the first stage can be varied (through negative feedback) to provide the correct amount of

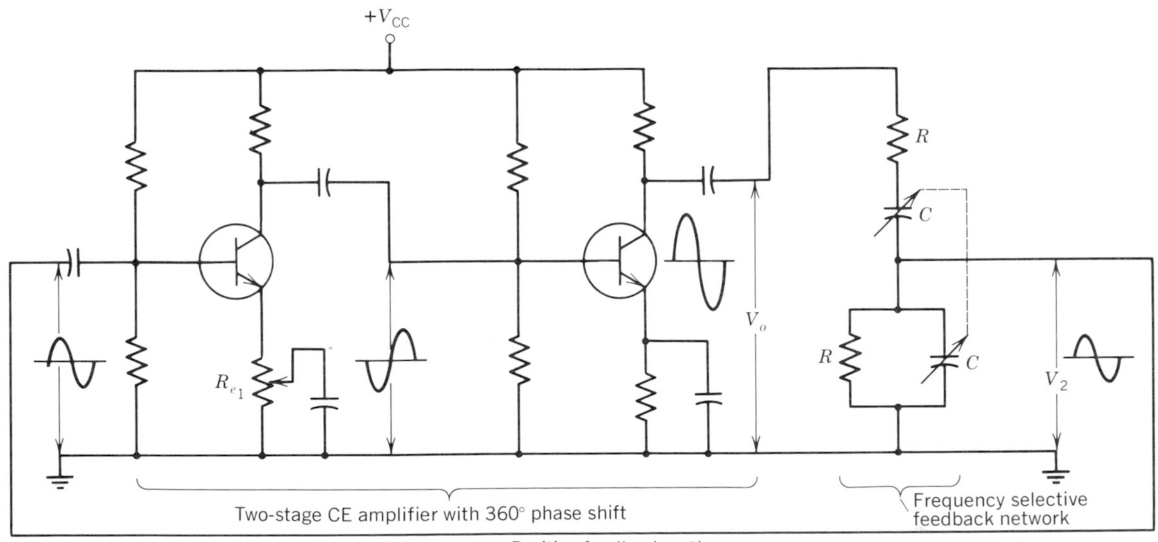

FIGURE 29-18
Variable frequency Wien bridge oscillator. An example of how positive feedback causes oscillations.

overall gain to make kA_v only slightly greater than one for minimum distortion.

Variable frequency operation from 20 Hz to 200 kHz is possible by the simultaneous adjustment of the two capacitors. The frequency of oscillation is given by

$$f = \frac{1}{2\pi RC} \tag{29-24}$$

where: f is the frequency of oscillation in hertz, Hz

R is the resistance in ohms, Ω

C is the capacitance in farads, F

EXAMPLE 29-9

What is the frequency of oscillation in a Wien bridge oscillator in which $R = 10\ \text{k}\Omega$ and $C = 200$ pF?

SOLUTION

$$f = \frac{1}{2\pi RC} \tag{29-24}$$

$$= \frac{1}{2\pi \times 10 \times 10^3\ \Omega \times 200 \times 10^{-12}\ \text{F}}$$

$$= \mathbf{79.6\ kHz}$$

Many different forms of oscillators exist. Some, designed to produce radio frequencies, use resonant LC tank circuits. Examples of these are the Hartley and Colpitts oscillators found in communications transmitters and receivers.

Nonsinusoidal oscillators often produce a square wave output. They are referred to as multivibrators and usually involve two interconnected transistors that are switched alternately full-on and then full-off. Timing is accomplished by means of the discharging rate of a capacitive circuit.

It should be noted that many of the amplifying and oscillating functions mentioned in this chapter can be found in encapsulated integrated circuits. (Discrete components are used, however, where the power levels are very high.) Nevertheless, it is often necessary to connect external components to these ICs for proper operation. This makes a working knowledge of each part of an amplifier a good preparation for using ICs.

SUMMARY

1. Bipolar junction transistors may be either PNP or NPN type.
2. The emitter-base and collector-base junctions of transistors can be checked with an ohmmeter (provided that the ohmmeter midranges are used so as not to damage the transistors).
3. In normal amplifying action the emitter-base junction is always forward-biased and the collector-base junction is always reverse-biased.
4. In either a CE or CB connection, majority carriers are swept across the narrow, lightly doped base to the collector. The resulting collector current is controlled by a relatively small base current that is the result of the forward-biased, emitter-base junction.
5. The forward current transfer ratio in a CE configuration is called h_{fe} or β (beta) and is the ratio of the collector current to the base current; in a CB, configuration, it is called h_{fb} or α (alpha) and is the ratio of the collector current to the emitter current.
6. The ratio β ranges from 10 for power transistors up to 500 for small-signal transistors; α is usually only slightly less than one.
7. Alpha can be obtained from beta using:

$$\alpha = \frac{\beta}{1 + \beta}$$

8. A CE amplifier involves superimposing an ac signal current on a dc base bias current. The corresponding variations in collector current (magnified by an amount equal to β) cause an ac voltage to be developed across the collector resistor. This output voltage is 180° out of phase with the input signal.

9. A CE amplifier can be operated from a single power supply and provides both current and voltage gain to give the largest power gain.

10. The power gain can be calculated in decibels using

$$\text{dB} = 10 \log \frac{P_o}{P_i}$$

11. Most practical CE amplifiers use emitter self-bias and a voltage-divider, forward-bias circuit to provide bias stabilization. This reduces the problems arising from temperature increases.

12. An amplifier's frequency response curve shows a drop in voltage gain at low frequencies due to circuit capacitances and at high frequencies due to transistor junction capacitances.

13. The voltage gain of an amplifier that uses voltage feedback is given by

$$A_{vf} = \frac{A_v}{1 - kA_v}$$

In the above equation:

a. If kA_v is *negative,* the gain is *reduced* due to *negative* feedback.

b. If kA_v is *positive,* the gain is *increased* and may cause oscillations.

14. The beneficial effects of negative feedback include a reduction in distortion, greater stability, and an increase in bandwidth.

15. Negative feedback can be introduced in a CE amplifier by means of an unbypassed resistor in the emitter lead.

16. The frequency of oscillation of a Wien bridge oscillator can be varied by changing the two capacitors in the frequency selective network;

$$f = \frac{1}{2\pi RC}$$

At this point the output is in phase with the input, thereby sustaining oscillations due to positive feedback.

SELF-EXAMINATION

Answer true or false.
(Answers at back of book)

29-1. The symbol for a *PNP* transistor has the arrow on the collector pointing inward. ________

29-2. The two-diode equivalent circuit for an *NPN* transistor has the base-emitter diode pointing out and the collector-base diode pointing in.

29-3. The base, emitter, and collector regions must be manufactured with exactly the same thicknesses. ________

29-4. For an *NPN* transistor an ohmmeter connected from collector to base, positive at the collector, will have a high reading. ________

29-5. A transistor manufactured using the gaseous diffusion technique has similar sizes of collector and emitter. ________

29-6. The reason for connecting the collector to the metal case in a power transistor is to help dissipate heat. ________

29-7. The location of the base terminal in a signal-type transistor is always in the middle between the collector and emitter. ________

29-8. In normal amplifying action, both the emitter-base junction and the collector-base junction must be forward-biased. ________

29-9. If the base is open in a CE amplifier, the current that flows in the collector is the leakage current called I_{CEO}. ________

29-10. A larger emitter-base forward bias causes more minority carriers to flow from the emitter to the collector for either the *NPN* or *PNP*. ________

29-11. The characteristic curves for a common emitter consist of graphs of the collector current against the collector-to-emitter voltage with the base current as the control. ________

29-12. If, in a common emitter, the collector current is 15 mA when the base current is 50 μA, the transistor's β is 30. ________

29-13. If a transistor's α is 0.99, the base current will be 0.1 mA when the emitter current is 10 mA. ________

29-14. The collector characteristic curves for a common-base are essentially horizontal lines with the emitter and collector currents practically equal. ________

29-15. If a transistor's $\alpha = 0.99$, its $\beta = 101$. ________

29-16. If a transistor's $\beta = 99$, its $\alpha = 0.99$. ________

29-17. In a CE amplifier the base dc bias current must be at least equal to the amplitude of the input ac signal current. ________

29-18. In a CE amplifier, as the base input current increases, the collector current also increases. ________

29-19. A single supply can be used in a CE amplifier by connecting a resistor from the collector supply voltage to the base. ________

29-20. Input and output capacitors are used for blocking direct current. ________

29-21. Phase inversion occurs between the input and output voltages in a common emitter. ________

29-22. A CE amplifier provides current and power gain but no voltage gain. ________

29-23. A CB amplifier provides both current and voltage gain. ________

29-24. A bypassed emitter resistor provides bias stabilization due to dc negative feedback. ________

29-25. Removing the bypass capacitor in a CE amplifier introduces ac negative feedback but still provides dc bias stabilization. ________

29-26. A frequency response curve shows how constant an amplifier's output is over a band of frequencies. ________

29-27. A reduction in gain at low and high frequencies is due to capacitive effects either in the amplifier circuit or within the transistor itself. ________

29-28. The CE amplifier is noted for its high power gain and the CB amplifier for its high-frequency capability. ________

29-29. Negative feedback is used in an amplifier to increase its voltage gain. ________

29-30. The advantages of negative feedback include reduced distortion and reduced bandwidth. ________

29-31. An oscillator involves positive feedback in which a signal is fed back from the output, in proper phase, to the input. ________

29-32. Oscillators can be made to produce outputs at different frequencies by varying components in the feedback network. ________

REVIEW QUESTIONS

1. Draw the symbol for an *NPN* transistor and its two-diode equivalent circuit, indicating the readings of ohmmeters connected across the three pairs of terminals for each ohmmeter polarity.

2. What is the advantage of the collector region's being physically much larger in area than the emitter?

3. How is a *PNP* type distinguished from an *NPN* type in many foreign transistors?

4. a. In what manner are the emitter-base and collector-base junctions biased in a *PNP* transistor when used in an amplifier?

b. Does it matter whether it is used in a collector-base or a common emitter?

5. a. Why are the leakage currents of a transistor very temperature-dependent?

b. How are they designated?

6. Explain how it is possible for current to flow from the emitter to the collector if the collector base junction is reverse-biased.

7. Why does an increase in base current cause an increase in collector current in a common emitter?

8. Of what is β a measure? How is it related to α?

9. a. Sketch a CE amplifier circuit with a *PNP* transistor and indicate clearly the polarities of the base and collector supply voltages.

b. Do the same with a CB amplifier, also using a *PNP* transistor.

c. Do you see why a CE amplifier is preferred over a CB as far as supply voltages are concerned?

10. Explain in your own words why phase inversion occurs between the input and output voltages in a CE amplifier.

11. Compare the magnitudes of the input currents in a CB and CE (for the same emitter-base voltage) and suggest which has the higher input resistance.

12. Distinguish between the following notations:

a. i_c, i_C, and I_C

b. $(I_E = I_C + I_B)$ and $(i_E = i_C + i_B)$

c. V_{BB} and V_{CC}

13. Describe in your own words the origin of the name "transistor."

14. Describe how you would experimentally obtain a transistor's

a. ac beta.

b. dc alpha.

15. If A_i is less than one in a CB amplifier, how is it possible to get any power gain?

16. a. Sketch the circuit for a *PNP* CE amplifier with emitter and voltage divider bias.

b. Indicate the polarities of the dc voltages across R_2, R_e, and the base-emitter junction.

c. Describe in your own words how this circuit is stabilized against any increases in ambient temperature.

17. In the circuit of Question 16, describe what happens to the collector current if the transistor is replaced with one having a much lower value of β.

18. What factors are responsible for the drop-off in voltage gain at low and high frequencies?

19. Why are f_1 and f_2 referred to as half-power frequencies in a frequency response curve?

20. a. Describe why negative feedback, which reduces the output voltage, *widens* the bandwidth of an amplifier.

b. Sketch the typical frequency response curves of an amplifier before and after adding negative feedback.

21. Why do you think kA_v is referred to as *loop gain* in Eq. 29-17?

22. Refer to Fig. 29-17. What happens to the following characteristics as the variable contact on R_e is moved upward?

a. Voltage gain.

b. Output distortion.

c. Frequency response.

d. Input resistance.

23. a. If an amplifier has a positive feedback loop, what condition can cause oscillations?

b. What are oscillations?

24. a. In Fig. 29-18, explain why V_o is in phase with the input to the first stage.

b. If a second stage was not used, what would the feedback network have to do?

25. Assume that the circuit of Fig. 29-18 is providing oscillations.

a. Describe what happens to V_o, in quantity and quality, as the moving contact on R_{e1} is moved downward.

b. What extreme condition could result?

26. a. What effect will a reduction of capacitance have on the frequency of oscillation in the Wien bridge oscillator?

b. Could this have been achieved by a similar reduction of resistance?

c. Why is it preferable to reduce the capacitance and not the resistance?

PROBLEMS

(Answers to odd-numbered problems at back of book)

29-1. Determine the unknown current given the following:

a. $I_B = 300\ \mu A$, $I_C = 15$ mA, $I_E = ?$

b. $I_E = 10.1$ mA, $I_C = 9.9$ mA, $I_B = ?$

c. $I_B = 250\ \mu A$, $I_E = 25$ mA, $I_C = ?$

29-2. Determine the unknown current given the following:

a. $I_E = 6.4$ A, $I_C = 5.9$ A, $I_B = ?$

b. $I_B = 100$ mA, $I_C = 2.5$ A, $I_E = ?$

c. $I_E = 250$ mA, $I_B = 2.5$ mA, $I_C = ?$

29-3. For the information given in the three parts of Problem 29-1, determine:

a. The value of β for each part.

b. The value of α for each part.

29-4. For the information given in the three parts of Problem 29-2, determine:

a. The value of β for each part.

b. The value of α for each part.

29-5. For the CE characteristic curves of a silicon NPN transistor given in Fig. 29-19, determine $\beta_{dc} = h_{FE}$ at the following points:

a. $I_B = 20\ \mu$A, $V_{CE} = 15$ V

b. $I_B = 100\ \mu$A, $V_{CE} = 15$ V

c. $I_B = 180\ \mu$A, $V_{CE} = 15$ V

d. Sketch a graph of β_{dc} versus I_C, for the three calculated values of β_{dc}.

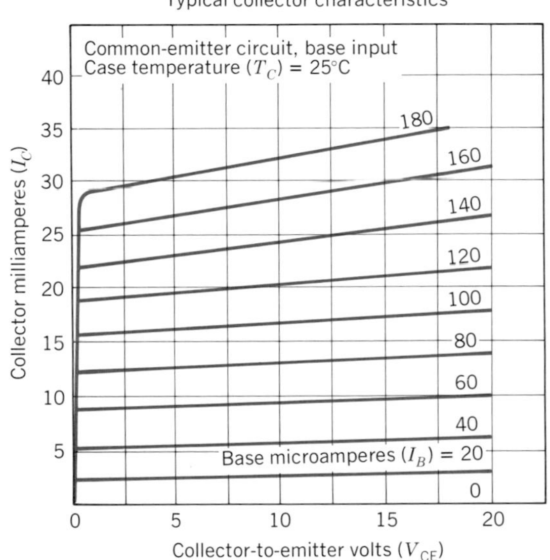

FIGURE 29-19
Collector characteristics for Problems 29-5 through 29-8.

29-6. Repeat Problem 5 using $V_{CE} = 2.5$ V instead of 15 V. Sketch the curve of β_{dc} versus I_C on the same axes as in Problem 29-5 and comment on the results.

29-7. Using the CE characteristic curves in Fig. 29-19 determine $\beta_{ac} = h_{fe}$ at $V_{CE} = 15$ V for various average collector currents from 2 to 35 mA. Sketch a curve of β_{ac} versus I_C.

29-8. Repeat Problem 29-7 using $V_{CE} = 2.5$ V instead of 15 V.

29-9. A transistor's $\beta = 150$. If its base current is 0.15 mA, calculate the collector and emitter currents.

29-10. A transistor's $\beta = 200$. If its collector current is 45 mA, calculate the base and emitter currents.

29-11. The emitter current in a transistor is 3 A. If the transistor's $\beta = 20$, calculate the base and collector currents.

29-12. Find the value of α for the transistors in:

a. Problem 29-9

b. Problem 29-10

c. Problem 29-11.

29-13. A transistor in the CB connection has its emitter current increased from 15 to 20 mA. This caused an increase in its base current from 0.32 to 0.48 mA.

a. Calculate α_{ac}.

b. Determine β_{ac} in two ways.

29-14. A transistor in the CE connection has its base current increased from 15 to 20 μA, causing an increase in the collector current from 6 to 8 mA.

a. Calculate h_{fe}.

b. Determine h_{fb} in two ways.

29-15. a. A germanium transistor's collector current is 51 mA when the base current is 0.4 mA. If $\beta_{dc} = 125$, calculate the transistor's collector cutoff current I_{CEO}.

b. Why would this be an impractical method for determining a silicon transistor's I_{CEO}?

29-16. A silicon power transistor has $\beta_{dc} = 180$ and $I_{CEO} = 40$ mA. Calculate the collector current for the following conditions:

a. $I_B = 0$.

b. $I_B = 10$ mA.

c. $I_B = 20$ mA.

29-17. A suitably biased CE amplifier circuit, as in Fig. 29-12*b*, has a transistor with $\beta = 125$ and $R_C = 2.2$ kΩ. An input voltage of 50 mV peak to peak causes a 40-μA peak-to-peak variation in the base current. Its 3-dB bandwidth is 15 kHz with a distortion of 4%. Calculate:

a. The peak-to-peak variation in the collector current.

b. The current gain, A_i

c. The peak-to-peak output voltage.

d. The voltage gain, A_v

e. The input resistance, $R_i = \dfrac{\text{input voltage}}{\text{base current}}$

f. The voltage gain, A_v, using Eq. 29-13. Compare with part (d).

g. The power gain, A_p, in dB.

29-18. Repeat Problem 17 using:

a. $R_C = 3.3$ kΩ.

b. $\beta = 100$ and $R_C = 1$ kΩ.

29-19. For the amplifier conditions in Problem 17, with $V_{BE} = 0.6$ V and $I_B = 40$ μA, determine:

a. V_{CE}.

b. V_{CB}.

c. The value of R_B if $V_{CC} = 30$ V.

29-20. Repeat Problem 29-19 using $R_C = 3.3$ kΩ.

29-21. To provide bias stabilization, a 100-Ω resistor (bypassed adequately with a capacitor) is placed in the emitter lead of the amplifier in Problem 29-17. Calculate the value of R_B to maintain $V_{BE} = 0.6$ V, $I_B = 40$ μA, with $V_{CC} = 30$ V. [Compare with Problem 19(c).]

29-22. If a voltage-divider, R_1–R_2, is used instead of R_B to provide forward bias for the amplifier in Problem 29-21, what voltage must exist across R_2?

29-23. An amplifier has a power output of 20 W and a voltage gain of 200 at a frequency of 1 kHz. When operated at the extremes of its −3-dB bandwidth calculate:

a. The power output of the amplifier.

b. The voltage gain of the amplifier.

c. The rms output current if the input voltage is 50 mV.

29-24. Repeat Problem 29-23 using −6 dB instead of −3 dB.

29-25. An amplifier has a voltage gain of 150. Calculate the new voltage gain for the following conditions:

a. 5% negative feedback introduced.

b. 10% negative feedback introduced.

c. 0.5% positive feedback introduced.

29-26. Repeat Problem 29-25 with voltage gain = 100.

29-27. For the amplifier in Problem 29-17 assume a 100-Ω unbypassed emitter resistor is added as in Fig. 29-17. Calculate:

a. The feedback factor, k.

b. The new voltage gain.

c. The new input resistance.

d. The new bandwidth.

e. The new distortion.

f. The gain-bandwidth product before and after adding the 100-Ω resistor.

g. The current gain, A_i

h. The new power gain, in dB.

29-28. Repeat Problem 29-27 using R_C = 3.3 kΩ.

29-29. Repeat Problem 29-27 assuming a 1-kΩ load resistor, R_L, is added, as in Fig. 29-17.

29-30. If C = 50 pF is the lowest practical value to use in the Wien bridge oscillator circuit of Fig. 29-18, determine the value of R and the maximum value of C to provide a range of frequencies from 20 to 200 kHz.

29-31. a. Determine the frequency of oscillation of a Wien bridge oscillator with R = 5.1 kΩ and C = 500 pF.

b. If V_2 in Fig. 29-18 is 200 mV, what is the output of the oscillator, V_o?

c. What must be the minimum voltage gain of the two-stage CE amplifier in Fig. 29-18?

APPENDIX

APPENDIX A-1

COMMON SYMBOLS USED IN SCHEMATIC DIAGRAMS

Conductor or wire

Crossing conductors not connected

Connected conductors

Shielded single conductor

Single cell

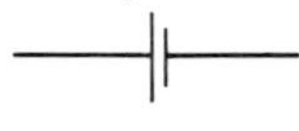

Multiple cell battery

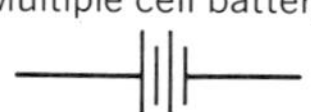

Alternating current source

Momentary contact push button

Single-pole single-throw switch, SPST

Single-pole double-throw switch, SPDT

Double-pole double-throw switch, DPDT

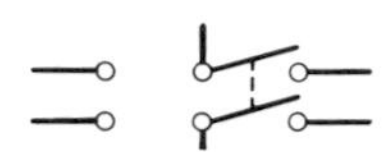

Incandescent lamp

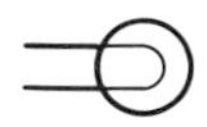

Neon lamp

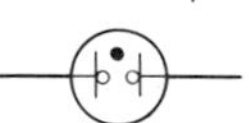

Buzzer

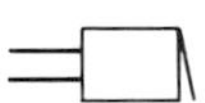

Bell

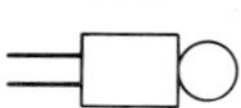

Loudspeaker

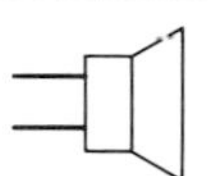

Motor

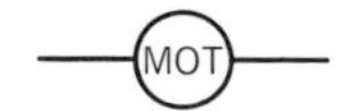

Generator

Fuse

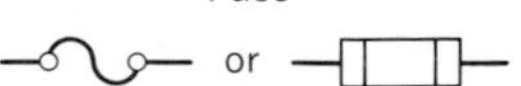

Circuit breaker

Ground (earth)

Chassis

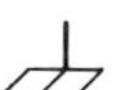

Common return path at same potential

Antenna

Microphone

Nonpolarized connector (plug)

Resistor (fixed)

Adjustable resistor (rheostat)

Potentiometer

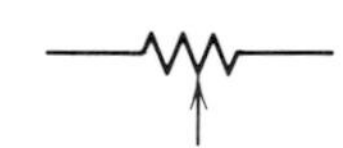

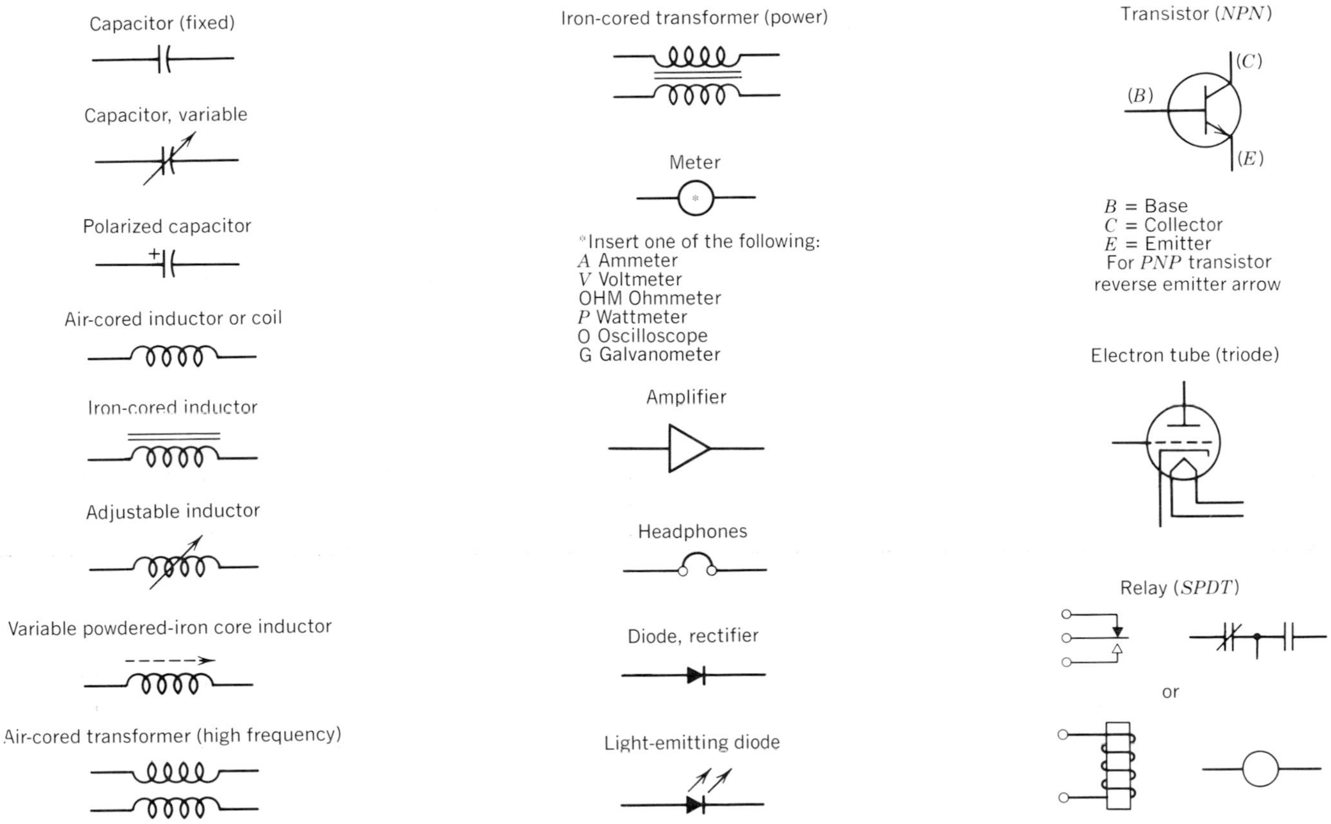

Capacitor (fixed)

Capacitor, variable

Polarized capacitor

Air-cored inductor or coil

Iron-cored inductor

Adjustable inductor

Variable powdered-iron core inductor

Air-cored transformer (high frequency)

Iron-cored transformer (power)

Meter

*Insert one of the following:
A Ammeter
V Voltmeter
OHM Ohmmeter
P Wattmeter
O Oscilloscope
G Galvanometer

Amplifier

Headphones

Diode, rectifier

Light-emitting diode

Transistor (*NPN*)

(*C*)
(*B*)
(*E*)

B = Base
C = Collector
E = Emitter
For *PNP* transistor reverse emitter arrow

Electron tube (triode)

Relay (*SPDT*)

or

APPENDIX A-2

STANDARD RESISTANCE VALUES FOR COMMERCIAL RESISTORS

20% Tolerance	10% Tolerance	5% Tolerance
10	10	10
		11
	12	12
		13
15	15	15
		16
	18	18
		20
22	22	22
		24
	27	27
		30

20% Tolerance	10% Tolerance	5% Tolerance
33	33	33
		36
	39	39
		43
47	47	47
		51
	56	56
		62
68	68	68
		75
	82	82
		91

Note: Commercial resistors are available also in submultiples, such as 2.2 Ω, 2.4 Ω, etc., as well as in multiples 220 Ω, 2.2 kΩ, 24 kΩ, etc.

APPENDIX A-3

AWG CONDUCTOR SIZES AND METRIC EQUIVALENTS

Gauge	Dia. (mil)	Resistance (Ω/1000 ft)	Dia. (mm)	Resistance (Ω/km)
0000	460	0.049	11.68	0.160
000	409.6	0.062	10.40	0.203
00	364.8	0.078	9.266	0.255
0	324.9	0.098	8.252	0.316
1	289.3	0.124	7.348	0.406
2	257.6	0.156	6.543	0.511
3	229.4	0.197	5.827	0.645
4	204.3	0.248	5.189	0.813
5	181.9	0.313	4.620	1.026
6	162	0.395	4.115	1.29
7	144.3	0.498	3.665	1.63
8	128.5	0.628	3.264	2.06
9	114.4	0.792	2.906	2.59
10	101.9	0.999	2.588	3.27
11	90.7	1.26	2.30	4.10
12	80.8	1.59	2.05	5.20
13	72	2	1.83	6.55
14	64.1	2.52	1.63	8.26
15	57.1	3.18	1.45	10.4
16	50.8	4.02	1.29	13.1
17	45.3	5.06	1.15	16.6
18	40.3	6.39	1.02	21.0
19	35.9	8.05	0.912	26.3
20	32	10.1	0.813	33.2

Gauge	Dia. (mil)	Resistance (Ω/1000 ft)	Dia. (mm)	Resistance (Ω/km)
21	28.5	12.8	0.723	41.9
22	25.3	16.1	0.644	52.8
23	22.6	20.3	0.573	66.7
24	20.1	25.7	0.511	83.9
25	17.9	32.4	0.455	106
26	15.9	41	0.405	134
27	14.2	51.4	0.361	168
28	12.6	64.9	0.321	213
29	11.3	81.4	0.286	267
30	10	103	0.255	337
31	8.9	130	0.227	425
32	8	164	0.202	537
33	7.1	206	0.180	676
34	6.3	261	0.160	855
35	5.6	329	0.143	1071
36	5	415	0.127	1360
37	4.5	523	0.113	1715
38	4	655	0.101	2147
39	3.5	832	0.090	2704
40	3.1	1044	0.080	3422

APPENDIX A-4

SOLVING LINEAR EQUATIONS USING DETERMINANTS

The process of solving a set of linear equations in two or more variables can be reduced to a purely mechanical process.

This process involves writing the coefficients of the unknowns in rows and columns in what is known as a *determinant*. The following is an example of a 2 × 2 determinant:

$$\begin{vmatrix} 4 & -5 \\ 3 & 7 \end{vmatrix}$$

A determinant can be "evaluated" as follows.

$$\begin{vmatrix} a_1 & b_1 \\ a_2 & b_2 \end{vmatrix} = a_1 \times b_2 - a_2 \times b_1$$

Thus
$$\begin{vmatrix} 4 & -5 \\ 3 & 7 \end{vmatrix} = 4 \times 7 - 3 \times (-5)$$
$$= 28 - (-15)$$
$$= 28 + 15 = \mathbf{43}$$

Consider two equations in x and y with *known* coefficients, written in the following form:

$$a_1x + b_1y = c_1$$
$$a_2x + b_2y = c_2$$

Cramer's rule gives the solution for x and y in terms of determinants:

$$x = \frac{\begin{vmatrix} c_1 & b_1 \\ c_2 & b_2 \end{vmatrix}}{\begin{vmatrix} a_1 & b_1 \\ a_2 & b_2 \end{vmatrix}} \qquad y = \frac{\begin{vmatrix} a_1 & c_1 \\ a_2 & c_2 \end{vmatrix}}{\begin{vmatrix} a_1 & b_1 \\ a_2 & b_2 \end{vmatrix}}$$

Note that the *denominator* determinant is the same for both x and y and is formed by writing down the coefficients of x and y in the same order as they appear in the original equations.

The *numerator* determinant is formed by repeating the denominator determinant except that the coefficients for the variable being sought are replaced by the constants. Thus, if we are solving for x, we replace a_1 and a_2 by c_1 and c_2, respectively. When solving for y, we replace b_1 and b_2 by c_1 and c_2, respectively.

EXAMPLE A4-1

Solve the following equations from Example 10-1.

$$2.5I_B + 2.0I_G = 13.2$$
$$2.0I_B + 2.1I_G = 14.5$$

SOLUTION

By Cramer's rule:

$$I_B = \frac{\begin{vmatrix} 13.2 & 2.0 \\ 14.5 & 2.1 \end{vmatrix}}{\begin{vmatrix} 2.5 & 2.0 \\ 2.0 & 2.1 \end{vmatrix}} = \frac{13.2 \times 2.1 - 2.0 \times 14.5}{2.5 \times 2.1 - 2.0 \times 2.0}$$
$$= \frac{27.72 - 29.0}{5.25 - 4.0}$$
$$= \frac{-1.28}{1.25} = \mathbf{-1.024}$$

$$I_G = \frac{\begin{vmatrix} 2.5 & 13.2 \\ 2.0 & 14.5 \end{vmatrix}}{1.25} = \frac{2.5 \times 14.5 - 2.0 \times 13.2}{1.25}$$
$$= \frac{36.25 - 26.4}{1.25}$$
$$= \frac{9.85}{1.25} = \mathbf{7.88}$$

Cramer's rule applies to any number of unknowns. We shall limit our application to a set of *three* linear equations, since a simple method of evaluating 3 × 3 determinants exists that does *not* apply to higher orders. Given a determinant:

$$\begin{vmatrix} a_1 & b_1 & c_1 \\ a_2 & b_2 & c_2 \\ a_3 & b_3 & c_3 \end{vmatrix}$$

its value can be found by repeating the first two columns and making the indicated products:

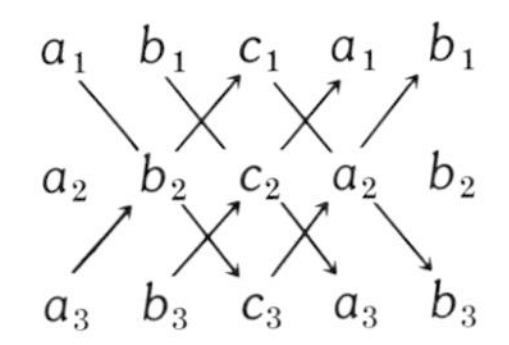

$$\text{Value} = a_1 \times b_2 \times c_3 + b_1 \times c_2 \times a_3 + c_1 \times a_2 \times b_3 - a_3 \times b_2 \times c_1 - b_3 \times c_2 \times a_1 - c_3 \times a_2 \times b_1$$

EXAMPLE A4-2

Solve the following equations from Example 10-2 for I_1:

$$3I_1 + 0I_2 + 2I_3 = 10$$
$$2I_1 - 4I_2 + 9I_3 = 0$$
$$0I_1 + 9I_2 - 4I_3 = -15$$

SOLUTION

$$I_1 = \frac{\begin{vmatrix} 10 & 0 & 2 \\ 0 & -4 & 9 \\ -15 & 9 & -4 \end{vmatrix}}{\begin{vmatrix} 3 & 0 & 2 \\ 2 & -4 & 9 \\ 0 & 9 & -4 \end{vmatrix}}$$

Numerator:

$$\begin{matrix} 10 & 0 & 2 & 10 & 0 \\ 0 & -4 & 9 & 0 & -4 \\ -15 & 9 & -4 & -15 & 9 \end{matrix}$$

$$\begin{aligned} \text{Value} &= 10 \times (-4) \times (-4) + 0 \times 9 \times (-15) + 2 \times 0 \times 9 \\ &\quad -(-15) \times (-4) \times 2 - 9 \times 9 \times 10 - (-4) \times 0 \times 0 \\ &= 160 + 0 + 0 - 120 - 810 - 0 \\ &= 160 - 930 = \mathbf{-770} \end{aligned}$$

Denominator:

$$\begin{matrix} 3 & 0 & 2 & 3 & 0 \\ 2 & -4 & 9 & 2 & -4 \\ 0 & 9 & -4 & 0 & 9 \end{matrix}$$

$$\begin{aligned} \text{Value} &= 3 \times (-4) \times (-4) + 0 + 2 \times 2 \times 9 \\ &\quad -0 - 9 \times 9 \times 3 - 0 \\ &= 48 + 36 - 243 = \mathbf{-159} \end{aligned}$$

$$I_1 = \frac{-770}{-159} = \mathbf{4.84}$$

For practice, solve for I_2 (−2.67) and I_3 (−2.26).

APPENDIX A-5

EFFECT OF ANGLE ON THE VOLTAGE INDUCED IN A MOVING CONDUCTOR

Consider a conductor moving at velocity v through a magnetic field B at some angle θ with the field, as shown in Fig. A5-1

After a period of time Δt, the conductor has moved a distance Δs. But the only part of this distance that contributes anything to the induced voltage is the horizontal component, $\Delta s'$, which actually cuts the field. The relation between $\Delta s'$ and Δs obviously has something to do with the angle θ. This relation (see Appendix A-6), is given by

$$\frac{\Delta s'}{\Delta s} = \sin\theta$$

For example, if θ is a 30° angle, $\sin\theta = 0.5$. This simply means that $\Delta s'$ is one-half the distance Δs. Thus

$$\Delta s' = \Delta s \times \sin\theta$$

But

$$v_{\text{ind}} = B \times l \times \frac{\Delta s'}{\Delta t}$$

$$= B \times l \times \frac{\Delta s}{\Delta t} \times \sin\theta$$

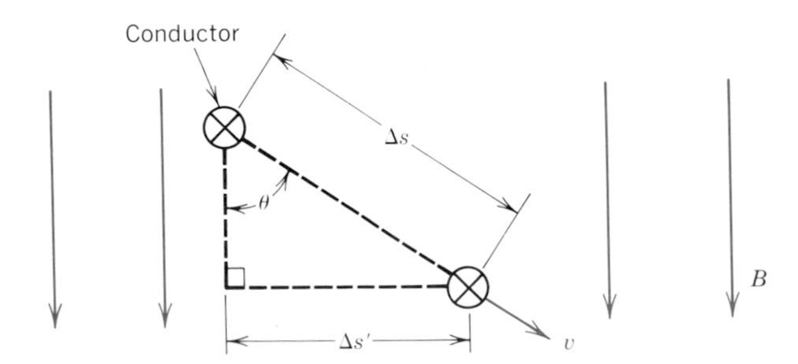

FIGURE A5-1
Conductor being moved at velocity v, through a magnetic field B at an angle θ.

And

$$\frac{\Delta s}{\Delta t} = v$$

Thus

$$\boldsymbol{v_{\text{ind}} = Blv \sin\theta = V_m \sin\theta} \qquad (14\text{-}5)$$

APPENDIX A-6
TRIGONOMETRIC RATIOS

It is found that in any right-angled triangle, no matter what its size, the ratio of one side to another has a definite value, dependent only upon the angle θ.

Consider the triangle shown in Fig. A6-1. The side opposite the right angle is called the hypotenuse. As far as the angle θ is concerned, side a is the opposite side and side b is the adjacent side.

A name is given to the ratio of the opposite side to the hypotenuse in terms of the angle θ. It is called sine θ.

$$\sin \theta = \frac{\text{opposite side}}{\text{hypotenuse}} = \frac{a}{c}$$

where $\sin \theta$ is an abbreviation for sine θ.

Similarly, two other common ratios are named:

$$\cos \theta = \text{cosine } \theta = \frac{\text{adjacent side}}{\text{hypotenuse}} = \frac{b}{c}$$

$$\tan \theta = \text{tangent } \theta = \frac{\text{opposite side}}{\text{adjacent side}} = \frac{a}{b}$$

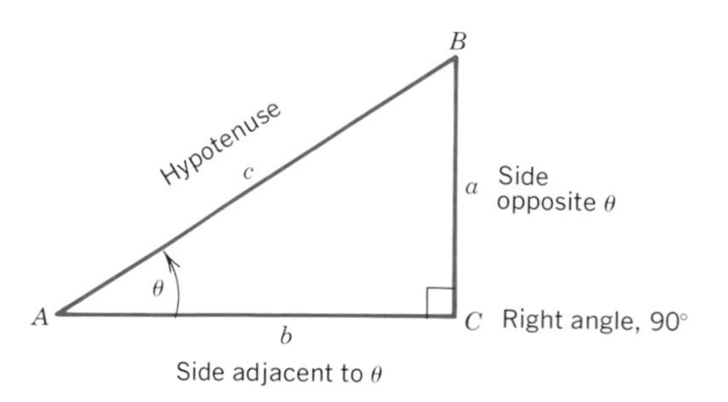

FIGURE A6-1
Relation of sides to θ in a right triangle.

Each of these three ratios, called trigonometric ratios or functions, has its own numerical value for any given angle. Trigonometric tables give values of these ratios for angles between 0 and 90°. Alternatively, they can be obtained on a scientific calculator similar to the one introduced in Chapter 1. Table A6-1 shows values for some common angles.

This table can be interpreted as follows: If, in a right triangle, θ is 30°, the side opposite θ is only half the length of the hypotenuse, since sin 30° = 0.5. Similarly, if θ is 45°, the adjacent and opposite sides are equal in length, since tan 45° = 1.

Angles between 0 and 90° are called *acute* angles. All their trigonometric ratios are positive. It is possible also to consider the trigonometric ratios of angles from 90 to

TABLE A6-1
Trigonometric Values of Some Common Angles

Angle, θ Degrees	Angle, θ Radians	Sin θ	Cos θ	Tan θ
0	0	0	1.000	0
30	$\frac{\pi}{6}$	0.500	0.866	0.577
45	$\frac{\pi}{4}$	0.707	0.707	1.000
60	$\frac{\pi}{3}$	0.866	0.500	1.732
90	$\frac{\pi}{2}$	1.000	0	∞

360°, called *obtuse* angles. Some of these ratios are negative and do not specifically refer to the side of a triangle. That is, although we defined the trigonometric ratios in terms of the sides of a right triangle, alternative definitions include angles of any size. For angles greater than 360°, the trigonometric ratios are a repeat of those from 0 to 360°.

A scientific calculator gives the trigonometric ratios of *any* angle, given in radians or degrees, both in magnitude and sign (positive or negative). Thus

$$\sin 270° = -1$$

$$\cos 390° = 0.866$$

$$\tan 120° = -1.732$$

$$\sin 2 \text{ rad} = 0.909$$

INVERSE TRIGONOMETRIC FUNCTIONS

When the sine (or any trigonometric ratio) of an angle is known, it is possible to find the angle itself using the *inverse* function.

For example, given $\sin \theta = 0.5$. We can say that θ is "an angle whose" sine is 0.5. Mathematically, we express this statement for θ using the following notation:

$$\theta = \text{arc} \sin 0.5$$

where "arc" stands for "an angle whose."

Another notation used is

$$\theta = \sin^{-1} 0.5$$

where $\sin^{-1}$ also means "an angle whose."

Both of these notations refer to the *inverse sine function*. Note that $\sin^{-1}$ does *not* refer to any reciprocal.

Since, for this example, we know that the sine of 30° is 0.5, then

$$\theta = \text{arc} \sin 0.5 = \sin^{-1} 0.5 = 30°$$

In general, we can obtain the value of an inverse trigonometric function using a scientific calculator.

For $\sin^{-1}$ 0.5, enter 0.5 into the calculator, press INV, press SIN. The result is 30° or 0.5236 rad. (Some calculators may have a SIN^{-1} button.) Note that the angle displayed is in the first quadrant. That is, sin 150° is also 0.5.

Other examples of inverse trigonometric functions are

$$\theta = \text{arc} \tan 1 = \tan^{-1} 1 = 45°$$

$$\theta = \text{arc} \cos 0.8 = \cos^{-1} 0.8 = 36.87°$$

APPENDIX A-7

FACTORS AFFECTING THE AMPLITUDE OF A GENERATOR'S SINE WAVE

Refer to Fig. 14-9.

$$\text{Voltage induced in each conductor} = Blv \sin\theta \quad (14\text{-}5)$$

For two conductors A and B in series, $v_{\text{ind}} = 2\,Blv \sin\theta$.

We need an expression for the velocity, v:

$$v = \frac{\text{distance traveled}}{\text{time taken}}$$

$$= \frac{\text{distance traveled by one conductor in one revolution}}{\text{time for one revolution}}$$

Let b = diameter of the coil.

$$\text{distance traveled} = \text{circumference of circle of diameter } b$$
$$= \pi b$$

Let n = the speed of rotation of the coil in rev/s.

$$\text{time for one revolution} = \frac{1}{n}\ \text{s}$$

Therefore,
$$v = \frac{\pi b}{1/n} = \pi bn$$

and
$$\begin{aligned} v_{\text{ind}} &= 2\,Blv \sin\theta \\ &= 2\,Bl\pi bn \sin\theta \\ &= 2\pi nBA \sin\theta \end{aligned}$$

since bl = area of the coil, A.

If the coil has N turns, the total voltage induced by the coil is

$$\boldsymbol{v_{\text{ind}} = 2\pi nBAN \sin\theta = V_m \sin\theta} \quad (14\text{-}7)$$

APPENDIX A-8

INDUCED VOLTAGE IN TERMS OF FREQUENCY AND TIME

A sine wave of voltage can be expressed by the equation $v = V_m \sin \theta$. The angle θ can be expressed in terms of the frequency, f, which is a more readily known quantity, as follows.

Consider a generator with one pair of poles. We know that one cycle of voltage corresponds to one revolution or 360°. (This is also true for any number of pairs of poles if we use electrical degrees.)

The angle "swept out" or passed through by a coil generating a voltage of f cycles per second equals $360 f$ degrees every second.

If at $t = 0$, the angle (and voltage induced) is zero, then at some later time t, in seconds, the angle that the coil has passed through, $\theta = 360ft$ degrees. Thus the induced voltage is now given by

$$v = V_m \sin (360\,ft)^\circ$$

But $360^\circ = 2\pi$ radians.

Thus
$$v = V_m \sin 2\pi ft \qquad (15\text{-}3)$$

If we let $\omega = 2\pi f$ (the angular frequency in rad/s), then

$$\boldsymbol{v = V_m \sin \omega t} \qquad (15\text{-}4)$$

APPENDIX A-9

ALGEBRAIC SOLUTIONS FOR INSTANTANEOUS AND TRANSIENT CURRENTS AND VOLTAGES IN A SERIES *RL* CIRCUIT

The universal time-constant curves are called *exponential* curves. This is because the algebraic solutions for the current and coil voltage are given by

$$i = \frac{V}{R}(1 - e^{-t/\tau})$$
$$= I_f[1 - e^{-t/(L/R)}]$$

or

$$\boldsymbol{i = \frac{V}{R}[1 - e^{-(Rt/L)}]} \quad \textbf{amperes (A)} \quad \text{(A9-1)}$$

and

$$v_L = Ve^{-t/\tau}$$
$$= Ve^{-t/(L/R)}$$

or

$$\boldsymbol{v_L = Ve^{-(Rt/L)}} \quad \textbf{volts(V)} \quad \text{(A9-2)}$$

where: i is the instantaneous value of the current in amperes, A

v_L is the instantaneous value of the coil voltage in volts, V

V is the applied dc emf in volts, V

R is the circuit resistance in ohms, Ω

L is the circuit inductance in henrys, H

t is the time in seconds, s

e is the constant, 2.718 (the base of natural logs)

EXAMPLE A9-1

Solve Example 19-4 using Eq. A9-1; $L = 8$ H, $R = 400$ Ω, and $V = 20$ V. That is, find:

a. The current 46 ms after closing the switch.
b. How long it takes for the current to reach 27.5 mA.

SOLUTION

a. $$i = \frac{V}{R}[1 - e^{-(Rt/L)}] \quad \text{(A9-1)}$$

$$= \frac{20\text{ V}}{400\ \Omega}(1 - e^{-\frac{400 \times 46 \times 10^{-3}}{8}})$$

$$= 50\text{ mA }(1 - e^{-2.3})$$

The evaluation of $e^{-2.3}$ utilizes the e^x button on most scientific calculators. Enter 2.3; obtain the negative, −2.3; and press the e^x button. The result is 0.100. Therefore,

$$i = 50\text{ mA }(1 - 0.1)$$
$$= 50\text{ mA} \times 0.9 = \mathbf{45\ mA}$$

b. $$i = \frac{V}{R}(1 - e^{-Rt/L}) \quad \text{(A9-1)}$$

$$27.5\text{ mA} = 50\text{ mA}(1 - e^{-400t/8})$$
$$\frac{27.5}{50} = 1 - e^{-50t}$$
$$e^{-50t} = 1 - 0.55 = 0.45$$

Take the natural log, $\log_e = \ln$, of both sides:

$$\ln e^{-50t} = \ln 0.45$$
$$-50t \ln e = \ln 0.45$$

The natural log can be readily obtained on a scientific calculator. Enter 0.45 and press the ln x button. The result is $-0.7985 \approx -0.8$. Similarly, $\ln e = 1$.

$$-50t = -0.8$$
$$t = \frac{0.8}{50}\text{ s}$$
$$= 0.016\text{ s} = \mathbf{16\ ms}$$

These answers are the same as the graphical solutions in Example 19-4.

EXAMPLE A9-2

Solve Example 19-5 using Eqs. A9-1 and A9-2.

SOLUTION

a. $\tau = \frac{L}{R}$ (19-2)

$= \frac{30 \times 10^{-3}\text{H}}{2 \times 10^{3}\Omega} = 15 \times 10^{-6}\text{ s} = \mathbf{15\ \mu s}$

b. $v_L = Ve^{-Rt/L}$ (A9-2)

Initially, at $t = 0^+$,

$v_L = 50\text{ V} \times e^{-0}$

$= 50\text{ V} \times 1 = \mathbf{50\ V}$

c. $v_L = Ve^{-Rt/L}$ (A9-2)

At $t = 7.5\ \mu s$,

$v_L = 50\text{ V} \times e^{-(2\times10^3\times7.5\times10^{-6})/(30\times10^{-3})}$

$= 50\text{ V} \times e^{-0.5}$

$= 50\text{ V} \times 0.607 = \mathbf{30.4\ V}$

d. $I_f = \frac{V}{R}$

$= \frac{50\text{ V}}{2 \times 10^3\ \Omega} = \mathbf{25\ mA}$

e. $i = \frac{V}{R}(1 - e^{-Rt/L})$ (A9-1)

At $t = 45\ \mu s$,

$i = \frac{50\text{ V}}{2 \times 10^3\ \Omega}[1 - e^{-(2\times10^3\times45\times10^{-6})/(30\times10^{-3})}]$

$= 25\text{ mA} \times (1 - e^{-3})$

$= 25\text{ mA} \times (1 - 0.05)$

$= 25\text{ mA} \times 0.95 = \mathbf{23.8\ mA}$

f. $v_R = iR$

$= \frac{V}{R}(1 - e^{-Rt/L})R$

$= V(1 - e^{-Rt/L})$

$37.5\text{ V} = 50\text{ V}(1 - e^{-Rt/L})$

Therefore $e^{-Rt/L} = 1 - \frac{37.5\text{ V}}{50\text{ V}} = 0.25$

$-\frac{Rt}{L} = \ln 0.25 = -1.39$

Therefore $t = 1.39 \times \frac{L}{R}$

$= 1.39 \times 15\ \mu s = \mathbf{20.8\ \mu s}$

g. $t = 5 \times \frac{L}{R} = 5 \times 15\ \mu s = \mathbf{75\ \mu s}$

The exponential equations for decaying current and voltage in a series RL circuit with a direct current being interrupted are

$$\boldsymbol{i = I_f e^{-t/\tau} = \frac{V}{R} e^{-Rt/L}} \quad \textbf{amperes (A)} \quad \text{(A9-3)}$$

and

$$\boldsymbol{v_L = -Ve^{-t/\tau} = -Ve^{-Rt/L}} \quad \textbf{volts (V)} \quad \text{(A9-4)}$$

with symbols as before.

These equations show why the shape of the falling current (and v_R) and the decaying voltage v_L are given by curve (b) in the universal time-constant curves of Fig. 19-4.

EXAMPLE A9-3

Solve Example 19-7 using Eqs. A9-3 and A9-4.

SOLUTION

a. $i = \frac{V}{R}e^{-Rt/L}$ (A9-3)

$= \frac{12\text{ V}}{500\ \Omega}e^{-(500\times0.1\times10^{-3})/(20\times10^{-3})}$

$= 24\text{ mA} \times e^{-2.5}$

$= 24\text{ mA} \times 0.082 = \mathbf{1.97\ mA}$

b. $v_L = -Ve^{-Rt/L}$ (A9-4)

$= -12\text{ V} \times e^{-(500\times20\times10^{-6})/(20\times10^{-3})}$

$= -12\text{ V} \times e^{-0.5}$

$= -12\text{ V} \times 0.607 = \mathbf{-7.28\ V}$

c. $i = 24\text{ mA} \times e^{-(500\times0.2\times10^{-3})/(20\times10^{-3})}$

$= 24\text{ mA} \times e^{-5}$

$= 24\text{ mA} \times 0.0067 = \mathbf{0.16\ mA}$

$v_L = -12\text{ V} \times e^{-5}$

$= -12\text{ V} \times 0.0067 = \mathbf{0.0804\ V}$

These answers should be compared with those from Example 19-7.

APPENDIX A-10

DERIVATION OF THE INDUCTIVE REACTANCE FORMULA, X_L

Refer to Fig. A10-1.

The voltage induced across the coil will be a maximum, V_m, when the current is changing at its maximum rate. Thus

since $$v_L = L\left(\frac{\Delta i}{\Delta t}\right)$$

$$V_m = L\left(\frac{\Delta i}{\Delta t}\right)_{\max} = L \times \text{initial slope of } i \text{ curve}$$

Let us find an expression for the initial slope of the current curve:

Given $i = I_m \sin 2\pi ft, \qquad i = 0, \text{ when } t = 0$

Assume that a short time interval, Δt, elapses. Then

$$i = I_m \sin 2\pi f\ \Delta t$$

But if Δt is small (near zero), $2\pi f\ \Delta t$ is also small (near zero). For this condition, it can be shown that

$$\sin 2\pi f\ \Delta t \approx 2\pi f\ \Delta t \qquad \text{in radians}$$

$$\left[\text{Try} \qquad \sin 1° = \sin\left(\frac{\pi}{180}\text{ rad}\right) \approx \frac{\pi}{180}\right]$$

Thus, for a very small Δt, near zero,

$$i = I_m \sin 2\pi f\ \Delta t = I_m\, 2\pi f\ \Delta t$$

The initial slope of the i curve is given (see Fig. A10-1) by

$$\text{slope} = \frac{\text{rise}}{\text{run}} = \frac{i}{\Delta t} = \frac{I_m\, 2\pi f\ \Delta t}{\Delta t} = 2\pi f I_m$$

$$V_m = L \times \text{initial slope of the } i \text{ curve}$$

Therefore, $V_m = L \times 2\pi f I_m = 2\pi f L I_m$

The effective voltage across the coil (equal to the supply voltage) is

$$V = \frac{V_m}{\sqrt{2}} = \frac{2\pi f L I_m}{\sqrt{2}}$$

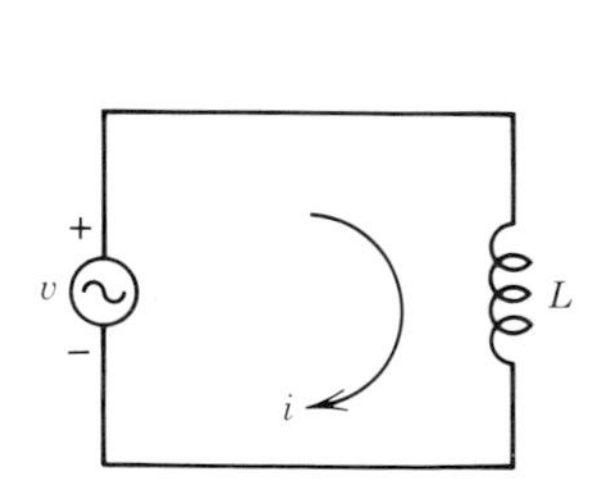

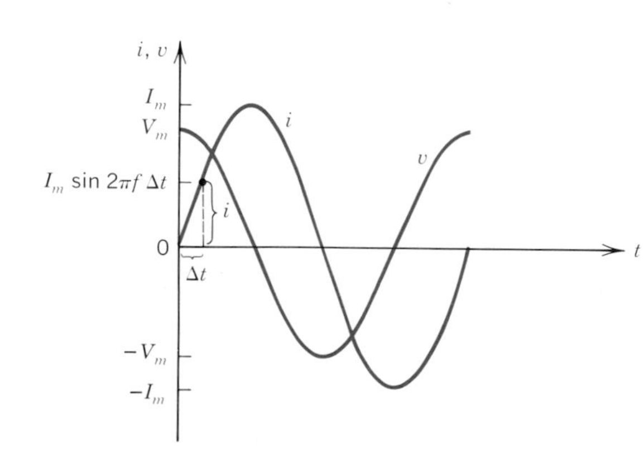

FIGURE A10-1
Initial slope of current curve equals $i/\Delta t$.

also, $$I = \frac{I_m}{\sqrt{2}}$$

Opposition to the ac current,

$$X_L = \frac{V}{I} = \frac{\dfrac{2\pi f L I_m}{\sqrt{2}}}{\dfrac{I_m}{\sqrt{2}}}$$

$$= \frac{2\pi f L I_m}{\sqrt{2}} \times \frac{\sqrt{2}}{I_m} = 2\pi f L$$

$$\mathbf{X_L = 2\pi f L = \omega L} \qquad (20\text{-}2)$$

APPENDIX A-11

DERIVATION OF THE CAPACITANCE EQUATION WITH AIR DIELECTRIC

The electric field intensity between the two parallel plates of a charged capacitor is given by

$$E = \frac{V}{d} \quad \text{volts/meter (V/m)}$$

where: E is the electric field strength in volts per meter, V/m

V is the potential difference across the plates in volts, V

d is the separation of the plates in meters, m

The electric field intensity is also given by

$$E = \frac{\sigma}{\epsilon_0} \quad \text{newtons/coulomb, (N/C or V/m)}$$

where: E is the electric field strength in newtons per coulomb, N/C (or V/m)

σ is the charge per unit area on the plates in coulombs per square meter, C/m^2

ϵ_0 is the permittivity of free space $= 8.85 \times 10^{-12}$ C^2/Nm^2

Since $\quad \sigma = \dfrac{\text{charge on one plate}}{\text{area of one plate}} = \dfrac{Q}{A}$

then $\quad E = \dfrac{\sigma}{\epsilon_0} = \dfrac{Q}{\epsilon_0 A} \qquad Q = \epsilon_0 AE$

Also, since $\quad E = \dfrac{V}{d}, \qquad V = dE$

thus $\quad C = \dfrac{Q}{V} = \dfrac{\epsilon_0 AE}{dE}$

and $\quad \mathbf{C = \epsilon_0 \dfrac{A}{d}}$

If the dielectric is a material other than air, having a dielectric constant K, the capacitance is

$$\mathbf{C = K \frac{\epsilon_0 A}{d}} \qquad (21\text{-}2)$$

APPENDIX A-12

ALGEBRAIC SOLUTIONS FOR INSTANTANEOUS CAPACITOR VOLTAGE AND CURRENT

We can calculate the capacitor voltage and current at any instant using the following exponential equations:

During charging:

$$v_C = V(1 - e^{-t/\tau}) = V(1 - e^{-t/RC}) \quad \textbf{volts (V)} \quad \text{(A12-1)}$$

and $$i_C = \frac{V}{R} e^{-t/\tau} = \frac{V}{R} e^{-t/RC} \quad \textbf{amperes (A)} \quad \text{(A12-2)}$$

During discharging:

$$v_C = V_{\text{initial}} \times e^{-t/RC} \quad \textbf{volts (V)} \quad \text{(A12-3)}$$

and $$i_C = \frac{V_{\text{initial}}}{R} \times e^{-t/RC} \quad \textbf{amperes (A)} \quad \text{(A12-4)}$$

where: v_C is the instantaneous value of the capacitor voltage in volts, V

i_C is the instantaneous value of the capacitor current in amperes, A

V is the applied dc voltage in volts, V

V_{initial} is the initial voltage of the charged capacitor in volts, V

R is the circuit resistance in ohms, Ω

C is the circuit capacitance in farads, F

t is the time in seconds, s

e is the constant 2.718 (the base of natural logs)

(These equations are the mathematical descriptions of the universal time constant curves in Fig. 22-8.)

Note the result of substituting $t = RC$ in Eq. A12-1:

$$\begin{aligned} v_C &= V(1 - e^{-t/RC}) \\ &= V\,(1 - e^{-RC/RC}) \\ &= V\,(1 - e^{-1}) \\ &= V\,(1 - 0.368) \\ &= 0.632\,V = 63.2\% \text{ of } V \end{aligned}$$

That is, after one time constant, $\tau = RC$, the capacitor voltage has reached approximately 63% of the final steady-state voltage. This is the usual definition of the time constant.

EXAMPLE A12-1

Refer to the circuit of Fig. A12-1 (as in Example 22-7). Calculate, using Eqs. A12-1 through A12-4:

a. The capacitor voltage 3.5 s after the switch is thrown to position 1.

b. The capacitor current at this instant.

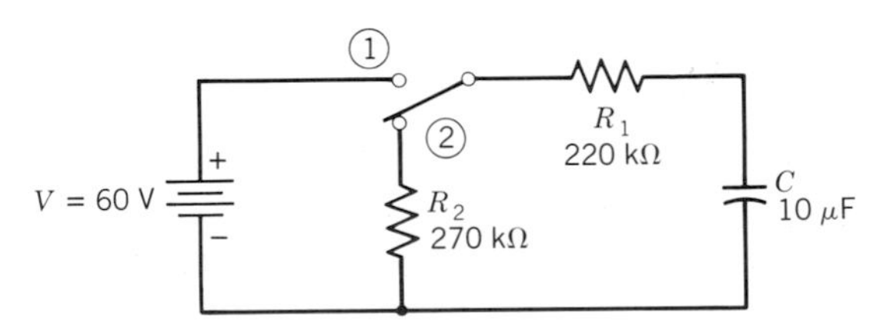

FIGURE A12-1
Circuit for Example A12-1.

c. The time required for the capacitor to discharge to 24 V after the switch is returned to position 2.

d. The capacitor current at this instant.

SOLUTION

a. Charging time constant

$$\tau_1 = R_1C \quad (22\text{-}3)$$

$$= 220 \times 10^3\ \Omega \times 10 \times 10^{-6}\ \text{F} = 2.2\ \text{s}$$

$$v_C = V(1 - e^{-t/RC}) \quad (\text{A12-1})$$

$$= 60\ \text{V}(1 - e^{-3.5\ \text{s}/2.2\ \text{s}})$$

$$= 60\ \text{V}(1 - e^{-1.591})$$

$$= 60\ \text{V}(1 - 0.204)$$

$$= 60\ \text{V} \times 0.796 = \mathbf{47.8\ V}$$

This compares favorably with 48 V using the graphical method. (See the solution to Example 22-7a.)

b. $$i_C = \frac{V}{R}e^{-t/RC} \quad (\text{A12-2})$$

$$= \frac{60\ \text{V}}{220 \times 10^3\ \Omega} \times e^{-3.5\ \text{s}/2.2\ \text{s}}$$

$$= 0.273\ \text{mA} \times e^{-1.591}$$

$$= 0.273\ \text{mA} \times 0.204 = \mathbf{0.056\ mA}$$

c. Discharging time constant

$$\tau_2 = (R_1 + R_2)C \quad (22\text{-}3)$$

$$= (220 \times 10^3 + 270 \times 10^3)\ \Omega \times 10 \times 10^{-6}\ \text{F}$$

$$= 4.9\ \text{s}$$

$$v_C = V_{\text{initial}} \times e^{-t/RC} \quad (\text{A12-3})$$

$$24\ \text{V} = 47.8\ \text{V} \times e^{-t/4.9\ \text{s}}$$

$$e^{-t/4.9\ \text{s}} = \frac{24\ \text{V}}{47.8\ \text{V}} = 0.502$$

Take the natural log of both sides:

$$\ln e^{-t/4.9\ \text{s}} = \ln 0.502$$

$$-\frac{t}{4.9\ \text{s}} = -0.689$$

Therefore, $t = 4.9\ \text{s} \times 0.689 = \mathbf{3.38\ s}$

This compares favorably with 3.43 s using the graphical method. (See the solution of Example 22-7b.)

d. $$i_C = -\frac{V_{\text{initial}}}{R} \times e^{-t/RC} \quad (\text{A12-4})$$

$$= -\frac{47.8\ \text{V}}{(220 + 270) \times 10^3\ \Omega} \times e^{-3.38\ \text{s}/4.9\ \text{s}}$$

$$= -0.098\ \text{mA} \times 0.502 = \mathbf{-0.049\ mA}$$

The negative sign indicates a discharging current.

APPENDIX A-13

OBSERVATION OF A TRANSFORMER CORE'S B-H HYSTERESIS LOOP USING AN INTEGRATING CIRCUIT

A *B-H* curve of a transformer's core can be observed on an oscilloscope using the circuit in Figure A13-1.

The current in the transformer primary develops a voltage across R_2 that is proportional to the magnetizing intensity, $H = \frac{NI}{l}$. This voltage, connected to the oscilloscope's horizontal input, deflects the beam proportionally to H, 60 times per second. Here N is the number of turns in the primary and l is the length of the magnetic circuit of the transformer's core. If we know the value of R_2 and the peak voltage, the horizontal axis can be calibrated in At/m.

On the transformer secondary,

$$i = C \frac{\Delta v_C}{\Delta t} \tag{22-1}$$

$$= C \frac{dv_C}{dt} \quad \text{(using the differential form)}$$

Thus $\quad i\,dt = C\,dv_C \quad$ and $\quad dv_C = \frac{1}{C} i\,dt$

or

$$v_C = \frac{1}{C} \int i\,dt$$

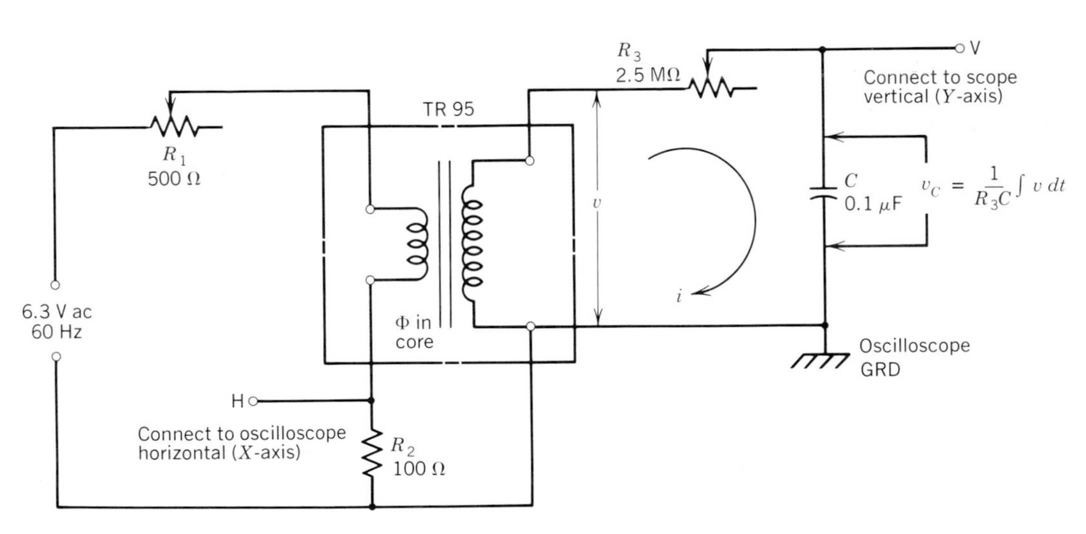

FIGURE A13-1
Circuit to observe a Lissajous figure of a transformer's *B-H* hysteresis loop.

The integral sign, $\int$, indicates the integrating process referred to in Section 22-8.1.

A minimum value of R_3 is approximately 1 MΩ. Since $X_C \approx 30$ kΩ, the secondary current is determined primarily by R_3:

$$i \approx \frac{v}{R_3} \quad \text{and} \quad v_C \approx \frac{1}{R_3C}\int v\,dt$$

But
$$v = N\frac{d\Phi}{dt} \tag{14-2}$$

Therefore,
$$\begin{aligned} v_C &\approx \frac{1}{R_3C}\int N\frac{d\Phi}{dt} \times dt \\ &= \frac{N}{R_3C}\int d\Phi \\ &= \frac{N\Phi}{R_3C} \end{aligned}$$

The capacitor voltage is the *integral* of the input voltage v. Since the secondary number of turns, N, is constant, as are R_3 and C, the capacitor voltage v_C is proportional to the core flux Φ. Also, since A, the cross-sectional area of the core, is constant, the flux density $B = \dfrac{\Phi}{A}$ is proportional to v_C.

Thus the oscilloscope beam is deflected vertically proportional to the flux density B, resulting in a Lissajous figure that is representative of the core's B-H characteristics.

If the horizontal and vertical axes are known for their volts/cm deflection, the two axes can be calibrated in AT/m and webers/m², respectively. Also, the core's permeability can be obtained quantitatively, using $\mu = \dfrac{B}{H}$.

APPENDIX A-14

DERIVATION OF THE CAPACITIVE REACTANCE FORMULA, X_C

The current will be a maximum, I_m, when the voltage is increasing at its maximum rate, at the origin (see Fig. A14-1). Thus

since $$i = C\left(\frac{\Delta v}{\Delta t}\right)$$

$$I_m = C\left(\frac{\Delta v}{\Delta t}\right)_{\max} = C \times \text{initial slope of the } v \text{ curve}$$

Given $v = V_m \sin 2\pi ft$, $\quad v = 0 \quad$ when $\quad t = 0$

Assume a short time interval, Δt.

$$v = V_m \sin 2\pi f\, \Delta t$$

But if Δt is small (near zero), $2\pi f\, \Delta t$ is also small (near zero). For this condition, it can be shown that $\sin 2\pi f\, \Delta t \approx 2\pi f\, \Delta t$ in radians. Thus, for a very small Δt, near zero,

$$v = V_m \sin 2\pi f\, \Delta t = V_m\, 2\pi f\, \Delta t$$

The maximum initial slope of v curve is given by

$$\text{slope} = \frac{\text{rise}}{\text{run}} = \frac{v}{\Delta t} = \frac{V_m\, 2\pi f\, \Delta t}{\Delta t} = 2\pi f V_m$$

Therefore, $$I_m = C \times \text{initial slope of the } v \text{ curve} = C \times 2\pi f V_m = 2\pi f C V_m$$

Effective capacitor current $$I = \frac{I_m}{\sqrt{2}} = \frac{2\pi f C V_m}{\sqrt{2}}$$

Also, $$V = \frac{V_m}{\sqrt{2}}$$

Opposition to the ac current,

$$X_C = \frac{V}{I} = \frac{V_m/\sqrt{2}}{2\pi f C V_m/\sqrt{2}} = \frac{V_m}{\sqrt{2}} \times \frac{\sqrt{2}}{2\pi f C V_m} = \frac{1}{2\pi f C}$$

Then $$\mathbf{X_C = \frac{1}{2\pi f C} = \frac{1}{\omega C}} \qquad (23\text{-}2)$$

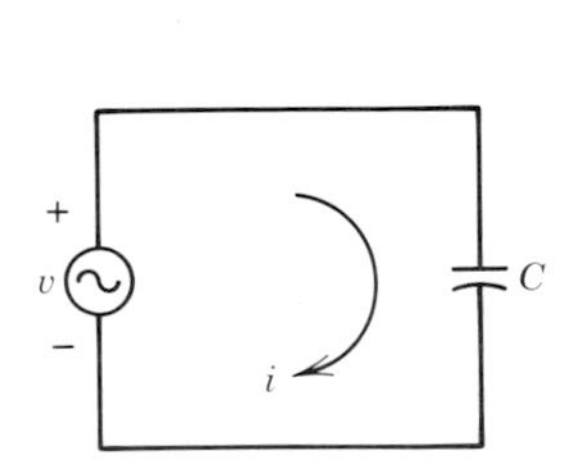

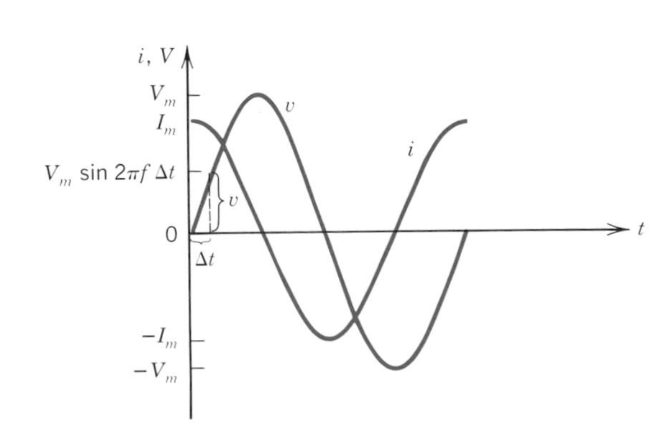

FIGURE A14-1
Initial slope of voltage curve equals $v/\Delta t$.

GLOSSARY

TERM	DEFINITION	SYMBOL OR ABBREVIATION	BASIC UNIT
Admittance	A measure of the ability of a reactive circuit to permit the flow of current. The reciprocal of impedance.	Y	siemens
Alternating current	A current that continuously reverses direction because of the continuously reversing polarity of an alternating voltage.	ac	ampere
Alternator	An ac generator.		
Ammeter	An instrument used to measure electrical current.	—(A)—	
Ampere	The basic unit of electrical current (coulomb per second).	A	
Ampere-hour	The basic unit of charge storage capacity of a cell or battery. A measure of the capacity of a battery to supply current for a given period of time.	Ah	
Ampere-turn	The basic unit of magnetizing force (magnetomotive force) caused by current flowing through a coil.	At	
Ampere-turn per meter	The basic MKS unit of magnetizing intensity. Magnetomotive force per unit length of magnetic circuit.	At/m	
Amplification	The process of increasing the voltage, current, or power of an electrical signal.		

TERM	DEFINITION	SYMBOL OR ABBREVIATION	BASIC UNIT
Amplifier	A device used to increase the voltage, current, or power of an electrical signal by means of an active device such as a transistor or electron tube.	(amplifier symbol: triangle marked A)	
Amplitude	The maximum instantaneous value of an alternating voltage or current. Peak value.	V_m, I_m	volt, ampere
Analog	A measurable quantity that takes on a *continuous* set of values. Nondigital.		
Anode	That terminal of a device (such as a battery or electron tube) that *loses* electrons in the external circuit during normal operation. In a battery the *negative* electrode is the anode.		
Apparent power	The power that *appears* to be delivered to a reactive circuit found by multiplying the applied voltage and current.	P_s	VA Volt-ampere
Aquadag	A graphite coating on the inside of a cathode-ray tube for collecting secondary electrons emitted by the screen.		
Arc	A spark or flash caused by a high voltage ionizing a gas or vapor.		
Arc tangent	An inverse trigonometric function meaning "the angle whose tangent is." Also called *inverse* tangent or $\tan^{-1}$.	$\tan^{-1}$	
Armature	In a relay, the moving part that is attracted to the electromagnet. In a generator, the windings in which a voltage is induced. In a motor, the rotating part or rotor.		

TERM	DEFINITION	SYMBOL OR ABBREVIATION	BASIC UNIT
Atom	The smallest particle of an element that combines with similar particles of other elements to form molecules and compounds.		
Attenuation	The process of reducing the strength of a signal.		
Audio	The range of sound frequencies that a human ear can detect, nominally from 20 Hz to 20 kHz.		
Autotransformer	A transformer that uses one coil for both primary and secondary.		
Average power	Average rate of energy consumption.	P_{av}	watt
AWG	American Wire Gauge. A table of standard wire sizes based on their circular mil areas.		
Ayrton shunt	A series interconnection of resistors placed in parallel with a meter movement to provide a multiple-range ammeter.		
Bandwidth (3 dB)	The range of frequencies over which the output of a device does not drop by more than 70.7% (or 3 dB) below its maximum output.	BW Δf	Hz
Base	The central semiconductor region in a bipolar junction transistor.		
Battery	Two or more interconnected cells that use a chemical reaction to produce electrical energy.	+ ⊣||⊢ −	
Beta	The ratio of collector current to base current in a bipolar junction transistor.	β	dimensionless
B-H curve	A graph showing the relation between flux density B and applied magnetizing intensity H for a magnetic material.		

TERM	DEFINITION	SYMBOL OR ABBREVIATION	BASIC UNIT
Bias	The application of a voltage or current to an active device (transistor, tube, etc.) to produce a desired mode of operation, such as in the linear region for an amplifier.		
Bimetal strip	Two different metals welded together, each having a different temperature coefficient of expansion. Bends when heated or cooled.		
Bipolar junction transistor (BJT)	A three-terminal semiconductor device with a base sandwiched between a collector and an emitter used for amplification or switching. May be PNP or NPN.		
Bohr model	A planetary model of the atom in which the central nucleus, made up of neutrons and protons, is surrounded by orbiting electrons.		
Branch	One of the paths for current in a parallel circuit.		
Bridge	Four components arranged in a diamond-shaped circuit with a voltage source connected to one pair of opposite corners, and a galvanometer connected to the other pair. Used for precise measurement of resistance, etc.		
Brush	A carbon or graphite structure used to conduct current between a stationary terminal and a rotating commutator or slip ring in a motor or generator.		
Capacitance	The property (of a capacitor) to store electrical charge and energy.	C	farad
Capacitive reactance	The opposition provided by capacitance to the flow of alternating current.	X_C	ohm

TERM	DEFINITION	SYMBOL OR ABBREVIATION	BASIC UNIT
Capacitor	An electrical device consisting of two metal plates separated by an insulating material called a dielectric.	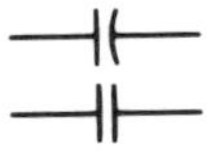	
Capacity	The ability of a battery to store electrical charge and energy.		Ah
Cell	A device that uses a chemical reaction to produce a dc voltage. The basic building block of a battery.	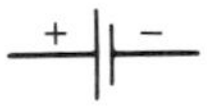 + −	
Cemf	Counter (or back) electromotive force, produced in a motor or by an inductance, that opposes the applied voltage.		volt
Center tap	A connection at the midpoint of a transformer's primary or secondary winding.	CT	
Charge	An electrical property of an electron (negative) or a proton (positive) that can cause a potential difference to exist between two points when excess charges accumulate.	Q	coulomb
Chassis	A metal case or frame upon which an electrical circuit or system is constructed.		
Choke	Another name for an inductor or coil, especially one that *chokes* or blocks high frequencies.	L 	
Circuit	An interconnection of components to form a complete path for the flow of electrical current.		
Circuit breaker	A resettable thermal or electromagnetic switch that opens a circuit when the current in the circuit exceeds a selected value.		
Circular mil	The unit of the cross-sectional area of a wire, found by squaring the diameter given in mils (thousandths of an inch).	CM	

TERM	DEFINITION	SYMBOL OR ABBREVIATION	BASIC UNIT
Clamp-on ammeter	An ammeter to measure current by clamping the meter around the wire instead of physically breaking the circuit. Uses the magnetic field *around* the current-carrying wire for its operation.		
Closed circuit	A complete path that permits the flow of current.		
Coefficient of coupling	The portion of magnetic flux set up by one coil that links a second coil.	k	dimensionless
Coercivity	A measure of the demagnetizing intensity (coercive force) required to reduce the residual magnetism of a material to zero.		
Coil	A common term for an inductor or the turns of wire making up the windings in a transformer, motor, generator, solenoid, and so on.		
Common	A term sometimes used for a common reference point or ground (neutral) in an electrical circuit or system.		
Common-base (CB) amplifier	A junction transistor amplifier circuit where the base is common to both the input and output.		
Common-emitter (CE) amplifier	A junction transistor amplifier circuit where the emitter is common to both the input and output.		
Commutator	Copper segments on the rotor of a motor or generator that switch current at the correct instant to generate direct current or permit a motor to run from a source of direct current.		
Complex number	A combination of a real and an imaginary number that may be given in rectangular or polar form.	$a + jb$ $M\angle\theta$	

TERM	DEFINITION	SYMBOL OR ABBREVIATION	BASIC UNIT
Condenser	Older name used for a capacitor	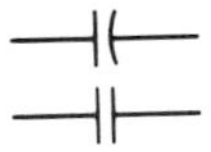	
Conductance	Ability of a circuit to allow the flow of current. Reciprocal of resistance.	G	siemens
Conductivity	A measure of the ease with which a material conducts electrical current. Reciprocal of resistivity.	σ	siemens/meter
Conductor	A material that allows electrical current to flow relatively easily, such as copper, aluminum, etc.		
Continuity	A complete path for current to flow. A resistance or impedance less than infinity.		
Core	A material (usually magnetic) inside a coil to concentrate the magnetic flux.		
Coulomb	The unit of electrical charge.	C	
Counter emf	See Cemf.		
CRT	Cathode Ray Tube.		
Curie temperature	The temperature at which a magnetic material loses all of its ferromagnetic properties (magnetism).		
Current	The rate of flow of electrical charge in a circuit.	I	ampere
Current transformer	A transformer used to extend the range of an ac ammeter by stepping down the primary line current by a known ratio	C.T. 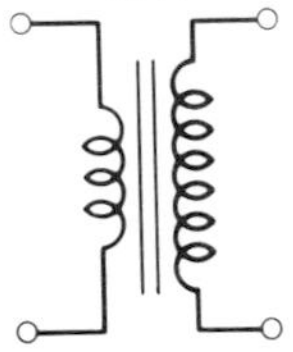	
Cycle	One complete positive and one complete negative alternation of a repetitive quantity such as voltage, current, etc.	~	
Decibel	A logarithmic unit used to show the power or voltage gain of an amplifier or the attenuation of a filter, etc.	dB	

TERM	DEFINITION	SYMBOL OR ABBREVIATION	BASIC UNIT
Degree	A unit of angular measure equal to 1/360th of a complete circle or revolution.	°	
Delta	A method of connecting a three-phase system so that the line and phase voltages are equal. A configuration of three connected components sometimes referred to as a π (pi) connection. Symbol meaning "a change in."	Δ	
Diamagnetic material	Those materials that have a permeability slightly less than that of free space (air), and slightly oppose the setting up of magnetic lines of force.		
Dielectric	An insulator; usually refers to the insulating material between the plates of a capacitor.		
Dielectric constant	The ratio of capacitance with a dielectric to the capacitance with air, all other variables held constant. A measure of the ability of an insulator in a capacitor to induce an opposing electric field.	K	dimensionless
Dielectric strength	The electric field intensity required to cause an insulator to break down and conduct.		V/m
Differentiator	A circuit that produces an output that is proportional to the slope (or derivative) of the input.		
Digital	A nonlinear process in which a variable takes on discrete values.		
Diode	An electronic device that allows current to flow through it in only one direction. May be a tube or a semiconductor.	—▶|—	
Direct current	An electric current that flows in only one direction.	dc	

TERM	DEFINITION	SYMBOL OR ABBREVIATION	BASIC UNIT
Dissipation factor	An indication of how much power is dissipated (lost) in a capacitor (or coil) at a given frequency. The reciprocal of Q.	D	dimensionless
Distortion	An undesired change in output waveform compared with input due to irregularities in amplitude, frequency, or phase.		
DMM	Digital multimeter.		
Domains	Groups of adjacent atoms ($10^{15}-10^{17}$) that respond collectively to an external magnetic field.		
DVM	Digital voltmeter.		
Eddy current	An induced current that circulates in a conductng core that carries a varying magnetic field.		
Effective value	A measure of the heating effect of a varying voltage or current. Also known as the rms or root-mean-square value.	V_{eff} I_{eff}	volt ampere
Efficiency	The ratio of output power (or work) to input power (or work), generally expressed as a percentage.	η	%
Electric field (intensity)	A region near a charged body where a force is exerted on another electrically charged body.	E	V/m N/C
Electricity	A form of energy due to electric charges at rest (static electricity) or in motion (current) that can be transmitted through wires to produce effects of light, heat, motion, sound, etc.		
Electrode	A conductor or terminal, usually metallic, through which an electric current enters or leaves.		

TERM	DEFINITION	SYMBOL OR ABBREVIATION	BASIC UNIT
Electrolyte	A solution or paste, capable of conducting an electric current because of its dissociation into positive and negative ions, as in a voltaic cell (battery).		
Electromagnet	A magnet produced by passing current through a coil of wire wound on a soft iron core.	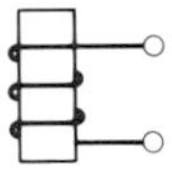	
Electromotive force	The open circuit voltage produced by an electrical source, such as a battery, photovoltaic cell, generator, or piezoelectric crystal, that causes current to flow in a complete circuit.	emf	volt
Electron	Negatively charged particle of an atom.		
Electronic	Related to the movement and control of electrical charges through semiconductor and vacuum tube devices.		
Electrostatic force	The force of attraction or repulsion between stationary charged bodies.		
Element	A substance composed entirely of one type of atom.		
Emitter	One of the three terminals of a BJT that emits the majority carriers.	(E)	
Energy	The ability or capacity to do work.	W	joules ft lb
Equivalent circuit	A circuit that represents a more complicated circuit but permits easier analysis.		
Exponent	The power of a number indicating how many times that number should be multiplied by itself.		
Farad	The unit of capacitance.	F	

TERM	DEFINITION	SYMBOL OR ABBREVIATION	BASIC UNIT
Feedback	The process of returning some of the output of a circuit (such as an amplifier) back to the input. May be positive or negative.		
Ferromagnetic materials	Those materials that have a permeability hundreds of times greater than that of free space, resembling the magnetic properties of iron.		
Field	A region in which a force is exerted on a static charge (electric field), or on a moving charge (magnetic field), or on a mass (gravitational field).		
Field winding	A coil in a motor or generator used to set up a magnetic field.		
Filter	A frequency selective circuit that passes certain frequencies and rejects all others.		
Flux	The lines of force in a magnetic or electric field.		
Flux density	The amount of flux per unit cross-sectional area.		
Free electron	A valence electron that has acquired enough energy to break away from its parent atom and thus is free to move at random through the atoms of the given material.		
Frequency	The number of times a varying quantity goes through a complete cycle of variation per second.	f	Hertz
Fuel cell	A chemical reaction, aided by a catalyst, in which electrical energy is produced directly using a process that is the reverse of electrolysis.		

TERM	DEFINITION	SYMBOL OR ABBREVIATION	BASIC UNIT
Full-wave rectifier circuit	A diode circuit that utilizes both positive and negative alternations of an alternating current to produce a direct current.		
Function generator	A piece of electronic equipment that is capable of producing several waveforms such as sine, square, triangular, and pulse.		
Fuse	A thermal protective device that opens when current becomes excessive to disconnect a circuit from the source.	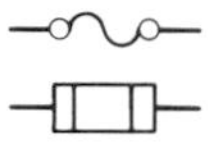	
Gain	The ratio of the output of an amplifier to the input, in terms of voltage, current, or power.	A_v A_i A_p	dimensionless
Galvanometer	An electrical instrument (ammeter) used to detect (measure) very small amounts of direct current.		
Gauss	The cgs unit of magnetic flux density.	G	
Gaussmeter	An instrument (fluxmeter) used to measure magnetic flux density.		
Generator	A machine that converts mechanical energy into electrical energy, either dc or ac.		
Germanium	A semiconductor material used to make diodes and transistors.	Ge	
Giga	A prefix used to designate 10^9.	G	
Ground	A common or reference point in an electrical circuit that can be a chassis or earth ground.		
Hall emf	The voltage produced across the edges of a semiconductor strip that is carrying a current and is located in a magnetic field.		

TERM	DEFINITION	SYMBOL OR ABBREVIATION	BASIC UNIT
Harmonic	An integral multiple of a fundamental frequency.		
Henry	The unit of inductance.	H	
Hertz	The unit of frequency. One cycle per second.	Hz	
Heterodyne	The process of beating or mixing two signals of different frequencies.		
Hole	A positive charge carrier (majority) in a *P* type semiconductor. The space in an atom left vacant by a departed electron.		
Horsepower	Unit of power used in the English system. 1 hp = 550 ft lb/s = 746 W.	hp	
Hydrometer	A device used to measure specific gravity.		
Hypotenuse	The longest side in a right triangle that is opposite the 90° right angle.		
Hysteresis	A lagging of the magnetic flux in a magnetic material behind the magnetizing force that is producing it.	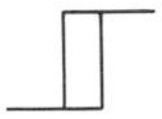	
Impedance	The total opposition to the flow of alternating current in a circuit containing resistance and reactance.	Z	ohm
Induced voltage	Voltage induced in a wire or coil as a result of a changing magnetic field.	v_{ind}	volt
Inductance	The property of an inductor to oppose a change of current through the inductor.	L	henry
Induction motor	An ac motor in which a rotating magnetic field, produced by current in the stator windings, induces current in the rotor and a subsequent turning motion.		
Inductive reactance	The opposition that inductance offers to an alternating current.	X_L	ohm
Inductor	A coil or choke having the property of inductance.		

TERM	DEFINITION	SYMBOL OR ABBREVIATION	BASIC UNIT
Infinite	Having no bounds or limits.	∞	
Input	The voltage, current, or power applied to a circuit.		
Instantaneous value	The value of a variable quantity at a given instant of time.		
Insulator	A material that does not conduct electrical current.		
Integrated circuit	A circuit in which all components and their interconnections are fabricated on the same chip of semiconductor material.	IC	
Integrating circuit	A circuit that produces an output that is proportional to the mathematical integral of the input. A summation device.	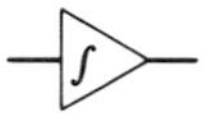	
Internal resistance	The resistance contained within a voltage or power source, such as a battery, power supply, or amplifier.	r	ohm
Ion	An atom that has an excess or deficiency of electrons.		
Iron-vane meter	An instrument that uses a moving iron-vane and a stationary coil to measure low-frequency ac voltage and current.		
Joule	The unit of work and energy in the MKS system.	J	
Kilo	A prefix used to designate 10^3.	k	
Kilowatt-hour	A common unit of energy used by power companies to compute electricity bills.	kWh	
Kirchhoff's laws	A set of laws that describes how voltage and current are distributed throughout a circuit, both ac and dc.		
Knee	An abrupt change in slope between two relatively straight segments of a curve.		

TERM	DEFINITION	SYMBOL OR ABBREVIATION	BASIC UNIT
Lag	A condition that describes the delay of one waveform behind another in time or phase.		
Laminated core	A core built up from thin sheets of metal (often insulated from each other) and used in transformers, relays, motors, and so on, operating from an ac supply.		
Lead	A condition that describes how one waveform is ahead of another in time or phase. Also, a wire or cable connection.		
Lenz's law	A law that states that the polarity of induced emf is always such as to oppose the cause that created it.		
Linear	Refers to a relation between two variables where one is directly proportional to the other. A straight-line relationship.		
Line of force	A line drawn to represent the direction of force in an electric or magnetic field.		
Load	A device that converts electrical energy into another form such as heat, light, or motion.		
Logarithm	The exponent to which a base number must be raised to equal a given number. ($\log_{10} 100 = \log_{10} 10^2 = 2$)	log	
Loop	A closed path in a circuit.		
Magnet	A magnetized material that aligns itself with the earth's magnetic field and attracts magnetic materials.		
Magnetic	Related to magnetism; north and south poles and their associated lines of force.		
Magnetic circuit	A complete circuit for the establishment of magnetic lines of force.		

TERM	DEFINITION	SYMBOL OR ABBREVIATION	BASIC UNIT
Magnetic core	A form or frame made from those materials (iron, nickel, cobalt, and certain alloys) that can be magnetized.		
Magnetic field	A region in which a force of attraction is exerted on a magnetic material or on a moving charge.		
Magnetic field strength	The magnetomotive force per unit length of magnetic circuit. Also known as magnetizing force, magnetic field intensity, or magnetizing intensity.	H	At/m
Magnetic flux	Magnetic lines of force.	Φ	Webers
Magnetic flux density	Magnetic lines of force per unit area. Also known as magnetic induction.	B	Wb/m^2 or Teslas
Magnetize	To make a magnetic material act as a magnet by aligning its magnetic domains.		
Magneto	A generator that produces alternating current using a permanent magnet.		
Magnetomotive force (mmf)	The magnetizing force produced by current flowing through a coil of wire wrapped around a magnetic core.	F_m	At
Magnitude	The size of some electrical quantity such as voltage or current without regard to its phase angle.		
Maximum power transfer	A circuit condition where the load resistance equals the source resistance (dc), or where the load impedance is the conjugate of the source impedance (ac), causing maximum power to be transferred from source to load.		
Maxwell	The cgs unit of magnetic flux.	Mx	

TERM	DEFINITION	SYMBOL OR ABBREVIATION	BASIC UNIT
Mega	A prefix designating 10^6 (1 million).	M	
Megger	A portable hand-operated, high-voltage dc generator used as an ohmmeter to measure insulation and other high resistances.		
Megohm	A million ohms.	MΩ	
Mesh	An arrangement of current loops in a circuit.		
Meter loading	A change in circuit conditions caused by the finite resistance of a meter when connected to the circuit to make measurements.		
Micro	A prefix designating 10^{-6} (1 millionth).	μ	
Milli	A prefix designating 10^{-3} (1 thousandth).	m	
Milliammeter	An ammeter that measures current in thousandths of an ampere.		
Modulation	The process of varying the amplitude (AM), frequency (FM), or phase (PM) of a carrier wave using a signal such as audio or video to convey intelligence.		
Multimeter	An instrument, often portable, designed to measure two or more electrical quantities.		
Multiplier	A resistor used to extend the voltage range of a meter movement.	R_s	ohm
Mutual inductance	The inductance between two separate coils that share a common magnetic field that links the two.	M	henry
Nano	A prefix designating 10^{-9}.	n	
Negative charge	A body or terminal that has an excess of electrons compared with a neutral state.		
Neutral	The common or grounded wire in a single or three-phase system.		

TERM	DEFINITION	SYMBOL OR ABBREVIATION	BASIC UNIT
Neutron	The electrically neutral particle of an atom that carries no electrical charge, but has a mass equal to that of a proton.		
Node	A junction in a circuit where current divides into different paths.		
Nonlinear	A scale that has unequal divisions for equal increments. A device that produces an output that is not directly proportional to the input.		
Nucleus	The central part of an atom that consists mainly of protons and neutrons.		
Null	A zero or minimum.		
Oersted	The cgs unit of magnetizing intensity.	Oe	At/cm
Ohm	The unit of resistance.	Ω	
Ohmmeter	An instrument that measures resistance.	—(Ω)—	
Open circuit	A circuit with an incomplete path for current. Infinite resistance.		
Oscillator	A circuit that uses positive feedback to produce an output without an external input signal.		
Oscilloscope	An instrument that displays on a screen the manner in which a waveform varies with time.	—(O)—	
Output	The voltage, current, or power developed by a circuit in response to an input.		
Overload	A load that draws more than the rated current or power.		
Parallel	A connection of two or more components that operate from the same voltage.		

TERM	DEFINITION	SYMBOL OR ABBREVIATION	BASIC UNIT
Parallel resonance	A condition in a parallel *RLC* circuit in which the circuit impedance is resistive and a maximum, causing the applied voltage and current to be in phase.		
Paramagnetic material	Those materials that have a permeability slightly greater than that of free space (air), and slightly aid the setting up of magnetic lines of force.		
Peak value	The maximum value of a waveform, equal to the amplitude for symmetrical alternating waveforms.	V_p I_p	volt ampere
Period	The time required for a periodic waveform to complete a full cycle.	T	s^{-1}
Periodic	A variation that repeats at definite time intervals.		
Permeability	A measure of the ease with which magnetic lines of force can be established in a material.	μ	Wb/At-m
Permittivity	A measure of the ability of a material to permit an electric field to be established in the material.	ϵ	C^2/Nm^2
Phase angle	An angle that shows the displacement of a time-varying waveform from some reference.	θ, ϕ	degree radian
Phasor	A line drawn to represent a time-varying waveform in magnitude and phase angle.		
Photovoltaic cell	A silicon *PN* junction that produces voltage from light energy. A solar cell.	λ	
Pico	A prefix designating 10^{-12}.	p	
Piezoelectric	The production of voltage by applying pressure to a crystal.		

TERM	DEFINITION	SYMBOL OR ABBREVIATION	BASIC UNIT
Polarity	Refers to the negative and positive terminals in an electric circuit or the north and south poles of a magnet.		
Polarization	The collection of hydrogen gas around the electrode of a discharging cell, or the distortion of the path of orbiting electrons in the dielectric of a charged capacitor.		
Pole	The regions of a magnet where the flux lines are concentrated. An electrode of a battery.		
Polyphase	A circuit or supply that utilizes more than one phase of alternating current.		
Positive charge	A body or terminal that has a deficiency of electrons compared with a neutral state.		
Potential	The ability to do work with respect to some reference.		volt
Potential difference	A measure of how much work may be done to move electrical charge between two points.	V	volt
Potentiometer	A three-terminal variable resistor device used to vary voltage. Also an instrument used to make accurate voltage measurements.		
Power	The rate of doing work or converting energy from one form to another.	P	watt
Power factor	The ratio of a circuit's true power in watts to the apparent power in volt-amperes.	$\cos\theta$	dimensionless
Power supply	A piece of equipment that supplies voltage and current, often variable, ac or dc, from the 120-V, 60-Hz line or other available sources.		

TERM	DEFINITION	SYMBOL OR ABBREVIATION	BASIC UNIT
Primary cell	A cell (battery) that cannot be recharged to its initial state.	[battery symbol: + ‖ −]	
Primary winding	The winding of a transformer that is connected to the input.		
Proton	The positively charged particle in the nucleus of an atom.		
Pulsating direct current	A varying direct current that periodically returns to zero.		
Quality factor	The ratio of reactive power to true power in a coil or resonant circuit. Used to show how much inductive reactance a coil has compared with its ac resistance.	Q	dimensionless
Radian	A unit of angular measurement. Approximately 57.3°.	rad	
Range	The maximum value that a meter is capable of measuring.		
Reactance	The opposition to the flow of alternating current offered by capacitance or inductance or both.	X	ohm
Reactive power	The power that is alternatively delivered to a reactive component or circuit and then returned to the source.	P_q	var
Reactor	Another name used for an inductor.	[inductor symbol]	henry
Real power	The power that is dissipated as heat in a resistor or converted to another form. Also known as true power.	P	watt
Rectification	The process of converting alternating current to direct current.		
Rectifier	An electronic device or circuit that converts alternating current to direct current.	[diode symbol]	

TERM	DEFINITION	SYMBOL OR ABBREVIATION	BASIC UNIT
Regulation	The change in terminal voltage of a source from no load to full load, expressed as a percentage of full load voltage.	V_{reg}	%
Relative permeability	The ratio of the permeability of a material to the permeability of free space (air).	μ_r	dimensionless
Relay	A switching device that uses electromagnetism to move an armature to open or close one or more contacts; often controlled from a remote location.	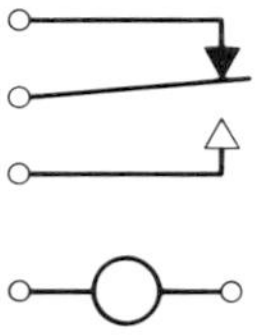	
Reluctance	The opposition that a material offers to the establishment of magnetic flux.	R_m	At/Wb
Residual magnetism	The magnetism that remains in a material after the magnetizing force has been removed.		Wb/m^2
Resistance	The opposition to the flow of current that converts electric energy into heat.	R	ohm
Resistivity	The resistance of a specific quantity of a given material.	ρ	$\Omega\cdot m$
Resistor	An electrical component whose property of resistance is used to oppose current.		
Resonance	A series or parallel *RLC* circuit which, at one frequency, has the applied voltage and current in phase with each other.		
Resonant frequency	That frequency that causes a series or parallel *RLC* circuit to be in a state of resonance.	f_r	hertz
Retentivity	The measure of the ability of a material to retain its magnetism after the magnetizing force is removed.		
Rheostat	A two-terminal, variable resistor, used to vary current.	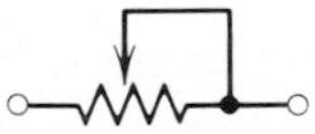	
Right angle	A 90° angle.		

TERM	DEFINITION	SYMBOL OR ABBREVIATION	BASIC UNIT
Root mean square	That value of direct current that is as effective in producing heat as the given time-varying waveform of current or voltage. Also known as the effective value.	rms	ampere volt
Saturation	A condition in which an increase in the driving force produces little or no effect, as in the saturation of a magnetic material.		
Secondary cell	A cell (battery) that is capable of being recharged to its initial state repeatedly.	+ ⊣⊢ −	
Secondary winding	The output winding of a transformer that is connected to the load.		
Selectivity	The ability of a (tuned) circuit to respond (develop an output) to a certain band of frequencies and reject others. Inversely related to bandwidth.		
Self-inductance	The property of a coil to induce a voltage within itself due to a changing current in the coil.	L	henry
Self-induction	The process of generating a voltage in a coil or circuit due to a changing magnetic field within the coil or circuit itself.		
Semiconductor	A material such as silicon or germanium that generally has four valence electrons and a resistivity between that of a conductor and an insulator, but closer to that of a conductor.		
Series	A connection of two or more components so that there is only one path for the flow of current.		

TERM	DEFINITION	SYMBOL OR ABBREVIATION	BASIC UNIT
Series resonance	A condition in a series *RLC* circuit where, at that one frequency, the impedance is a minimum and pure resistance due to equal inductive and capacitive reactances.		
Short	A path of zero or very low resistance between two points.		
Shunt	A parallel-connected component used, as in an ammeter, to shunt or divert current around a meter movement.		
Siemen	The unit used for conductance, susceptance, and admittance. The reciprocal of an ohm.	S	
Signal	A low-level voltage or current that is generally time varying.		
Silicon	A semiconductor material used to fabricate electronic devices such as diodes, transistors, integrated circuits, and photovoltaic devices.	Si	
Sine	In a right triangle the name given to the ratio of the side opposite a given acute angle to the hypotenuse.	sin	dimensionless
Sine wave	An alternating waveform whose instantaneous value is related to the trigonometric sine function of time or angle.		
Sinusoidal	A waveform having the shape of a sine wave.		
Skin effect	The effect that takes place at high frequencies due to induced voltage causing an alternating current to flow near the surface of a conductor, which results in an effective increase in the resistance of the conductor.		

TERM	DEFINITION	SYMBOL OR ABBREVIATION	BASIC UNIT
Slope	The steepness of a line or curve given by the ratio of rise to run of a tangent line drawn to the curve.	m $\frac{\Delta y}{\Delta x}$	
Solar cell	A semiconductor device that converts light energy directly into electrical energy.		
Solenoid	A coil with a movable plunger that operates because of electromagnetism.		
Source	A device that produces or converts energy.		
Specific gravity	The ratio of the weight of a given volume of a substance to the weight of the same volume of water.		
Static	A fixed nonvarying condition; without motion.		
Steady state	A condition following a transient period where no sudden changes take place any more.		
Susceptance	The ability of a purely reactive component to permit the flow of alternating current. The reciprocal of reactance.	B	siemens
Switch	An electrical or electronic device used to control the flow of current in a circuit.	—o⁄ o—	
Tangent	In a right triangle the name given to the ratio of the side opposite a given acute angle to the adjacent side.	tan	dimensionless
Tank	A parallel resonant LC circuit.		
Temperature coefficient	A constant specifying the amount of change in the value of a temperature dependent variable per unit change of temperature.	α	per °C
Tesla	The MKS unit of magnetic flux density. $1\ \text{T} = 1\ \text{Wb/m}^2$	T	

TERM	DEFINITION	SYMBOL OR ABBREVIATION	BASIC UNIT
Thermistor	A semiconductor device that has a large change in resistance (usually a decrease) when its temperature increases.	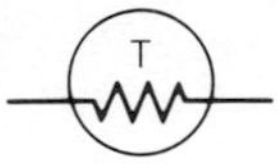	
Thermocouple	A pair of dissimilar metals joined together so that an emf is generated at the open ends when the junction is heated or cooled.		
Theta	A Greek letter used to represent an angle.	θ	degree radian
Three-phase	Three sinusoidal voltages or currents displaced from each other by 120 electrical degrees.	3ϕ	
Time constant	The time required for the voltage or current in an *RC* or *RL* circuit to change 63.2% from the initial value to the final value.	τ	second
Tolerance	The upper and lower limits of variation from a component's nominal value.		
Transducer	In electricity and electronics, a device that converts a variable such as temperature, pressure, or flow to an electrical signal; or a device that converts an electrical signal to a different form of energy such as sound.		
Transformer	An electrical device that uses electromagnetic induction to transfer electrical energy from a primary winding to a secondary winding causing an increase or decrease in the ac voltage.	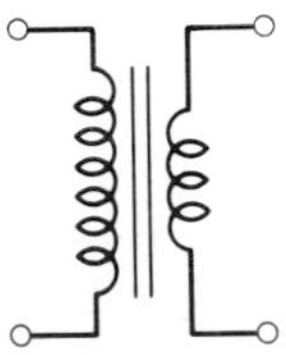	
Transient	A short-lived change in circuit conditions from one steady state to another.		
Transistor	A three-terminal semiconductor device used for amplification or switching. May be of the bipolar or field-effect type.		

TERM	DEFINITION	SYMBOL OR ABBREVIATION	BASIC UNIT
Tuned circuit	A resonant circuit that responds to a narrow band of frequencies.		
Turns ratio	The ratio of the number of turns in a transformer's primary winding to the number of turns in the secondary winding.	a	dimensionless
Unity power factor	A condition in an ac circuit where the inductive and capacitive reactive powers are equal, causing the applied voltage and current to be in phase with each other.	$\cos \theta = 1$	dimensionless
Valence electron	The electron that is in the outermost orbit of an atom.		
Vector	A line drawn to represent a quantity in magnitude and direction.		
Volt	The unit of voltage, electromotive force, or electrical potential.	V	
Voltage	The potential difference between two points. A measure of the potential energy available to move charge or a measure of the amount of work that must be done to move charge between two points.	V	volt
Volt-ampere	The unit of apparent power in an ac circuit.	VA	
Volt-ampere reactive	The unit of reactive power in an ac circuit.	var	
Voltmeter	An electrical instrument used to measure the voltage between two points.	—(V)—	
Voltmeter sensitivity	The ohms per volt rating of a voltmeter equal to the ratio of the total resistance of a voltmeter on a given range to the FSD voltage of that range. Also equal to the reciprocal of the meter movement's full-scale current for analog instruments.	S	Ω/V

TERM	DEFINITION	SYMBOL OR ABBREVIATION	BASIC UNIT
Watt	The MKS unit of power.	W	
Watt-hour	A unit of energy.	Wh	
Wattmeter	An electrical instrument used to measure power.	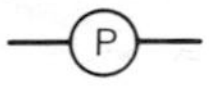 —(P)—	
Waveform	The shape of a wave obtained when instantaneous values of a varying quantity are plotted against time.		
Wavelength	The distance, usually expressed in meters, traveled by a wave during the time interval of one complete cycle.	λ	meter
Weber	The MKS unit of magnetic flux.	Wb	
Wheatstone bridge	See Bridge.		
Winding	The coil or turns of wire on an inductor or transformer.		
Wiper	The movable contact in a potentiometer or other variable device.		
Work	The product of an applied force and the distance moved by the force.		
Wye	A method of connecting a three-phase system using a common point so that the line and phase currents are equal. A configuration of three connected components sometimes referred to as a T (tee) connection.	Y	
Zener diode	A diode used with reverse bias in the breakdown mode where the voltage is relatively independent of current. Used for voltage regulation.		

ANSWERS TO SELF-EXAMINATIONS

CHAPTER 1

1. T 2. F 3. T 4. T 5. T 6. F 7. T 8. T 9. T 10. F 11. T 12. F 13. F 14. T 15. T 16. T 17. F 18. T 19. T 20. T 21. F 22. T 23. T 24. F 25. F 26. F 27. T 28. T

CHAPTER 2

1. T 2. F 3. T 4. T 5. F 6. T 7. T 8. T 9. F 10. T 11. T 12. T 13. T 14. T 15. F 16. T 17. F 18. T 19. T 20. T 21. F 22. T 23. T 24. F 25. T 26. T 27. T 28. F 29. T 30. T 31. T 32. F 33. T 34. T

CHAPTER 3

1. T 2. T 3. T 4. F 5. T 6. F 7. T 8. F 9. T 10. F 11. T 12. T 13. T 14. T 15. F 16. T 17. F 18. T 19. F 20. T 21. 10 Ω 22. 20 A 23. 50 V 24. 3 A 25. 220 V 26. 40 W 27. 22 Ω 28. 1.2 kWh 29. $2.00 30. YVYG

CHAPTER 4

1. d 2. b 3. T 4. F 5. b 6. a 7. c 8. T 9. F 10. b 11. T 12. T

CHAPTER 5

1. T 2. T 3. F 4. T 5. F 6. T 7. c 8. c 9. b 10. a 11. d 12. a 13. b 14. c 15. d 16. d 17. T 18. T 19. T 20. F

CHAPTER 6

1. T 2. F 3. T 4. F 5. c 6. a 7. b 8. d 9. c 10. c 11. c 12. d 13. b 14. a 15. b

CHAPTER 7

1. T 2. F 3. F 4. F 5. d 6. c 7. b 8. T 9. F 10. T 11. T 12. F 13. T 14. T 15. T 16. F

CHAPTER 8

1. F 2. c 3. b 4. T 5. T 6. F 7. b 8. d 9. F 10. c 11. F 12. T 13. c 14. a 15. d 16. T 17. T 18. T 19. F

CHAPTER 9

1. T 2. c 3. F 4. b 5. c 6. F 7. d 8. T 9. T 10. F 11. d 12. T 13. F 14. b 15. F 16. F

CHAPTER 10

1. T 2. T 3. F 4. F 5. T 6. T 7. T 8. T 9. T 10. T 11. F 12. T 13. F 14. T 15. T 16. F 17. T 18. T 19. T 20. F 21. F 22. T

CHAPTER 11

1. T 2. F 3. T 4. F 5. T 6. T 7. T 8. T 9. F 10. F 11. T 12. F 13. T 14. T 15. F 16. T 17. F 18. F 19. T 20. T 21. F 22. T 23. T 24. F 25. F 26. F 27. T 28. T 29. F 30. T

CHAPTER 12

1. T 2. F 3. T 4. F 5. T 6. T 7. F 8. d 9. T 10. T 11. T 12. F 13. T 14. b 15. T 16. T 17. T 18. a 19. c 20. T 21. c 22. F 23. T 24. b 25. d

CHAPTER 13

1. T 2. T 3. F 4. T 5. F 6. F 7. T 8. F 9. c 10. F 11. T 12. T 13. F 14. T 15. T 16. T 17. F 18. T 19. T 20. T 21. F 22. T

CHAPTER 14

1. F 2. T 3. F 4. T 5. T 6. F 7. T 8. c 9. c 10. b 11. c 12. F 13. T 14. T 15. F 16. a 17. T 18. T 19. F 20. T 21. d 22. b 23. b 24. T 25. F 26. T 27. b 28. T

CHAPTER 15

1. c 2. d 3. F 4. T 5. T 6. a 7. d 8. a 9. b 10. c

CHAPTER 16

1. T 2. T 3. F 4. F 5. T 6. T 7. F 8. F 9. T 10. b 11. c 12. T 13. F 14. T 15. F 16. T 17. F 18. T

CHAPTER 17

1. T 2. F 3. b 4. d 5. F 6. T 7. F 8. T 9. F 10. c 11. d

CHAPTER 18

1. T 2. T 3. F 4. a 5. c 6. F 7. b 8. T 9. F 10. F 11. F 12. T 13. F

CHAPTER 19

1. T 2. T 3. F 4. F 5. T 6. F 7. c 8. b 9. F 10. F 11. T 12. T 13. F 14. T 15. T

CHAPTER 20

1. T 2. T 3. T 4. F 5. d 6. a 7. T 8. T 9. F 10. T 11. F 12. F 13. T 14. F 15. T 16. T 17. T

CHAPTER 21

1. T 2. F 3. F 4. c 5. d 6. b 7. c 8. T 9. b 10. T 11. T 12. F 13. T 14. T 15. F 16. F 17. T 18. F 19. T 20. F 21. c 22. a 23. F 24. F 25. F

CHAPTER 22

1. T 2. T 3. F 4. F 5. b 6. c 7. e 8. a 9. T 10. c 11. F 12. T 13. F 14. d 15. T 16. F 17. T

CHAPTER 23

1. T 2. T 3. T 4. F 5. c 6. b 7. F 8. F 9. F 10. T 11. T 12. F 13. T 14. F 15. T

CHAPTER 24

1. T 2. F 3. T 4. T 5. c 6. T 7. a 8. d 9. F 10. F 11. b 12. F 13. F 14. T 15. F 16. F 17. F 18. F 19. T

CHAPTER 25

1. T 2. T 3. T 4. F 5. c 6. a 7. b 8. c 9. F 10. T 11. T 12. F 13. T 14. d 15. T 16. T 17. F 18. T 19. F 20. d

CHAPTER 26

1. T 2. F 3. T 4. F 5. T 6. b 7. a 8. d 9. F 10. b 11. F 12. T 13. F 14. c 15. b 16. T 17. F 18. T 19. T 20. T

CHAPTER 27

1. f 2. c 3. s 4. h 5. b 6. d 7. e 8. k 9. m 10. i 11. g 12. j 13. n 14. p 15. l 16. q 17. o 18. t 19. r 20. a

CHAPTER 28

1. T 2. T 3. F 4. T 5. T 6. T 7. F 8. F 9. T 10. F 11. T 12. F 13. T 14. T 15. T 16. F 17. T 18. T 19. T 20. F 21. T 22. T 23. F 24. T 25. T 26. F 27. T 28. T 29. F 30. F 31. T 32. T 33. T 34. F 35. T 36. T 37. F

CHAPTER 29

1. F 2. F 3. F 4. T 5. F 6. T 7. F 8. F 9. T 10. F 11. T 12. F 13. T 14. T 15. F 16. T 17. T 18. T 19. T 20. T 21. T 22. F 23. F 24. T 25. T 26. T 27. T 28. T 29. F 30. F 31. T 32. T

ANSWERS TO ODD-NUMBERED PROBLEMS

CHAPTER 1

1–1 a. $5.710,00 \times 10^5$
b. 6.45×10^{-5}
c. 2.306×10^3

1–3 a. 5.7×10^5
b. 6.5×10^{-5}
c. 2.3×10^3

1–5 a. 6×10^5
b. 6×10^{-5}
c. 2×10^3

1–7 a. 479,000
b. 0.001,135
c. 907.9

1–9 a. 9.84×10^1
b. 1.87×10^6
c. 2.22×10^6

1–11 a. 34.3 cm
b. 0.881 m
c. 201 km
d. 33.1 l

1–13 a. 0.050 A
b. 50,000 μA

1–15 a. 2.5 kV
b. 350 mV

1–17 a. 1.2 kΩ
b. 750 mΩ
c. 1.5 MΩ

1–19 a. 50,000 ps
b. 0.050 μs

1–21 a. 0.213 ns
b. 1000 μF
c. 500 μH

1–23 a. 0.23 mA, 0.54 mA, 0.87 mA
b. 5.75 mA, 13.5 mA, 21.75 mA
c. 230 mA, 540 mA, 870 mA

1–25 a. 0.35 V, 0.9 V, >1.5 V ac
b. 0.23 V, 0.75 V, 1.36 V dc
c. 7.2 V, 23.8 V, 43 V dc
d. 23 V, 75 V, 136 V ac

1–27 a. 1.7 Ω, 7.5 Ω, 35 Ω
b. 1.7 kΩ, 7.5 kΩ, 35 kΩ

CHAPTER 2

2–1 a. 5.63×10^{-4} N
b. 1.27×10^{-4} lb

2–3 1.56×10^{11} electrons

2–5 1.5 V

2–7 0.33 A

2–9 12,000 C

2–11 16,200 J

2–13 500 s

2–15 6.25×10^{14} electrons

2–17 a. 2.5 mA
b. 75 μA
c. 5 A

2–19 a. 9.37×10^{-4} m/s
b. 2.34×10^{-4} m/s
c. 5.86×10^{-5} m/s

2–21 a. 5.62 cm
b. 1.41 cm
c. 0.35 cm

2–23 a. 1.5 kΩ
b. 4.7 MΩ
c. 45 mΩ

2–25 a. 0.667 mS
b. 0.213 μS
c. 22.2 S

2–27 a. 25 Ω
b. 2.2 MΩ
c. 1 Ω

CHAPTER 3

3–1 146.7 Ω
3–3 6 kΩ
3–5 2.2 mA
3–7 306 V
3–9 3×10^7 ft-lb
3–11 30.3 hp
3–13 a. 3.6×10^5 J
b. 0.1 kWh
3–15 83.3%
3–17 1¢
3–19 a. 3.6×10^6 J
b. 2.78×10^{-7} kWh
3–21 a. 1.44 kW
b. 1.93 hp
3–23 117 V
3–25 39.8 A
3–27 7.07 mA
3–29 26.1 V
3–31 2 W
3–33 a. 33 kΩ ± 10%
b. 510 Ω ± 5%
c. 2.2 MΩ ± 20%
d. 6.8 Ω ± 5%
e. 10 Ω ± 10%
f. 4.7 kΩ ± 5%
g. 750 Ω ± 5%
h. 3.9 MΩ ± 10%
3–35 a. 15 kΩ
b. 270 kΩ
c. 5.6 MΩ

CHAPTER 4

4–1 a. 5.43×10^{-2} Ω
b. 0.136 V
c. No. 16
4–3 2.92×10^3 m
4–5 9.69×10^{-6} m²
No. 7 min. gauge
4–7 9.89 cm
4–9 a. 0.625 mm
b. 0.0417 mm
4–11 0.239 Ω
4–13 18.7 Ω
4–15 363.8°C
4–17 a. 923 Ω
b. No cold inrush current
4–19 1003.6 Ω

CHAPTER 5

5–1 500 Ω
5–3 a. 180 mA
b. 27 V, 54 V, 9 V
5–5 a. 4.86 W, 9.72 W, 1.62 W
b. 16.2 W
5–7 a. 11.4 V
b. −8.6 V
5–9 4 kΩ
5–11 1.25 kΩ, 1.25 kΩ, 2.5 kΩ, 1.25 kΩ
5–13 36.1 V

5–15 a. 0.6 Ω
b. 43.3 A

5–17 100-W lamp: 90 V, 56.25 W
60-W lamp: 150 V, 93.75 W

5–19 a. 30.1 mV
b. 0 V
c. 15.1 mV

5–21 6.02 μA

5–23 181 mW

5–25 a. 15 V
b. 90 Ω, 60 Ω
c. −15 V
d. −9 V

5–27 a. 0.8 mA
b. +42 V
c. −38 V

CHAPTER 6

6–1 56.88 Ω

6–3 4.03 kΩ

6–5 19.70 Ω

6–7 253 W

6–9 4.7 kΩ

6–11 8 kΩ, 8 kΩ

6–13 a. $I_{47\,k} = 6.24$ mA
$I_{51\,k} = 5.76$ mA
b. $P_{47\,k} = 1.83$ W (2 W)
$P_{51\,k} = 1.69$ W (2 W)

6–15 14.1 V

6–17 $I_1 = 1.707$ mA
$I_2 = 0.776$ mA
$I_3 = 0.517$ mA

CHAPTER 7

7–1 a. 7.02 kΩ
b. 5.7 mA
c. 2.28 mA
d. 26.8 V

7–3 a. 2.21 mA
b. 0.98 mA
c. 4.88 mW

7–5 a. No. 5
b. 541 W
c. 6762 W

7–7 a. 680 Ω
b. 529.4 Ω
c. 600 Ω

7–9 a. 4.16 mA; 10.4 mW
b. The first 800 Ω

7–11 a. 30 Ω, 1.2 W
(30 Ω, 5%, 2 W)
b. 6.6 V
c. 13.2 V

7–13 $R_B = 1.5$ MΩ, 1.5 W
(1.5 MΩ, 5%, 2 W)
$R_S = 83.3$ kΩ, 3 W
(82 kΩ, 5%, 5 W)
1.895 kV

7–15 a. 8.87 kΩ
(9.1 kΩ, 5%)
b. $P_B = 7.1$ W (10 W)
$P_S = 10.2$ W (15 W)
c. 322 V

7–17 a. 36 Ω, 5%
b. 4 Ω
(3.9 Ω, 5%)
c. 2.25 W each
d. 17.9 V
e. 13.43 V

7–19 $R_S = 2.5$ kΩ, 4 W
(2.4 kΩ, 5%, 5 W)
$R_1 = 8$ kΩ, 3.2 W
(8.2 kΩ, 5%, 5 W)
$R_B = 18$ kΩ, 0.45 W
(18 kΩ, 5%, 1 W)

7–21 $R_1 = R_4 = 100$ Ω, 0.64 W
(100 Ω, 5%, 1 W)
$R_2 = R_3 = 400$ Ω, 0.36 W
(390 Ω, 5%, 1 W)

CHAPTER 8

8–1 a. 9.38 lb
b. 10.62 lb

8–3 a. 16.5 h
b. 53 h
c. 87 h

8–5 a. 0.49 Ω, 283 W
b. 31.2 Ω
c. 5 min
d. No. Insufficient current

8–7 a. 30 A
b. 0.05 Ω

8–9 a. 6.2 A
b. 0.24 Ω

CHAPTER 9

9–1 a. 10 mA
b. 25 V
c. 75 V
d. 0.44 W
e. 40 mA

9–3 a. 0.25 Ω
b. 1.25 Ω
c. 0.5 Ω

9–5 a. 2.256 V
b. 0.256 Ω

9–7 20%

9–9 11.5 V

9–11 12 A

9–13 a. 24 V, 10 A, 0.04 Ω
b. 6 V, 40 A, 0.0025 Ω
c. 12 V, 20 A, 0.01 Ω

9–15 a. 12.68 V
b. 0.024 Ω
c. 2.92 V
d. 252 A
e. 3.81 kW, 2.54 kW

9–17 a. 0.02 Ω
b. 300 A
c. 1800 W
d. 1.7 hp
e. 1800 W
f. 6 V
g. 50%

9–19 a. 1.375 Ω, 46.625 Ω
b. 14.7%, 85.4%
c. 46.6 Ω

CHAPTER 10

10–1 a. 1.55 A recharging
b. 14.23 A
c. 12.68 A
d. 12.68 V

10–3 a. 180.4 A discharging
b. 176.8 A
c. 3.6 A opposite direction
d. 3.6 V opposite polarity

10–5 $I_{R_1} = 3\frac{1}{3}$ mA
$I_{R_2} = 3\frac{1}{3}$ mA
$I_{R_3} = 0$
$I_{R_4} = 1\frac{2}{3}$ mA
$I_{R_5} = 1\frac{2}{3}$ mA

10–7 a. $I_{R_1} = 2.04$ mA
$I_{R_2} = 2.51$ mA
$I_{R_3} = 1.39$ mA
$I_{R_4} = 3.16$ mA
$I_{R_5} = 0.65$ mA
b. 1.43 V
c. 3.3 kΩ

10–9 a. 3.04 kΩ
b. 4.03 kΩ

10–11 $I_{R_1} = 3.33$ mA
$I_{R_2} = 3.33$ mA
$I_{R_3} = 0.01 \text{ mA} \approx 0$
$I_{R_4} = 1.66$ mA
$I_{R_5} = 1.66$ mA

10–13 a. 1.52 A recharging
b. 14.2 A
c. 12.68 A
d. 12.68 V

10–15 3.12 mA

10–17 0.65 mA

10–19 3.08 mA

10–21 a. 0.186 mA
b. 0.243 mA

10–23 5.7 mA

10–25 3.33 mA

10–27 All 30 kΩ

10–29 2.6 kΩ

10–31 0.5 mA, 150 mV

CHAPTER 11

11–1 0.6 T

11–3 1.6×10^{-4} Wb

11–5 2.43 A

11–7 2125 At/m

11–9 a. 1.35 T
b. 750 At
c. 1705 At/m
d. 7.92×10^{-4} Wb/At-m
e. 630

11–11 9.66 A

11–13 a. 1.07×10^{6} At/Wb
b. 1.4×10^{-4} Wb

11–15 a. 3.75 N
b. 0.563 N · m

11–17 a. 0.5 A
b. 3 A
c. 6 V
d. 114 V
e. 78 W
f. 420 W
g. 282 W, 0.38 hp
h. 67%

CHAPTER 12

12–1 a. 25 μA
b. 40 kΩ

12–3 a. 40.8 Ω
b. 0.2 V
c. 40 Ω

12–5 a. 222.2 Ω, 40.8 Ω, 8.03 Ω, 2.002 Ω, 0.400 Ω
b. 0.2 V
c. 200 Ω, 40 Ω, 8 Ω, 2 Ω, 0.4 Ω

12–7 $R_1 = 1600\ \Omega$, $R_2 = 320\ \Omega$, $R_3 = 80\ \Omega$

12–9 a. $R_1 = 142.1\ \Omega$, $R_2 = 15.8\ \Omega$
b. 15.7 Ω
c. 15.1 Ω
d. 15 Ω

12–11 a. ±1 mA

b.	c.
4 to 6 mA	20%
14 to 16 mA	6.7%
24 to 26 mA	4%

12–13 a. 0.682 mA, 1.02 mW
b. 0.469 mA, 0.48 mW
c. 0.439 to 0.499 mA
d. 0.652 mA
e. 0.352 to 0.952 mA

12–15 a. 10 kΩ, 100 kΩ, 500 kΩ
b. 8 kΩ, 98 kΩ, 498 kΩ
c. 10 kΩ/V
d. 500 kΩ

12–17 a. 10 kΩ, 100 kΩ, 500 kΩ
b. 8 kΩ, 90 kΩ, 400 kΩ
c. 10 kΩ/V
d. 500 kΩ

12–19 a. 40 μA
b. 6 V
c. 8.33 V
d. 9.91 V

12–21 a. 4 V
b. 3.88 V

c. 0.92 V
d. 0.92 V
e. 2.93 V

12–23 a. 4.902 V
b. 4.898 V
c. 4.46 V
d. 4.46 V
e. 4.85 V

12–25 a. 9.32 kΩ/V
b. 13.87 to 16.37 V
c. 245 μA
d. 9%

CHAPTER 13

13–1 a. 1.253 V
b. −0.047 V
c. 3.1%

13–3 a. 113.24 Ω
b. 1.1%

13–5 a. 19.9 kΩ
b. 12.5%

13–7 a. 50.1 Ω
b. 10%

13–9 0, 7.5 kΩ, 20 kΩ, 45 kΩ, 120 kΩ, ∞

13–11 a. 18.7 kΩ
b. 1.3 kΩ too high
c. 0.2 V

13–13 a. 0
b. 57 kΩ
c. 0.25 A
d. 125 mA, 25 μA
e. 12.5 mA, 25 μA
f. 36 Ω, 4 Ω

13–15 ×3 kΩ

13–17 a. 45.2 Ω
b. ±0.6%
c. 0.181 A

13–19 a. 104.5 kΩ
b. ±3.1%
c. 55 μA
d. 320 μW

13–21 a. 104.5 kΩ
b. ±3.1%
c. 18.5 μA
d. 35.8 μW

13–23 a. 1 ± 0.0003 V, ±0.03%
b. 1 ± 0.0012 V, ±0.12%

13–25 a. 2 ± 0.003 mA, ±0.15%
b. 9 ± 0.01 mA, ±0.11%

13–27 a. ±0.264 kΩ
b. ±0.032%

CHAPTER 14

14–1 a. Negative
b. Positive

14–3 a. Negative
b. Positive

14–5 a. Negative
b. Positive

14–7 a. 3.3×10^{-3} Wb/s
b. 0.5 Wb/s
c. 0.9 Wb/s
d. 4×10^{3} Wb/s

14–9 5×10^{-3} Wb/s

14–11 1 T

14–13 b. +1.5 Wb/s, 0, +2 Wb/s, −5 Wb/s
c. 75 V, 0, 100 V, −250 V

14–15 15 V

14–17 a. 7.5 V
b. 10.6 V
c. 14.8 V
d. 15 V

14–19 a. 12.5 V
b. 12.5 V, −11.9 V, 10.8 V

14–21 471 V

14–23 a. 60 Hz
b. 16.7 ms

14–25 a. 5.77 V
b. 199 V

CHAPTER 15

15–1 a. $v = 5 \sin 40\pi \times 10^3 t$ volts
b. 2.94 V

15–3 a. 70 V
b. 140 V
c. 50 rad/s
d. 7.96 Hz
e. 0.126 s
f. −18.4 V

15–5 a. 1.82 A
b. 400 Hz
c. $v = 60 \sin 800\pi t$ volts
d. $i = 1.82 \sin 800\pi t$ amperes

15–7 a. 109 W
b. 54.5 W
c. 800 Hz

15–9 a. 5.3 V
b. 4.95 mA
c. 7.07 mV
d. 12 A

15–11 a. 339 V
b. 679 V
c. 42.4 A
d. 4.5 μV

15–13 a. 127.6 V
b. 74 W

15–15 0.25 W

15–17 a. 11 mW
b. 2.27 kΩ

15–19 a. 5 Ω, 3.08 Ω, 1.47 Ω
b. 28.8 W, 17.7 W, 8.5 W
c. 22.9 V

CHAPTER 16

16–1 a. 26.5 mV
b. 0.53 V
c. 13.25 V

16–3 a. 1.25 Hz
b. 312.5 Hz
c. 6.25 kHz
d. 125 kHz

16–5 600 Hz

16–7 a. 0.98 mA
b. 2.5 kHz
c. $v = 4.6 \sin 5000\pi t$ volts
d. $i = 0.98 \sin 5000\pi t$ mA

16–9 2.65 kΩ

16–11 a. 42.4 V
b. 120 V_{p-p}
c. 1.93 mA
d. 0.2 ms

16–13 a. 28 V
b. 9.33 kΩ
c. 5 kΩ
d. 3 kΩ

16–15 a. 2 mA
b. 3.43 mA, 0 V
c. 0 mA, 0 V, 48.6 V_{p-p}

CHAPTER 17

17–1 a. 14 A/s
b. 98 V

17–3 600 A/s

17–5 a. 0.13 mH
b. 23.4 mH

17–7 0.35 V, 0 V, −3 V

17–9 a. 120 mH
b. 10.9 mH
c. 20 mH in series with a parallel combination of 40 and 60 mH

17–11 a. 21 H, 9 H
b. 0.4

17–13 419 μH

CHAPTER 18

18–1 16.3 V

18–3 a. 32.6 A
b. 7.1 Ω

18–5 a. 9.58
b. 1.6 A
c. 0.167 A
d. 38.4 VA

18–7 a. 4.8 A
b. 5.3 A

18–9 a. 2.83
b. 17.7 V, 0.277 A
c. 6.3 V, 0.782 A
d. 4.9 W
e. 1.9 W

18–11 a. 755 A
b. 2523 A
c. 5000 kW

18–13 a. 1210 W
b. 11 A
c. 10.1 A
d. 0.9 A
e. 8.2%

18–15 a. 40 A
b. 1725 V
c. 52.5 kW
d. 69 kVA

CHAPTER 19

19–1 a. 1500 A/s
b. 2 mA
c. 1.33 μs
d. 6.7 μs

19–3 a. 48 Ω
b. 30 ms
c. 1.44 H

19–5 a. 0.33 A
b. 2.4 V
c. 41.3 ms
d. 17.5 ms

19–7 0.563 J

19–9 b. 46 mA
c. −12 V
d. 2.8 μs

19–11 a. 64 μs
b. 11 mH
c. 1.94 V

19–13 a. 2.5 A/s
b. 2.42 ms
c. 6.1 mA
d. 12.1 ms
e. 20 V
f. 122 V
g. 15.3 A/s
h. 0.34 ms
i. 1.7 ms

19–15 a. 0.324 A
b. 2.2 V
c. 40.2 ms
d. 17.3 ms

19–17 b. 22.3 mA
c. −6.07 V
d. 1.39 μs

CHAPTER 20

20–1 2.4 kΩ

20–3 1.19 H

20–5 3.37 MHz

20–7 a. 0.8 mA
b. $i = 1.13 \sin 2\pi \times 10^4 t$ mA

20–9 a. 12 V dc
b. 44 mV_{p-p}

20–11 a. 1 mA
b. 10 V, 20 V, 30 V

20–13 a. 892 Ω
b. 11.2 mA
c. 4.3 V, 5.7 V

20–15 0.36 H, 0.72 H

20–17 a. 0.128 H
b. 0.191 H
c. 0.67

20–19 a. 5.48 kΩ
b. 10.9 mA

20–21 a. 1.78 kΩ
b. 580 Ω

CHAPTER 21

21–1 5 μF

21–3 20 V

21–5 a. 0.02 μF
b. 1600 V

21–7 a. 1.51×10^{-3} m^2
b. 330 V

21–9 5.08×10^{-3} m^2

21–11 6.7×10^6 V/m

21–13 a. 122 μF
b. 16 V
c. 1.95 mC

21–15 a. 6.25 μF
b. 312.5 μC
c. 12.5 V
d. 100 V

21–17 a. 0.00857 μF
b. 114.3 V, 85.7 V

21–19 60 μF

21–21 0.0156 J

21–23 32 μF

21–25 a. 484 pF
b. 6 μF

CHAPTER 22

22–1 a. 5000 V/s
b. 1 mA
c. 10 μC

22–3 45 mA, 0, −37.5 mA

22–5 a. 1818 V/s
b. 0.27 mA
c. 6 V

22–7 a. 25 s
b. 0.2 V/s
c. 3.15 V
d. 125 s
e. 50 mA

22–9 a. 205 μA
b. 75.7 μA
c. 167 V
d. 440 s

22–11 a. 6 s
b. 100 mA
c. 15 ms
d. 40 A
e. 24 J
f. 1600 W

22–13 a. 34.4 V
b. 3.1 s

22–15 a. 131 V
b. 4 mA
c. 2.3 ms
d. −2.5 mA

22–17 a. 120 μs
b. 1.2 μF
c. 833 Hz

22–19 a. 13.3 kHz
b. 0.33 V/μs

22–21 a. 8.8 V
b. 0.2 V
c. 8.9 V, 0.1 V

22–23 a. 35.3 V
b. 3.36 s

22–25 a. 132 V
b. 3.8 mA
c. 2.26 ms
d. −2.5 mA

CHAPTER 23

23–1 575 Ω

23–3 19.9 V

23–5 1.8 μF

23–7 a. 1.19 mA
b. 119 mA
c. 11.9 A

23–9 82.9 Hz

23–11 a. 8.62 kΩ
b. 2.78 mA
c. 11.1 V, 7.4 V, 5.5 V
d. 0.046 μF

23–13 a. 5.89 kΩ
b. 1.81 mA, 2.71 mA, 3.61 mA
c. 8.15 mA
d. 0.45 μF

23–15 a. 100 Ω
b. 0.33

CHAPTER 24

24–5 a. 89.4 V
b. 63.4°
c. 2:1

24–7 a. 21.8 V
b. 24.6°
c. 9.9 mA
d. 0.4 H

24–9 a. 1.2 kΩ
b. 537 Ω
c. 1074 Ω
d. 10.74 V
e. 21.48 V
f. 2.85 H

24–11 a. 48.7 Ω
b. 145 mA
c. 15°

24–13 a. 1050 Ω
b. 81°
c. 159.6 Ω
d. 1037 Ω
e. 6.5
f. 2.75 H

24–15 a. 119 V
b. −62.8°
c. 41.3 μF

24–17 a. 1.7 kΩ
b. −25.1°

24–19 a. 45.7 kΩ
b. 76.6 μA
c. V_R = 0.17 V, V_L = 2.31 V, V_C = 5.81 V
d. −87.2°

24–21 a. 22.6 kΩ
b. 154.9 μA
c. V_R = 0.34 V, V_L = 9.36 V, V_C = 5.87 V
d. 84.4°

24–23 a. 56.4°
b. 19.9 V
c. 36 V

24–25 a. 55.3 mA
b. 868 Ω
c. −43.7°

24–27 a. 2.36 mA
b. 2.12 kΩ
c. −16°

CHAPTER 25

25–1 11.9 Ω

25–3 a. 19.1 vars
b. 19.1 VA

25–5 175.7 μF

25–7 69 VA

25–9 946 VA

25–11 a. 107.8 W
b. 73.9 vars
c. 130.7 VA

25–13 a. 21.3 W
b. 26.5 vars
c. 34 VA
d. 0.63

25–15 a. 1.1 W
b. 1.8 vars
c. 2.1 VA
d. 0.52

25–17 a. 306 W
b. 111 vars (inductive)
c. 326 VA
d. 2.7 A
e. 0.94 (lagging)

25–19 a. 53.1°
b. 1870 VA
c. 1122 W
d. 1496 vars

25–21 a. 11.8 A
b. 9.2 A
c. 8.3 A
d. 9.2 A

25–23 a. 82 μF
b. 5.1 A

25–25 a. 0.248 A
b. 29.8 VA
c. 12.3 W
d. 3.1 W
e. 0.52
f. 25.5 vars
g. 4.7 μF
h. 0.13 A, Dimmer lamp

25–27 a. 0.87
b. 86.7%

25–29 150 Ω resistance in series with an inductive reactance of 531 Ω; (L = 42 μH)

CHAPTER 26

26–1 a. 31.8 Ω capacitive
12.04 Ω resistive
29.7 Ω inductive
b. −67.8°, −4.8°, 66.2°
c. 0.629 mA, 1.66 mA, 0.673 mA
d. 7.55 mV, 880 mV, 899 mV
19.9 mV, 2.35 V, 2.35 V
8.1 mV, 961 mV, 943 mV

26–3 1.5005 MHz

26–5 0.25 μF

26–7 a. 240 kHz, 260 kHz
b. 3 mW
c. 12.5

26–9 a. 118
b. 12.7 kHz

26–11 54.4

26–13 a. 37.1 V (rms)
b. 0.02 μF
c. 476 Hz
d. 83.3 mW

26–15 60 to 258 pF

26–17 a. 142.4 kHz
b. $I_R = 0.2\ \mu A$
$I_L = I_C = 22.4\ \mu A$
c. 0.2 μA

26–19 a. 112
b. 50 kΩ
c. 22.4 μA
d. 1.27 kHz

26–21 a. 5.6 mV
b. 1.1 mV

26–23 145.5 kHz

26–25 a. 140.1 kHz
b. 5.5
c. 25.5 kHz

26–27 a. 18.4 Ω
b. 2.3 Ω, 136.7 kHz

26–29 1.8 to 2.6 pF

CHAPTER 27

27–1 $80 + j40$ V
89.4 V $\angle 26.6^\circ$

27–3 $25 - j40\ \Omega$
47.2 Ω $\angle -58^\circ$

27–5 $68 + j75.4\ \Omega$
101.5 Ω $\angle 48^\circ$

27–7 a. 200 $\angle 35^\circ$
b. 12.5 $\angle 85^\circ$
c. 0.08 $\angle -85^\circ$

27–9 a. 2.54 A $\angle 58^\circ$
b. 63.5 V $\angle 58^\circ$
c. 101.6 V $\angle -32^\circ$

27–11 a. 0.47 A $\angle -48^\circ$
b. 32.2 V $\angle -48^\circ$
c. 35.7 V $\angle 42^\circ$

27–13 a. Capacitive
b. 240 Ω $\angle -30^\circ$
c. 208 Ω, 120 Ω

27–15 a. $11 + j4$
b. $-5 - j18$
c. $5 + j18$

27–17 $50 + j100\ \Omega$,
112 Ω $\angle 63.4^\circ$

27–19 a. $68 - j24$
b. $-0.308 - j0.462$
c. $-1 + j1.5$

27–21 a. $6.1 + j13.7\ \Omega$,
14.95 Ω $\angle 66^\circ$
b. 1.6 A $\angle -66^\circ$
c. 0.89 A $\angle -68.2^\circ$,
0.72 A $\angle -63.4^\circ$

27–23 a. 2.89 kΩ $\angle 4.8^\circ$
b. 34.6 μA $\angle -4.8^\circ$
c. 92.8 μA $\angle -68.2^\circ$,
83.3 μA $\angle 90^\circ$

27–25 a. 11.7 $\angle 59^\circ$
b. 7.6 $\angle -66.8^\circ$
c. 10 $\angle -143^\circ$

27–27 14.93 Ω $\angle 66^\circ$

27–29 a. $-37.3 + j53.2$
b. $3.7 - j11.4$
c. $-18.8 - j6.8$

27–31 10.9 Ω, 127 μH

27–33 a. 118.5 mA
b. 400 Ω $\angle 50^\circ$
c. 3.7 W

27–35 a. 382.7 Ω $\angle 45.9^\circ$
b. 62.7 mA $\angle -45.9^\circ$
c. 20.7 V $\angle 26.4^\circ$
d. 63.1 mA $\angle -31.2^\circ$

27–37 a. 4.86 kΩ, 15.9 H
b. 4.85 kΩ, 38.2 mH

27–39 0.57 μA $\angle -41.6^\circ$

27–41 $R = 232$ kΩ, $C = 0.043\ \mu F$

27–43 $R_x = \dfrac{R_2 R_3}{R_1}$, $L_x = C_s R_2 R_3$

CHAPTER 28

28–1 a. 10.8 V
b. 72.2 mA
c. 23 mA
d. 0.25 W
e. 34 V

28–3 a. 20 V
b. 56.6 Ω

28–5 a. 108 V
b. 108 mA
c. 11.7 W
d. 339 V
e. 120 Hz

28–7 a. 40 V
b. 0.57 A

28–9 a. 216 V
b. 216 mA
c. 46.7 W
d. 339 V
e. 120 Hz

28–11 a. 281 V
b. 1.58 kW
c. 294 V
d. 360 Hz

28–13 a. 1.2 V
b. 3.4 V

28–15 b. 8.8%

28–17 a. 2.41%
b. 8.55 V

28–19 1205 μF

28–21 2.5 H

28–23 1.3 H, 130 μF

CHAPTER 29

29–1 a. 15.3 mA
b. 0.2 mA
c. 24.75 mA

29–3 a. 50, 49.5, 99
b. 0.98, 0.98, 0.99

29–5 a. 125
b. 172
c. 190

29–7 At I_C = 2.5 mA, β_{ac} = 146
At I_C = 17 mA, β_{ac} = 181
At I_C = 32 mA, β_{ac} = 223

29–9 22.5 mA, 22.65 mA

29–11 0.143 A, 2.86 A

29–13 a. 0.968
b. 30.3

29–15 a. 1 mA
b. I_{CEO} is too small

29–17 a. 5 mA
b. −125
c. 11 V
d. −220
e. 1.25 kΩ
f. −220
g. 44.4 dB

29–19 a. 19 V
b. 18.4 V
c. 740 kΩ

29–21 722.5 kΩ

29–23 a. 10 W
b. 141
c. 1.42 A

29–25 a. 17.6
b. 9.4
c. 600

29–27 a. 0.045
b. −20
c. 13.75 kΩ
d. 165 kHz
e. 0.36%
f. −3300 kHz
g. −125
h. 34 dB

29–29 a. 0.145
b. −6.3
c. 13.8 kΩ
d. 165 kHz
e. 0.36%
f. −1035 kHz
g. −125
h. 29 dB

29–31 a. 62.4 kHz
b. 600 mV
c. 3

INDEX

Accelerating anode, 290
Acceptor atom, 546
Accuracy:
  FSD, 211, 229
  numerical, 8
Ac resistance, 372
Active devices, 10
Active networks, 156–171
Addition:
  by geometrical construction, 440
  of instantaneous values, 439
  of sine waves, 439
  vector, 440, 520
Admittance, 529, 621
Air-core coils, 310
Air gap, 198, 206, 352
Algebra, vector, 513, 532
Alternating current:
  average value, 278
  effective value, 278
  instantaneous value, 257, 274, 276, 294
  peak value, 276, 293
  sine-wave, 258–261, 274–282, 293
  waveforms, 275
Alternating-current bridges, 539
Alternating-current circuits, 435–453
  impedance networks, 531
  parallel, 282, 450–453, 529
  power in, 277, 461–478
  series, 282, 437–450
  simple, 275, 438
  three-phase, 261, 476
Alternating emf:
  average value, 278
  effective value, 279
  generation of, 253, 257, 261
  instantaneous value, 258, 274, 294
  nature of, 257
  peak value, 258, 276, 293
  sine-wave, 258–261, 274–282, 293
Alternator, 261, 556
Amalgamation, 121
American wire gauge, 63, 109
Ammeter:
  ac, 298, 336
  clamp-on, 212, 336, 626
  dc, 44, 207–213
  multirange, 16, 208
Ammeter calibration, 228
Ammeter loading, 212
Ammeter shunts, 207
Ampere, definition, 33, 621
Ampere-hour, 123, 128, 146, 621
Ampere-turn, 186, 621
Amplification, 621
Amplifier, 577–586, 622
  bandwidth, 582, 584
  bias, 577, 580
  bypass capacitor, 582
  cascaded, 580
  CE, ac, 577
  complementary symmetry, 574
  coupling capacitor, 577
  current gain, 579
  distortion, 579, 584
  feedback, 585
  frequency response, 582
  gain, 579
  gain-bandwidth product, 584
  input resistance, 579, 584
  power gain, 579
  voltage gain, 579
  waveforms, 579
Amplitude, of sine wave, 258, 608, 622
Amplitude modulation, 498
Angle:
  effect of, 256
  impedance, 442
  phase, 364, 438
  power factor, 470
  in radians, 274
Angular frequency, 275
Angular velocity, 275
Anode, 121, 290, 550, 622
Antiresonance, 497
Apparent power, 324, 465, 622
Aquadag, 210, 622
Arcing, 352
Armature, 183, 184, 196, 622
Atom, 623
  acceptor, 546
  Bohr, 30
  donor, 546
  structure, 31
Audio generator, 295, 350
Auto transformer, 333, 623
Average value:
  current, 298, 553
  power, 277
  voltage, 263, 553, 556
Ayrton shunt, 209, 623

Back-off scale, 232
Balanced three-phase load, 477
Ballast, 354, 368
Bandwidth, 623
  amplifier, 582, 584
  resonant circuit, 491, 500

Battery:
  alkaline-manganese, 122
  charging, 126, 128
  construction, 121, 122, 127, 128
  dry-cell, 11, 121
  lead-acid storage, 125–128
  lithium sulfide, 131
  mercury, 124
  nickel-cadmium, 128–130
  paper-thin, 124
  solar, 133
  temperature effects, 123, 125
  Weston standard, 226–228
  wet-cell, 120, 125
  zinc-nickel oxide, 130
Bell, 183
BH curves, 188, 623
Bias, 624
  fixed, 581
  stabilization, 581
  voltage divider with emitter, 581
Bifilar winding, 356
Bimetallic switch, 184, 624
BJT, *see* Transistors
Bleeder resistor, 112
Bonding:
  covalent, 31, 544
  incomplete, 546
Branch current method, 156
Bridge, general equation, 535
Bridge rectifier, 335, 554
Bridges:
  ac, 535
  impedance, 374
  Maxwell, 539
  Wheatstone, 159, 172, 234–237
Brownout, 261
Brushes, 196, 257, 261, 263, 624
Buzzer, 183
Bypass capacitor, 582

Calculator, 6
Calibration, 228, 231, 293
Capacitance, 381–394, 624
  in ac circuits, 423–429
  in dc circuits, 401–414
  definition, 382
  derivation of formula, 614
  factors governing, 383
  farad, 383
  measurement, 391, 411
  stray, 411
Capacitive, 448, 452
Capacitive reactance:
  definition, 425, 624
  derivation of formula, 619
  factors governing, 425
Capacitive susceptance, 529
Capacitors:
  basic action, 382
  charging, 382, 403, 615
  checking, ohmmeter, 406
  construction, 388
  definition, 382, 625
  dielectrics, 384–388
  discharging, 382, 407, 615
  electrolytic, 388
  energy stored, 394
  filter, 412
  graphic symbol, 11, 384
  leaky, 389
  parallel, 390, 428
  power factor improving, 475
  series, 392, 427
  smoothing, 412, 552
  types, 389
  typical, 12, 390
  working voltage, 386, 393
Capacity, 123, 128, 146
Cathode, 121, 290, 550
Cell:
  alkaline-manganese, 122
  carbon-zinc, 121
  copper-zinc, 120
  fuel, 132, 631
  gelled-electrolyte, 130
  heavy duty, 124
  lead-acid, 125–128
  Leclanché, 121
  lithium, 124
  lithium sulfide, 131
  maintenance-free, 127
  mercury, 124
  nickel-cadmium, 128–130
  parallel-connected, 145
  photovoltaic, 133, 639
  Planté, 126
  primary, 120–124
  secondary, 125–131
  series-connected, 145
  silver cadmium, 130
  silver oxide, 124
  silver zinc, 130
  sodium-sulfur, 131
  solar cell, 133
  standard cell, 226–228
  voltaic, 120
  zinc-nickel oxide, 130
Characteristic curves:
  diode, 549
  transistor, 574
Charge, electric:
  bound, 548
  in capacitors, 383
  carriers, 28, 544
  coulomb, 27
  definition, 27
  induced, 387
  positive and negative, 26, 31
Charging current, 126, 128, 382, 403
Chassis, 625
Chassis ground, 83
Chemical energy, 120, 125, 129, 132
Choke, filter, 367, 559
Choke coil, 311, 367, 625
Circuit:
  ac, 10, 277, 282, 435–453
  capacitive dc, 402–414
  coupled, 427
  CR, 401–414
  dc, 10, 76–173
  differentiating, 412
  electric, 10
  electronic, 10
  equivalent, 78, 94, 161, 166, 172, 390, 530
  integrating, 414
  LR, 344–357
  magnetic, 197
  magnetically coupled, 324
  main parts, 10
  parallel, 11, 94–100, 282, 390, 428, 450, 529
  RC, delay, 414
  rectifier, 552–556
  resonant, 488, 497
  series, 11, 76–86, 282, 392, 427, 437–450, 516
  series-parallel, 106–113, 527
  three-phase, 262, 476
Circuit breaker, 330, 625
Circuit diagram, 13
Circular mil, 63, 625
Coefficient:
  of coupling, 314, 626
  temperature, 64, 390

Coercivity, 188, 626
Coils:
  air-core, 310
  choke, 311, 367
  degaussing, 191
  iron-core, 310
Commutator, 196, 263, 626
Complex numbers, 513–532, 626
Compound motor, 197
Conductance, 36, 45, 95, 529, 627
Conductivity, 627
Conductors:
  electric, 10, 30, 627
  semi-, 31, 64, 195, 643
Conjugate quantity, 477, 522
Constant-current source, 164
Control grid, 290
Conventional current direction, 33
Conversion:
  polar-to-rectangular, 526
  rectangular-to-polar, 524
Conversion factors, 8
Conversion of units, 9
Coordinates:
  polar, 514
  rectangular, 519
Coulomb, definition, 27
Coulomb's law, 26
Counter emf, 308, 344, 627
Coupled circuit, 427
Coupling:
  coefficient of, 314, 626
  magnetic, 324
  mutual, 314
Coupling capacitors, 427, 577
Covalent bonding, 31, 544
Critical damping, 503
CRO, 290–297
CRT, 290, 627
CR time constant, 404
CR wave shaping, 411
Curie temperature, 191, 627
Current:
  alternating, 257–261, 274–282
  ampere, 33
  branch, 94, 156
  capacitive, 382, 424, 452
  charging, 126, 382, 402
  collapse, 350
  conventional, 33
  definition, 32
  dependent variable, 45
  direction, 33
  discharging, 29, 126, 382, 407
  displacement, 385
  eddy, 325
  electric, 10, 28, 32
  electron flow, 33
  exciting, 323
  hole flow, 544
  inductive, 345, 364, 451, 452
  inrush, 67, 404
  instantaneous, 258, 276, 279
  lagging, 437, 440, 451
  leading, 437, 445
  leakage, 389
  line, 110
  loop, 158
  magnetizing, 323
  polar form, 517
  pulsating, 263, 552
  rate of change, 345
  resonant rise of, 500
  rise, 344
  short-circuit, 99, 125, 164
  sine-wave, 258, 276
  steady-state, 345
  tank, 500
  total, 94, 106, 112, 450
Current carriers, 546
Current divider principle, 98, 283
Current ratio, 323
Current sensitivity, 207, 216
Current source, 164
Current transformer, 334, 627
Cycle, 258, 627

Damping, resonant circuits, 503
D'Arsonval movement, 206
Decay, current, 347
Decibels, 580, 627
Declination, 192
Deflecting plates, 290
Degaussing, 191
Degrees, electrical and mechanical, 259
Delay circuit, 414
Delta-connected three-phase system, 262
Delta-wye transformation, 172
Demagnetization, 190
Demodulation, 506
Dependent variable, 45
Depletion region, 548
Depolarizer, batteries, 121
Determinants, 603
Diagram:
  circuit, 11
  impedance, 441
  phasor, 438–452
  pictorial, 11, 13
  schematic, 11, 13
  symbols, 11, 598
Diamagnetic materials, 180, 628
Dielectric, 384–388, 628
Dielectric constant, 384, 628
Dielectric hysteresis, 429
Dielectric strength, 64, 386, 628
Difference in potential, 28, 84
Differentiating circuit, 412, 628
Digital, *see* Meters
Diode:
  avalanche effect, 550
  characteristic *VI* curves, 549
  construction, 547, 551
  forward-biased, 548
  free-wheeling, 356
  germanium, 551
  light emitting, 543
  ohmmeter checking, 550
  peak inverse voltage, 550
  rectifiers, 552–556
  reverse-biased, 548
  silicon, 547–552
  symbol, 548
  Zener, 550
Direct current, 10, 32, 628
Direct current motors, 195
Direct probe, 293
Discharge:
  battery, 124, 126, 131
  capacitor, 382, 407
  displacement current, 385
  resistor, 356
Dissipation factor, 428, 629
Distortion, 579, 584, 679
Divider:
  current, 98, 283
  voltage, 81, 282, 393
Domains, magnetic, 189, 629
Donor atoms, 546
Doorbell, 183
Door chimes, 183
Dot notation, 313
Drift, electron, 34, 548
Drop of potential, 76, 109

Dry cell, 11, 121, 128
Dry charge, 127
Dynamic microphone, 253

Earth's magnetic field, 192
Eddy current, 325, 629
Edge frequency, 492
Edison three-wire system, 330
Effective resistance, 373
Effective value, 278, 629
Efficiency, 47, 49, 147, 324, 629
Electrical degrees, 259
Electric charge, 27
Electric energy, 10, 47, 51, 125, 349, 394
Electric field, 385, 387, 629
Electricity:
  definition, 629
  static, 26
Electric lines of force, 385
Electric potential, 28, 84
Electric power, 49, 148
Electric quantity, 27
Electric shock, 14, 394
Electrode, 30, 121, 629
Electrodynamometer, 298
Electrolysis, 132
Electrolyte, 121, 128, 630
Electromagnet, 182, 630
Electromagnetic induction, 250
Electromagnetism, 182
Electromotive force:
  alternating, 257–261, 274–282
  counter, 308, 344
  definition, 30, 630
  direct, 10, 30, 120–134, 263
  Hall, 194, 213, 336
  induced, 250, 255, 308
  source, 120–134, 250
  volt, 29
Electron gun, 290
Electron-hole pair, 544
Electronic calculator, 6
Electrons:
  charge, 27
  drift, 34
  energy levels, 368
  flow, 28, 33
  free, 32, 544
  mass, 31
  shells, 31
  theory, 27
  valence, 31, 368, 544
  velocity, 34
Electron trace, 290
Electrostatic field, 385, 387
Electrostatic induction, 387
Electrostatics, 28, 630
Energy:
  chemical, 120, 125, 129, 132
  cost, 51
  definition, 47, 630
  electric, 10, 47, 126, 146, 195
  heat, 373
  joule, 29, 47
  levels in atom, 368
  mechanical, 46, 196
  storage, 349, 394
  work, 47
Engineering prefixes, 9
Equations:
  general bridge, 535
  Kirchhoff's law, 77, 94, 156
  loop, 158
  mesh, 158
  Wheatstone bridge, 235
Equivalent circuits:
  capacitor, 390, 429
  circuit simplification, 166
  delta-wye, 172
  Norton, 166
  parallel, 94, 97, 145, 502
  series, 78, 86, 145
  series-parallel, 106, 145
  Thévenin, 161, 531
Equivalent impedance, 529
Equivalent reactance, 530
Equivalent resistance, 78, 95, 97, 373, 530
Equivalent susceptance, 530
Errors, 211, 229
Exciting current, 323
Exponential curves, 347, 409
Exponential equations, 610, 615

Factor:
  conversion, 8
  dissipation, 428
  power, 470
  reactive, 471
Fall of current, 350
Farad, definition, 383
Faraday's law, 251, 322
Feedback:
  advantage, 584
  loop gain, 583
  negative, 583
  positive, 583, 586
Ferrites, 189, 325
Ferromagnetic materials, 180, 631
Field:
  electric, 385, 387
  magnetic, 180
Field intensity:
  electric, 385
  magnetic, 186
Field windings, 196, 261, 264, 631
Filter capacitors, 412, 557
Filter chokes, 367, 559
Filter networks, 367, 412, 557–561
Fluorescent:
  lamp, 368
  screen, 290
Flux:
  electric, 631
  magnetic:
    definition, 180
    density, 186
    leakage, 310
    lines of force, 180
    linkages, 310, 314
    mutual, 314
    weber, 252
Flux density, magnetic, 186
Focusing anode, 290
Force:
  coercive, 188
  electromotive, 30, 120–134
  electrostatic, 26
  gravitational, 47
  magnetic, 192
  magnetizing, 186
  magnetomotive, 186
  mechanical, 46
Forward-current transfer ratio, 574, 576
Four-way switch, 330
Four-wire three-phase system, 262
Free electrons, 32, 544, 631
Frequency:
  angular, 274
  conversion, 505
  definition, 259, 631
  edge, 492
  fundamental, 501

harmonic, 501
intermediate, 505
measurement, 293, 295, 300
resonant, 490, 502
Fuel cell, 132, 631
Full scale deflection, 207, 214
Full-wave rectification, 262, 298, 553, 632
Function generator, 296
Function switch, 15
Fundamental, 501
Fuse, 99, 330, 632

Galvanometer, 226, 235, 253, 632
Gauss, 191, 632
Gaussmeter, 194, 632
General bridge equation, 535
Generation of ac and dc, 249–264
Generator:
ac, 257–261
dc, 263
definition, 257, 632
four-pole, 260
three-phase, 261
two-pole, 257
Geometric construction, vector diagrams, 440
Germanium diode, 551
Graphic symbols, 11, 598
Graticule, 293
Ground, 83, 113, 300, 632
Ground fault circuit interrupter, 331

Half-cycle average, 553
Half-power points, 492, 582
Half-wave rectification, 552
Hall effect, 194, 213, 336, 632
Hand-rule, magnetic field direction, 182
Harmonics, 501, 633
Heat energy, 373
Henry, 309
Hertz, 259, 633
Heterodyne, 505, 633
Hole flow, 544, 633
Horsepower, 48, 633
Hydrometer, 128, 633
Hysteresis:
dielectric, 429
magnetic, 188, 324, 633
Hysteresis loop, 188, 617
Hysteresis loss, 189, 324, 429

IR drop, 77
Ignition system, automotive, 354
Imaginary component, 518
Impedance:
conjugate, 477
definition, 441, 633
internal, 477
parallel, 451, 523
polar form, 517
rectangular form, 519
reflected, 328, 477
series, 441–448, 521
series-parallel, 527
Impedance angle, 442, 514
Impedance bridge, 374
Impedance diagram, 442
Impedance matching, 327, 477
Impedance networks, 531
Impedance transformation, 327
Impedance triangle, 441
Impurity materials, 545
acceptor, 546
donor, 546
pentavalent, 545
trivalent, 546
Independent variable, 45
Induced electric field, 387
Induced emf, 250, 255, 308, 633
Inductance, 306–315
in ac circuits, 363–374
in dc circuits, 343–357, 610
definition, 309
factors governing, 310
henry, 309
measurement, 351, 374, 391, 443, 466
mutual, 312, 370
self-, 308, 633
total, 313
Induction:
electromagnetic, 250
electrostatic, 387
magnetic, 181
mutual, 312, 370
self-, 308, 605
Induction heating, 325
Induction motor, 633
Inductive, 448, 452, 518
Inductive reactance:
definition, 365, 633
derivation of formula, 612
factors governing, 366

Inductive susceptance, 529
Inductors:
definition, 311, 633
energy stored in, 349
filter, 367, 559
graphic symbol, 11, 309
parallel, 311, 371
practical, 311
series, 311, 370
smoothing, 367, 559
typical, 12
Input resistance, 579, 584
Inrush current, 67, 404
Instantaneous value:
addition, 439
current, 276
power, 277
voltage, 257
Instrument, measuring, 206–218, 225–239, 289–301
Instrument transformers, 334
Insulation resistance, 389
Insulator breakdown, 64, 352, 385
Insulators, 30, 64, 634
Integrated circuit, 634
Integrating circuit, 414, 617, 634
Intensity:
electric field, 385
magnetic field, 186
Intermediate frequency, 505
Internal resistance, 96, 109, 121, 142–146, 327
International system of units, 8
Ionization, 352, 368
Ions, 120, 634
Iron, magnetic properties, 189
Iron-core coils, 309
Iron-core transformers, 322–334
Iron-vane meter, 300, 634
Isolation transformer, 323, 332

j operator, 518
Joule, definition, 29, 47

Kilowatthour, definition, 51, 634
Kirchhoff law equations, 77, 94
Kirchoff's laws:
in ac circuits, 439
current law, 94, 156
in dc circuits, 77, 156, 344, 404
voltage law, 77, 158

Laminations, 325
Lamp:
  fluorescent, 368
  incandescent, 67
  neon, 353, 414
Laws:
  Coulomb's, 26
  Faraday's, 251, 322
  Kirchhoff's, *see* Kirchoff's laws
  Lenz's, 251, 308, 344, 635
  Ohm's, 44, 197, 281
LC filter, 560
LC ratio, 494
Leakage current, 389
Leakage flux, 310
Leclanché cell, 121
Lenz's law, 251, 308, 344, 635
Linear impedance network, 531
Linear resistor, 45, 66
Line current, 110
Lines of force:
  electric, 385
  magnetic, 180
Line voltage, 109, 330
Linkages, flux, 310
Lissajous figures, 295, 448
Load:
  balanced, 477
  delta-connected, 173
  electric, 10, 46, 143
  unbalanced, 476
  wye-connected, 173
Loading, 212, 216
Load resistance, 46, 143
Lodestone, 179
Loop, tracing, 156, 158
Loop current, 158
Loop equations, 159
Loss:
  eddy current, 325
  hysteresis, 189, 324, 429
  winding, 324
Loudspeaker, 195
LR time constant, 345

Magnet:
  bar, 180
  electro-, 182, 630
  permanent, 180
  temporary, 181
Magnetic circuits, 197, 635
Magnetic core memory, 190
Magnetic declination, 192
Magnetic domains, 189
Magnetic field, 180, 192, 636
  intensity, 186
  strength, 186, 636
Magnetic flux, 180, 636
  density, 180, 186, 636
Magnetic force, 192
Magnetic induction, 181
Magnetic lines of force, 180
Magnetic materials:
  diamagnetic, 180
  ferrites, 189, 325
  ferromagnetic, 180, 189, 325
  nonmagnetic, 180
  paramagnetic, 180
"Magnetic Ohm's law," 197
Magnetic permeability, 187
Magnetic poles, 180, 259
Magnetic relay, 184
Magnetic reluctance, 197
Magnetic retentivity, 181, 188
Magnetic saturation, 188
Magnetic shielding, 325
Magnetic tape, 190
Magnetic variation, 192
Magnetism, 179–199
  residual, 181, 188, 264
Magnetite, 179
Magnetization curves, 188
Magnetizing current, 323
Magnetizing force, 186
Magnetizing intensity, 186
Magnetomotive force, 186, 636
Majority current carriers, 546, 548
Matching, 147, 328
Maximum power transfer, 147, 327, 477, 636
Maxwell bridge, 539
Mechanical energy, 46
Mechanical force, 46
Mechanical power, 48
Memory effect, 130
Mesh equations, 158
Meter movement:
  D'Arsonval, 206, 297
  electrodynamometer, 298
  moving-coil, 206, 297
  moving-iron, 300
  Weston, 206
Meter multiplier resistor, 213
Meters:
  ammeter, 16, 207–213, 299, 336
  capacitance, 391
  connection of, 16, 44
  digital, 213, 237, 300, 336
  frequency, 300
  galvanometer, 226, 235
  gaussmeter, 194
  impedance, 515
  multimeter, 15, 237, 336
  ohmmeter, 16, 231–234
  oscilloscope, 290–297
  potentiometer, 226
  power factor, 472
  true reading rms, 281, 299, 300
  varmeter, 465
  vector, 515
  voltmeter, 16, 213–218, 515
  VTVM, 15, 216
  wattmeter, 148, 299
Meter shunt resistor, 207
Microphone, 253
Millman's theorem, 169
Minority current carriers, 546, 548
Mixer, 505
Motor action, 192, 251
Motors, dc, 195
Moving-coil movement, 206, 297
Moving-iron movement, 300
Multimeter, 15, 237, 336
Multiplication, vector, 515, 522
Multiplier, voltmeter, 213
Multirange meters, 15, 208, 214, 233
Mutual coupling, 314
Mutual flux, 312, 314
Mutual inductance, 312, 314, 370, 637
Mutual induction, 312, 370
Mutual reactance, 370

Negative charge, 26
Negative potential, 29, 84
Network analysis, 154–173
  branch current, 156
  delta-wye, 172
  loop current, 158
  mesh current, 158
  Norton's, 166
  superposition, 159
  Thévenin's, 161
Networks:
  active, 10, 156–171
  attenuator, 177
  bridge, 159, 172

bridged-T, 177
delta, 172, 262
Edison three-wire, 330
multisource, 156, 159
passive, 10, 172
pi-, 173
power distribution, 328
resistance, 156–173
T-, 173
three-terminal, 172
two-terminal, 161
Wheatstone bridge, 159, 172, 234–237
wye, 172, 262
Network theorems:
Millman's, 169
Norton's, 166
superposition, 159
Thévenin's, 161, 531
Neutral, 262, 300, 330, 637
Neutron, 30, 638
Newton, 26
Nodes, 94, 638
Noninductive resistors, 356
Nonlinear magnetic circuits, 198
Nonlinear resistors, 45, 67
Nonlinear scales, 232, 299
Nonmagnetic materials, 180
Normal magnetization curve, 188
Norton's theorem, 166
Notation, scientific, 6
N-type semiconductor, 545
current in, 546
Nucleus, 31, 637
Null condition, 226, 235, 638
Numerical accuracy, 8

Oersted, 181, 191, 638
Ohm, definition, 35
Ohmmeter, 16, 46, 231–234, 406
definition, 62
Ohm's law, 44, 197, 281
Ohms per volt rating, 216
Open circuits, 80
Open-circuit voltage, 142, 161
Operator *j*, 518
Orbits, electron, 31
Oscillator:
local, 505
Wien bridge, 587
Oscilloscope, 290–297, 638
calibration, 293
construction, 290
controls, 291–293
dual-beam storage, 297
measurement of frequency, 293, 295
measurement of phase angle, 449
measurement of voltage, 293
operation, 291
probe, 293
probe compensation, 293
single beam, 292
Overload, 99, 638

Parallax, 211
Parallel capacitors, 390, 428
Parallel cells, 145
Parallel circuits, 94–100, 282, 390, 428, 450
Parallel impedances, 523
Parallel inductors, 311, 371
Parallel resistors, 94–100, 282
Parallel resonance, 452, 497–504, 639
Paramagnetic materials, 180, 639
Passive networks, 10, 172
Peak value of sine wave, 258, 639
Percentage error, 211
Period, 260, 300, 639
Permanent magnet, 180
moving-coil, 206, 297
Permeability:
definition, 187, 639
free space, 187
relative, 187
Permittivity:
absolute, 388, 639
free space, 384
relative, 387
Phase angle, 364, 424, 437, 445, 639
measurement, 448
Phase inversion, 578
Phasor, 436, 639
see also Vector
Phasor diagram, 438–452
Phosphor, 290
Photoflash, capacitor, 408
Photovoltaic cell, 133, 639
Pictorial diagram, 11
Piezoelectric effect, 134, 639
Pi-network, 173
PN junction, 547–552
breakdown voltage, 550
capacitance, 549
characteristic curves, 549
construction, 547
cut in voltage, 549
depletion region, 548
forward-biased, 548
potential barrier, 548
reverse-biased, 548
reverse current, 549
temperature dependence, 552
Zener voltage, 550
Polar coordinates, 514
Polarity markings, 79, 277, 300, 308, 551
Polarity of voltage drop, 76
Polarization:
atom, 385, 387, 640
battery, 121, 646
capacitor, 388
generator, 264
Poles, magnetic, 180, 259
Polyphase systems, 261, 640
Positive charge, 26, 640
Positive potential, 29, 84, 112
Potential:
barrier, 548
depletion layer, 548
ionization, 352, 368
Potential difference, 10, 28, 84
instantaneous, 257
Potential drop, 77
Potential rise, 77
Potential transformer, 334
Potentiometer, 53, 82, 226, 640
Power, 461–478
apparent, 324, 465
average, 277, 463
definition, 48, 640
electrical, 49
factor, 324, 470, 640
instantaneous, 277, 462
lagging reactive, 465
leading reactive, 465
maximum transfer of, 147, 327, 477
measurement, 148, 299, 476
mechanical, 48
net reactive power, 469
parallel circuit, 95
peak, 277
in pure capacitance, 464
in pure inductance, 462

in pure resistance, 462
reactive, 463
series circuit, 79
series-parallel, 109
three-phase, 476
true, 281, 462
var, 463
voltampere, 324, 465
watt, 48, 462
Power distribution, 328
Power factor:
angle, 470
correction, 473
definition, 324, 470
lagging, 472
leading, 472
meter, 472
unity, 473
Power supply, 85, 561, 640
Power triangle, 466
Prefixes, 9
Primary, transformer, 322
Primary cell, 120–124, 641
Probes, 293
Propagation velocity, 35
Proton, 30, 641
P-type semiconductor, 546
Pulsating current, 263
Pulsating power, 277
Pythagoras, 440, 467, 468

Q factor, resonant circuits, 492, 499, 641
Quadrature power, 463
Quality of coil, 372, 444
Quantity of electric charge, 27

Radian, 274, 641
Radio frequencies, 505
Range switch, 15, 208, 214
Rate of change, 251, 254, 345, 402, 424
Reactance:
capacitive, 424, 619
inductive, 365, 612
mutual, 370
net, 447
Reactive factor, 471
Reactive power, 463, 471, 641
Real component, 518
Real power, 462, 641
Recombination in semiconductors, 546, 548
Rectangular coordinates, 519
Rectifier, 261, 298, 552–556, 641
Rectifier circuits:
bridge, 554
full-wave, 553
half-wave, 552
three-phase, 556
Rectifier filters, 557–561
Reference phasor, 436, 450, 517
Reference signs, 277
Reflected impedance, 328, 477
Regulation, voltage, 112, 143, 642
Regulator, voltage, 261, 550
Relative permeability, 187, 642
Relative permittivity, 387
Relay, magnetic, 184, 262, 642
Reluctance, 197, 642
Residential wiring, 330
Residual magnetism, 181, 188, 264, 642
Resistance:
ac, 372
definition, 35, 642
effective, 373
equivalent, 78, 94, 161, 373
factors governing, 36, 62
insulation, 389
internal, 96, 109, 121, 142–146, 327
load, 10, 46, 143
measurement, 16, 229
nature of, 36
ohm, 35
ohmic, 373
reflected, 328
specific, 62
substitution box, 55
temperature coefficient, 65
total, 78, 216
Resistance networks, 156–173
Resistive, 448, 451, 452, 497
Resistivity, 62, 642
Resistors:
bifilar, 357
bleeder, 112
carbon composition, 52
color code, 52
construction, 52
decade, 55
definition, 51
DIP, 53
discharge, 356
fixed, 52
graphic symbol, 11
linear, 45, 66
metal film, 66
multiplier, 213
nonlinear, 45, 67
parallel, 94–100
series, 76–86
series-dropping, 111
series-parallel, 106–113
shunt, meter, 207
standard values, 600
thyrite, 69
tolerance, 52
typical, 12
variable, 53
wirewound, 53
Resonance, 487–506
definition, 488, 502, 642
parallel, 452, 497–504
series, 448, 488–496
Resonant circuit, 488, 497
Resonant frequency, 490, 502, 642
Resonant rise of current, 500
Resonant rise of voltage, 494
Retentivity, 181, 188, 642
Reverse breakdown voltage, 550
RF generator, 295, 499
Rheostat, 53, 254, 642
Right-hand rule, flux direction, 182, 253
Ripple factor, 558
Ripple voltage, 368, 412, 558
Rise:
current, 344
potential, 77
rms value of sine wave, 278, 643
Rotating magnetic field, 261
Rotating vector, 436
Rotor, 196, 261
Rounding-off, 8

Safety, 14
Saturation, magnetic, 188, 643
Sawtooth wave, 291
Scales, reading, 16
Schematic diagram, 11
Scientific notation, 6
Secondary, transformer, 322, 643
Secondary cell, 125–131, 643

Selectivity, resonant circuit, 491, 500, 643
Self-excited, 264
Self-inductance:
 definition, 308, 643
 factors governing, 310
 graphic symbol, 11, 302
Self-induction, 308, 643
Semiconductors, 31, 64, 195, 643
 P-type, 546
 recombination in, 546, 548
Sensitivity:
 meter movement, 207
 resonant circuit, 496
 voltmeter, 216
Separately excited, 264
Series capacitors, 392, 427
Series cells, 145
Series circuit:
 electric, 11, 76–86, 282, 392, 427, 437–450, 516
 magnetic, 198
Series-dropping resistor, 111
Series impedances, 521
Series inductors, 311, 370
Series motor, 196
Series-parallel circuits, 106–113, 527
Series resistors, 76–86, 282
Series resonance, 448, 488–496, 644
Shelf-life, 123, 130
Shells, electronic, 31
Short-circuit current, 99, 125
Short circuits, 80
Shunt:
 ammeter, 207, 644
 Ayrton, 209
Shunt motor, 196
Siemens, definition, 36, 529, 644
Signal generators, 295
Significant figures, 8
Silicon:
 current in, 544
 diode, 547–551
 electron-hole pairs, 544
 extrinsic, 545
 holes in, 544
 intrinsic, 545
 N-type, 545
 P-type, 546
Sine wave:
 addition, 430
 amplitude, 258, 608
 average value, 278
 cycle, 258, 294
 definition, 258, 644
 effective value, 278
 half-wave rectified, 263
 instantaneous value, 258, 279
 peak value, 258, 293
 period, 260, 294
 rms value, 278
SI system, 8
Skin effect, 373, 644
Slip rings, 257
Smoothing capacitor, 412, 557
Smoothing choke, 367, 559
Solar battery, 133, 645
Solenoid, 182, 645
Solenoid valve, 183
Source:
 constant current, 164
 of emf, 10, 119–134, 250
 energy, 10
 three-phase, 261
Specific gravity, 128
Specific resistance, 62
Split rings, 263
Square wave, 350, 411
Standard cell, 226–228
Star-connected system, 172, 262
Starter, fluorescent, 368
Static electricity, 26
Stator, 196, 261
Steady-state values, 345, 404
Storage battery, 125–131
Superconductivity, 64, 66
Superheterodyne receiver, 504
Superposition theorem, 159
Susceptance, 529, 645
Switch, electric, 11, 645
Symbols, graphic, 11, 598

Tables:
 atomic structure, 31
 capacitor, types, 389
 color codes, 52
 conversion of units, 9
 dielectric constants, 384
 dielectric strengths, 64, 386
 filter circuits, 561
 graphic symbols, 11, 598
 meter movement resistances, 207
 multimeter characteristics, 238
 prefixes, 9
 primary cells, 125
 resistivities, 62
 secondary cells, 131
 standard resistance values, 600
 temperature coefficient of resistance, 65
 trigonometric functions, 606
 wire, 601
Tank current, 500
Temperature, 64
 Curie, 191
Temperature coefficient of resistance, 64
Temperature rise, 66
Temporary magnet, 181
Terminal voltage, 97, 121, 124, 131, 142
Tesla, definition, 186, 645
Theorems:
 maximum power transfer, 147
 Millman's, 169
 Norton's, 166
 superposition, 159
 Thévenin's, 161, 531
Theoretical parallel resonance, 497
Thermal runaway, 581
Thermal stability, 581
Thermistor, 67, 646
Thermocouple, 134, 183, 646
Thermostat, 184
Thévenin's theorem, 161, 531
Three-phase systems:
 definition, 261, 646
 delta, 262
 four-wire wye, 262
 power, 262, 328, 476
 rectifier, 556
Three-terminal networks, 172
Three-way switch, 330
Three-wire system, 330
Thyrite, 69
Time constant:
 CR networks, 404, 615, 646
 LR networks, 345, 610, 646
T-network, 173
Toroid, 198
Torque, 196, 206
Tracing loops, 157
Transducer, 237, 646
Transformation ratio, 323
Transformations:
 delta-wye, 172
 impedance, 327

Transformer:
audio, 328
auto-, 333
center-tapped, 330, 553
current, 334
efficiency, 324
instrument, 334
iron-core, 322–334, 646
isolation, 323, 332
loading, 323
potential, 334
power, 325
radio frequency, 325
step-up and step-down, 323, 354
tuned, 505
Transformer action, 322
Transformer rating, 324, 475
Transient, 345, 404, 610, 615
Transistors, 570–577, 646
ac current gain, 575, 576
alpha, 576
amplifying action, 572
beta, 574
bias, 572, 577, 580
collector characteristic curves, 574, 576
common base, 576
common emitter, 572
construction, 571
curve tracer, 575
cutoff frequency, 582
dc current gain, 574, 576
diode analogy, 570
dissipation, power, 571
forward-current transfer ratio:
CB, 576
CE, 574
lead identification, 571
leakage current, 572
NPN, 570
ohmmeter checking, 570
packaging, 571
PNP, 570
switch, 573
symbol, 570
types, 570
Trigonometric ratios, 606
True power, 462
True-reading rms, 281, 299, 300
Tuned circuits, 496
Tuned transformers, 505
Turns ratio, 322, 647
Two-terminal networks, 161, 531
Two wattmeter power measurement, 476

Uncovered charges, 548
Units:
conversion, 8
international system, 8
MKS, 9
prefixes, 9
Universal motor, 197
Universal shunt, 209
Universal time constant curves:
CR circuits, 409
LR circuits, 347

Vacuum-tube voltmeter, 15, 216
Valence electrons, 31, 647
Var, 463
Variation, 192
Varistor, 69
Varmeter, 465
Vector:
addition, 440, 520
definition, 436, 647
division, 515, 522
multiplication, 515, 522
rotating, 436
subtraction, 520
Vector algebra, 513–532
Volt, definition, 29, 647
Voltage:
breakdown, 64, 385
definition, 30, 647
Hall, 194, 213, 336
induced, 250, 255, 308, 609
line, 109, 330
open circuit, 142, 161
polar form, 514
rate of change, 402, 424
rectangular form, 519
resonant rise of, 494
terminal, 97, 121, 124, 131, 142
Voltage divider, 81, 111–113, 226, 282, 393, 500
Voltage divider principle, 81
Voltage drop, 46, 76, 109, 142
Voltage ratio, 322
Voltage regulation, 112, 143
Voltage regulator, 261, 550
Voltage rise, 77
Voltage triangle, 441
Voltaic cell, 120
Voltampere, 324, 465, 647
reactive, 463, 647
Volt-box, 228
Voltmeter, 15, 44, 213–218, 226, 647
calibration, 228
loading effect, 216
multipliers, 213
sensitivity, 214, 216, 647
VOM, 15, 216, 234
VTVM, 15, 216

Watt, definition, 48
Watthour meter, 330
Wattmeter, 148, 299, 476, 648
Wave:
sawtooth, 291
sine, 258, 274
square, 350, 411
Wavelength, 648
Weber, definition, 186, 252
Weston standard cell, 226–228
Wet-cell battery, 120, 125
Wet charge, 127
Wheatstone bridge, 159, 172, 234–237
Window, 156, 158
Wire table, 63, 601
Work, 29, 46, 648
Wye-connected three-phase system, 262, 648
Wye-delta transformation, 172

Yoke, 290

Zener diode, 550, 648

NOTES

NOTES

NOTES

NOTES

NOTES